THE DYNAMICS OF THE EARTH

An Introduction to Physical Geology

Other Books by
Edgar Winston Spencer

BASIC CONCEPTS OF HISTORICAL GEOLOGY

GEOLOGY: A SURVEY OF EARTH SCIENCE

THE DYNAMICS OF THE EARTH

An Introduction to Physical Geology

by
EDGAR WINSTON SPENCER
Washington and Lee University

with drawings by
Elizabeth Humphris Spencer

THOMAS Y. CROWELL COMPANY
New York Established 1834

tyc

Cover photograph courtesy of the United States Geological Survey.

Published simultaneously in Canada by Fitzhenry & Whiteside Limited, Toronto.

L. C. Card 74-179776
ISBN 0-690-24844-X

Design by Elliott Epstein
Supplementary drawings by Vantage Art Inc.

Manufactured in the United States of America

For Liz

PREFACE

In the past decade, geology and related earth sciences such as oceanography and geophysics have undergone dramatic changes. New data have accumulated at an extremely rapid and accelerating rate. Our present understanding of the earth is far more advanced than it was even ten years ago. Very few branches of geology have remained significantly unchanged, but in some these changes are fundamental and almost revolutionary. This is particularly evident in our understanding of the ocean basins and structure of the upper mantle and crust. We have seen the theory of continental drift return to a place of eminence and the idea that the ocean basins are ancient and stable fall into disrepute. New theories of sea-floor spreading and plate tectonics have captured our imagination; these theories are being applied to old problems of crustal instability, orogeny, changes in sea level, and other concepts of large-scale earth structure. New questions, new problems, and new fields of inquiry are developing so rapidly that we can hardly escape the conclusion that the earth sciences today are among the most dynamic of all fields of scientific endeavor.

Toward the end of the 1960's, the effects of urbanization, population growth, and environmental pollution focused attention on the character of the earth's surface and the processes acting there. The general public is at last aware of the pressures that expanding populations and economic activity place on limited resources of minerals, land, air, and water; geologists everywhere must carefully re-examine the relevance of their field of study to these problems. Geology is indeed in a unique position to contribute to the understanding of the total environment and to the solution of many of the problems we face in maintaining its fitness for man.

New discoveries, the increasing importance of environmental problems, and students' increasing awareness of the significance of the earth sciences have been primary considerations in the development of this text. Considerable attention has been given to oceanography, slope stability, stream action, and ground water. Geophysics is recognized as playing an increasingly significant role in geology, and a full chapter is devoted to the evolution of our ideas about tectonics. Major topics have been given the fullest treatment commensurate with the introductory course in geology, and a considerable amount of descriptive material makes the book useful as a reference without burdening the reader with unnecessary detail.

The student with normal preparation in elementary mathematics (exclusive of calculus) and some exposure to secondary school science should have little difficulty in understanding the text; the student with lesser preparation can use the text with a minimal review of secondary school mathematics and science, especially the elements of chemistry. At the ends of the chapters, review is facilitated by chapter outlines, lists of new terms, review questions, and, for the especially interested student, suggested readings. On the endpapers are a modern geologic time scale, a comparison of metric and English measure, and a partial periodic chart containing the elements discussed in the text and used in various mineral formulas.

For the most part, the United States government style manual *Suggestions to Authors of the Reports of the United States Geological Survey* has been followed in the preparation of the text. As far as practicable, measurements in the text have been expressed in metric units, but English units have been retained in some source materials such as a number of the diagrams and tables. The conversion table on the front endpapers will help the reader translate from one measure to the other.

Five appendices will assist the reader in reading topographic and geologic maps, in understanding the basic physics of forces in the earth, in reviewing chemical symbolism, and in understanding the special problems of oil and gas exploration. The appendix containing, in one place, the primary sources of geological information should prove useful to both beginning and advanced students. Following the appendices is a glossary of terms commonly employed in geology, useful in a general review of the topics covered by the text. After the appendices and preceding the general index, the full references cited in the text are given.

The original work of a great many scientists is presented in this book. The contributions of many are acknowledged in the text, bibliography, cited references, and captions of illustrations; the work of others has become part of our general knowledge. A debt of gratitude is acknowledged to all whose original contributions are recorded in this book.

Special thanks are extended to Donald Coates, Jacob Freedman, William Bonini, Robert Boyer, Grant Gross, Francis Wu, Don Kissling, Rosemary Vidale, and William MacDonald for their comments on the manuscript for this book. I appreciate the advice and help of my colleagues Donald Doehring, Sam Kozak, Odell McGuire, Fred Schwab, and Hugh Buchanan. I am indebted to those students who read parts of the manuscript and helped in preparation of materials for it, and to Mrs. Stuart Brewbaker, Patricia McMahon, and Mata McGuire for help in the preparation and typing of the manuscript.

I am particularly grateful to my wife for her encouragement and for the preparation of many of the line drawings.

Edgar Winston Spencer

CONTENTS

I
MATERIALS OF THE EARTH'S CRUST

6
WEATHERING AND SOIL DEVELOPMENT, 137

7
SEDIMENTS AND SEDIMENTARY ROCKS, 157

II
THE OCEANS

8
THE OCEAN BASINS, 183

9
PHYSICAL OCEANOGRAPHY, 203

10
ENVIRONMENTS AT THE MARGINS OF THE OCEANS, 229

11
DEEP-SEA ENVIRONMENTS, 259

III
THE CHANGING LANDSCAPE

12
GRADATION OF THE EARTH'S CRUST, 275

13
MOVEMENT OF SURFICIAL MATERIALS BY GRAVITY, 283

14
STREAMS AND THEIR ROLE IN LANDSCAPE DEVELOPMENT, 303

15
WIND, 345

16
GROUND WATER, 361

V
THE STRUCTURE OF THE EARTH

THE DYNAMICS OF THE EARTH

An Introduction to Physical Geology

I

MATERIALS OF THE EARTH'S CRUST

1
INTRODUCTION

1 INTRODUCTION

GEOLOGY—DIMENSIONS OF THE FIELD

The term *geology* means the study of the earth, and it was initially used by Richard de Bury in 1473 to apply to earthly science as opposed to *theology*, or heavenly science. Such a broad definition implies that geology includes the other natural sciences, which it does, in that other sciences are freely drawn upon to solve geological problems. But in the twentieth century the various scientific disciplines have become increasingly specialized even though often they are concerned with the same subject matter.

Recently, the physical sciences that deal most directly with the earth are often referred to collectively as the *earth* sciences. These may be grouped in various ways; one such *grouping* is the following:

EARTH SCIENCES[1]

Geological Sciences

Geochemistry is the study of the relative and absolute abundances of the elements and of the atomic species, isotopes, in the earth, and the distribution and migration of the individual elements in the various parts of the earth.

Geomorphology is the systematic examination and description of land forms, study of the processes of their origin and alteration, and the interpretation of them as records of past history.

Glacial geology is the study of the cause, distribution, erosion, transportation, deposition, and other effects of the formation of large masses of ice on the continents.

Glaciology is the study of form, physics, and regimens of modern glaciers.

Mineralogy is the study of minerals, their composition, the ways in which they occur, their crystal structure, and their behavior.

Paleontology is the study of fossils, the remains and traces of the life of the past. Fossil assemblages are used to correlate rock units, and some assemblages are excellent indicators of the environments in which they lived.

Petrology is the study of the origin of rocks, their structures, textures, and their alteration.

Petrography is the systematic description and classification of rocks.

Sedimentation is a specialized branch of petrology that deals with the way in which sedimentary rocks are deposited in water, air, and ice.

Stratigraphy is the description and study of stratified rocks. It is concerned with the distribution, composition, thickness, age, variations, and correlation of rock strata.

Structural geology is the study of the shape and configuration of the rocks in the crust of the earth and their causes.

Geophysical Sciences

Geodesy is the investigation of the shape and dimensions of the earth. This field includes analysis of the earth's gravitational field.

Seismology is concerned with the detection and interpretation of earthquake waves and the study of the elastic properties of the earth.

Terrestrial magnetism is the study of the earth's magnetic field, its causes, variations, and the history of changes that have taken place in the position of the field and the poles.

Oceanography is broadly defined to include the study of the oceans and the ocean basins including *physical oceanography*, which en-

[1] The following definitions are based on the American Geological Institute's *Glossary of Geology and Related Sciences* (Washington, D.C.: The American Geological Institute, 1957).

TABLE 1–1
Earth sciences

Geological sciences	Geophysical sciences	Oceanography	Meteorology
Geochemistry	Geodesy	Chemical oceanography	Climatology
Geomorphology	Seismology	Marine biology	
Glaciology and Glacial geology	Terrestrial magnetism	Marine geology and geophysics	
Mineralogy		Physical oceanography	
Paleontology			
Petrology			
Stratigraphy			
Structural geology			

compasses study of tides, waves, and currents; *marine geology* and *geophysics; chemical oceanography* (chemistry of sea water); and *marine biology*.

Meteorology is the study of the atmosphere, atmospheric phenomena, and weather.

GEOLOGY—KEY TO EARTH'S PHYSICAL ENVIRONMENT

A dramatic increase in public awareness of the importance of understanding our environment has taken place in recent years. The consequences of long-continued and virtually unrestrained population growth coupled with intensive industrialization, wasteful use of natural resources, and lack of understanding of relationships among the various components of our environment have now combined to produce what many scientists consider an environmental crisis. Some predict that the very survival of our species depends on our recognition and solution of environmental problems—and soon.

The environmental problems we face have many dimensions. Explosive population expansion places a strain on food and natural resource supplies, as well as on our political and social institutions. In an industrial society such as ours, increased urbanization and demand for goods pose an immediate threat to the environment by making necessary more intensive use of land, placing a greater demand on our natural resources. Expedient methods of goods production have been undertaken without regard to environmental considerations, resulting in increasing amounts of pollution of air, water, and land. We have at last become aware, even if only vaguely so, of some of the consequences of pollution and uninformed, intensive use of land. Part of the price is contaminated water supply, lakes and streams transformed into open sewers, air filled with deadly fumes, despoiled beaches, flooding, and billions of dollars lost in property damage. Perhaps even more costly is the damage done to the life on our planet. And still uncertain are the possible long-term effects on our weather and the consequences this will have on the ice–water balance and sea level. The quantity of ice on land is sufficient to raise sea level about 60 meters if it should melt.

The earth sciences are concerned with the physical and, to some extent, biological aspects of the earth's environment. The goals of these sciences are to describe the earth and to gain understanding of the composition and properties of the materials that make up the earth, the origin of these materials, the processes acting on and within the earth that cause changes on earth, and the ways in which the atmosphere, lithosphere, hydrosphere, and, to some extent, biosphere are interrelated.

THE ROLE OF GEOLOGY IN SOCIETY

The subject matter of most geological research is such that the results are directly applicable to human problems. Most geologists are employed in the search for and development of natural resources, but as society becomes increasingly aware of the importance of understanding geological processes and earth materials, geologists are required to contribute their knowledge to such fields as civil engineering, regional planning, and urban development. Increased demand for energy, natural resources, and more intensive land use is a direct consequence of population growth. Demand for electric power, for example, has doubled about every decade. So far, we have met this demand through generation of electric energy by the burning of fossil fuels (coal, oil, and gas), through hydroelectric generation, and through the use of nuclear power generators. Fossil fuels are limited in quantity; known reserves are insufficient to meet demands projected for the next thirty years. Unless other satisfactory energy sources can be developed quickly, even more intensive exploration for oil and gas, especially in the oceans, will be necessary. Natural gas especially is sought because it can be burned with a minimum of atmospheric pollution. Hydroelectric power development is limited by the availability of dam sites. We may, however, expect more pump storage systems in which water is pumped into a high reservoir during periods of low demand (at night) and used to generate power at times of peak demand. Use of reactors to generate power has developed much more slowly than earlier predicted, owing to the problems of disposal of radioactive waste products, thermal pollution of water used to cool reactors, and dangers inherent in the presence of a reactor in a population center.

Geological knowledge is applicable to many aspects of energy-supply problems. For example, geology is used in the exploration for coal, oil, gas, and radioactive minerals, and a knowledge of foundation stability is critical in the choice of location and in the construction of dams and reactors. An understanding of the dangers from natural disasters such as floods, earthquakes, and volcanic activity is obviously important. And waste disposal in the earth, in streams, or in the oceans involves many geological considerations.

Relatively little attention has been given to geological considerations in our society's use of land. The price in lost property and lives has been great and will become even greater as people crowd into land areas where natural hazards exist. Most zoning laws give scant recognition to possible hazards from earthquakes, floods, slope instability, tsunami, volcanic eruption, land subsidence, storm waves, or submergence. Most people are not even aware these dangers exist; warnings are inadequate, protection by law is rare, and insurance usually does not cover natural disasters, often called "acts of God."

The value of water, however, is gaining increased recognition. Many areas in the Western States are now sites of overdraft—more water is being withdrawn from the ground than is being naturally replenished. Withdrawal of water in coastal areas is causing intrusion of salt water into water-bearing strata. In other areas ground water is being contaminated by improper disposal of sewage and other waste while in still others permanent loss of porosity and storage capacity is being induced by withdrawal of water. And now plans are being laid to transport large quantities of water from one drainage basin to another. Our knowledge has advanced far enough for us to predict some of the consequences of particular actions, but often this knowledge is not applied. Much greater public concern is being shown for the pollution and destruction of our natural surface-drainage systems.

We can thus summarize areas of special application of geology to society as follows:

1. **Natural Resources**
   a. Exploration for natural resources, broadly classified as fuels, metals, and nonmetals, which include our primary sources of energy, raw materials for industry, and building.
   b. Evolution and development of natural resources.
   c. Planning of resource development.
2. **Geological engineering.** The application of geology to civil engineering problems, especially foundations for dams, large buildings, reactors, bridges; evacuations such as tunnels and road cuts; coastal engineering projects.
3. **Land use.** Analysis of the suitability of terrain for various types of development. Consideration must be given to the various hazards arising from earthquakes, volcanic activity, slope instability, and so on.
4. **Water resource development.** Analysis of ground-water supplies, movements, contamination, and interchanges with streams and the ocean.
5. **Streams.** The character of the stream system and processes affecting it must be understood if we are to control flooding, prevent or reduce pollution, and avoid undesirable side effects resulting from artificial modification of streams.

Study of the character of earth materials, the processes acting on and within the earth, and the history of the earth provides a unique perspective on our physical environment. One of the objectives of this text is to develop this perspective.

GEOLOGY—A PREVIEW

Earth scientists are directly concerned with the solid earth, its interior and surface, the great bodies of water that cover nearly three-fourths of that surface, and the surrounding envelope of gases, the atmosphere. Plants and animals are an integral part of many earth processes and cannot be disregarded in a study of the earth. The remains of plants and animals are an important constituent of many sediments, and fossils are an important part of the rock record; they have provided the basis for unraveling a large portion of the earth's history.

The geologist sees the earth in terms of the materials that compose it, the shape of its surface, the nature of its interior, the natural processes acting in and upon it, and its history. It is the historical aspect of geology that sets it apart most distinctly from other physical sciences. Mathematics, physics, and chemistry, as well as biology, are applied to the study of the earth, but time is frequently the key element in natural processes acting on the earth. The geologist is continually confronted with a historical perspective in his concern for the origins and the evolution of changes that have led to the present state of the earth. The earth must be considered in terms of complex interrelations of particular configurations of materials and natural processes leading to changes through time. Because of the complexity of nature, the large number of variables, and the size and time factors, rigorous quantitative analysis is at best difficult and sometimes impossible. Because the various fields of earth science are so interrelated, none can be satisfactorily treated independently of all others. Ideally, the student needs a background in the other natural sciences and in other fields of geology before starting the study of any earth problem. However, this is clearly impossible. The purpose of this introduction is to provide an overview, a framework, to which the following chapters can be related.

MATERIALS

Chemical elements occur naturally on earth as organic compounds in animal and plant tissue, oil, and coal, or as inorganic compounds called *minerals*. Minerals occur sometimes in a pure form, but more generally as *rocks*, which are aggregates of minerals. The most common chemical elements in the earth are oxygen, silicon, aluminum, iron, magnesium, calcium, sodium, and potassium. These are combined in such common minerals as *feldspar* ($KAlSi_3O_8$), *quartz* (SiO_2), *mica* ($KAl_2(OH,F)_2AlSi_3O_{10}$), *hornblende* ($Ca_2$-$(Mg,Fe)_5(OH)_2(Al,Si)_8O_{22}$), *calcite* ($CaCO_3$), and *olivine* ($(Mg,Fe)_2SiO_4$).

Rocks and their component minerals are known to form in three main ways: (1) through crystallization from molten masses within the earth (*igneous rocks*); (2) through settling out (*sedimentation*) and consolidation from water, ice, or air, as is the case with sediment in oceans and lakes, sand dunes, volcanic dust, and glacial deposits (*sedimentary rocks*); and (3) through *metamorphism*, a process that alters igneous or sedimentary rocks by high temperature, stress, pressure, or chemically active fluids (forming *metamorphic rocks*). Igneous rocks composed of minerals rich in silicon and aluminum are called *sialic* rocks; an example is *granite*, whose component minerals are quartz, mica, and feldspar. When iron and magnesium are more prominent, the igneous rock is called *simatic;* an example is *basalt*, composed of olivine, hornblende, and feldspar. Most sedimentary rocks are shale, sandstone, limestone, or mixtures of these. Metamorphism of limestone produces marble; sandstone is transformed to quartzite; and shale under high stress is transformed into slate or, under other conditions, into schist.

GRADATION

Weathering

Natural processes acting on and within the earth are constantly shifting, altering, and rearranging the materials of the earth's crust. Where rocks come in contact with the elements of the atmosphere, mechanical and chemical reactions take place that lead to the decomposition and breakdown of the rocks and the formation of soil. Taken as a group, these processes acting on the earth's crust are called *weathering*.

Degradation

Both weathering products, soil and the consolidated or semiconsolidated rock beneath, are subject to the effects of various geomorphic agents such as running water, glaciers, wave action,

FIGURE 1–1
Stream dissection of flat-lying beds of a plateau. This is an early stage in the dissection of this plateau. Later the divides will be narrowed and the valley will become wider.
(U.S. Air Force)

FIGURE 1–2
Glaciers on the flanks of Mount McKinley, Alaska.
(U.S. Air Force)

ground water, and wind. Each of these agents brings about erosion, transportation, and eventually deposition of earth materials through a complex of natural processes peculiar to itself. The overall effect of these agents is a shifting of materials from higher to lower places on the earth. These forces collectively can be called processes of *degradation*. Each of the geomorphic agents brings about erosion through different processes, and the methods of transportation and deposition differ. Thus, where one of these dominates, the land surface gradually assumes a shape that is characteristic of that geomorphic agent. Because each agent is different, it becomes possible to recognize the land forms and the character of the deposits formed through each agent, even though that agent is no longer acting.

Igneous Activity and Diastrophism

If the earth were affected only by processes that tended to reduce its surface, it would long since have obtained a nearly flat and featureless surface. The main factors in keeping portions of the earth high are *igneous activity*, the production of rock melts, and *diastrophism*, movements of the crust. Igneous activity includes *volcanism*, through which volcanoes are built and vast outpourings of lava take place. Approximately 200 active volcanoes are known, and one of the most extensive of the lava plateaus is the Columbia River Plateau of the northwestern United States.

Carefully surveyed level lines across large areas of continents reveal that parts of the continents are even now being elevated in relation to other parts. Diastrophism may act slowly, as in this almost imperceptible uplifting, or violently, as in the Alaskan earthquake of 1964. Rock strata near the top of the Grand Canyon are composed of sediments that today form only in the oceans, and these sediments contain fossils of animal groups that today live only in the oceans. Marine fossils are found in the rock units near the top of Mount Everest, nearly 9,000 meters above sea level. These are but a few bits of the mass of information indicating that the crust of the earth is not stable. The crust is sometimes warped over a large area, elevating some portions and depressing others. In addition, certain long belts of the crust have been strongly deformed so that the rocks are folded, broken by faults, and uplifted to form high mountain belts. This deformation is such that it could be produced only by great stresses within the crust. Rocks subjected to great stress and lying at great depths behave quite differently from rocks at the earth's surface. Under the high confining pressures and at the high temperatures of the earth's interior, rock loses its brittleness and flows. This flowage is apparent in the deformed rocks found in the cores of uplifted mountains that have been dissected by weathering and erosion. The behavior of rocks under such conditions gives us an indication of the nature of the earth's interior.

FIGURE 1–3 (*above*) Kilauea flank eruption, Hawaii, 1960. Fountains about 60 meters high throwing new spatter. (U.S. Geological Survey)

THE INTERIOR OF THE EARTH

Most of what we know of the earth's interior comes from indirect observations. In exposed cores of mountain ranges we can examine rocks that have been at depths of several kilometers; a few mines penetrate 1.6 kilometers or more deep, and a few oil wells have been drilled to depths of 7 or 8 kilometers. However, these distances are slight when compared to the earth's diameter, and we have no direct information about the interior of the earth below the crust. Such information is gained chiefly through *seismology*, the study of shock waves that have traveled through the earth from earthquake foci, and through measurements of the gravity and magnetic fields of the earth. Based on the known depths within the earth at which changes occur in the velocity and character of shock waves, a number of important zones have been recognized. The closest of these to the surface is called the *Mohorovičić discontinuity (M-discontinuity)*, and it is used to delineate the base of the earth's *crust*. The crust is deepest under mountain ranges, and deeper under continents than under ocean basins. Velocities of the shock waves through the continents suggest that the composition of the continents is rich in silicon and aluminum, termed *sial*. The ocean basins appear to be underlain by *simatic* rocks, rocks rich in iron and magnesium, like basalt. The differences in composition and level of the continents and ocean basins seem to be related to the density of the rock types. Sialic rocks are less dense than simatic rocks. This suggests that the crust is more or less "floating" on

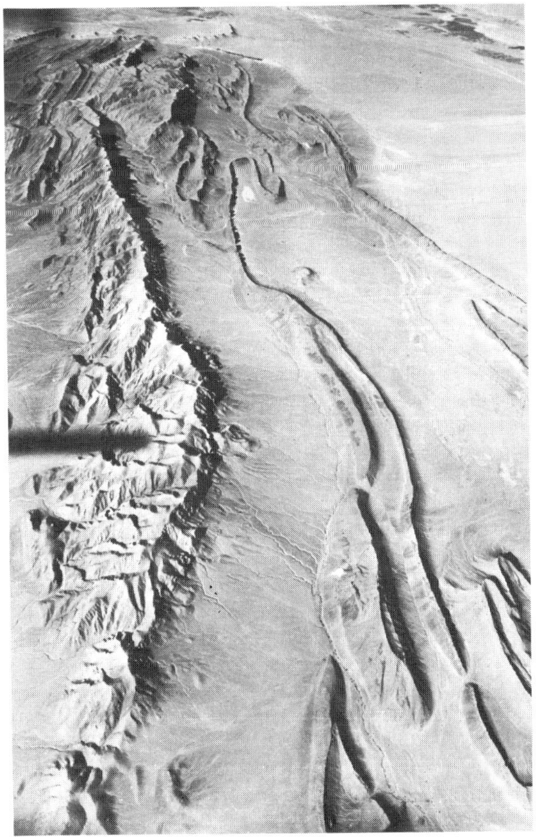

FIGURE 1–4 Belt of folded mountains in Africa (Atlas Mountains). Deformation and uplift of the crust of the earth is one of the main factors in maintaining parts of the crust high. Already these mountains have been deeply dissected. Erosion products of the mountains lie on the flanks of the ridges. (U.S. Air Force)

the top layer of the interior of the earth—the *mantle.*

If such flotation is indeed the case (and it is borne out by many lines of evidence), then the rocks of the mantle must be able to *flow.* Of course this flowage is not like that of most familiar liquids, but more like that of asphalt, which appears to be a solid but behaves as a plastic or highly viscous liquid. A zone or shell of high plasticity in the upper mantle at a depth of about 100 kilometers was discovered by Beno Gutenberg and is named in his honor. The lower limit of the mantle is defined by another zone of seismic discontinuities. Below the level of about 2,280 kilometers, the outer boundary of the core, only those shock waves are transmitted that are capable of moving through liquid. This has led to the conclusion that the earth has a liquid core. The density of this core is so great that only a few materials as we know them at the surface of the earth could form it. The most probable of these is a nickel-iron compound like that commonly found in meteorites.

LARGE-SCALE FEATURES OF THE EARTH

Continents and Ocean Basins

The most obvious as well as the most important divisions of the crust are the continents and the ocean basins. These differ in many respects and are separated by zones of transition that have different characteristics from those of either the continents or the ocean basins. Among the most striking differences between the continents and the ocean basins are the following:

1. The elevation of the continents is, on the average, almost 0.8 kilometer above the level of the present-day seas, but the average depth of the ocean floors is almost 5 kilometers beneath the surface of the waters —a difference of about 5.8 kilometers in surficial elevation.
2. The thickness of the crust is strikingly different under oceans and under continents. Measured from the ocean's surface, the crust in the ocean basins ranges between 8 and 10 kilometers in thickness whereas under the continents it is about 20 kilometers thick, reaching a maximum under some mountains, where its thickness may be as great as 40 kilometers.
3. The crust under the oceans has a bulk composition that differs from that under the continents. Almost the entire continental crust is thought to be sialic, but the oceanic crust contains a thin layer of sialic material and in some places none at all. The minerals of the oceanic crust are those that make up basalt and are simatic.
4. The products of igneous activity under the oceans are different from those under the continents in that extrusions and intrusions under the oceans tend to be basaltic in composition, whereas those in and on the continents tend to be granitic in composition.
5. Even the processes of erosion and deposition are different, due to the various properties of the fluids directly in contact with the lithosphere on the continents and under the oceans.
6. In general, the continents may be considered the sites of the most rapid erosion, whereas the ocean basins are primarily sites of deposition.
7. The variety of factors involved in the erosion and decay of rocks is greater on continents than in oceans.

It is possible to make further subdivisions of the crust, distinguishing a large number of unique crustal segments, but for the purpose of general discussion the definition of six types of crustal segments is sufficient.

Continental shields. Each continent contains a large and usually centrally located segment composed primarily of crystalline igneous and metamorphic rocks more than 600 million years old. Although these regions, known as continental shields, were unstable in their early history, today they are the most stable portions of the continental crust, having few earthquakes and no young, folded mountain belts.

Folded mountain belts. The youngest of the folded mountain belts are physiographically high regions like the Alps, Rockies, Andes, and Himalayas. All belts that have been deformed in the last 600 million years fall into the category of folded mountain belts even if they are no longer high. They are characterized by folded and faulted strata, the presence of igneous intrusions along their axial zones, past or present earthquakes, unusually thick sequences of sedimentary rocks, linear patterns, and long-term instability.

Continental shelves. The classic example of a continental shelf is the eastern margin of the United States. Where exposed above sea level, the continental margin is relatively flat and is called the *coastal plain.* This nearly level topography extends out under the ocean as far as 160 kilometers in a few places before it breaks away in a steeper slope leading down into the depths of the ocean basins. The shelves are transition zones between the continents and the ocean basins. Most of the major streams extend onto the shelves in the form of submarine canyons. Sediments carried across these shelves are deposited on the slopes leading into deeper water. During the ice ages of the past 2 million years, these shelves were periodically exposed as a result of lowered sea level.

Island arc systems. These are the least stable portions of the crust. Here earthquakes of great intensity are concentrated, and here also most of the earth's volcanic activity occurs. Where the islands have been cut by erosion, great intrusions are found. Many of the island arcs have arcuate patterns, with the convex side toward the ocean basins; the Aleutian Islands are the classic example. Off the islands are the deep and narrow submarine trenches that contain the greatest deeps of the oceans. Most of the island arcs today are located along the western margin of the Pacific Ocean. These island arcs appear to be places where oceanic crust sinks down into the mantle.

Deep oceanic crust. At depths of around 5,400 meters are found the flattest portions of the crust, large expanses known as the *abyssal plains.* Beneath these plains the structure of the crust, as now known, is rather simple. There is a thin veneer of modern unconsolidated sediments, or, in some places, a thin section of consolidated sediments. Beneath these sediments are, first, layers of basalt and, then, at a shallow depth of about 10 kilometers, the base of the crust.

Midocean ridges. The ocean basins present a highly varied topography. One of the most prominent features is the *midocean ridges.* These ridges are comparable in relief to the Rocky Mountains, and they are culminated by a few peaks that rise as islands above the sea. Of these ridges, the Mid-Atlantic Ridge is the best known. It has a relief of over 6,000 meters above the adjacent abyssal plains. Most of the samples collected from this ridge show it to be volcanic material of basaltic composition. This is borne out by the fact that all of the islands projecting above water are volcanoes. One notable feature of the midocean ridges is a rift zone along their crests, where apparently the ridges have been pulled apart by stresses within the crust.

CONTINENTAL DRIFT

Many lines of evidence support the theory that the continents have drifted. Excellent geographic fits are found, for example, between South America and Africa and between Antarctica and Africa; and many other reasonable fits have been suggested. Reassembly of the continents into a single plate or perhaps two plates explains the distribution of ancient (over 200 million years old) glacial deposits found in India, Africa, South America, and Antarctica. Reassembly also provides an explanation of the distribution of many fossil plants and animals, and it allows connection of some folded mountain belts now located on opposite sides of the ocean. Interest in continental drift was greatly intensified by the discovery of *sea-floor spreading,* and the continental drift theory has also been supported by many discoveries about the structure of the crust and upper mantle.

Sea-Floor Spreading

New crust is thought to originate by injection of molten rock from the mantle into the crust near the crest of the midocean ridge, causing the sea floor to spread away from the ridge crest. As this melt cools, the polarity of the magnetic field becomes frozen into the rock. The concept of sea-floor spreading is supported by the discovery of nearly symmetrically disposed bands of variations in the intensity of the magnetic field on either side of the midocean ridges. These variations can be correlated with changes in the polarity of the magnetic field, and the age of these changes can be determined by radiometric dating techniques. Thus the distance of a particular band from the crest of the ridge is a measure of the amount of new oceanic crust formed since that band was formed.

This theory provides an excellent mechanism for continental drift. The continents could be moved passively on the underlying mantle rock as the sea floor spreads. The theory also serves to explain characteristics of the island arcs. The arcs are places where the oceanic crust moves down into the mantle, thus accounting for the great depth and pattern of earthquakes in the arcs and the presence of deep-sea trenches on the oceanic side of the arcs.

GEOLOGIC CONCEPT OF TIME

Geologists must learn to think in terms of lengths of time that far exceed our common experience. It is a simple matter to be aware of the fact that the earth is thousands of millions of years old, and to know the approximate time of many events in the earth's history. It is not so easy to comprehend or appreciate fully the implications that these vast spans of time hold for the processes acting on the surface of the earth and within its interior. For example, rises in sea level too small to be detected within a lifespan could flood most of the continents if continued over a period of millions of years. Although this example is purely hypothetical, seas have invaded the continents in the past and have, in fact, covered almost all the land surface of the world. Some areas are being slowly submerged today, whereas others are emerging, perhaps to rise eventually as mountains. The importance of time can also be illustrated by changes in the land. How could a canyon like the Grand Canyon, several kilometers wide and as much as 1.6 kilometers deep, have been formed? Without a consideration of the time involved, it is a perplexing problem. But for those who know how streams erode and transport rock debris it is not at all difficult to visualize the formation of the canyon over a period of hundreds of thousands of years as the rocks exposed in the walls of the canyon gradually crumble, break away, and fall into the Colorado River, which transports them to the sea. The canyon is being enlarged today in about the same way and perhaps as fast as it ever had been in the past. The face of the earth has never before looked exactly the way it does right now; it will never be the same again. It is not even the same from day to day, but the changes are so slow that we generally fail to perceive them at all. In the short term, these changes seem insignificant; in the long term, they may convert areas occupied by seas into mountains, and mountains into plains. It is the cumulative effects of natural processes over periods of geologic time that have brought about the record of change that is the history of the earth.

MEASURING GEOLOGIC TIME

Two basic techniques are used to measure geologic time. The first, and most important technique, is the observation of some feature of a rock that gives evidence of the existence in the past of a process that was *unidirectional* and *nonreversible*. The movement of the hands of a clock is such a process, although it can hardly be used for the geologic past. The clock would be of no use if one could not depend on the hands to move in one direction all the time. It has the added advantage of moving at a constant rate in that one direction. Some "geologic clocks" do not proceed at constant rates; therefore they cannot be used to tell absolute time. Nevertheless, they are useful in fixing the time of one event *in relation* to the time of some second event. The primary unidirectional, nonreversible processes in nature include:

1. *Decay of radioactive substances.* This is the best "geologic clock" because the rate of decay is constant and dates obtained are absolute dates (see Chapter 2).
2. *Evolution of life.* Evolution has proceeded since the first forms of life appeared on earth. Evidence of the process is found in the fossil record. Many of the same phyla or types of animals are represented through long periods of time, but, as a group, fossils of any one time are a unique assemblage. These assemblages provide a reference system for determining the order of succession of geologic events. Fossils are

FIGURE 1–5
The Grand Canyon of the Colorado River. The Colorado River and many of its tributaries are deeply entrenched in nearly flat-lying strata. These strata provide a record of the history of this region for a period of more than a billion years of earth history. Some of the rock units exposed near the rim of the canyon contain fossils of marine invertebrate animals. That the rim is now about 1.6 kilometers above sea level indicates that the region has been uplifted in relation to the sea. In the distance the extinct volcanoes of the San Francisco Mountains of central Arizona are visible.
(Union Pacific Railroad)

by far the most widely used means of determining sequences of events.

3. *Accumulation of sediments.*
4. *Erosion of the land surface.*
5. *Growth of plant matter as exemplified by tree rings.*

A second technique by which geologic time is measured is analysis of the record of events that have resulted from processes that recur with a fixed periodicity. Such processes are not uncommon within small areas, but geologists disagree on the question of the existence of such processes on a world-wide scale. Locally, seasonal variations in climate influence the types of sediments deposited in lakes and the rate of their deposition. This results in the formation of finely banded, or *varved,* deposits. Some geologists believe evidence supports the existence of periodic changes of greater extent. Among the suggested pulsations are the following:

1. Periodic changes in the world climate, bringing periods of glaciation in high latitudes.
2. Periodic movements within the earth's crust, causing deformation and formation of mountain ranges.

DETERMINATION OF SEQUENCE OF EVENTS IN ROCKS

Understanding the sequence of events that led to the present configuration of the rocks, minerals, and land forms of the earth is fundamental to the science of geology, and it is this historical perspective that distinguishes the methods of approach of geologists most clearly from those of other physical scientists. Several branches of geology, notably stratigraphy and paleontology, are primarily concerned with reconstruction of the life and physical history of the earth, and these are normally treated in courses in historical geology, but some sense of geologic time is important in trying to comprehend the significance of modern processes.

Relative time relationships of events represented in rock masses were well established long before development of the techniques of radiometric dating that determine age in years. Relative time relationships are based on such assumptions as the following:

1. Younger beds lie above older beds in undisturbed sedimentary sequences (Fig. 1–5).
2. An ancient erosion surface in a sequence of rocks is younger than the rocks below that erosion surface.

FIGURE 1–6
Felsic dike; the dikes must be younger than the rock they intrude.
(Crown Copyright)

3. An intrusion is younger than the rocks intruded (Fig. 1–6).
4. Rocks cut by a fault are older than the fault.
5. Rocks folded are older than the deformation.

It was possible to reconstruct the sequence of events in much of the earth's history by use of the simple logic outlined above, and the geologic time scale was developed on this basis before we had access to radiometric dating procedures.

HISTORICAL DEVELOPMENT OF FUNDAMENTAL GEOLOGICAL CONCEPTS OF TIME

Law of Superposition

Nikolaus Steno (1638–87) made a number of astute observations about the formation of sedimentary rock strata, and from these he formulated a group of laws governing the formation of the stratigraphical succession. These laws include the following:

1. Any one stratum must either cover the whole earth or be limited laterally by other solid deposits.
2. During the period of accumulation of a deposit there is above it only the water from which it is precipitated; therefore, the lower layers in a series of strata must be older than the upper.

The second of these laws is one of the underlying principles of historical interpretation in geology, known as the *law of superposition*—in a normal, undisturbed sequence the oldest stratum lies below younger strata. Steno recognized that stratigraphic sequences are not always found undisturbed. He cited a number of examples of local *crust-inthrow,* places where the strata had been thrown into a perpendicular position or arched upward in folds. Thus he recognized the possibility that sequences could be turned upside down. He also suggested that the arching and faulting might originate from uplifting resulting from volcanic activity.

Uniformitarianism

Georges Fushsel (1722–73), a French physician, made one of the greatest contributions to the young science of geology in his statement of the *principle of uniformitarianism:* "Indeed, the manner in which nature at the present time is still acting and producing things must be assumed as the rule in our explanation; we know no other."

The concept of uniformitarianism is that the same processes and principles that govern natural changes on the earth today have acted in the past to produce similar changes. In other words, *the present is a key to the past.*

A controversy about the origin of crystalline rocks arose in the eighteenth century; Fushsel's principle of uniformitarianism, which had been published in Latin, was ignored. Meanwhile, a German geologist, Abraham Werner, succeeded in convincing the scientific community that crystalline rocks, including most igneous and metamorphic rocks, were chemical precipitates. Ultimately, however, the view of a Scotsman, James Hutton, that most crystalline rocks are of volcanic (igneous) origin, prevailed.

Hutton's observations not only led him to the conclusion that many crystalline rocks are of igneous origin, but also enabled him to state clearly the principle of superposition and the principle of uniformitarianism. Though both of these important ideas had been advanced earlier, they were not widely accepted during Hutton's life.

Charles Lyell, uniformitarian theorist. One of the foremost exponents of the theory of uniformitarianism was Charles Lyell (1797–1875). The following quote from his textbook (1833) summarizes his reasoning regarding the uniformity of earth processes.

> By these researches into the state of the earth and its inhabitants at former periods, we acquire a more perfect knowledge of its present condition, and more comprehensive views concerning the laws now governing its animate and inanimate productions. . . .
>
> If we reflect on the history of the progress of geology, we perceive that there have been great fluctuations of opinion respecting the nature of the causes to which all former changes of the earth's surface are referable. The first observers conceived the monuments which the geologist endeavors to decipher to relate to an original state of the earth, or to a period when there were causes in activity, distinct, in kind and degree, from those now constituting the economy of nature. These views were gradually modified, and some of them entirely abandoned, in proportion as observations were multiplied, and the signs of former mutations more skillfully interpreted. Many appearances, which had for a long time been regarded as indicating mysterious and extraordinary agency, were finally recognized as the necessary result of the laws now governing the material world; and the discovery of this unlooked-for conformity has at length induced some philosophers to infer, that during the ages contemplated in geology, there has never been any interruption to the agency of the same uniform laws of change. The same assemblage of general causes, they conceive, may have been sufficient to produce, by their various combinations, the endless diversity of effects, of which the shell of the earth has preserved the memorials; and, consistently with these principles, the recurrence of analogous changes is expected by them in time to come.

Catastrophism

Before the time of Hutton's teachings most students of the earth subscribed to the idea that the earth was no more than a few thousand years old, and that its history had been punctuated by catastrophes during which all animals had been wiped out and after which completely new creatures appeared. Such catastrophic events were widely accepted as the explanation of the formation of mountains and other natural features. This view survived into the nineteenth century, as is shown in the following passage by the prominent American geologist Amos Eaton (1818):

> In whatever manner the ancient world was supported, it is evident that when the wickedness of man drew down the vengeance of the Almighty, its foundations gave way and it sank to the bottom of the ocean never to be again uplifted. . . .
>
> While this tremendous crash of Nature was going on, scales of various thicknesses from the various strata were shot up, detached and broken, which gave formation to our surrounding hills, the ragged cliffs of the Catskill and the bleak brow of the Andes. Some were formed at the bottom of the sea by volcanic fires; others have arisen from various causes since the great deep retired.

EVIDENCE FOR UNIFORMITARIANISM

Natural processes. Since there is no way for us actually to observe that uniformity has prevailed, this concept must be considered an assumption. It is one of the most important, fundamental assumptions of geology. There are two reasons for its general acceptance among geologists.

First, in the time that man has been actively observing and recording the nature of natural processes, there is no evidence of any change in the way in which they operate. The principles governing physical and chemical reactions and interactions have remained constant, and apparently they may be applied as well to events of the past as to those observable in the present. We assume that limestone exposed at the ground surface in the past was dissolved by rainwater carrying carbon dioxide in solution just as it is today. This does not mean that the same amount of carbon dioxide in the rainwater now was present then, or that limestones were as widely distributed then as they are now. Further evidence of the validity of uniformitarianism is found in the ease with which the geologic record of the past can be interpreted by its use.

Reconstruction of the past. It is possible to make remarkable restorations of conditions long past by using uniformitarianism. Past climatic conditions may be postulated. Positions of coast lines and mountains as well as depth and even temperature of water at a given time in the past may be judged. What can be learned of past geologic ages must be obtained from the rocks formed during those ages and preserved since that time. This imposes several restrictions on our interpretations and makes the problem a much greater challenge. How can we possibly discover whether or not there was a mountain range in a certain area 100 million years ago, if the mountain range itself disappeared from the face of the earth 90 million years ago? The answer is that the material removed from the mountain was deposited on its flanks and in the areas around it. If it was located near a sea, it is probable that the sediment accumulated there. The presence of the mountains will have to be inferred from the character of the deposits in the sea. Thus the nature of sedimentary deposits is extremely important, and in order to understand them we must know the conditions under which they were formed.

Past environments. By studying environments in which present-day sediments are deposited, we learn how to interpret those of the past. It is worthwhile to study not only the type of sediment that is formed in a certain place but also the animals that live there. These animals die, and their remains become incorporated in the sediment. Just as different animals live in different climates or regions on land, so certain marine organisms live in particular marine environments. The fauna and flora of a swamp are strikingly different from those of the continental shelf or those of the deep-sea floor. Some marine organisms live only in warm water, others only in cold water. Some require currents to bring them food, others move about and collect their food. Thus a valuable means of interpreting past environments lies in knowing the environments in which similar animals and plants live today.

LIMITATIONS OF UNIFORMITARIANISM

In the strict sense, uniformitarianism means that the nature of processes and physical and chemical laws have remained constant throughout time. This does not imply that because ice caps occur at the poles today they have always been there, nor does it suggest that sea level has always tended to rise, or that climatic variations have always been as great as they are now. An example of how the concept of uniformitarianism may be stretched is the use of rates of sedimentation to judge the length of time involved in the formation of a sedimentary rock. The present rate of accumulation for a particular sediment type is measured, the thickness of an ancient deposit of similar nature is measured, and, by assuming that the same rates of sedimentation applied in the past, the length of time required to form the ancient deposit is calculated. The adherent to a strict interpretation of the concept of uniformitarianism will not accept this line of reasoning. Uniformitarianism may be used to evaluate the mechanism or mode of deposition, but not its rate. The rate would depend on many other factors.

One of the first objections to uniformitarianism was the amount of time required for completion of all the work accomplished by natural processes. In the eighteenth century, most people believed that the earth was no more than 6,000 years old. This would hardly leave time enough for a feature the size of the Grand Canyon to be formed by a stream cutting in the manner and at the rate of the Colorado River. Even if we accept the possibility that the river might have cut more rapidly in the past than it is now cutting, 6,000 years is not conceivably enough time. One apparent solution to the problem was the aforementioned theory of catastrophism. It has even been proposed that the Grand Canyon is a huge rift in the earth's crust, torn apart in as short a time as a day or a week, and that the rift has simply been modified by subsequent erosion. However, no field evidence indicates such a catastrophic event.

Periods of mountain formation mark breaks between some of the major divisions of geologic time. Many have envisioned these mountains as rising rapidly, in a matter of a few hundreds of years. With the advent of accurate dating methods and with a greatly improved idea of the age of the earth, it is unnecessary to call on catastrophic events to explain these. Uplifts of as little as a fraction of an inch a year may eventually create mountains. Instead of uplifting suddenly, mountains rise slowly.

The Role of Minor Catastrophes

Minor catastrophes are common on earth today. The earthquakes that struck Agadir, Africa, and

Chile in 1960 and Peru in 1970, the 1963 landslide in Italy, the typhoons that hit Japan in 1959 and Pakistan in 1970, the Alaskan earthquake of 1964, and the southern California mudflows of 1969 are examples. The effects of such catastrophes are pronounced, as the following examples show:

1. Every year rivers flood and wash away huge amounts of topsoil; hurricanes hit coasts, storm beaches are formed, and ordinary beaches are washed away; torrential rainfall causes alteration of desert landscapes.
2. Strain is built up on opposite sides of faults for periods of many years; then release of this energy and subsequent displacement takes place in a matter of a few minutes.
3. Volcanoes may stand dormant for many years and suddenly begin to eject great quantities of lava, ash, or other pyroclastic material.
4. Tsunami break up rocks along the sea cliffs around the oceans of the world and inundate lowland areas in short periods of time.
5. Masses of loose sediment or decayed rock stand on steep slopes for years until they become unstable; then the slightest disturbance sets huge landslides in motion. The counterpart of this process in the oceans is the turbidity current. Some of these currents occur with the seasons, after large quantities of sediment have been deposited on the steep submarine slopes off the mouths of rivers.

These examples serve to illustrate the importance of minor "catastrophic" events in accomplishing erosion and deposition. The cumulative effects of the constant downslope creep of soil, the gradual decay of rocks in the atmosphere, and the gradual removal of material by streams over millions of years are probably greater than the effects of these minor catastrophes. Nevertheless, minor catastrophes should not be discounted.

THE LAW OF FAUNAL SUCCESSION

The fifteenth century marked the start of a long debate about the origin of fossils. Some scientists believed that they grew from seeds, others interpreted them as marine organisms laid on the surface of the earth during Noah's Flood, and still others suggested that they were placed in rocks by the Devil to confuse man. Leonardo da Vinci was one of the first to recognize fossils as remnants of marine organisms deposited in the seas at a time when the seas were higher or the land lower. Robert Hooke, who is famous for his work in physics, is credited with being the first to suggest that fossils might be used in the study of the earth's history.

John Woodward (1665–1722), an English paleontologist, strongly opposed the religious dogma of the period, which held that all fossils were remnants of the Flood. Woodward considered fossils to be remnants of flora and fauna that lived in the distant past. His writings reveal that he correctly identified and interpreted widespread sedimentary strata.

> The Result of this was that in time I was abundantly assured that . . . there were enclosed in the Stone, and all the other denser kinds of terrestrial Matter, great numbers of Shells, and other Productions of the Sea.

The importance of this insight lies in that it opens the way to consider the possibility of correlating events in historical perspective.

The most significant conclusions regarding the value and use of fossils were offered by William Smith. Smith (1769–1839) was a man with an insatiable curiosity and a habit of taking notes on nearly everything he observed. His life's work was engineering, and he brought an immense amount of imagination and skill to it. Employed to drain swamps, locate canals, and restore springs, he realized that an understanding of rock units was of great value in accomplishing these tasks. His success in these projects put him in such great demand that he was called to all parts of England. Often he traveled as much as 16,000 kilometers a year, and on all these trips he kept careful notes of the types of rocks he saw. Probably his most important contribution was the observation that certain rock units can be identified by the particular assemblages of fossils they contain. This led to the formulation of the *law of faunal succession.*

This law states in effect that fauna and flora vary with time and that for this reason the fossils found in rock layers of different age will be different. Once the fossil faunas of a given age are recognized and differentiated from those that are younger or older, it becomes possible to identify the age of the rock unit by use of the fossil assemblage in it (Fig. 1–7).

DEVELOPMENT OF A GEOLOGIC TIME SCALE

Not all of the strata recognized by Smith occur at the ground surface in any single locality; therefore, it took some time before the correct positions of some units within the sequence could be established. Studies of the fossil fauna and flora paralleled the studies of rock lithologies. Smith had recognized that different rock groups contain different fossil assemblages. The observed differences were great enough in some cases to foster the idea that a number of catastrophes in the past had destroyed most forms of life, and this misconception was used as a basis for subdividing the rock record. It was noted that these marked changes in the forms of life occurred where the rocks of one group or series were deposited on those of an earlier sequence that had been folded or warped, causing angular unconformity (Figs. 1–8 and 1–9) between strata of the two groups. Other breaks were found to be related to *disconformities,* erosional or nondepositional breaks in a stratigraphic succession where layers above and below the break are parallel.

FIGURE 1–7
Law of faunal succession. William "Strata" Smith learned to recognize various rock units by the particular assemblages of fossils that they contained. In this case it would be possible to recognize the angular unconformity by the difference in the inclination of the beds above and below the break in the record.

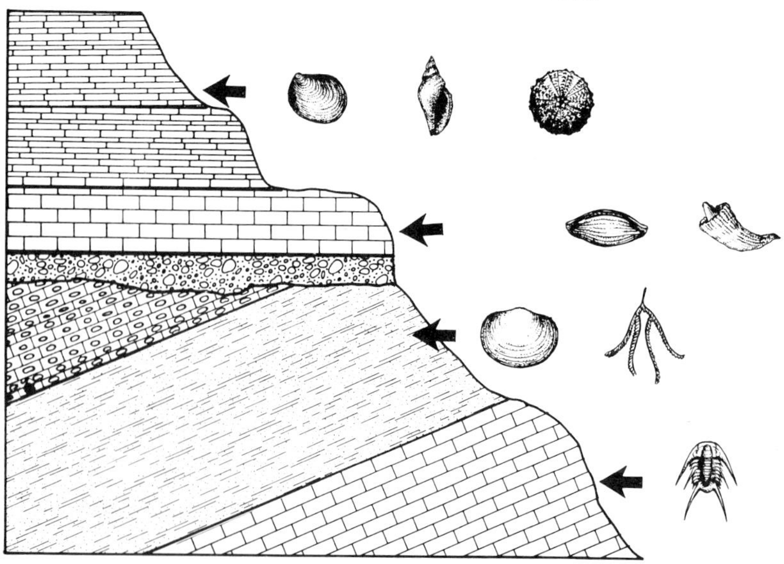

It was soon discovered that it was possible to correlate groups of strata by their position in relation to the unconformities. The same rock units containing almost identical fossils were found immediately above an angular unconformity in many localities in Europe and England. Eventually it was shown that similar units and fossils had identical relations to unconformities in North America. The evidence suggested that there had been a number of periods in the geologic past when the earth's crust was strongly deformed.

Many geologists consider such widespread deformations to be the ultimate basis for correlation of geologic events throughout the world. But the question of whether periodic earth movements affect large parts of the earth's crust almost simultaneously is still subject to debate. We do know that world-wide correlation on this basis is not a simple process. Complications are found as stratigraphic study is pursued in greater detail. At the present time geologists are working on problems related to this question. They are making more exact correlations between rock units, fossil assemblages, and geologic time, and their relation to deformation of the crust (Fig. 1–10). In general, their findings seem to point to major errors in earlier concepts based on the belief that geologic time may be divided into several major eras separated by almost catastrophic periods of deformation. Mountains are found to develop slowly—over periods of millions of years. Even changes in forms of life characteristic of different eras seem to be much less drastic than formerly thought.

One of the first attempts to classify the rocks of the earth's crust in the order of their development was made by Johann Lehmann in 1756. Lehmann divided the various types of rock units he observed into three groups:

Primary rocks. Crystalline rocks containing no fossils—"formed at the beginning of the earth."

FIGURE 1–8
Lower Old Red sandstone conglomerate resting unconformably on folded and eroded slates and limestones, Argyll, England.
(Crown Copyright)

Secondary rocks. Consolidated and stratified rocks in which fossils could be found—"formed during the Flood."

Alluvial rocks. Unconsolidated sediments and loosely consolidated material—"formed since the time of Noah's Flood."

The first appearance of a geologic time scale similar to that used today was in 1833 in a textbook by Charles Lyell, *Principles of Geology*. Lyell's classification has a threefold subdivision, Primary (or Paleozoic), Secondary (or Mesozoic), and Tertiary (or Cenozoic), but the meaning given to these terms in 1833 was completely different from the meaning initially assigned to the same terms by Lehmann. Lyell's time scale included the sedimentary rock units identified by William Smith.

THE MODERN GEOLOGIC TIME SCALE

Eras

Prominent unconformities separate Primary, Secondary, and Tertiary groups of Lyell's classification in many places. Primary rocks were found to consist of two parts: (1) unfossiliferous crystalline rocks, and (2) sedimentary rocks characterized by fossils of invertebrates and fishes. The Secondary group was characterized by remains of certain mollusks (cephalopods, ammonites) and

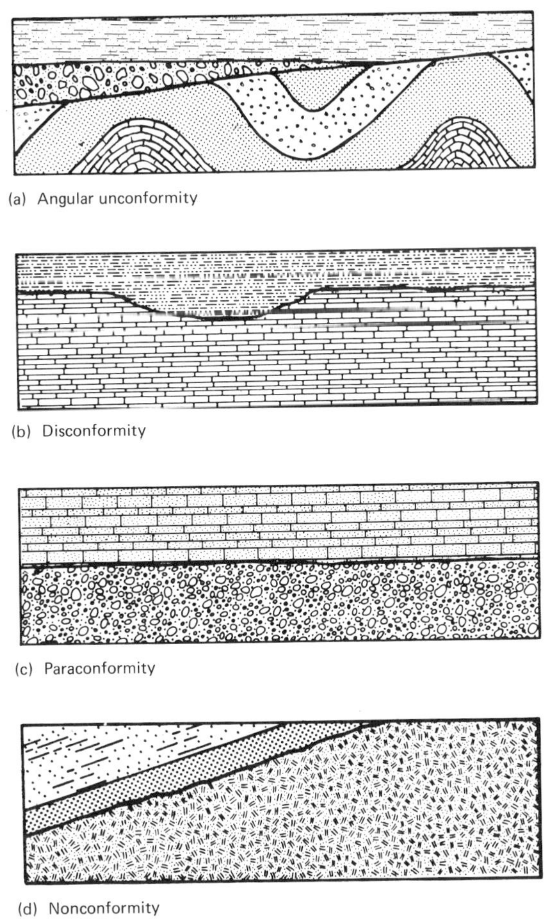

FIGURE 1–9
Unconformities. Four types of breaks in the rock record are depicted. (a) An angular discordance occurs between the strata above and below the unconformity. The older folded strata have been subjected to erosion and were planed off before the younger strata above the erosion surface were deposited. (b and c) Similar types of breaks appear in b and c except that part of the topography was buried in b whereas the contacts between the units in c are almost perfectly parallel. (d) The nonconformity is the break between the older igneous mass and the overlying strata that have been tilted. The igneous mass was intruded, solidified, uplifted, and eroded before the deposition of the sandstone unit over it.

FIGURE 1–10
Correlation between outcrops may be done by matching the rock types in the layers, the sequence of layers, and the fossils present in the layers.

FIGURE 1–11
Trilobites. These fossils of the species *Olenoides serratus* were collected from the Burgess shale, which is of Middle Cambrian age.
(U.S. National Museum)

reptiles, including dinosaurs; the Tertiary group by mollusk and mammal remains. The prominence of unconformities and impressive changes in faunas between these divisions led to their acceptance as *eras* of geologic time. They were given names that reflect the stages of the development of life in each—*Paleozoic* (old life), *Mesozoic* (middle life), and *Cenozoic* (modern life).

Periods

Other unconformities and noteworthy changes in faunas and floras are found within rock units of each era. Many of these breaks can be recognized in different parts of the world and have come to be generally accepted. The largest of these subdivisions of time is called a *period.* Rock units formed within periods are called *systems,* and systems and periods are usually given the same names. These are names of the locality in which the system was first recognized and described. Ages of other rocks that are classified in the same period must be correlated with the rocks of the type section.

The Paleozoic era is divided into seven periods (see Fig. 1–12): Cambrian, Ordovician, Silurian, Devonian, Mississippian, Pennsylvanian, and Permian. The Pennsylvanian and Mississippian are combined into what is known as the Carboniferous period in Europe. The Mesozoic era is divided into three periods: Triassic, Jurassic, and Cretaceous. The Cenozoic era is divided into Tertiary and Quaternary. Younger rock units generally overlie older strata and tend to cover them. Younger units are more widespread, are better exposed, and can be studied in greater detail than older units, and thus they can be subdivided more readily.

The lower part of the Primary series in the classification of Lyell is composed largely of igneous and metamorphic rocks that contain few identifiable fossils. These rocks represent more than three-quarters of the earth's history. It is probable that they include as much as seven-eighths of the record, yet the exact position of the division in relation to overlying divisions is often unclear. This, the largest interval of earth history, is called the *Precambrian.* It began with the formation of the first rocks on earth about 4 to 5 billion years ago and lasted until the fossil animals found in the Cambrian sedimentary units came into existence. These fossils, which serve as guides to the start of the Cambrian, are largely trilobites (Fig. 1–11), and they are advanced invertebrate animals; therefore we can conclude that life had a long history of development in the Precambrian. So far, relatively few of the older Precambrian fossils have been found. Most of the Precambrian fossils are remains of algae, worms, and very similar low-level life forms. This difference between Cambrian and Precambrian fossils is due in part to the great time gap so frequently found between the ages of Cambrian and Precambrian rocks where they are in contact.

Epochs

The periods are subdivided into *epochs* of time and the corresponding rock units are called *stages.* The periods and epochs of the Cenozoic era are shown in Fig. 1–12. Although the Quaternary period, in which we now live, is the shortest of all, it is highly significant because its history is known in great detail and it is the one with which we deal most often in the study of physical geology. (Also see front endpapers for time scale.)

The beginning of the Recent epoch is set at a time about 11,000 years ago when there was an abrupt change in sea level as it began rising, accompanied by warming of the surface layers of the ocean and rapid retreat of the continental

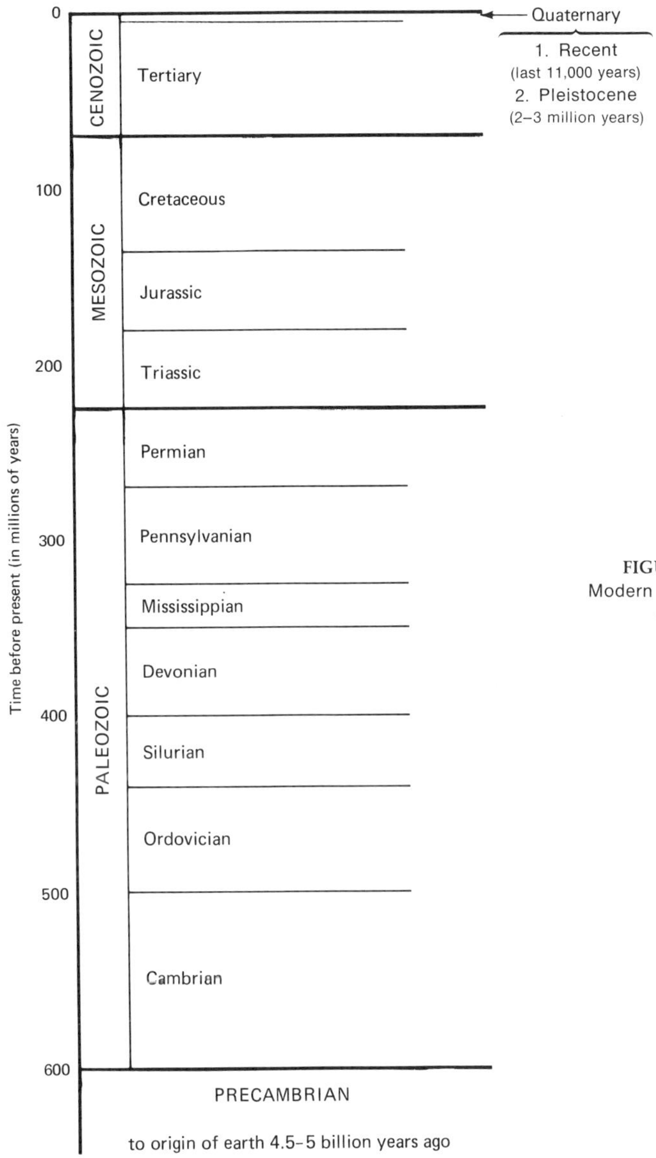

FIGURE 1–12
Modern geologic time scale.

glaciers that occupied the continents in the Northern Hemisphere.

The start of the Pleistocene is still not fixed precisely in absolute dates, but it is selected as the time when the ice ages began, when sea level fell, the surface of the earth cooled, and continental glaciers grew. These glaciers advanced and retreated a number of times during the Pleistocene; thus it may be that the Recent is nothing more than an interglacial interval in the Pleistocene.

CHAPTER OUTLINE

Geology—Dimensions of the Field
- Geological Sciences
- Geophysical Sciences

Geology—Key to Earth's Physical Environment

The Role of Geology in Society

Geology—A Preview

Materials

Gradation
- Weathering
- Degradation
- Igneous Activity and Diastrophism

The Interior of the Earth

Large-Scale Features of the Earth
- Continents and Ocean Basins
  - *Continental shields*
  - *Folded mountain belts*
  - *Continental shelves*
  - *Island arc systems*
  - *Deep oceanic crust*
  - *Midocean ridges*

Continental Drift
- Sea-Floor Spreading

Geologic Concept of Time

Measuring Geologic Time

Determination of Sequence of Events in Rocks

Historical Development of Fundamental Geological Concepts of Time
- Law of Superposition
- Uniformitarianism
  - *Charles Lyell, uniformitarian theorist*
- Catastrophism
- Evidence for Uniformitarianism
  - *Natural processes*
  - *Reconstruction of the Past*
  - *Past environments*

Limitations of Uniformitarianism
- The Role of Minor Catastrophes

The Law of Faunal Succession

Development of a Geologic Time Scale

The Modern Geologic Time Scale
- Eras
- Periods
- Epochs

TERMS

abyssal plains
alluvial rocks
basalt
catastrophism
chemical oceanography
climatology
coastal plain
continental drift
continental shelf
continental shield
crust
degradation
diastrophism
disconformities
epoch
era
flow
geochemistry
geodesy
geologic time scale
geology
geomorphology
geophysics
glacial geology
glaciology
granite
igneous activity
igneous rocks
island arc systems
law of faunal succession
law of superposition
mantle
marine biology
marine geology
metamorphic rocks
metamorphism
meteorology
midocean ridges
mineralogy
minerals
Mohorovičić discontinuity (*M*-discontinuity)
oceanography
paleontology
period
petrography
petrology
physical oceanography
primary rocks
rocks
sea-floor spreading
secondary rocks
sedimentation
seismology
sialic
simatic
stages
stratigraphy
structural geology
systems
terrestrial magnetism
tsunami
uniformitarianism
varved
volcanism
weathering

REVIEW QUESTIONS

1. Give several reasons why the more recent subdivisions of the geologic time scale cover shorter time intervals than the older subdivisions.
2. What accounts for the shape of the topography shown in Fig. 1–4. Explain the causes of the differences you observe in different parts of this area.
3. What basic types of data are used to distinguish the various subdivisions of the earth's crust?
4. Cite several reasons why geologists do not abandon the geologic time scale in favor of an absolute (numerical) scale.
5. Examine Fig. 1–7 and outline the sequence of events indicated by this exposure. The brick pattern indicates limestone, dashes indicate shale, and dots and circles indicate gravel.
6. Early geologists thought the earth was very ancient even before radiometric dating techniques were available. What evidence was available to them?
7. How might you determine that a sedimentary sequence is upside down?
8. Why is the study of ecology so important to stratigraphers?
9. How might you decide the type of environment a fossil of a now-extinct group inhabited if you found the fossil embedded in a rock layer?
10. Explain why the laws of faunal succession and superposition were necessary before a geologic time scale could be constructed.

SUGGESTED READINGS

Albritton, C. C., Jr., ed., 1963, The Fabric of Geology (prepared under the directions of a committee of the Geological Society of America, in commemoration of the Society's 75th anniversary): Reading, Mass., Addison-Wesley, 372 p.

Berry, W. B. N., 1968, Growth of a Prehistoric Time Scale, Based on Organic Evolution: San Francisco, W. H. Freeman, 158 p.

Cattermole, Peter, 1969, The World of Geology: Fleet Street, London, E.C.S., Frederick Muller, Ltd., 174 p.

Croneis, C., and **Krumbein, W. C.,** 1961, Down to Earth, An Introduction to Geology (Phoenix Science Series): Chicago, Ill., University of Chicago Press, 499 p.

Engel, A. E. J., 1969, Time and the earth: American Scientist, v. 57, no. 4, p. 458–483.

Faul, Henry, 1966, Ages of Rocks, Planets, and Stars: New York, McGraw-Hill, 109 p.

Flawn, P. T., 1970, Environmental Geology, Conservation, Land-Use Planning, and Resource Management (Harper's Geoscience Series): New York, Harper & Row, 313 p.

Harbaugh, J. W., 1968, Stratigraphy and Geologic Time (Brown Foundations of Earth Science Series): Dubuque, Iowa, Wm. C. Brown, 113 p.

Hurley, P. M., 1959, How Old Is the Earth? (Science Study Series, Anchor Books): Garden City, N.Y., Doubleday, 160 p.

Leet, L. D., and **Leet, F. J.,** 1961, The World of Geology (McGraw-Hill Paperbacks in Science, Mathematics, and Engineering): New York, McGraw-Hill, 262 p.

2 MINERALS: THE BUILDING BLOCKS OF THE EARTH'S CRUST

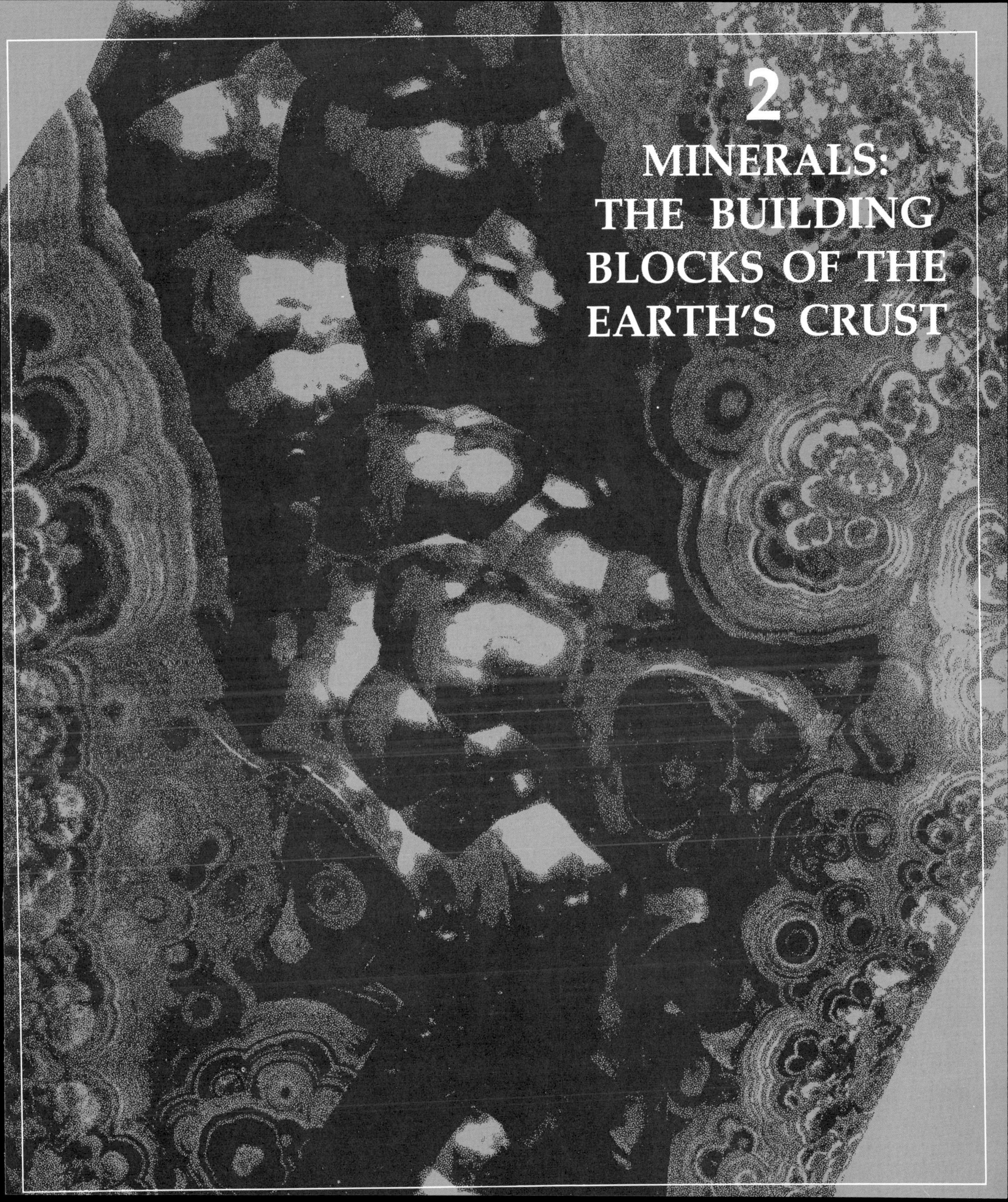

2
MINERALS: THE BUILDING BLOCKS OF THE EARTH'S CRUST

The study of materials is one of the oldest areas of scientific inquiry; yet in spite of exciting modern discoveries many important questions remain. Indeed, rock and soil samples from the moon have recently pointed up the significance of this field. But it was long recognized that to understand the earth, its moon, or the other planets, we must first know the materials of which they are composed, the conditions under which these materials were formed, and the way one material reacts with another. And although the rock and soil exposed at the earth's surface have been carefully scrutinized, the earth's crust remains a subject of intense study. Beyond that, we are only beginning to investigate the materials buried at depths just beneath the earth's crust. These more deeply buried materials have played an important role in processes affecting the crust —processes of mountain-building, of continental drift. And we must not lose sight of the importance of crustal materials to man. They are the source of nutrients that support life, the foundation of our civilization's structures, the raw materials of our industry, a key element of our environment.

Rocks and minerals are composed of atoms; thus it is with atomic structure that we begin our study of the earth.

MODELS OF ATOMIC STRUCTURE

Lord Rutherford (1871–1937) studied atomic physics shortly after the discovery of radioactivity, before its behavior was explained. The three types of radiation from radioactive elements—alpha, beta, and gamma rays—were known, but Rutherford was the first to realize that the phenomenon of radioactivity is caused by the spontaneous disintegration of heavy, unstable atoms. He recognized the potential usefulness of bombarding atoms of various elements with radiation from radioactive materials, particularly alpha (positive) rays. A beam composed of these rays was directed through a small slit and allowed to pass to a fluorescent screen where tiny flashes would appear when the rays hit. The flashes were concentrated in a small point. When a piece of thin foil was placed in their path, the rays were scattered, hitting over all parts of the screen and even being deflected back toward the viewer.

At the time of Rutherford's experiment, the accepted model of the atom was one of negative electrons floating in a positive electric fluid. However, the scattering observed by Rutherford would not be possible in such a fluid because negative electrons in the positive fluid would be so closely intermingled that large deflections of an alpha ray would not occur (Fig. 2–1). That large angles of deflection did occur led Rutherford to postulate that the atom has its positive charge concentrated at the center in a nucleus and that the electrons are at a distance from the nucleus. In order for them to retain this distance, he reasoned, they would have to be moving in orbits at great speed, similar to the way in which the planets of the solar system revolve around the sun. Thus started the modern *planetary model* of the atom. Rutherford's model recognized electrons as carriers of negative charge and postulated the existence of some other particle or particles in the nucleus as carriers of positive charge. The carriers of this positive charge were later recognized as *protons*, each of which carries a charge equal in magnitude to that of an electron.

Bohr Model of Atomic Structure

One of the greatest differences between the atom as visualized by Democritus (460–357 B.C.) and Lucretius (99–55 B.C.) and the atom as analyzed by modern atomic physicists lies in its internal structure. The Greek philosophers thought of atoms as indivisible particles, but they are now known to be composed of electrons, protons, neutrons, and a great many other smaller particles. And unlike the Greeks, we now know the size of atoms. Although they are too small to be seen in microscopes of the highest power, the mass, electrical charge, and other characteristics of atoms and many of their component parts have been successfully measured. If an atom were enlarged 100 million times it would be about the size of a pea, and even at this size the electrons, protons, and neutrons still would not be visible with a high-powered microscope. The atomic weight scale now in use employs the most abundant type of *carbon atom* as its standard. This carbon atom is arbitrarily defined as having a weight of exactly *12 mass units.* All other weights are expressed in terms relative to the weight of *carbon-12,* a stable isotope (p. 37) of carbon.

The *electron* is the smallest of the three fundamental particles. Its mass, 0.00055 mass unit, was determined by experiments of J. J. Thomson, and the value of its negative charge was measured by R. A. Millikan's oil-drop method. The *proton* has a single positive charge, but it is much larger than the electron. It has a mass of 1.00758 mass units, which makes it about 1,800 times as massive as the electron. The *neutron* represents a compounding of a single electron and a single proton. Its net resultant electrical charge is zero. The mass is almost equivalent to the addition of one electron and one proton, or 1.00896 mass units.

Protons, electrons, and neutrons combine in various ways to form complex structures called *atoms.* Atoms formed by the various combinations are stable and cannot be broken down by ordinary chemical methods; these are *elements.* We know of 103 of these atomic configurations, which differ in the number of protons, electrons, and neutrons in each. The number of protons and electrons present ranges from 1 to 103 apiece. Each element possesses unique physical and chemical characteristics. An element's *atomic number* corresponds to both its total number of electrons and the number of protons in its nucleus; chlorine, for example, is element No. 17.

The model of atomic structure that lends itself to easiest explanation is that devised by the Danish physicist, Niels Bohr. His model is an extension of the model devised by Rutherford.

Radiation is given off by atoms when they are excited by high temperature. This radiation consists of certain frequencies that are characteristic of the particular atoms from which it is emitted. The frequencies can be determined by means of line spectra, which have provided a most valuable source of information about atomic structure. It was the line spectrum of hydrogen that Niels Bohr, in 1913, first explained successfully by applying the quantum theory to atomic structure. In the model of the hydrogen atom, a single electron orbits a single proton in the nucleus, the two being held together by an electrical attractive force. The movement of the system is analogous to the movement of a planet around the sun,

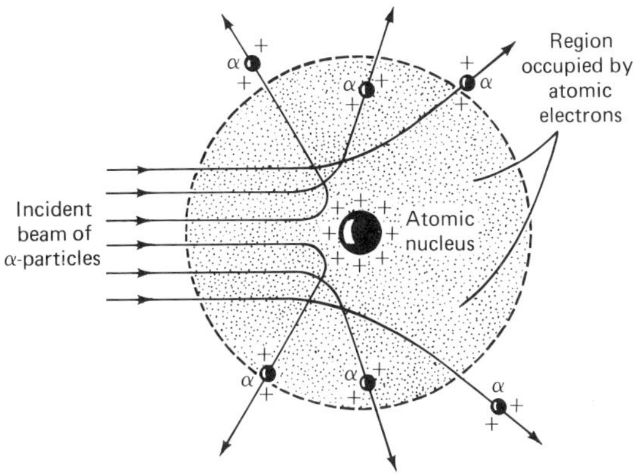

FIGURE 2–1 The discovery of protons was made by directing a beam of α-particles through a foil. Resulting scattering was explained by concentrations of positive charge in the nucleus.

which Newton's Laws of Motion have described so well. If we apply Newton's Laws of Motion to the hydrogen atom, we are led to predict that the orbit is either circular or elliptical and that its size is determined by the energy level of the system. According to electromagnetic theory, the light produced by an electron revolving about a nucleus should have a frequency equal to that of the revolution. Thus the line spectrum of hydrogen should show a continuous spread but instead shows a few specific frequencies. This discrepancy between the observed discrete frequencies and the continuous spectrum predicted by Newton's Laws of Motion was resolved by Bohr when he applied quantum theory to the problem. Planck and Einstein had developed the theory that light energy is emitted and absorbed only in multiples of certain amounts, called *quanta.* Bohr then theorized that the electrons in an atom can exist only in certain *energy states,* the lowest, or normal, state being the *ground state.* The atom can change from one state to another only by the emission or absorption of a specific amount of radiant energy. Bohr also succeeded in developing a method to calculate the energy of the various states and to derive the line spectrum of hydrogen. According to Bohr's conclusions concerning electron motion within atoms, the following rules hold:

1. Electrons are restricted to only a few circular and elliptical orbits around the nucleus; selection of these orbits is carried out according to well-defined rules. The amount

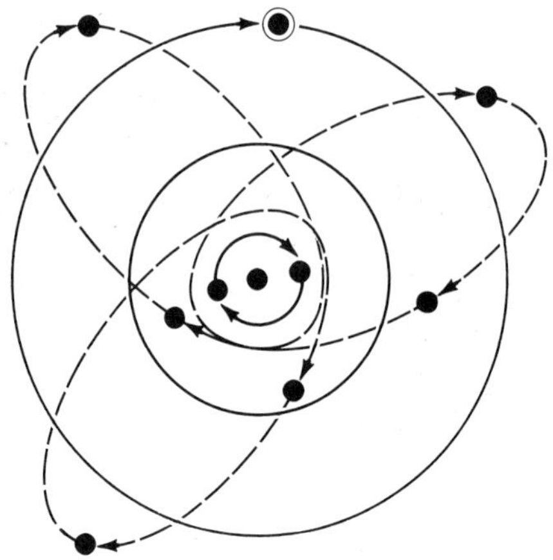

FIGURE 2–2
Bohr's model of an atom.

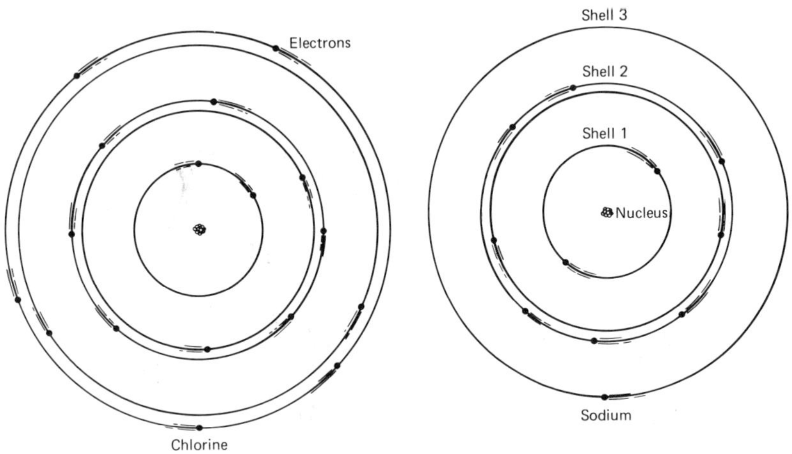

FIGURE 2–3
Hypothetical model of the atomic structure of sodium and chlorine. This model shows the distribution of electrons in various shells.

of energy an electron has is related to the distance of its orbit from the nucleus. The closer the electron is to the nucleus, the lower its energy.

2. While an electron is moving in one of these orbits, no electromagnetic waves are given off.
3. Under certain conditions an electron may jump from one orbit to another. When it jumps from an outer orbit to one closer to the nucleus, it does so with the emission of radiant energy. The longer the jump, the greater the amount of energy emitted.

The Bohr model of hydrogen was so successful that his approach was quickly applied to more complex atoms.

According to the Bohr model, a positive charge is concentrated in the nucleus, about which negative electrons move in circular and elliptical paths at relatively great distances from the nucleus (Fig. 2–2). Electrons occupy elliptical orbits only at certain specific distances from the nucleus. The number of electrons in each of these orbits is strictly limited. For example, no more than two electrons may occupy the innermost orbit, which is called the *first shell.* When atoms have more than two electrons, the additional electrons are located in other shells. The second shell may be considered to have two subshells (Fig. 2–3). The inner subshell may contain up to two electrons, the outer subshell up to six, giving a total of eight electrons in the second shell. The element that has eight electrons in its second shell is neon, one of the *inert gases;* neon has a total of ten electrons, two in its first shell plus eight in the two subshells of its second shell. The third shell may have up to three subshells, which are occupied by a maximum of two, six, and ten electrons, respectively. The fourth shell may have up to four subshells, occupied by a maximum of two, six, ten, and fourteen electrons, respectively.

Certain of the elements, known as the *noble* or *inert gases,* are remarkably stable, to the extent that, until recently, no compounds of these elements were known. They are helium, element No. 2 ($_{2}$He); neon, No. 10 ($_{10}$Ne); argon, No. 18 ($_{18}$Ar); krypton, No. 36 ($_{36}$Kr); xenon, No. 54 ($_{54}$Xe); and radon, No. 86 ($_{86}$Ra). Because these gases are inert, the grouping of the electrons in these elements must be very stable. With the exception of helium with two electrons, the most stable electron configuration contains eight electrons in the outermost orbit. Elements that have this configuration do not lose electrons from their outer orbits. Other elements tend either to lose or gain electrons until there are eight electrons in the outer shell.

Ions

The presence of eight electrons in the outer shell, or orbit, of an atom makes it particularly stable. Atoms have a tendency to attain this stable form.

When an atom loses an electron it has a net positive charge. Similarly, if it should lose two electrons, it will have an excess of two positive charges. Charged atoms are called *ions.* Ions may be either positive or negative, depending on whether they gain or lose electrons. Which atoms are most likely to become positive ions and which negative ions is easily predicted. If a given atom has eight electrons in the next to the outermost shell and one in the outermost shell, it will lose that one electron more readily than it will gain seven additional ones to achieve the stability of eight in the outer shell. Likewise, an atom with two electrons in its outer shell is more likely to lose these than it is to gain six. On the other hand, an atom with six or seven electrons in its outer orbit is more likely to gain one or two than it is to lose six or seven.

FIGURE 2–4
Quartz crystals.
(U.S. National Museum)

MINERALS

Geologists deal with elements or compounds in the form of *minerals* and *rocks.* A *mineral* is defined as a naturally occurring element or compound formed by inorganic natural processes; it generally has a definite chemical composition and a characteristic atomic structure that is expressed in an external crystalline form as well as in other physical properties (see p. 44). Most *rocks* are aggregates of minerals. These are the raw materials with which we must become thoroughly acquainted if we seek a basic understanding of this and other planets.

FIGURE 2–5
Crystal of microcline feldspar.
(U.S. National Museum)

Crystals

The internal orderly arrangement of atoms in minerals is illustrated by the symmetry and geometrically perfect external form of a crystal. A *crystal* is a homogeneous substance having surfaces that formed into smooth planes (Figs. 2–4 and 2–5). If a mineral in the process of growth or formation is unrestricted by the shape of the space in which it can grow, it may achieve the perfect shape of a crystal. Such crystals are thus found where they have freedom of growth, as in a cavity or along an open fracture. Some crystals form from solutions or rock melts. In the late stages of a melt's crystallization, before solidification is complete, many minerals are forced to crystallize in

FIGURE 2–6
Dendritic growth. (Scale in inches.)

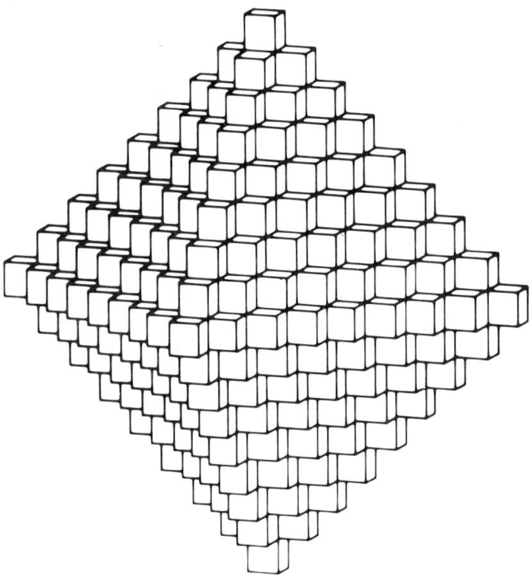

FIGURE 2–7
Haüy's model of crystal structure of an octahedral crystal composed of small cubes.

restricted spaces. Such minerals still have an *orderly internal atomic structure* and therefore are said to be *crystalline* although no crystals of the minerals may be seen. When the individual minerals of a rock can be distinguished only with the aid of a microscope, the mineral is said to be *microcrystalline,* and when matter is detectably crystalline but individual crystals are not distinguishable, it is *cryptocrystalline.* The crystalline nature of cryptocrystalline minerals may be determined by means of electron probes or X-ray analysis. Materials lacking an ordered internal atomic arrangement (e.g., glass) are *amorphous.*

CRYSTALLOGRAPHY[1]

History of Crystallography

Crystals have always been admired for their beauty, but they were not subjected to scientific scrutiny until the seventeenth century. Although they were known much earlier, their nature was very poorly understood. Even so eminent a scientist as William Gilbert, who is famous for his work with magnetism, suggested in his book *De Magneti* (1600) that gems are made of water that is somehow "concreted." Nicolaus Steno (1669) was able to establish that mineral crystals grow in a manner different from that of living organisms. At this point crystallography became a science. The history of crystallography may be divided into three stages of development. (1) Research in the seventeenth and eighteenth centuries was devoted to observation, classification, description, and speculation; (2) in the nineteenth century, the theoretical foundations of crystallography on a mathematical basis were laid; and (3) in the twentieth century, theory has been verified by experiments, and new observations and speculations have developed.

The scientists of the seventeenth century had great difficulty in understanding the composition of crystals and how they grew, since they lacked knowledge of chemical elements and how these are bound together in compounds. But nevertheless the alchemists observed that certain chemical substances combined to form crystals, and the artificial production of crystals was begun. Robert Hooke (1665) was one of the earliest to report *dendritic* crystal growths (skeletal crystals), which resemble the trunk and branches of a tree (Fig. 2–6) and may occur on the surfaces of some metals and on fracture planes in rock masses. The alchemists concluded that metal crystals formed directly from the liquid melt as it cooled, but that mineral crystals formed by some other process.

[1] This section follows a treatment given by Earle Ryba, 1968.

Gradually the notion became established that crystals grew by the addition of primary clusters of particles that attached together systematically. This was a simple extension of the contemporary observations on the growth of living tissues. Johann Kepler (1611) and René Descartes (1596–1650) suggested that the primary clusters for snowflakes were spheres packed together closely but systematically to outline the flake. This idea, used by Robert Hooke (1665) to account for the shapes of mineral crystals, is almost identical with the modern view that minerals are composed of close-packed spherical atoms. The sphere-packing theory was later modified to hold that primary clusters are arranged into polyhedra that possess shapes related to the shape of the crystal, a view supported by Robert Boyle (1667) and Sir Isaac Newton (1730).

The first attempts to classify crystalline materials according to their crystal form were made in 1772 by J. B. L. Romé de L'Isle. He listed 110 basic crystal forms and classified many crystals. The mechanics of using his classification was greatly simplified when the contact goniometer was invented by Arnould Carangeot. It is essentially a simple protractor with a rotating arm that can be used to measure the angle between crystal faces.

René Just Haüy (1784–1822), one of the leading crystallographers, developed a technique for determining the primary shapes of crystals by breaking them down mechanically until no changes in shape could be seen. He concluded that these small fragments represented the "primary clusters," or building blocks, for the crystal. These building blocks could be formed by fitting together only three basic shapes—the parallelopiped, the triangular prism, and the tetrahedron, and these building blocks, when placed together and arranged into layers, make up all crystals (Fig. 2–7). The steplike patter on a crystal's surface was thought to be too small to be visible. This idea was widely accepted for many years.

Crystal Systems

A dramatic change in crystallography was introduced by Christian Weiss, who revised Haüy's hypothesis. Weiss found that he could relate each crystal to a set of three properly oriented reference axes. The cube provides a simple example (Fig. 2–8). The cube has three mutually perpendicular axes and the faces of the cube intersect the axes at the same distance from the center. By 1815, Weiss had successfully identified the cubic (later isometric), tetragonal, hexagonal, and orthorhombic crystal systems. The monoclinic and triclinic systems were recognized by Friedrich Mohs in 1825. A crystal can be classified as belonging to one of these systems by the lengths of the axes and the angles between them. Weiss noticed that crystals do exhibit a certain symmetry as they are rotated about these axes, but he did not develop this idea. He did, however, point out that the position of a crystal face can be specified by giving the intercept distance on the crystal axes in terms of unit lengths of the axes.

Point and Space Groups

Weiss' recognition of crystal axes led to several very important developments in crystallography. One of the first was the concept of three-dimensional space lattices in which the crystallographic axes provide the reference lines. Following Haüy's construction of polyhedral building blocks, it was found useful to represent each polyhedron by a point. The configuration of these points in space is a *space lattice*. These space lattices can be classified into crystal systems using the same criteria (length and angular relationships of the axes) established by Weiss for crystals.

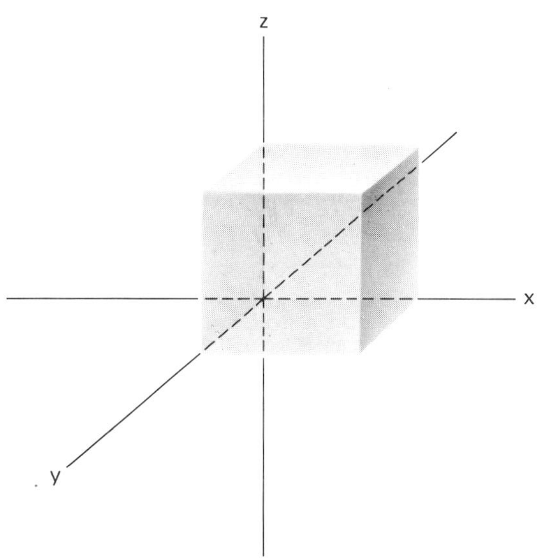

FIGURE 2–8
Cube showing coordinate axes. $ox = oy = oz$.

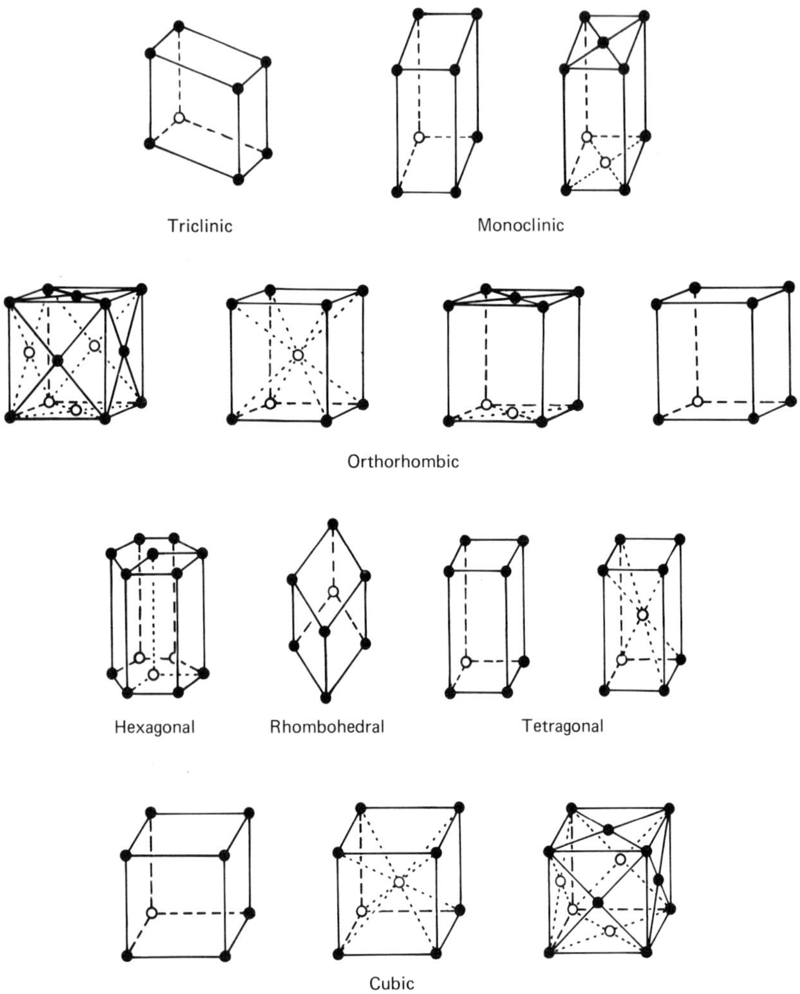

FIGURE 2–9 *(above)* Unit cells of Bravais' fourteen-space lattices.

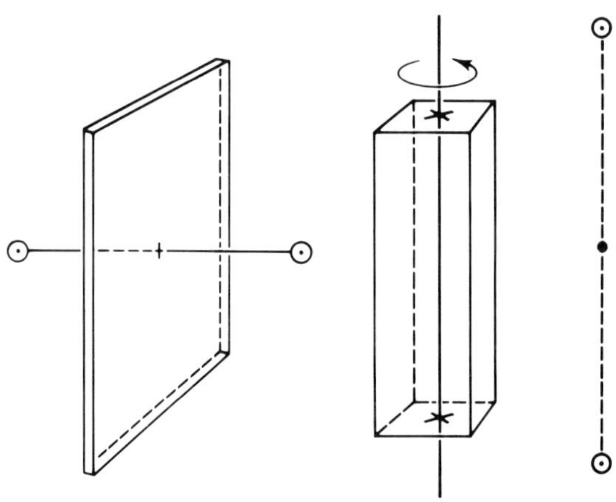

FIGURE 2–10 Three types of symmetry relationships. Mirror image or plane of symmetry, left; an axis of fourfold symmetry, center; and symmetry about a point.

Auguste Bravais (1848) demonstrated that there are only fourteen ways to make up a space lattice, as illustrated in Fig. 2–9.

Crystal Symmetry

Weiss had noticed that crystals exhibit certain types of *symmetry,* and this concept was greatly extended by Johann Hessel (1830). We are all familiar with the type of symmetry represented by a mirror plane: all points on one side of the plane have corresponding mirror images on the other side of the plane. Other symmetry operations can be related to an axis of rotation and a center of symmetry. An axis of rotation can be envisioned if we imagine a solid body, such as a cube or hexagonal prism, with an axis as shown in Fig. 2–10. As the solid is rotated about the axis, the shape of the body will have the same appearance in several positions during one complete rotation—the cube in four positions, the hexagonal prism in six, etc. Thus, in these shapes, two-, three-, four- and sixfold axes are found.

A *center of symmetry* is present when points on the surface of a solid have similar points on the opposite side of the crystal at an equal distance from the center along a straight line.

Symmetry is defined by the number of *symmetry planes,* the number of two-, three-, four- and sixfold axes of symmetry, and the presence or absence of a center of symmetry. The nine symmetry planes of a cube are illustrated in Fig. 2–11.

Hessel concluded that there are only thirty-two different combinations of the symmetry operations, each called a *class,* and all crystals must fall into one of these. Thus any given crystal will belong to one of seven crystal systems, depending on the length and angular relationship of the crystallographic axes, to one of thirty-two classes or point groups, depending on its symmetry, and to one of the fourteen Bravais lattices, depending on the unit cell (polyhedra, building blocks) of which the crystal is composed.

X-Ray Crystallography

When Dr. Max von Laue (1912) was working with diffraction gratings, to separate beams of radiant energy into component wave lengths, and he needed a finer grating than was available at that time, he turned to natural crystals. Crystals were thought to contain closely spaced layers of atoms that would constitute a three-dimensional grating. He hypothesized that the waves would be diffracted within a crystal in such a way that separate beams of radiant energy would emanate at specific angles. His experiments consisted of placing a crystal in the path of a beam of X-rays. The X-rays passed through the crystal and their paths were recorded on a photographic plate placed behind the crystal. When developed, the plate showed a dark spot in the center, where the X-rays had traveled directly through the crystal, and a large number of small dark spots arranged around the center in a regular geometrical pattern (Fig. 2–12), located where X-rays reflected from various layers were reinforced. Thus Laue proved that crystals do possess an internal orderly arrangement of atoms.

W. H. Bragg and W. L. Bragg, a father and son who worked at the Royal Institution of Great Britain, set to work with the X-ray methods devised by Laue and began to determine the internal structure of minerals by X-ray methods, using the idea that the X-rays are reflected from planes of atoms in the crystal, as shown in Fig. 2–13. Atoms are shown to be regularly arranged in layers of constant thickness. Radiant energy directed on such a material will be scattered by the layers so that waves will be reinforced in some directions and canceled out in others. The use of X-ray techniques developed rapidly and has proven to be a powerful tool for analysis of the internal structure of crystals.

When a crystal is bombarded by X-rays, the rays interact with the planes of atoms, causing intense radiation to be reflected out in certain directions. These directions are indicated by the following relations, called Bragg's Law:

$$n\lambda = 2d \sin \theta$$

where n is a small integer, the order of diffraction; λ is the wave length of the X-ray; d is the spacing between planes of atoms; and θ is half the angle between the incident and emitted radiation.

The X-rays used are produced by bombarding a metal target with high-voltage electrons. The amount of energy the X-rays have is different for various metals. By using different targets in the X-ray tubes, variations in the wave-length distributions of X-rays coming from the tubes can be produced. This allows a greater range of experimental studies of crystal structures. When the wave length of X-rays used is known, the angle θ is measured experimentally, and the d-spacing

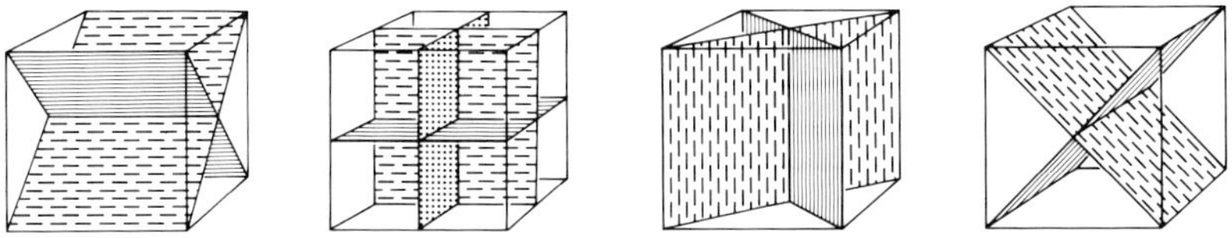

FIGURE 2–11
Planes of symmetry of a cube.

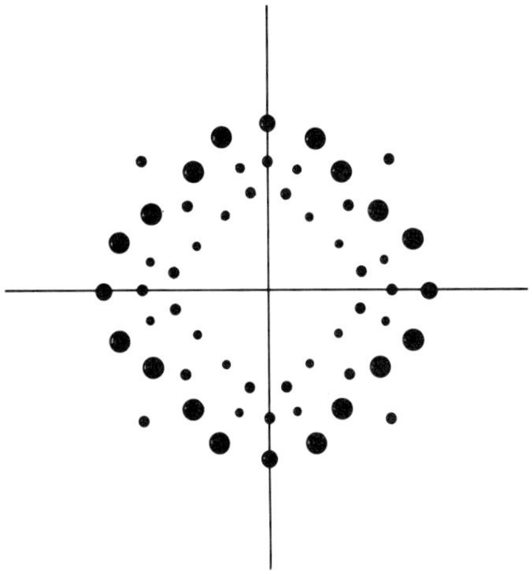

FIGURE 2–12
Pattern formed by X-rays that passed through a piece of halite.

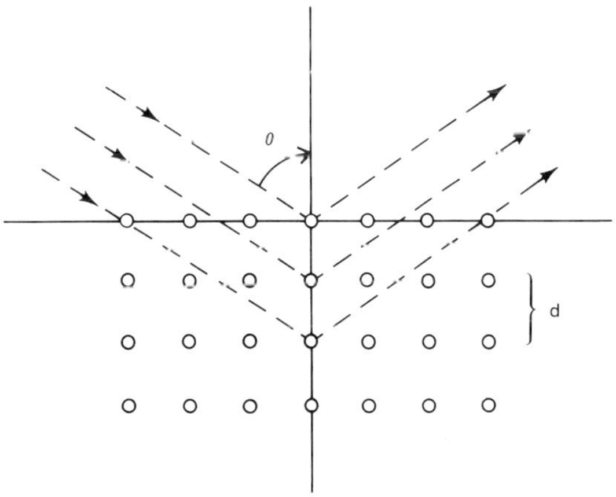

FIGURE 2–13
Schematic illustration of reflection of X-rays on passing into a crystalline solid. The circles are atoms.

Element	Symbol	Ion	Radius of ion	Coordination number
Oxygen	O	O^{-2}	1.40 A	
Chlorine	Cl	Cl^{-}	1.81 A	
Sulfur	S	S^{-2}	1.85 A	
Sodium	Na	Na^{+}	0.97 A	6 or 8
Magnesium	Mg	Mg^{+2}	0.66 A	6
Aluminum	Al	Al^{+3}	0.51 A	4 or 6
Silicon	Si	Si^{+4}	0.42 A	4
Potassium	K	K^{+}	1.35 A	12
Calcium	Ca	Ca^{+2}	0.99 A	8

TABLE 2–1 *(above)* The radii of the ions are important in determining the way in which the ions will pack together. Given are the ionic radii of several common ions, the ionic charge, and the predicted number of oxygen ions that can be packed around them, the coordination number.

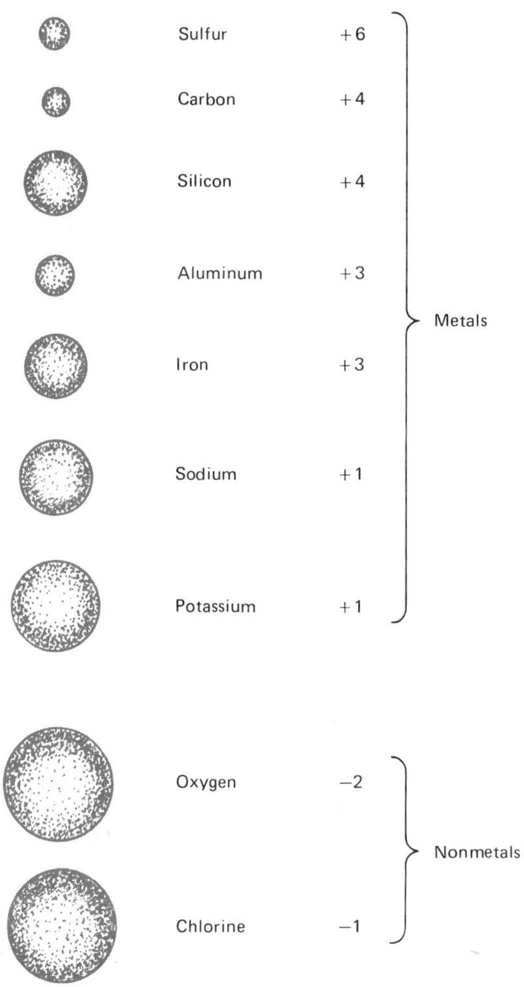

FIGURE 2–14 Relative sizes of some common ions and the normal ionic charge.

can be calculated from the above equation. Every mineral has a different *d*-spacing which may be used to identify the mineral.

A technique for X-ray microanalysis with a device called an *electron probe* has come into widespread use in recent years. A beam of electrons is focused on a point on the surface of a sample. Electron bombardment causes the sample to emit X-rays. The spectrum of X-rays emitted is characteristic of the elements present and the intensity of various wave lengths is a measure of the concentration of the elements. The electron probe makes it possible to determine chemical composition within very small areas (e.g., 5 microns), which allows the probe to be used to study zoning within individual crystals, contact zones between crystals or phases, and other small-scale chemical variations.

MINERALS AS CHEMICAL COMPOUNDS

Compounds are formed by the combination of elements through organic and inorganic processes. With the exception of some of the "native" metals, such as gold, silver, platinum, and copper, and a few other substances, such as sulfur and pure carbon, few elements occur in pure, uncombined form. Most minerals and rocks are compounds of two or more elements held together by electrical forces. Compounds have physical and chemical properties that differ from those of their component elements. The bonds holding elements together in the compounds are of four types—ionic, covalent, metallic, and Van der Waals bonds.

Ionic Bonds

Ions form when an atom gains or loses one or more electrons and becomes charged. The net charge is the difference between the number of protons in the nucleus and the number of electrons in the orbits of the ion. Ions obey the laws of static electricity and are attracted to ions with an opposite charge and repelled by those of a like charge. If a large number of ions, some positive and some negative, are placed in solution in a liquid through which they can move freely, they will tend to form an ionic compound. The negative ion will attract positive ions, and they in turn will attract more negative ions. In this manner a solid crystal may be built up.

The shape that this crystal will assume is dependent on the size of the ions of which it is composed. Atoms and ions, visualized as spherical or nearly spherical, differ in diameter. Ions with more shells or electrons may have diameters many times greater than those of elements having a lower number. The sizes of some of the common ions (see Table 2–1) are illustrated in Fig. 2–14. The forces acting on the ions tend to pull them as close together as possible. The number of positive ions, a *cation,* that can fit around one negative ion, an *anion,* is dependent on the relative sizes of the two. Therefore, the most important factor in determining the shape of the configuration is the size of the ions involved. The most common arrangements of ions are depicted in Table 2–2.

The ratio of the radius of a cation to that of the anion with which it combines is a convenient way to express the relative sizes of ions. This ratio is significant, too, because it serves as a guide to possible configurations a given cation/anion combination may have. The number of anions that are in contact with a cation in any given packing arrangement is called the *coordination number.* Anions and cations of any relative radius can be packed together if they are packed in coordination of 1 or 2, but three anions can fit around one cation on the corners of an equilateral triangle only if the radius ratio falls in a certain range (0.155–0.225). Coordination of 4 to form a tetrahedron, of 6 to form an octahedron, of 8 to form a cube, and of 12 to form a cube are possible only within certain limits of radius ratios (see Table 2–2). Thus knowledge of the ionic radius of any cation and anion should enable one to predict probable coordination and configuration of the two.

One of the most common ionic compounds is sodium chloride, the mineral halite. As sodium is element No. 11, it will have two electrons in the first shell, eight in the second, and one in the third. Therefore, it will have a tendency to lose the single electron in the outer shell in order to obtain a stable outer shell and become a sodium ion with a positive charge (Na^+). Chlorine, $_{17}Cl$, has two electrons in its first shell, eight in its second, and seven in its third shell. It tends to gain one electron in the third shell to become a chlorine ion with a negative charge (Cl^-). The Na^+ ion and the Cl^- ion, having different charges, are attracted toward one another and pack as closely as possible. Their sizes are such that each negative ion has six positive ions adjacent to it in the configuration shown in Fig. 2–15. The resulting crystal is cubic in shape.

TABLE 2–2

Coordination number	Radius ratio R cation/R anion	Configuration of atoms	
1	0–∝	side by side	
2	0–∝	on two sides	
3	0.15–0.22	corners of an equilateral triangle	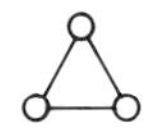
4	0.22–0.41	corners of a tetrahedron	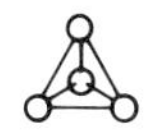
6	0.41–0.73	corners of an octahedron	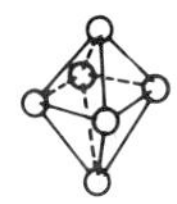
8	0.73–1	corners of a cube	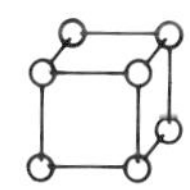
12	1	center of cube edges	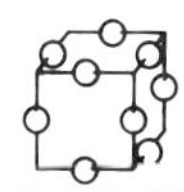

FIGURE 2–15
Packing of sodium and chlorine ions in the ionic compound sodium chloride, NaCl. The structure is enlarged at right. This is a face-centered cubic structure.

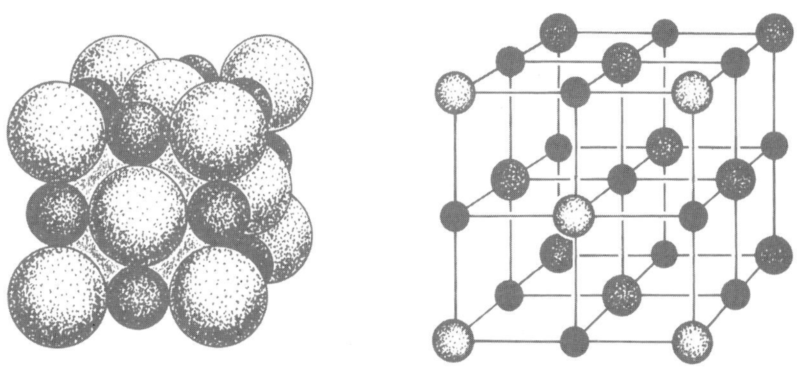

Most metals, like sodium, tend to lose their outer electrons, whereas nonmetals tend to gain electrons. If the metal involved loses only one electron but the nonmetal picks up two, the ionic compound probably will contain two metal ions to each nonmetal ion (e.g., Na_2S). Many ionic compounds are complex structures in which electrons from metals are given up to groups of nonmetals (e.g., CO_3^{-2}, SO_4^{-2}) (Fig. 2–16), rather than to single atoms.

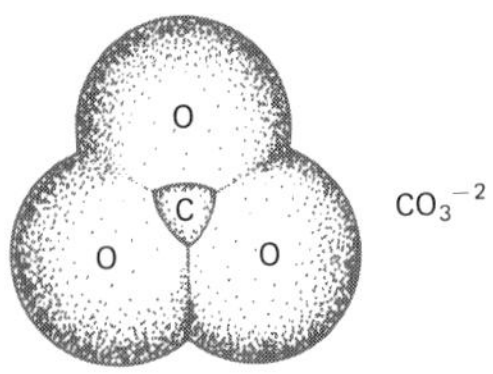

FIGURE 2–16
Structure of the carbonate ion.

The amount of energy needed to remove an electron from an atom is called the *ionization energy* of that atom. The lower this energy level, the easier it is for the atom to give up an electron. The alkali metals [Group I in the Periodic Table (Appendix B)—Li, Na, K, Rb, and Cs] form positive ions readily. Ionic bonds thus can be expected between two atoms when one has a low ionization energy and the other a high electron affinity.

Covalent Bonds

A second type of bond is formed when elements simply share electrons. No transfer of electrons from one atom to another occurs and the shared electrons tend to remain between the two atoms. It is thus possible for each of the atoms to achieve the stability of having eight electrons in its outer orbit by sharing those of one or more other atoms. Diamond (Fig. 2–18), essentially a single large molecule of carbon atoms, is a good example of a mineral with covalent bonding.

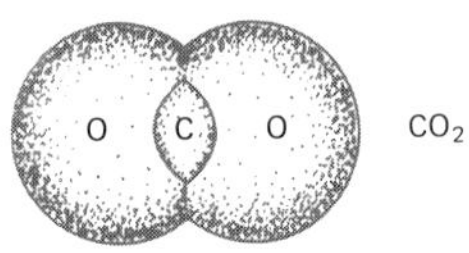

FIGURE 2–17
Structure of carbon dioxide.

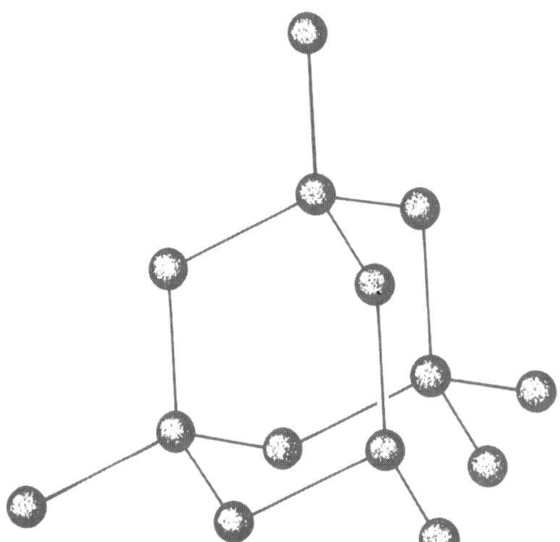

FIGURE 2–18
Schematic arrangement of carbon atoms in diamond.

Methane, marsh gas, which is formed in swamps from the decay of organic matter, exhibits covalent bonding. Methane is a compound of carbon and hydrogen. Hydrogen ($_1H$) has one electron in its orbit. Carbon ($_6C$) has two electrons in its first orbit and four in its outer orbit. A stable compound is formed when the carbon is able to share the electrons of four hydrogen atoms. Thus methane is expressed as CH_4.

```
    H
    |
H—C—H
    |
    H
```

Another common compound held together by covalent bonds is water, H_2O. Hydrogen has one electron and oxygen ($_8O$) needs two electrons in its outer orbit to have eight. Thus one oxygen atom combines with two hydrogen atoms to form the compound H_2O.

The distinction between ionic and covalent bonds lies in whether the electrons are shared by two atoms or held mainly by one atom. Bonding in some compounds is not wholly ionic or covalent but is of a mixed character.

Some covalent compounds contain molecules that behave somewhat like small magnets; these are called *polar covalent* compounds. Water is one such molecule. Its magnetic qualities are caused by the way the two hydrogen atoms are attracted to the oxygen atom (Fig. 2–19). When the electron density on one side of the molecule is increased, the molecule acts as though one end is slightly positive and the other slightly negative. This *polar nature* of the water molecule is the reason it is such an effective solvent.

Consider a grain of salt or another ionic compound placed in water. Water molecules acting as small magnets become oriented with their positive ends toward negative ions in the solid and their negative ends toward positive ions. The effect of the attractions set up between water molecules and ions is to pull the ions out of the solid structure as soon as enough molecules are oriented to counterbalance the electrical forces within the crystal. The amount of solution that takes place is limited, however, for when a large number of ions have gone into solution there will not be enough free water molecules left to attract ions of the solid effectively. Hydrochloric acid (HCl) is also a polar covalent compound.

Covalent bonds are important in many complex minerals. Silicon and oxygen are held together by covalent bonds in silicon-oxygen tetrahedra, described later, and aluminum, sulfur, and carbon also form covalent groups.

Metallic Bonding

Metallic bonding, which is characteristic of pure metals, is different from ionic and covalent bonding in that electrons are shared throughout the crystal lattice. The atoms, all of the same size, are packed tightly together so that each has twelve others around it. Twelve spheres of a single size can always be packed around a thirteenth sphere of the same size (coordination number 12) regardless of what that size is. The atoms of the metals lose their outer electrons easily. The electrons in this case can move through the crystal lattice, making it possible for metals to conduct electricity. Although the electrons are in a sense shared by the whole crystal structure, they should not be thought of as completely free.

Because of the nature of metallic bonding, metals are weaker than perfect ionic or covalent minerals. Since atoms can be rearranged without the metals losing strength, and since they do not have strong directional preferences in bond strength, metals can be deformed easily and tend to be ductile.

Van der Waals Bonds

The weakest bonds are those postulated nearly one hundred years ago by the Dutch physicist Van der Waals. These forces exist even among the otherwise electrically neutral inert gases. The forces arise from the constantly changing positions of the electrons in the atoms, which result in charge imbalances. At any given moment these forces may be thought of as somewhat like the water molecule, which has a *polar character* as a result of the grouping of the hydrogens on one side of the oxygen. Unlike in the polar molecules, however, the position of the electrical imbalances changes from moment to moment in the case of Van der Waals forces.

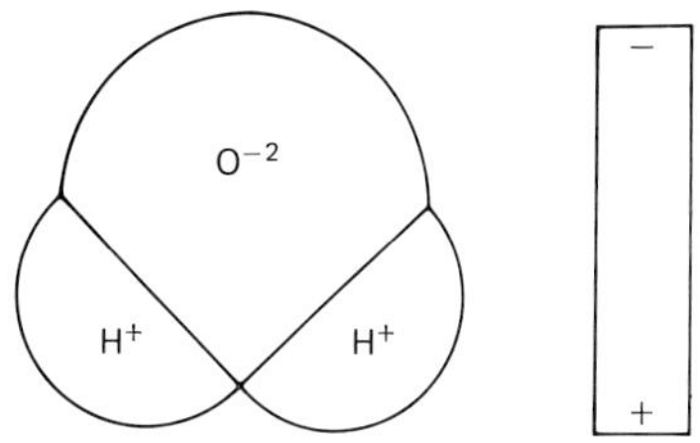

FIGURE 2–19
Water molecule. Because the hydrogen atoms are close together the molecule acts somewhat like a small bar magnet.

These bonds are important in compounds whose lattices are composed of whole molecules instead of atoms or ions. Crystals of the inert gases are examples. Such crystals usually have low melting and boiling points and low mechanical strength. Van der Waals forces also act in many more complex structures to bind ionic or covalently bonded groups together. For example, the carbon in graphite is bound into sheets by covalent linkages, but the sheets are held together by Van der Waals bonds. These bonds are rare among common minerals, however.

ISOTOPES

Isotopes are useful tools in the solution of geological problems as diverse as determining the age of rocks or identifying and tracing the movement of water in the deep oceans. Usually each element has a nucleus composed of equal numbers of neutrons and protons; however, elements also have alternate forms in which there are different numbers of neutrons in the nucleus. Forms having chemical and physical properties so similar that they are almost identical to the element are called *isotopes* of that element. Thus an isotope has the same atomic number as its element but a different mass. A mass spectroscope is used to distinguish isotopes. Some isotopes are unstable and break down spontaneously by the emission of radiant energy. Such isotopes are said to be *radioactive*. The energy they radiate is composed of protons and neutrons, which compose the nucleus, and electrons. Through the emission of these particles, the isotope disintegrates to daughter products. The rate of disintegration is constant for each isotope. Disintegration rates are such that a constant amount of time is required for one-half the mass of the isotope to break down. This amount of time is called the *half life* of the isotope. As a result of the breakdown of an iso-

Element	Symbol	Atomic number	Percentage of volume of earth's crust	Percentage of weight of earth's crust
Oxygen	O	8	91.97	46.60
Silicon	Si	14	.80	27.72
Aluminum	Al	13	.77	8.13
Iron	Fe	26	.68	5.00
Magnesium	Mg	12	.56	2.09
Calcium	Ca	20	1.48	3.63
Sodium	Na	11	1.60	2.83
Potassium	K	19	2.14	2.59

TABLE 2–3 *(above)* Composition of the earth. Only the more abundant elements are listed.
(from Mason, 1952)

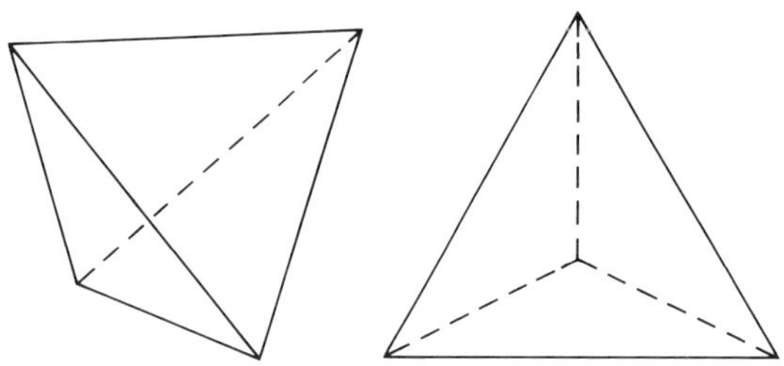

FIGURE 2–20 A tetrahedron seen from two different angles.

tope, new isotopes or elements may form. The new ones may be radioactive also and break down again, but eventually a stable isotope or element is formed. Because the half life of a radioactive isotope is a constant, and because a stable element is formed as an end product, it is possible to use radioactive isotopes as geologic clocks. This is possible only if the half life (see Appendix B) is known, if the steps in the radioactive breakdown and the half lives of each intermediate isotope are known, if decay has not proceeded too far, and if the stable end product is present in measurable amounts.

MINERAL GROUPS

Because so many different minerals have been identified, about 2,000, classification is necessary. Minerals are classified on the basis of their chemical composition as well as their crystal symmetry. In the chemical method, a negative ion or group of ions is used to define the class to which the mineral belongs. Negative ions are used because they are bigger and control properties of the mineral. The principal classes according to Dana's classification are listed with examples of each.

Native elements: gold, silver, copper, platinum, arsenic, sulfur, diamonds.

Sulfides: sulfur compounds, i.e., galena (PbS), sphalerite (ZnS), pyrite (FeS_2).

Oxides: compounds of oxygen, i.e., ice (H_2O), hematite (Fe_2O_3), corundum (Al_2O_3), magnetite (Fe_3O_4).

Halides: compounds of chlorine or fluorine, i.e., halite (NaCl), fluorite (CaF_2).

Carbonates: compounds of carbonate (CO_3), i.e., calcite ($CaCO_3$), dolomite ($CaMg(CO_3)_2$).

Sulfates: compounds of sulfate (SO_4), i.e., gypsum ($CaSO_4 \cdot 2H_2O$).

Phosphates: compounds of phosphate (PO_4), i.e., apatite ($Ca_5(PO_4)_3(F)$).

Silicates: compounds of silicon and oxygen, i.e., quartz (SiO_2), feldspar ($KAlSi_3O_8$), kaolinite ($Al_2(Si_2O_5)(OH)_4$).

IMPORTANT ROCK-FORMING MINERALS

Estimates of the chemical composition of the earth's crust (see Table 2–3) have been made on the basis of the approximate volumes of different rock types and their compositions. The estimated distribution of rock types on the continents is reasonably accurate and can be considered to include a depth of a few miles on the basis of deep drilling and seismic data.

Only a few of the earth's known elements make up a large proportion of the crust, both by weight and by volume. Oxygen and silicon are far more abundant than any other elements, with aluminum, iron, calcium, sodium, potassium, and magnesium following in sizable quantity; these eight elements account for about 98 percent of the crust by weight.

These abundances alone would lead us to expect compounds of silicon and oxygen to be important constituents of rocks. The silicates,

FIGURE 2–21
Various ways of representing the silicon-oxygen tetrahedral grouping. The silicon atom is located in the center as shown in the enlarged view at left. The four oxygens occupy the corners of a tetrahedron.

compounds of silicon and oxygen, by far the most important group of rock-forming minerals, deserve our special attention. The metal ions most abundant in the silicates are Al, Fe, Ca, Mg, K, and Na, as also might be expected. These are the main constituents of such minerals as feldspar, mica, clay, amphibole, pyroxene, olivine, and garnet.

The chief structural component of the unit cells among the silicates is the silicon-oxygen tetrahedron. Silicon ($_{14}Si^{+4}$) has an ion with four positive charges. It is intermediate in character between metals and nonmetals. Oxygen ($_{8}O^{-2}$), on the other hand, has an ion with two negative charges. The size of the silicon ion is such that four oxygen ions can cluster around it to form a four-sided tetrahedron (Figs. 2–20, 2–21, and 2–22). Each silicon and oxygen are bound together by a single electron-pair bond, leaving each of the four oxygens with a single valence electron free to be bound to another atom. Thus the SiO_4^{-4} tetrahedron has a net negative charge of minus four.

Each of the four oxygens is linked to another silicon atom in a continuous three-dimensional network in quartz, SiO_2. This network is electrically neutral. Because the bonds are strong in every direction in this structure, they produce a mineral that (1) is hard, with no tendency to break in any preferred directions, and (2) has a high melting point, 1,470°C.

The silicon-oxygen tetrahedron is the basic building block of all the silicates. The way these tetrahedra are related in the crystal structure has become the basis for classification of silicates. The classes included are the following (see Fig. 2–23):

1. **Three-dimensional network.** Described above for quartz, the simplest of these silicates (e.g., quartz, feldspar, zeolites).
2. **Independent tetrahedral groups $(SiO_4)^{-4}$.** These tetrahedra are not directly linked together but are bonded to metal ions as in Mg_2SiO_4 and Fe_2SiO_4 (e.g., olivine, garnet, aluminum silicates).
3. **Double tetrahedra.** Two tetrahedra linked together by sharing only one oxygen, as in $(Si_2O_7)^{-6}$ (e.g., epidote).
4. **Ring structure.** Produced when the oxygens of adjacent tetrahedra are shared in such a way that a ring of three $(Si_3O_9)^{-6}$, four $(Si_4O_{12})^{-8}$, or six $(Si_6O_{18})^{-12}$ tetrahedra is produced (e.g., beryl, tourmaline).
5. **Chain structures.** Single chains are produced by linking tetrahedra in such a way that the ratio of Si:O is 1:3 (e.g., pyroxene). Double chains consist of two cross-linked chains with a Si:O ratio of 4:11 (e.g., amphibole).

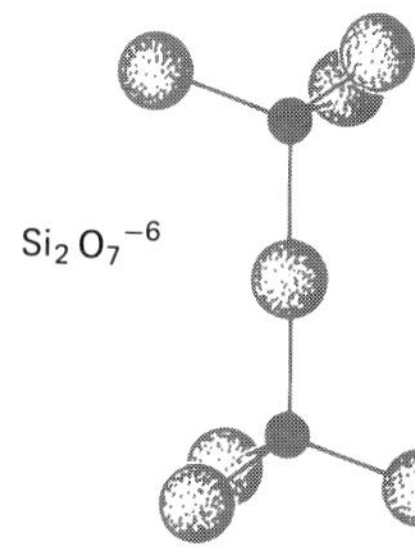

FIGURE 2–22
Two silicon-oxygen tetrahedra linked.

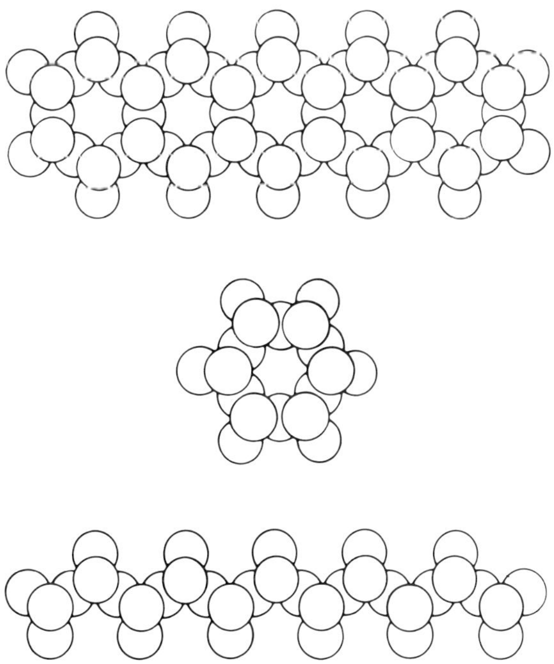

FIGURE 2–23

(a)
A double-chain structure of silicon-oxygen tetrahedra.

(b)
A ring structure formed by silicon-oxygen tetrahedra.

(c)
A chain formed by sharing of one oxygen atom by two tetrahedra.

6. **Sheet structures.** These are produced by the sharing of each of the three oxygens with another tetrahedron to produce an extensive sheet (e.g., mica, clay).

Metal ions fit into the silicate structures to satisfy the net electrical charges on the above tetrahedral grouping arrangements and to bind the sheets, rings, chains, and so forth, together. Water molecules, hydroxyl groups (OH), and additional oxygen atoms also are found with some of the silicate tetrahedral groupings.

Aluminum is of particular interest because of its abundance and because it can replace silicon in the SiO_4 tetrahedra. Note that aluminum ($_{13}Al^{+3}$) has three positive charges and its size is such that it can have either four or six oxygens grouped around it. As aluminum can exist in either 4- or 6-coordination groups, it can substitute for Si^{+4}, Mg^{+2}, or Fe^{+2}. $(AlO_4)^{-5}$ has a tetrahedral shape whereas $(AlO_6)^{-9}$ is octahedral.

Expressing the Chemical Formulae of Silicates

The chemical formulae of minerals may be written in several ways. For example, silicate compositions may be expressed as oxides:

$CaO \cdot SiO_2$	instead of $CaSiO_3$;
$CaO \cdot Al_2O_3 \cdot 2SiO_2$	instead of $CaAl_2Si_2O_8$;
$2MgO \cdot SiO_2$	instead of Mg_2SiO_4.

Unlike in the carbonates or sulfates where CO_3 and SO_4 are repeatedly used and are the naturally occurring forms in solution, silica is infrequent in solution. It is likely to combine in several ways with the same metal ions as, for example, $MgSiO_3$, Mg_2SiO_4, or Na_2SiO_3, Na_4SiO_4, $Na_4Si_2O_6$. The reactions between silicates are often most easily interpreted as interactions between oxides of the components.

When parentheses are used in a chemical formula, they indicate that either or both of the elements within the parentheses may be present in the crystal structure.

Olivine

Olivine is one of the minerals with independent silicon-oxygen tetrahedra. Aluminum does not often substitute for silicon in this type of structure. The general formula for olivine is $(Mg,Fe)_2SiO_4$, which indicates that either magnesium or iron or a mixture of the two may be present in the structure. In other words, Mg^{+2} and Fe^{+2} ions can be freely substituted for one another. This is an example of a *continuous reaction series,* the end members of which are the following:

forsterite------------------------fayalite
Mg_2SiO_4-----$(Mg_{50}Fe_{50})SiO_4$-----Fe_2SiO_4

The physical properties in such a series change from one end to the other.

Feldspar Group

The feldspar group includes the following minerals:

potash feldspar
- orthoclase (monoclinic) — $KAlSi_3O_8$
- microcline (triclinic) — $KAlSi_3O_8$

plagioclase feldspar (triclinic)
- general formula—$(Ca,Na)(Al_{1-2}Si_{2-3}O_8)$
- end member—anorthite $(CaAl_2Si_2O_8)$
- oligoclase
- andesine
- labradorite
- bytownite
- end member— albite $(NaAlSi_3O_8)$

The feldspars constitute about 60 percent of all silicates and thus rank among the most important rock-forming minerals. Aluminum substitutes for silicon in the feldspar structure. The structure is a continuous crystal network composed of SiO_4 and AlO_4 tetrahedra with positive ions of sodium, calcium, and potassium located in the spaces. The potash feldspars have several different forms, but the plagioclase feldspars constitute an *isomorphous solid-solution series,* all members of which are triclinic. The physical properties of the plagioclase series exhibit a continuous change as the composition changes. In hand specimens, the main distinction between the potash and plagioclase feldspars is the presence of distinctive gratinglike lines, called *twinning,* on the plagioclase (Fig. 2–24). The various plagioclase feldspars can be distinguished by density or other precise measurements of physical or optical properties.

In the plagioclase solid-solution series (Fig. 2–25), calcium ions, Ca^{+2}, can be substituted for sodium ions, Na^{+}, because they have about the same radius, but aluminum, Al^{+3}, also must be substituted for silicon, Si^{+4}, at the same time in order to maintain a neutral charge on the network. The top curve shows the temperature at which a plagioclase melt of any given composition (e.g., a) will begin to form its first crystals. These crystals will have a composition (e.g., b) shown on the lower curve. As crystals at b are formed, the composition of the melt changes and will shift toward the left. A continuous reaction takes place between the melt and crystals as crystallization continues. Eventually if all crystals undergo reaction completely, the final solid rock will be composed of crystals that have the composition of the initial melt. If reactions are not completed, as might happen if a crystal were removed from the melt, zoned plagioclase may result.

FIGURE 2–24
Twinning striations in labradorite feldspar.

Pyroxenes and Amphiboles

The pyroxenes are the most common members of the *single-chain* silicates. They closely resemble the double-chain amphiboles in physical and chemical properties as well as in crystal structure.

The orthorhombic pyroxenes are called orthopyroxenes, and their general formula is $(Mg,Fe)SiO_3$. The end members of the series are the following:

enstatite-------------------hypersthene
$MgSiO_3$--------------------$(Mg,Fe)SiO_3$

Monoclinic pyroxenes are called clinopyroxenes. The most abundant member of the group is augite, which has the general formula $(Ca,Mg,Fe)SiO_3$.

The chains of silica tetrahedra in both kinds of pyroxenes are linked by ions (usually Ca,Mg,Fe) in 6-coordination. Both pyroxenes and amphiboles have good cleavages in two directions, outlining prisms, because the bonds between the chains are weaker than the bonds within the chains. The cleavages in the two have different angles of intersection (Fig. 2–26), which is the most important key to differentiating them in hand specimens. The most common amphibole is hornblende, $NaCa_2(Mg,Fe,Al)_5(Si,Al)_8O_{22}(OH)_2$.

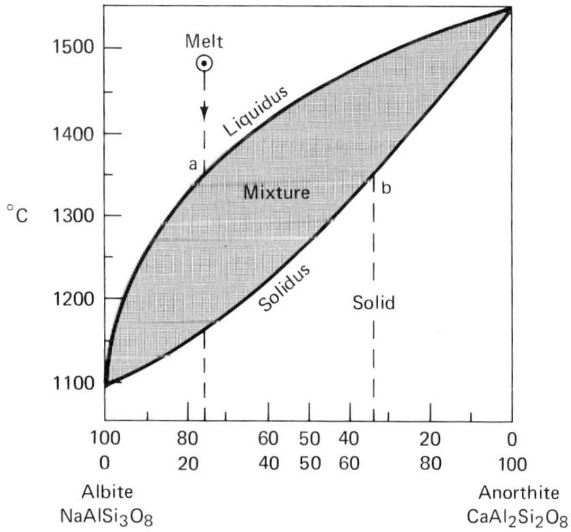

FIGURE 2–25
Phase diagram for the solid-solution series typical of the plagioclase feldspars.

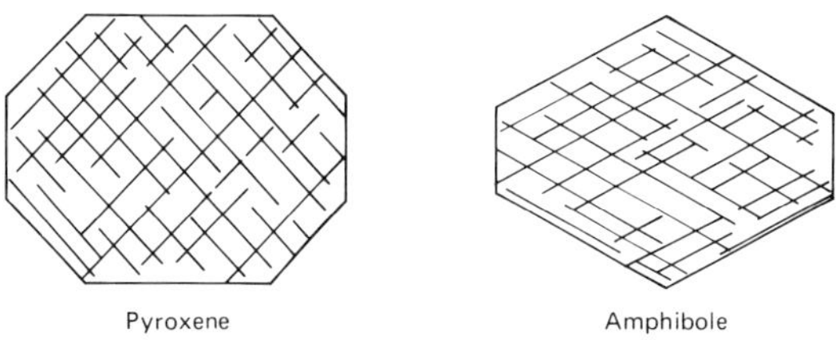

FIGURE 2–26
Cleavages of pyroxene and amphibole.

Mica

The single perfect cleavage common to micas makes them one of the most distinctive of all mineral groups. This cleavage arises from the sheet arrangement of the silica tetrahedra (Fig. 2–27).

Examples include the following:

muscovite: $KAl_2AlSi_3O_{10}(OH,F)_2$;

biotites: $K(Mg_3Fe)_3AlSi_3O_{10}(OH,F)_2$.

RADIOMETRIC AGE DETERMINATION

A number of naturally occurring radioactive isotopes are known to undergo spontaneous disintegration. The original isotope, which we will call the parent nuclide, breaks down to form daughter products which may or may not be radioactive and radiate energy. If the daughter product is radioactive, it disintegrates, eventually yielding a stable isotope. The rate of decay is such that the time required for one-half of the radioactive nuclide to disintegrate is constant; this time is designated the *half life,* and its measurement is now well known for most isotopes. The rate of disintegration is, in fact, so precisely known that the ratio of the daughter to the parent isotope can be used to determine the length of time the process of disintegration has been going on. Among the isotopes (see Table 2–4) used for age determination are uranium-lead, lead-thorium, lead-lead, potassium-argon, rubidium-strontium, ionium-thorium, and carbon-14.

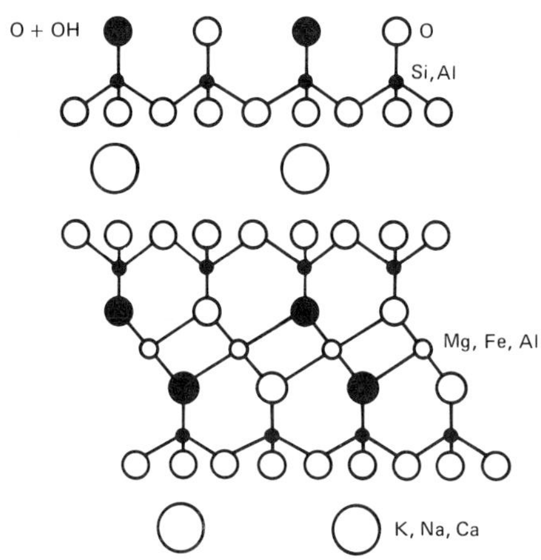

FIGURE 2–27
Side view of the atomic structure of mica.

Reliable ages can be determined if the following conditions hold:

1. The parent element content of the mineral has been changed only by radioactive decay.
2. The steps in the decay process are correctly understood.
3. The parent element was separated completely from daughter products at the time the mineral formed.
4. All of the daughter products have been retained since the mineral formed.
5. The amounts of both parent and daughter are sufficient in the sample to permit the necessary processing.
6. No errors are introduced by the preparation of samples and in the analytical procedure.

Decay of Uranium to Lead

Uranium-238, like most other radioactive substances, is transformed into a stable substance, lead-206, only after passing through a number of other radioactive isotopes. The half life of uranium-238 is 4.56×10^9 years. One million grams of uranium-238 will produce $\frac{1}{7600}$ gram of lead-206 per year.

During a period of diastrophism, magma may be intruded into the deformed rocks. Let us suppose that one of those minerals that crystallizes from the magma contains uranium-238. As soon as it has crystallized, nuclei of uranium-238 atoms begin to disintegrate. A certain percentage of the atoms disintegrate every minute. After 4,560 million years, half of them will have disintegrated. As the uranium-238 disintegrates, helium and lead-206, both of which are stable and undergo no further change, are formed as end products of decay. Thus the ratio of helium and lead-206 to the amount of uranium-238 bears a definite relationship to the age of the mineral in which it is found and, therefore, to the time of its formation. In our example, the rock was formed during a

Radioactive nuclide	Half life (in years)	Range (in years)	Kind of sample
Uranium-Lead U^{238}—Pb^{206}	4.5×10^9	10 million +	Uraninite, pitchblende, zircon minerals that occur in igneous and metamorphic rocks
U^{235}—Pb^{207}	0.7×10^9	10 million +	Same as above
Rubidium-Strontium Rb^{87}—Sr^{87}	4.7×10^{10}	10 million +	Mica, microcline feldspar, and whole rock in metamorphic rocks Glauconite in sediments
Potassium-Argon K^{40}—Ar^{40}	1.3×10^9	100,000 +	Mica, feldspar, volcanic rock Glauconite and sylvite in sediments
C^{14}	5710	0–50,000	Wood, charcoal, shells, water, grain, bone, peat

TABLE 2–4
Principal methods used in geochronology.

period of deformation; thus the age of the rock is the age of the deformation.

Lead-207 is the stable daughter product of uranium-235, which is also present in such minerals as uraninite and pitchblende. Since both U-235 and U-238 occur together, age for a sample can be determined by two methods, allowing a good check. The age equations used are:

$$T_{238} = 6.50 \times 10^9 \ln\left(1 + \frac{\text{Pb-206}}{\text{U-238}}\right)$$

$$T_{235} - 1.03 \times 10^9 \ln\left(1 + \frac{\text{Pb-207}}{\text{U-235}}\right)$$

where T = age in millions of years.

Potassium-Argon Dating

Because potassium is present in several of the most abundant minerals on earth, notably the micas, clay, and feldspar, the time it takes potassium to decay to the inert gas argon is now one of the most widely used standards for radiometric age determination. The formula for age is:

$$T = 1885 \ln\left(1 + 9.10 \frac{\text{Ar-40}}{\text{K-40}}\right)$$

The major problem with this method is that there is a loss of argon by diffusion at high temperatures. Some argon is lost from glauconite at 50°C, from biotite at 150°C, and from muscovite at 200°C. Thus heating after the minerals crystallize may cause dates to be wrong.

Carbon-14

Well known for its application to archeological findings, this method of dating depends on the presence of radioactive carbon, carbon-14, in organic matter (Fig. 2–28). Carbon-14 is formed in the upper atmosphere when nitrogen is bombarded by cosmic rays. Nitrogen, when hit by a cosmic ray, emits a proton and becomes carbon-14. Carbon-14 combines with oxygen to form carbon dioxide, which is absorbed by all living matter. When a plant or animal dies, it stops absorbing carbon-14, and carbon-14 present in the organism begins to decay, forming nitrogen again. The half life of carbon-14 is about 5,710 years. In order to date the remains of a plant or animal, the amount of carbon-14 in the organic matter is measured and compared with the amount found in living matter. The ratio of the carbon-14 content in the dead and living matter is a measure of the length of time carbon-14 in the dead matter has been decaying, and therefore of its age. For example, if exactly half as much carbon-14 is found in a piece of wood that has been buried in an old river deposit as in a living tree, then the age of the old wood can be estimated. Assuming that the old wood had the same concentration of carbon-14 in it when it was living as the present-day tree has, then one-half has decayed. Thus the age is equal to the half life of carbon-14, or 5,710 years.

Although radiometric methods of dating are the best we have for determining absolute ages of rocks, there are certain limitations to their use-

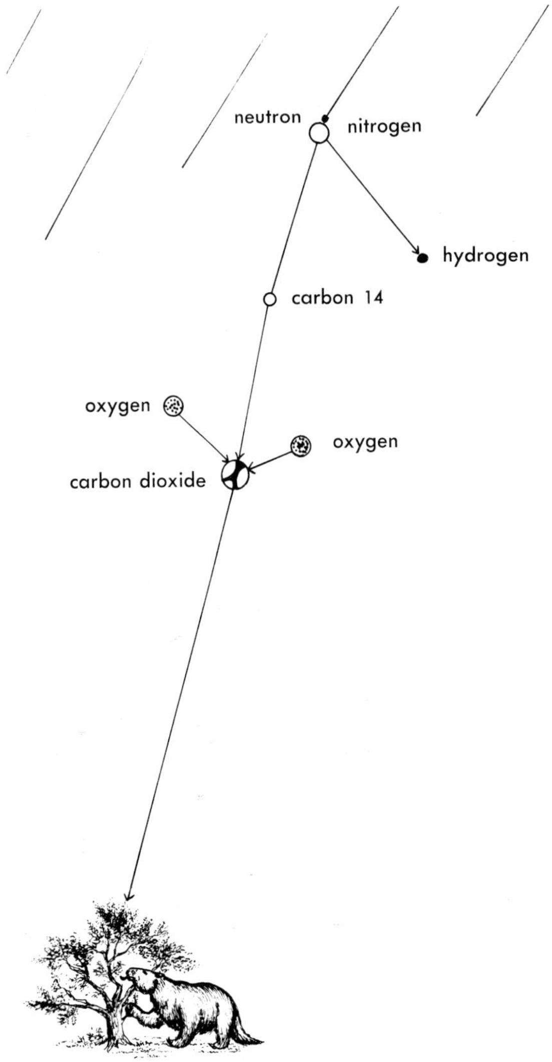

FIGURE 2–28
Carbon-14.
(See p. 43)

fulness—limitations implicit in the mode of occurrence and nature of the radioactive isotopes.

1. The half life of carbon-14 is so short that after about 50,000 years too little remains to measure it accurately. Therefore it is of no value in dating older forms of organic matter.
2. Carbon-14 is confined to organic matter; thus it cannot be used to date rocks that do not contain organic matter.
3. Uranium-238 is largely confined to igneous rocks, particularly pegmatites. Potassium-argon dating is useful in dating micas of metamorphic rocks. But incidence of uranium is not sufficiently widespread to solve more than a small fraction of all dating problems. Micas are largely confined to igneous and metamorphic rocks, so there is little opportunity to date the most commonly found rocks, those of sedimentary origin, by these means.
4. Where uranium is found in igneous rocks and dated, we learn the age of the igneous rock. But if that igneous body has intruded a sedimentary rock unit, we have only the minimum age for the unit.
5. All radioactive dating methods are time-consuming, difficult, and expensive. They require special laboratory equipment and cannot be used in the field.
6. Radioactive decay can be interrupted by recrystallization of the rock. This amounts to resetting the "radioactive clocks."
7. Radiometric methods of age determination are based on the assumption of a closed system, that is, that neither parent nor daughter nuclides are lost. However, some of the daughter nuclides are gases that are readily lost through natural processes such as decomposition or disintegration of the rock. Losses may also occur in the process of preparing the sample. If part of the parent element is lost, the determined age of the specimen is greater than it should be; if the daughter product is lost, the age is less.

PHYSICAL PROPERTIES OF MINERALS

Specific gravity. The density of a mineral is the ratio of its weight to that of an equal volume of water. Graphite is about twice as dense as water; that is, two cubic centimeters of water would weigh the same as one cubic centimeter of graphite. The specific gravity of graphite is therefore 2. Some idea of the range of values found for the specific gravity of various minerals is gained from comparing graphite, 2, with hematite, the most commonly used ore of iron, 5.2, and gold, 19.

The density of a mineral is determined by the weight, size, and crystal structure (packing) of the chemical components of the mineral. For example, magnesium olivine, Mg_2SiO_4, has a specific gravity of 3.22, but iron olivine, Fe_2SiO_4, has a specific gravity of 4.41.

Cleavage. Some minerals split along definite smooth planes; this property is referred to as *cleavage* (see Figs. 2–29, 2–30, and 2–31). These planes are a simple indication of the difference in bond strength between atoms in the crystal. If bonds are strong between atoms in one plane and relatively weak between those in other planes, the mineral will break along the weaker planes. In some minerals differences in bond strength are great; in others, there is little difference. Cleavage planes are always parallel to a crystal face or a possible crystal face. Because they differ in several ways in different minerals, they are very useful in mineral identification.

1. *In number.* There may be one or as many as six directions of cleavage.
2. *In angles between them.* Two may be at right angles or oblique to one another; three may be mutually perpendicular, or all may intersect at acute angles.
3. *In degree of development.* Some minerals break to form very smooth cleavage surfaces, in which case the cleavage is said to be perfect; others that do not break so easily have imperfect cleavage. Still others may be almost impossible to break along cleavage.

Fracture. It is sometimes helpful to note the way in which a mineral has *fractured.* Some break into splinters; others fracture smoothly and evenly. Some break in rough, irregular surfaces; and a few, such as quartz, break with a smooth, curved fracture known as a *conchoidal fracture.*

Luster. The *luster* of a mineral is the appearance of its surface in reflected light. Lusters fall into two large groups, metallic and nonmetallic. Luster depends on the structure of the surface of the mineral, its transparency, and the way light is reflected or refracted from it. There is great variation in the nonmetallic lusters, and there is no substitute for direct observation in learning to

FIGURE 2–29 (*left*) Gypsum. The crystal variety on the left showing cleavages is called selenite; that on the right is satin spar.

FIGURE 2–30 (*below*) Cleavage rhombs of calcite. (Scale in inches.)

FIGURE 2–31 (*bottom*) Fragments of halite showing cubic cleavage.

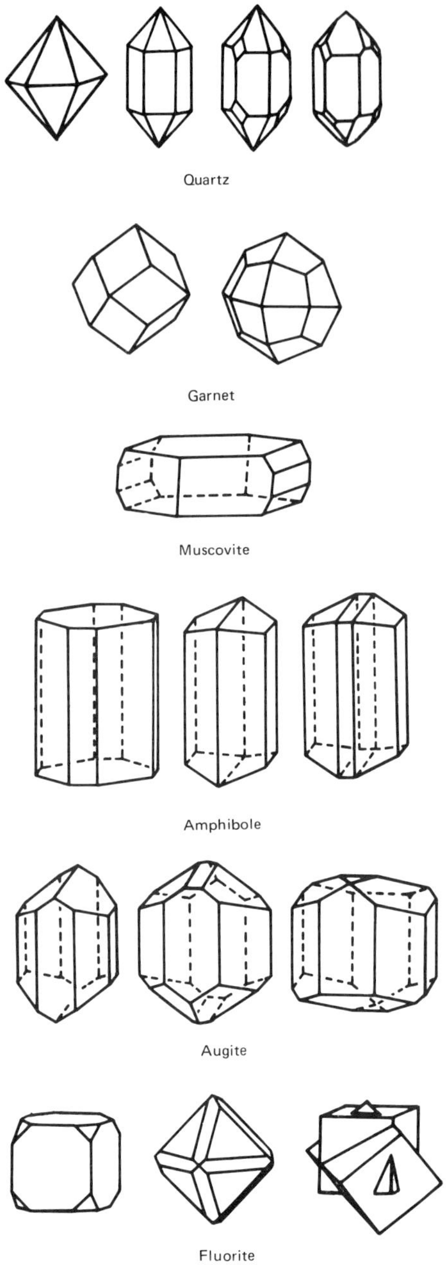

FIGURE 2–32
Crystal shapes of some typical crystals of quartz, garnet, muscovite, amphibole, augite, and fluorite.

identify these. The most common are the following:

> Adamantine: brilliant luster of diamond.
> Vitreous: luster of glass.
> Resinous: luster of yellow resins.
> Greasy: luster of oily glass.
> Pearly: luster of pearl.
> Silky: like silk, usually resulting from fibrous structures.

Color. Of all physical properties, color is frequently the most impressive and least definitive. Color in nonmetallic minerals may be due to impurities. Very small quantities of some elements are sufficient to give large masses of other minerals bright colors. It is not uncommon for a single mineral to occur in as many as five or six different colors. Do not depend on color in identification of most nonmetallic minerals.

Streak. The color of the powder of a mineral is called its *streak.* Although the mineral may occur in several different colors, the color of the powder will always be the same. Streak is tested by rubbing a piece of the mineral on an unglazed porcelain plate, or crushing a fragment.

Hardness. Hardness in the sense used here does not mean breaking strength, but the ability of one mineral to scratch another. A scale of ten relatively common minerals is used as a basis for hardness comparisons. The scale is nonlinear (i.e., No. 5 is not five times harder than No. 1; actually it is many times harder).

Mohs' Hardness Scale

1. Talc
2. Gypsum (fingernail)
3. Calcite (a copper penny)
4. Fluorite
5. Apatite (a steel knife) (plate glass)
6. Feldspar (a file)
7. Quartz
8. Topaz
9. Corundum
10. Diamond

A mineral with a hardness of 6.5 will scratch feldspar, but it will not scratch quartz. In addition, it can be scratched by quartz, but not by feldspar. The hardness of an unknown mineral may be found by scratching it with minerals or other materials of known hardness.

Optical Properties

Variations in optical properties of minerals are widely used as means of identification by geologists. Light is bent, refracted, as it passes from one medium (e.g., air) into another (e.g., mineral). According to Snell's Law, the ratio of the sine of the angle of incident light (i) to the sine of the angle of the refracted light (R) is a constant, N, called the index of refraction. It can be shown that this is equal to the ratio of the velocity (V) of light in one medium to that in the other.

$$\frac{\sin i}{\sin R} = N = \frac{V \text{ air}}{V \text{ mineral}}$$

The index of refraction is related to the crystal structure and composition of the mineral or substance. The velocity of propagation is constant in all directions through some materials, for example, those that are isometric and noncrystalline. The velocity varies in different directions through all other substances, which are referred to as optically *anisotropic*. Light in anisotropic materials may be thought of as splitting into two rays that vibrate at right angles to one another and have different indices of refraction. When light with one of these directions of vibration is removed the remaining light is said to be *plane polarized*.

These optical properties are not often used in hand-specimen identification, but optical calcite can be used to demonstrate the splitting of the light rays through the crystal. Double images can be seen through calcite.

Luminescence. Some minerals exhibit fluorescence when placed under an ultraviolet light. These minerals absorb ultraviolet light and emit visible light, light of longer wave length. Short- and long-wave ultraviolet light both produce fluorescence, but often in different minerals. Unfortunately, fluorescence is not often diagnostic because it is produced by even small amounts of minor or trace elements in the host crystal. *Phosphorescence* is a somewhat related phenomenon through which absorbed radiant energy is released in the form of light after the excitation is past.

Radioactivity

Radiation from radioactive minerals containing uranium or thorium can be detected readily by Geiger counters and scintillometers, which are now widely available. Radioactivity plus other physical properties are often diagnostic, and some radioactive minerals may be distinguished by the intensity of the radiation.

SHAPES OF CRYSTALS AND AGGREGATES

Only rarely is a perfect crystal found. Aggregates of crystals grown together irregularly or more massive forms without crystal faces are much more common. Although crystals with equally developed faces and perfect or near-perfect symmetry are unusual, the angles between adjacent crystal faces are constant, and these interfacial angles provide a reliable key to identification. Single crystal shapes (Figs. 2–32 and 2–33), are commonly described as columnar, equant or stubby, tabular, bladed, acicular, or fibrous.

Columnar. A crystal that is nearly equidimensional in cross section but elongate in one direction.

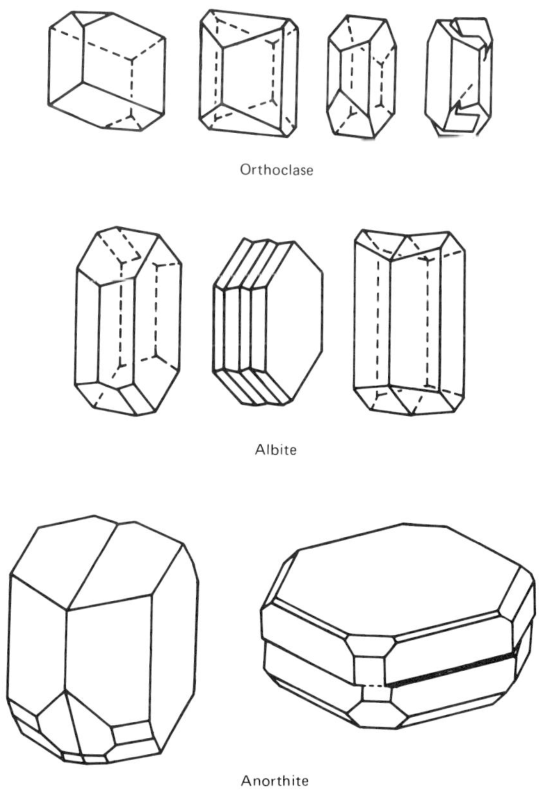

FIGURE 2–33
Crystal shapes of feldspars.

FIGURE 2–34
Natural asbestos, variety chrysotile.
(U.S. National Museum)

Equant (stubby). A crystal that is nearly equidimensional.

Tabular. A crystal that has two dimensions appreciably longer than the third.

Bladed. A tabular crystal with bladelike form.

Acicular. Needlelike crystals.

Fibrous. Individual crystals similar to fibers.

Crystal aggregates may consist of individual crystals of any of the above types, but other special patterns and shapes are also formed as a result of the way individual crystals have grown together. Some of these are diagnostic keys, typical of certain mineral species. Examples of terms used follow:

Granular Aggregates (coarse, medium, or fine-grained). Terms applied when the individual minerals are nearly equidimensional and show no strong preferred orientations.

Radiating. A term applied to crystals that diverge from a common center.

Micaceous (lamellar, foliated). Minerals that consist of thin flat plates (e.g., mica). This is usually due to the presence of a single perfect cleavage.

Stalactitic. The elongate calcium carbonate deposits that hang from cave roofs initially as straw-shaped projections that later become more tapering and rounded at the end are called stalactites. Other minerals may assume similar form.

Dendritic. This form has a branchlike outline. Dendrites are skeletal crystals that occur as growths on surfaces such as a fracture plane.

Colloform. Spherical or rounded masses of various sizes and shapes resulting from the radiating growth of individual crystals (e.g., *botryoidal,* grape size; *reniform,* kidney shaped; *mammillary,* shaped like mammillary glands).

Oölites. Round egglike bodies usually having concentric internal layering. The term *pisolite* is applied if the bodies exceed 2 millimeters in diameter.

FIGURE 2–35
Malachite, copper carbonate.
(U.S. National Museum)

FIGURE 2–36
Native copper from Bolivia.
(U.S. National Museum)

TABLE 2–5 MINERAL IDENTIFICATION

(Hardness)	Color	Streak	Mineral
	I Metallic Luster		
1	gray	gray	graphite
2½	copper red	copper red	native copper
2½	gray	black	galena
2½	gold	gold	gold
2½	bronze (purple iridescence)	gray black	bornite
3	brass yellow	greenish black	chalcopyrite
3½	red to black	brownish red	cuprite
4	reddish bronze	black	pyrrhotite
5½	black	brown-gray	chromite
6	red or black	red	hematite
6	black	black	magnetite
6	brass yellow	greenish black	pyrite
6½	red-brown to black	white to brownish	cassiterite
	II Nonmetallic Luster		
1	variable	white	talc
2	yellow	yellow	sulfur
2	white	white	gypsum
2½	white	white	halite
2½	black	dark	biotite
2½	white	white	muscovite
2½	greenish	pale green	chlorite
2–5	green or black	white	serpentine
3	white	white	calcite
3	pink	white	lepidolite
3½	blue	blue	azurite
3½	brown yellow	light brown	sphalerite
3½	varies	white or gray	dolomite
	variable	variable	bauxite
4	green	green	malachite
4	variable	white	fluorite
4–7	pale blue	white	kyanite
5	rust	yellow brown	limonite
5	variable	white	apatite
5½	black	greenish black	hornblende
5½	black	gray green	augite
6	varies	white	feldspar group
6	red to black	red	hematite
6½	green, brown, or yellow	white	epidote
6½	variable	white	garnet
6½	green	white	olivine
6½	gray brown	white	sillimanite
7	variable	white	quartz
7	variable	grayish	chert
7	variable	white	staurolite
8	white, clear	white	topaz
9	variable	variable	corundum
10	variable	white	diamond

TABLE 2–6 MINERAL GROUPS

Native elements

Copper (native) / $2\frac{1}{2}$ [*copper red*] **Cu**	It occurs in irregularly shaped masses in pore spaces in lavas and gravels. It is ductile.
Diamond / 10 [*colorless*] **C**	Hardness of diamond is its most distinguishing feature. Diamond crystals are most commonly octahedra, cubes, and slight modifications of these shapes, but they are often rounded and distorted. Diamond has octahedral cleavages that are perfect. In addition to the colorless varieties there are yellowish, red, green, blue, and black diamonds. The brilliance of diamond is due to its high dispersion of light.
Gold (native) / $2\frac{1}{2}$ [*gold*] **Au**	It usually is found as rolled scales, grains, or nuggets. Its high specific gravity and color distinguish it from most other metals. It is malleable and ductile, has no cleavage.
Graphite / 1 [*gray*] **C**	Massive forms are most common. These are scaly, foliated, granular, or earthy. The color, dark gray, hardness of 1, and one perfect cleavage make it easy to identify. It feels greasy.
Sulfur / 2 [*yellow*] **S**	Granular, fibrous, crust, or compact forms are most common. Straw-yellow color is characteristic. There is no cleavage.

Sulfides

Bornite / $2\frac{1}{2}$ [*bronze*] $\mathbf{Cu_5FeS_4}$	This important copper-bearing mineral occurs in massive forms that are usually granular or compact. It is opaque. The peacock-colored tarnish is most characteristic. It is found associated with chalcopyrite, malachite, and other copper-bearing minerals.
Chalcopyrite / 3 [*brass yellow*] $\mathbf{CuFeS_2}$	Crystals are usually tetrahedra, but it also occurs in compact and granular forms. It may be distinguished from pyrite by its deeper color and greater softness. Common associates are pyrite, bornite, galena, sphalerite, and chalcocite.
Galena / $2\frac{1}{2}$ [*gray*] **PbS**	Cubic crystals and lead-gray color combined with its high specific gravity make galena easy to identify. But it also occurs in granular aggregates. It has perfect cubic cleavage.
Pyrite / 6 [*brass yellow*] $\mathbf{FeS_2}$	Cubic crystals are commonly found. These crystals usually are striated. It may also be found in massive granular forms. When weathered, pyrite may turn into limonite.
Pyrrhotite / 4 [*bronze*] **FeS**	Pyrrhotite occurs in massive granular form frequently mixed with chalcopyrite. It tarnishes readily and is slightly magnetic. It can be distinguished from pyrite and chalcopyrite by its color, tarnish, and magnetism. Nickel is often associated.
Sphalerite / $3\frac{1}{2}$ [*brown yellow*] **ZnS**	Resinous luster is one of the important characteristics. Yellowish colors are also typical. When in crystals they are tetrahedral. Cleavage is prominent in six directions.

Carbonates

Azurite / $3\frac{1}{2}$ [*dark blue*] $\mathbf{Cu_3(OH)_2(CO_3)_2}$	The azure-blue color is characteristic. It will effervesce in hydrochloric acid. It is a minor ore of copper, most commonly found with malachite.
Calcite / 3 [*variable*] $\mathbf{CaCO_3}$	Crystals are found in many shapes. Crystals or granular aggregates are most common, but it is also found in stalactitic form in caves. Color is highly variable. In clear specimens double refraction of light may be observed—two images of objects below the calcite will be seen. There are three perfect cleavages not at right angles. Calcite will effervesce in dilute hydrochloric acid.
Dolomite / $3\frac{1}{2}$ [*variable*] $\mathbf{(Ca,Mg)(CO_3)}$	It is similar in appearance to calcite, but crystals of rhombohedral shape have curved faces. It also occurs in massive forms. It is distinguished from calcite by its failure to effervesce in weak hydrochloric acid.
Malachite / 4 [*green*] $\mathbf{Cu_2(OH)_2CO_3}$	Crystals are needlelike and form in groups. More commonly it is found in massive fibrous, stalactitic, kidney-shaped masses with internal banding, as velvety crusts or as earthy masses. The silky luster and bright green color are characteristic.

Oxides

Cassiterite / $6\frac{1}{2}$ [*black*] $\mathbf{SnO_2}$	It occurs as short prismatic crystals or in granular forms, sometimes with radial fibrous structures. Cleavage is indistinct, and the color is highly variable, but usually it is black.
Chromite / $5\frac{1}{2}$ [*black*] $\mathbf{FeCr_2O_4}$	It is usually found in granular masses. Grains are cubic. It is opaque, has a metallic luster, and is sometimes very slightly magnetic. It is an important ore mineral of chromium.
Corundum / 9 [*variable*] $\mathbf{Al_2O_3}$	It is frequently found in hexagonal and barrel-shaped prismatic crystals, but it also occurs in massive granular forms. Color is variable, but the most common colors are gray, green, and blue. Hardness is the most distinguishing feature, but care must be exercised in determining the hardness because chlorite is often associated with corundum and makes it appear softer than it is.
Cuprite / $3\frac{1}{2}$ [*red to black*] $\mathbf{Cu_2O}$	Crystals are usually octahedral but massive and earthy forms also occur. The red color, brownish red streak, hardness, and translucence make cuprite distinctive. It is mined for copper.

Hematite / 6 *[red or black]* **Fe_2O_3**	There are several important varieties of hematite. *Specularite* is characterized by its metallic luster, steel-gray or iron-black color, and reddish streak. It occurs as crystals, in a micaceous form, and as granular masses. *Oölitic* or *fossil hematite* is characterized by very small egg-shaped bodies. Hematite may occur as a cement in sandstone, as oölites, or as a replacement of fossils. Compact hematite occurs as kidney-shaped masses that have fibrous radial internal structures. The luster is submetallic, and the color is iron black or brownish red. All hematite has a red streak.
Limonite / 5 *[rust]* **$FeO(OH) \cdot H_2O$**	It is usually in massive forms that are nodular, compact, earthy, or stalactitic. The rusty color is characteristic of the surface. The compact forms are fibrous, but earthy limonite is more common.
Magnetite / 6 *[black]* **Fe_3O_4**	Crystals belong to the cubic system and are often octahedra. Magnetite also is found in compact, granular, and lamellar forms. Cleavage is indistinct. Unlike most other minerals it is strongly magnetic.
Bauxite / *[variable]* **Al hydroxides**	It is found in a massive, earthy or claylike form. Rounded pea-shaped concretions are characteristic. It is the main ore of aluminum. Hardness is variable.

Silicates

Amphibole Group: **Hornblende** / $5\frac{1}{2}$ *[black]* **$Ca_2(Mg, Fe)_5$-$(OH)_2(Al, Si)_8O_{22}$***	It occurs in long, prismatic crystals and in massive forms of small bladed, fibrous, granular, or compact grains. There are two good cleavages at angles of 56° and 124°.
Chlorite / $2\frac{1}{2}$ *[greenish]* **$Mg_5Al(OH)_8Al$-Si_3O_{10}**	Crystals are tabular, hexagonal, and often bent. It is micaceous in appearance in most occurrences. Colors are shades of green. One perfect cleavage is responsible for the micaceous appearance. It occurs in metamorphic rocks, and as a scaly coating on other minerals. The flakes are flexible but not elastic.
Epidote / $6\frac{1}{2}$ *[green, brown, or yellow]* **(Ca, Al, Fe, Al, Si, O)**	It occurs as elongated prismatic crystals, but is more commonly found in massive forms that may be columnar, fibrous, or granular. The greenish brown, greenish yellow, and yellow colors are characteristic. Common associates are feldspars, amphiboles, and pyroxenes.
Feldspar Group: **Microcline** / 6 *[green]* **$KAlSi_3O_8$** **Orthoclase** / 6 *[pink, yellow, or brown]* **$KAlSi_3O_8$**	Both orthoclase and microcline contain potassium. They occur as thick tabular crystals and in massive forms that may be granular or cleavage masses. Common colors are white, gray, pink, and, for microcline, green. They may be distinguished from other feldspars by rectangular cleavage, and by absence of fine striations on cleavage surfaces that are the result of twinning. Albite, labradorite, and anorthite are *plagioclase* feldspars. These contain
Feldspar Group (cont.) **Plagioclase Series** **Albite** / 6 *[white or gray]* **$NaAlSi_3O_8$** **Labradorite** / 6 *[gray or green gray]* **$(Na, Ca)(AlSi_3O_8)$** **Anorthite** / 6 *[colorless or white]* **$CaAlSi_3O_8$**	various amounts of sodium and calcium. Albite is one end member of the group. It contains only sodium. Anorthite, the other end member, contains only calcium. Other members of the group contain varying amounts of sodium and calcium, and the physical properties vary with the relative amount of each of these constituents. The plagioclase feldspars are all twinned. There are fine striations on crystal and cleavage faces. Labradorite is the most easily identifiable member of the group because it is usually gray and has a play of colors as it is turned in light.
Garnet / $6\frac{1}{2}$ *[variable]* **(Ca, Mg, Fe, Al)-(SiO_4)**	Crystals are common and are cubes or dodecahedra, but it occurs in massive granular forms also. Color is highly variable, but deep red colors are characteristic of some varieties. Cleavage is not distinct. Hardness is one of the best ways of distinguishing it from similar minerals.
Kyanite / 4–7 *[blue]* **Al_2SiO_5**	Long, bladed crystals that are curved and radially grouped are characteristic of kyanite as are the bluish streaks or spots in it. The hardness varies with direction, being 4 in one direction and as much as 7 in the other. It occurs in metamorphic rocks.
Mica Group: **Biotite** / $2\frac{1}{2}$ *[black]* **$K(Mg,Fe)_3$-$(OH,F)_2AlSi_3O_{10}$** **Lepidolite** / 3 *[pink]* **$K_2Li_3Al_3(F, OH)_4$-$(AlSi_3O_{10})_2$** **Muscovite** / $2\frac{1}{2}$ *[white]* **$KAl_2(OH,F)_2Al$-Si_3O_{10}**	Members of this group are hexagonal in shape. Soaplike sheets formed as a result of the one perfect cleavage are the most characteristic occurrence in rocks. Biotite is black, brownish, or blackish green. Muscovite is colorless, white, or yellowish. Lepidolite is pink or lavender. The colors are characteristic of these three.
Olivine / $6\frac{1}{2}$ *[green]* **$(Mg,Fe)_2SiO_4$** **Fayalite** / *[brown and black]* **Fe_2SiO_4** **Forsterite** *[white]* **Mg_2SiO_4**	It is usually found in granular masses. The green color is typical, and striations of the crystal faces may be apparent in larger crystals, but are difficult to see in granular aggregates.
Pyroxene Group: **Augite** / $5\frac{1}{2}$ *[black]* **$Ca(Mg,Fe,Al)$-$(Al, Si)_2O_6$**	Augite is the most common member of this group. It is most easily confused with hornblende, but the two can be distinguished by the angle between the cleavage faces. The acute angle between cleavages in augite is 87° compared with an angle of 56° in hornblende. Most augite is altered to hornblende and both cleavage angles may be observed on one specimen.

* Some single, long chemical expressions are broken with a hyphen into two lines in this table.

TABLE 2–6 MINERAL GROUPS (cont.)

Quartz / 7 [*variable*] **SiO_2**	Crystalline varieties are hexagonal prisms. Crystal faces are horizontally striated. This group includes rock crystal quartz and milky quartz. The conchoidal fracture, glassy luster, and hardness are characteristic. Cryptocrystalline varieties are also very common. In these the crystal structure is not apparent. They appear as massive forms that may be nodular, kidney-shaped, banded, concretionary, stalactitic, or compact masses. The luster of these forms is waxy or vitreous. *Chalcedony* has a waxy luster, hardness of 7, and conchoidal fracture. *Agate* is banded chalcedony. *Onyx* is agate in which the lines or bands are even and straight. *Jasper* is distinguished by its red color which is derived from specks of hematite in the quartz. Opal is an amorphous silica mineral that contains water.
Serpentine / 2–5 [*green or black*] **$Mg_6\ (OH)_8\ Si_4O_{10}$**	It is found only in massive forms that may be compact, columnar, fibrous, or granular. The luster is greasy or waxy. It is often spotted, clouded, or multicolored. *Asbestos* is a variety.
Sillimanite / $6\frac{1}{2}$ [*gray brown*] **Al_2SiO_5**	It occurs as long, thin needlelike forms or as radiating fibrous masses in metamorphic rocks. Silky luster is an important characteristic of most occurrences.
Staurolite / 7 [*gray, brown, or black*] **Fe $(OH_2\cdot 2Al_2O\text{-}SiO_4$**	Prismatic crystals are common. These are usually in a crosslike shape. It is found in metamorphic rocks.
Talc / 1 [*variable*] **$Mg_3\ (OH)_2\ Si_4O_{10}$**	When in crystals it is thin and tabular. More common are the foliated massive forms or fibrous, granular, or compact masses. Luster is pearly, and it feels greasy.
Tourmaline / $7\frac{1}{2}$ [*black, pink, or brown*] A complex silicate of boron and aluminum	Crystals are common as long or short hexagonal prisms characterized by a triangular outline in cross section. Color is sometimes distributed zonally along the crystals, which may vary from transparent to nearly opaque.

Others

Anhydrite / $3\frac{1}{2}$ [*colorless to gray*] **$CaSO_4$**	Anhydrite usually occurs in massive, fine granular or fibrous forms; rarely as crystals. There are three mutually perpendicular cleavages that distinguish it from gypsum with which it is frequently associated. The mineral is translucent and has pearly to vitreous luster.
Apatite / 5 [*highly variable*] **$CaPO_4$ (F,(OH),Cl)**	Crystals are hexagonal prisms. It also occurs in nodular or earthy massive form in crystalline limestone, ore deposits, and in igneous rocks. Color is highly variable. The single cleavage is imperfect.
Carnotite / — [*bright yellow*] **(K, U, O, V, H_2O)**	It occurs as a powdery incrustation in sand or sandstone. Canary-yellow color and radioactivity are the most distinctive characteristics.
Fluorite / 4 [*variable*] **CaF_2**	It occurs in cubic crystals and in cleavage masses or in granular form. Color is variable, but purples and greens are most common. The cleavage is octahedral and is perfect. Fluorite is most often confused with calcite, but differs in that it has a hardness of 4.
Gypsum / 2 [*white*] **$CaSO_4\cdot 2H_2O$**	Crystals are tabular or prismatic. Massive forms may be laminated, granular, fibrous, or earthy. It is usually colorless, white, or gray. Colorless transparent crystals or cleavage plates are called *Selenite*. Fibrous forms that have silky luster are called *satin spar*. Granular forms are called *alabaster*.
Halite / $2\frac{1}{2}$ [*colorless*] **NaCl**	Crystals are cubic in shape. Massive forms may be cleavable masses, granular, fibrous, stalactitic, or crusts. Colors are white, gray, or colorless. Because halite can absorb water it may feel damp and slick. The salty taste is characteristic, as is its perfect cubic cleavage. It is common table salt.

Clay Minerals

Kaolinite / 2 [*white*] **$Al_4Si_4O_{10}(OH)_8$**	Clay minerals cannot be distinguished from one another in hand specimens. X-ray diffraction is normally used to determine clay minerals. Clay has a greasier feel than chalk with which it may be confused. Clay will not effervesce in dilute acid; chalk will. Kaolinite is used in manufacture of paper.
Montmorillonite / 2 [*white*] **$Al_2Si_4O_{10}(OH)_2\cdot H_2O$**	See above. Montmorillonite is used in drilling muds. The mineral swells by absorbing water, producing a gel-like substance.

CHAPTER OUTLINE

TERMS

acicular (crystal shape)
amorphous
anisotropic
atomic number
atoms
bladed (crystal shape)
botryoidal
cleavage (mineral)
colloform
columnar (crystal shape)
conchoidal fractures
coordination number
covalent bond
cryptocrystalline
crystal
crystal class
crystal systems
d-spacing
dendritic
electrons
elements
equant (crystal shape)
fibrous (crystal shape)
fracture
ground state
half life
hardness
inert gases
ionic bonds
ionization energy
ions
isomorphous solid-solution series
isotopes
isotropic
luminescence
luster
mammillary
metallic bonds
micaceous
microcrystalline
mineral
mineral groups
neutron
oölites
polar covalent compounds
proton
pyroxene
radioactive isotopes
reniform
rock
silicon-oxygen tetrahedron
space lattice
specific gravity
stalactitic
streak
symmetry
tabular (crystal shape)
Van der Waals bonds

REVIEW QUESTIONS

1. Why do some elements (e.g., Na and Al) have two possible coordination numbers while other elements have only one?

2. How can you predict what elements will form ions with a single, double, etc., charge? (Refer to Appendix B.)

3. What types of symmetry are shown by the cubic and tetragonal forms in Fig. 2–9?

4. How can twinning striations be distinguished from mineral cleavage.

5. Study Fig. 2–25. What effect would the removal of the early-formed crystals of anorthite from the melt have on the composition of the final rock?

6. Explain the origin of the cleavage of mica in terms of the cross-sectional view of mica (Fig. 2–27) and the ionic charges on the various elements shown.

7. What conditions must be fulfilled if a radiometric date determined from a sample is to be a correct indication of the age of the sample?

8. Explain why some minerals have a single chemical equation while others (e.g., plagioclase) have a range of compositions.

9. Most physical properties of minerals can be explained in terms of atomic structure. How can hardness, cleavage, and twinning be explained?

10. What is the ionic configuration of potassium and oxygen?

SUGGESTED READINGS

Berry, L. G., and **Mason, Brian,** 1959, Mineralogy: Concepts, Descriptions, Determinations: San Francisco and London, W. H. Freeman, 630 p.

Ernst, W. G., 1969, Earth Materials (Foundations of Earth Science Series): Englewood Cliffs, N.J., Prentice-Hall, 150 p.

Holden, A., and **Singer, P.,** 1969, Crystals and Crystal Growing (Science Study Series, Anchor Books): Garden City, N.Y., Doubleday, 320 p.

Mason, B., and **Gerry, L. G.,** 1968, Elements of Mineralogy (a series of books in geology, ed. by Gilluly, J., and Woodford, A. O.): San Francisco, W. H. Freeman, 550 p.

Pearl, Richard M., 1963, Rocks and Minerals: New York, Barnes & Noble, 275 p.

Ryba, Earle, 1968, The history of crystallography; Earth and Mineral Sciences, v. 37, no. 5, p. 49–54.

3
VOLCANIC ACTIVITY

3 VOLCANIC ACTIVITY

VOLCANIC ERUPTIONS

Krakatoa

One Sunday afternoon in August of 1883 a few mild explosions rocked the island of Krakatoa, located in the Sunda Strait between Java and Sumatra. The next morning the cone of this volcano was ripped apart by an explosion that sent more than 4 cubic kilometers of rock into the air. A cloud of dust, gases, and debris rose nearly 27.2 kilometers into the atmosphere. Heavier debris fell back to earth, but smaller particles of dust were caught in upper air currents and were blown around the earth. For the next two years, before this great quantity of dust settled, it colored sunsets. Krakatoa and a neighboring island were literally blown to bits. A sounding was made over the original peak of the volcano, which had been 792.5 meters above sea level; the sounding was −304.8 meters. Few people lived on Krakatoa, but in the lowlands of the southwest Pacific islands thousands were killed by a *tsunami*, or *seismic wave* (commonly and erroneously called a "tidal wave"), created by the explosion.

The site on which Krakatoa was built was one previously occupied by volcanoes. A fringe of islands marks the rim of a former volcano that collapsed or exploded before recorded history.

Volcanic eruptions, whether violent explosions like Krakatoa's or relatively quiet, effusive outpourings of lava like those in Hawaii, are dramatic evidence of the continued activity of natural processes on and within the earth. Volcanism is proof that there is molten rock in the earth's crust. Most modern volcanoes are found in well-defined belts around the perimeter of the Pacific, near the equator, and along the crest of midoceanic ridges in all oceans. These same belts are sites of earthquake activity and deformation of the earth's crust. Many intriguing questions are raised by the phenomenon of volcanic eruptions as well as by the fact that volcanoes are associated with other types of crustal activity. Where does the molten rock originate? From what depths does it come? Is it being generated now? Why are there so many different types of volcanic activity? What causes volcanoes to be located where they are? Are they in earthquake belts because there is a connection between the generation of molten rock and earthquakes, or do earthquakes open conduits through which the melt can move? Are igneous activity and earthquakes related to movement of crustal plates?

ORIGIN OF MOLTEN ROCK

There are two principal schools of thought regarding the origin of the earth. The prevalent view today is that the earth was formed by the coalescence of a mass of small, cool particles. As these collected and were pulled together by gravitational forces, the interior of the earth heated and eventually the earth became molten. These fluids cooled, becoming liquid melts composed largely of silicates called *magma*. This primitive magma eventually crystallized as minerals, forming rocks. According to the other view, the earth originated from the sun as a mass of fluids. In both theories the earth was molten in its early stages of development. Thus, all rocks have been derived ultimately from magmas.

Early students of volcanism thought that the earth's molten interior was covered by a thin crust through which fiery masses from below broke, causing volcanic eruptions. Through seismology it has been discovered that the earth's interior is rigid enough to transmit seismic waves that depend on rigidity to a depth of about 2,900 kilometers. It is highly unlikely that magmas are coming from such a depth. The modern view is that these magmas are generated within the crust or in the upper part of the mantle, at depths ranging from 10 to 100 kilometers. A likely source is a plastic layer at a depth of about a hundred kilometers, described in Chapter 21. Once generated, magma moves in response to stresses and as a result of volume increases that accompany the melting process. The expanding volume of molten matter is forced upward along fractures and faults, if they are present, or it pushes the overlying strata upward into domes through which it may eventually break along newly formed cracks. If magma finds its way to the surface, a volcanic eruption takes place. If there are great quantities of gas held under pressure, the eruption may be an explosion; if gases have been allowed to escape, the eruption is usually quiet. When magma reaches the ground surface it is called *lava*. Lava, gases, and fragments of rocks through which the

FIGURE 3–1
An aerial view showing Mount Ngauruhoe in eruption. Tongariro National Park, New Zealand.
(National Publicity Studios, New Zealand)

magma has moved are the principal products of volcanism. Magmas do not always break through to the surface. Many of them cool and crystallize beneath the ground surface, giving rise to many different types of *igneous rocks*.

TYPES OF VOLCANIC ACTIVITY

The type of volcanic activity that occurs at any given time or place depends on many different factors, such as the following:

1. The temperature of the magma.
2. The relative amounts of gases, lava, and solid material present.
3. The types of potential openings through which release is possible.
4. The fluidity of the magma.
5. The amount of pressure that can be built up in the magma and feeder system before an opening to the surface is formed.
6. The composition of the magma.
7. Where the eruption takes place.

The above factors are complexly interrelated. For example, basaltic lavas are generally much more fluid than the quartz-rich rhyolite lavas. For this reason rhyolite lavas are much more likely to build up bulbous masses when they are extruded whereas basalts flow out laterally to fill depres-

FIGURE 3–2
Cinder cone at Craters of the Moon National Monument, Arco, Cutte County, Idaho.
(H. R. Cornwall, U.S. Geological Survey)

sions in the topography. Most of the lava plateaus and shield volcanoes are composed of basalt flows. This difference in fluidity is partly due to compositional effects. Silicon-oxygen tetrahedra tend to become linked and build up large three-dimensional structures. Linkages of this type take place more quickly when silicon and oxygen are the primary constituents of the lava. As these structures grow, they impede the movement of the lava. But fluidity (the inverse of viscosity) of most liquids is highly temperature-sensitive. Viscosity is directly proportional to the logarithm of the pressure and inversely proportional to the logarithm of the temperature, as indicated by the following relationship:

$$\eta = \eta_0 \, e^{aP} \, e^{(b + cP)/T}$$

where a, b, and c are constants, and P = pressure, T = temperature.

Viscosity measurements of dry silicate melts of olivine basalt indicate changes from 900 poises at 1,150°C to 200 poises at 1,300°C. For comparison, the viscosity of glycerine is 10 poises and that of water 0.01 poise at room temperature. Dissolved gases in basaltic magma could reduce the viscosity to a value close to that of water. An orthoclase melt has a viscosity of 10 million poises at 1,300°C, which is not very different from that of asphalt at room temperature; however, it is much more viscous than olivine basalt. Viscosity is also closely related to the amount of gases present. When temperature and pressure reach very high levels and when rhyolitic magma contains a lot of volatiles, the eruptions may take the form of incandescent gaseous clouds, glowing clouds or *nuée ardente,* which are extremely fluid and cover large areas rapidly. Thus one type of eruption starts with explosive release of pressure and gases followed by the relatively quiet extrusion of highly viscous lava.

The environment in which the eruption takes place also has a marked effect on the type of activity. Eruptions under water or ice are very different from eruptions in air.

Feeder systems through which magma reaches the surface are also highly variable. Some eruptions occur along faults in the earth's crust that have been caused by crustal deformation. Other eruptions force their way through systems of faults, fractures, and explosive craters generated by the upward movement of the magma from below. In extreme cases, pressures build up in the magma chamber until the only release is an explosion that can destroy entire volcanoes, leaving large depressions, but more frequently the pressure is relieved slowly with the pipe and vent functioning as a valve. High gas pressures below are relieved through the eruption of lava, ash, and gas. When the vent becomes filled, pressure builds up again until the next eruption clears it out. If the pipe or vent becomes too tightly filled by chilled lava, an eruption may occur on the flank of the volcano or an explosion may result in the vent.

The most familiar types of volcanic activity are two: (1) eruptions whereby large clouds of dust, ash, and larger fragments are ejected and accumulate around a central vent to build a steep-sided cinder cone (Fig. 3–2); and (2) explosions whereby lava as well as ash, dust, and larger fragmented debris are extruded from the central vent to build large composite cones (Fig. 3–3). However, lava plateaus formed by the relatively quiet eruption of vast quantities of lava from long fissures account for a high percentage of all volcanic materials. Examples include the Columbia River Plateau of Washington and Oregon, the Deccan Plateau of India, and the shield volcanoes in the Hawaiian Islands.

Still another type of volcanic activity is characterized by violent explosions followed primarily by the exhalation of gases. This usually produces a circular or oval zone of thoroughly broken rock, since the rock breaks up not only in the explosion accompanying the release of gas but also in its collapse into the chamber from which the gas escaped.

Central-vent volcanoes grow above a pipelike feeder system or over a fissure in a zone where the materials can move upward; other eruptions may occur simultaneously along fissures that are typically many kilometers long. Eruptions from central vents and open fissures may be prolonged, may occur intermittently over many years, or may be sudden, violent, and explosive.

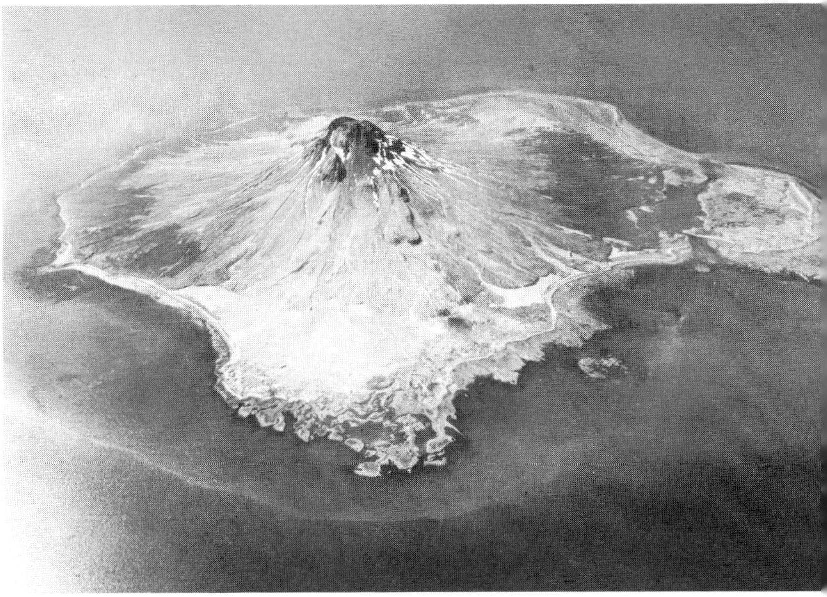

FIGURE 3–3
An extinct composite cone surrounded by its own erosion materials. This recently active volcano is located in Alaska.

Classifying Types of Volcanic Activity

There have been several attempts to classify types of volcanic activity. One system uses the eruption patterns of well-known volcanoes as a standard—for example, Hawaiian, Strombolian, and Peléan activity. Each of these is described later. This system is not entirely successful because the course of events is so rarely alike in different volcanoes. The International Association of Volcanology has attempted to characterize types of volcanic activity by indicating which of the following features or types of activity are present:

1. *Type of conduit.* Summit crater, parasitic craters, radial fissures, or regional fissures.
2. *Location of volcano.* Submarine, sublacustrine (under lakes), subglacial, or subaerial eruption.
3. *Type of eruption.* Normal explosion, incandescent cloud, lava stream, or phreatic eruption (a steam explosion)

PRODUCTS OF VOLCANIC ACTIVITY

Volcanoes differ in the amount, size, and type of materials they eject, and in the shape these materials take after they have been ejected. A volcano may erupt the following: (1) solid fragmental materials, called *pyroclastics;* (2) lava; and (3) gases composed largely of steam.

Pyroclastics

The fragmental materials erupted include pieces of rock torn from the sides of the pipe or feeder system through which the lava moved, pieces of the cone previously ejected, and pieces of lava in a liquid state but cooled and at least partially solidified in the air. These pyroclastic materials are classified primarily according to their size and shape. *Blocks* are the largest masses of rock blown out of volcanoes; violent explosive eruptions can move huge blocks. A 2-ton block was blown a distance of about 3.2 kilometers during an explosion of the Italian volcano Stromboli, and blocks weighing tens of tons have been moved short distances. Smaller fragments such as *bombs* (32 millimeters in diameter or larger), *lapilli* (4–32 millimeters in diameter), *ash* ($\frac{1}{16}$–4 millimeters in diameter), and *dust* (less than $\frac{1}{16}$ millimeter in diameter) are much more common. Bombs and lapilli are formed from bits of lava blown into the air, where the outer part is chilled and partially solidified before the fragment hits the ground. They have an oval center but are often turned in

FIGURE 3–4 (*right*) Pyroclastics. Shown here are ribbon and spindle bombs from the Craters of the Moon National Monument, Idaho.
(H. T. Stearns, U.S. Geological Survey)

TABLE 3–1

Lava	Plutonic compositional equivalent	Principal minerals in lava
Rhyolite	Granite	Silica-rich minerals (e.g., quartz) and alkalic feldspar
Dacite	Granodiorite Quartz diorite	Plagioclase, quartz, pyroxene, or hornblende
Trachyte	Syenite	Alkalic feldspar (minor biotite, hornblende, or pyroxene)
Andesite	Diorite	Plagioclase (andesine) (minor mafic constituents)
Basalt	Gabbro	Plagioclase (calcic), pyroxene, locally olivine
Phonolite	Nepheline syenite	Orthoclase (sodic), nepheline, diopside
Ultramafic (rare)	Dunite Peridotite	Olivine, pyroxene, hornblende, and biotite

the air, producing twisted ends (Fig. 3–4). When the principal materials are fragmental, steep-sided cones are produced (Fig. 3–2). Finer sized fragments may be blown great distances and volcanic dust can be moved hundreds, even thousands of kilometers. Dust from Italian volcanoes can be found in North Africa and some dust may even reach the upper atmosphere and circulate around the earth, causing brilliant sunsets. If there is a strong wind blowing, the small fragments may be very well sorted; the finest fraction will be carried far away, whereas the coarser materials will lie near the vent. Large areas of relatively uniform particles are produced in this way.

A light-colored, porous, glassy, highly vesicular rock called *pumice* is formed when gases froth through a highly viscous rhyolitic lava. When gases that have been prevented from escaping are suddenly released in a violent eruption, pumice may be the end product. Vesicles are created in the melt while it is being ejected. As a fragment enters the air, the outside pressure on the fragment is reduced, causing gases in the lava to expand and form bubbles, which in turn grow to form cells separated by membranelike walls of glass. The city of Pompeii was buried under pumice when Vesuvius erupted in A.D. 79. *Scoria* is another type of vesicular volcanic rock formed by the expansion of gases in lava as it is ejected, but scoria (which contains magnesium and iron) is heavier and darker in color because of its more mafic composition. If the fragments of scoria are still plastic when they hit the ground they may weld together. This is particularly likely near a lava fountain where layer after layer may be built up, producing mounds or towerlike structures called *spatter cones.*

When ash or dust becomes compacted to form a rock, the term *tuff* is applied even though a broad range of different-sized particles may be included. When medium to large sizes dominate, the rock is an *agglomerate.* Fragmental eruptives that contain large angular particles are called volcanic *breccias. Welded tuffs* or *ignimbrites,* in which the ash particles are fused together by intense heat, form when a magma rich in silica is violently erupted from fissures to form very hot incandescent clouds. The resulting rock may resemble a lava flow in that it is massive and continuous over a long distance.

Lava

The principal mineral constituents of lavas are the same as those of plutonic rocks. Quartz, potash and soda feldspars, plagioclase, biotite, pyroxene, hornblende, and olivine are the most common minerals in volcanic rocks. Their presence or absence and relative abundance are the accepted bases for identifying and classifying solidified lava (see Table 3–1).

The range of textures in rocks crystallized from lava is great. When chilled lava is frozen to form glass, it can be identified best by making an analysis of the bulk chemical composition. A less expensive technique involves the calibration of the glass's index of refraction with respect to its chemical composition. Melts that cool rapidly crystallize into rocks with small grains, but as the time it takes to cool the melt increases, crystals grow, grain size increases, and perfect crystals may form in the melt. If the melt is cooled above the crystallization temperature of some constituents while it is in the magma chamber or in the feeder system, the early-formed crystals may grow quite large. If this melt is then extruded and the fraction that remains liquid is cooled rapidly, the resulting rock is composed of large crystals (called *phenocrysts*) embedded in a fine-grained groundmass. Such rocks are called *porphyry*. Depending upon its cooling history each of the lava types may occur as glass or as porphyry.

Fluidity, the amount of gas in the lava, and the rate at which the magma is erupted and cooled are important determinants of the form the lava takes. *Áá,* or blocky lava (Fig. 3–5), and *pahoéhoé,* or ropy lava (Figs. 3–6 and 3–7), are two common forms found in Hawaii as well as other places. The blocky lava begins to take shape where a crust forms over the surface of a moving lava flow. The crust contains numerous openings caused by gases that were trapped in the cooling lava. This crust breaks up because the lava contracts as it cools and because the liquid center of the flow moves, dragging and deforming the crust with it. Such a flow advances as the outer, cooled, edge of the flow is overridden by the slowly advancing lava. Blocks fall forward off the steep front of the flow and are buried as the flow advances. If the rate of advance of the lava increases, the flow may break out and proceed as a river of lava. Ropy

FIGURE 3–5
Áá flow top at Sunset Crater, Coconino County, Arizona.
(H. R. Cornwall, U.S. Geological Survey)

FIGURE 3–6
Flow patterns were frozen in this pahoéhoé lava as it poured back into a fissure. Note lava-encased trees.
(U.S. Geological Survey)

FIGURE 3–7
Ropy (pahoéhoé) lava. A recent flow covers slightly older irregular surface formed as a result of splattering lava that originated from the small cone in the right background.
(U.S. Geological Observatory)

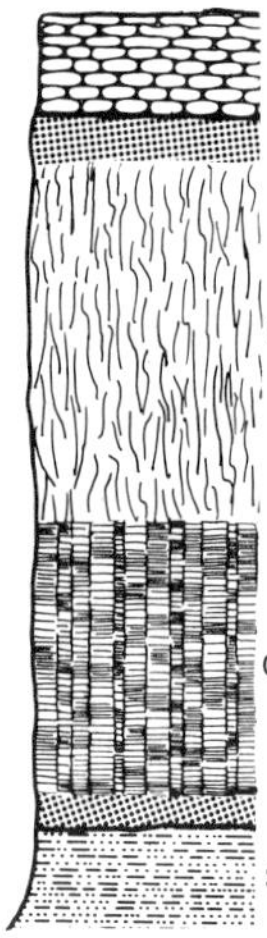

FIGURE 3–8
A cross section of one of the lava flows that compose the Watchung Mountains of New Jersey. At the base is Triassic sandstone. The lava flow may be divided into several structural layers. The bottom is composed of hexagonal columnar structures. Above these the columns become somewhat less well defined. Near the top is a massive layer of basalt and pillow lava, which is usually formed where lava is extruded underwater.

lava is characterized by the twisted, ropelike form of lava that is slowly emitted from a hole or crack and flows forward as a thin sheet or rope. It cools and begins to twist as chilling of the outer layer of the lava slows it down.

If lava crystallizes as a sheet several meters or more thick in a lava lake, in the vent, in a flow that has stopped moving, or in an intrusion between sedimentary beds, the rock may be relatively uniform and dense. The top of the sheet may be charged with gas bubbles and the margins most likely will be chilled and fine-grained. Such lavas often have a uniform columnar jointing, the columns being polygonal-shaped in cross section and oriented perpendicular to the surfaces through which heat was lost as cooling took place (Figs. 3–8 and 3–9).

Gases

Several different types of activity are associated with the near-surface emplacement and extrusion of magma. Many hot springs and all geysers are thought to arise as a result of the heating of ground water in the rocks over and around magmas or cooling igneous rocks. Ground water is heated to form steam in geysers, but violent expulsions of steam and water are relatively rare compared with the widespread abundance of volcanoes. *Fumaroles,* holes or vents through which fumes or vapors issue (Fig. 3–10), are far more common than geysers. Fumaroles are formed under a great variety of conditions in volcanic areas: they are found where cracks or holes, developed through the chilled surface layer of a lava flow, lead into a hot or molten center; they are found in the central vent of cinder cones where gases are making their way upward through the ash and cinder fill in the conduits; they are particularly abundant where explosion or collapse of a volcano has led to the formation of intensive fracturing of the rock over the magma; they form in the crust that develops over lava lakes when the vent or feeder system of a volcano fills with lava. Water vapor is the most abundant gaseous emanation (see Table 3–2) but hydrogen sulfide also frequently is given off at fumaroles. The ground around fumaroles is often yellow from sulfur crystallized from the gas. Sometimes sulfur and sulfur-bearing minerals such as lead and pyrite crystals form an encrustation around the vent of a fumarole. Sulfuric and hydrochloric acid also may be abundant gaseous constituents.

FIGURE 3–9
Polygonal jointing due to cooling and contraction in basalt.
(Crown copyright)

THE FORMS OF VOLCANIC PRODUCTS

Volcanic activity is one of the important processes whereby the surface of the earth is built up and new land forms are created (Fig. 3–11). The volume of material extruded and ejected is often measured in cubic kilometers and a few volcanic provinces such as the Columbia Plateau contain volcanic products estimated to exceed 417,000 cubic kilometers (km^3). The form a particular volcano develops depends on the following:

1. The pre-existing topography.
2. The type and nature of material ejected.
3. The forcefulness of the eruption.
4. The length of time the activity goes on.
5. The history of events at the volcano.

Where there has been a long period of activity, the land forms have changed, often dramatically and unpredictably.

FIGURE 3–10 (*above*)
An aerial view of White Island, Bay of Plenty, New Zealand.
(National Publicity Studios, New Zealand)

Cinder cone. The *cinder cone* (Fig. 3–2) is one of the simplest of volcanic forms. The pyroclastics pile up around the vent, forming steep slopes away from the vent area. The steepness of the slopes is determined by the angle of repose of the pyroclastics, usually between 30–40°. As the cone is built, an inverted cone-shaped depression called the *crater* normally develops around the vent so that materials that do not clear the rim of the crater are funneled back into the vent. Cones such as this are found in many volcanic fields. Beautiful examples can be seen in Craters of the Moon Park in Idaho and in Oregon, Arizona, and Mexico, where the cones often stand isolated from one another, but they may be found also in Hawaii, Iceland, New Zealand, and other places

TABLE 3–2
Composition of volcanic gases from Kilauea Volcano, Hawaii.

Constituent	Percent of total volume
H_2O	67.7
CO_2	12.7
N_2	7.65
SO_2	7.03
SO_3	1.86
S_2	1.04
H_2	.75
CO	.67
Cl_2	.41
A	.20

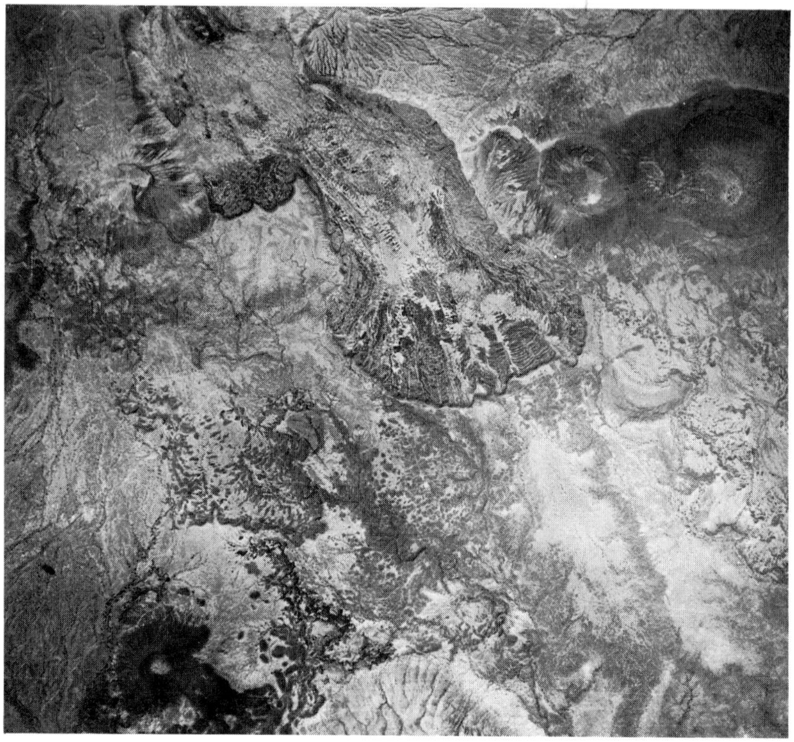

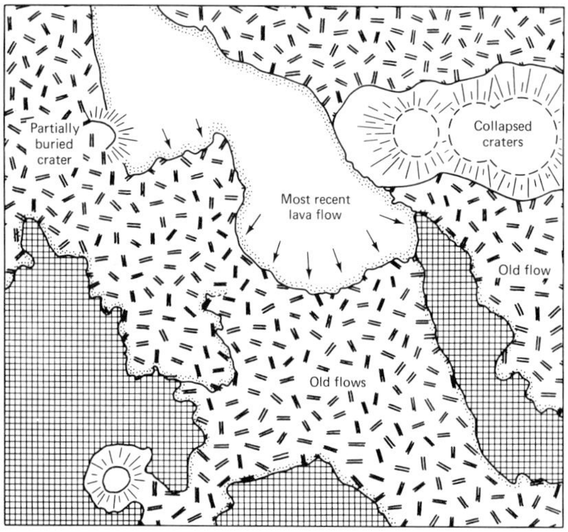

FIGURE 3–11
Vertical photograph of lava flows and cinder cones in Africa. Several flows of different ages are shown. The margins of the flows and the cones are outlined on the sketch map at right.
(U.S. Air Force)

where the volcanism is dominated by extrusion of other types of material.

Composite cone. *Composite cones* are the most beautiful volcanic cones. They are circular in plan and concave upward in profile. Composite cones are built around a central vent when lavas as well as pyroclastics are erupted (Figs. 3–12 and 3–13). Frequently extrusion of lava will alternate with ejection of pyroclastics. Lavas may form a lava lake in the crater before spilling over the rim and flowing down the sides of the cone, or lava may break through the flanks of the cone and spill out in fluid streams down the slope. As lava fills the feeder system and forms a lake, great pressure is built up in the magma chamber. This pressure may be released by flank eruptions. Explosive pressures may be built up when the lava is highly viscous or when the lava in the pipe crystallizes to form a *plug* (Fig. 3–14), as it did at Mount Pelée. The lavas tend to expand the base of composite cones whereas the pyroclastics build the rims higher, leading to the characteristic profiles illustrated in Figs. 3–12 and 3–13. Many of the most famous volcanoes, such as Mount Fuji in Japan and Mount Mayon in the Philippines, are of this type.

Shield volcano. The name *shield volcano* is applied to the largest type of cones, which resemble a shield or low-sloping dome in profile; Mauna Loa in Hawaii is a classic example. The eruption of large quantities of low viscosity lava gives rise to such volcanoes. The lava may come from a central vent part of the time, but usually there are flank eruptions from fissures or rifts commonly oriented in a radial pattern out from the top of the volcano.

When eruption takes place primarily along fissures and no central high area is created, the lava field may become a plateau. The flows that form such areas are very fluid, usually basaltic lava, and they fill low depressions in the topography as they have in the Columbia River Plateau and Iceland.

Caldera. The eruption of large quantities of lava or pyroclastic debris from any type of volcano may lead to the development of a void under the volcano or lava field. Voids may be created by the expulsion of material driven by gas pressures that are released during eruption, or by the contraction of the magma as it cools and crystallizes. The loose volcanic debris and lava are rarely strong enough to stand up over a shallow void and

FIGURE 3–12 *(above)*
A composite volcano located in Japan.
(Japan Air Lines)

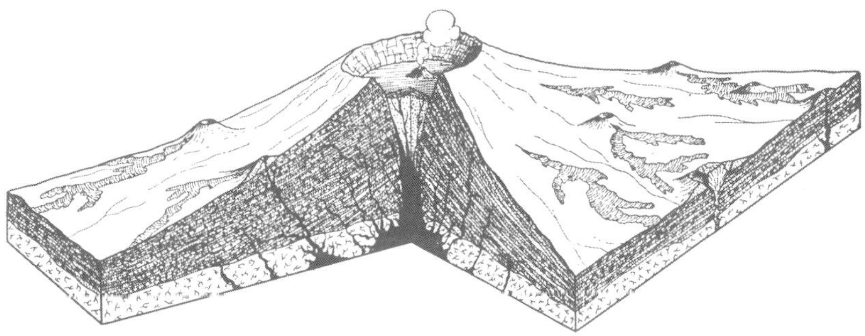

FIGURE 3–13
Diagram of a composite volcano. Main cone built of lavas and fragmental ejecta. Five parasitic cinder cones on the flanks; also a plug dome with a parasitic cone, shown in section on the right. Many dikes cut the interior of the main cone, and one serves as a feeder to a parasitic cone.
(from Howell Williams, 1962)

FIGURE 3–14
Shiprock, N. Mex. This volcanic plug of lava once occupied the pipe of a large volcano that has been eroded away, leaving only the erosion-resistant plug and dikes that radiate out from the pipe.
(New Mexico State Tourist Bureau)

FIGURE 3–15
Oblique view into the crater of Kilauea volcano; part of the shield volcano has collapsed and subsided. Mauna Loa is visible at top of picture.
(U.S. Geological Survey)

usually will *subside* or *collapse* into the magma chamber, creating an enclosed or partially enclosed depression called a *caldera* (Fig. 3–15). Crater Lake, Ore. (Fig. 3–16), occupies a caldera developed over a central vent, but calderas may be found in most regions of volcanic activity. Some volcanoes are situated within the calderas of older volcanoes, as is the case of Wizard Island in Crater Lake (Fig. 3–17), a small cone that rises as an island. Calderas may measure up to tens of kilometers across. Usually a caldera is oval or circular in map view with steep inward-sloping walls formed by the concentric breaks along which the collapse occurred. Broader regional subsidence undoubtedly occurs in some volcanic provinces but these are not termed calderas.

A few calderas result from violent explosions that destroy the cone by blowing it away, as in the case of Krakatoa, but even in such instances a large part of the material falls back into the area around the cone.

Lava domes. When the magma coming out of a vent or fissure is highly viscous, as is the case with most rhyolite lavas, the flow does not spread quickly and usually tends to form steep slopes near the edge of the flow and assume a bulbous or domed form. The bulbous mass may fill a crater of a volcano, grow as an injected mass under the older volcanic products as an intrusion, or develop into a lava flow like those of the Pitchstone Plateau in Yellowstone Park.

Welded tuffs. Particularly disastrous and violent eruptions occur when the ejected mass is composed of incandescent gases charged with droplets of lava (usually rhyolite). These clouds of red-hot dust and ash give off incandescent gases that lubricate the solid particles and can flow at rates of 96 kilometers per hour. These intensely hot clouds spread rapidly and as the material, initially carried up in the cloud, condenses and falls, it forms a deposit, in which the ash and glassy particles are welded together. This deposit is called *welded tuff*. Such deposits often resemble lava flows and have been mistaken for them.

Mudflows. When gas and water are ejected from a volcano or when eruptions disrupt lakes in craters or melt ice, the water is likely to mix with pyroclastics and other debris to form *mudflows* (called *lahars* when of volcanic origin). They assume the form of lobate masses that move downslope, following drainage or low topography whenever possible.

Pillow lava. Eruption of lava underwater may produce flows characterized by pillow-shaped masses, *pillow lava*. Typical pillow lava is composed of oval-shaped masses with glassy, chilled margins formed when the lava encounters water. Pillow lava contains marine sediments and sometimes chert formed from the siliceous remains of single-celled organisms known as radiolarians. These siliceous deposits are frequently found in very deep water.

SELECTED CASE HISTORIES

Tamboro

The most violent explosion of recorded history occurred on a small island in the East Indies in 1815 when the volcano Tamboro blew about 200 cubic kilometers of the crust into the air. The debris created by this tremendous explosion covered islands for hundreds of kilometers around. Explosions of this sort are rare, and most of them occur in volcanoes that have been dormant for long periods of time. Explosion is a means of release of pressures built up beneath the volcano's vent in the *pipe,* the connection between the volcano and the magma, or within the magma itself. Magmas contain large quantities of gases. Gases are compressible and may continue to build up pressure until the pressure exceeds the strength of the plug of frozen lava in the pipe. If the gases can be released slowly, violent explosion is unlikely.

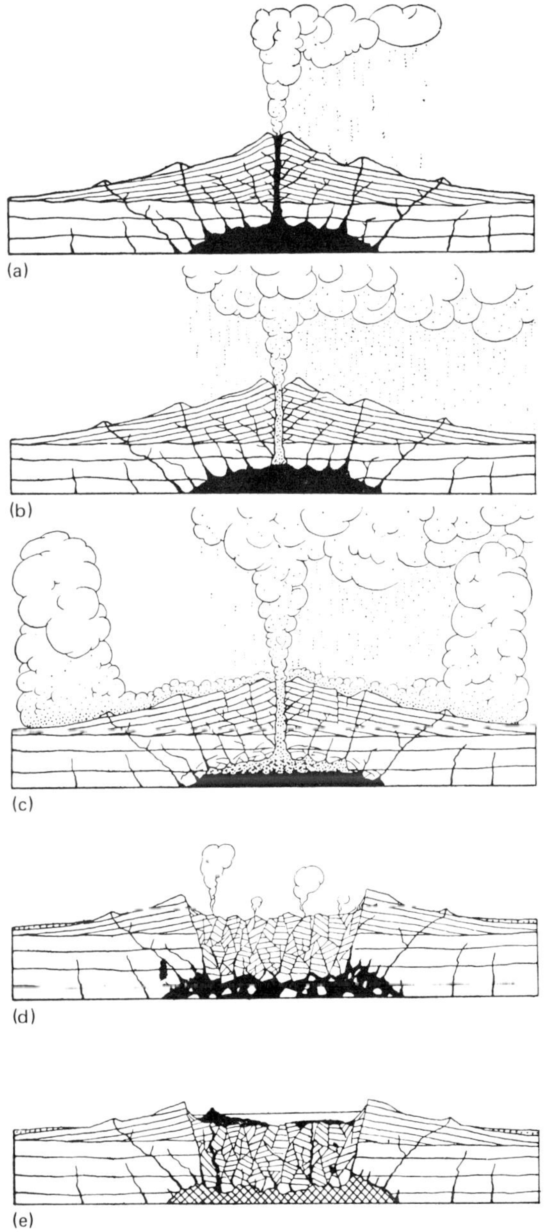

FIGURE 3–16 The evolution of Crater Lake. (a) Beginning of the great eruption. (b) Eruptions increase in violence and showers of pumice become heavier. Lava-level sinks in the conduit. (c) Climax of activity. Glowing avalanches of pumice and scoria sweep down the mountainsides. Magma reservoir is rapidly drained. (d) Top of volcano collapses into the caldron. (e) Crater Lake today, showing Wizard Island. Magma in the reservoir is largely, if not entirely, solidified.
(from H. Williams, 1962)

Mount Pelée

Mount Pelée, situated on the slopes above the city of St. Pierre on the island of Martinique in the West Indies, is famous for an eruption that took place in 1902. This eruption was so extraordinary and its effects so disastrous that the name *Peléan* is used to describe this particular type of violent eruption. Mount Pelée stands over 1,341 meters above sea level. It was known to have been active, but eruptions were rare and never caused serious damage. In April of 1902 activity started. Fumaroles, holes from which gases and boiling water issue, formed in the upper valley of the volcano. A few weeks later, ash began to be ejected and sulfur gases could be smelled. Late in April explosions began and increasing amounts of ash fell on St. Pierre. The population of the island became uneasy, but officials assured them there was no danger and troops were used to prevent people from leaving the city. By the morning of May 8, activity had subsided and only a thin

FIGURE 3–17 (*below*) Profile through Crater Lake including the Watchman, Wizard Island, and Mount Scott.
(from F. E. Matthes, U.S. Geological Survey, Crater Lake quadrangle)

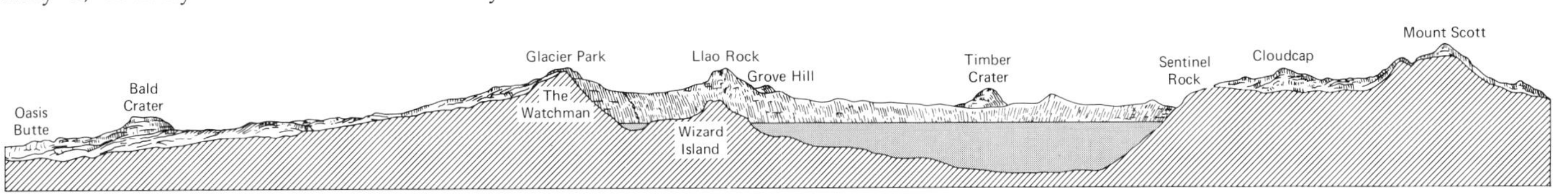

cloud of smoke was rising from the crater when suddenly, at about 8:00 A.M., the volcano exploded in four great bursts. A huge black cloud spread out from the volcano and rolled down the slope, engulfing the city and port in two minutes. The cloud consisted of hot gases and dust. The population of the city, estimated to be about 30,000, was wiped out in a matter of minutes. The only survivors were on two of the eighteen ships in the harbor and a prisoner who was confined in a dungeon. Most people were killed by heat and poisonous gases; only a small percentage were buried under ash and lava.

Parícutin

On February 20, 1943, a Mexican farmer was plowing his corn field when he noticed a thin cloud of steam rising from the ground. Within a day frequent small explosions were taking place in that part of the field; clouds of smoke and dust were boiling up from a small hole. Before the end of that week a volcano almost 140 meters high had been built of pyroclastics, and a flow of basalt lava had issued from a fissure located near the cone. Within the first year the cone rose to a height of 325 meters. Activity continued until 1952 when the volcano, Parícutin, now 410 meters high, became one of many inactive volcanoes in the western part of Mexico.

Vesuvius

The destruction of the city of Pompeii is one of the classic incidents in ancient Roman history. In A.D. 79 Pompeii was buried to a depth of 15 meters by a hot, thick mud of dust and ash mixed with steam.

Vesuvius now stands about 1,220 meters high in the center of the collapsed crater of an older and larger volcano, Mount Somma. All that is left of Mount Somma is a ridge that makes a half-circle around Vesuvius. Varied and periodic activity has been the character of Vesuvius' history in modern time. The volcano has been active frequently since the explosion that destroyed Pompeii. Activity seems to follow a pattern. After a period of violent eruption and explosion the remains of its composite cone begin to collapse where it is unstable or oversteepened. The crater catches much of this debris. Within a few years lava begins to flow into the crater and may eventually fill it. Weight of material in the crater allows pressure to build up within the pipe. Lava may break through the sides of the cone, and some may flow over the rim of the crater. Eventually pressures within become too great to be contained, and a major eruption begins. This may take the form of incandescent ash, bombs, and dust blown into the air, or activity may be in the form of a quiet extrusion of glowing rivers of red-hot lava. On steep slopes these flow rapidly downhill, carrying partially cooled blocks along with them. At the foot of the volcano, where the slopes break, the motion is slowed and the lava begins to consolidate. A crust forms over the top of the flow, but internal heat keeps the center fluid, and it flows, breaking and cracking the crust as it moves. The front margins of the flow look like a pile of smoldering broken blocks. The top blocks keep breaking off and falling in front as the margin gradually moves forward.

Katmai

Located on the Alaskan peninsula, Katmai has been a source of much information about volcanoes since its eruption in 1912. There was no previous record of activity at this deeply eroded and dormant volcano until it exploded, sending great quantities of dust and bits of *pumice* (a natural frothy glass) into the air. Fine dust was carried around the world while the larger particles were dropped out within a few hundred kilometers. So much material was ejected that parts of Kodiak Island, 96 kilometers away, were covered with over a meter of dust. Robert Griggs described the activity at Katmai in his report of 1922:

> Over an area of fifty square miles the ground is all broken open, and hot gases from the molten material below are even now everywhere pouring out, forming the several millions of fumaroles that constitute the Valley of Ten Thousand Smokes. . . . If such an eruption should occur on Manhattan Island, the column of steam would be conspicuous as far as Albany. The sounds of the explosions would be plainly audible in Chicago. The fumes would sweep over all the states east of the Rocky Mountains. . . . As far away as Toronto the acid raindrops would cause

stinging burns wherever they fell on face or hands. . . . Ash would accumulate in Philadelphia a foot deep. . . . The whole of Manhattan Island, and an equal area besides, would open in great yawning chasms, and fiery fountains of molten lava would issue from every crack. In its deepest parts the near-molten sand would probably overtop the tallest skyscrapers, though the tip of the Woolworth tower might protrude.

Studies of the Valley of Ten Thousand Smokes have given us much information about volcanoes, their gaseous emanations, and their contents. One of the most interesting of the observations made there is that metallic minerals are dispersed in the porous *tuffs* (ash deposits) around fumaroles. Magnetite, specular hematite, molybdenite, pyrite, galena, sphalerite, covellite, and other metallic minerals are found. This is taken as strong evidence that gaseous emanations can collect, transport, and deposit metals, an important consideration in the origin of ore deposits. It was also discovered that fumaroles around Katmai yielded approximately one and one quarter million tons of hydrochloric acid and 200,000 tons of hydrofluoric acid each year. Such acids undoubtedly play an important role in the alteration of rocks near large intrusions.

Stromboli

Stromboli, located at the northern end of the Aeolian Islands in the Mediterranean Sea, typifies the volcanic activity called *Strombolian*. Stromboli is so consistently active that it has been called the "lighthouse of the Mediterranean" because of the red glow that follows each explosion. Its activity has been persistent, moderate, and uniform since the earliest records were made, at about the time of Aristotle. Incandescent fragments of lava and white clouds of steam with small quantities of ash are emitted in a typical eruption.

Stromboli has a near-perfect-shaped cone that rises about 3,500 meters from the sea floor and stands 915 meters above sea level. The crater has steep-sided walls around most of its perimeter, but one side is lower than the others; this may be the result of an eruption that destroyed one side of the cone. Now during eruptions, lava and ash flow down a steep slope on this breached side of the crater. There are three vents in the crater, two located toward the sides and a third in the center. Large blocks that have fallen from the cliffs around the crater protrude from the crater floor. Lava in the vents rises and falls by as much as 6 meters. As the lava rises, large bubbles form and burst, throwing lava fragments into the air; then the lava subsides. Sometimes basaltic lava flows almost continuously from one vent and cascades down the slope toward the sea.

The activity described is the normal and prevailing condition as long as the vents remain open, but when the vents become temporarily blocked, more violent explosions occur. These break up the material and give rise to ash-laden clouds.

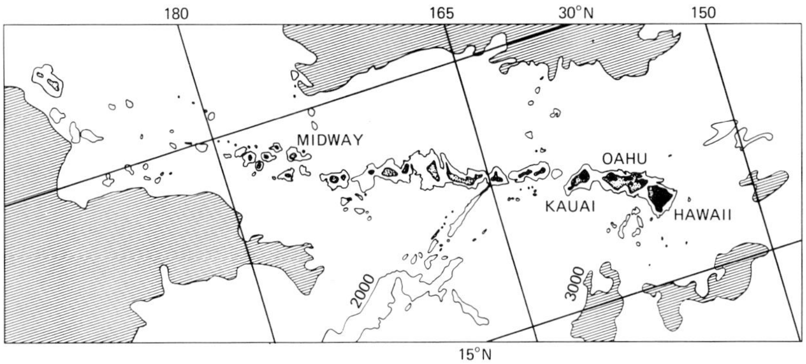

FIGURE 3–18
Map of Hawaiian Islands. Ruled area is below 3,000 fathoms (5.4 kilometers); shallow banks and seamounts are shown dotted, and major islands are black.
(after U.S. Navy Hydrographic Office map)

Volcanic Activity of the Hawaiian Islands

The Hawaiian Islands are a classic region for study of volcanism. A volcano observatory on the rim crater Kilauea has kept precise records of the activity of this volcano. This chain of islands extending nearly 2,560 kilometers along a northwest-southeast line in the mid-Pacific is formed almost entirely of volcanic rocks (Fig. 3–18), and the sediments derived from them. The islands rise nearly 4,800 meters from the sea floor and reach peaks as high as Mauna Loa (pronounced Mahōō'nah Lowah), 9,725 meters above the sea floor (Fig. 3–19). This enormous pile of lava and ash was built up by thousands of flows and extrusions that are being added to today in much the same way as they have been for millions of years. The highly fluid lava flows have resulted in formation of volcanoes that have the profile of a broad shield—hence the name *shield volcano* (Figs. 3–15

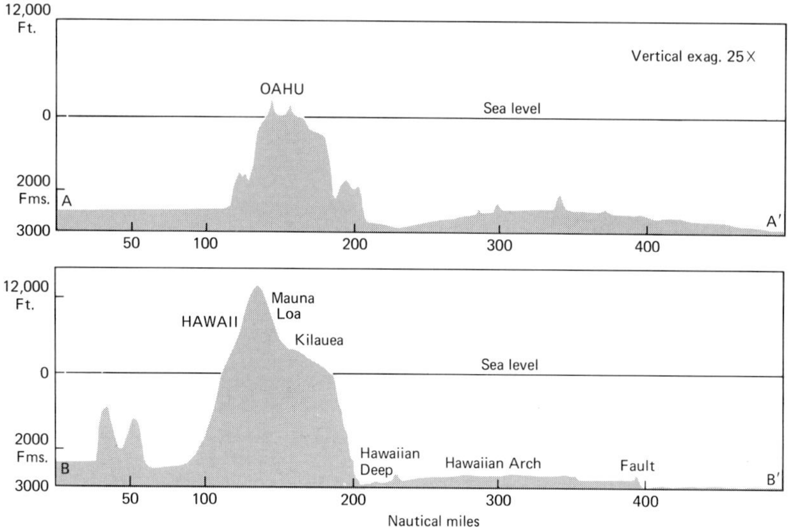

FIGURE 3–19
Profiles across Oahu and Hawaii, showing the bathymetry of the sea floor to the east and to the west of those islands. Fms = fathoms = 6 feet = 1.8 meters.
(from Dietz and Menard, 1953)

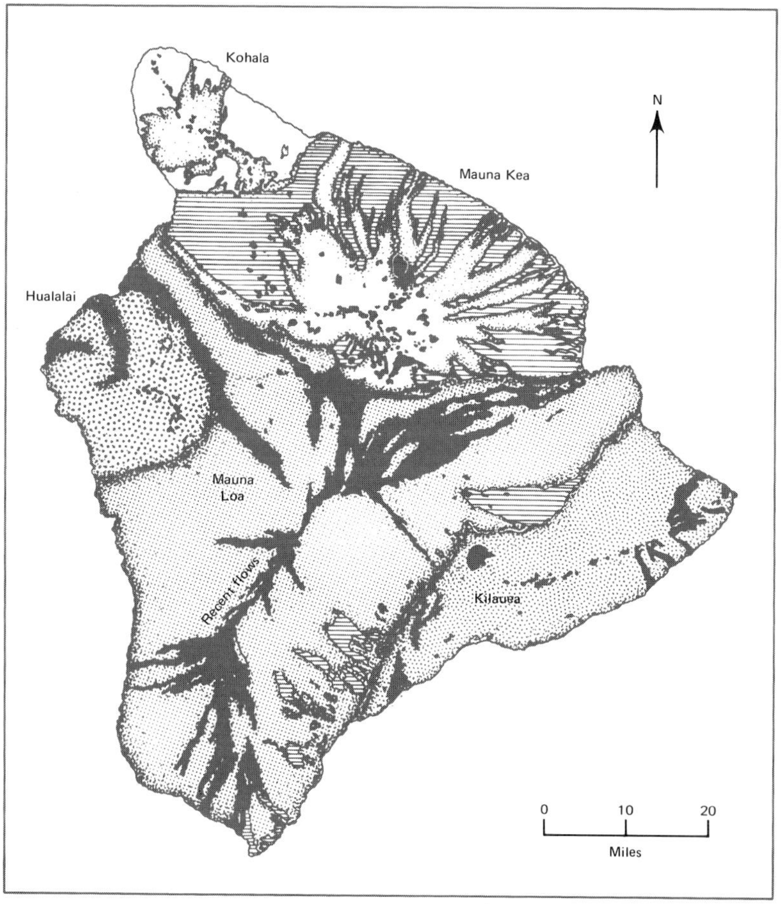

FIGURE 3–20
Geologic map of Hawaii showing the five main centers of eruption. The oldest flows are shown without pattern and with horizontal ruling. Dot patterns define younger flows from each of the five centers. Recent flows are shown black.
(modified after a map by Macdonald, 1961)

and 3–21). Orientation of the islands along a nearly straight line supports the idea that they formed along one of the great fracture or rift zones of the oceanic crust. Such a fracture provides a zone of low resistance to the upward movement of magma generated below the crust. Movements of magma in the islands may be traced by sensitive seismographs that record earth tremors produced by faulting and by movement of magma in pipes. Magma starts from a depth of about 60 kilometers (Fig. 3–21). The magma, having a lower density than the rocks that compose the oceanic crust, creates an unstable condition, facilitating upward movement of the magma. When magma reaches the surface it flows out at temperatures of about 1,000°C. Generally there is no explosive eruption; rather, the magma flows through central vents and along fissures that open in the flanks of the volcanoes.

Most of the lava erupted in the Hawaiian Islands is olivine basalt made up largely of plagioclase, pyroxenes, euhedral or subhedral olivine crystals, and minor amounts of magnetite and ilmenite; a variation that does occur is andesite. The bulk chemical compositions of these two in oxides are shown in Table 3–3.

Because the parent rock from which all the Hawaiian lavas are derived is oceanic crust or upper-mantle material, which is thought to be relatively uniform in composition, differences in composition of lavas can be explained in terms of processes acting within the magma chamber and during the magma's ascent. Samples of lava have been taken during eruptions of Kilauea in order to study variations in composition over a period of time. The chronological changes in composition of the principal oxides are illustrated in Fig. 3–22 for flank eruptions in 1955 and 1960. The most notable changes are in the magnesium oxides, which increased, and silicon and aluminum oxides, which decreased, in February of 1960. This can be related directly to the amount of olivine, $(Mg,Fe)_2SiO_4$, in the lava. Olivine is found in small euhedral crystals, suggesting that it crystallizes very early and at high temperature, a fact substantiated by laboratory experiments. Thus the variation in olivine content must be related to the history of the magma. If a magma is produced from a homogeneous parent rock at temperatures above the melting point of olivine and brought to

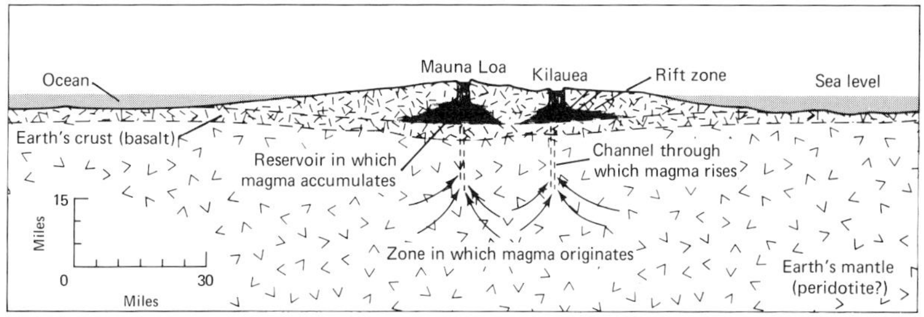

FIGURE 3–21 *(above)*
Diagram illustrating the probable internal structure of Kilauea volcano. Note that although the walls of the magma reservoir within the cone appear sharp in the diagram, they probably are actually gradational. Fissures tapping the magma chamber lead upward to the caldera and laterally into the rift zones.
(after Eaton and Murata, 1960, and Macdonald, 1961)

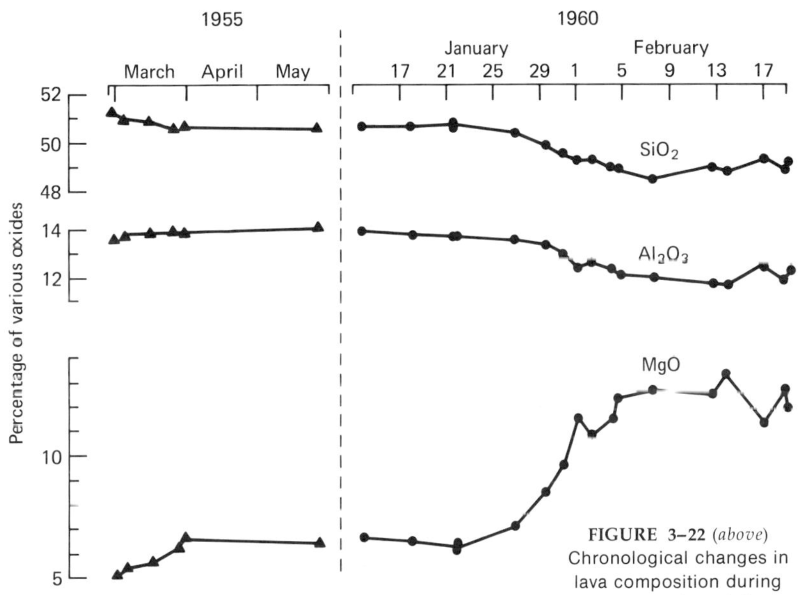

FIGURE 3–22 *(above)*
Chronological changes in lava composition during the 1955 and 1960 flank eruptions in the Puna district, Hawaii.
(modified after Murata and Richter, 1966)

Constituent	Basalt	Andesite
SiO_2	48.3 %	48.7 %
Al_2O_3	13.18	15.8
Fe_2O_3	2.35	4.10
FeO	9.08	7.53
MgO	9.72	4.74
CaO	10.34	7.99
Na_2O	2.42	4.50
K_2O	0.58	1.58

TABLE 3–3
Composition of basalt and andesite from Hawaii.

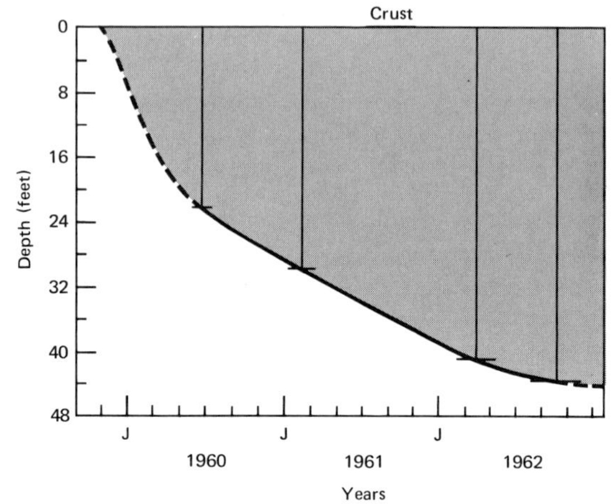

FIGURE 3-23
Growth of the crust (dotted) of basalt over the lava at Kilauea Iki crater. The vertical lines are drill holes.
(after Richter and Moore, 1966)

FIGURE 3-24 *(below)*
Maps and cross sections of Kilauea caldera in 1825 (after Malden) (A), and in 1960 (B) The large central pit of 1825 has been entirely filled with new lava.
(modified after Stearns and Macdonald, 1946)

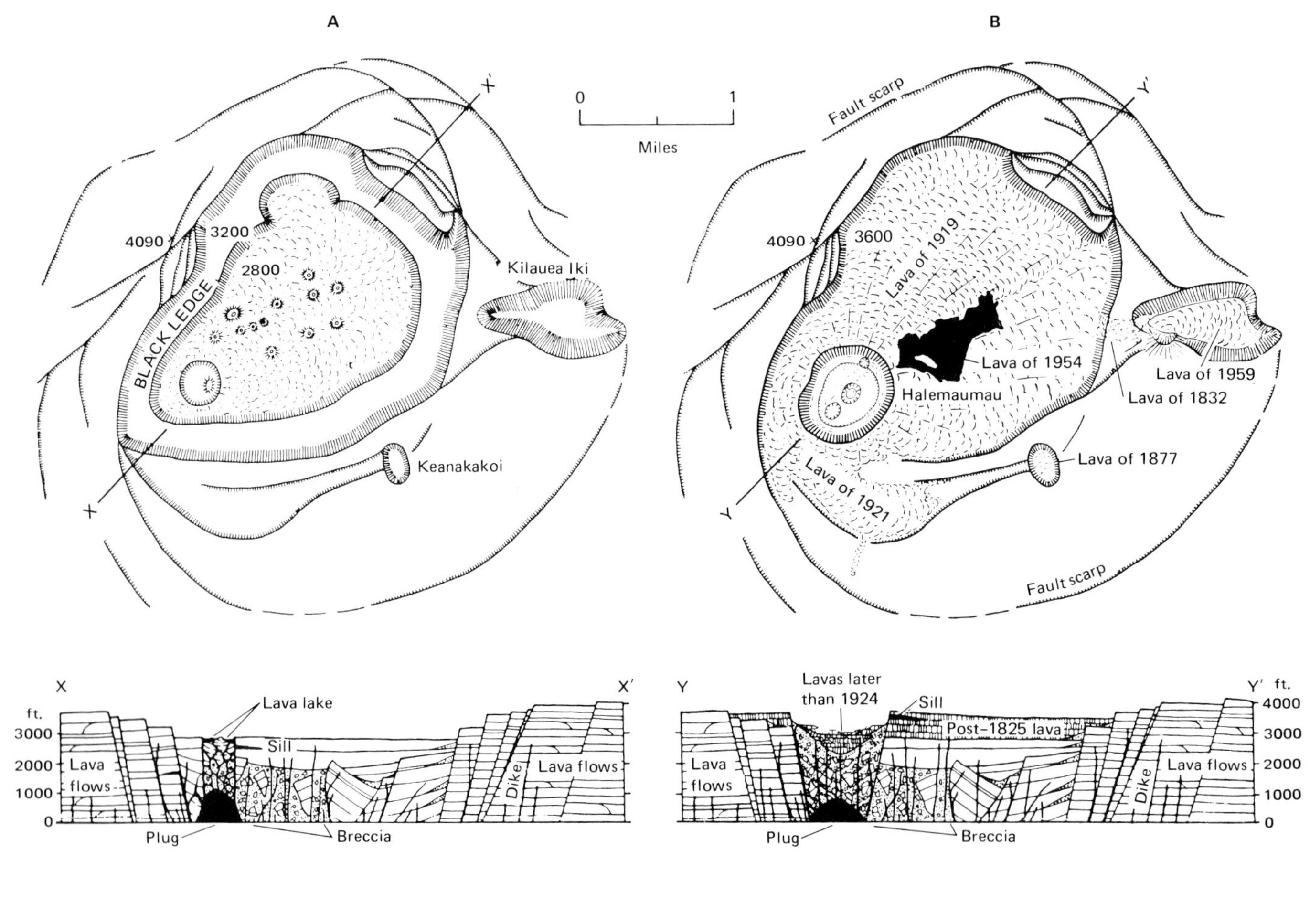

the surface rapidly, a normal content of olivine will be present in the lava. If the magma cools enough in the magma chamber or during ascent, olivine may crystallize out and because its density is higher than that of the lava, it will sink. If it sinks faster than the magma rises or if the olivine settles out in the magma chamber, the lava that reaches the surface will be deficient in olivine (hence in MgO). Then if at a later date olivine that has settled out is picked up and brought to the surface, an olivine-rich lava is produced. This appears to be the case at Kilauea in February 1960 (Fig. 3–22). This is one type of *magmatic differentiation*—one due to gravity separation.

Following the 1960 eruption of Kilauea a series of cores were taken from the lava lake at Kilauea Iki. From these samples the rate of crystallization of the basalt in the lake could be measured (Fig. 3–23). Almost 1.2 meters (4 feet) crystallized during the first month but only 10 centimeters (4 inches) more had crystallized by June of 1962, two and a half years later.

One of the most notable eruptions in the islands took place in 1959–60 (Figs. 3–25, 3–26, and 3–27). That eruption was described as follows by members of the staff of the United States Geological Survey's Hawaiian Volcano Observatory (Eaton and Murata, 1960).

> The first suspicious sign appeared during September 1959, when a series of very shallow, tiny earthquakes began recording on the North Pit seismograph. . . . By the first of November, quakes of this swarm exceeded 1,000 per day. . . . A hurried remeasurement of tilting (tiltmeters are sensitive leveling devices used to detect changes in the slope of the volcano sides) at bases around the caldera . . . revealed that dramatic changes were in progress: the summit of Kilauea was swelling at least three times faster than during previous months. In mid-afternoon on 14 November earthquakes emanating from the caldera suddenly increased about tenfold in number and intensity. At frequent intervals during the next five hours the entire summit region shuddered as earthquakes marked the rending of the crust by the eruptive fissure splitting toward the surface. Then, at 8:08 P.M., the lava broke through in a half-mile-long fissure about halfway up the south wall of Kilauea Iki crater, just east of Kilauea caldera. Abruptly the swarm of earthquakes stopped, and seismographs around the caldera began to record the strong harmonic tremor characteristic of lava outpouring from Hawaiian volcanoes.
>
> During the next 24 hours the erupting fissure gradually shortened until only one fountain remained active. But then the rate of lava outpouring, which had decreased as the erupting fissure shortened, began to increase again, and it continued to increase steadily until the fountain died out suddenly on 21 November. The 40 million cubic yards of lava poured into Kilauea Iki crater filled it to a depth of 335 feet, slightly above the level of the vent. . . .
>
> During the following three weeks 14 more eruptive phases of shorter and shorter duration but with increasingly vigorous fountaining took place at the Kilauea Iki vent. The highest fountain was measured during the 15th phase, on 19 December, when a column of incandescent gas-inflated lava jetted to 1,900 feet, by far the greatest fountain height yet measured in Hawaii. At its highest stand, at the end of the eighth phase, the lava pond was 414 feet deep and contained 58 million cubic yards of lava. At the end of each phase the fountain died abruptly, and from the 2nd to the 16th phase, a mighty river of lava surged back down the vent as soon as the fountaining stopped. . . .
>
> Tiltmeters (Fig. 3–27) around Kilauea caldera showed that the volcano was swelling rapidly as phase after phase of the eruption delivered lava to the surface and then swallowed it up again. When surface activity ceased at Kilauea Iki on 21 December, far more lava was stored in the shallow reservoir beneath the caldera than when the eruption began. It appeared that Kilauea was in an unstable state and that further activity was very likely.
>
> During the last week of December a swarm of small earthquakes began to record on the seismographs at Pahoa. . . . The source of these earthquakes was soon traced to the east rift zone of Kilauea, about 25 miles east of the caldera. . . . The magma that inflated the summit region most probably exerted pressure on the plastic core of the rift zone, and earthquakes revealed where the rift zone yielded and where dikes began to extend toward the surface. . . .

FIGURE 3–25 Kilauea flank eruption (eruption on rift zone) near the town of Kapoho in the Puna district, Hawaii, 1960.
(Volcano Observatory, U.S. Geological Survey)

(a)
Rift of fountains which have built a spatter rampart. Subsequently the trees were destroyed and the fountains built a pumice and spatter cone over 90 meters high.

(b)
Steam cloud formed by pahoéhoé lava entering the ocean. In less than two days after the eruption started, the 0.8-kilometer-wide lava front had flowed 3.2 kilometers from an 18-meter elevation to the ocean and had begun to extend the Hawaiian shore line.

(c)
Fountains about 60 meters high throwing new spatter onto the ramparts they are building around the erupting rift.

(d)
Áá lava front creeping slowly into the ocean with its still hot fragments heating the near-shore water and killing the fish.

(e)
A moving áá lava flow burning the Kapoho school building and covering the spot a few minutes later. The galvanized sheets of roofing mark the position of another school building site just covered.

(f, g)
Fountain of molten lava reaching over 300 meters above the previously flat sugar-cane fields on which it has built an extended cone over 90 meters high. High fountains produce a dark cloud of light-weight pumice fragments that are scattered over distances of several kilometers by the winds. Lava flows extend toward the observer and the cane and cane roads are covered with over 15 centimeters of pumice.

(h)
Pahoéhoé apron at the ocean front. One method by which lava flows extend the island boundaries is shown here as countless little tongues of pahoéhoé lava form the patchwork apron. At the ocean front are several new black-sand areas formed as the molten lava is blown into glassy, sand-sized fragments by jets of steam produced at the littoral zone. Steam is seen rising from many places over the still hot lava field.

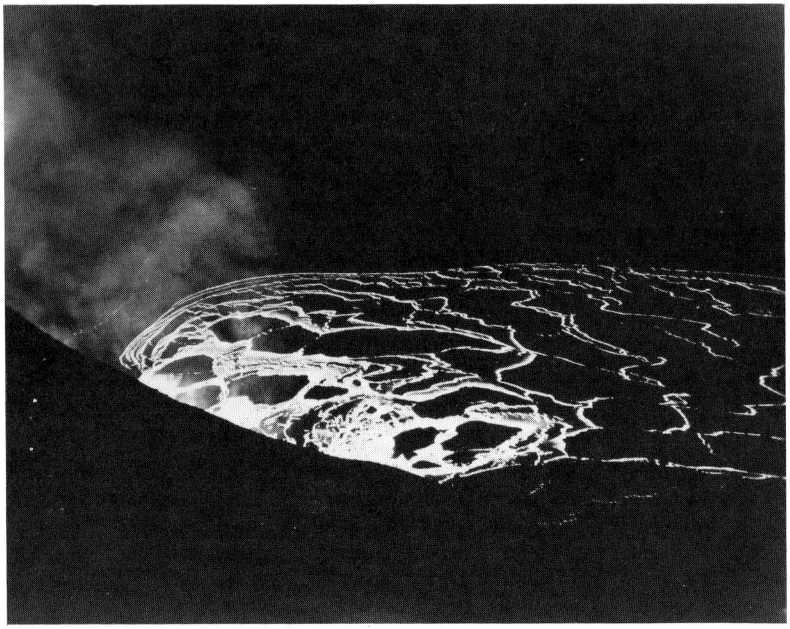

FIGURE 3–26 (*above*) Lava lake at Kilauea Iki. Dark areas are called lava crust. Lava is seen flowing back into the feeder. (U.S. Geological Survey)

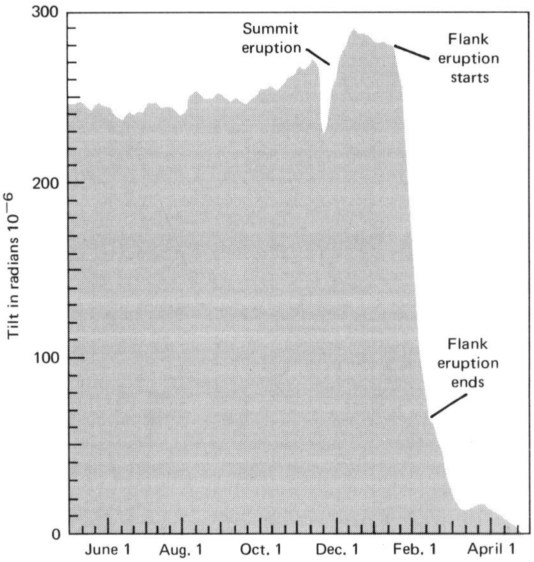

FIGURE 3–27 East-west component of tilting at a point west of Kilauea crater during the 1959–60 eruption. Swelling preceded the eruption and collapse followed. (after Eaton, 1963)

On 13 January the village of Kapoho was rocked by frequent, very shallow earthquakes, and by nightfall a graben (a grave-like fault zone) 0.5 mile wide and 2 miles long that contained about half of the town had subsided several feet. At 7:30 P.M. the earthquake swarm gave way to harmonic tremor, and the flank eruption broke out along a fissure 0.75 mile long near the center of the subsiding graben, a few hundred yards north of Kapoho and nearly 30 miles east of the summit of Kilauea.

During the next five weeks nearly 160 million cubic yards of lava poured out of the vent north of Kapoho and reshaped the topography of the eastern tip of Hawaii. . . .

On 17 January, only four days after the flank eruption began, the summit of Kilauea began to subside precipitously as lava began to drain from beneath the caldera and to move through the rift zone toward the Kapoho vent.

Activity at Kilauea continued until April.

Rhyolitic Deposits of New Zealand

An area in the central part of North Island, New Zealand (Fig. 3–28), contains one of the best known and largest rhyolitic deposits in the world. Several violent explosive eruptions have occurred in this area in recorded history, there are several magnificent active composite volcanoes (Ruapehu, Tongariro, and Ngauruhoe) at the southern end of the central volcanic region, and one of the world's three major geyser basins is located here at Rotorua. But these features are dwarfed in significance when we consider the eruptions that gave rise to the blanketlike deposits of rhyolite flows, welded tuffs (ignimbrites), and pumice that cover thousands of square kilometers. Many of these deposits appear to have been associated with violent eruptions in which vast clouds of incandescent particles covered the central part of North Island. The continuity of these deposits over thousands of square kilometers as well as indications that the particles were air-borne suggest how fast these clouds may have covered the region. An eruption of comparable magnitude now would be an extreme disaster.

Ignimbrites are generally light colored (white, buff to brown in color) and take the form of hard, massive sheets, often tens of meters thick. The rock is composed of fine glassy shards (fragments), dust of pumice, and sometimes large fragments of pumice. Plagioclase is a prominent mineral constituent with quartz and minor magnetite. Many of the minerals and fragments are welded together, indicating how extremely hot these materials were when they were deposited.

Layers of pumiceous ash, welded glassy ignimbrite, and loose ash are frequently interbedded. The pumice fragments are often lenticular in shape, having flowed laterally. These features indicate that the ignimbrites formed at extremely high temperatures and that they settled out of the air. From the thickness (up to 304 meters in places) and extent (one continuous sheet covers about 4,404 square kilometers) of the ignimbrite sheets we get an impression of the magnitude and violence of such an eruption.

Pumice deposits are close associates of the ignimbrites. These cover over 20,725 square kilometers near Taupo, New Zealand. Pumice ash, lapilli, and blocks varying in thickness from less than 0.3 meter on steep slopes to several meters cover older deposits. The last of these huge explosive outbursts is dated about A.D. 131.

The centers of eruption are obscured by the deposits, but the most recent eruptions have occurred along rift zones. A number of long faults are present in the region. We might envision the eruptions as consisting of violent explosions accompanied by the outpouring of clouds of gases and incandescent particles of siliceous composition. These clouds must have contained tremendous volumes of material including incandescent particles that became welded together, ash that went higher in the air and settled out slowly, and pumice. As the clouds expanded rapidly over the area, they produced a steady rain of ash, pumice, and welded tuff. Layering then resulted from variations in the settling rates and the type of materials ejected from the eruptive centers.

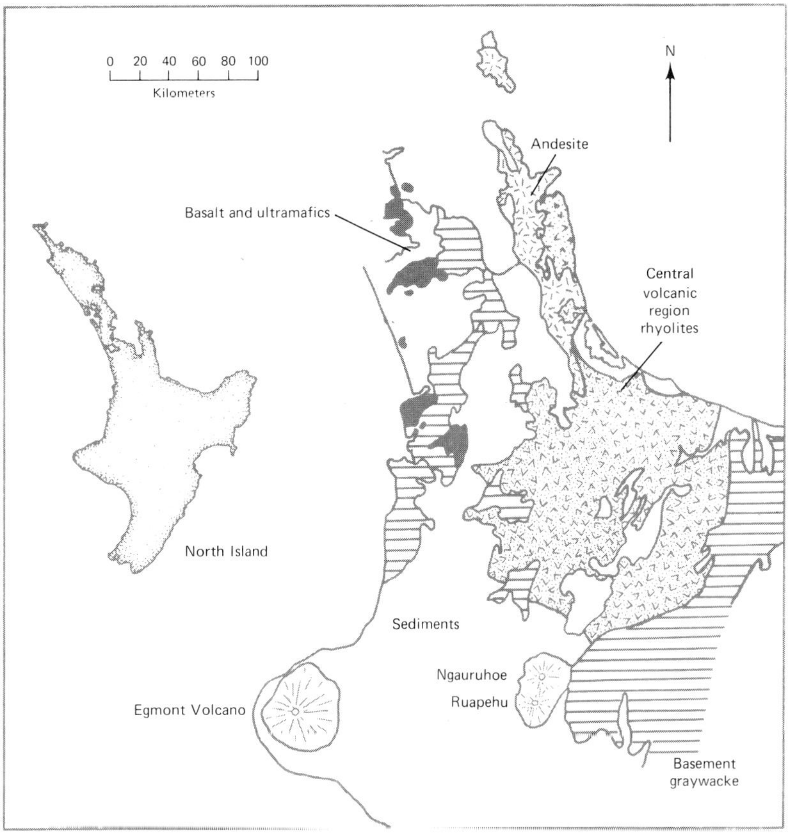

FIGURE 3–28
Map of North Island, New Zealand, showing location of central volcanic region rhyolites and some of the andesite volcanic centers.
(after Thompson and others, 1965)

Central Volcanic Region of New Zealand

Rhyolite flows, welded tuff (ignimbrite), and pumice are found in an area that is separated from mountains on either side by high angle faults. The mountains on both sides of the lower volcanic region give it the general appearance of a down-dropped block partly filled in by extrusive rocks (Fig. 3–29). The bulk chemical composition of the sedimentary rocks that make up the mountains (mainly quartz and feldspar) is so similar to that of the volcanic products that it seems probable the lavas were produced when parts of the sediment melted at different depths. This vol-

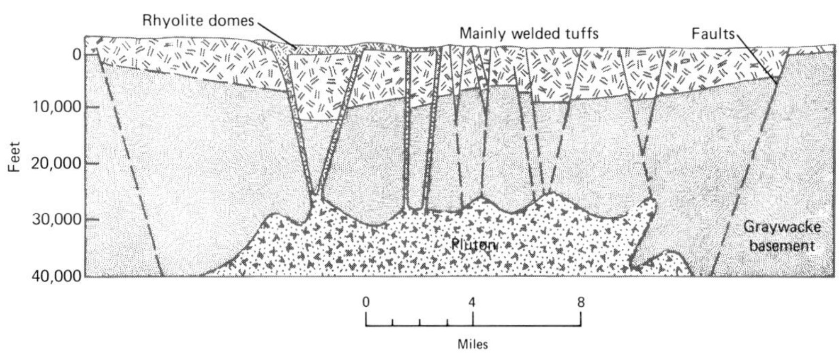

FIGURE 3–29
Simplified cross section (east to west) of the central volcanic region of New Zealand.
(after Grindley, 1960)

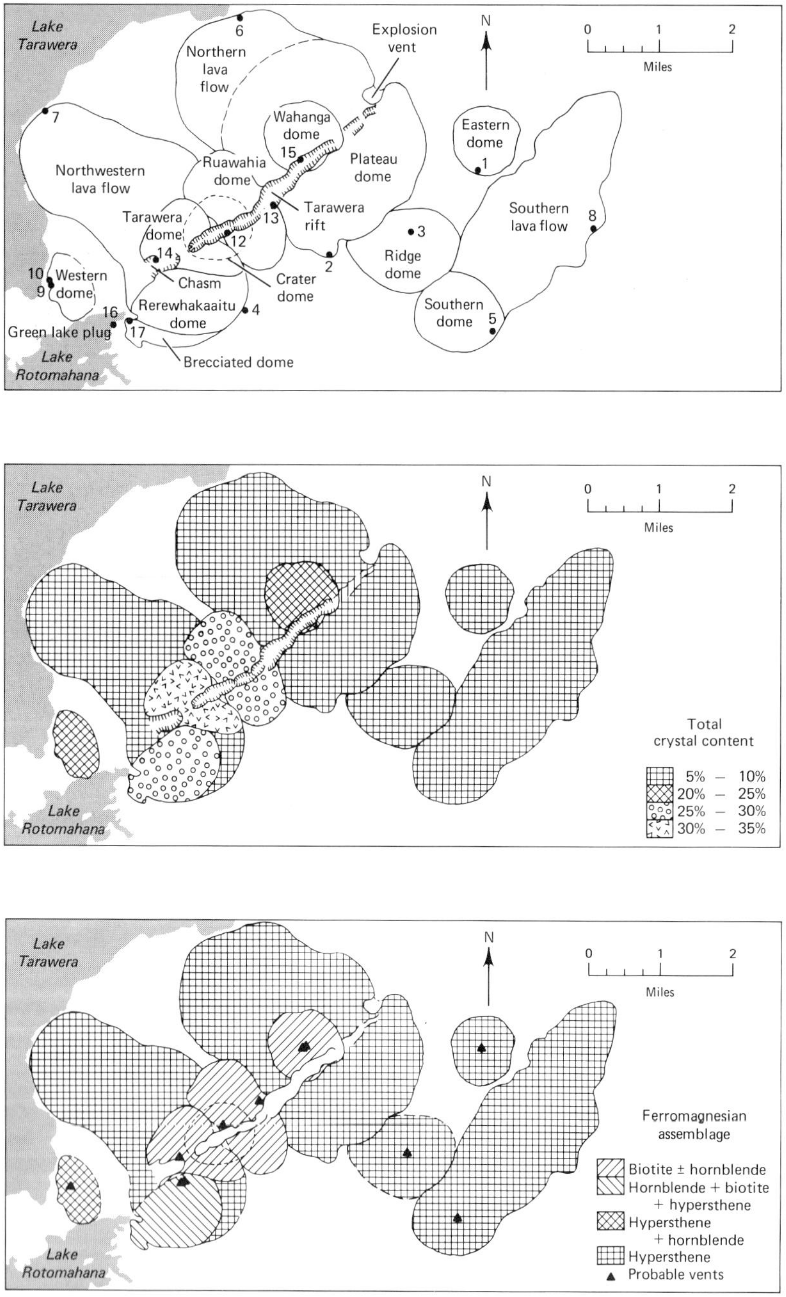

FIGURE 3–30
Domes and lava flows of the Tarawera volcanic complex. The rift is that of the 1886 eruption. Also total crystal content of the rhyolite domes of the Tarawera complex, exclusive of vesicle space and ferromagnesian assemblage of the rhyolite domes of the Tarawera complex. (Magnetite occurs in all the domes and flows.)
(from J. W. Cole, 1965)

canic region, located on continental crust about 20 kilometers thick, is composed largely of rocks that contain quartz and feldspar. Thus this region stands in sharp contrast with the Hawaiian Islands where the rock in and beneath the volcanoes is basaltic or *ultramafic* in composition.

Many different types of volcanic features are found in the central volcanic region of New Zealand. One of particular interest is the Tarawera rhyolite domes.

Tarawera rhyolite domes. During recorded history the Tarawera complex in the central volcanic region of New Zealand has been the focal point of rhyolite eruptions. Characterizing the recent activity is the formation of large lava domes (Fig. 3–30) from lava flows and related pyroclastic rocks (both ash falls and ash flows). There were violent eruptions in 1920, when rhyolite domes were built, and in 1886, when a basaltic dike reached the ground surface. Explosions and the eruption of ash and debris over a large area accompanied emplacement of the dike and a spectacular rift formed in the older dome complex (Fig. 3–31). Accounts of the 1886 eruption describe a series of earthquakes that started about an hour before the first eruption, which occurred when a small dike broke out in an older crater; scoria and rhyolite were erupted. This was followed by the breaking out of a 10-meter-wide dike and a series of explosions that produced large mushroom-shaped clouds. The scoria was so hot that it welded together and formed small flows. Later explosions, caused by the "flashing" (an explosion of water heated above the boiling point) of superheated ground water, took place in lakes, blowing out soft lake sediments and volcanic ash to form mud. This mud buried a Maori village located nearby, killing an estimated one hundred people. Five hours after it started, the eruption was over, leaving a series of interconnected craters along the fissure for a distance of about 14.4 kilometers.

Volcanic Provinces of the Northwestern United States

Products of volcanic activity, particularly lava flows, are exposed or lie under the soil of large areas of Washington and Oregon; these flows and associated volcanoes extend into northern California, and across southern Idaho (Fig. 3–32). The high central part of this region, covering well over 250,000 square kilometers, most of it in Washington and Oregon, is called the Columbia River Plateau. It is high and relatively uniform in level over much of its area. A number of young volcanoes stand out as prominent topographic features west of this plateau. These mountains, the Cascades, extend from Lassen Peak in California to Mount Rainier near Seattle, Wash. The Columbia Plateau province includes a number of smaller plateaus that have been broken by faults, locally tilted, and dropped down. The southeastern portion of this province, the Snake River Plateau, is covered by numerous small craters and volcanic peaks that are very recent, though none is now active.

Columbia Plateau. The Columbia Plateau has been the site of volcanic activity over a very long time. The oldest extensive lavas are of Eocene age (about 40–60 million years old) but most of the area is covered by younger flows (Miocene age, 11–25 million years old). Deposits and flows less than 2 million years old (Pleistocene to Recent) cover most of the Cascade region, the Snake River, and Yellowstone areas. Despite their great age, the older flows are nearly horizontal over vast areas and are only locally folded or broken by faults. The thickness of individual flows is about 5 meters but the aggregate thickness is great. Thicknesses of 2 kilometers can be measured in many river canyons and thicknesses of 4 kilometers are known. Total volume estimates of the lava exceed 417,000 cubic kilometers. Subsidence of the surface following the extrusion of such quantities of lava must account for some of the folding and faulting observable in the flows, although other deformational processes also have been important. The extent of these lava flows (individual flows can be traced 160 kilometers or more), their thickness, and their general uniformity of composition and horizontality (particularly in Washington and Idaho) indicate that they resulted from fissure eruptions of great quantities of highly fluid lava. The flows filled topographic lows and extended toward the edge of the regional basin as flow after flow erupted.

FIGURE 3–31
An aerial view of the craters of Mount Tarawera, Rotorua, New Zealand.
(National Publicity Studios, New Zealand)

Cascade Mountains. The region west of the Columbia Plateau was uplifted as a broad and long arch several million years ago. The older flows were folded into long upfolds, particularly in Oregon, and the older deposits were covered by the building of large composite cones such as Mount Hood, Mount Lassen, Crater Lake, Three Sisters, Mount Adams, and Mount Rainier. All of these are so perfectly shaped that they cannot be very old.

Yellowstone Park

Yellowstone Park (Fig. 3–33) is located east of the great lava fields of the Columbia River Plateau and the Snake River region. Although comparable in age to the volcanic field farther west, Yellowstone has quite different rock types. Most of the park is situated on a plateau underlain by rhyolite. Mountains surround the park, giving it a basinlike form. The lava flows along the eastern

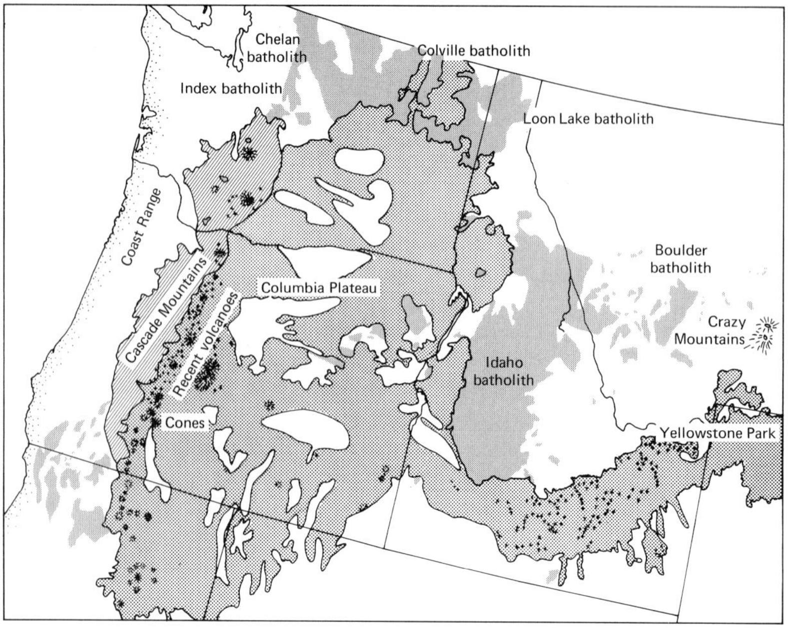

FIGURE 3–32

Distribution of volcanic and igneous intrusions in the northwestern United States.

(after the Tectonic Map of the United States, American Association of Petroleum Geologists)

FIGURE 3–33

Sketch map of the Yellowstone Park region. The areas undesigned are mainly outcrop areas of rock that predate the volcanic activity. The youngest lavas are designated Qb (Quaternary basalt) and Qr (rhyolite). Slightly older volcanics are indicated by Tv (Tertiary volcanics); Teb (Tertiary early basic lavas and breccias); and Tlb (Tertiary late basic breccias). The stars mark craters in the Snake River basalt lavas.

(after maps of Montana and Wyoming in Hamilton, 1960)

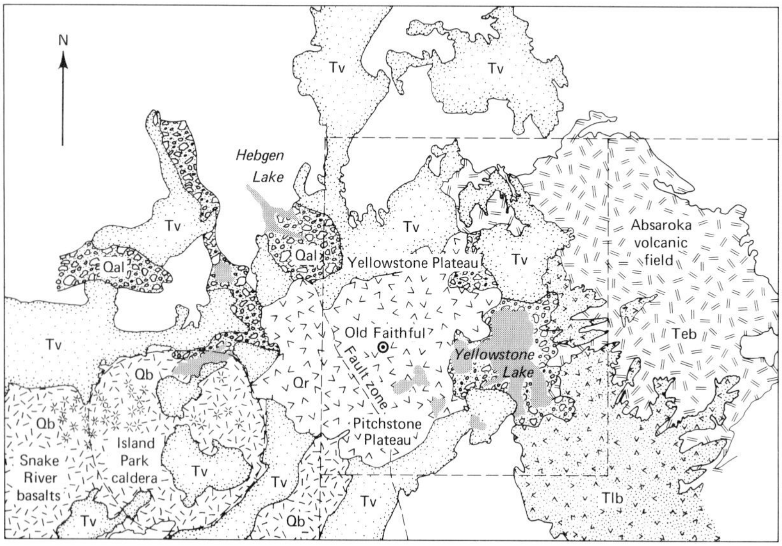

edge of the park are beautifully exposed in the deeply dissected mountains of the Absaroka Range where stream action and glaciation have cut valleys as much as 1.6 kilometers deep into relatively flat-lying flows.

Volcanic activity has been going on for many millions of years in the Yellowstone area. Eruptive centers of Eocene age have been found in the high mountains both east (Absaroka Range) and west (Gallatin Range) of Yellowstone Park. These older eruptives are basaltic and andesitic in composition and large quantities were ejected in the form of breccias, some of which spread as huge mudflows. Welded rhyolite tuff is the dominant older volcanic rock in some surrounding areas. All of these older volcanic rocks have been faulted and displaced, sometimes thousands of meters.

More recent volcanic activity involved the Yellowstone Plateau and the Snake River Plateau. Fluid basaltic lavas poured over the Snake River Plateau during the ice ages, but the flows of the same age on the Yellowstone Plateau are mainly rhyolitic, and they are surrounded by slightly older, welded rhyolite tuff. The rhyolites lie over glacial deposits, showing that this volcanism took place during or after the advance of glaciers into the Yellowstone Park region. Welded tuffs may have been formed during a particularly violent eruption like those in the central volcanic region of New Zealand, but the rhyolite flows were very viscous masses erupted from fissure zones along the crest of the Yellowstone Plateau. These flows extend about 48 kilometers along former river valleys and contain about 400 cubic kilometers of material. The surface of the flows is marked by flow ridges that are concentric and lobate; their terminal edges, still well preserved and moderately steep (20° or more), indicate how viscous the lava was. The upper 30 meters of the flow is composed of obsidian flow breccia that passes into a rhyolite zone in which flow patterns are over 100 meters thick and grade into more massive rhyolite, finally terminating in a base of obsidian breccia. Some of these flows, those located on the Pitchstone Plateau, are very fresh in appearance.

Rhyolite flows in the Yellowstone area are thought to occupy a collapse caldera about 48 kilometers in diameter that formed about 2 million years ago when the roof of the older lava and tuff accumulations subsided or collapsed into the underlying magma chamber (Hamilton, 1960). The caldera now has been filled by more recent rhyolite flows, tuffs, and minor amounts of basalt. The cooling rock melt at depth under Yellowstone Park is responsible for the high thermal gradient and geyser activity described in Chapter 16.

CHAPTER OUTLINE

Volcanic Eruptions
- Krakatoa

Origin of Molten Rock

Types of Volcanic Activity
- Classifying Types of Volcanic Activity

Products of Volcanic Activity
- Pyroclastics
- Lava
- Gases

The Forms of Volcanic Products
- *Cinder cone*
- *Composite cone*
- *Shield volcano*
- *Caldera*
- *Lava domes*
- *Welded tuffs*
- *Mudflows*
- *Pillow lava*

Selected Case Histories
- Tamboro
- Mount Pelée
- Parícutin
- Vesuvius
- Katmai
- Stromboli
- Volcanic Activity of the Hawaiian Islands
- Rhyolitic Deposits of New Zealand
- Central Volcanic Region of New Zealand
  - *Tarawera rhyolite domes*
- Volcanic Provinces of the Northwestern United States
  - *Columbia Plateau*
  - *Cascade Mountains*
- Yellowstone Park

TERMS

agglomerate
ash
bombs
caldera
cinder cone
composite cones
crater
fluidity
fumarole
lapilli
lava
magma
magmatic differentiation
mudflows (lahars)
phenocrysts
pillow lava
pipe
plug
poise
porphyry
pumice
scoria
seismic sea wave
shield volcano
spatter cones
tsunami
tuff
vent
volcanic blocks
volcanic breccia
volcanic dust
welded tuff (or ignimbrite)

REVIEW QUESTIONS

1. What are the most significant differences and similarities between volcanism in the ocean basins and that on continents? Compare the compositions (both mineralogic and bulk chemical), distribution, and forms.
2. If you lived in an area near active or dormant volcanoes, what types of precautions would you like to see the local government take to insure public safety from volcanic eruption?
3. On the basis of what geological information would you predict the type of activity you should expect from a dormant volcano?
4. What are the most important determinants of lava viscosity and how does viscosity affect the form of volcanic features?
5. What does the presence of the broad troughlike depression called the Hawaiian deep, which almost

encircles the Hawaiian Islands, imply about the strength of the earth's crust and mantle?

6. What features would you look for to determine the top and bottom of an ancient basalt lava flow that had been tilted into a vertical position?

7. Describe the types of movement indicated for the blocks of country rock in the Halemaumau crater (Fig. 3–24) and compare these types of movement with those in the central volcanic region of New Zealand, Fig. 3–29. Do the pattern of these movements suggest compression or extension in the area?

8. Heavy rains often accompany volcanic eruptions. What types of deposits would you expect to form as a result?

9. What features would help you identify an ancient volcanic feeder pipe in an area where the volcano had been so changed by erosion that you could no longer recognize the form of the volcano?

10. Make up an outline of the various types of volcanic activity. Include every type described in this chapter.

SUGGESTED READINGS

Bullard, Fred M., 1962, Volcanoes, In History, In Theory, In Eruption: Austin, Tex., University of Texas Press, 441 p.

Eaton, J. P., and **Murata, K. J.,** 1960, How volcanoes grow: Science, v. 132, no. 3432, p. 925–938.

Griggs, Robert F., 1922, The Valley of Ten Thousand Smokes: National Geographic Society.

Herbert, Don, and **Bardossi, Fulvio,** 1968, Kilauea: Case History of a Volcano: New York and Evanston, Ill., Harper and Row, 191 p.

MacDonald, G. A., and **Abbott, A. T.,** 1970, Volcanoes in the Sea, The Geology of Hawaii: Honolulu, Hawaii, University of Hawaii Press, 441 p.

MacDonald, Gordon A., and **Hubbard, Douglass H.,** 1961, Volcanoes of the National Parks in Hawaii: Honolulu, Hawaii, Mid-Pacific Press, 52 p.

Rittmann, A., 1962, Volcanoes and Their Activity: New York, John Wiley and Sons, 305 p.

Sterns, Harold T., 1966, Geology of the State of Hawaii: Palo Alto, Calif., Pacific Books, 266 p.

Williams, Howell, 1962, The Ancient Volcanoes of Oregon (Condon Lectures; Oregon State System of Higher Education, 3rd ed): Eugene, Ore., University of Oregon Press.

4 IGNEOUS ROCKS

4 IGNEOUS ROCKS

NATURE OF MAGMA

The term *igneous* means born from fire. It is used to describe rocks that originate in and on the earth's crust from the solidification of molten masses of rock, called *magma*. Igneous applies to both volcanic rocks and rocks that crystallize slowly within the earth's crust. These two groups comprise a large part of the crust of the earth—90 percent of all rock down to a depth of 16 kilometers is igneous rock.

The origin of igneous rocks from melts was not firmly established until the late 1700's when James Hutton, a Scotsman, succeeded in proving that the basaltic dikes and sills exposed in Scotland are very similar in composition, texture, and occurrence to igneous rocks found near and within active volcanoes in the Mediterranean region. The notion that "igneous rocks" such as granite and basalt were sediments had been in vogue for many years. This idea had been taught by a great German geologist, Abraham Werner (1750–1817), who had little opportunity to study volcanoes and attributed them to local effects related to the burning of coal seams underground. He argued that the basalts were interbedded with sandstones and other sediments and could be seen to grade into other sediments. It is hard to imagine such an origin now, but very little was known then of the volcanoes around the Pacific or in the Mediterranean where the principal types of magma can be studied at active volcanoes. Other types of magma never come to the earth's surface in a molten form, but much can be learned about these from igneous rocks that crystallized at depth and were subsequently uplifted and laid bare by erosion.

Magmas consist of mixtures of solids, fluids, and dissolved gases. They are hot silicate melts containing water and varying amounts of highly reactive fluids and gases in solution. These reactive fluids, including hydrochloric acid and hydrofluoric acid, are responsible for much of the alteration that is found where magma comes in contact with limestone and other carbonate rocks. The temperature of magma can be obtained from measurements of lava temperature. Such measurements have been made at many places, notably near Katmai Volcano in Alaska and in the Hawaiian Islands. Direct measurements made in basaltic lava show that it has a temperature of 1,100–1,200°C as it flows out on the surface. Since it has been forced up from a considerable depth, it may be cooler than when it was first generated, or it may be hotter as a result of the burning of volatiles at the surface. Temperatures of 1,400°C have been measured in the lava lake at Vesuvius.

GENERATION OF MAGMA

Magmas are generated by partial or complete melting of a part of the earth's crust or upper mantle. We have no way of making direct observations of the formation of a magma. Thus we must rely on our knowledge of the materials involved, the processes acting in the earth, and laboratory simulation of pressure and temperatures at depth. Temperature and pressure are obviously important influences on melting. No mineral will remain solid under a given pressure after the temperature reaches its fusion point. The fusion point is increased by higher pressure. The temperature range for fusion of basalt under atmospheric pressure is 900–1,100°C. Confining pressure on rocks within the earth is great, and it increases with depth, so to generate a magma in the crust, either the temperature must be raised high enough to reach the fusion point of the rock, or the pressure on hot rock must be reduced to achieve a lower fusion point.

Thermal Gradient

That the earth's interior is hot is unquestioned, but how that heat originated and how high the temperature is at depth are still debatable. Some heat may be residual heat from a once-molten earth trapped by the insulating cover of the crust. Heat is now being liberated as radioactive minerals decay.

Temperature increase with depth may be observed in deep wells and mines. Temperature is usually found to increase at a rate of about 20°C/km (about 1°C/100 feet). This is a very steep gradient and would indicate 1,000°C temperatures at depths of 40 kilometers, which would mean that all materials below 100 kilometers would be molten. Studies of shock waves generated during earthquakes show that the mantle reacts as a rigid body and cannot be molten at such shallow depths; therefore, the geothermal gradient must lessen at depth and probably lies within the temperature estimates shown in Fig. 4–1. Note that pressure increases in a nearly linear fashion with depth, but all estimates of temperature show a decreasing gradient with depth. Although we do not expect a molten interior or mantle to exist close to the surface of the earth, rocks are near the fusion temperature within the outer 100 kilometers, and it seems probable that most primary magma is generated at depths ranging from 100 to 200 kilometers. The temperature and pressure conditions under which granite and basalt will be solid, partly molten, and liquid are shown in Fig. 4–2. Compare the conditions favorable for melting with the estimated temperature and pressure curves. For example, granite begins to melt at considerably lower temperatures than basalt within the upper 40 kilometers of the crust.

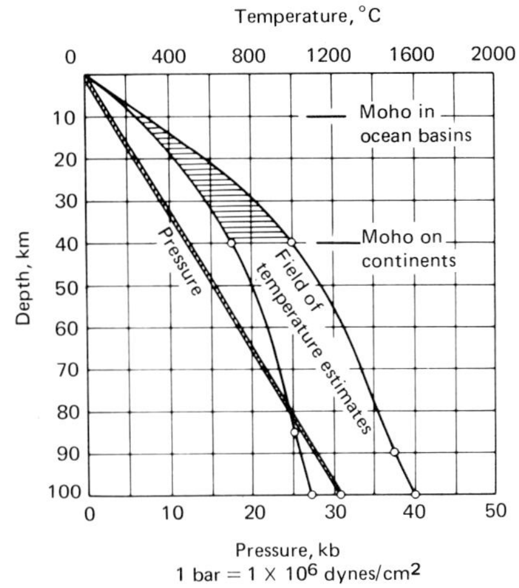

FIGURE 4–1
Temperature and confining pressure variation with depth in the crust and upper mantle.
(after Gutenberg, 1951)

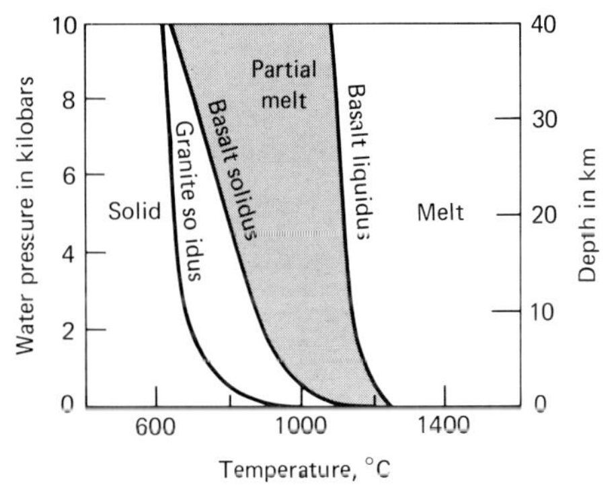

FIGURE 4–2
Plot of pressure vs. temperature showing conditions under which basalt and granite are solids and melts.
(after Wyllie, 1966)

Mechanisms to Increase Heat

Curves showing the increase of pressure and temperature with depth in the earth's crust and upper mantle (Fig. 4–1) indicate that temperatures are not high enough in the crust of either continents or oceans for granitic or basaltic melts to form. A complete melt of basalt appears most probable at depths of 70 kilometers or more. If melts are generated at shallow depths there must be some process by which temperature can be increased. Heat liberated by radioactive decay, frictional heat accompanying movements, and heat rising from greater depths are among the most likely sources of abnormally high temperatures.

Radioactive minerals occur in many rocks in the earth's crust. They are more abundant in the granitic rocks found on continents than in the basalts of the oceanic crust. Heat is liberated in radioactive decay processes, and because rock is such a poor conductor the heat accumulates. Concentrations of radioactive minerals are believed to be sufficient to melt rock locally, but little is known about the amount of radioactive minerals in subcrustal rocks, and it is from the upper mantle that most magma appears to originate. Furthermore, much modern volcanism takes place in the ocean basins where the crustal rocks have low amounts of radioactive minerals. The total heat produced by radioactive decay in basalt is only about 15 percent that in granites.

Heat is generated in the zone of movement between masses of material held together under pressure. The amount of frictional heat produced when two surfaces move over one another depends on the texture of those surfaces and on the force holding them together. The surfaces or zones in the earth along which movements occur are called *faults*. At depth the crustal blocks separated by these zones are forced together by great pressures from the weight of overlying rocks. Thus deep faults or zones of movement may be sites of abnormally high temperatures during and following movement. Volcanoes are found in is-

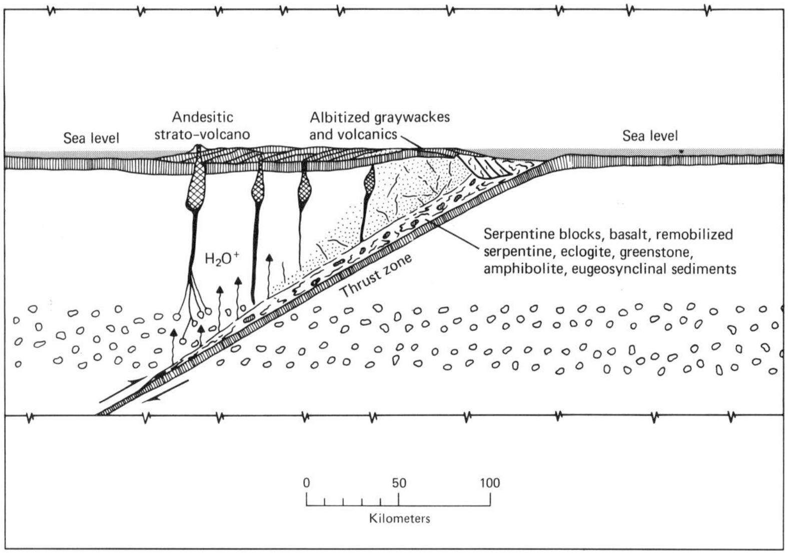

FIGURE 4–3 *(above)* Generalized cross section across an island arc showing hypothetical development of magma from sediments dragged down along a thrust fault to a depth where they begin to fuse and mix with basaltic magma derived from the mantle.
(from Coats, 1962)

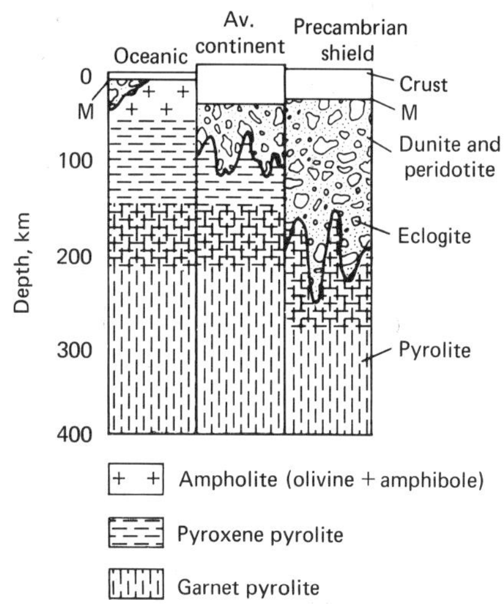

FIGURE 4–4 Petrological model for the upper mantle.
(from Clark and Ringwood, 1964)

land arcs and along the midocean ridges where faulting is active, but no volcanoes are found along many other faults. It may be that the breaks simply provide lines of easy access to the surface for magmas generated by other mechanisms.

The upward movement of hot gases from the liquid interior of the earth might raise temperatures near the surface to the fusion point.

Decrease of Pressure

Before a decrease in confining pressure can be effective in melting parts of the earth's interior, temperatures must be very high (i.e., in excess of the temperatures of molten rock at the surface). Decreasing pressure lowers the fusion point. Thus a reduction in pressure may cause the melting of rocks that were relatively close to the fusion point at a higher pressure.

The sites of most current igneous activity are in mobile belts of the earth's crust—the circum-Pacific belt, midocean ridges, and mountain belts. These are places where crustal deformation is in process, where uplift and faulting are active. If uplift in these places is caused by abnormally high pressures at depth, then pressure drops might be expected where faults or fissures begin to open. These would be particularly effective if the fissure were generated at depth and propagated upward. Fissures of this type are most likely to form where the rocks above the magma are being extended (as in tension). This appears to be the case in the midocean ridges (see Chapters 8 and 25) but extension around the Pacific and in young folded mountains is not as clear.

PARENT MAGMA

Most magmas are thought to originate in the upper mantle at depths in the range of 40 to 100 kilometers. It is from this depth, for example, that the Hawaiian lavas rise. Thus the composition of the upper mantle is very important. Unfortunately there are few ways to obtain direct evidence about the mantle's composition. Seismic wave velocities passing through the mantle are compatible with compositions of mafic and ultramafic rocks such as peridotite (olivine, amphibole, or pyroxene rock); dunite (all olivine); and eclogite (garnet and pyroxene rock).

This information agrees with field observations. Peridotite and other ultramafic bodies, especially serpentines, are found in the cores of the folded mountain belts throughout the world. The peridotite intrusions appear to have been emplaced at depth. Also ultramafic blocks are occasionally erupted from deep-seated volcanoes and they comprise the deep diamond pipes. We hope to find out the composition of the mantle by drilling a hole through the crust someday, but in the meantime all the best evidence points toward an ultramafic composition for these rocks.

The upper mantle is the subject of extensive study and debate because it appears to be very important in processes such as continental drift and the formation of mountains. Several hypotheses about the transition from the crust to the mantle, a boundary now drawn on the basis of a change in seismic velocity, have been advanced. Some think the boundary corresponds to a physical discontinuity caused where a phase change from eclogite to basalt or gabbro occurs. The two are very similar in composition but gabbro differs from eclogite in density. An alternative suggestion is that the boundary marks a chemical change to ultrabasic rocks such as peridotite or dunite in the mantle. According to this view, partial melting of the mantle has yielded basaltic magma. A petrological model of this hypothesis showing differences in the mantle under continents and oceans is depicted in Fig. 4–4. *Pyrolite* is the name given to mixtures of olivine, pyroxene, plagioclase, and garnet that are thought to make up the mantle.

In light of the suggested composition of the mantle and the proposed depth for the generation of primary magma, we may wonder why ultramafic rocks are not more plentiful in the crust. The answer seems to lie in the processes of magmatic differentiation, through which some constituents of the magma are selectively removed (as when heavy crystals sink through the melt), and partial melting, which results from selective fusion. These processes make possible selected concentration of certain ions and elements in a magma.

Basalt is the main type of lava produced by volcanoes in the ocean basins all over the world. This is taken with seismic and other geophysical evidence to support the idea that the oceans are generally underlain by basalt. By contrast, continents are granitic in bulk composition. The earth's crust is thin (about 10 kilometers average) in the oceans. Basalts extruded in the oceans almost certainly have come from magmas generated in the upper mantle, as in Hawaii. Rhyolites and corresponding plutonic rocks like granite are rare in the ocean basins, but they are far more common than basalt on continents. This raises the question of whether or not it is possible that most magmas are derived from basalt or peridotite with granitic magmas representing a more highly differentiated form. N. L. Bowen explored this possibility through a series of now classic experiments described in the following sections. The origin of much granite is highly controversial. Some suggest granite magma is produced by the melting of old granite or sedimentary rock with the bulk chemical composition of granite; and some propose that granite may be a differentiation product of basaltic magma, but much evidence supports the idea that granites can be produced by an alteration process in which there is no melt at all (see page 134).

CRYSTALLIZATION OF A MAGMA

Geologists must deal primarily with the products of igneous activity, the crystallized rock. Our task is to infer from the rock as much as we can about its history, and in order to do this we must ask the right questions. What is the meaning of the mosaic intergrowth of crystals in igneous rocks? Why are some rocks composed of large crystals and others of small crystals or glass? What is the meaning of mixtures of sizes? Why do some minerals in an igneous rock have perfect geometric shapes and others irregular forms? Why do certain mineral configurations occur together in the same rock and others never? Some of the answers to these questions have been revealed through field observations of rocks, microscopic studies, chemical analyses, and experimental studies of simple chemical systems.

Magmas are multicomponent silicate melts that usually contain solid minerals, liquids at high temperature, and gases under pressure. The processes by which a magma becomes a solid igneous rock are complex; they combine precipitation of saturated components in a solution with freezing of a liquid. The crystallization of magma is affected by changes in temperature, rate of cooling, amounts of gases present, pressure, viscosity, and composition.

Careful examination of the crystals that comprise an igneous rock may lead to the conclusion

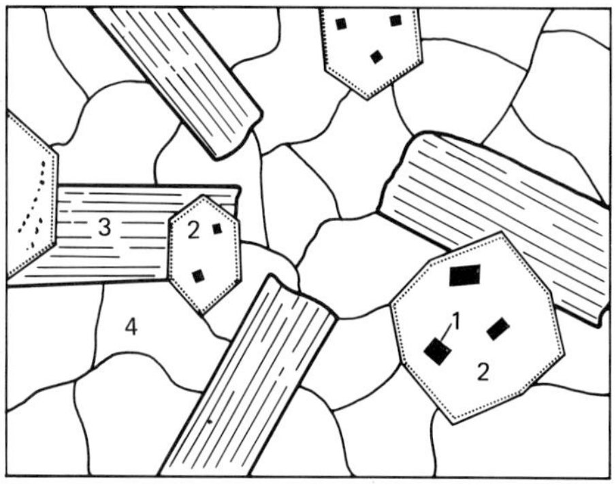

FIGURE 4–5
An enlarged cross section of an igneous rock. This section demonstrates how the sequence of crystallization may be determined. In general, larger crystals form before smaller ones, euhedral crystals form before anhedral or subhedral crystals, and enclosed crystals form before the minerals that enclose them.

that the minerals did not all form at the same time. Notice, for example, that in Fig. 4–5 crystals 2 and 3 are larger than crystal 1; crystals 1, 2, and 3 are *euhedral* (perfect geometric forms), whereas crystal 4 is *subhedral* (some crystal faces) or *anhedral* (no crystal faces); and some small minerals, crystal 1, are included (completely surrounded) within other minerals. The larger crystals probably formed in the early stages of crystallization when there was plenty of room and an abundance of the elements needed for them to grow. The euhedral crystals probably formed before subhedral and anhedral crystals. The first-formed crystals grew into perfect forms in a liquid host, but those crystallizing later had to fit into the spaces left among previously formed crystals before the process of solidification was complete. Any mineral that is included within another one formed before the including mineral. Using these techniques it is possible to make a good estimation of the order of crystallization of the various mineral constituents of an igneous rock.

Laboratory studies of simple rock melts have contributed greatly to our understanding of igneous rocks and their crystallization. N. L. Bowen was among the first to show that the principles of physical chemistry can be applied successfully to problems of rock genesis. Rocks can be melted and then allowed to recrystallize in the laboratory. The temperature, pressure, and composition of the solid end products can be observed, and crystallization can be stopped at any stage by quenching the melt.

The temperature at which most minerals begin to crystallize is known from laboratory studies. But it is not easy to approximate the actual conditions within magmas because magmas are not simple melts containing only the necessary elements in the proper amounts to form a certain mineral. Magmas are generated by the melting of rock that contains a number of minerals, each of which has certain elements within it. As melting proceeds, the minerals break down. Some elements are freed and others remain grouped as complexes. Once the rock is melted, the elements are free to form new compounds with other elements. What new minerals result from crystallization will depend on factors such as the following:

1. The composition of the magma and the natural affinities of certain ions and elements for one another.
2. The rate at which the magma is cooled.
3. The impurities present.
4. The presence of any chance fragments or early-formed solid nuclei (e.g., small crystals).
5. The events during crystallization of the magma that bring about differentiation of the constituents. (See crystallization differentiation, p. 95.)

Crystallization of Simple Melts—Phase Rule

The crystallization of even a simple rock melt is complicated by the presence of liquid, solid, and gaseous phases at various times in the cooling history of the magma. *Phase* is used here to describe a part of a system that is physically different from other parts—a part that can be mechanically separated. Gases are completely miscible (able to be mixed) so only one gaseous phase is present, but liquids may or may not be miscible with one another, and any number of crystalline phases may coexist. Willard Gibbs (1874) established a relationship between the number of phases that can coexist in a solution that is in equilibrium and the number of components of the system.

$$p = c + 2 - f$$

where p = the number of phases in equilibrium, c = the number of components (e.g., the smallest number of chemical species needed to make up the phases), and f = the number of degrees of freedom of the system.

A system has no degrees of freedom if the number of phases present can coexist at only a single temperature and pressure (e.g., water–water vapor–ice). The system has one degree of freedom if one of the variables (temperature or pressure) is selected arbitrarily (e.g., ice–water). The system has two degrees of freedom if both temperature and pressure are chosen arbitrarily.

The phase rule is useful because it indicates how many minerals can occur in equilibrium in a given rock. Pressure and temperature are not constant in the crystallization of magma; therefore melts have two degrees of freedom and the phase rule reduces to the following,

$$p = c$$

indicating that the number of minerals in a rock should not exceed the number of components present.

Examination of the crystallization or changes of state in simple systems will help clarify the nature of the more complex history of a magma. Changes of phase in a single-component system such as ice or a single mineral are shown in Fig. 4–6, in which the various states are functions of temperature and pressure. As the temperature is increased, a mineral may first undergo a change in crystal form, called an inversion, followed by the production of a liquid melt. (Not all minerals have multiple crystal forms such as those in the case illustrated.) Quartz (Fig. 4–7) undergoes inversions of this nature (α-quartz–β-quartz–tridymite–cristobalite–melt). Further increase in temperature or reduction of pressure may initiate boiling and produce a gaseous phase. At still higher temperature a point is reached above which the gaseous and liquid phases become indistinguishable. This point is called the *critical temperature*.

Sublimation, transition directly from the solid phase to the vapor phase without passing through the liquid phase, is not very important in the crystallization of most magmas because the necessary drop in pressure rarely occurs in magmas. The sublimation of dry ice is a familiar example of the process.

When melting occurs in a one-phase system the temperature remains nearly constant as the heat energy is used to melt the solids. If heat energy is supplied rapidly, the rate of melting increases. Temperature tends to remain constant until melting is complete. When the temperature drops, crystallization takes place. The texture of the resulting crystalline form depends on how fast cooling proceeds.

Two-Component Systems

Single-component systems such as water or quartz are the simple equilibrium systems. When two compounds (minerals) are in a system the behavior of the melt is more complex because the solid compounds may react with their own melt. The most common types of behavior found in two-component systems are *simple eutectic* and *solid solution*.

A system composed of the minerals diopside (a pyroxene, $CaMgSi_2O_6$) and anorthite (a plagioclase feldspar, $CaAl_2Si_2O_8$), two igneous rock-forming minerals (Fig. 4–8), is an example of a simple eutectic system. This system deserves special attention because its composition is close to that of basalts and gabbros. Pressure effects are

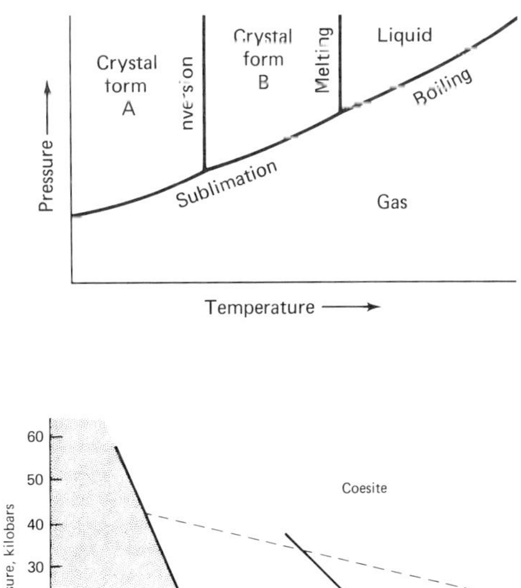

FIGURE 4–6
Change of state. A solid, such as crystal form A may change to a different form of the same material, crystal form B, by an increase of temperature. The solid will melt if the temperature continues to rise, and eventually the liquid will boil to form a gas. Likewise, a solid may go into the gaseous state through variation of the temperature or pressure conditions. A generalized scheme for these changes is shown.

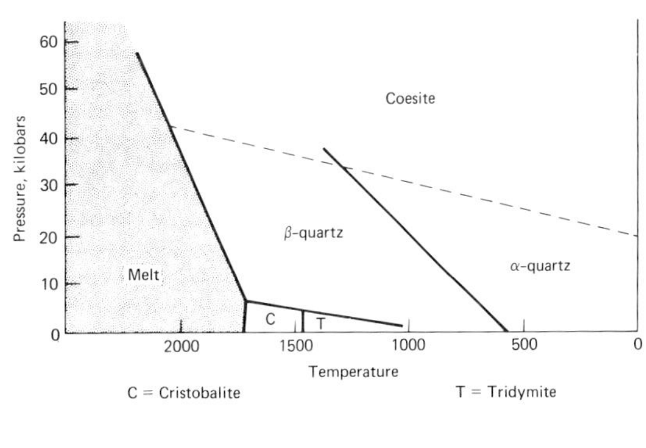

FIGURE 4–7
Pressure-temperature (°C) stability fields for various forms of SiO_2.

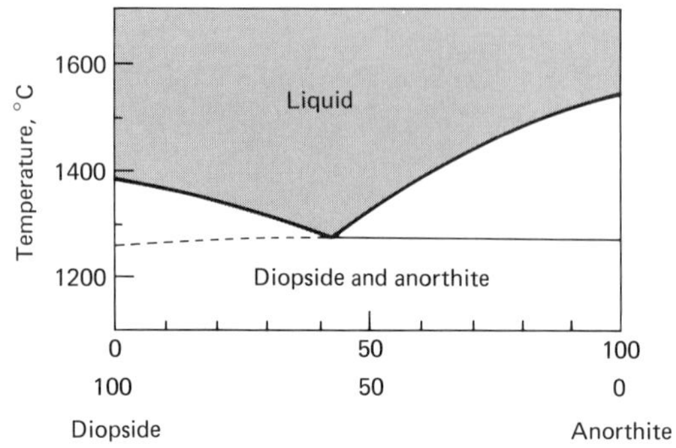

FIGURE 4–8
Simple eutectic system diopside-anorthite.
(after Bowen, 1915)

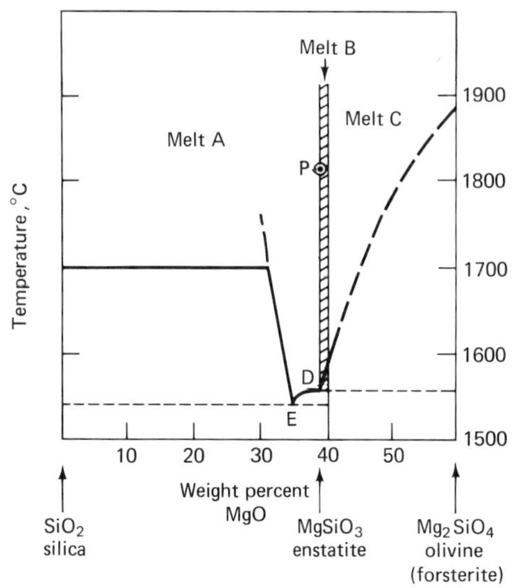

FIGURE 4–9
Equilibrium diagram of the system forsterite-silica.
(after Bowen, Andersen, and Greig)

neglected in this example, temperature is shown on the ordinate increasing upward, and the compositional variations from 100 percent diopside at the left to 100 percent anorthite on the right are shown on the abscissa. A system composed entirely of diopside would have a melting point at about 1,390°C, as shown on the left ordinate; an anorthite system melts at about 1,550°C. Any point (defined by a composition and temperature combination) that falls above the two curves lies in a field with a liquid phase. The various resulting combinations of the end members exhibit characteristics taken from both minerals. If a melt of any composition is cooled, the composition will remain constant until the experimentally determined curve of crystallization is reached. Assume, for example, that the temperature of a rock containing 25 percent anorthite and 75 percent diopside is raised until all solid crystals are melted, and then the melt is allowed to cool slowly. As the temperature drops, the melt reaches a point at which the first solid appears. At a temperature of 1,310°C, diopside starts to crystallize. This temperature is unique for a magma composed of 25 percent anorthite and 75 percent diopside. As new crystals of diopside continue to form and earlier formed crystals grow, the net composition of the melt is shifted along the crystallization curve progressively toward anorthite. The temperature necessary to crystallize this composition is slightly lower, so more diopside forms, and again the composition of the melt shifts; and the temperature of crystallization is lowered correspondingly. This process continues until a temperature of 1,265°C, unique for these two minerals, is reached; this temperature is called the *eutectic* point. At the eutectic point both diopside and anorthite crystallize, so the relative amounts of the two minerals in the melt do not change and crystallization continues until the entire melt is a solid of the original bulk composition. The resulting rock is composed of large crystals of diopside, which formed early, embedded in a groundmass of small crystals of diopside and anorthite, which formed last. If the composition of the melt lies on the anorthite side of the eutectic, the first crystals to form will be anorthite; diopside will crystallize only after the eutectic is reached.

Plagioclase feldspar provides a good example of a *solid-solution series* (Fig. 2–25). Albite ($NaAlSi_3O_8$) and anorthite ($CaAl_2Si_2O_8$) are the end members of this series, and the temperature and compositional representations are the same as those for the simple eutectic. At any point above the experimentally determined upper curve, called the *liquidus,* only the melt phase exists; below the experimentally determined lower curve, the *solidus,* only crystals exist; and between the two, crystals and melt occur together. If a melt of composition 25 percent An, 75 percent Ab is cooled, the first crystals will form when the melt is cooled to about 1,350°C (point *a,* Fig. 2–25) and these crystals will have a composition indicated by point *b* (i.e., a liquid at point *a* is in equilibrium with crystals at point *b*). Crystals use more Ca than Na, thus enriching the liquid in albite. Therefore, as the liquid cools and crystallizes its composition shifts toward albite. However, unlike in the eutectic system, plagioclase crystals will react with the Na-enriched melt as Na is substituted for Ca in the feldspar crystal structure. The composition of the solid plagioclase shifts along the solidus curve as the composition of the liquid plagioclase moves along the liquidus curve. If the reactions go to completion, the final plagioclase will have the same relative amounts of Ca and Na as the melt had initially.

Discontinuous Reaction Series

When several minerals can be formed from an essentially two-component system, as exemplified by olivine (forsterite, Mg_2SiO_4) and silica (cristobalite, SiO_2) (Fig. 4–9), the system is called a *discontinuous reaction series.* Which minerals crystallize depends on the composition of the melt. If the composition lies to the left of a vertical line drawn through *P* and *D*, melt A, the system behaves as a simple eutectic composed of silica and enstatite. If the composition of the melt lies in the field labeled melt C, olivine (forsterite) is crystallized when the temperature reaches the solidus curve. Further crystallization causes the composition of the melt to shift along the curve until it reaches point *D*. At this point the forsterite reacts with the melt to form enstatite. The final rock is composed of olivine and enstatite. If the melt has composition in zone B, the initial crystals are olivine and the composition moves along the solidus to point *D* where the olivine reacts with the melt to form enstatite; however, some melt will remain from which silica will be formed.

Three-Component Systems

Three-component systems can be depicted by means of triangular (ternary) diagrams (Fig. 4–10). Each corner of the triangle corresponds to a single "pure" component; and each side corresponds to a two-component system such as a solid-solution series, a simple eutectic, or a discontinuous series. A point within the triangle corresponds to a three-component system. An artificial system with three components comes much closer to representing natural melts than single- or two-component systems.

Temperature is indicated on the triangular diagram by means of *isotherms*—contours connecting points of equal temperature. The simplest type of three-component system is one in which each side is a simple eutectic (Fig. 4–11). When a three-component melt begins to crystallize, one of the three components usually crystallizes first. If this melt's composition is represented by *P* in Fig. 4–11, crystals of A first form when the temperature of the melt reaches that shown by the isotherms at *P*. The composition of the melt shifts progressively down the isotherms until a point *X*

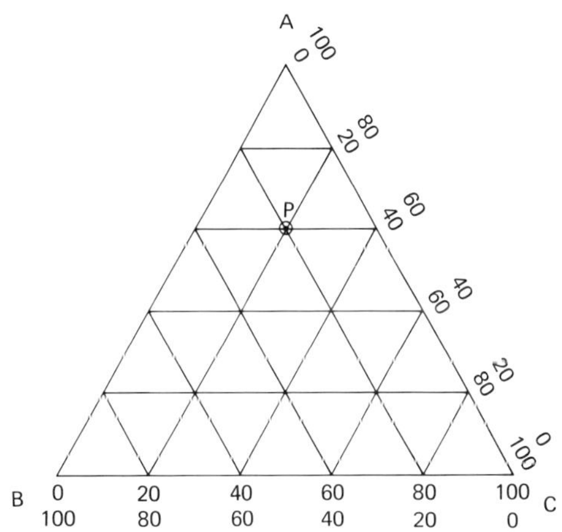

FIGURE 4–10 Coordinate system used to plot compositions on a three-component diagram. The composition at *P* is 20 percent B, 20 percent C, 60 percent A.

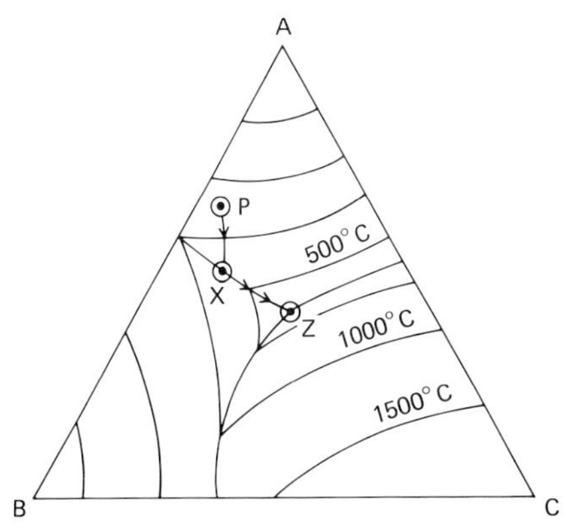

FIGURE 4–11 Triangular diagram depicting a three-component system in which each of the components forms a simple eutectic with the other components.

is reached on the line between two components (called a *cotectic line*). Here the two components on either side of the line crystallize together and the composition shifts along the cotectic line to point Z (called the *ternary eutectic*) where all three components crystallize together.

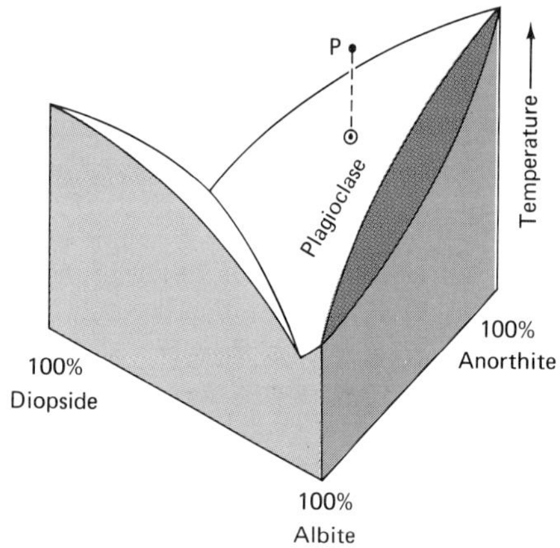

FIGURE 4–12
Block diagram of a three-component system.
(simplified after Bowen, 1915)

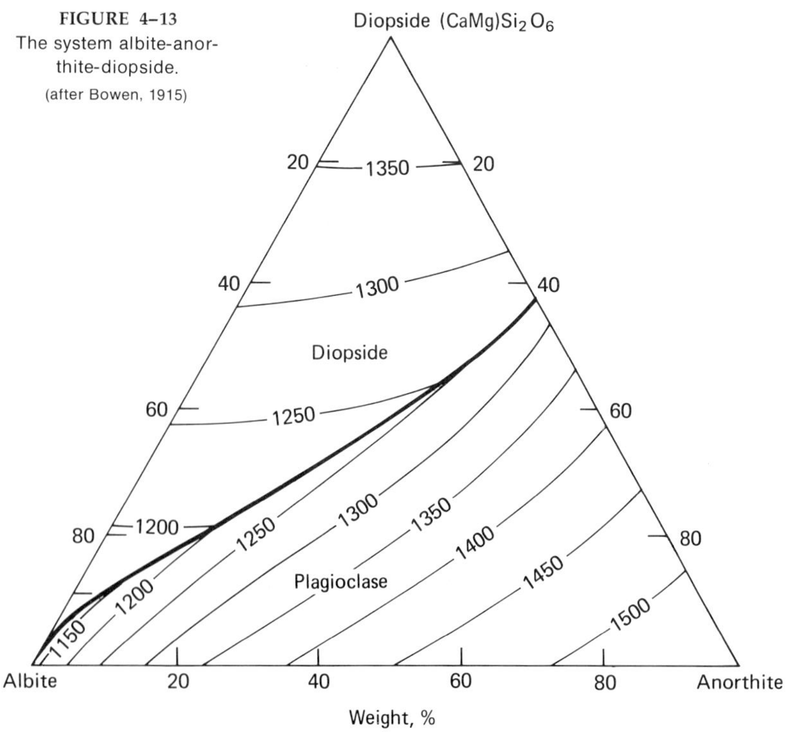

FIGURE 4–13
The system albite-anorthite-diopside.
(after Bowen, 1915)

The system diopside-albite-anorthite is compositionally similar to basalt magma. Albite-anorthite forms a solid-solution series and diopside-anorthite and diopside-albite have simple eutectics. Thus this system is more complex than one with three simple eutectics. A single cotectic line separates the field into two parts. On one side is the field under which the plagioclase solid solution is found. The other is a field in which diopside will crystallize first. If a melt has a composition such as that at point *P* in Fig. 4–12, the first crystals are plagioclase. These react with the melt and undergo continual changes in composition just as in the solid-solution series. But as the composition of the melt is shifted down the isotherms it eventually reaches the cotectic line, and from that point on diopside is also crystallized.

Four-component systems can be depicted on a tetrahedron; however, the representation is complex. The cotectic lines on the three-component diagrams can be represented by planes within the tetrahedron. Obviously the complexity of the crystallization process increases with each new component. Most igneous rocks contain at least three minerals in large quantity, and usually a number of accessory minerals as well as traces of others; so the analysis of natural magmas is highly complicated. The artificial system albite-anorthite-diopside has only five oxides. Actual basaltic magma contains five important components and numerous minor ones.

Bowen's Reaction Series

N. L. Bowen studied the crystallization and chemical reactions that take place in a basaltic magma as it cools. Figure 4–14 is a diagrammatic representation of the reactions. The first signs of crystallization in a basaltic magma start to take place at a temperature of about 1,100°C. The minerals that begin to appear are olivine, the pyroxene augite, and calcium-rich plagioclase feldspar. As soon as the first solids form in a magma the original composition of the melt is changed, being diluted or depleted in elements of the crystallizing minerals (Mg, Fe, Ca, Cr) and enriched in the remainder (Na, K, SiO_2, H_2O). Bowen found that each of these early-formed crystals tends to react with the elements of the melt to form other minerals. As the temperature decreases, the olivine crystals react with the melt to form augite. By the time the magma reaches a temperature of 700°C, therefore, an initially large olivine crystal may have only a nucleus of olivine left surrounded by a shell of augite. At still lower temperatures, the augite reacts with the magma to form hornblende. Likewise, the calcium-rich plagioclase feldspar reacts with the melt to form plagioclase feldspar increasingly rich in Na. By the time the magma has been reduced in temperature to 573°C quartz is formed, and as the temperature falls toward 200°C such minerals as sericite, chlorite, epidote, and zeolite may be formed through reactions between the nearly solid igneous rock and the water-rich solutions or residues of the melt. These water-rich solutions can move into the country rock along fractures and between crystals to ultimately form pegmatites, which are frequently composed of quartz and potash feldspar, and quartz veins.

Note that the reaction series (Fig. 4–14) includes a continuous solid-solution series, plagioclase feldspar, discussed earlier, and a discontinuous series, olivine-biotite. The mineral composition of rock formed by quick chilling of a basaltic magma at any stage in the crystallization can be predicted from Bowen's reaction series. For example, if the magma cools to the point where amphibole and alkalic plagioclase begin to form and then it is chilled, the resulting rock will not contain minerals shown below that position in the series except as small crystals in the groundmass. It also becomes apparent that different rock types could be produced by separation of the liquid phase of a magma from earlier formed crystals in the magma. The composition would be a function of the stage at which separation takes place.

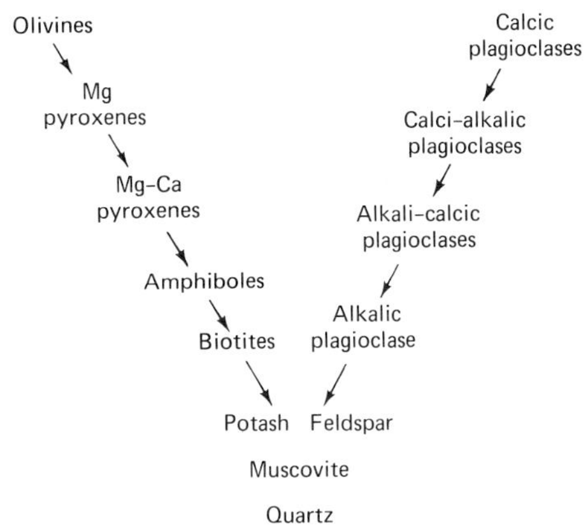

FIGURE 4–14
Bowen's reaction series.

RATE OF CRYSTALLIZATION, TEXTURE

Crystal sizes in igneous rocks range from amorphous glasses to crystals tens of meters long found in pegmatite veins. Most igneous rocks contain crystals that range in size from those in basalt, which are just a little too small to be seen with the naked eye, to crystals several millimeters across. These crystals are generally grown into an interlocking mosaic.

The rate at which a magma cools has a controlling effect on the texture of a rock. *Texture* refers to the size, shape, and arrangement of the minerals in a rock. Extremes in cooling rates may be compared between the crystallization of a lava extruded from a volcano and flowing rapidly down the flanks of the cone and into the sea with crystallization of a magma that is intruded 16 kilometers beneath the surface and cools over a period of tens of thousands or even millions of years. The lava that flows into the sea is frozen as the temperature drops from nearly 1,000°C to the boiling point of water in a few minutes, whereas the intruded mass has enormous amounts of time for crystals to grow.

Crystal nuclei. Pure melts, devoid of any solids, may be cooled below the temperature at which they would ordinarily crystallize if the cooling proceeds without agitation or any other disturbance. If, while this process known as *supercooling* is taking place, the melt is agitated, or particularly if small mineral grains are introduced, crystallization starts immediately. The mineral fragments act as *nuclei* on which the mineral matter from the melt accumulates. The most favorable nuclei are fragments of the minerals that would ordinarily be crystallizing out of the melt. Nuclei form spontaneously even from a pure melt, but minerals differ in their ability to form the initial grain or crystal seed. The necessary ions must come together in the proper form for a crystal to start growth. The valence forces in such embryonic crystals are not satisfied so the crystals

FIGURE 4–15 (*above*) Orthoclase phenocrysts in a felsic matrix. (Scale in inches.)

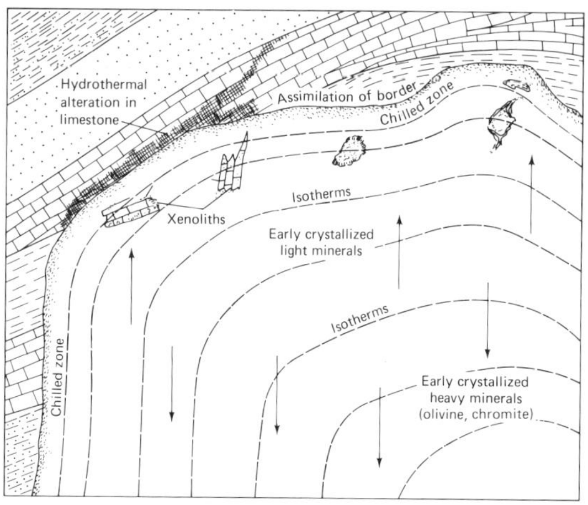

FIGURE 4–16 Magmatic differentiation. Many different types of igneous rocks may originate from a single magma. The differences arise from rates of cooling, settling of early crystallized heavy minerals, contamination of the margins of the magma by the assimilation or melting of country rock, as well as other processes.

are highly soluble, and they tend to dissolve unless they can grow rapidly. Once the nucleus reaches a critical size it can survive and is less likely to dissolve. The critical size is related to the degree of supersaturation of the solution. The higher the degree of supersaturation, the more easily nuclei form. Thus coarse-grained rocks may indicate crystallization from a slightly saturated solution.

The rate of cooling plays an important role in crystal growth because it determines how long a crystal grows. As soon as a crystal begins to grow, it pulls to it from the immediate vicinity those ions or elements of the type required for its structure. This leaves atoms and ions around it that cannot fit into its crystal structure. Thus time is required for other necessary ions and atoms to be brought to the crystal. This is accomplished by convection within the melt or by movement of the crystal through the melt. Keeping these two variables (rate of cooling and production of crystal nuclei) in mind, consider what happens in a magma that is cooling very slowly and that contains only a small number of available nuclei. The rock crystallizing from this magma will be composed of few but very large crystals. As the rate of cooling and the rate of nuclei production increase, more medium-sized and small crystals form. If a magma cools very rapidly and has a large number of nuclei, the resulting rock will be composed entirely of crystals too small to see. Finally, a magma may cool so rapidly that the rock is in effect frozen. It cools too fast for nucleation and few if any crystals have time to form. Such a rock is a natural glass. It differs from minerals in that it contains no ordered crystalline structure; it is amorphous.

Porphyritic texture. An igneous rock is called *porphyritic* when it is composed of large crystals, called *phenocrysts,* embedded in a groundmass of very small crystals or glass (Fig. 4–15). A few scattered large crystals form while the magma is cooling slowly at depth. Then the cooling rate is increased as the magma is forced toward or to the surface. Rapid cooling of the remainder of the magma produces a matrix of crystals or, if cooling is very fast, glass. A porphyritic rock contains two or more distinct grain sizes.

MAGMATIC DIFFERENTIATION

Magmas of basaltic composition are found predominantly in the ocean basins and granites are predominant on continents, but igneous rocks of other compositions are found throughout the world. Why do we find such diversity, especially in view of the likelihood that most igneous rocks may have evolved from essentially one type of parent magma?

Differences in composition and texture may be found within a single body or in numerous spatially related bodies of igneous rock, all of which crystallized from a single magma. Some of these differences are of a gradational nature, the composition of a particular rock being only slightly different from that of those nearby. *Crystallization differentiation, gravity separation, filter pressing, mixing of magmas, liquid immiscibility,* and *assimilation* are among the most important processes of differentiation (Fig. 4–16).

Crystallization differentiation. This process, also known as *fractional crystallization,* includes all processes that bring about a separation of the magma into two phases (a solid and a liquid), resulting, after crystallization, in two or more different types of rocks. This may be accomplished by complete or partial isolation of the liquid from the solid, or the migration of certain constituents of the magma into zones where they become concentrated. Cooling is much faster near the margins of a magma where it contacts cooler country rock than it is within the magma chamber. As the margins cool, a sort of insulating mat for the rest of the magma forms. Because cooling is proceeding more rapidly in one part of the chamber, convection currents develop and ions migrate, aided by diffusion. As these ions are incorporated in the crystals forming at the margins, the net composition of the magma is changed, and differentiation continues.

Gravity separation.[1] If the first crystals that form in a magma are heavier or lighter than the melt, they will sink or rise through the melt and become segregated either at the bottom or as a raft of floating minerals near the top of the magma chamber. This may give the resulting igneous rocks a banded appearance (Fig. 4–17). Olivine, the first silicate mineral to crystallize from a basaltic magma, is heavy and is found concentrated near the bottom of basaltic intrusions such as the Palisades sill (Fig. 4–17), along the west shore of the Hudson River in New Jersey. This process may also account for some ore bodies such as chromite, another early-forming high-density mineral. Chromite occurs in layers near the bottom of a number of large intrusions such as the Stillwater complex of Montana and the Bushveld complex of South Africa.

[1] Stokes' Law, which governs settling velocities, is described on page 159.

Filter pressing. As crystallization proceeds, a point is reached at which the once-molten mass is a mesh of crystals loosely interlocked and connected. Residual liquids surround these crystals and fill the pore spaces. If the magma chamber is deformed residual liquids are likely to be squeezed away from the compacting crystal mesh and become isolated. When the residual liquid solidifies it will be quite different from the earlier formed rocks in composition and possibly also in texture, because the residual fluids are likely to be deficient in those components required for the early-formed minerals.

Mixing of magmas. Some igneous rocks do not represent the systematic crystallization of a single magma. A magma may be intruded by another magma of a different composition. Thus all the rocks formed after this time will have a strikingly different composition from those formed earlier. Studies of Hawaiian magmas demonstrate that they become mixed with differentiation products

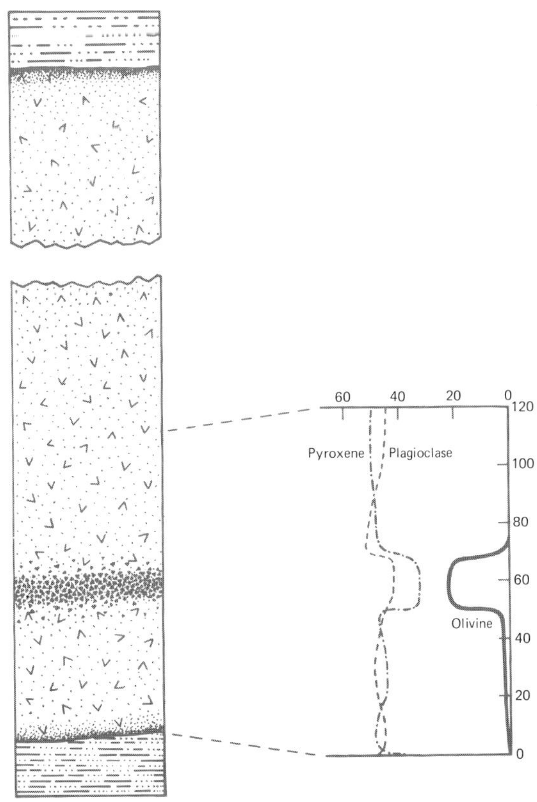

FIGURE 4 17
Section through Palisades sill (left), showing chilled margins. The graph at right shows the percentage of the principal minerals in the bottom part of the sill.
(from Walker, 1940)

of earlier magmas as they ascend, producing olivine-rich and olivine-deficient lavas. A much greater range of compositional variation would result from the mixing of a magma generated below the crust (basaltic in composition) with one produced by melting of sediments.

Liquid immiscibility. One liquid is said to be *immiscible* in another if it will not mix with it or will do so to only a very limited degree. Oil and water are immiscible liquids. The two become separated by the difference in their specific gravity. There is little evidence of immiscibility in silicate magmas, but experimental work with sulfide melts indicates that some of these are immiscible.

Assimilation (modification of a magma by the digestion of its margins). If two magmas, one 800°C and one 1,200°C, are intruded into the same type of country rock, the effects at the margins may be very different. Any minerals that melt at relatively low temperatures will melt where they come in contact with either magma, but those minerals with a fusion point above the temperature of the cooler magma will not melt. The hotter magma may cause pure melting of a thick zone surrounding the magma chamber and this new melt will mix with the magma and alter its composition, particularly around the margins. Even partial melting will cause a localized change in the composition of the magma.

Bowen's reaction series leads us to predict another type of assimilation process—reaction of the melt with minerals that formed earlier at a higher temperature.

		Oversaturated			Saturated			Undersaturated
		Quartz	Quartz and feldspar		Feldspars			Mafic minerals predominate
			Orthoclase	Plagioclase	Orthoclase	Na-rich plagioclase	Ca-rich plagioclase	
Deep intrusions	Coarse-grained phaneritic	Vein quartz	Granite		Syenite	Diorite	Gabbro	Peridotite
		← Porphyry (mixed textures) →						
Extrusions and shallow intrusions	Fine-grained aphanitic		Rhyolite	Dacite		Andesite	Basalt	Olivine basalt
	Glass		Pumice ← Pitchstone → Scoria; ← Obsidian →					
		LIGHT COLORS						DARK COLORS

FIGURE 4–18 *(above)* Classification of igneous rocks.

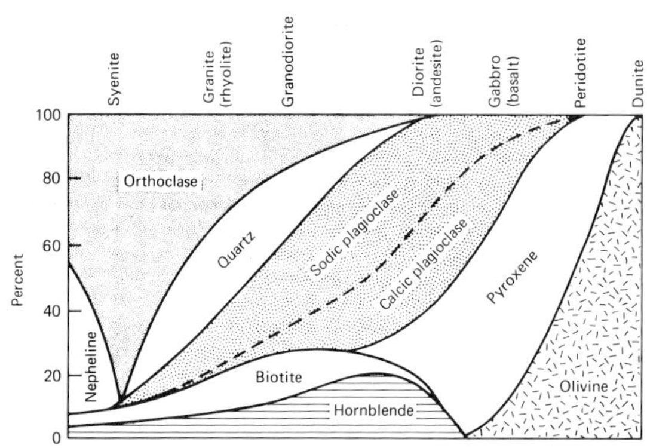

FIGURE 4–19 Diagrammatic representation of the composition, in terms of minerals, of the most important rock types. As indicated at the top of the diagram, a given assemblage of minerals may have two or more names, the particular rock name being determined by the texture.
(from Gutenberg, 1951)

CLASSIFICATION OF IGNEOUS ROCKS

Igneous rocks are classified on the basis of their composition and texture. Both of these range in an almost continuous fashion between extremes. The purpose of classification is to subdivide this continuum into meaningful and manageable categories of rock types that can be recognized easily and provide a useful and convenient basis for discussion and study. Specialists may use whole-rock chemical analyses and statistical counts of the mineral constituents to obtain a precise determination of a rock's composition. But the layman, the field geologist, and the student will find it more convenient to classify hand specimens by identifying mineral constituents and estimating their abundance. Only an approximate classification is possible with these techniques, of course. Minerals can be positively identified by optical techniques involving use of a petrographic microscope, by electron probes, and by use of X-ray techniques described earlier (p. 32).

Rock-type descriptions frequently specify that certain minerals are *essential*—that is, they are always present in a given rock by definition. Other minerals that are commonly present in small quantities may be termed *accessory minerals.*

A classification chart (Figs. 4–18 and 4–19) can be organized to reflect the relative abundance or degree of saturation of the magma with silica.

Certain minerals do not occur together (e.g., quartz and olivine) because, as in this sample, quartz forms only in the presence of an excess of silica whereas olivine is a silica-deficient mineral that will not form if free silica is present.

The term *texture* may apply to structural features and fabric, the way the individual grains fit together, but it is used most frequently in reference to size of the crystals. For hand-specimen identification the following textural terms are in use.

Aphanitic. Applies when the rock is crystalline but the individual crystals are too small to be seen or identified by the unaided eye. The extremes of this size are *microcrystalline* and *cryptocrystalline* (crystallinity is evident only in optical or X-ray study, respectively).

Phaneritic. Applied when the crystals are large enough to be seen with the unaided eye. Phaneritic textures may be subdivided into coarse-, medium-, and fine-grained.

Glassy. Applied to the texture when the material is amorphous, shows no crystals, and is like glass.

Porphyritic. Applied to textures in which two or more sizes are present, as when large crystals, *phenocrysts*, are in a groundmass of fine crystals or glass.

Pegmatitic. Applied to rocks that have extremely coarse crystals. The extremes of crystal size are indicated by the 12-meter crystal found in a pegmatite in the Black Hills. Individual crystals over an inch in length are much more common in pegmatites. Many pegmatites are dikelike bodies.

DESCRIPTIONS OF IGNEOUS ROCKS

Coarse-Grained Intrusive Rocks

Granite. A coarse- to medium-grained equigranular rock composed of potassium feldspars (orthoclase and microcline) and quartz, with some biotite mica and hornblende, usually present in small amounts. Many other minerals may be present but in small amounts. These include black specks of magnetite, honey-yellow crystals of sphene, red grains of garnet, and other less common minerals.

Porphyry composed of fine ground-mass and large crystals.

Graphic granite.

A coarse-grained granite.

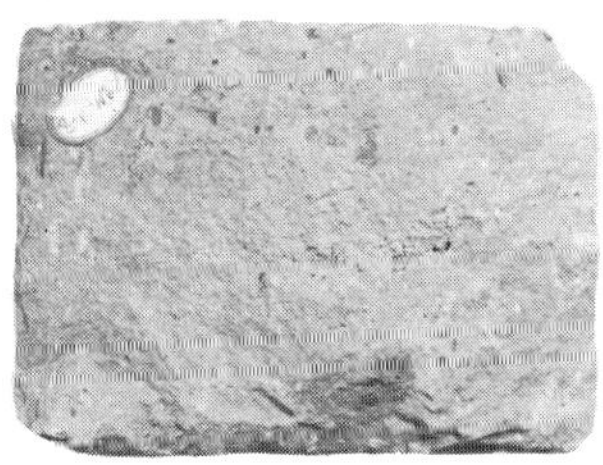

A fine-grained felsite.

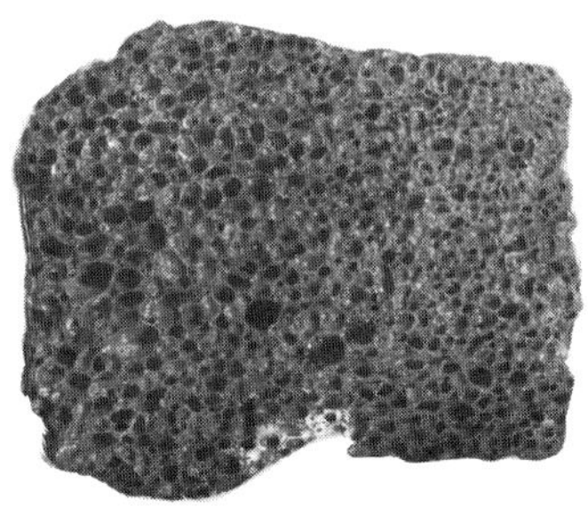

A porous, vesicular rock called scoria.

Obsidian, natural glass.

FIGURE 4–20
Textural variations in igneous rocks.
(U.S. National Museum)

Migmatite. Rocks of mixed nature containing metamorphic- and igneous-appearing rock are called *migmatite*. Sometimes the term is restricted to injection of granite into schist along the layering or schistosity of the schist.

Graphic granite. Rocks of granitic composition that have a texture consisting of an intergrowth of orthoclase and quartz in such a way that cross sections resemble certain hieroglyphics (Fig. 4–20).

Granite pegmatite. A very coarse-grained rock with granitic composition. It generally contains microcline, quartz, and mica (biotite or muscovite). A great variety of rare minerals may be present. Pegmatites are usually found near the borders of large plutons composed of rocks of finer grain size. Water-rich solutions may have been present in sufficient volume to permit ready migration of ions to develop large crystals.

Syenite. A coarse- to medium-grained equigranular rock composed primarily of orthoclase. Some hornblende, biotite, and pyroxene are present. Either a small amount of quartz or nepheline is usually present depending on degree of silica saturation.

Granodiorite. A medium- to coarse-grained plutonic rock that contains quartz, calcic plagioclase (oligoclase-andesine), and orthoclase as light-colored constituents and biotite, hornblende, or pyroxene as mafic constituents. It contains at least twice as much sodic plagioclase as orthoclase.

Diorite. A coarse- to medium-grained rock composed of intermediate plagioclase and biotite, hornblende, or pyroxene. Small quantities of quartz may be present. The amount of dark minerals ranges from 12 to 36 percent of the total.

Gabbro. A rock composed of coarse- to medium-grained pyroxene, hornblende, and biotite. The amount of these may exceed the amount of feldspar, which is calcic plagioclase. The mineral olivine is usually present, and quartz is absent. Most gabbros are dark because dark minerals predominate. They are equigranular. It is not uncommon to find gabbro pegmatites. Like the granite pegmatites, these are simply extremely coarse-grained equivalents of the gabbro probably formed from residual patches of water-rich magma.

Peridotite. A dark, coarse-grained, equigranular rock containing a large quantity of olivine and/or pyroxene or hornblende, but no quartz or feldspar.

Dunite. A rock composed almost completely of the mineral olivine; accessory pyroxene and chromite may be present.

Pyroxenite. A dark, coarsely crystalline rock composed mostly of pyroxenes.

Hornblendite. An igneous rock of coarse texture and nearly equigranular fabric composed mostly of hornblende.

Fine-Grained Igneous Rocks

Felsite. A general term applied to igneous rocks of fine-grained texture and light color indicating a granitic composition. If large crystals are present in a felsitic groundmass, it is called a felsite porphyry. Colors of felsites may range from white, gray, pink, yellow, brown, to purple and light green. When phenocrysts of quartz are present the rock is termed a quartz felsite or, more commonly, a quartz porphyry.

Basalt. A dark-colored, fine-grained rock composed primarily of plagioclase and pyroxene, with or without olivine. Basalts are dark because they contain large percentages of hornblende, pyroxene, biotite, olivine, or other dark minerals. It is often an extrusive igneous rock and may have cavities formed by bubbles of gas.

Diabase (Dolerite). A rock of basaltic composition, consisting mainly of labradorite (a plagioclase feldspar) and pyroxene, and with a texture characterized by euhedral or subhedral crystals of plagioclase embedded in a very fine-grained matrix of pyroxene. The term diabase is restricted by English geologists to altered dolerites.

Glasses

Obsidian. A glassy rock compositionally equivalent to granite or felsite; a solid, natural glass containing no crystals. The name has in recent years been somewhat restricted to glasses having a very low water content as contrasted with pitchstones and perlites. Most obsidian is black, but it also occurs in green and brown. It breaks with a *conchoidal* (curved) fracture. Natural glasses frequently contain round or spherical bodies composed of incipient crystals of feldspar and silica of white, gray, or red color, with a radiating, fibrous internal structure. These bodies are called *spherulites* and form from less rapid cooling of magma than glass. Also found in the natural glasses are the collapsed remains of glass bubbles, *lithophysae*. They form as gases escape from the glass. The bubbles break, and the hot glass flows back down but does not completely melt again.

Pitchstone. A variety of obsidian with a luster like that of a resin instead of glass. Pitchstone differs in composition from obsidian in that it contains 5 percent or more of water. Obsidian contains less than 1 percent. Colors include black, gray, red, brown, and green.

NOTE: The extrusive equivalents of the plutonic rocks are indicated in Table 3–1 and described in Chapter 3.

COMPARISON OF LUNAR AND EARTH ROCKS[2]

The rock samples brought back from the moon during the Apollo 11 mission bear a strong resemblance to mafic igneous rocks found on earth. The Tranquillity Base rocks are basaltic and gabbroic igneous rocks thought to be derived from near-surface melting of the crust of the moon. Microbreccia, a mechanical mixture of soil and rock fragments, and lunar soil composed largely of dust (see Chapter 6) make up the surface layer on which the crystalline rocks were found scattered as fragments. The basalts and gabbros have textures typical of their earth analogs. Some are very fine-grained with vesicles like vesicular basalt on earth, a few millimeters across, others contain larger cavities, and still others are medium-grained equigranular rocks. The minerals present in abundance include pyroxene, plagioclase feldspar, ilmenite (titanium-rich), olivine, and cristobalite (SiO_2). Three additional minerals not found on earth are present in these samples. Moon rocks differ in several significant respects from their earth counterparts: they show no effects of hydrothermal alteration; they are anhydrous; they contain free metallic iron as a common accessory mineral but no ferric iron; and some of them have unusually high titanium contents. Chemical analyses of some Apollo samples and a comparison of these analyses with some similar earth rocks are given in Tables 4–1 and 4–2.

[2] See *Science*, January 30, 1970, v. 167, for a full issue devoted to findings about Apollo 11 samples.

TABLE 4–1
Chemical compositions (weight percent) of Apollo II regolith and two rocks. Analyst, C. G. Engel. Al_2O_3 values corrected after X-ray fluorescence analyses for Mn and Cr by J. S. Wahlberg (U.S. Geological Survey, Denver, Colo.).
(from Engel and Engel, 1970)

	Apollo 11 Gabbro	Apollo 11 Vesicular diabase (basalt)	Apollo 11 Regolith (dust)
SiO_2	42.01	39.79	41.50
TiO_2	8.81	11.44	7.50
Al_2O_3	11.67	10.84	14.31
Fe_2O_3	.00*	.00 ?	.06 ?
FeO	17.98	19.35	15.62
MnO	.24	.20	.22
MgO	6.25	7.65	7.95
CaO	12.18	10.08	11.84
Na_2O	.48	.54	.48
K_2O	.11	.32	.16
H_2O^+	.00	.00	.00
H_2O^-	.00	.01	.01
P_2O_5	.08	.17	.10
Total	99.81	100.30	99.75

* Actual Value,–0.08.

TABLE 4–2 (*below*)
Compositions (weight percent) of lunar gabbro, average basaltic achondrite, and common terrestrial basalts. The H_2O^+ content in terrestrial oceanic basalt and alkali-olivine basalt due in part to secondary alteration.
(from Engel and Engel, 1970)

	Apollo 11 gabbro	Basaltic achondrite	Oceanic basalt	Alkali-olivine basalt
SiO_2	42.01	48.51	50.01	48.01
TiO_2	8.81	.48	1.37	2.92
Al_2O_3	11.67	13.04	16.18	15.97
Fe_2O_3	.00	1.11 ?	2.32	3.87
FeO	17.98	15.90	7.07	7.56
MgO	6.25	7.87	7.71	5.26
CaO	12.18	11.00	11.33	9.04
Na_2O	.48	.50 ?	2.79	3.73
K_2O	.11	.08 ?	.22	1.89
H_2O^+	.00	.07	.87	1.33
P_2O_5	.08	.19	.13	.42

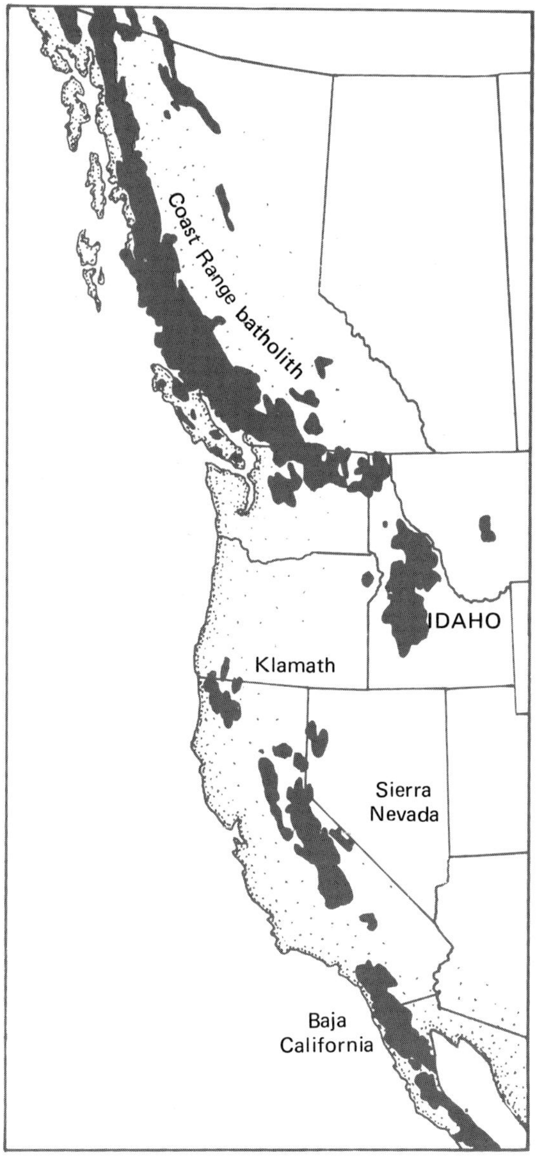

FIGURE 4–21
Location of major batholithic intrusions in western North America.

FIGURE 4–22
Xenoliths. This glacial boulder from North Sudbury contains basic xenoliths in a matrix of quartz.
(W. R. Hansen, U.S. Geological Survey)

Experiments with lunar samples indicate that the basalts and gabbros crystallized from igneous melts in the range of 1,060°C to 1,210°C and that crystallization took place in a highly reducing (oxygen deficient) system. The oxidation level was very much lower than for earth-derived basalts. Calculations based on the observed lunar compositions also indicate that the lunar magma had a viscosity an order of magnitude lower than earth basalts. This may account for the vast extent and flatness of the surface of the lunar maria.

RECOGNITION OF PLUTONIC BODIES

Igneous and metamorphic rocks contain many of the same minerals, and they may have similar textures and structural features. We need to establish criteria by which igneous plutons can be distinguished from metamorphic country rock and by which one intrusion can be distinguished from another of different age where the two are together. The controversy over the origin of granite has been mentioned. One approach to this problem is to set criteria by which a granite pluton can be identified. Igneous rocks are formed from a melt; so the high temperature and chemical effects associated with a rock melt may logically provide a key to the identification of plutons.

Because heat is conducted away from an injection at its contact with the surrounding country rock, a fine-grained or chilled margin may be formed. Coarser grains will be found at a distance from such a margin.

When large quantities of magma are intruded the heat and chemically active fluids in the magma cause alteration of the country rock. The degree of alteration dies out away from the contact zone as the heat decreases. These alteration effects may produce a baked contact zone followed by concentric zones of alteration around the pluton. Such effects are called *contact metamorphic aureoles.* Blocks of country rock become dislodged and sink into the magma as xenoliths. As they sink they are tilted or rotated and partially melted and assimilated by the magma. After erosion these are recognizable as rotated xenoliths—one of the better proofs of a melt.

Sharp cross-cutting boundaries are used to

identify igneous bodies. A dike, for example, is often intruded along a fracture or fault that cuts across layering and pre-existing structures. Thus the igneous rock may appear as a clearly defined cross-cutting mass (Fig. 4–24). The relative age of the rock masses can be determined readily since the younger rock cuts across older rock. Where multiple injections take place, their relative age can be determined from cross-cutting relationships (Fig. 4–25).

The internal structure of a pluton is distinctly different from that of the country rock, and the contact marks the boundary between the two more or less sharply. Within a pluton, elongate or platy minerals, drawn out clusters, xenoliths, and plates of minerals called *schlieren* will become aligned in the flow pattern (Fig. 4–23). The pattern of such flowage may reflect internal eddy currents but will almost always be parallel to the contact zone at the margin of the magma.

FIGURE 4–23 (*above*) Flow structure in granitic rock.

THE FORM OF PLUTONS

Factors Affecting the Emplacement of Intrusions

The regularity and symmetry of some plutons contrasted with the irregular form of others poses intriguing questions about how these bodies are emplaced. Why are some plutons injected as thin layers whereas others form bulbous masses or appear to engulf large areas of country rock? We can study the shape of plutons by mapping their aerial distribution, by drawing cross sections where they have been uplifted and dissected by erosion, and by examining their relationship to the enclosing country rock. And we can obtain some understanding of the mechanics of intrusion by considering the physical properties of rock melt.

The pre-existing structure and strength of the rock into which the magma is injected, the depth at which intrusion takes place, and the relation of the timing of intrusion to deformation that may accompany the intrusion, all play a role in shaping the form of intrusions. Factors related to the physical properties of the magma include temperature, composition, and, in particular, viscosity. The rate at which the intrusion takes

FIGURE 4–24 Basalt dike. (Crown Copyright)

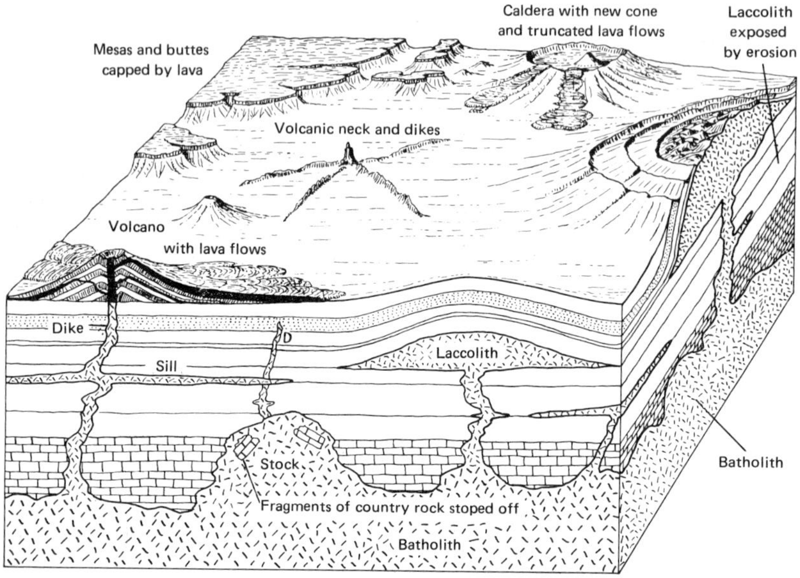

FIGURE 4–25 Occurrence of igneous rocks. (from Frederick Young)

place and the rate of cooling are also important determinants of structure and form.

Some igneous rock bodies cool and crystallize where the magma was generated; they form *in situ,* but magma tends to be mobile because of its fluidity, high pressure, and because rock melt has a lower density than rock. All of these factors tend to produce an unstable condition. Fluids normally move in the directions of least resistance; thus most magmas rise toward the surface following lines of weakness in the overlying country rock. The magma may be injected along bedding planes between strata and along fractures. Large bodies of magma must move by forcing the overlying rock aside or upward into domes, by melting the country rock, or by enveloping blocks of country rock. When blocks over the magma are wedged loose and sink into the magma, the process is referred to as *magmatic stoping.*

The viscosity of magma is determined by its composition, temperature, and the amount of water in the magma. Low-viscosity (highly fluid) magmas transmit pressure rapidly and readily enter any zone of weakness, such as fractures or bedding planes, in the rock adjacent to the magma. Highly viscous magmas do not spread so readily, and pressure within a highly viscous magma is increasingly directed upward as the body spreads laterally. As a result, high-viscosity magmas tend to develop bulbous forms whereas low-viscosity magmas extend farther from their source and along smaller passageways.

Viscosity also determines the capacity of the magma to slow the settling velocity of blocks stoped from the roof and sides of the magma body and thus accounts for the presence of xenoliths near the margins and roof of intrusions. A number of related factors, particularly higher temperature and increased fluidity, combine to reduce the number of xenoliths preserved in low-viscosity magmas. High temperatures favor melting and assimilation as well as more rapid sinking of xenoliths through the magma. Fluid intrusions are also more likely to become differentiated by gravity separation of early-formed high-density minerals.

If a magma is intruded slowly so that cooling along the edges substantially increases the viscosity at these edges, the portion of the magma over the feeder system will grow increasingly thicker. Pressure in this central area will accentuate the doming over the feeder and the flanks will become increasingly steeper. Ultimately the flanks will approach being vertical. If at this point the pressure in the remaining highly viscous magma is sufficient to move the overlying rock, a break roughly circular in plan is likely and further movement occurs as a more or less cylindrical mass is forced upward.

CLASSIFICATION OF PLUTONS

The shape or form of a pluton and its structural relationship to the enclosing rock provides a useful basis for classification of intrusions[3] (Fig. 4–25). Despite this, a large number of bodies have irregular shapes that cannot be readily classified.

A. **Injected masses.** Masses entirely enclosed by the invaded formations, except along relatively narrow channels.

I. **Concordant** (injected along planes of stratification, layering, or schistosity).

[3] The classification that follows (with brief definitions) is one modified and simplified after a classification proposed by R. A. Daly in 1933.

FIGURE 4–26
Columnar structure in lava flows exposed in a cliff near Tower Falls in Yellowstone Park.

1. **Sill.** Sheetlike body with large lateral extent relative to its thickness. Transitions occur between sills and laccoliths. These are termed multiple if there are successive injections of one kind of magma, and composite if there is a compound intrusion of successive injections of more than one kind of magma.

2. **Laccolith.** A planoconvex or doubly convex lens, flattened in the bedding plane of the invaded formation. The roof is arched over the laccolith. These may also be multiple and composite intrusions.

3. **Lopolith.** A lenticular concordant intrusive mass in which the thickness is approximately one-tenth to one-twentieth of its width. The central portion is concave upward.

4. **Phacolith.** Intrusion of lensoid shape in cross sections, located at the hinge of folds. It may extend parallel to the axis of the fold.

II. **Discordant** (injected across planes of stratification or schistosity).

1. **Dike.** Injected body with parallel or subparallel walls that is narrow relative to its lateral extent.
2. **Dike swarm.** Many dikes of similar trend or orientation occurring together.
3. **Intrusive vein.** When the path of a discordant injected body is less regularly defined than is true of dikes the wavy threadlike protrusion is called a vein.
4. **Apophyses or tongues.** Dikes or veins that can be traced to larger intrusive bodies as the source of magmatic supply.
5. **Ring dike.** A dike of arcuate to circular outcrop.
6. **Cone sheet.** A dike of arcuate outcrop and regular inclination toward a focus.
7. **Volcanic neck.** Solid lava occupying a volcanic vent.

B. **Subjacent masses.** Bodies that are not visibly floored. In many cases the mode of emplacement is uncertain.

I. **Batholith.** "A batholith is a stock-shaped or shield-shaped mass intruded as the result of the fusion of older formations. On the removal of its rock cover and continued denudation, the mass either holds its diameter or grows broader to unknown depth" (Suess, 1918). (Usage confines batholith to masses of more than 103.6 square kilometers surface exposure.)

II. **Stocks and bosses.** These two terms are used almost synonymously. They refer to masses similar to batholiths except that they are smaller. Bosses are stocks of nearly circular ground plan.

FIGURE 4–27 (*above*) Dike swarms at the Spanish Peaks of Colorado. Stocks are located near the center of the swarm. (redrawn after Johnson, 1961)

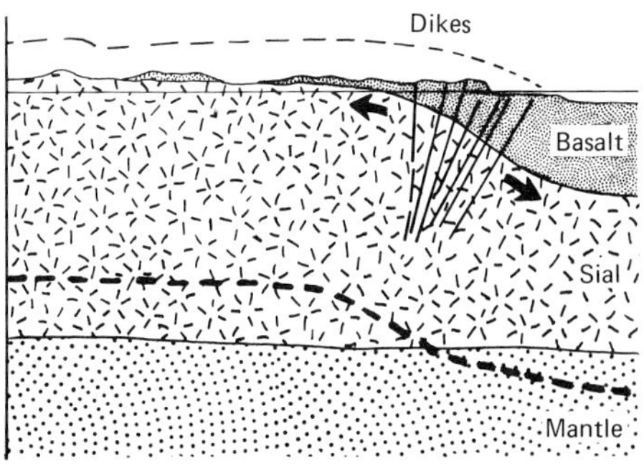

FIGURE 4–28 Vertical section along longitude 77°25′ showing the observed thickness of the lavas and the relation of the dike swarm to the crustal flexure. (from Wager and Deer, 1938)

Dikes

Dikes are often useful keys to geologic history and structure. Because they extend outward and upward from large bodies, at least a maximum age can be determined for the igneous activity by observing the youngest strata into which they intrude. Because they follow joints and faults, dikes give useful information about the existing patterns at the time of intrusion. Sometimes these are early planes of weakness opened during the intrusion; others, identified by their unique patterns, appear related to stresses that accompanied the igneous activity itself. The ring dikes and cone sheets are examples of this, and the radiating dike swarms sometimes found around centers of eruption and intrusion, as at Ship Rock, N. Mex.,

Spanish Peaks, Colo. (Fig. 4–27), and the Crazy Mountains, Mont. Another example is the dike swarm located in the zone of flexure (bending) along the coast of Greenland (Fig. 4–28). Dikes follow fractures formed as a result of extension of the country rock during flexure. That the most common dike patterns are identical with fracture and fault patterns emphasizes that dikes are intruded along the lines of least resistance.

Thin long dikes must have been formed of low-viscosity magma, and they are usually interpreted as having been injected rapidly. Dikes are usually continuous for many miles; sources of most basaltic dikes cannot be seen even in deeply dissected mountains where erosion relief in excess of 1.6 kilometers allows them to be traced at some depth.

The largest dike known is the Great Dike of Rhodesia, an ultrabasic body that is about 500 kilometers long and ranges from 3 to 12 kilometers in thickness. It is unusual in other respects, notably in that it has a lensoid layering that has been explained as having originated as a result of intrusion from several centers along its length.

Swarms of dikes occur in many places. Notable examples are found in Scotland where dikes have been separated into three major age groups, Tertiary dikes of predominant northwest trend, Late Paleozoic dikes of east–west trend, and Caledonian (Middle Paleozoic) dikes of east-west and northeast trend (Fig. 4–29). The large numbers of dikes and the consistency of trends make this one of the most unusual dike systems. No comparable system is known in North America.

FIGURE 4–29
Trends of Tertiary dikes in Scotland.
(from Richey, 1939)

Laccoliths

The Henry Mountains of south-central Utah afford one of the best known examples of intrusions of laccoliths. It was in this area that G. K. Gilbert first recognized and named this type of lensoid intrusion in which the magma has moved along bedding surfaces and locally accumulated, forcing the overlying strata to arch. The Henry Mountains are located on the Colorado Plateau in a region where flat-lying Mesozoic sedimentary strata cover large areas. Local warps and elongate, broad, open folds are broken by regional fracture and fault systems. The intrusions consist of a number of large stocks with associated laccoliths.

Location of the stocks shows a distinct align-

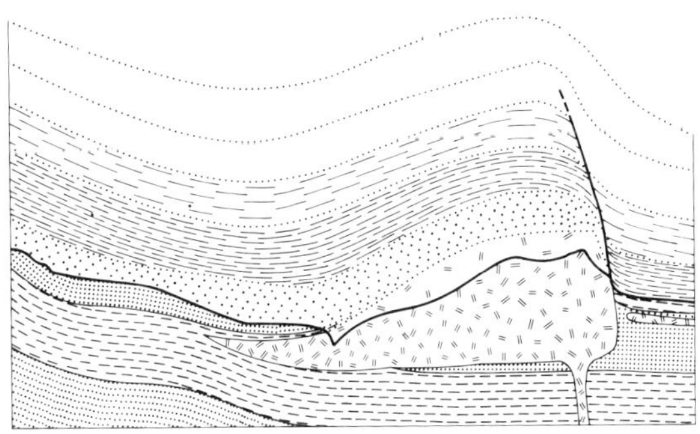

FIGURE 4–30
(a)
Cross section through Mount Marcellina, West Elk Mountains, Colo. Section is about 11.2 kilometers long.
(after Cross)

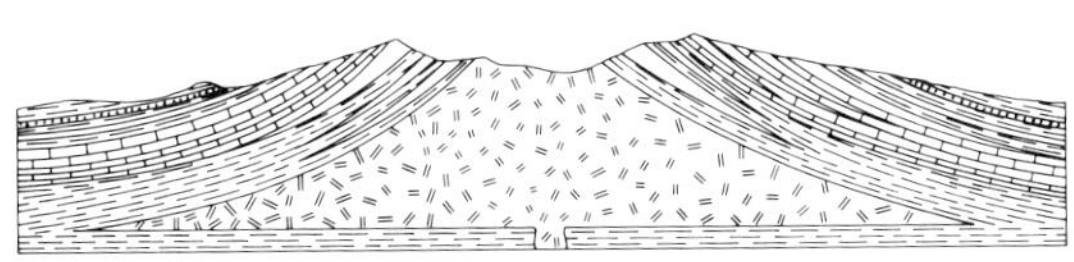

FIGURE 4–30
(b)
Section through the Alpine laccolith, Judith Mountains, Mont.
(after Weed and Pirsson; from Hunt, 1953)

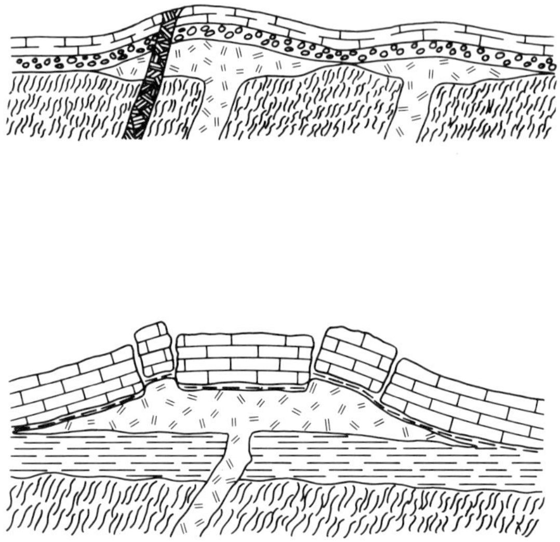

FIGURE 4–31
Diagrammatic cross sections of some laccoliths in the Black Hills at Deadwood Gulch (above) and Ragged Top laccolith (below).
(after Irving; from Hunt, 1953)

ment along regional fractures to which the feeder systems are probably related. The intrusions consist of a central stock that is roughly circular in plan and that has been the source of supply for magma that has moved laterally into surrounding sedimentary units, forming the laccoliths. In some instances the magma has broken across strata, moved upward in the section, and formed other laccoliths.

The sedimentary formations are turned up at the margins around each of the stocks, but they are well enough preserved to allow calculated restorations, which show that if they were restored to their original horizontal position the space now occupied by igneous rock would be closed. Thus the intrusions must have been emplaced by forceful injection and not by stoping or partial melting. By making a comparative study of the dimensions of the various laccoliths, Hunt (1953) was able to determine that at least 800 meters of uplift in an area 9.6 kilometers in diameter could be accomplished by stretching of the strata without creating space for cross-cutting intrusions. Central stocks are found in all larger domes, and there is greater shattering of country rock, more contact metamorphism, a greater variety of magma types, and much greater volume of satellitic intrusions around the larger stocks.

The laccoliths formed around the stocks primarily by lateral movement of magma from the stocks; this magma moved out to form tongue-shaped masses. But some magmas developed a more ideal laccolithic form where a feeder system broke up and through strata to provide a central feeder. Others formed laccoliths with the shape of linear bulges. There exists a great variety of forms reflecting local structure and conditions, but the laccoliths are all of similar origin (Fig. 4–31).

Intrusion of magma from the central stocks into the strata is clearly influenced by the character of the country rock. The greatest number of laccoliths and the largest volume of magma are located in the weak shale and clay units and particularly at the contact of sandstone beds within the shales.

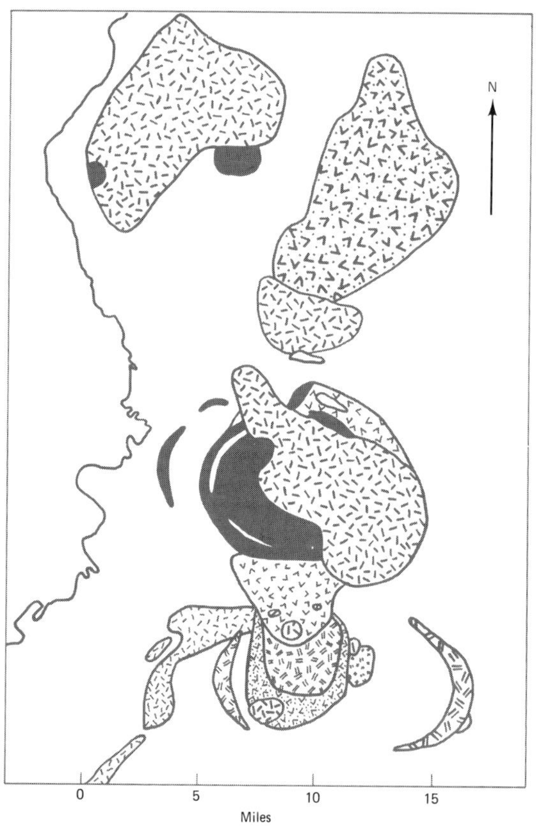

FIGURE 4–32
Stocks and dikes with arcuate and circular outcrop in New Hampshire.
(after the geologic map of New Hampshire compiled by M. P. Billings, 1955)

Ring Complexes

Circular or oval intrusive and extrusive complexes are known in many parts of the world. Those of Scotland were the first to be studied intensively, but similar bodies occur in New Hampshire, and other good examples are found in Nigeria and South West Africa. These bodies have the following general characteristics:

1. The outcrop patterns are circular, oval, or arcuate in form.
2. The bodies are frequently compound and/or composite. A great variety of igneous rock types may be represented.
3. Outcrops indicate a variety of structural forms, often closely associated, including: (a) circular, near-vertical dikes, ring dikes; (b) circular and arcuate dikes that dip toward the center and that, if projected at depth, would form cones, cone sheets; (c) dikes that dip steeply away from the center; (d) central plugs that appear to have the shape of steep-sided cylinders; (e) plugs that have outward-dipping margins.
4. The complexes sometimes are covered partially by lavas or tuffs.
5. Platy minerals and other planar structures are often oriented so that they are inclined inward.

A geologic map of the ring dikes in New Hampshire (Fig. 4–32) illustrates the complex nature of these structures. Because they are dissected deeply only in mountainous terrain and because there is little subsurface information, the nature of the structures at depth is largely hypothetical. The ring patterns are thought to originate over shallow intrusions as a result of forceful injection of a more or less circular stock or collapse of the roof over a shallow intrusion. The failure patterns are cylindrical, conical, and parabolic fractures. Further development of the structure depends on what happens to the central blocks. Depending on the density contrast between the country rock and magma, viscosity of the magma, and pressure within the magma, as well as geological factors such as the presence of a floor to the magma chamber at shallow depth, the central blocks may rise, subside, or tilt. A series of simplified models (Fig. 4–33) depicts a wide range of possibilities.

Subjacent Plutons

Subjacent plutons vastly exceed all other types of igneous rocks in area of outcrop and in volume.

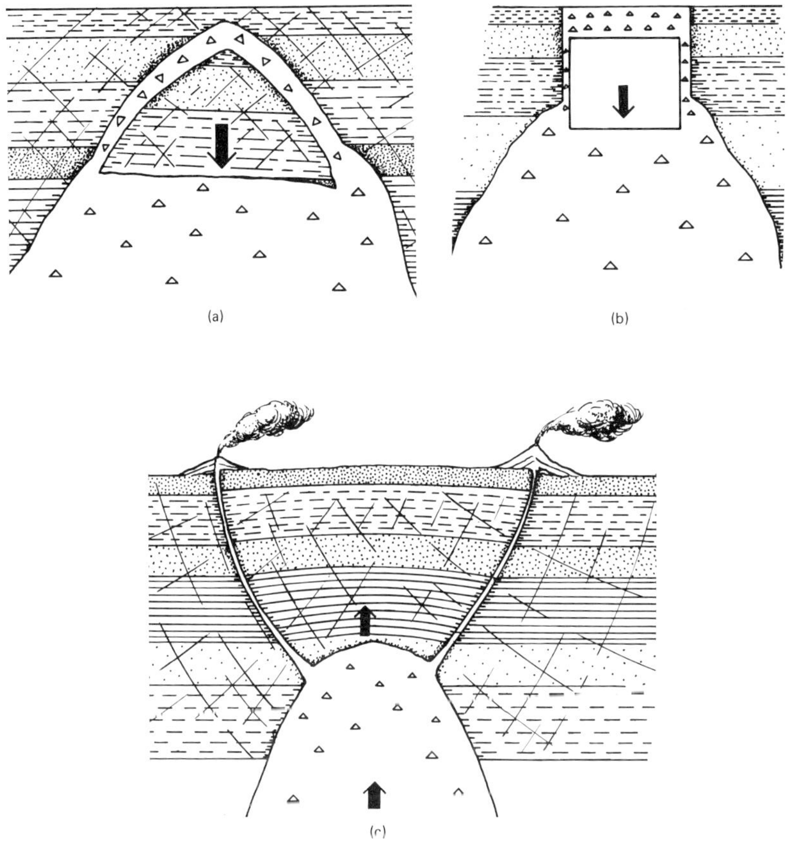

FIGURE 4–33
Cross sections illustrating the formation of cone-sheet (a and c) and ring-dike (b) intrusions.
(modified after Chapman, 1966)

However, a number of problems concerning batholiths remain unsolved, and the chief of these is the determination of their mode of origin and emplacement.

Most of the subjacent plutons are located on continents or marginal islands like Japan. Typically they are coarse-grained rocks consisting of quartz and potash feldspar, with some dark minerals such as biotite and hornblende. Most such bodies are thus granitic in composition. It has been estimated that more than 90 percent of all known plutons are of this character whereas basic rocks, basalt and the like, are the most abundant extrusives. However, some gabbro plutons and many small ultramafic bodies are found on continents in the folded mountain belts, and large ultramafic intrusions are known in the circum-Pacific belt.

ORIGIN OF PLUTONS

Large granitic masses can originate in the following different ways:

1. Magmatic origin and emplacement.
2. Metasomatic origin—products of granitization.
3. Combinations of magmatic and metasomatic origin.

Masses of the first class, magmatic, are igneous plutons in which a granitic magma was generated by one of the following:

1. Melting of a granitic layer in the crust.
2. Melting of pre-existing sediments.
3. Differentiation from an original basaltic magma.

These magmas are emplaced by forceful injection into the crust, by magmatic stoping, or by solidification in place. Magma injected forcefully must displace country rock by causing it to be domed or uplifted, or by forcing it aside by folding.

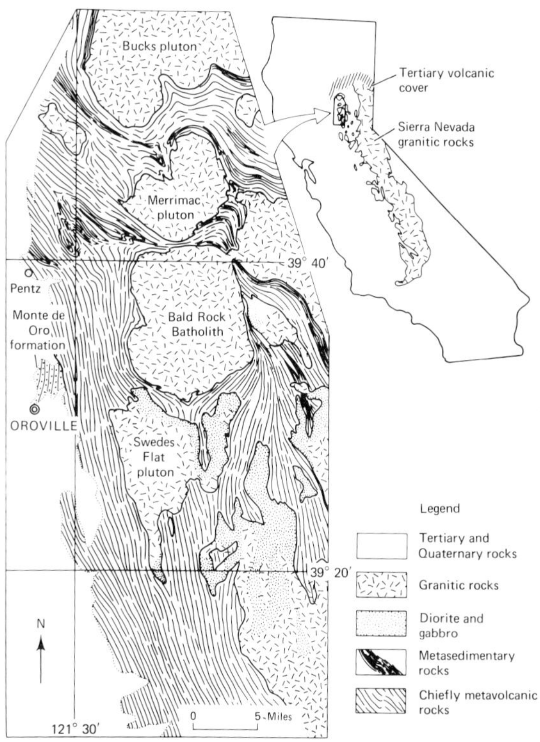

FIGURE 4–34 Geologic map of a part of the northwestern Sierra Nevada Mountains. Lined pattern shows general trends of bedding and foliation in metamorphic rocks; outline map of California indicates position of area with regard to main Sierra Nevada pluton. (from Compton, 1955)

The process of magmatic stoping is accomplished by the magma's invasion of the country rock along fractures, bedding surfaces, or other lines of weakness accompanied by foundering of the detached blocks of country rock in the magma. This process would be accelerated by doming, which would open fractures and cause new ones to form as a result of stretching of the country rock over the dome. Fractures that permit stoping can form only at shallow depths in the crust.

Many granites once classed as subjacent plutons are now considered to be products of a metasomatic process called *granitization*. Granitization converts original sedimentary rocks into granites by liquid or gaseous diffusion. This conversion requires some changes in the bulk chemical composition of the sediments, notably addition of silica and alkalis and removal of calcium, iron, and magnesium. Primary and secondary sedimentary structures preserved in some of these granites indicate that they are not of igneous origin.

The third class consists of plutons of magmatic origin with altered margins and metasomatic granites with localized partial melting.

Daly (1933) summarized the characteristics of subjacent plutons as follows:

1. Location in orogenic belts.
2. Elongation parallel to the tectonic axis of the mountain range.
3. Date of intrusion following, more or less closely, an antecedent period of mountain building.
4. Cross-cutting relations to the country rocks.
5. An irregularly domical roof, commonly with projections downward and re-entrants upward.
6. Steeply inclined walls.
7. Relative smoothness of the walls.
8. Downward enlargement; no floor visible.
9. The appearance of having replaced the invaded formation during the intrusion.
10. Composition usually felsic; relatively homogeneous to visible depth.
11. Great volume with corresponding long magmatic life, slow cooling, considerable ultimate contraction, and resulting necessity of prolonged readjustments of the

material of both batholith and intruded terrain.

Bald Rock Batholith, California

The Bald Rock Batholith is one of a chain of small plutons in the northern part of the Sierra Nevada Mountains intruded into Late Paleozoic rocks that had been deformed and metamorphosed. The intrusion appears to have been the result of forceful injection accompanied by stoping. There is local cross-cutting, but foliation and layering in the country rock is concordant with the edge of the batholith around three-fourths of the contact. Evidence of stoping is found in the large concentric outliers and a "hull" of migmatite in the contact zone (Fig. 4–34).

That the country rock at the edge of the batholith was forced aside during injection is supported by the observation that rock layers near the east and west contact were thinned, presumably by stretching. Stretched pebbles and hornblende prisms parallel to the direction of stretching can be found.

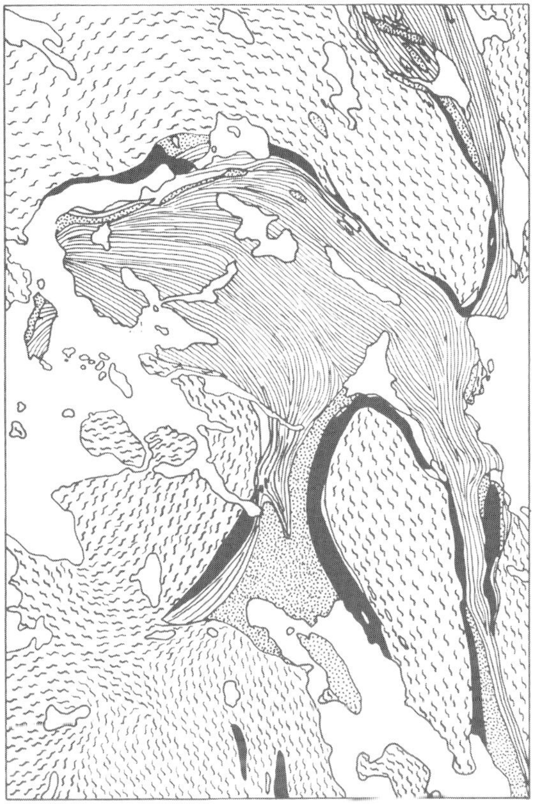

FIGURE 4–35
Mantled gneiss domes near Kuopio, Finland. Wavy pattern indicates granite and gneiss; black shows distribution of quartzite that borders many of the domes; solid lines indicate schist; dots show amphibolite; white indicates water.
(simplified after Eskola, 1949)

MANTLED GNEISS DOMES

Domes composed of layered granitic rock and more or less encased in a mantle of sedimentary rocks in such a way that the layering or foliation of the gneiss is parallel to stratification at the marginal contact occur in many folded mountain belts.

A number of sedimentary materials are found mantling the domes in Finland (Fig. 4–35), but the most diagnostic are those containing a basal conglomerate with boulders composed of the same gneiss that makes up the dome. Thus the gneiss dome has been intruded into a rock produced by the erosion of the gneiss. In some places the layering is truncated at the contact, but more frequently the layering is parallel to the contact and the contact is a zone of migmatite with a strongly deformed and veined structure. These contacts are also zones of granitization (Eskola, 1949).

Granite cuts across the contact locally, indicating that the gneisses became mobilized or that a magma was formed by partial melting during formation of the migmatite.

The mantled gneiss domes appear to represent an earlier granite intrusion or metasedimentary rock that was truncated by erosion, received a cover of sediment, and then was remobilized in a subsequent period of deformation. Similar domes are found in many folded mountain belts. The gneisses at Baltimore, Md., are of this nature.

ECONOMICALLY IMPORTANT DEPOSITS

Igneous rocks contain many of the largest ore bodies. *Ore* is defined as a mineral that may be used to obtain one or more metals at a profit. It must be economically feasible to mine an ore before the deposit qualifies as an ore legally. Many ore deposits originate from solutions that emanate from magmas and move into metamorphic or sedimentary rocks where the ore minerals are deposited. The deposits are generally called *hydrothermal deposits.* The close association between ore-forming solutions and magma is well established. Most of the famous mining camps of the western United States are located in or very near igneous intrusions. In addition, magnetite, galena, and sphalerite have been found growing around fumaroles at Katmai Volcano, proving that gases and solutions in volcanoes have a capacity for carrying metal ions.

Ore minerals normally constitute a minor part of the whole of an igneous rock and are economically important only when concentrated. However, a few minerals (e.g., diamonds) are of such great value (or are maintained at a high price by a cartel that limits supply) that it is economical to mine even large volumes of rock in order to recover the disseminated minerals. Ore minerals may be *disseminated* throughout the igneous rock as diamonds are in the *pipes* of South Africa and southwestern Arkansas. These pipes are carrot-shaped bodies of ultramafic rock (peridotite, kimberlite, or dunite) hundreds of meters in diameter and 1,000 meters or more deep. The ultramafic rocks are intrusions from very great depth—perhaps from the earth's mantle where the pressures and temperatures are sufficient to generate diamonds.

Ores may be concentrated in magma chambers by processes of magmatic segregation and magmatic concentration, as when early-formed heavy minerals sink to the floor of the chamber or when early-formed light minerals rise to the top. The segregated minerals may also be injected into the country rock as dikes, sills, or veins. The chromite deposits of the Stillwater complex, Montana, and the Bushveld, South Africa, are good examples of gravity-induced segregation. Chromite crystallizes early from a magma and the chromite is found in layers of both these intrusive complexes near their bottoms. The nickel deposits of the Sudbury, Ontario, lopolith have also been interpreted as segregation and injection deposits.

When a magma is mostly crystallized, the last water-rich silicate solutions may form pegmatites that often contain rare minerals.

The hydrothermal solutions left after pegmatite injection move upward carrying metals as well as other elements and losing heat as they go. Composed primarily of water, these solutions also carry many mineralizing agents. The dilute solutions can move through fractures and fault zones, through pore spaces, cavities, along bedding planes, and even through crystal lattices. Volcanic pipes, breccia, and cooling cracks also afford the necessary openings. Cooling of the solutions as they move causes deposition because solubility is decreased by decreasing temperature and pressure. Chemical reactions between the solutions and country rock, especially carbonates and andesites, also account for ore deposits.

Hydrothermal deposits are classified according to the depth at which they have formed.

1. **Hypothermal deposits.** The deepest are associated with metamorphic rocks, intensely folded rocks; they are lenticular in shape and commonly contain *native gold,* chalcopyrite, tin, and molybdenite.
2. **Mesothermal deposits.** Intermediate depth deposits tend to be in more continuous veins; there is low-grade metamorphism, if any, and they are often associated with granites and diorites. These deposits may contain lead, zinc, gold, and copper.
3. **Epithermal deposits.** Shallow depths of formation; they occur in planar veins, fractures, and usually have simple mineralogy.

Deposits from hydrothermal solutions may occur as a direct precipitation of the ore mineral from solution in a cavity as a result of reduced solubility, lowered temperature, changes in pressure, and reaction of the solution on contact with other solutions, especially ground water. Metasomatic replacement, a process by which there is simultaneous solution and deposition as new minerals grow at the expense of existing minerals in a rock, is also very important. The size and texture of the original minerals may be preserved as the new mineral grows to assume the space and shape of the replaced mineral, creating a *pseudomorph.*

CHAPTER OUTLINE

TERMS

accessory minerals
aphanitic
apophyses (or tongues)
assimilation
basalt
batholiths
cotectic line
concordant injections
cone sheets
contact metamorphic aureoles
critical temperature
crystal nuclei
crystallization differentiation
diabase
dike swarms
dikes
diorite
discontinuous reaction series
discordant
dunite
epithermal deposits
fault
felsite
filter pressing
flexure
fractional crystallization
gabbro
glassy
granite
granite pegmatite
granitization
granodiorite
graphic granite
gravity separation
hornblendite
hypothermal deposits
igneous
isotherms
laccoliths
liquid immiscibility
liquidus
lithophysae
lopoliths
magmas
magmatic stoping
mesothermal deposits
metasomatic replacement
migmatite
miscible
mixing of magmas
multiple dike
obsidian
ore
pegmatitic
peridotite
phacoliths
phaneritic
phase
phase rule
pitchstone
pseudomorph
pyrolite
pyroxenite
ring dikes
schlieren
sills
simple eutectic system
solid-solution series
solidus
spherulites
stocks and bosses
subjacent masses
sublimation
supercooling
syenite
ternary eutectic
volcanic necks
xenoliths

REVIEW QUESTIONS

1. How would you distinguish a lava flow from a sill of similar composition?
2. What effect does increasing water pressure have on the temperature at which granite and basalt melt? See Fig. 4–2.
3. Describe the crystallization of a magma formed by melting rock composed of 80 percent diopside and 20 percent anorthite. Assume that crystallization is slow. Describe the composition of early-formed crystals and the groundmass. Refer to Fig. 4–8.
4. Summarize evidence that supports the idea that most crustal rocks have been derived from the mantle and outline the processes by which this is possible.
5. Explain why igneous rocks do not normally contain both olivine and quartz. Refer to Fig. 4–9.

6. Refer to Fig. 4–11 and describe the crystallization of a melt of composition 60% C, 20% B, and 20% A.

7. What would be the effect according to Bowen's reaction series if the reactions do not go to completion in the alteration of an early-formed olivine to pyroxene and hornblende?

8. What processes of magnetic differentiation were thought to be important in causing the observed variations in composition of Hawaiian basalts (see Fig. 3–22) described in the preceding chapter?

9. How does the composition of lunar rocks compare with that of earth rocks?

10. Summarize the criteria used to recognize igneous intrusions.

SUGGESTED READINGS

Bayly, Brian, 1968, Introduction to Petrology: Englewood Cliffs, N.J., Prentice-Hall, 371 p.

Daly, R. A., 1933, Igneous Rocks and the Depths of the Earth: New York, Hafner Publishing Co., 598 p.

Engel, A. E. J., and **Engel, C. G.,** 1970, Lunar rock compositions and some interpretations: Science, v. 167, no. 3918, p. 527–528.

Ernst, W. G., 1969, Earth Materials: New York, Prentice-Hall, 149 p

Hamilton, W., and **Myers, W. B.,** 1967, The nature of batholiths: U.S. Geol. Survey Prof. Paper 554-C, 30 p.

Krauskopf, Konrad B., 1967, Introduction to Geochemistry: New York, McGraw-Hill, 721 p.

Mason, Brian, 1952, Principles of Geochemistry (3rd ed., 1966): New York, London, Sydney, John Wiley and Sons, 329 p.

Simpson, Brian, 1966, Rocks and Minerals: Oxford, Eng., Pergamon Press, 302 p.

5
METAMORPHISM

5 META-MORPHISM

WHAT IS METAMORPHISM?

The term metamorphic means literally "change of form." Sir Charles Lyell (1833) introduced the concept of *metamorphic theory* by pointing out that fossiliferous formations of various ages, some as young as Tertiary in the Alps, have been converted to crystalline rocks such as gneiss, mica schist, and marble.

> The transmutation has been effected by the influence of subterranean heat acting under great pressure, and aided by thermal water or steam and other gases permeating the porous rocks, and giving rise to various chemical decompositions and new combinations. . . . To this . . . action the fusion of granite . . . as well as the development of the metamorphic texture in sedimentary strata, may be attributed; and in accordance with these views the age of each metamorphic formation may be said to be twofold, for we have first to consider the period when it originated, as an aqueous deposit, in the form of mud, sand, marl, or limestone; secondly, the date at which it acquired a crystalline texture.

More recently, Turner and Verhoogen (1960) have defined metamorphism as:

> The mineralogical and structural adjustments of solid rocks to physical and chemical conditions which have been imposed at depths below the surface zones of weathering and cementation and which differ from the conditions under which the rocks in question originated.

whereas Arie Poldervaart defines it as

> The response of rocks to changes in energy environment.

Metamorphic rocks are the most enigmatic of all rocks. Although we can observe directly the initial stages in the formation of sedimentary rocks as well as the crystallization of rock melts, nowhere can we observe directly the development of metamorphic rocks. They can be seen around the edges of plutons and in the cores of mountain ranges. They may contain minerals that do not occur in either sediments or igneous rocks. Sometimes they have structural features that suggest strong deformation. How do these unusual mineral assemblages develop? What connection is there between these rocks' structural features and their process of formation? How can we account for their unusual textural features?

Metamorphic changes may be thought of as adjustments oriented toward production of a chemically and physically stable rock under conditions different from those under which the rock formed. For example, clay is stable under surface conditions but at higher temperatures and pressures both physical and chemical changes occur. Water is pressed out of the clay and the elements present become involved in chemical reactions that produce new minerals. Since these processes cannot be observed directly, we must deduce the processes and changes from relationships we see in the field (e.g., proximity of certain mineral suites to the contact of a pluton). The processes can be further clarified by use of controlled experiments.

Field Evidence That Metamorphic Changes Do Take Place

Metamorphic rocks consist of an interlocking network of crystals much larger than those of most sedimentary materials; yet field observations indicate that many coarsely crystalline metamorphic rocks were once sediments. Fossils and primary sedimentary structures such as ripple marks, cross-bedding, and graded bedding can be recognized in some metamorphic rocks. Such primary features are called *relics.* It is possible to trace the transition of an unmetamorphosed sequence of sedimentary units through a progressively recrystallized metamorphic sequence in the contact zones of some dikes and sills and around the margins of large stocks and batholiths.

METAMORPHIC TEXTURES

Metamorphic rocks may be classed in one of two textural groups, those that are foliated and those that are unfoliated. The terms *foliation* and *schistosity* are applied to any parallel or nearly parallel planar or linear structure or mineral fabric (Fig. 5–1). A number of structures may be responsible

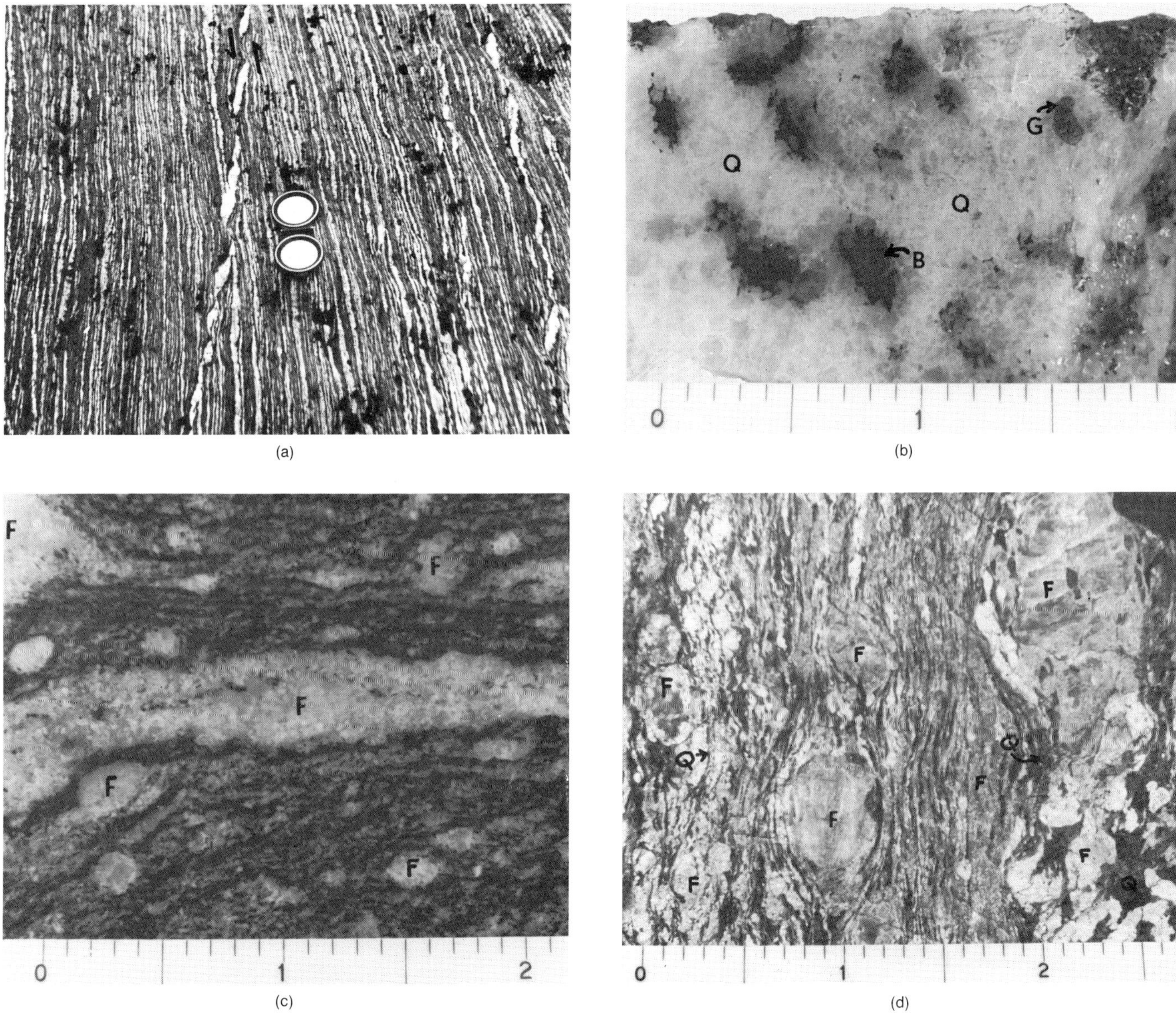

(Scales in inches)

FIGURE 5–1
Textures of metamorphic rocks. Banded texture typical of gneiss is shown in (a) and (c). Note the offsets in the cross-cutting quartz vein in (a). Feldspars are seen growing in (c) and (d). In (b) garnets are growing in a quartzite. The dark minerals in (a), (c), and (d) are predominantly biotite and hornblende. Q = quartz, F = feldspar, G = garnet, B = biotite.

for this planar *fissility* (a property of splitting along closely spaced parallel planes).

1. The parallel orientation of masses of tabular, platy, or prismatic minerals such as micas, chlorite, sericite, or hornblende. If all of these are oriented so that they lie in parallel planes, the rock produced will break much more easily along that plane.
2. Finely spaced planes of slippage, fracture, or shear produced by directed stress.
3. Bands or layers of different minerals that give the rock a laminated structure. These may be produced as a result of segregation of mineral assemblages during recrystallization and recombination.
4. Crystals flattened or drawn out as a result of deformation.

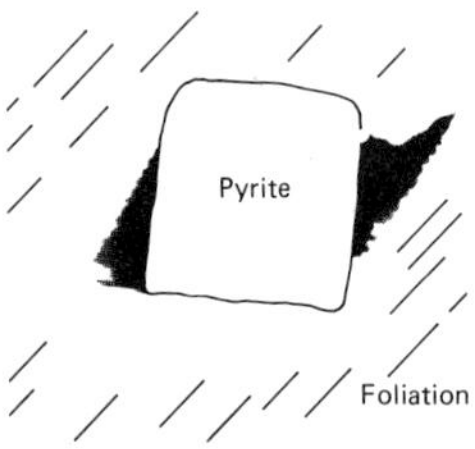

FIGURE 5–2
Pressure shadow.

The most perfect foliation is that caused by near-perfect alignment of micaceous minerals in slate, called *slaty cleavage* (rock cleavage is *not* synonymous with and does not originate in the same way as mineral cleavage).

Both foliated and unfoliated metamorphic rocks are crystalline. Unlike in a magma, where the early crystals grow in a liquid melt, metamorphic crystals must compete for space in an already solid rock. Nevertheless, many metamorphic rocks contain large, perfectly shaped (euhedral) crystals.

One special metamorphic texture resembles the porphyritic texture of igneous rocks. It is called *porphyroblastic.* Although the two have the same appearance, they are formed in entirely different ways. Porphyroblastic texture consists of large, well-formed crystals in a matrix of finely crystalline material, but instead of being an early-formed crystal or having cooled at different temperatures, the porphyroblast grows at the expense of its surrounding matter.

Most regionally metamorphosed rocks are foliated at least slightly. Nonfoliated metamorphic rocks have mosaiclike textures that consist of crystals growing together to form an interlocking pattern, or granulose textures such as the fused sand grains of a quartzite.

The texture of a metamorphic rock depends on (1) the original texture, (2) the way the rock is deformed, and (3) the processes of crystallization. The original texture may be that of igneous rocks, sedimentary rocks with compositional stratification, fossils, primary sedimentary structures, and so on, or metamorphic rocks with strongly developed foliation, layering, and lineations.

When metamorphism is accompanied by strong deformations the original rock may be brecciated, crushed, rolled out, and eventually emerge as a fine crush powder called *mylonite.* The rock in hand specimens appears compact, chertlike, and shows parallel streaks or layers. *Pressure shadows* can develop where fragments that do not get crushed are rotated, or bent (Fig. 5–2).

A number of special textures may be formed as a result of recrystallization. Large lenticular-shaped crystals or crystal-aggregates have *augen* (eyelike) texture (Fig. 5–3). If early-formed crystals are partially replaced, a rim of replaced material results. If the entire crystal is replaced, a *pseudomorph* is formed. This is a crystal that has been completely replaced by another mineral in such a way that the original crystal form is preserved.

FIGURE 5–3
Finely layered quartz feldspar gneiss showing large feldspar augen (eyelike forms). The pencil indicates scale.

DESCRIPTION OF THE COMMON METAMORPHIC ROCKS

1. **Slates** are the most perfectly foliated metamorphic rocks. Blackboard slates and roof-

ing shingles are split out of large blocks of slate. The foliations or slaty cleavages are so closely spaced that a single piece of slate may be split into thin sheets. Theoretically, one could split them into sheets no thicker than a piece of paper. The origin of these foliations is still being debated. Foliations may develop when clay becomes altered to micaceous minerals as the original shale is folded and tightly compressed. The minerals tend to become aligned in planes that almost bisect the folds. The mineral constituents are so small that they cannot be identified megascopically.

2. **Phyllites** are fine-grained schistose rocks identified by the lustrous, silky sheen that characterizes light reflected from chlorite and muscovite micas of which they are composed. Quartz and albite are often present. The grain size is larger than that of slates but finer than that of schists. They are usually greenish or red and may show the initial stages of segregation of some mineral constituents into layers.

3. **Schists** are strongly foliated rocks of medium to coarse crystalline texture. Most of the mineral constituents are easily identified without the use of a microscope (this is not true of phyllites). Foliation of schistosity is caused by parallel or nearly parallel alignment of micaceous minerals. The most common minerals in schists are quartz, feldspars, and micas. If one of the constituents makes up 50 percent or more of the rock, the name of that constituent is attached as a modifier (e.g., mica schist, quartz schist, hornblende schist). If no constituent comprises 50 percent, the names of the two most abundant constituents are used (e.g., garnetiferous-mica schist).

4. **Gneisses** have as their characteristic feature the compositional layering that produces a banded appearance (Fig. 5–4). Gneiss is medium- to coarse-grained; it usually contains quartz and feldspar interlayered with thin layers rich in hornblende or mica that induce planar weakness in the rock. The quartz, feldspar, and other constituents usually have interlocking boundaries. The layering is thought to develop as a result of segregation of mineral constituents during metamorphism and it often corresponds to original bedding of metasedimentary rocks. When gneisses are known to be derived from igneous rocks they are called *orthogneisses*. When the parent material is sedimentary, the name *paragneisses* is applied.

FIGURE 5–4
A small isoclinal fold is seen in one layer of this quartz-feldspar-amphibole gneiss.

Schists are distinguished from *gneisses* by the way schists break into thin platy slabs (1 to 10 millimeters thick) or into pencil-like columns, whereas gneisses break into much thicker slabs or even across the layering. Most gneisses also have a much higher feldspar content. Twenty percent feldspar is used to distinguish the two rock types in some classifications.

5. **Marble** is the metamorphic equivalent of calcite, limestone, or dolomite. Marbles often contain micaceous impurities that give them a foliation. The texture is usually that of an interlocking mosaic growth of calcite or dolomite crystals (Fig. 5–1). Marbles of each locality tend to be slightly diferent either in color or in structures and patterns.

6. **Quartzite** is the metamorphic equivalent of quartz sandstones, and it is composed of about 80 percent or more quartz. Tightly cemented sandstones, called *orthoquartzites*, may resemble *metaquartzites* in hand specimens. The grain boundaries of metaquartzites are interlocked and the rock

often has some foliation. Orthoquartzites form as a result of cementation of sandstone by silica from ground water in the absence of true metamorphism. The boundaries of the original grains of sand are usually visible. The cementing material is silica, and the bonds are so strong that, when quartzite is broken, fractures cut indiscriminately across sand grains and cement. In contrast, when a sandstone is fractured, the fracture cuts through the cement and around the grains of sand. Orthoquartzite may be interbedded with unmetamorphosed sedimentary rocks.

7. **Hornfelses** are fine-grained, nonfoliated, dense, usually dark-colored rocks formed near the contacts of igneous intrusions. The minerals form a mosaic of interlocking grains. They commonly break into splintery fragments that have translucent edges, like horn. The term is most frequently applied to baked shales.
8. **Amphibolites** contain mostly plagioclase and hornblende and often some biotite. The prismatic hornblende crystals lie in the plane of foliation and may induce a planar fissility or a linear schistosity. Plagioclase is often strongly aligned also. Quartz-free amphibolite is usually derived from basaltic or mafic tuffaceous rocks. When quartz is present it suggests the possibility of derivation from marls or other sedimentary rocks.

NATURE OF METAMORPHIC CHANGES

Metamorphic changes take place in solid rock. If the temperature becomes high enough to melt the rock, forming a silicate melt, it will solidify into an igneous rock. This is the distinction between igneous activity and metamorphic activity. The two phenomena grade so imperceptibly into one another that it may be difficult to determine which name, igneous or metamorphic, is applicable. The problem centers on determination of whether a rock remains "solid" at very high temperatures. Igneous rocks pass through stages of gradual crystallization so that when the magma is nearly all crystallized a fluid fraction remains. Similarly, as a sedimentary rock is heated, the rock is gradually recrystallized and a fluid fraction composed of liquids and gases permeates the whole rock. The temperatures attained, the proportion of fluids used, the structures, the textures, and even the compositions of the rocks produced in these two cases may be very similar. Fortunately we do not run into this problem of identification for most metamorphic rocks. When metamorphism takes place at lower temperatures, end products are less likely to be confused with those of igneous activity.

A new rock of different texture and often different mineralogical composition is produced through metamorphism. New minerals form by recombination of elements present, old minerals may be recrystallized as larger crystals, new textures develop, and the internal structure and fabric of the rock may change. Such changes may take place without the addition or removal of elements, called *isochemical changes*. This type of system is described as *closed*. The addition and removal of elements is sometimes necessary to explain the compositional changes. Such systems are *open*.

FIGURE 5–5
Cross section of sulfide ore bodies at MacKay, Idaho, formed by contact metamorphism of large limestone inclusions in a granite porphyry.
(after Umpleby)

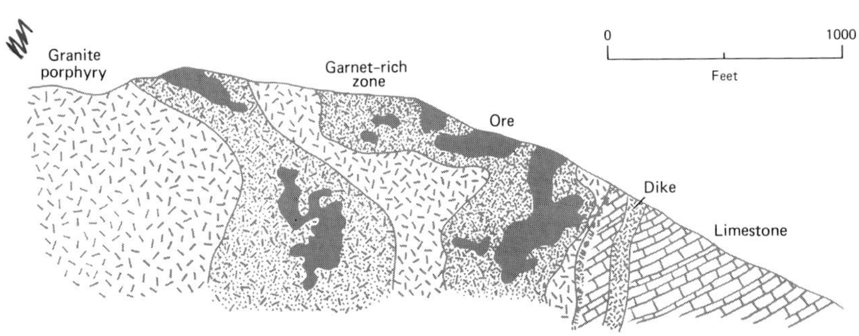

OCCURRENCE AND TYPES OF METAMORPHISM

The special geological circumstances that bring about metamorphic changes can be deduced from studying the mode of occurrence of metamorphic rocks. Experimental approaches have also proved valuable to those seeking qualitative and quantitative conclusions about the processes in action.

Metamorphism is caused by heat and fluids from igneous intrusions, by heat at depth in the earth's crust, and by processes of rock deformation. The country rock around plutons is usually altered (Fig. 5–5), producing a succession of concentric aureoles in which the variations in heat and fluid emanations from the magma are reflected in the variations in rock type at increasingly greater distances from the margin of the pluton, called *contact metamorphism.* Recrystallization of minerals and recombination of chemical constituents of rock are often found in conjunction with rock deformation. When deformation is significant in the transformation of the rock the metamorphism is called *dynamic metamorphism.* Deformation that imparts a new fabric in the rock is often accompanied by the third type of metamorphism, *regional metamorphism.* Most of the metamorphic rocks exposed on the earth's surface are found in Precambrian shields (Fig. 24–1) and in orogenic belts (Fig. 24–2). Even the shields appear to have been parts of orogenic belts in which folding and faulting were closely associated with igneous and metamorphic processes. A distinction is made between dynamic and regional metamorphism because some metamorphic rocks do not show evidence of strong deformation. Regional metamorphism may be caused by the burial of rock at depths over 10,000 meters (where temperature exceeds 300°C) in one of the deeply subsiding belts of the earth's crust, or by ascension of hot gaseous solutions into rock at shallower depths.

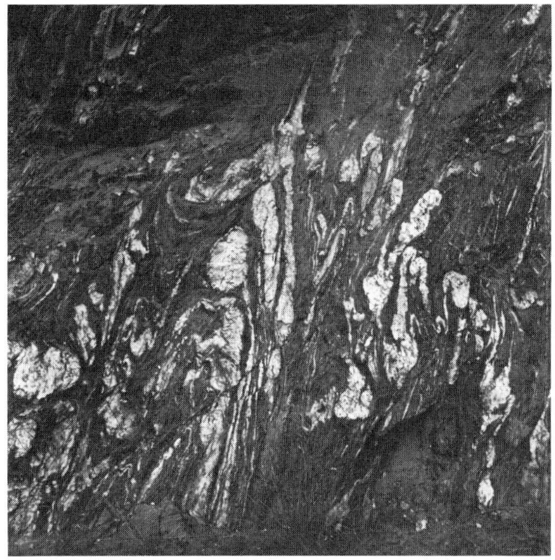

FIGURE 5–6
Highly deformed metamorphic rock. Quartz layers show the extent of folding.

FIGURE 5–7
Migmatite (mixed metamorphic rock). Note hammer for scale.

PROCESSES OF METAMORPHISM

Four processes taking place during metamorphism are responsible for most of the observed features of metamorphic rock texture, composition, and structure. These are the following:

1. Rock flowage } Mechanical Transformations
2. Granulation } Mechanical Transformations
3. Recrystallization } Chemical Transformations
4. Recombination } Chemical Transformations

The particular geologic circumstance causing metamorphism to occur and the particular group of factors that control the metamorphism determine which of the four are most effective.

Rock flowage.[1] That it is possible for rock to flow is shown by the flowage-type folds in gneisses, schists, and other metamorphic rocks (Figs. 5–6 and 5–7). Flow in the sense used here is not the same as the viscous flow associated with fluids. It is like the flowage of glacier ice in which the material remains solid yet undergoes continuous deformation. Flowage of this type results from movements along planes of slip within the component crystals of the rock, and the grains of the rock may be rotated and displaced relative to each

[1] The theory and experimental evidence concerning rock flowage are presented in Chapter 23.

other (*intergranular rotation*); usually it is accompanied by recrystallization of minerals. Rock subjected to this process can undergo large-scale deformation without developing fractures and losing cohesion within its mass. This type of rock flowage is known as *pseudoviscous flow;* the rock behaves as a ductile material. Flow is most likely to take place at high temperatures and high confining pressures (great depth of burial). Rock flowage that accompanies metamorphism may be accelerated by movement of solutions and migration of elements in the solids toward points of less stress.

Granulation. Granulation is the crushing of rock under such conditions that no visible openings result. The directed stress acting on the rock must be great enough to exceed the crushing strength of the rock, and usually this sort of action is favored by intense and rapidly applied stress of short duration. Granulation is most likely to occur in shallow rocks that are composed of hard, brittle, insoluble minerals; otherwise the rock would be likely to undergo pseudoviscous flow or recrystallization. The rock produced by granulation is likely to be characterized by mineral grains that become flattened as though smeared between two hard surfaces pressed together. The grains may be rotated as they break up, crushing off the irregular corners or eventually forming a powdery substance resembling mortar. When the crystals are reduced to this fine powder, the rock may be said to have a *mylonite structure.* Such rock is often (1) flinty, banded, or streaked in appearance, (2) tends to be localized along fault surfaces and zones of intense shearing, and (3) is a product of what is often termed *cataclastic deformation.*

Recrystallization. A complete reorganization of the elements of the original minerals in a rock may take place when the rock is subjected to high temperature and pressure, and particularly when water is present. It is not necessary that any new elements be introduced or that any original elements be removed, although such additions and removals may also occur. Unstable minerals may break down slowly into their constituent parts, while at the same time these elements recombine to form new minerals that are stable. A reaction of this sort might be represented as follows: $AB + CD \rightarrow A^+, B^-, C^+,$ and D^- ions that recombine to form $AD + BC$. The compounds AB and CD were unstable, whereas AD and BC are stable under the metamorphic conditions. The alternative reaction may be that the original minerals recrystallize into larger crystals or slightly different forms of the same minerals. Chemical changes produce some minerals that are unique in metamorphic rocks. Where metamorphic rocks grade gradually into the sedimentary rock from which they were derived, it is possible to trace the changes accurately and in detail.

FACTORS IN METAMORPHIC PROCESSES

A number of factors affect the operation of metamorphic processes, influencing the metamorphic changes that take place and the type of new rock that is produced. These factors include the following:

1. The original composition of the rock.
2. The depth at which metamorphism takes place (expressed in terms of the weight of the overlying rock mass in kilobars).
3. The temperature.
4. Directed stress.
5. The presence of chemically active fluids.
6. The amount of water present.
7. The size of the grains in the rock before metamorphism.
8. The porosity and permeability of the original rock.
9. The length of time the processes act, whether or not the processes act over a long enough time for the chemical changes to go to completion and for the rock to reach a new state of equilibrium under the existing conditions.

Grain Size

Chemical reactions occur on surfaces; thus, the surface area exposed to solutions is a factor in the rate of reaction between the solid and the solution. The total surface area of a given volume of rock increases as the rock is broken into smaller and smaller pieces. Thus fine-grained materials go into solution faster than coarser grained materials, and reactions occur faster in finer materials; so rocks like shale, composed of microscopic micaceous flakes, are more sensitive to metamorphic changes than are coarser sediments.

The fragmental character of many sediments frequently gives rise to surfaces along which chemical reactions may take place. Pore spaces between fragments allow solutions to move freely along grain surfaces, but even igneous and metamorphic rocks have some porosity due to fractures, cracks through crystals, and minute separations along grain boundaries.

Confining Pressure

Confining pressure and *lithostatic pressure* are terms applied to nondirected pressure in rocks. The pressure acts as a compression directed from all sides on the surface contacts of the material. The amount of this pressure at any given depth is equal to the weight of the overlying volume of rock. The pressure on the top and bottom of a small piece of material is essentially equal, and pressures from opposite directions tend to balance one another. Lithostatic pressure increases rapidly with depth and is on the order of 250–300 bars per kilometer of depth (1 bar = 1 atmosphere or 14.7 pounds per square inch). Most metamorphic phenomena take place under lithostatic pressure in the range of 0–10 kilobars (0–10,000 bars). Few of the rocks now exposed at the surface of the earth have ever been at depths of much more than 16 kilometers. Rocks that were once deeply buried are now exposed most often in folded mountain belts.

Directed Pressure

The response of rocks subjected to *directed pressure* is either to move or to change shape (become distorted). If the magnitude of the directed pressure is low, the rock may behave simply as an elastic material. Most rocks behave as brittle, elastic materials at low temperatures and low confining pressures. As the temperature or confining pressure is increased, rock undergoes a transition from brittle to ductile behavior. If the amount of directed pressure is high and temperature and confining pressure are low, the crushing strength of the rock may be exceeded, and shearing, crushing, and granulation will occur in the brittle rock. If conditions in the environment (e.g., high temperature and confining pressure) favor ductile behavior, directed stress may induce rock flowage.

Directed pressure is also an important factor in bringing about recrystallization. Riecke's principle, that solubility increases under pressure, applies in such cases. Granular rocks have points of greatest stress where the individual grains touch one another. The points of least stress are the pore spaces around the grains. Under stress the grains may go into solution at the points of contact; the material moves into the pores where it recrystallizes, giving rise to minerals elongated in the direction of minimum stress. In stratified rocks the lines of easiest movement are parallel to the beds, not across them. Thus when a stratum is metamorphosed under pressure, bands of new minerals may form parallel to the old bedding surfaces.

Temperature

Temperature is an important factor in both mechanical and chemical processes of metamorphism for the following reasons:

1. It accelerates most chemical reactions.
2. It facilitates recrystallization and recombination by increasing molecular activity.
3. It favors dehydration and endothermic (heat-absorbing) reactions.
4. It makes most rocks more ductile and accelerates pseudoviscous flow.

In addition, heat is important because it induces increased activity in water and other solutions. Heat is the most important factor in metamorphism that takes place close to igneous intrusions. The country rock may be baked, without melting. This type of alteration, called *pyrometamorphism,* is found at the contacts of batholiths, dikes, sills, and in the fragments of country rock that break off the sides and roof of the intrusion. Basaltic lavas attain temperatures of 1,200°C. Temperatures in the range of 100–300°C may be expected as normal

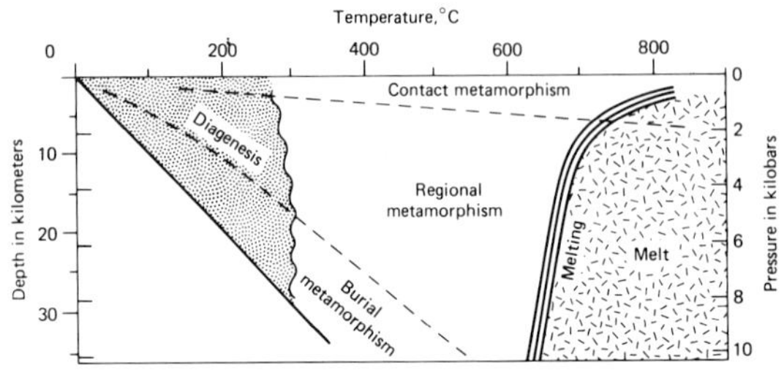

FIGURE 5–8
Pressure-temperature fields of various types of metamorphism. The term *diagenesis* refers to processes of compaction and lithification.
(after H. G. F. Winkler, 1965)

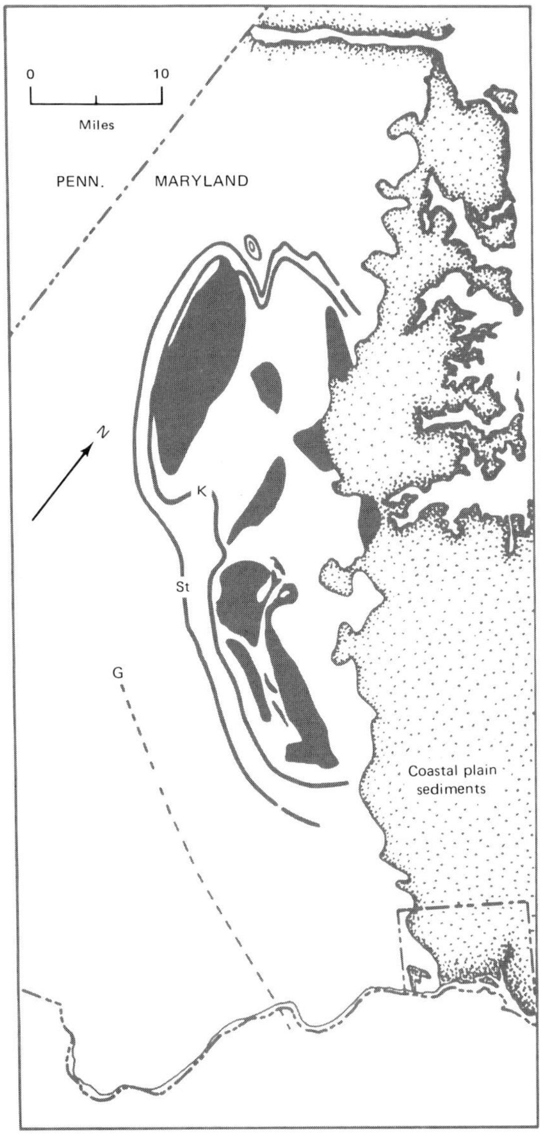

FIGURE 5–9
Gneiss domes in the northern Piedmont and metamorphic zones. G = garnet isograde; St = staurolite isograde; K = kyanite isograde.
(from the State Geological Map of Maryland, 1968)

at the depths of burial in folded mountain belts, that is, 10–20 kilometers (Fig. 5–8). Experimental evidence indicates that temperatures in the range of 200–800°C are common in regional metamorphism. Above 800°C, granitic magmas may evolve at high pressures.

CONCEPTS OF METAMORPHIC ZONES AND GRADES

Much of the early field work in metamorphic terrains was done in the Scottish Highlands, where George Barrow recognized and mapped zones of progressive metamorphism in the early 1900's. He perceived that progressive changes in mineralogy and texture (e.g., slates to coarse-grained sillimanite–garnet–mica schists) can be traced in rocks that have almost the same bulk chemical composition. A *metamorphic zone* is an area that was exposed to similar metamorphic conditions. The boundaries are recognized by and the zone is named for selected *index minerals*. The boundaries between zones were later called *isograds* (Fig.5–9), lines or planes connecting points of equal metamorphic intensity. Each zone or isograd represents a particular degree or grade of metamorphism.

The zones recognized by Barrow (1912) are the following:

1. **Chlorite zone.** The typical mineral assemblage is quartz–muscovite–albite.
2. **Biotite zone.** Recognized by the appearance of red-brown biotite.
3. **Garnet zone.** The typical mineral assemblage is quartz–muscovite–biotite–almandine garnet–albite.
4. **Staurolite zone.** The typical mineral assemblage is quartz–biotite–muscovite–plagioclase almandine–staurolite.
5. **Kyanite zone.** Contains quartz–biotite–muscovite–plagioclase–almandine–kyanite.
6. **Sillimanite zone.** Contains quartz–biotite–muscovite – plagioclase – almandine – sillimanite.

Role of Water

Water may be retained in sedimentary rocks from the time they were deposited. Water influences chemical reactions and the rate at which they proceed; it is also the medium in which materials move through the rock.

The effect of water is temperature-dependent. Silica is almost insoluble in water at surface temperatures, but at elevated temperatures and with small quantities of alkalis present, water will dissolve significant amounts of silica. As temperature rises, the mobility of water molecules is increased. At 374°C the critical point is reached. Above this temperature the gaseous and liquid conditions of water merge. The fluid is so active at these temperatures that it can penetrate even the most minute opening in rocks and crystals.

The role of water in reactions depends on the confining pressure, water pressure, and amount of water present as well as on the temperature. Earlier it had been thought that the particular assemblage of minerals present in a metamorphic rock was related almost entirely to temperature and confining pressure. Experimental work at the Geophysical Laboratory (Yoder, 1955) revealed that the minerals typical of any metamorphic grade can be produced at a constant temperature simply by varying the amount of water concentration. The lower the water concentration, the higher the metamorphic grade produced. It was concluded that metamorphic grade can be indicative of temperature or water vapor pressure (Fig. 5–10).

The chemical equilibrium of any reaction in which water appears as a reactant or a product is dependent on the partial pressure of water. (Partial pressure of water is that part of the total vapor pressure that is due to water vapor.) Other gases such as CO_2 may also be present and contribute to the total vapor pressure. An example is seen in the following reaction:

muscovite + quartz ⇌
potash feldspar + sillimanite + water

This reaction can proceed from left to right only so long as water-vapor pressure remains low. If water-vapor pressure rises, the reaction is driven from right to left.

The presence of fluids may be extremely important in initiating chemical reactions and in providing a means by which the products of these reactions can be moved through the rock. Fluids that move through solid rocks and achieve recrystallization must move along grain or crystal boundaries and through interconnected pores. Fluids may be introduced in an open system when fluids from a rock melt escape into the surrounding country rock or when pore fluids enter from an adjacent rock mass of different composition. This changes the net chemical composition of the rock and increases the chances for formation of new minerals. The old minerals may be replaced in a process of simultaneous solution and deposition called *metasomatism*. The new minerals may be partly or wholly different in composition from the ones they replace.

PRINCIPLE OF CHEMICAL EQUILIBRIUM IN METAMORPHISM

Metamorphic chemical reactions tend toward establishing an assemblage of minerals that is stable under the existing physical conditions. Temperature, pressure, and bulk composition are of primary importance. Fluids initially present, those that evolve during reactions, and any introduced materials are also highly significant. When equilibrium is not reached because metamorphic processes are interrupted or reactions fail to go to completion because they are too slow, a number of characteristic features may be produced. Plagioclase feldspar may show a compositional zoning (Fig. 5–11) or crystals of one mineral may be partly altered to another mineral (e.g., a relict augite crystal with a rim of chlorite).

The theory that a tendency to reach an equilibrium condition is a characteristic of metamorphism can be tested by comparison of field observations and experimental evidence. An

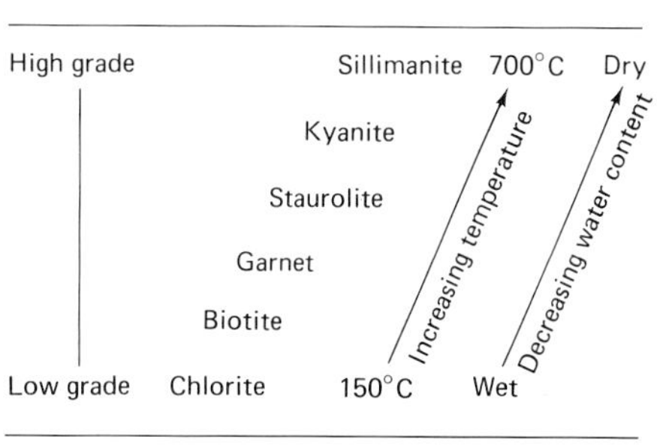

FIGURE 5–10
Mineral grades. Metamorphic grades are recognized by the appearance of certain diagnostic mineral assemblages. Some of the important key minerals are given in this chart. Their occurrence depends on the temperature, pressure, and amount of water present.

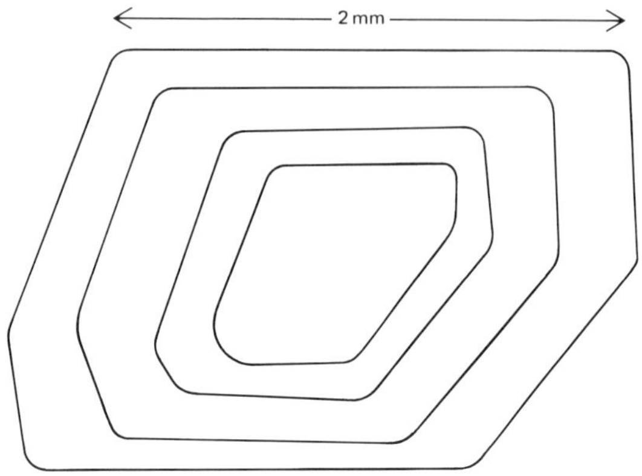

FIGURE 5–11
Zoned plagioclase.

TABLE 5–1 (*below*)
Metamorphic equivalents of common sediments.

Sediment	Sedimentary rock	Metamorphic rock
sand	sandstone	quartzite
limy mud	limestone	marble
marl	sandy limestone	siliceous marble
calcareous sand	calcareous sandstone	quartzite
clay, mud	shale	slate (dynamic metamorphism)
		phyllite, schist, gneiss (regional metamorphism)
tuffs	tuff	slate (dynamic metamorphism)
		phyllite, schist, gneiss (regional metamorphism)
graywacke	graywacke	schist, gneiss (regional metamorphism)

unaltered rock unit is traced into an area where progressive changes in mineral assemblages can be observed. The original unaltered rock is then altered experimentally under controlled conditions approximating those thought to be responsible for the metamorphism. If the same mineral assemblages observed in the field can be reproduced in the experiments, the conditions under which metamorphism occurred can be inferred.

Gibbs' phase rule (see p. 88) is applicable to both igneous and metamorphic rocks. According to the phase rule, a system with three components (e.g., CaO, MgO, SiO_2) can yield a rock with no more than three minerals. Those three may be any one of a group of chemically equivalent minerals (e.g., quartz-diopside-enstatite or diopside-forsterite-periclase). Additional variations are possible, depending on whether the system contains gases such as carbon dioxide or water vapor. When a system is open and contains a mobile component like CO_2, the phase rule is altered to p max $= c - N$ (number of mobile components).

Original Composition

The composition of the original rock places definite limitations on the possible metamorphic reactions (see Table 5–1). The composition of a rock may be expressed either in terms of its bulk chemical content or in terms of the amount of each mineral present. Only those new minerals can be produced that contain as part of their essential composition elements that are originally present or added during metamorphism. Despite this, some rocks can be metamorphosed to produce a wide variety of metamorphic rocks. A shale, for example, initially composed of quartz, mica, chlorite, hydrous aluminum silicates, and amorphous iron oxides, will become a dense, hard, fine-grained rock called a hornfels if it comes in contact with a very hot magma. The new rock is composed of the minerals quartz, andalusite, cordierite, biotite, and feldspar. If the same shale is subjected to higher pressure and deformation, it might be changed to a rock composed of quartz, muscovite, biotite, and garnet, called a *garnetiferous-mica schist.* Under conditions of lower temperature and pressure the shale could be transformed into a slate.

Often it is difficult to determine which of several possible original rock types was metamorphosed to produce a given metamorphic rock. For example, amphibolite (a hornblende-plagioclase-garnet rock) could form from metamorphism of such diverse rocks as basalt, gabbro, basaltic tuff, or an impure shaly dolomite because all may have the same chemical composition.

Rocks may be broadly grouped into the following four compositional classes to study their metamorphism.

1. **Argillaceous (pelitic) rocks** (derived from clay) are the finest grained products of decomposition of the materials that are under equilibrium at surface conditions. They are exceptionally responsive to temperature and pressure increases and are thus ideally suited for establishment of metamorphic zones (see p. 124).
2. **Carbonate rocks**–limestone, dolomite.
3. **Siliceous rocks**–quartz-rich igneous rocks, tuffs, schists, and gneisses.
4. **Intermediate and basic igneous rocks**–volcanic tuffs, and shaly and sandy dolomites.

ACF and AKF diagrams. It is useful to examine metamorphic processes in terms of both bulk chemical composition and mineralogic composition. Triangular phase diagrams, called ACF and AKF diagrams, have been developed to allow graphic representation of mineralogic data as a function of the chemical composition of rocks. Chemical compositions of rocks are expressed in terms of the percentages of the various oxides present. If a rock's composition can be represented in terms of three oxides, it can be shown on a triangular diagram. The oxides Al_2O_3, CaO, and (Mg,Fe)O are used as corners for the ACF diagram. Diagrams are also constructed to show rocks with excessive amounts of other oxides, and corrections are applied so the sum of A, C, and F is very close to 100 percent.

$$A = Al_2O_3 + Fe_2O_3 - (Na_2O + K_2O)$$

$$C = CaO$$

$$F = MgO + FeO + MnO$$

The AKF diagram is used to supplement ACF plots for rocks that contain a single calcic phase:

$$A = Al_2O_3 - (CaO + Na_2O + K_2O)$$

$$K = K_2O$$

$$F = FeO + MgO + MnO$$

The compositions of a number of rock types are shown plotted on ACF and AKF diagrams in Fig. 5–12.

ACF diagrams provide an excellent way to depict the possible mineral assemblages for any three-component system. The phase rule indicates such a system will contain three minerals. Such a representation for the mineral assemblages found in quartzo-feldspathic gneisses and amphibolites in Quebec is shown in Fig. 5–13. The possible assemblages are the following:

1. calcite–diopside–plagioclase
2. diopside–plagioclase–hornblende
3. hornblende–plagioclase–garnet
4. plagioclase–garnet–sillimanite
5. garnet–sillimanite–biotite

Experimental Approaches to Metamorphic Reactions

Controlled experiments help to define probable physical conditions (temperature, pressure, water-vapor pressure) under which certain critical metamorphic reactions may occur. Usually

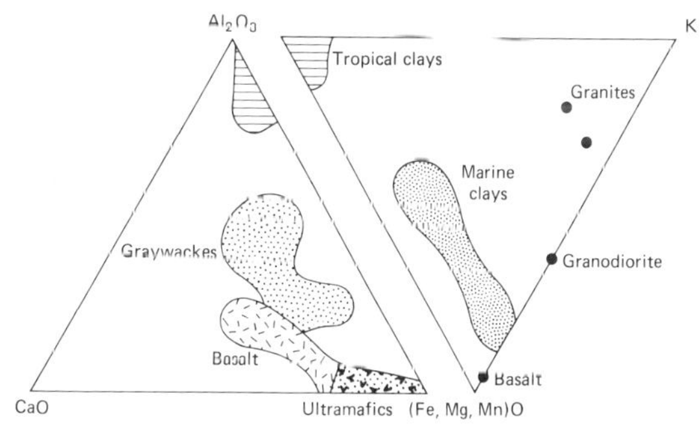

FIGURE 5–12
Chemical composition of various rocks generalized in terms of three components.
(modified after H. G. F. Winkler, 1965)

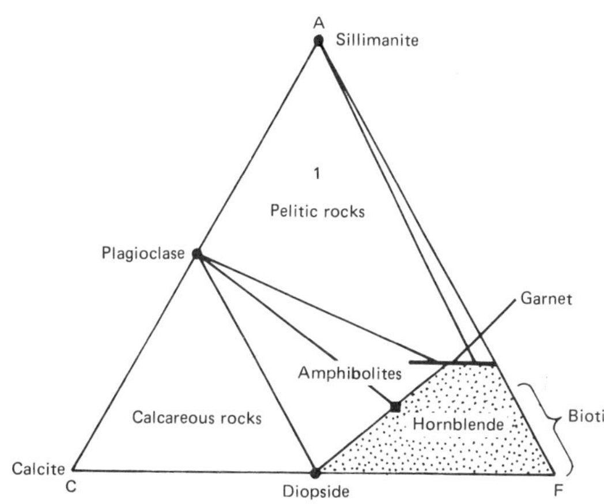

FIGURE 5–13
ACF diagram for mineral assemblages in quartzo-feldspathic gneisses and associated amphibolites in an area of southwestern Quebec. Stippled field not recorded.
(data from R. Kretz, 1959, 1963; after Turner, 1968)

the compositions used in laboratory experiments are less complex than those encountered in the field, and when water-vapor phases or carbonates are investigated, the fluid pressure P_{H_2O} or P_{CO_2} is set equal to the load pressure. *Univariant equilibrium curves* have been established for a number of metamorphic reactions. These show the combinations of temperature and pressure at which an equilibrium condition exists for the reaction. Many other experiments define only an upper temperature limit for a given reaction.

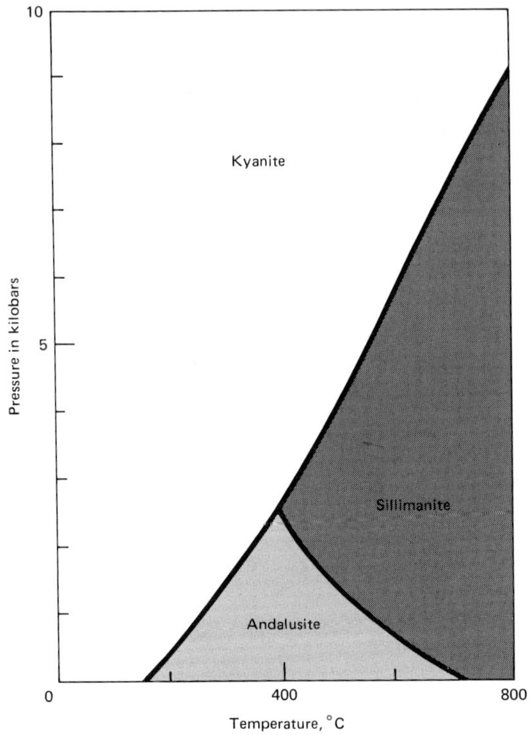

FIGURE 5–14
Phase diagram showing stability fields of the polymorphs of Al_2SiO_5. (data from W. S. Fyfe; after Turner, 1968)

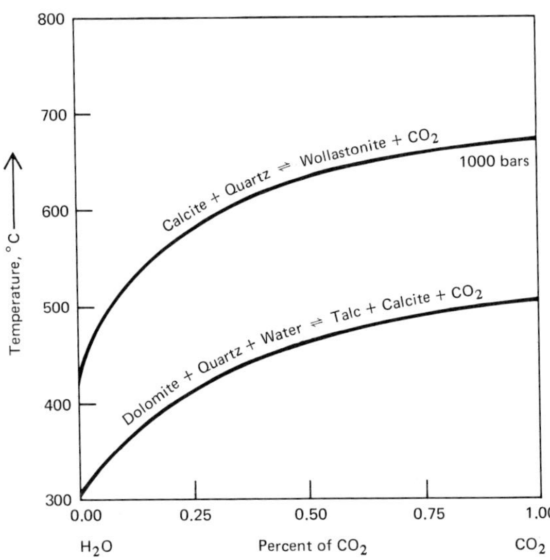

FIGURE 5–15
Equilibrium curves for two metamorphic reactions plotted for variations in temperature and partial pressure of carbon dioxide.

As examples of the experiments used, we will consider the stability fields of andalusite-kyanite-sillimanite (Fig. 5–14) and the univariant equilibrium curves for reactions important in the metamorphism of impure dolomites (Fig. 5–15).

Andalusite, kyanite, and sillimanite were used by Barrow (1912) to recognize high-grade zones of metamorphism. These minerals all have the same chemical composition, Al_2SiO_5, but different atomic structure. The stability fields of the three minerals (Fig. 5–14) are consistent with field observations that suggest that sillimanite occurs in rocks subjected to higher grade metamorphism than rocks containing andalusite or kyanite.

Metamorphism of serpentine takes on special significance in light of unusual modes of its occurrence. Serpentine is found in places along the crest of the midocean ridges and in the core of folded mountain belts. Alteration of ultrabasic rocks to serpentine plays an important role in hypotheses regarding the origin of mountains and continental drift (see Chapter 25). Ultramafic rocks such as peridotites (olivine + pyroxene) and dunite (olivine) are found altered to serpentine, possibly according to the following reaction:

$$\text{olivine} + \text{water} \rightleftharpoons \text{serpentine} + \text{brucite}$$

$$4Mg_2SiO_4 + 6H_2O \rightleftharpoons Mg_6Si_4O_{10}(OH)_8 + 2Mg(OH)_2$$

The metamorphism of serpentine involves a progressive dehydration of the minerals formed.

METAMORPHIC FACIES

Metamorphic changes are such that the original mineral assemblage, no longer in equilibrium with its environment, is altered to produce an assemblage of minerals that is stable under the new environmental conditions. If the metamorphic reactions have proceeded to completion or near-completion, the new minerals will have arrived approximately at an equilibrium state and thus will tend to undergo no further change until the conditions are changed. Therefore it should be possible to relate certain metamorphic rocks or assemblages of minerals to the environmental conditions of their formation, such as confining pressure, water pressure, temperature, and original composition.

At about the turn of this century, two now famous geologists, V. M. Goldschmidt and P. Eskola, studied the mineral assemblages in metamorphic rocks in the regions of Oslo, Norway, and Orijarvi, Finland. Goldschmidt examined the contact zones around a number of plutons that had invaded a series of shaley, sandy, and calcareous sedimentary rocks. He found that the mineral assemblages in the various contact aureoles were remarkably simple in mineralogical composition despite the great differences in bulk chemical composition of the country rock around different plutons. In each case the rock was composed primarily of four or five minerals. These minerals were present in abundance, but chemically equivalent pairs (pairs with the same bulk chemical composition) did not appear together. Eskola found the same effects in Finland and attributed the differences in mineralogy to differences in the physical conditions accompanying the metamorphism. The concept of facies was suggested by Eskola (1915) and later defined by him as follows:

> A certain *metamorphic facies* comprises all rocks exhibiting a unique and characteristic correlation between chemical and mineralogical composition, in such a way that rocks of a given chemical composition have always the same mineralogical composition, and differences in chemical composition from rock to rock are reflected in systematic differences of their mineralogical composition.

Thus the mineralogical composition of any metamorphic rock in a given metamorphic facies depends on the chemical composition of the rock as well as on the environmental conditions. A group of minerals, a facies, will always result if original

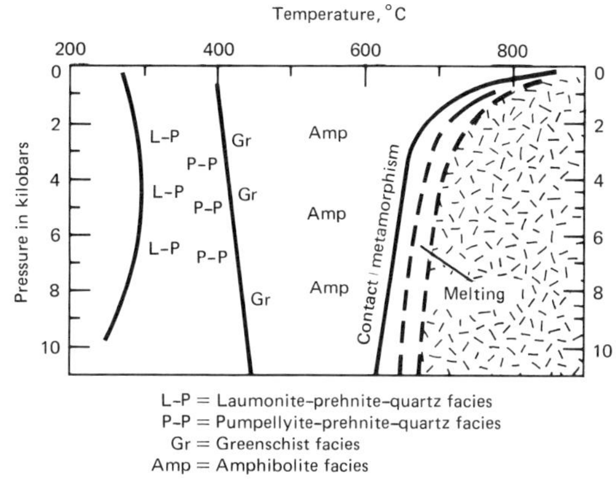

FIGURE 5–16
Pressure-temperature conditions for various facies of metamorphism.
(modified after Winkler, 1965)

rocks of the same bulk chemical composition are metamorphosed under the same environmental conditions, no matter what their original mineralogical composition was. A single facies can consist of a number of different metamorphic rocks, each derived from original rocks of different chemical composition. The name of the facies is thus an expression of an environmental condition in which certain mineral phases will be present if the rocks in the environment contain the necessary constituents.

If an assemblage of minerals happens to be stable over a wide range of temperature, pressure, and water content, then that assemblage may belong to several metamorphic facies. A precise definition of a facies is possible when a critically sensitive mineral assemblage is stable only within relatively narrow limits of environmental changes. The environmental conditions of some of the more important facies are illustrated in Fig. 5–16 and summarized in Table 5–2.

Greenschist Facies

Metamorphic effects produced by burial metamorphism begin to disappear as temperatures and pressures rise and the original sedimentary minerals, including clay and zeolites, are altered. The rocks tend to assume a greenish appearance because of the formation of chlorite and epidote. The sedimentary character of the rock has been significantly altered by the time this facies is formed. Original primary structures may be pre-

Temperature	Shale	Mafic rocks
200–500°C	**Greenschist facies** quartz–albite–chlorite–muscovite	albite–epidote–actinolite–chlorite
300–600°C	**Amphibolite facies** quartz–plagioclase–muscovite–biotite–almandine garnet	plagioclase–hornblende
500°C+	**Granulite facies** quartz–plagioclase–orthoclase garnet–sillimanite	plagioclase–garnet–hypersthene–quartz
600°C+	**Eclogite facies** none known	omphacite–garnet

TABLE 5–2 Metamorphic facies formed at high pressures. Pressures in the range of 3,000–12,000 atmospheres, depths of 10–40 kilometers. (after Fyfe, Turner, and Verhoogen, 1958)

served, but the rock has a new texture, schistosity, and original mineral grains are largely replaced with new minerals.

The mineral assemblages present depend on the temperature-pressure-water vapor and original composition of the rock from which they are derived. *Albite–biotite–muscovite–chlorite–epidote–tremolite–actinolite* are characteristic. Calcite and dolomite may be present if the original rocks contained excessive amounts of calcium or magnesium (e.g., dolomite-impure limestones), and other minerals may appear when rocks containing excessive amounts of particular elements are altered. The temperature-pressure field of the greenschist facies is illustrated in Fig. 5–16.

Amphibolite Facies

The mineral assemblages of the greenschist facies undergo additional changes as temperature and pressure are increased. Such increases occur during regional metamorphism at deeper levels in the crust. The most characteristic minerals of the amphibolite facies are hornblende and almandine garnet, although garnet is not always present. Staurolite, kyanite, muscovite or sillimanite, and orthoclase are also often present.

The original composition is a controlling influence on which assemblages are produced. Studies in the Scottish Highlands show the following typical mineral assemblages:

Original rock	Amphibolite facies
shale	quartz–muscovite–almandine–plagioclase
quartz–feldspar-rich sediments	quartz–muscovite–plagioclase
mafic (basalt, mafic tuff)	hornblende–plagioclase–almandine
calcareous	calcite–diopside–grossularite garnet

Higher grade assemblages are indicated by the appearance of new minerals in the following order:

chlorite → biotite → garnet → staurolite → kyanite → sillimanite

Granulite Facies

High-grade metamorphic rocks in which micaceous minerals are altered to garnet and kyanite or sillimanite, and in which hypersthene and diopside are produced from hornblende, belong to the granulite facies. They are thought to originate under high confining pressures, temperatures in the range of 700–800°C, and under dry conditions. Experiments demonstrate that amphibolites will remain unaltered until they melt if water is present in excess. Several students of granulites have concluded that their origin is best explained in terms of a secondary metamorphism of an earlier gneiss-amphibolite complex from which the liberated water has had time and opportunity to escape.

The characteristic mineral assemblages used to recognize the granulite facies where amphibolite is metamorphosed are plagioclase-hypersthene-hornblende-diopside (pyroxene). Almandine garnet is prominent where quartz-feldspar rich sediments are metamorphosed to this facies, and diopside-calcite-quartz assemblages are found where the original rocks were calcareous.

Eclogite Facies

Eclogite is a metamorphic rock composed primarily of a red garnet and a green pyroxene (omphacite) and has a bulk chemical composition that could only have formed from rocks of basaltic or gabbroic composition. Hornblende, rutile, and sphene are among the common accessory minerals. Eclogites are among the rarest of metamorphic rocks and their mode of occurrence usually suggests that they were formed under conditions different from those that affected the rock in which they are found. Most eclogites occur as blocks in the diamond pipes, in blocks erupted from deep-seated volcanoes, and as masses that appear to have been squeezed into place from greater depth in rocks of granulite facies. The reactions required to form the minerals in eclogite proceed only under very high confining pressure; the minimum pressure of about 10 kilobars would suggest that eclogite may originate in the mantle. This conclusion is verified by laboratory experiments in which basalt is altered to eclogite at 17 kilobars confining pressure and 1,000°C in a dry test.

Eclogites are also found locally in rocks of amphibolite facies and they are even known to occur interlayered with glaucophane schists. Such eclogites must have originated under conditions very different from those described above, for glaucophane schist is indicative of much lower temperature conditions.

CONTACT METAMORPHISM

The alteration of country rock around an intrusion is known as *contact metamorphism.* If the magma is relatively dry—if few volatile materials are present—little change in the bulk chemical composition of the country rock can occur and the elevated temperatures near the contact are the most important agent of metamorphism. Usually some volatiles are present that will act as catalysts and react with the country rock. Alteration by chemically active emanations is most evident above the pluton. The magma may invade or partially melt the country rock, creating a rim of intimately mixed rock, called *migmatite,* around the pluton.

The degree of metamorphism varies with the size of the pluton, the initial temperature, the rate at which it cools, the amounts and types of volatiles present in the magma, and the texture and composition of the country rock. Rocks are not very good conductors of heat; therefore, contact effects are localized near the margins of the magma. If the country rock is fractured or permeable, or if the magma contains much gas and liquid, alteration may extend far beyond the margins. Since temperature is greatest at the marginal contact and decreases outward, zones of different metamorphic effects are established. Characteristic zonal effects are usually found where granite plutons are intruded into shale. At the marginal contact the shale is metamorphosed to a *hornfels,* a thoroughly recrystallized, fine-grained aggregate of andalusite, biotite, muscovite, and quartz. This grades into a second zone characterized by hornfelses with spots composed of incipient crystals of andalusite in them. The groundmass is a recrystallized mass of mica and quartz. This zone grades into a spotted slate in which the spots are segregations of carbonaceous matter. Farther away, the rock is practically unaltered.

Carbonate rocks are particularly susceptible to contact metamorphism. Calcite and dolomite are highly reactive and very easily recrystallized. When calcite is heated, it becomes recrystallized. The reaction is continuous unless the system is open and may be represented by the following reaction:

$$CaCO_3 + \text{heat} \rightarrow$$

$$\begin{pmatrix} CO_2\uparrow + CaO \\ \text{open system} \end{pmatrix} \rightarrow \begin{pmatrix} CaCO_3 \\ \text{closed system} \end{pmatrix}$$

If the carbon dioxide cannot escape, it will recombine with the CaO upon cooling to form a recrystallized calcite or marble with increased grain size. Many unusual metamorphic minerals occur in marbles when impurities are present. Limestone or dolomite may combine with silicate minerals in the limestone to form such minerals as garnet, wollastonite, diopside, tremolite, and epidote. A number of chemical reactions may occur in country rock composed initially of dolomite and quartz, but which reactions these are depends on the temperature. As the temperature increases, the following series of reactions (stated in terms of mineral content) take place (equilib-

Temperature	Mineral assemblages derived from shale	Mineral assemblages derived from mafic rocks
200–500°C	**Albite–epidote hornfels facies**	
	quartz–albite–muscovite–biotite	albite–epidote–actinolite–chlorite
300–600°C	**Hornblende hornfels facies**	
	quartz–plagioclase–microcline–biotite–muscovite	plagioclase–hornblende
500°C+	**Pyroxene hornfels facies**	
	quartz–plagioclase–orthoclase–cordierite–andalusite	plagioclase–diopside–hypersthene
600°C+	**Sanidinite facies**	
	tridymite–cordierite–mullite–glass	plagioclase–diopside–hypersthene

TABLE 5–3
Metamorphic facies formed at low pressures. Examples of typical mineral assemblages that would be derived from shale and mafic igneous rock under low pressure (depth less than 10 km).
(after Fyfe, Turner, and Verhoogen, 1958)

rium curves for two of these reactions are illustrated in Fig. 5–15):

dolomite + quartz ⇌
calcite + tremolite + carbon dioxide

$$(Ca,Mg)(CO_3)_2 + SiO_2 \rightleftharpoons CaCO_3 + Ca_2Mg_5(Si_8O_{22})(OH)_2 + CO_2$$

dolomite + tremolite ⇌
forsterite + calcite + carbon dioxide

$$(Ca,Mg)(CO_3)_2 + Ca_2Mg_5(Si_8O_{22})(OH)_2 \rightleftharpoons Mg_2SiO_4 + CaCO_3 + CO_2$$

calcite + tremolite + quartz ⇌
diopside + carbon dioxide

$$CaCO_3 + Ca_2Mg_5(Si_8O_{22})(OH)_2 + SiO_2 \rightleftharpoons CaMg(Si_2O_6) + CO_2$$

calcite + tremolite ⇌
diopside + forsterite + carbon dioxide

$$CaCO_3 + Ca_2Mg_5(Si_8O_{22})(OH)_2 \rightleftharpoons CaMg(Si_2O_6) + Mg_2SiO_4 + CO_2$$

dolomite ⇌
periclase + calcite + carbon dioxide

$$(Ca,Mg)(CO_3)_2 \rightleftharpoons MgO + CaCO_3 + CO_2$$

calcite + quartz ⇌
wollastonite + carbon dioxide

$$CaCO_3 + SiO_2 \rightleftharpoons CaSiO_3 + CO_2$$

The facies of contact metamorphism are listed below in order of increasing temperature (if water pressure is constant) or decreasing water pressure (if temperature is constant).

1. Albite, epidote hornfels
2. Hornblende hornfels
3. Pyroxene hornfels
4. Sanidinite

DYNAMIC METAMORPHISM

The name *dynamic metamorphism* is applied when effects of directed stress are important in metamorphic processes. The most obvious effects of dynamic metamorphism are seen in the texture, fabric, or structural features of the rock. Directed stress is responsible for folding, fracturing, faulting, shearing, and other types of rock deformation. Mylonite, shattering of mineral grains, and rotation of crystals are evidence that shearing accompanied metamorphism. Deformed oölites, primary structural features, fossils, and fold and fault structures are also positive evidence of directed stress. Foliation (Fig. 5–1) takes a number of forms:

1. Compositional banding seen in gneisses.
2. Approximate planar alignment of platy minerals (such as micas) and linear parallelism of elongate minerals (such as hornblende or sillimanite) typical of schists.
3. Almost perfect alignment of small platy minerals as seen in slaty cleavage.
4. Crystals flattened or drawn out as a result of deformation.

Foliations often bear definite and consistent relationships to larger structural features with which they are associated. Sometimes the foliation is parallel to original sedimentary layering. This can be proved in low-grade metamorphic rocks in which primary sedimentary structures formed during deposition of the sediment are still preserved after recrystallization. Such alignments can be explained readily in terms of lateral migration of fluids during recrystallization and flattening of clay minerals during compaction and burial. Well-developed alignment of clay minerals is found in unconsolidated sediment subjected to compression. The clay minerals assume orientations approximately perpendicular to the direction of the applied directed stress; the clay colloids become flattened. Subsequent recrystallization will produce a rock with strongly aligned chlorite or mica. But not all foliation is parallel to original bedding. This is evident in metamorphic rocks in which both original compositional layering and foliation of another orientation can be seen. Also some metamorphic rocks contain more than one direction of foliation. Foliations that do not lie in the bedding have been attributed to a number of mechanisms. Foliation is frequently found aligned parallel to the plane that bisects a fold symmetrically (Fig. 5–17); slaty cleavage often has this orientation. Many geologists have interpreted this as another example of formation of platy minerals aligned perpendicular to the direction of maximum stress. Slippage may be found along these foliations, indicating shearing along the closely spaced parallel foliations (Fig. 5–17).

The origin of slaty cleavage is still debated, particularly because thin layers of silt and sand have been found parallel to slaty cleavages (Fig. 5–18). The most satisfactory explanation of these stringers is that they were injected while the rock was still unconsolidated and water was present in the sandy layers. If this is the case, the cleavage originally must have been present in the sediment and thus it cannot be attributed to metamorphism. This interpretation does not rule out later recrystallization of minerals parallel to the original closely spaced fractures or subsequent shearing along this direction.

Deformation processes, particularly shearing, twinning, and granulation, may be important in facilitating the chemical reactions involved in metamorphism. These processes should make the movement of fluid in the rock easier and accelerate solid-solid reactions.

REGIONAL METAMORPHISM

The name *regional metamorphism* is applied to those metamorphic alterations that affect rocks over large areas and are indicative of widespread environmental changes rather than localized deformation or magmatism. Although the effects of dynamic and contact metamorphism are often

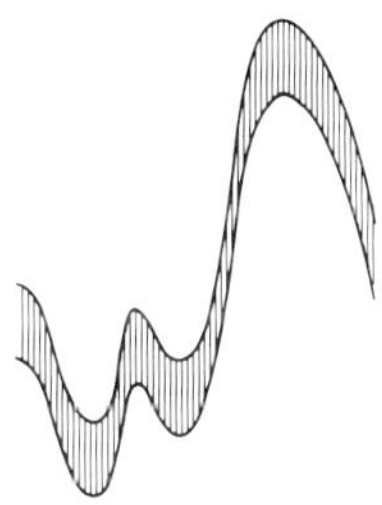

FIGURE 5–17 *(above and left)*
Folds caused by slip on closely spaced cleavages parallel to lines shown (see diagram at left). The dotted lines in the photograph above indicate such folds, with cleavages exposed.

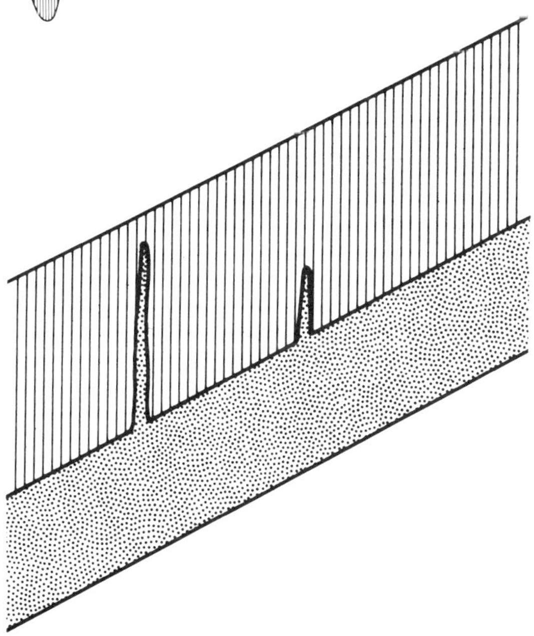

FIGURE 5–18
Sand stringers injected into a shaly bed with slaty cleavage.

combined with regional effects, they are not necessary for regional metamorphism. Some metamorphism may occur simply as a result of depth of burial. Such effects have been observed in rock buried at depths of 10,000 meters or more.

Certain belts of the earth's crust have long histories of instability. They have been sites of deep subsidence in which piles of shallow-water sediment have accumulated, achieving thicknesses of 10,000–15,000 meters. These belts have also been sites of uplift, strong deformation, metamorphism, and concentrated igneous activity.

The metamorphism in these mountain belts may be due primarily to depth of burial in one place, whereas in other parts of the belt high temperatures and emanations of magma or strong directed stresses may be the major cause. The common denominators of regional metamorphism are the increased confining pressures and elevated temperatures that are a consequence of depth of burial. For this reason early attempts were made to relate regional metamorphic effects to shallow, intermediate, or deep-seated conditions. But judging from the complex structural features of many regionally metamorphosed rocks, they must have originated under conditions in which dynamic and thermal metamorphic processes played an important role.

It is convenient to consider metamorphism as taking place when temperatures exceed 300°C, a temperature level at which mineral assemblages begin to form that cannot originate in near-surface sedimentary environments. This lowest temperature range is represented by the *zeolite facies*. The zeolites are a group of minerals that may co-exist with normal sedimentary minerals such as clay. Higher confining pressures and increases in temperature and pressure lead to the development of greenschist facies, amphibolite facies, and granulite facies.

THE ORIGIN OF GRANITES—GRANITIZATION

Rocks of granitic composition constitute the largest plutons exposed on continents. Many such plutons are now exposed in mountain belts where parts of the crust have been uplifted and deeply dissected. The Sierra Nevada and Idaho batholiths and the vast body that extends along the west coast of British Columbia (Fig. 4–21) are examples. Similar bodies are found in the Appalachians in the north-central states of Minnesota and Wisconsin and in central Canada. A considerable part of the region of central Canada is composed of granitic rocks, and granites are often encountered in deep drill holes in the central United States and in uplifted areas such as the Blue Ridge, the Adirondacks, and the Green Mountains, in the central Ozarks, the Llano Mountains of Texas, the Colorado, New Mexico, Wyoming, and Montana Rockies, and in the southwestern portion of Arizona. If all of these granitic rocks are true plutons, the continental basement rock consists mainly of a mosaic of plutonic masses.

Many granitic rocks are of unquestioned igneous origin. The true granitic pluton has fine-grained borders where the melt cooled more rapidly. Fragments of the country rock, called *xenoliths* (Fig. 4–22), may be found in the granite near the margins of the pluton. These xenoliths are altered by heat and solutions, partially melted perhaps, and rotated out of parallelism with the country rock as they move into the magma. Dikes and sills related in composition to the granite may be traced back into the pluton. The surrounding country rock may be forced aside or uplifted over the pluton as it is forced into place. Finally, some minerals, such as zircon, that may crystallize at high temperatures in the magma are euhedral.

Field relations at the margins of some granites cast doubt on a plutonic origin. Locally it is possible to walk along the bedding of sedimentary rocks through a transition zone into rocks that are massive granites. Sedimentary features such as cross-bedding can be seen in rocks that would appear, according to their mineral composition and texture, to be granites. In this case the xenoliths have not been moved out of parallelism with the country rock. The inference is that the sediments have been transformed into granite without the development of a magma. Granitic rocks such as this are often banded where the original sedimentary layering has been partially preserved. Zircons in such granites are rounded as they are in sediments. Thus granites may be produced in at least two different ways: (1) from magma and (2) by transformation of sediments, called *granitization*.

The idea of the transformist origin of granite is not new; it was proposed by a group of French geologists in the nineteenth century and maintained and elaborated by the Finnish geologist Sederholm in the early part of this century. They theorized that some granitic magma was produced and that the rocks above this were transformed by emanations from the melt. One of the most persuasive early arguments of the transformists was the "room problem." How can one account for the colossal amount of displacement that would be needed to allow for the emplacement of the known granitic bodies? The country rock around many granite terrains shows no folding or uparching. Another argument derives from Bowen's experimental work. According to Bowen, about 10 percent of a basaltic parent magma could differentiate to form granite; yet almost all of the plutons on continents are granitic in composition. Where has the 90 percent of the basaltic magma gone?

The most significant support for the transformists' views comes from field observations; the most serious problem for the transformists is the explanation of the geochemical processes that must take place for transformation of sediment to granite. The transformation is envisioned as taking place in solid rock by processes of ionic migration and reaction. The introduction of some ions is usually necessary because sediments do not often contain enough *sodium* to account for that found in the feldspars of granite. The ions can move in minute amounts of solution through the pore spaces, around grains in the sediment, along grain boundaries, through spaces in the crystal lattice, and along crystal defects. Ionic migration and reaction would be greatly facilitated if water were present and if temperatures were high even though below fusion temperature. The transformist view is that under proper conditions sediment can be altered to form the crystalline rock, granite, by the progressive growth of quartz and feldspars at the expense of the clay, silt, sand, and feldspar fragments in the sediment.

CHAPTER OUTLINE

What is Metamorphism?
- Field Evidence That Metamorphic Changes Do Take Place

Metamorphic Textures
Description of the Common Metamorphic Rocks
Nature of Metamorphic Changes
Occurrence and Types of Metamorphism
Processes of Metamorphism
- *Rock flowage*
- *Granulation*
- *Recrystallization*

Factors in Metamorphic Processes
- Grain Size
- Confining Pressure
- Directed Pressure
- Temperature

Concepts of Metamorphic Zones and Grades
- Role of Water

Principle of Chemical Equilibrium in Metamorphism
- Original Composition
- Experimental Approaches to Metamorphic Reactions

Metamorphic Facies
- Greenschist Facies
- Amphibolite Facies
- Granulite Facies
- Eclogite Facies

Contact Metamorphism
Dynamic Metamorphism
Regional Metamorphism
The Origin of Granites—Granitization

TERMS

amphibolite facies
amphibolite
cataclastic deformation
closed system
confining pressure
contact metamorphism
directed pressure
dynamic metamorphism
eclogite facies
fissility
foliation
gneiss

TERMS (cont.)

granulation
granulite facies
greenschist facies
hornfels
isograds
intergranular rotation
lithostatic pressure
marble
metamorphic facies
metamorphic zone
metaquartzite
metasomatism
mortar structure
mylonite
open system
orthogneiss
orthoquartzite
paragneiss
phyllite
porphyroblastic
pressure shadows
pseudomorph
pseudoviscous flow
pyrometamorphism
quartzite
recrystallization
relics
regional metamorphism
rock flowage
schist
schistosity
slate
slaty cleavage
univariant equilibrium curves
xenoblastic
zeolite facies

REVIEW QUESTIONS

1. Metamorphic textures are similar to as well as different from the textures of igneous and sedimentary rocks. What are some of the more significant similarities and differences in distinguishing metamorphic rocks from the other two rock types?
2. What are some of the more important metamorphic textures and how do they differ in the basic ways they were formed?
3. What important features distinguish a schist from a gneiss? A schist from a phyllite?
4. Some sandstones and quartzites can be similar in appearance, especially if the original grain boundaries are still evident in the quartzite. How could you distinguish a metaquartzite from a tightly cemented sandstone?
5. Dynamic and regional metamorphism are two of the three types of metamorphism. In certain ways the conditions under which they are formed are similar. Briefly compare and contrast.
6. Both rock flowage and recrystallization are most likely to occur at high temperatures and high confining pressure. Both could occur in the same rock. What factors might favor one process over the other?
7. Temperature and confining pressure are two important factors determining the type, extent, and so on, of metamorphism that takes place. In what way or ways do the two act together to produce a change in the behavior of a pre-existing rock body?
8. Water and water vapor were not thought to have much effect on metamorphism until experimental studies in the 1950's proved otherwise. In what ways does water, particularly at high temperature and pressure, affect metamorphism?
9. The concept of facies in metamorphic rocks is important, particularly in classification. Yet a single facies may contain several different metamorphic rocks that were derived from rocks of different chemical composition. Why?
10. The single most important factor in contact metamorphism is temperature. However, several other factors are almost as important in controlling the extent to which the metamorphism takes place. Discuss these briefly.

SUGGESTED READINGS

Krauskopf, K. B., 1967, Introduction to Geochemistry: New York, McGraw-Hill Co., 721 p. Refer to chapter 19.

Read, H. H., 1957, The Granite Controversy, Geological Addresses Illustrating the Evolution of a Disputant: London, Thomas Murby, 430 p. Refer to parts I, VII.

Turner, F. J., 1968, Metamorphic Petrology, Mineralogical and Field Aspects: New York, McGraw-Hill, 403 p.

Tyrrell, G. W., 1958, The Principles of Petrology, An Introduction to the Science of Rocks: London, Methuen, 347 p. Refer to chapters XVI–XXI.

6 WEATHERING AND SOIL DEVELOPMENT

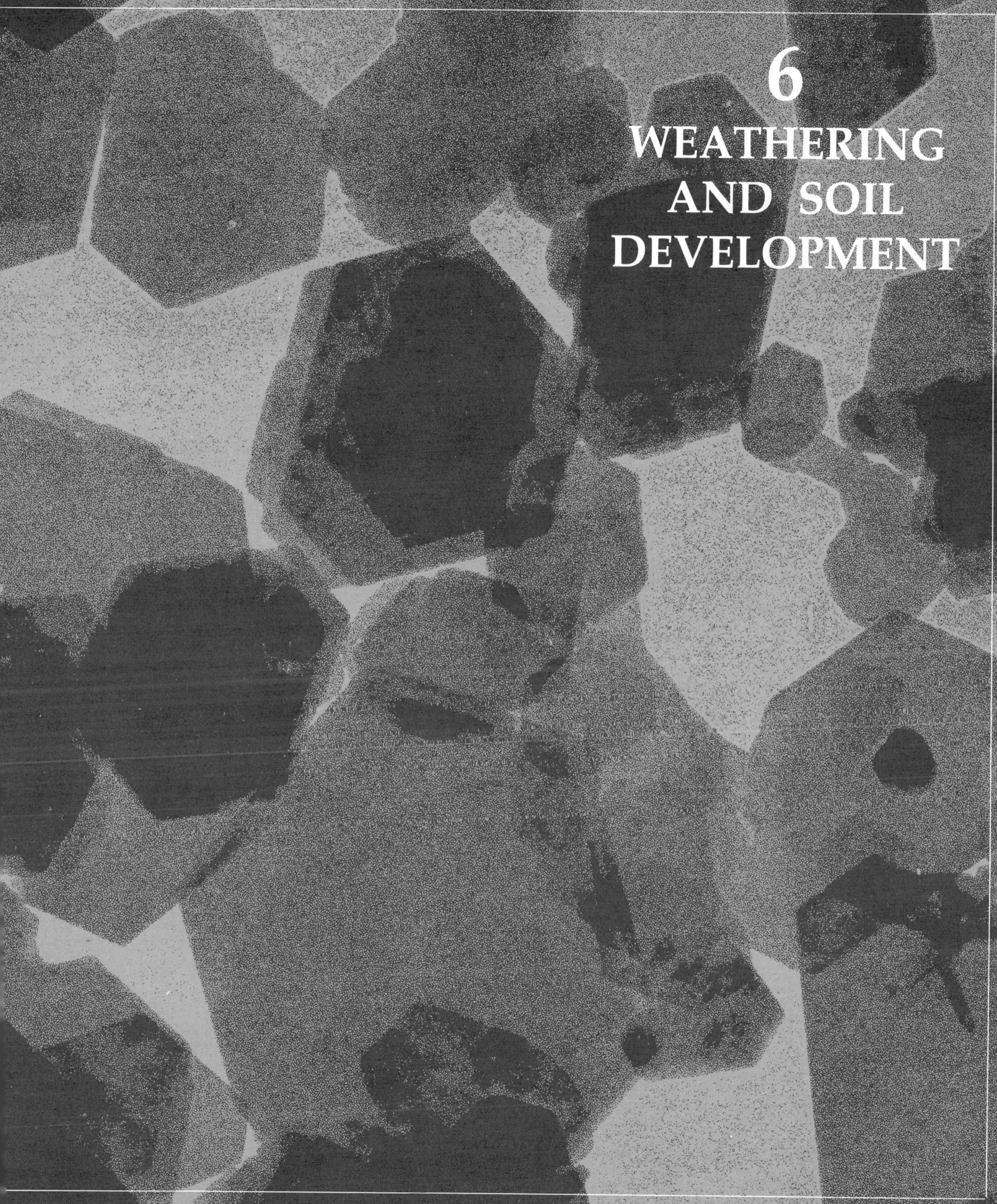

6 WEATHERING AND SOIL DEVELOPMENT

INTRODUCTION

In the sixteenth century many philosophers thought that rocks and minerals were living substances, that they were lower forms of life than plants. This belief was founded mainly on the observation that rocks appear to be in various stages of life, that is, youth, maturity, and old age. A particular type of rock in one outcrop seen to be a solid, strong, fresh, and youthful substance might be recognized elsewhere as a decaying mass, obviously "dead" and starting to disintegrate. The philosophers were observing what today we would call effects of weathering. *Weathering* changes occur where rocks and minerals come in contact with the atmosphere, surficial waters, and organic life under conditions normally found at the surface of the earth. Alteration by chemical changes produces new minerals that are stable at the surface of the earth from minerals originally formed under quite different environmental conditions.

The surface of the earth continually undergoes *gradation,* a result of the shifting of material. Weathering of rocks is not directly a process of gradation, but is related to it in that weathering causes decay and disintegration of rocks and facilitates their removal by geomorphic agents. It is usually an important—but not always necessary—initial step in degradation.

Soil is a product of weathering. Soils are usually composed of the products of decay, decomposition, and disintegration of rock. Some soils may be derived in place from solid rock or unconsolidated sediment; others may originate elsewhere and be transported into their present location. When a rock is weathered there may be a chemical interchange between the air or water and the mineral matter; therefore soil may contain components not present in the rock from which it is derived. In addition, some ions may go into solution during weathering and thus be removed in solution by the movement of water through the rock and soil.

Evidences of weathering are all about us. Sidewalks and concrete highways are built in separated sections to prevent them from buckling when heated and they expand or from cracking when temperatures fall and they contract. Houses in moist climates must be painted frequently to prevent decay and disintegration of wood. A knife left outside in the rain will react with oxygen and water to form rust. The surface of the earth's crust is constantly exposed to attack by these and other weathering agents. The effects are not confined to the surface; they penetrate into the crust along fractures and faults, between crystals, and along grain boundaries within rocks. Gradually they cause loosening, decay, and disintegration of even the hardest and most resistant rocks. Weathering processes bring about disintegration and decomposition; massive rocks can be, indeed often are, reduced to fragmental materials through these processes. One of the most important results of weathering is production of soils. Because soils form slowly, they are a valuable natural resource that must be conserved.

PROCESSES OF WEATHERING

Breakdown and decomposition of solid rock into soil vary from one climatic region to another and depend on the composition of the rock itself. Aside from climate and rock composition, it is possible to isolate a number of naturally occurring processes that produce weathering. Weathering processes may be grouped into mechanical processes that result in disintegration and chemical processes that result in chemical decomposition. Mechanical processes include freezing of water, frost action, unloading, crystal growth in soil, and organic activity. Chemical processes include oxidation, carbonation, hydrolysis, and hydration. It is possible to generalize the relationship between temperature, precipitation, and type of weathering, as shown in Fig. 6–1, because weathering phenomena are so closely related to climate, and climate is largely determined by temperature and precipitation considerations. Mechanical weathering dominates where there are temperature extremes but low moisture content. Chemical weathering is most effective where moisture and temperature are uniformly high, although chemical salts do produce much weathering in the Arctic. But over most of the earth mechanical and chemical processes combine forces to destroy rock coherence.

MECHANICAL WEATHERING

Freezing

Freezing water is effective in breaking down rocks because water expands when it crystallizes. In this respect it is different from most other liquids. Normally liquids contract when they crystallize, but water increases in volume by about 9 percent. As expansion takes place, high pressures are exerted on all sides of the space confining the water. Experiments have shown that when water is frozen in an enclosed vessel, pressures up to 2,100 kilograms per square centimeter (about 29,500 lb/in^2) are developed at temperatures of −22°C. Such pressures far exceed the strength of even the hardest and most durable rocks. The development of this pressure is of great significance, because rainwater seeps into fractures in rock not only where it is exposed but also beneath the soil cover. If freezing takes place while water is in a fracture, the fracture may be spread apart and extended in depth, or new fractures may be formed. The process is most effective where daily temperatures are above freezing and nightly temperatures below freezing. Once fractures have been opened, new surfaces become available for other types of mechanical and chemical weathering. The effects of freezing and thawing are cumulative. It is the repeated thaw and freeze cycle, day after day, year after year, that makes them so effective.

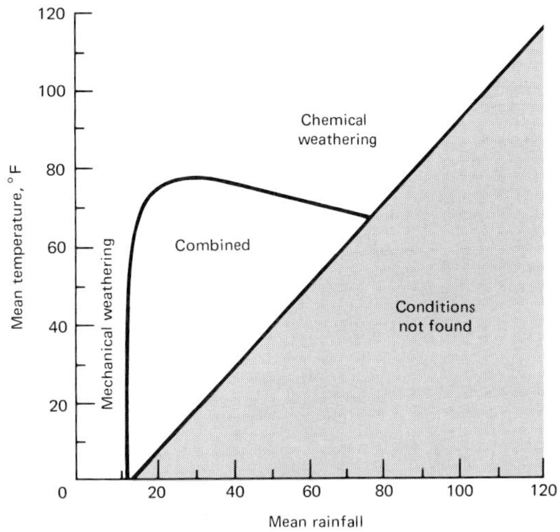

FIGURE 6–1
Relationship between rainfall, temperature, and the effectiveness of weathering processes.
(modified after Leopold, Wolman, and Miller, 1964)

FIGURE 6–2
Ice vein in muck, Seward, Alaska.
(S. Taber, U.S. Geological Survey)

Frost Action

Frost heaving is a process that causes rocks to be moved upward through soil, mud, or other loose sediment as a result of the formation of layers of ice in the soil (Fig. 6–2). Even when frost heaving occurs only rarely, or is restricted in degree or size of the area it involves, it loosens soil and allows air to pass into it. This process may be healthful for the soil, but it may also cause destruction through heaving up building foundations or through loosening soil on steep slopes where it can creep downhill or be washed away by rain. Layers of ice are most likely to form in soils that have differences of composition or texture. Once ice starts to crystallize, water is pulled into the surface of the layer by *capillary*

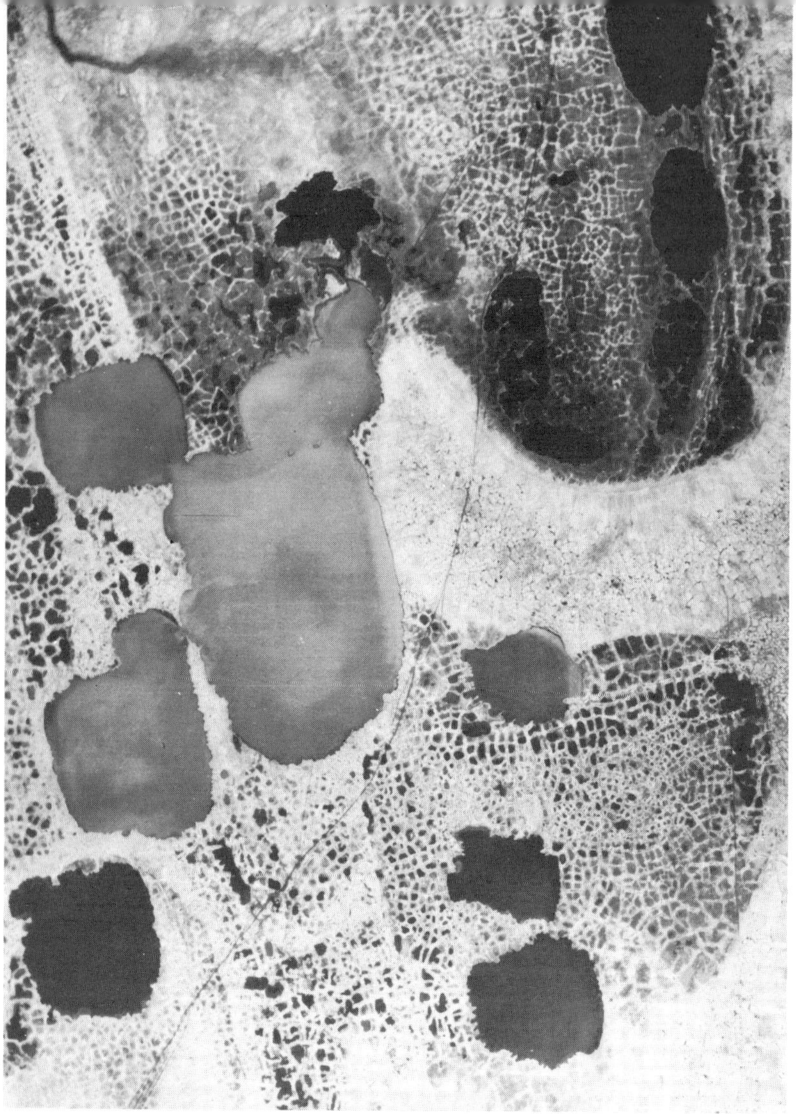

FIGURE 6–3
Ice-wedge polygons near Point Barrow shown in vertical aerial view. Thawed lakes surrounded by ice-wedge polygons developed in silt. Scale 1:20,000.
(U.S. Geological Survey)

action. This is possible because intergranular spaces in soil or rock act as *capillary tubes*. When crystallization takes place, a slight amount of heat is liberated, and this keeps a film of water on the ice. The film of water is connected to water in pore spaces and thus the process continues: water freezes on the ice surface; the ice slowly expands; more water is drawn in to freeze; and layers of considerable thickness may form eventually. The overlying soil is forced ever upward as the layer of ice thickens.

Rock fragments forced up through the soil as a result of frost heaving must be removed from plowed fields year after year. Frost heaving of stones begins when water immediately beneath a stone cools to the freezing point before the water on the sides of the stone. The stone is lifted as ice forms in the small capillary-sized opening under the stone. When there are large numbers of stones, a stone field may form at the surface. The stones are frequently arranged in polygonal-shaped patterns (Fig. 6–3), a phenomenon attributed by some geologists to near-surface expansion and contraction when the water in the soil freezes in the winter and expands. The center of each polygon is the point of greatest upward bulging, and the edges of the polygons are correspondingly lower. Stone polygons tend to merge and form stone stripes where the ground-surface slope increases.

Permafrost [1]

The name *permafrost* is applied to permanently frozen ground. Nearly 20 percent of the land area of the world has permafrost, including Antarctica, most of the land above latitude 50°N, and 85 percent of Alaska. Some ground within these areas is underlain by continuous and thick permafrost (400 meters of permafrost is recorded near Barrow, Alaska, and 550 meters occurs on Melville Island, Northwest Territories, Canada). Permafrost may be thin, discontinuous, or occur as isolated masses. Such areas present unique problems for those who would upset the thermal regime by construction of engineering works such as roads, airstrips, buildings, and the like.

Areas of thick, uniformly frozen ground are found where air temperatures rarely rise above freezing. Among areas inhabited by man, seasonal variations with temperatures above freezing in summer are more frequent. Unfrozen water may be found below the permafrost, as active layers within the permafrost, and above the frozen ground. The depth at which frozen ground is found may be much greater beneath a stream or lake or where the ground is saturated with sea water (Fig. 6–4).

Polygonal-shaped patterns frequently are found in areas of permafrost. The edges of these polygons in polar regions are usually underlain by wedges of ice (Fig. 6–3) that have filled deep cracks, which probably formed as a result of con-

[1] See Ferrians, Kachadoorian, and Greene (1969) for additional details.

traction during winter months. Serious foundation problems are encountered in these areas because the ice may melt, causing the ground to subside. Differential subsidence has been a serious problem in construction of railroads, highways, buildings, and airstrips over permafrost. Removal of vegetation that acts as a thermal cover is sufficient to trigger melting in many areas.

Pingos, small dome-shaped hills 10–100 meters in diameter and sometimes over 30 meters high formed by the growth and intrusion of ice mounds, are found in regions of permafrost. The ice mass is shaped like a laccolith (Fig. 4–31). Pingos originate in several different circumstances, but all result from the freezing and expansion of water in areas of permafrost. Some form where ground water under pressure moves from high ground into and between layers of frozen ground; others are found on the sites of former lakes. Frozen ground around the lake surrounds water-saturated sediments under the lake. Ground around the lake freezes progressively closer to the center of the former lake. Water is forced toward centers where it freezes, expands, and forces the former lake sediments upward.

Crystal Growth

Water soaks into the ground rapidly after a light rain. It may dissolve limestone or other soluble constituents of the rocks or soils directly beneath the surface as it percolates downward, but when the sun comes out and evaporation begins, water near the surface is drawn back to the surface as though it were in a capillary tube. Connected pore spaces in the soil are about the same size as capillary tubes and act in the same way. Water evaporates at the surface, and material in solution is precipitated to form crystals or crystalline aggregates in the soil. As crystals grow they can exert large expansion stresses, breaking up soil and even some rocks. This is the mode of formation of *caliche,* which is a layer or nodular mass commonly found in soils underlain by limestone or other rocks that contain calcium carbonate in semiarid plains or deserts. Other compounds that grow in this manner include some sulfates, phosphates, nitrates, gypsum, calcite, alum, epsom salts, and saltpeter.

Unloading

In quarrying granite for building stones it is sometimes possible to utilize natural forces to split the granite. A set of joints or fractures, called *sheeting,* may develop nearly parallel to the surface of the ground (Figs. 6–5 and 6–6) as a response to the release of confining pressure when the weight of overlying rock is removed. The process also occurs when glaciers have laid bare massive rocks like granite.

Insolation

The effect of the sun's heat on rocks exposed at the surface is called *insolation.* Daily variation in temperature is among the most universal processes, but its importance in bringing about

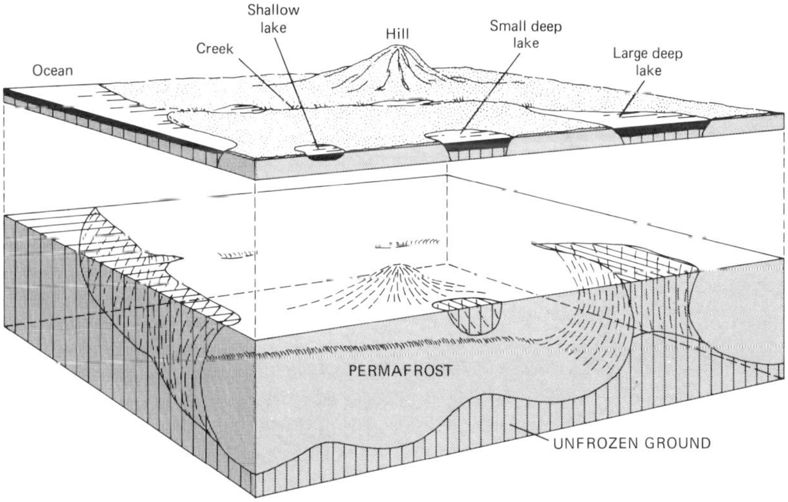

FIGURE 6–4
The effect of surface features on the distribution of permafrost in the continuous permafrost zone.
(after Lachenbruch, 1968)

FIGURE 6–5
Sheeting in granite at the Mount Airy quarry, North Carolina.

FIGURE 6–6
Half Dome in Yosemite National Park, Calif. Note the disintegration of the rock. Exfoliation and frost action are the dominant processes promoting the disintegration now. Unloading due to glacial erosion has also been an important factor.
(U.S. Geological Survey)

mechanical breakdown of rocks still is not clearly evaluated. Variations of 16°C or more may take place in desert regions where exceedingly high temperatures reached in the day are followed by rapid cooling at night. When solids are heated they expand, when they cool they contract. However, although the amount of expansion and contraction depends on the type of material, it is usually small. The question is whether or not repeated expansion and contraction will, over a long period of time, lead to mechanical breakdown of a rock. In laboratory experiments, repeated heating and cooling of rocks through wide ranges of temperature have failed to produce mechanical disintegration. But there is a time factor in nature that cannot be duplicated readily by short-term experiments. Possibly very slight fractures are formed. Differential expansion of different minerals in the rock would allow gradual breakdown between the grains. Once grains are loosened, solutions can get into them and bring about chemical decomposition. The development of scalelike sheets on exposed rocks has been attributed by some to insolation effects; such scaling is called *exfoliation*. Scales form on massive granites and other igneous rocks. Because they only form parallel to the surface of the exposed portion, the scales are thought to result from some surficial process. Many have thought that exfoliation is related to insolation, but universally accepted convincing evidence has yet to be found.

WEATHERING PROCESSES ON THE MOON

Weathering on the moon is of special interest because the moon does not have an atmosphere; both water and oxygen, which are so important on earth in chemical-weathering processes, are lacking; and no living organisms appear to be present. Weathering, then, is almost exclusively mechanical and results largely from impact of meteors on a lunar surface underlain by basaltic and related igneous rocks.

The soil brought back from Tranquillity Base is a mixture of fragments of basalt and gabbro, glassy fragments, and small bits of iron meteorites. This soil ranges from about 3 to 6 meters in thickness at the landing site. Soil there also contains fragments of microbreccia mechanically mixed as a result of high-velocity impacts of meteorites. The glasses appear to have been formed as a result of local melting with splashing, evaporation, and condensation following impact. They occur as small spheres, as stringers of glass in intensely fractured mineral fragments, as frothy ropy glass spattered on mineral fragments, and as glass linings on depressions. Some glass forms show evidence of later impact, resulting in the shattering of spheres and other forms. Many of the moon rocks contain pits and vesicles formed as a result of high-velocity impact.

Much of the moon's surface apparently is covered by a thin coating of lunar dust of the sorts described above. The chemical composition of the soil is very similar to that of crystalline rocks (page 99).

CHEMICAL WEATHERING

How rapidly a rock undergoes chemical decomposition depends largely on the climate. The amount of rainfall, its distribution through the year, the average temperature, and the temperature range are all important factors in determining rate and extent of chemical decomposition. The atmosphere contains a number of constituents that can react with minerals. Most important of these are water, carbon dioxide, and oxygen. Effectiveness of these chemical agents depends on the composition of the rocks and the size of their constituent particles. Quartz, for example, is almost insoluble in rainwater and is unaffected by carbon dioxide and other constituents of the atmosphere. Thus rocks composed primarily of quartz decompose slowly. Most other minerals are highly susceptible to water and show marked signs of reaction within a few months of exposure to the atmosphere. Four processes are notably responsible for chemical weathering. They are (1) oxidation, (2) hydration, (3) hydrolysis, (4) and carbonation. Almost all chemical-weathering processes cause expansion of the rock mass, thereby initiating stresses. Density is usually decreased.

Oxidation

This is the process by which oxygen combines with elements or compounds. Combination is aided by the presence of water and high temperatures. Water facilitates oxidation and may enter into other chemical reactions. Oxygen has a particular affinity for sulfur, manganese, and iron compounds, and these are among the most commonly oxidized materials. Pyroxenes, hornblende, biotite, and olivine all contain iron. As minerals are broken down they become oxidized, and iron is converted to ferric oxide (hematite, Fe_2O_3) or ferric hydroxide (limonite).

$$2FeS_2 + \frac{15}{2}O_2 + 4H_2O \rightarrow Fe_2O_3 + 4SO_4^{-2} + 8H^+$$

These processes are accompanied by color changes in the minerals and in the soils. Green and black soils change to red, yellow, or brown as limonite forms. Soils of these colors are most common in warm, moist climates. Reduction, the process by which oxygen is freed from its compounds, is not common in weathering.

Hydration, Hydrolysis, and Solution

Solution most commonly involves the dissolving of a mineral into water by the process of ioniza-

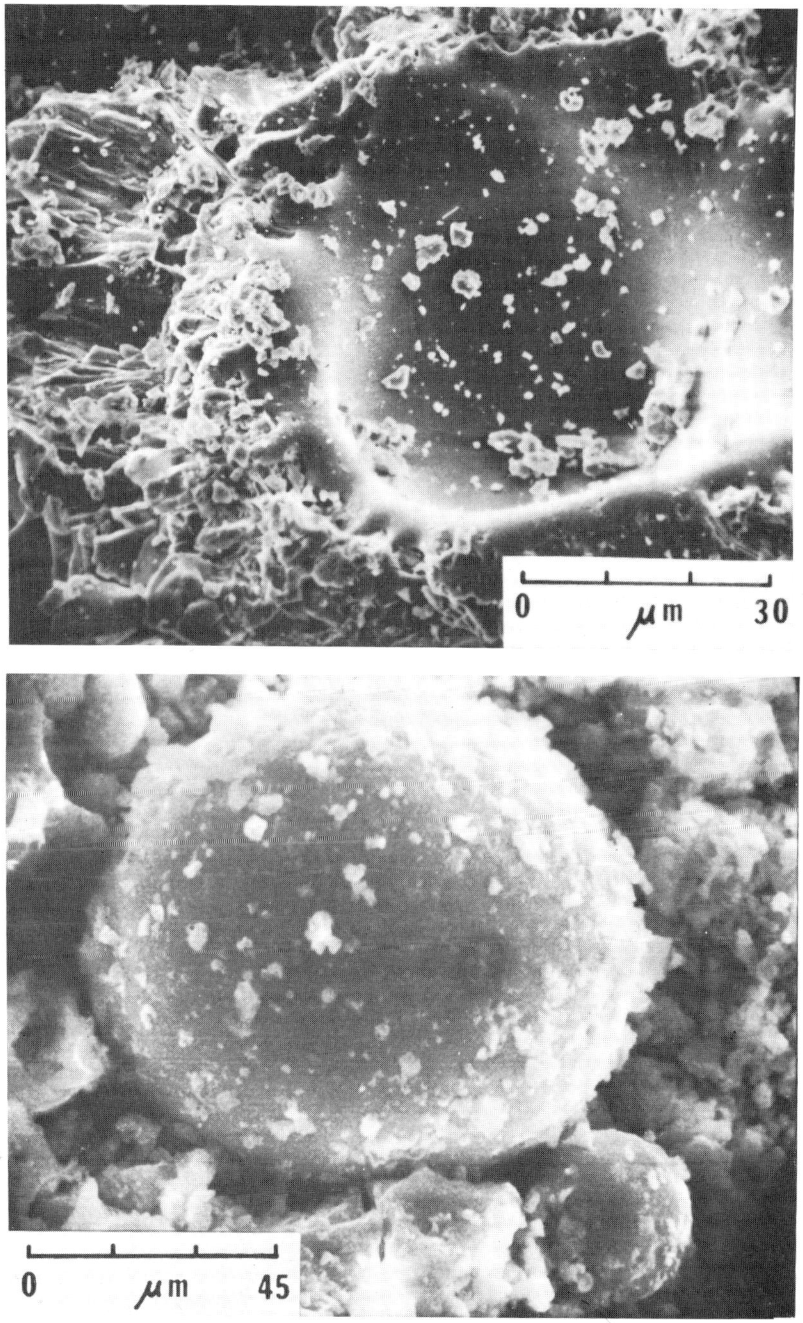

FIGURE 6–7
Scanning electron micrographs of lunar samples collected on Apollo 11 mission. At top is shown a glass-lined hole with splashed rim and glass fragments; at bottom, a sphere with adhering dust fragments in an angular microbreccia matrix. 1 μm = .000001 m

(photos courtesy of James L. Carter)

tion. Most silicate minerals dissolve so slowly in pure water that solution is negligible, but salt, gypsum, and, to a lesser extent, carbonate minerals may go into solution in appreciable quantity. Salt and gypsum may be present near the ground surface in outcrops or in the soil in arid climates. Some sand dunes, like those in the White Sands National Monument, N. Mex., are composed of gypsum, but these minerals are rare in the soils of humid climates.

Hydration refers to absorption and combination of water with other compounds, as when anhydrite absorbs water to become gypsum. *Hydrolysis* is a reaction between water and another compound. These processes, especially when combined with carbonation, are extremely effective in weathering of common rocks and minerals.

Hydrolysis is particularly important because it is the process by which most silicate minerals, the most abundant minerals in igneous and metamorphic rocks, are chemically weathered. The exact mechanism of feldspar weathering is not yet adequately defined, but the reaction may be expressed as follows (Krauskopf, 1967):

orthoclase feldspar + water ⇌
potassium + clay (kaolinite) + silicic acid

$$4KAlSi_3O_8 + 22H_2O \rightleftharpoons 4K^+ + 4OH^- + Al_4Si_4O_{10}(OH)_8 + 8H_4SiO_4$$

This equation proceeds in the laboratory only at temperatures over 200°C. It would be very slow at surface temperatures, and it is possible that the above reaction symbolizes a series of steps rather than a single reaction. It has also been suggested that the *alumina* and silica may be separated from the feldspar as *colloidal* particles that later combine to form the clay mineral. When feldspar or other silicates are found partially weathered, a discolored weathered rind remains from which some cations have been preferentially removed, leaving an outer protective shell around the original mineral grain.

The hydrolysis of silicates results in the production of a more basic solution. The acidity of carbonic acid solution is reduced, neutralized, or even made alkaline. The silica that is set free goes into solution as silicic acid (H_4SiO_4). If this is present in high concentration, some silica may come out of solution as colloidal or *amorphous* SiO_2. The silica ultimately may do one of three things: (1) it may go into the formation of clay; (2) it may be taken up by plants; or (3) it may be recrystallized into small quartz crystals.

Carbonation

In this process carbon dioxide is added to oxides of calcium, magnesium, sodium, and potassium to form carbonates of these metals. Carbon dioxide is a common constituent of the earth's atmosphere. It combines with water to form a weak acid, called *carbonic acid* (H_2CO_3), which has a *p*H of about 5.7. Acidity is expressed as *p*H. A neutral solution has a *p*H of 7. Acid solutions range from 0 to 6 *p*H, alkaline solutions from 8 to 14 *p*H. Thus rainwater, which contains minute quantities of carbon dioxide removed from the atmosphere, is a weak acid and is the most common solvent acting on the crust. Its action is evident in the formation of green copper carbonate on exposed copper roofs.

Natural waters may contain other acids such as the *humic acids* derived from decaying organic matter; these may lower the *p*H to 4.5–5.0. Nitrogen acids, although of minor importance, also may be formed through bacterial action and decay of organic matter.

Sulfuric acid can be found naturally in areas around volcanoes or where sulfide minerals are oxidized. These acids may be quite strong, with *p*H near 1.

Weathering of carbonate minerals (e.g., calcite, limestone) results primarily from the reaction of acids with carbonate (CO_3^{-2}) to form stable bicarbonate ions, HCO_3^-, which combine with H^+ to form carbonic acid. We can summarize the reactions in the solution of limestone by rainwater as follows:

1. Rainwater contains pure water and carbon dioxide, which combine to form carbonic acid, which in turn breaks down in an equilibrium situation to form hydrogen and bicarbonate ions.

$$H_2O + CO_2 \rightleftharpoons H_2CO_3 \rightleftharpoons H^+ + HCO_3^-$$

2. When the acid is weak, calcite combines with hydrogen ions to form bicarbonate and calcium ions.

$$CaCO_3 + H^+ \rightleftharpoons Ca^{+2} + HCO_3^-$$

3. The two above reactions can be summarized as follows:

$$CaCO_3 + H_2CO_3 \rightleftharpoons Ca^{+2} + 2HCO_3^-$$

When the reaction proceeds from left to right, calcium carbonate is removed, and when it proceeds from right to left, calcium carbonate is deposited.

SUSCEPTIBILITY OF ROCKS TO WEATHERING

Texture, structure, and composition of rocks make them of varying susceptibility to weathering. The surface area exposed is a function of the size of the particles that comprise the mass—the smaller the particles, the greater the surface area. Because chemical reactions take place on surfaces, fine-grained rocks or intensively fractured rocks open to solutions are more easily broken down than those that are coarse-grained and impervious. Fractures, bedding planes, and pore spaces all provide lines along which solutions can move. In addition, between crystals there are sometimes fine openings of capillary size along which solutions can be drawn.

Chemical weathering is often concentrated along bedding planes in sedimentary rocks and along fractures. Chemical reactions occur at the surface of contact between solutions and solid rock. Corners of blocks defined by bedding and fractures or intersecting fracture systems become rounded. Eventually the rock is reduced to a spherical or oval-shaped mass (Figs. 6–8 and 6–9).

Composition is of particular importance in chemical weathering. Some minerals react slowly with solutions and gases normally found in the atmosphere and in the ground; others break down rapidly in their presence. The most common rock-forming minerals include quartz, the feldspar group, micas, pyroxene, amphibole, olivine, and calcite. Of these, quartz is virtually unaffected by chemical weathering. Salt, gypsum, and calcite can be dissolved, and the others undergo complex chemical reactions. Carbonate water is very important as a naturally occurring acid, but acids may also be produced from the oxidation of sulfur compounds, and silicic acid is derived from hydrolysis of silicates. The silicates tend to yield

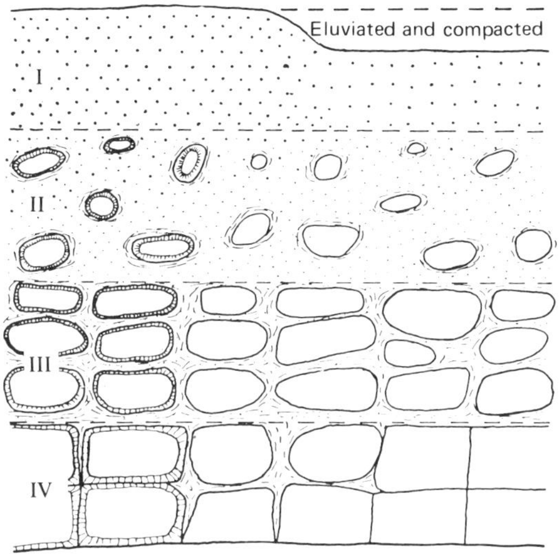

FIGURE 6–8
Zones in a profile of a granite that is being weathered chemically. Note that the alteration of the granite is concentrated in the fracture zones within the massive rock. Nearly spherical masses may form as successively deeper shells of the rock are attacked and decayed. The soil profile grades upward from fresh unaltered rock into zones of more and more complete decay until near the surface none of the original structure remains.
(from Ruxton and Berry, 1957)

FIGURE 6–9
Spherical weathering in a shale-siltstone sequence.

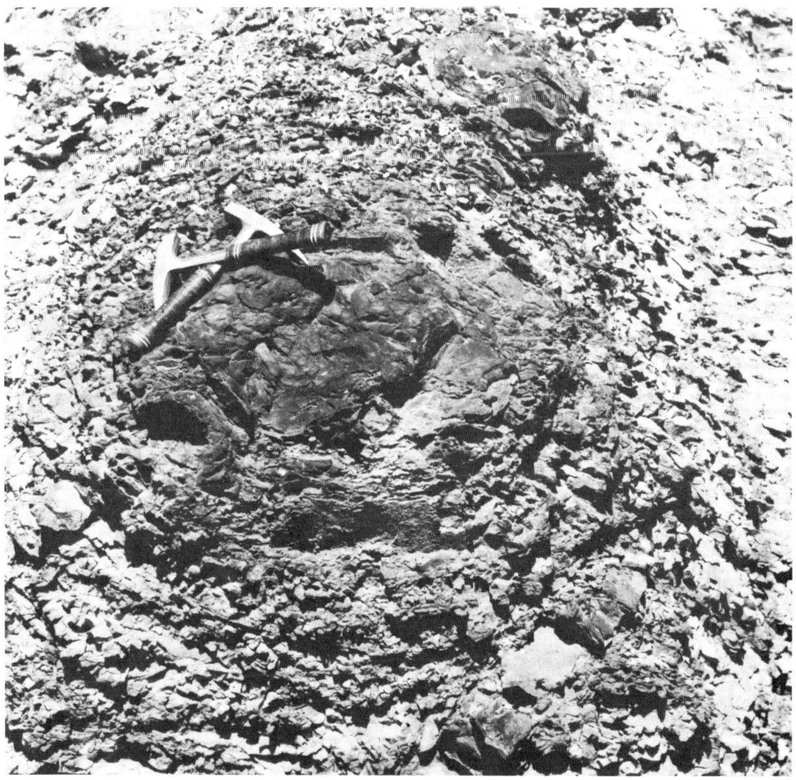

cations in solution, silicic acid, and residues composed of clay minerals. Iron, sulfur, and manganese-bearing minerals are subject to oxidation.

The rates of weathering of minerals can be observed and compared, and on this basis minerals can be placed in a series according to the rate at which they weather. The series thus constructed for common rock-forming minerals of igneous rocks is almost identical with the series constructed on the basis of the order of crystallization from melts (Bowen's reaction series, Fig. 4–14). The minerals that crystallize at the highest temperatures are most susceptible to weathering.

Animals as Weathering Agents

Animals and plants contribute significantly to the weathering and erosion of the land. Worms and other burrowing organisms produce holes in the soil, loosening it and providing passageways for atmospheric gases and solutions. Bacteria play an important role in breaking down organic compounds and liberating humic acids. Even larger animals, especially man, contribute to the breakdown and movement of rocks exposed at the surface. Man is particularly important in influencing the rates of weathering through activities that accelerate the erosion of the soil and affect the capacity of the soil to hold water. Certain crops remove particular chemical constituents from the soil, upsetting the chemical balance in such a way as to influence the direction of chemical reactions taking place in the soil and underlying rock. Several marine invertebrates, especially gastropods and echinoids, contribute to the weathering of exposed rocks along shore lines, particularly in the intertidal zone. These animals actually bore into the rock, producing holes or pits in which they live. Individual holes commonly become coalesced into a complex of channels as the rock surface is increasingly attacked.

Plants in Rock Weathering

Plant growth affects rock weathering in a number of ways. Roots force their way into the soil, opening it and leaving a cavity when the plant dies. This type of mechanical process also operates on solid rocks when roots grow down along fractures in the rock, forcing the cracks open and extending them farther into the rock mass. As the system of roots becomes interwoven, a mat of root matter is produced that effectively protects the underlying soil and rock from erosion. Plants greatly effect the rate of weathering in this way. The amount and type of plant growth influence the effect of precipitation on the soil or rock surface because plants may intercept raindrops, reducing their impact on the ground surface. Plants also effectively reduce the amount of moisture that reaches the ground, especially where rainfall is of short duration.

The biochemical processes brought into play by plants to work on the soil and underlying rock are probably far more important than any of these. One such process is the removal of certain elements from the ground by *accumulator plants,* those that take up certain elements to a greater degree than most plants (Lovering, 1958). We will consider the removal of silica as an example of this action.

A number of plants, including horsetail, corn, palm trees, bamboo, reeds, and other grasses, take up large amounts of silica. Silicon dioxide comprises about 77 percent of the ash in Giant reed, 47 percent in bluegrass, and 70 percent in China laurel and palms. The significance of these silica contents becomes apparent when we consider the amount of silica that could form in tropical forests where many of these accumulators are abundant. Lovering (1958) estimates that such a forest would take 0.4 ton of silica per acre per year out of the soil. If the underlying rock is basalt, containing about 49 percent SiO_2, one acre foot of the basalt would contain about 2,000 tons of silica that could be removed in about 5,000 years. Laterite is a common tropical soil and it is notably deficient in silica. Of course not all silica removed by plants would be lost because the ash is returned to the ground surface when a plant dies and decays, but the silica-rich ash is vulnerable to rapid removal by erosion resulting from stream action. With the surficial removal of silica, the soil is left deficient in silica and relatively enriched by other constituents.

Not only do plants contribute to weathering on land; they also make their contribution in the ocean where algae are present on rocks along the shore. Algae extend fine roots into the rock to

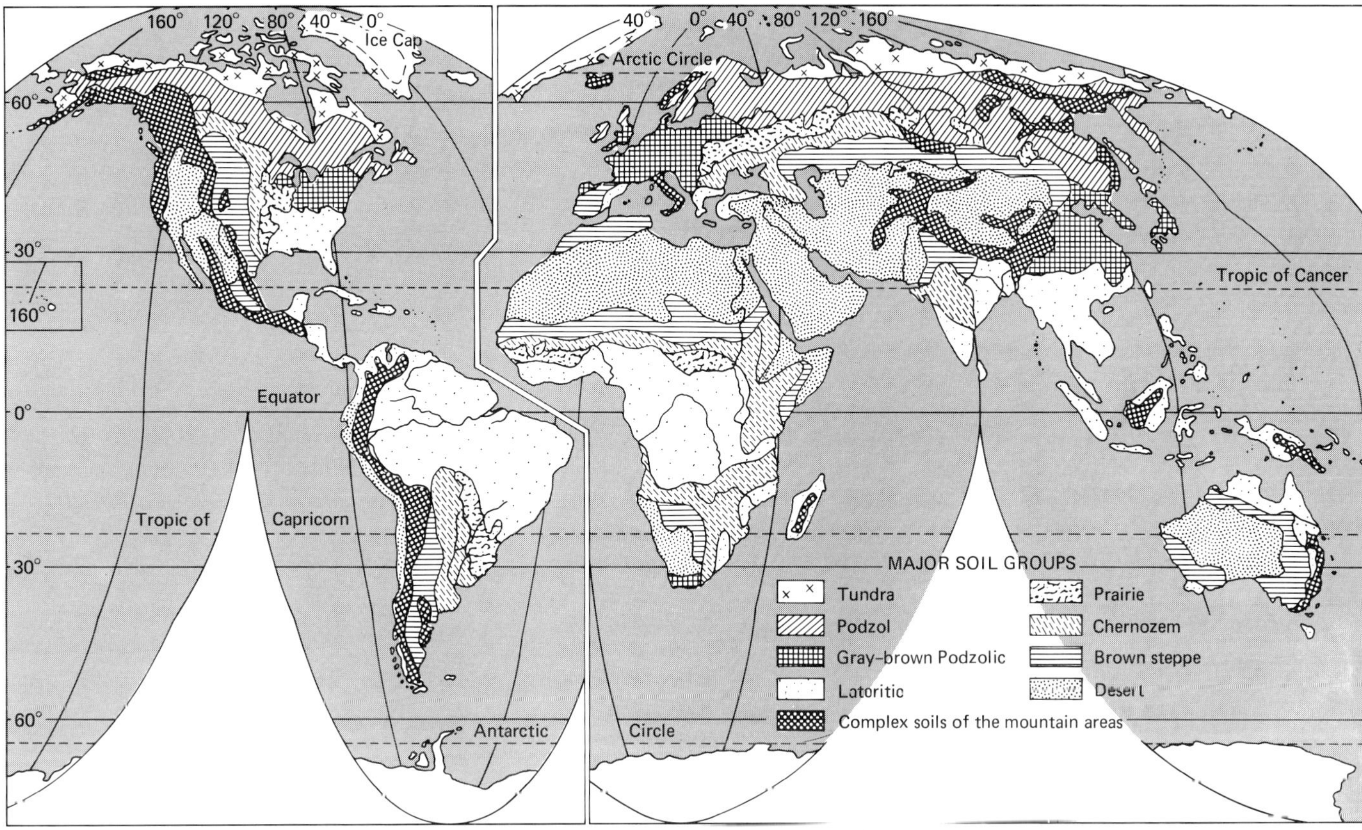

FIGURE 6–10
World distribution of great soil groups.
(from Miller and Langdon, 1964)

which they attach, loosening grains and altering the chemistry of the solutions around the roots.

SOIL

Most of the land surface of the earth is covered with soil (Figs. 6–10 and 6–11). In areas where it has been carefully studied and the processes governing its fertility, formation, destruction, and value are understood, soil has been conserved and continues to provide the riches of agricultural production. In other areas where study of soil has not been given careful attention, it has generally been depleted, eroded, and gradually abandoned as a depleted natural resource. At a time when overpopulation and starvation are among the most significant problems facing mankind, a thorough understanding of the soil is most important. Soil science, *pedology*, has expanded and is now a field apart from geology in much the same way as is meteorology.

Because soil is a product of rock decay as well as of organic processes, it is apparent that the type of rock from which a given soil is derived will be an important factor in its formation and in determining its qualities. In the light of the preceding discussion of weathering processes, it is also obvious that more than one type of soil may be derived from a given rock type. Among the most important factors governing soil formation are the following:

1. Rock from which the soil was derived.
2. Climate (particularly temperature, precipitation, and the distribution of these). (See Fig. 6–12.)

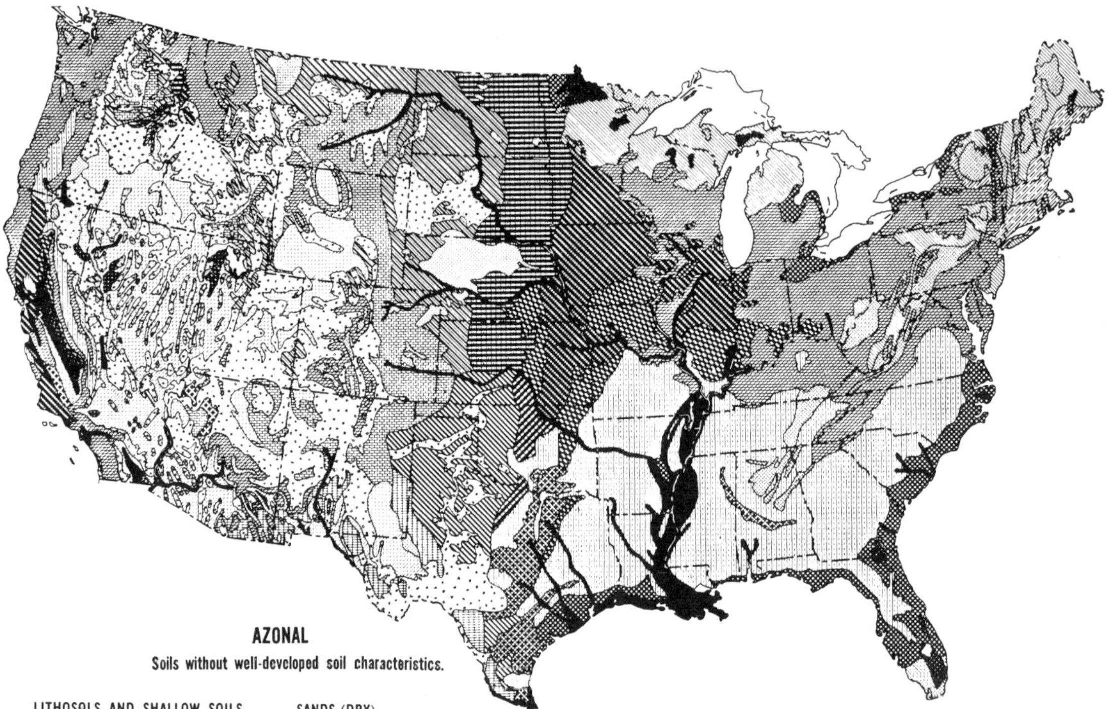

FIGURE 6–11
Soils in the United States.
(from U.S. Dept. of Agriculture)

AZONAL

Soils without well-developed soil characteristics.

LITHOSOLS AND SHALLOW SOILS
(ARID-SUBHUMID)
Shallow soils consisting largely of an imperfectly weathered mass of
(HUMID) rock fragments, largely but not exclusively on steep slopes.

SANDS (DRY)
Very sandy soils.

ALLUVIAL SOILS
Soils developing from recently deposited alluvium that have had little or no modification by processes of soil formation.

INTRAZONAL

Great groups of soils with more or less well-developed characteristics reflecting the dominating influence of some local factor of relief, parent material, or age over the normal effect of climate and vegetation.

RENDZINA SOILS
Dark grayish brown to black soils developed from soft limy materials in cool to warm, humid to subhumid regions, mostly under grass vegetation.

SOLONCHAK (1) SOLONETZ (2) SOILS
(1) Light-colored soils with high concentration of soluble salts, in subhumid to arid regions, under salt-loving plants.

(2) Dark-colored soils with hard prismatic subsoils, usually strongly alkaline, in subhumid or semiarid regions under grass or shrub vegetation.

BOG SOILS
Poorly drained dark peat or muck soils underlain by peat, mostly in humid regions, under swamp or marsh types of vegetation.

PLANOSOLS
Soils with strongly leached surface horizons over claypans on nearly flat land in cool to warm, humid to subhumid regions, under grass or forest vegetation.

WIESENBODEN (1), GROUND WATER PODZOL (2), AND HALF-BOG SOILS (3)
(1) Dark brown to black soils developed with poor drainage under grasses in humid and subhumid regions.

(2) Gray sandy soils with brown cemented sandy subsoils developed under forests from nearly level imperfectly drained sand in humid regions.

(3) Poorly drained, shallow, dark peaty or mucky soils underlain by gray mineral soil, in humid regions, under swamp-forests.

ZONAL

Great groups of soils with well-developed characteristics, reflecting the dominating influence of climate and vegetation.

CHESTNUT SOILS
Dark brown soils of cool and temperate, subhumid to semiarid grasslands.

REDDISH CHESTNUT SOILS
Dark reddish brown soils of warm-temperate, semiarid regions under mixed shrub and grass vegetation.

REDDISH BROWN SOILS
Reddish brown soils of warm-temperate to hot, semiarid to arid regions, under mixed shrub and grass vegetation.

BROWN SOILS
Brown soils of cool and temperate, semiarid grasslands.

NONCALCIC BROWN SOILS
Brown or light reddish brown soils of warm-temperate, wet-dry, semiarid regions, under mixed forest, shrub, and grass vegetation.

SIEROZEM OR GRAY DESERT SOILS
Gray soils of cool to temperate, arid regions, under shrub and grass vegetation.

RED DESERT SOILS
Light reddish brown soils of warm-temperate to hot, arid regions, under shrub vegetation.

PODZOL SOILS
Light-colored leached soils of cool, humid forested regions.

BROWN PODZOLIC SOILS
Brown leached soils of cool-temperate, humid forested regions.

GRAY-BROWN PODZOLIC SOILS
Grayish brown leached soils of temperate, humid forested regions.

RED AND YELLOW PODZOLIC SOILS
Red or yellow leached soils of warm-temperate, humid forested regions.

PRAIRIE SOILS
Very dark brown soils of cool and temperate, relatively humid grasslands.

REDDISH PRAIRIE SOILS
Dark reddish brown soils of warm-temperate, relatively humid grasslands.

CHERNOZEM SOILS
Dark brown to nearly black soils of cool and temperate, subhumid grasslands.

3. Shape of the surface on which soil develops (i.e., slope, relief, orientation of the slope).
4. Biological factors (amount and types of plants).
5. The length of time during which these factors have been acting.

The great number of combinations of different parent materials, climates, and other factors is reflected in the variety of soil types. Soil scientists differentiate 40 major types of soil, and for purposes of agricultural research these 40 are subdivided into hundreds of subtypes. In this discussion we will consider the gross characteristics of only the largest categories of soil types.

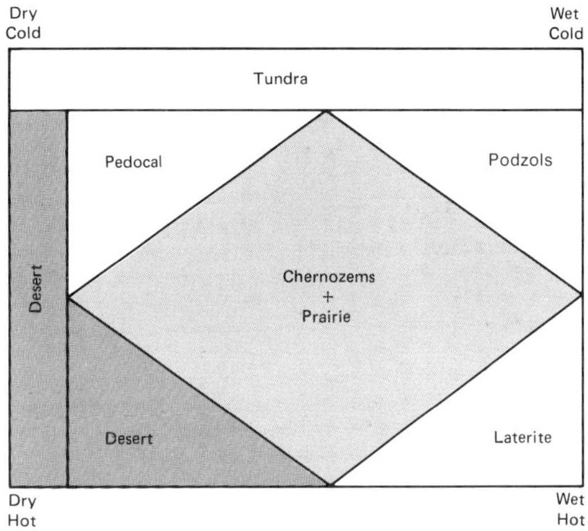

FIGURE 6–12 Schematic diagram indicating the relationship between soil type, temperature, and moisture. (simplified after Millar, Turk, and Foth, 1966)

FIGURE 6–13 *(below)* Humus-rich top layer in soil developed on Coastal Plain sediments, Mississippi.

Soil Horizons

Where a cut is made through soil to underlying rock it is possible to see well-defined zones, called A, B, and C *horizons* (Figs. 6–13 and 6–14), in some types of soils. Below the C horizon lies unweathered parent material such as solid rock or sediment. The C horizon contains weathered parent material, loose and partly decayed rock. It is from this material that the overlying soil is derived. Soils sometimes develop on material that is no longer located in its place of origin; thus rock in the C horizon may have been transported and the overlying soil may not be related to the bedrock below.

The B horizon is called the *subsoil*. It normally contains fewer organisms than the overlying layers, but more than the C horizon. Weathering has reduced all rock to fine material in the B horizon. It often contains more clay than either the A or C layers, and consequently it is harder when dry and stickier when wet than either of the others. Iron and aluminum hydroxide tend to be concentrated, particularly in humid and tropical climates, whereas calcium is concentrated in this layer in arid climates. Essentially, the B horizon is the site of accumulation of suspended materials from the overlying soil. Clayey material often has a blocky or prismatic structure.

The top layer of soil, the A horizon, is the site of maximum organic activity (Fig. 6–14). Processes of removal of soluble constituents in water are most active here. From the top down, through an idealized A horizon there would be leaves and loose organic debris that are largely undecomposed, then matted and partially decomposed organic material, next a mixture of organic and mineral matter, and finally a transition into clayey subsoil (Fig. 6–14).

The variability of soil has been emphasized. Variability also characterizes the degree of development and presence of the three horizons. Some soils show no zones at all, others have a zone missing, and still others have subzones of the general zones we have outlined. Texture and structure of soil are also highly variable. Soil constituents can be divided into sand, silt, and clay sizes, and these occur in all degrees of mixture. Four types of soil structure are recognized: (1) *granular,* (2) *blocky,* (3) *columnar* (also prismatic), and (4) *platy.* Granular and blocky soils have the best structure for most plants. Through proper soil management it is possible to influence structure.

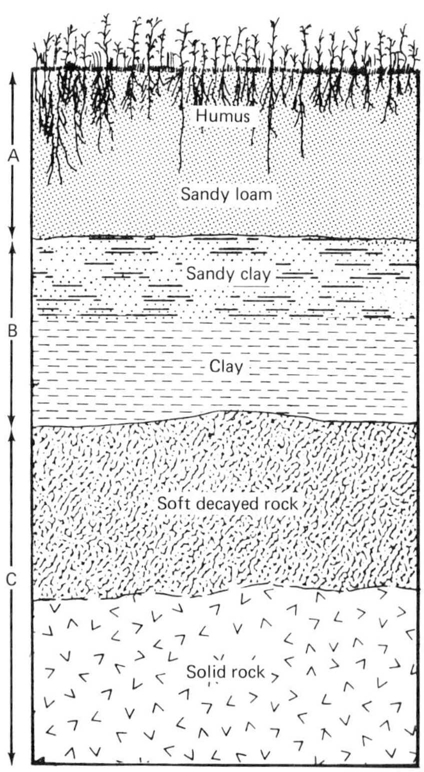

FIGURE 6–14
A soil profile of residual soil. The A horizon is characterized by decayed organic matter and sandy loam. The B horizon contains clay, and the C horizon is composed of the decayed rock from which B and A are derived.

Soil Chemistry

Chemical decomposition of a rock's constituent minerals is of particular importance in soils. Igneous and metamorphic rocks (granite, basalt gneiss, and schists) are composed primarily of quartz, feldspar, micas, pyroxene, amphibole, and olivine. Of these, quartz does not decompose rapidly. The others break down into clay minerals, iron compounds, silicon dioxide, and other compounds in solution. Thus of the over one hundred elements, only a relatively small number are common in soils, but many more may be present in trace amounts. Sediments and sedimentary rocks are products of breakdown and accumulation of sedimentary, igneous, and metamorphic rock constituents. Sedimentary rocks form the underlying rock beneath soil over nearly three-quarters of the earth's land surface. Common sediments are clay, sand (particularly quartz sand), lime, mud, and mixtures of these that occur in varying degrees of consolidation.

Of the various minerals present, clay minerals are of particular significance because they contain a number of different elements and also because they are so small that they are more readily involved in chemical reactions in soil. For this reason, clay plays a particularly significant role in chemical reactions taking place in soil. Clay, at least in small quantities, is present in most sediments and sedimentary rocks, and clay is also one of the most common weathering products of igneous and metamorphic rocks (Fig. 6–15).

Typical soil contains clay colloids, fine mineral fragments, organic matter, and some moisture. Mineral fragments undergo chemical decomposition very slowly. The effect of the process is to dissolve the minerals, reducing them to their ionic constituents. These go into solution in soil moisture. Some of the ions stay in solution, but because clay colloids are negatively charged they can hold some of the positive ions on their surface. This is particularly true of hydrogen, calcium, and sodium. In addition, smaller quantities of aluminum, magnesium, and potassium are held in this way. Concentration of hydrogen ions in a solution determines the acidity of the soil. The pH range for soils is between 4 and 10. The most common ions in acid soils are hydrogen and aluminum; calcium and magnesium are common in neutral soils; and sodium is most common in alkaline soils. Each type of plant grows best in soils of a particular acidity.

The chemistry of soils and soil solutions is complex, but it may be viewed simply in terms of an equilibrium condition. Positive ions (e.g., H^+, Ca^{+2}, and Na^+) are exchanged between the solid and liquid phases in the soil. An equilibrium condition is established in which the concentration of ions in soil water is maintained. As ions are removed from the soil by plants for use as food, they are replaced by ions liberated from minerals and organic matter and from clay colloid surfaces. Continued application of lime is necessary to keep many acid soils neutral, which is desirable because most plants thrive in soils with a pH of 6 or 7. When hydrogen ions in an acid soil solution are neutralized by lime, more hydrogen ions are released (or exchanged for calcium ions) from the surfaces of the colloids. The properties of soils that control their ability to exchange ions and maintain a supply of needed ions determine their desirability as a medium in which plants may grow. Some sixteen elements are necessary for plant growth: carbon, hydrogen, oxygen, nitrogen, phosphorous, sulfur, potassium, calcium, magnesium, iron, manganese, zinc, copper, molybdenum, boron, and chlorine.

In most soils the total supply of plant nutrients is high compared with the need of plants supported by it. When a plant dies in an uncultivated area, most of the ions return to the soil as decaying organic matter. If plants are crops, however, the soil loses these constituents because the crops are removed. The same thing happens when plants are burned off, a common practice in the tropics, and to an even greater extent if the ashes are washed away.

MAJOR SOIL TYPES

Desert Soil

"A zonal group that have light-colored surface soils and usually are underlain by calcareous material and frequently by hard layers. They are developed under extremely scanty scrub vegetation in warm to cool arid climates" (*Soil,* U. S. Dept. of Agriculture yearbook).

Large portions of some deserts do not have soil covers at all. Chemical weathering is particularly important in causing soil to form. Where there is little or no free moisture, bare rock, sand, and lag gravel are likely to predominate. Lag gravel is the name applied to boulders, cobbles, or pebbles left as a residual cover on deserts when finer sediment is blown away; it is not a soil. When soil is present in a desert, it is likely to be thin, loose, and friable. The low rainfall prevents leaching of soil and favors accumulation of salts. Salt content is exceedingly high in low portions of desert basins. Soil here often has salt crystals growing near the surface as a result of evaporation of water that has dissolved the salts from deeper in the soil. A hard layer containing calcium carbonate is sometimes found in the soil. This layer is referred to as *hard pan.*

FIGURE 6–15
Well-crystallized Georgia kaolinite clay. The enlargement (resolution) is × 90,000.
(K. M. Towe, Smithsonian Institution)

Tundra Soil

Tundra soil may be found on high mountains at all latitudes, but the vast expanse of it is within the Arctic Circle. It forms where there is very cold weather most of the year. Over large areas in the Arctic there is permanently frozen ground at a depth of a foot (0.3 meter) or more. The upper foot of soil freezes in the winter. As freezing takes place, water crystallizes out in layers of irregular size. The soil is forced up and disrupted, forming small mounds. These mounds are usually covered by a dense growth of moss, sedges, and flowering plants. Over much of the Arctic the land is low and poorly drained, giving rise to swamps and ponds. This poorly developed drainage is due to low slopes, permafrost, and the sheet of glacial deposits dropped there as ice sheets melted away. The soil is poorly drained in summer when ice in upper portions of the soil melts. Though there is abundant moisture, the rate of evaporation is low and there is little rainfall because of low temperatures. Low temperatures tend to slow down chemical reactions and inhibit growth of most plants, so the soils tend to be very thin.

Podzol Soil

"A zonal group having surface organic mats and thin, organic mineral horizons above gray leached horizons that rest upon illuvial [formed of colloids and soluble salts from above] dark-brown horizons developed under coniferous or mixed forests or under heath vegetation [low evergreen shrubs having whorls of needlelike leaves] in a cool-temperate, moist climate" (from *Soil*).

Podzolic soils have been the traditional agricultural soils in Europe and North America. They are widely developed in the central and eastern United States, which has a humid climate and large areas of hardwood forest cover. One common variety of podzolic soil has a gray-brown color and is crumbly and composed of silty loam. A loam such as this is composed of a mixture of clay, silt, sand, and organic matter. In wooded areas it has a matted layer of decaying leaves near the surface. The A horizon supports abundant plant growth and animal activity. Often there is a lower layer that has a light gray color and is partially leached. This layer tends to crumble readily and it is a silty loam also. The B horizon is richer in clay, coarser in texture, and darker in color than the A layer. The two, A and B, normally have an aggregate thickness of three feet (0.9 meter). The C horizon is composed of blocky clay. The soil profile is very similar, although it develops on rocks as different as limestone, shale, and glacial deposits. The soil is acidic and requires liming, but it supports a wide variety of plants, including cereals, legumes, root crops, vegetables, and fruits. Outstanding crop production can be expected when adequate care is taken to fertilize the soil, but even where this is not done the soil will produce for many years.

Chernozem Soil

"A zonal group having deep, dark to nearly black surface horizons and rich in organic matter, which grades into lighter colored soil below. At 1.5 to 4 feet, these soils have layers of accumulated calcium carbonate. They develop under tall and mixed grasses in a temperate to cool subhumid (semiarid) climate" (*Soil*).

These soils develop in climates where moisture is not abundant or consistent. A large expanse of this soil group is found in central western North America. The soil is very dark in color, almost black. Cereals are best adapted to it, and this region is the great wheat-producing area of North America. The low rainfall does not favor removal of ions from soil water by leaching. For this reason there is little need for fertilizer. Where calcium carbonate is dissolved near surface and moved down, the soil often has accumulated a layer of lime in the subsoil, but not to any great depth.

Tropical Soil (Latosolic)

There are possibly more different soil types in tropical and subtropical climates than in all other climates combined. This is due in part to the great variety of combinations of high temperature, variable amounts and distribution of rainfall, and topography. Tropical climates generally have heavy rainfall, but it is evenly distributed throughout the year in some areas while in others there are hot, humid periods of high rainfall separated by warm dry spells. Portions of the tropics have up to 1,000 centimeters (400 inches) of rainfall per year.

Soils in this climatic belt are almost always very thick. Conditions are ideal for chemical weathering. Plant matter tends to be abundant, and it usually quickly decomposes into soil. Plant nutrients are liberated rapidly from decomposing minerals, but the high rainfall soon washes these out of the soil. For this reason the soils become exhausted in a period of a few years if the forest cover is stripped and crops are cultivated. The usual farming practice is to cut a strip out of the forest, cultivate it for four or five years, then let it grow up in forest cover again and cut another strip. It takes about fifteen years for soil to become fertile again after farming is stopped. Because leaching is so extensive, soils are richest where they are continually rejuvenated by addition of new mineral supplies. These can come from volcanic ash, from stream deposits, or from the continued erosion of soil parent material at a slow but continuous rate. Because fires are so common during dry seasons, minerals from plants are often lost before they are returned to the soil.

Laterite is a special type of clay-rich soil that has a tendency to harden into a bricklike solid. It is found where the rainfall is highly variable, causing a large range of fluctuation in the level of the water table from within half a meter of the ground surface to depths of many meters. Silica is extensively leached from such soil and there is a concentration of hydrous aluminum oxides. In many instances these soils contain so much aluminum that they are potentially valuable sources of the metal. It is thought that further alteration of some laterites leads to the formation of *bauxite,* the principal ore of aluminum (Fig. 6–16).

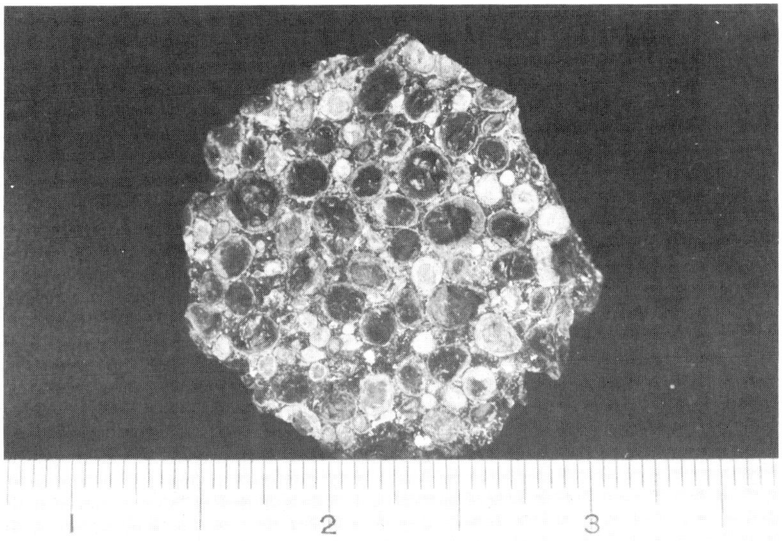

FIGURE 6–16 *(above)* Bauxite showing pisolitic structure. Scale is in inches.

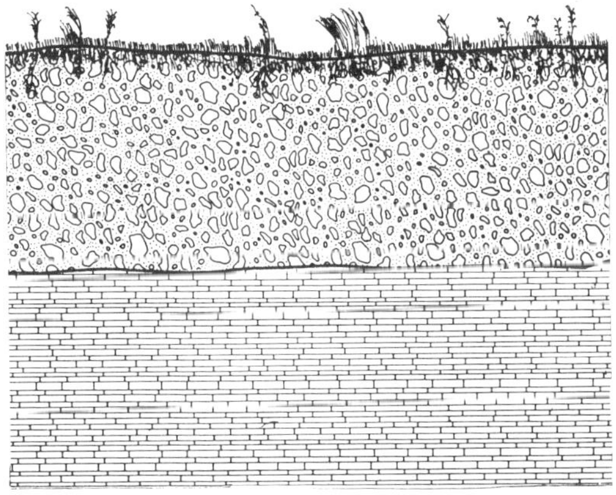

FIGURE 6–17 A transported soil. Note the sharp discontinuity between the bedrock and the soil. The pattern at bottom is for limestone whereas that above is alluvium and soil.

Alluvial Soil

Material transported by streams and deposited in stream valleys is called *alluvium.* Large streams flowing on low slopes usually have wide valleys, and during flood stages waters cover valley floors extensively. Soils developed on this transported alluvium are often rich and provide valuable farm land. Where soils are formed along modern rivers, the soil is young and original materials have not been modified very much by soil-forming processes (Fig. 6–17).

Mountain Soil

All the other types of soil can be found in mountains, and often several types occur close together, for climatic conditions on a mountain depend greatly on elevation. Usually mountain soils contain fragments of broken rock in slow transport down the mountainsides.

PRACTICAL VALUE OF A FAMILIARITY WITH WEATHERING

In addition to their importance in determining the concentration of ore deposits, both mechanical and chemical weathering affect critically the stability of rock and must be carefully considered in foundation evaluation and construction. A number of tragic mistakes have been made where foundations for dams have been built into rock that was either already weakened by weathering or subject to rapid weathering effects in water. For example, the 1926 failure of the 62.5-meter-high concrete St. Francis Dam was attributed to the weakness of a conglomerate in which one end of the dam was founded. It was discovered that the conglomerate had begun to flake and crumble when submerged in water due to the presence of veinlets of gypsum, which is soluble in water. Careful analysis of foundation rock must be made where large and heavy buildings, highways, dams, bridge abutments, tunnels, reactors, and similar structures are to be built to insure safety and prevent foundation failure. Degree of weathering and susceptibility to weathering are among the most important considerations.

Residual Concentrates

Weathering has been very important in the formation of economically important deposits of ore (metals) and nonmetals through processes of *residual concentration* and *mechanical concentration*. Residual concentration is a process by which some valuable mineral constituents accumulate while undesirable constituents are removed as a result of weathering and erosion. Among the residual concentrates are the following:

1. Iron (e.g., in the Great Lakes region).
2. Manganese (e.g., in India, the Gold Coast, Brazil, and Egypt).
3. Bauxite (e.g., in Arkansas).
4. Residual clay (for use in ceramics).
5. Nickel (e.g., in New Caledonia developed in laterite).
6. Phosphates (e.g., land pebble deposits of Florida).
7. Zinc (e.g., oxidized deposits of Virginia and Tennessee).

Many of the above residual concentrations are of minerals or rocks that owe their origin to chemical reactions that occurred during weathering; therefore weathering in these cases is not only responsible for concentration but also for the origin of the ore or nonmetallic minerals themselves. This is clearly the case with regard to bauxite and clay, which were described in the discussion of chemical weathering. The nickel deposits of New Caledonia, the world's largest source of nickel, provide another example. The nickel occurs in silicate minerals in large bodies of serpentinized peridotite that have been weathered under tropical conditions to form a lateritic soil. The peridotite has decomposed, silicate minerals have broken down, and hydrous silicates of magnesium and nickel have formed. These are concentrated in ore bodies in the lateritic soils.

Iron

The famous iron ores of the Great Lakes region that have been a major source of supply for the American steel industry are thought to be products of residual concentration through weathering of iron-bearing siliceous sediments. The ore is a hematite body that is very rich near the surface but becomes lower grade ore, called *taconite,* at depth. The taconite was originally derived from colloidal particles including silica and iron that were weathered and eroded from exposed rock, probably of igneous or metamorphic origin. These sediments were deposited as beds of iron hydroxide, iron silicate, silica, and carbonates. The deposits were later covered by a thick mud accumulation, after which the region became the site of extensive volcanic activity. Then the deposits were lithified and later folded. By this time the taconite, a banded cherty iron-bearing rock, was formed. The hematitic iron was later derived from the taconite, probably by weathering and ground water. Iron-bearing minerals of the taconite were oxidized and silica and carbonate were leached out and carried off by ground water. The above interpretation is consistent with the observed facts but it is hypothetical.

Supergene Enrichment

Oxidation is very important in the origin of certain types of near-surface deposits, particularly copper, which result from oxidation and *supergene enrichment.* These deposits originate as a result of the chemical weathering of the primary ore minerals by near-surface waters that cause some of the minerals (especially pyrite) to oxidize and yield acids that dissolve or react with other minerals. Thus the ore body is oxidized and leached of valuable elements that move down in solutions to the water table where secondary enrichment occurs. Pyrite and chalcopyrite are important because they oxidize to yield sulfuric acid, which is highly reactive. Most deposits of this nature are characterized by a limonite-rich leached cap, called *gossan,* and an enriched zone in which carbonates of the metal are concentrated at depth (copper and silver deposits may be enriched in this manner). Gossans are easily recognized in regions of sparse soil and thin plant cover, and they have been successfully used by prospectors in exploration for copper and other deposits, particularly in the southwestern United States.

Mechanical Concentration

Some deposits are formed primarily by concentration of a valuable mineral, for example, native gold or cassiterite (tin ore mineral), that occurs in the rock through weathering processes that break down the enclosing rock but do not destroy the ore mineral. These conditions are favorable for mechanical concentration, which involves first freeing the stable mineral through weathering and then concentrating it through mechanical processes such as gravity separation in streams, along coasts, or by wind. Only those minerals that are heavy, resistant to chemical weathering, and are hard, malleable, or durable are likely to survive the concentration process. Deposits formed in this way are called *placer deposits;* among the minerals found in them are gold, platinum, diamond, cassiterite (tin), magnetite (iron), chromite, ilmenite (titanium), copper, and monazite (phosphate). The stream placers of gold were particularly important in the development of the American West. Gold, which is malleable, weathered from gold-bearing quartz veins and accumulated in alluvial gravels along streams flowing out of the mountains where the gold veins cropped out. Gold accumulated behind boulders and other natural barriers across the streams and in gravels in stream channels.

CHAPTER OUTLINE

TERMS

accumulator plants
alluvial soil
alluvium
alumina
amorphous
bauxite
caliche
capillary action
capillary tube
carbonation
carbonic acid
chernozem soil
colloidal particles
desert soil
exfoliation
frost action
frost heaving
gossan
gradation
hard pan
humic acid
hydration
hydrolysis
illuvial
insolation
lag gravel
laterite
mechanical concentration
mountain soil
oxidation
pedology
permafrost
*p*H
pingos
pisolitic structure
placer deposits
podzol soil
residual concentration
sheeting
soil
subsoil
supergene enrichment
taconite
tropical soil (latosolic)
tundra soil
weathering

REVIEW QUESTIONS

1. Prepare a chart of igneous and metamorphic rocks showing probable susceptibility to weathering in a humid climate.
2. How could you use a knowledge of weathering if you were involved in locating industrial plants?
3. Outline the criteria by which you could differentiate a transported soil from a residual soil.
4. What weathering process are most active in the area where you live? What environmental conditions favor each?
5. Compare weathering processes on the moon with those in an arid climate on earth. How do they differ?
6. Why are clay minerals so important in soil-chemistry studies?
7. Explain differences that may be found in the soils developed on rocks of granite composition as a result of variations in climate.
8. What field evidence would prove that feldspar does weather to form clay minerals?
9. What special precautions must be taken in constructing bridge abutments and highways in regions of permafrost?
10. Refer to Fig. 6–11 and describe the location of azonal soils and zonal soils. Then explain some probable causes of this distribution.

SUGGESTED READINGS

Goldich, S. S., 1938, A study of rock weathering: Jour. Geol., 46.

Keller, W. D., 1955, Principles of Chemical Weathering: Columbia, Mo., Lucas Bros., 88 p.

Leopold, L. B., Wolman, M. G., and **Miller, J. P.,** 1964, Fluvial Processes in Geomorphology: San Francisco, W. H. Freeman, 552 p. Refer to chapter 4.

Ollier, Cliff, 1969, Weathering: New York, American Elsevier, 304 p.

Reiche, P., 1950, A Survey of Weathering Processes and Products, rev. ed.: Albuquerque, N. Mex., University of New Mexico Press, 95 p.

Stefferud, Alfred, 1957, Soil, The Yearbook of Agriculture: Washington, D.C., United States Government Printing Office, 784 p.

7 SEDIMENTS AND SEDIMENTARY ROCKS

7 SEDIMENTS AND SEDIMENTARY ROCKS

Sedimentary rocks cover about three-quarters of the land surface and are spread almost continuously across the ocean floors. They contain most of the world's fuel and water resources; they average almost 2,300 meters in thickness on continents; and it is largely from these rocks that the history of the earth has been deciphered. The term *sediment* applies to materials that settle out of suspension in water, but it applies more generally to particles that settle out of water, wind, or ice as well as to materials precipitated from solution and to deposits of organic origin. When sediments become consolidated to form a hard, solid rock, they are termed *sedimentary rocks.* Rocks are usually composed of a number of different minerals, but some minerals that occur in large masses (e.g., halite, calcite, gypsum) are also considered to be rocks.

Sands and gravels found on the bottoms of streams; sands, silts, or cobbles that make up beaches; muds on lake bottoms; decaying plants and mud in swamps; salts in deserts; sands, dust, and ashes transported by the wind; debris that melts out of glaciers; and calcite deposited from water dripping in a cave—all are sediments. If conditions prove favorable for their preservation, they may become sedimentary rocks; but most sediments on land are not likely to be preserved long. Forces of erosion continue to act upon them, and they are likely to be transported again toward the sea, where chances of burial and preservation are much greater.

SOURCE OF SEDIMENTS

The particles of soil, rocks, and minerals that comprise sediments are mostly derived from breakdown or decay of rocks. The great majority of the sediments and sedimentary rocks that form the shallow veneer over the face of the earth simply represent a reworking of materials in the earth's crust by the hydrosphere, atmosphere, and biosphere, with which they are in contact; but some sediments are composed in part of meteoritic particles from outside the earth, and other sediments are derived through volcanic action from sources deep within the earth.

Products of Decay

Particles produced by decay and disintegration of the crust and freed to form sediments are primarily clastic particles, ions, colloids, and organic matter.

Clastic particles. The term *clastic* means "broken off." Clastic particles are fragmental materials such as sand and gravel that are produced by the mechanical disintegration of all types of pre-existing rocks. A clastic sediment may consist of fragments of any type of rock.

Ions. These charged particles are formed through solution of minerals in water. When many minerals dissolve, they break down into their component ions. The process of solution may have been aided by the presence of acids or alkalis in the water. Solution is accelerated by mechanical breakdown of material that is being dissolved. Since solution takes place at the surface of a solid, the greater the exposed surface area, the more rapidly solution can take place. Once a solid has gone into solution, either it is available for formation of a sediment by direct chemical precipitation, or it may be removed from solution by animals or plants that use it to build their skeletons or shells. These in turn may become sediment after the animals die.

Colloids. This term refers to a particular size of matter—anything in the range of 10^{-5} to 10^{-7} centimeters (.00001—.0000001 cm) in diameter. Colloidal particles make up mud and clay. These particles are so small that we cannot distinguish them individually even with a high-powered optical microscope. If we stir clay in water until particles move in suspension through the solution, larger aggregates and particles will separate from the colloids. The heavier and larger aggregates will settle out, but colloids of clay will remain in suspension and move more or less at random through the solution. The dimensions of colloids are about 10 times or more the size of an atom.

Organic matter. This term is used in a very general sense to include excrement and other waste products of organisms, solid parts of an animal—all of which may become part of the sediment—and remains of plants, which may be converted into coal. Some sediments are made almost entirely of such remains of plants and animals.

PROCESSES OF SEDIMENTATION

Many sediments are chemical or biochemical precipitates that form under conditions in which settling is not always directly involved. Examples are the calcareous deposits formed around springs and in caves, silica deposited as petrified wood, flint, and as cavity fillings from ground water, and small, spherical masses of calcite precipitated from sea water where waves break over the Bahama Banks.

Settling of Solid Particles in Fluids

Many sediments are deposited as solid particles that settle out of a fluid medium in which they are temporarily held in suspension. The solid particles may originate in different ways, for example:

1. Sand and dust lifted as solid particles from the ground surface.
2. Grains and fragments freed by weathering and moved in suspension and by current action in streams.
3. Large grains that bounce along the bottom of a stream in the direction of current.
4. Extra-terrestrial material that falls through the atmosphere, then settles through ocean waters.
5. Volcanic products that settle through air and/or water.
6. Clay and other colloids that come together when they are introduced into salt water and lose their charge.
7. Chemical precipitates that form in water.
8. Shells and other remains of animals or plants living in the upper layers of water that settle when the organism dies.
9. Fecal pellets, mainly invertebrate excreta, that settle through water.

A solid particle is held in suspension when the forces holding it up are equal to the force of gravity pulling it down through the fluid in which it is suspended. If the fluid is at rest, the primary upward-directed force on the particle is that due to buoyancy. A fragment in water is buoyed up by a force that is equal to the weight of the volume of the fluid that is displaced. Thus a particle will have greater buoyancy in sea water than in fresh water, which is less dense. If gases or bubbles are attached to the body or within it, they also contribute to its buoyancy. If the weight of the particle exceeds the buoyancy, the particle settles.

Stokes' Law. Sir George Stokes studied the settling velocity of a solid sphere in a fluid and demonstrated that the frictional force on a solid sphere settling through a viscous liquid is proportional to the viscosity of the fluid, η, the radius of the sphere, R, and the velocity of settling, V. The friction, F, increases as the velocity increases.

$$F = 6\pi\eta RV$$

The other forces on the sphere are its weight ($mg = \frac{4}{3}\pi R^3\rho g$) and the buoyant force ($\frac{4}{3}\pi R^3\rho_0 g$) where ρ = density of the sphere, ρ_0 = density of the fluid, and g = gravitational acceleration. The net force acting downward is the mass of the sphere times its acceleration. As the sphere moves down it begins to experience a retarding force given by Stokes' equation and eventually achieves a constant velocity. This velocity is given by.

$$v = \frac{2}{9}\frac{R^2 g}{\eta}(\rho - \rho_0)$$

This equation applies provided turbulence does not set in. If it does, a much greater retarding force is experienced by the sphere. This equation is important geologically because it can be used to predict settling velocities of many sediments. Silt-size particles come closest to following this equation. The density and radius of the particle and the density and viscosity of the fluid determine settling velocity in such cases.

The equation does not hold when the fluid is in turbulent motion or when turbulence is set up around the particle, and the greater the departure of the shape of the particle from that of a sphere, the less valid the equation becomes. A general formula has been developed that incorporates what is called fluid impact, and it applies beyond the limiting velocities of Stokes' Law (Fig. 7–1). (Note that it applies to large sand-size material.)

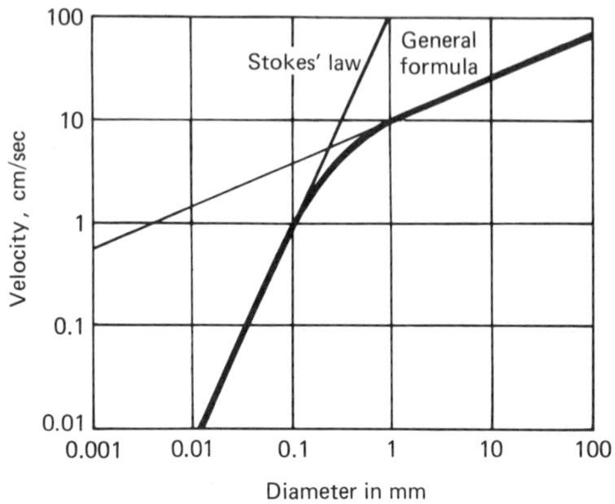

FIGURE 7–1
General formula for settling velocities of sediments of various sizes. The velocity predicted by Stokes' Law is shown also. The heavy line given by the general formulae agrees well with observed data.
(after W. W. Rubey, 1933)

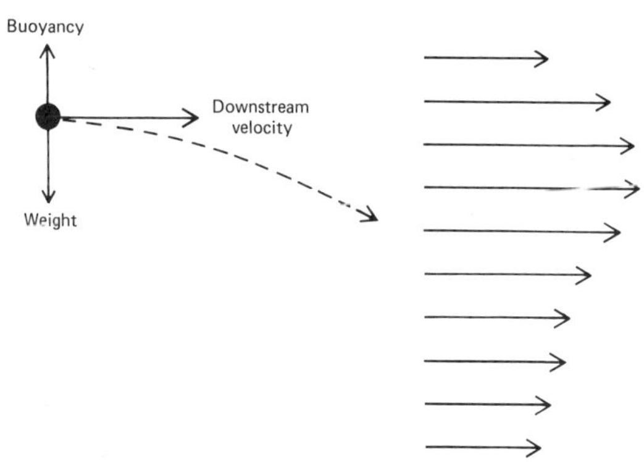

FIGURE 7–2
Schematic vertical cross section of a stream showing a particle and the forces acting on it. Its path is determined by its weight, buoyancy, and the downstream momentum.

Tabular or prismatic particles settle with different velocities, depending on how they are aligned in the fluid. When a flat surface on a tabular particle is aligned perpendicular to the direction of movement, resistance is greatly increased.

When the particle is settling through a fluid that is moving subhorizontally, like a stream (Fig. 7–2), the forces acting on it are frictional resistance to downward movement (Stokes' Law), buoyancy upward, gravity (weight) directed downward, and a lateral component in the direction of flow. The particle will settle along a curved path.

Study of settling velocities reveals the way sediments are sorted according to size.

Process of Chemical Precipitation

Precipitation is the process by which a solid may be separated from a solution. The solution initially may be a homogeneous system in which the ions (anions and cations) of some solid are held in solution. Water is the most common solvent and the substances dispersed in it may be gases, solids (ions), or liquids, but we will be most concerned with ionic solutions.

A solution is saturated (with respect to a particular ionic compound) when no more of the solute can be dissolved in the solvent. The maximum amount of the ionic compound dissolvable in a given amount of solvent under a given set of conditions of temperature and pressure is called the *solubility* of the compound. (High temperatures tend to increase most solubilities. Solubility of ionic compounds in water is more sensitive to temperature than to pressure variations. Most natural solutions are not single-component systems; they contain many ions in solution, and the chemical reactions involved are often complex. Solubility of most compounds is particularly affected by the acidity of the solution.

Precipitation can take place as a result of any of the following:

1. Evaporation of a solution until crystallization occurs.
2. Cooling of a solution until it is saturated.
3. Addition of a reagent that forms a new insoluble residue.
4. Addition or removal of a substance in the solution that has the effect of decreasing the solubility of another substance (e.g., if the solubility of one compound is increased by the presence of acid in the solution, then neutralization of the acid will lower the solubility of the initial compound, causing precipitation).

Biochemical Processes

Plants and animals are very important in the formation of most marine sediments. Some sediments are composed largely of the shells or skeletons of animals. Calcium carbonate (as calcite or aragonite) is the most common constituent of shells, but microscopic plants, diatoms (Fig. 7–3), and animals, radiolarians (Fig. 11–4), take silica out of sea water and build their hard parts of opal. The excreta of marine invertebrate animals, fecal pellets, are usually composed of calcium carbonate with impurities held together by a mucus material. The chemistry of sea water is affected by the presence of plants and animals. Slight changes in acidity are found in water that is immediately in contact with certain plants and animals. Some sedimentologists think most precipitation in sea water is a result of biochemical processes. Certainly important changes result where animals take in oxygen and give off carbon dioxide in the water and where plants take in carbon dioxide and give off oxygen as a result of photosynthesis.

LITHIFICATION

Most sediments originate as unconsolidated materials. All clastic sediments first must be deposited in unconsolidated form. When *chemical precipitation* occurs, ions come together, forming ionic compounds that either settle as a mass of tiny particles or crystallize as a hard, consolidated layer. Colloids often are deposited when fresh water carrying them becomes salty, causing the colloids to settle in little lumps. At other times colloids coagulate, forming a *gel. Lithification* is the process by which unconsolidated sediments become consolidated rocks. Some sediments undergo internal reorganization of the material. In addition, reactions with water and alterations due to solutions that may flow through the sediment occur.

One of the most universal processes affecting sediments is *compaction.* This is particularly important where sediments are deposited in an environment in which long-continued sedimentation is possible, as in the sea. Weight of the overlying sediment compresses the deeper sediment, forcing water out, eliminating void and open spaces, and pressing soft materials together. Colloids are particularly compressible. The volume of clay can be reduced by 40 percent if it is subjected to a pressure equal to that found at a depth of several thousand meters. Sands and other clastic sediments are only slightly compressible. Pressure due to weight of overburden may facilitate recrystallization of some minerals. When this happens, the compounds may dissolve at points of high pressure, where grains are in contact, and recrystallize in spaces of lower pressure, as in pore spaces.

Many sediments are mixtures of different sizes

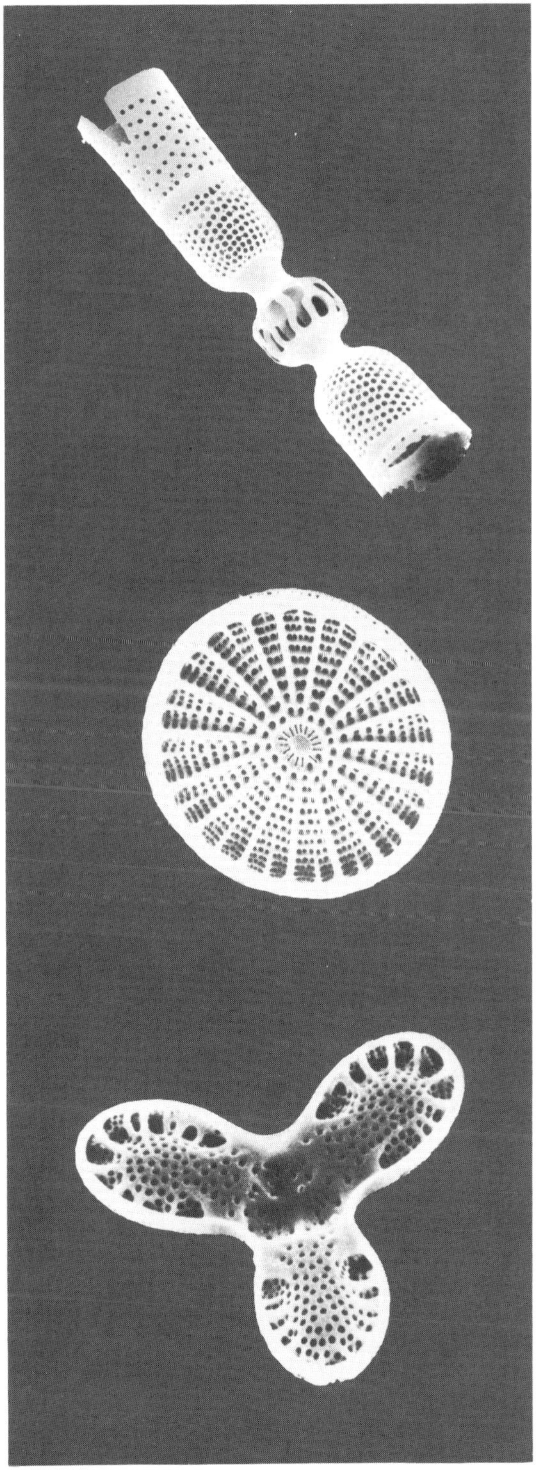

FIGURE 7–3
Diatoms. Top approximately ×2,000; middle ×480; bottom ×480.
(photos courtesy of H. C. Culver, Engis Equipment Co.)

and types of minerals. Mud may act as a cementing material. When solutions passing through the sediment carry ions, these ions may precipitate in pore spaces and cement the particles of sediment. The most common cementing materials are calcite, quartz, iron carbonate (siderite), and iron oxide (hematite). These may be derived from the sediment itself, or they may be introduced from outside by water flowing through it. Some cements are important economically. Iron oxide cement in the Clinton formation (Silurian age) in the Appalachians is rich enough to make the sandstone an iron ore. It is the principal ore for the Alabama iron industry.

FIGURE 7–4
Stratification in lake-bed deposits, Bryce Canyon.
(Union Pacific R. R.)

Recrystallization and reactions between some minerals and sea water, which often saturates marine sediments even after they have been buried deeply, bring about further alteration of sediment. During lithification a number of unusual structures may develop in sedimentary rocks. Nodular masses, concretionary structures, chert stringers, and radiating and fibrous textures are among the features formed during lithification.

STRATIFICATION

One of the most characteristic features of sedimentary rocks is the layering—*stratification*—that usually results from processes of sedimentation (Fig. 7–4). Consider a lake in which sediment is accumulating. If this lake is fed by streams, coarse sediments enter the lake at inlets. These sediments will fan out into the lake. The finest particles may be evenly distributed throughout the lake, but larger, heavy particles are dropped near the inlets. The fine sediments tend to form a thin layer over most of the lake bottom, but they are mixed with coarse sediment near the mouths of the streams. Any change in the type of material brought into the lake by the streams produces variations in the layers deposited, and these may be used to identify stratification.

Stratification may be the result of variations in the following:

1. Composition
2. Color of layers
3. Texture
4. Porosity
5. Structure

The layers are called *strata* or *beds,* but if they are very thin they are called *laminae.*

On continental shelves and in ocean basins, platelike or wedge-shaped sedimentary units cover thousands of square kilometers. Some of these units are uniform in thickness and composition. Others vary greatly in both composition and thickness, but stratification is usually present (Fig. 7–5).

Environments of Deposition

Although most sediments are ultimately deposited in the ocean basins or around margins of oceans, sediments accumulate in many other places as well, and even within oceans many variations in environment are found. Sediments deposited on land are influenced by the character of the surface on which they are deposited, composition of the atmosphere, temperature, humidity, precipitation, soil type, and organisms. Environmental factors in marine environments include temperature, depth, circulation, organisms, types of solids and gases present in the water, and turbulence and movement of the water. These environmental factors affect texture, composition, and structure of sediments that form. Most hard, consolidated sedimentary rocks now exposed on the flanks of mountain ranges hundreds of kilometers inland originated in the ocean. Because environmental conditions leave their imprint on sediment, it is possible to tell how deep the water was, the approximate temperature of the water, and, in short, to reconstruct the environmental conditions that prevailed during sedimentation.

The three major sedimentary environments are (1) marine, (2) transitional, and (3) continental. Some of the important subdivisions of each are the following:

1. Marine environments (see Chapters 10 and 11)
   a. Shallow water (neritic)
   b. Intermediate-depth water (bathyal)
   c. Deep water (abyssal)
2. Transitional environments (see Chapter 10)
   a. Deltas
   b. Lagoons
   c. Beaches
   d. Tidal swamps
   e. Estuaries
3. Continental environments (see Chapters 12–17)
   a. Deserts
   b. Glacial environments
   c. Streams
   d. Lakes
   e. Swamps
   f. Caves

Characteristics of deposits formed in these environments will be discussed as we consider the natural processes acting in each.

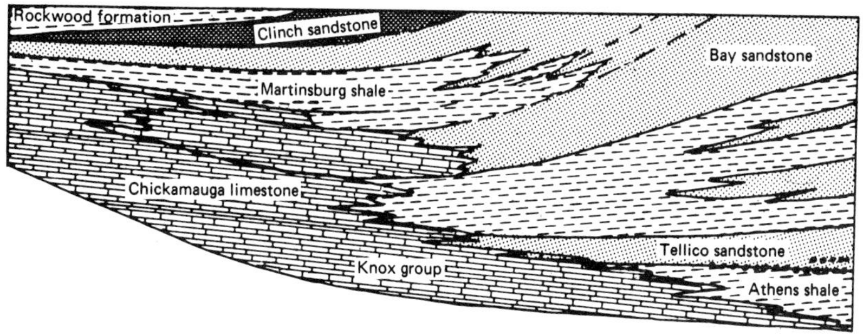

FIGURE 7–5
Cross section across eastern Tennessee. The coarser sediments of the east (right) grade westward into finer sediments. Limestones were laid down across the entire section early in the Middle Ordovician but by the end of the period coarse sediments were deposited almost to the westernmost end. The section is a little over 48 kilometers long.
(after P. B. King)

TEXTURES

Clastic

Clastic rocks are composed of fragments. The size of fragments is their most apparent feature, which makes them easy to identify. Everything from colloids up to boulders hundreds of meters across can be found, as well as every intermediate size. The difference in size between the smallest and largest fragments is extreme. To make analysis of sediments relatively easy, a system of sizes has been arbitrarily selected. This scale, known as the *Wentworth classification*, defines the limits of each type of clastic particle (also see Fig. 17–14):

Boulder. Anything above 256 mm in diameter.

Cobble. 64 mm–256 mm.

Pebble. 4 mm–64 mm.

Granule. 2 mm—4 mm.

Sand. $\frac{1}{16}$ mm–2 mm.

Silt. $\frac{1}{256}$ mm–$\frac{1}{16}$ mm.

Clay. $\frac{1}{256}$ mm or less.

FIGURE 7–6
Variation in roundness of grains of sand from very angular at bottom to well-rounded at top.
(after M. C. Powers, 1953)

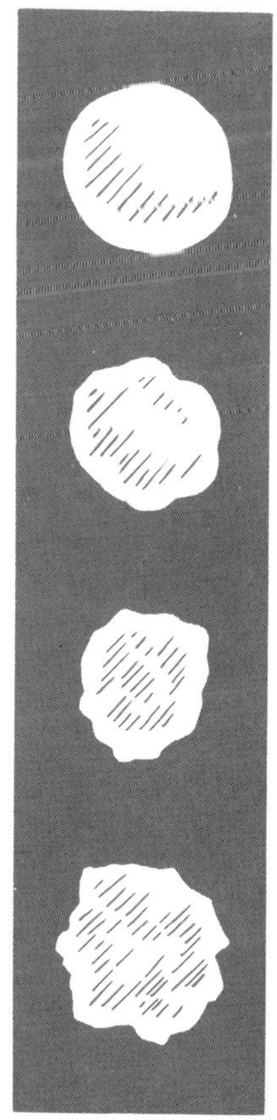

The shape of the grains in clastic sediments is also important in describing texture. Grains generally may be described as equidimensional, elongate, tabular, angular, subangular, subrounded, or rounded (Fig. 7–6). Angularity or roundness is often an important indicator of the distance the particles have been transported. Experiments with cubes of quartz subjected to repeated abrasion and impacts both in water and in a wind tunnel have demonstrated that the cubes are gradually reduced to spheres. A similar process should act where abrasion and impact are important on beaches, in streams, and in wind-blown sand.

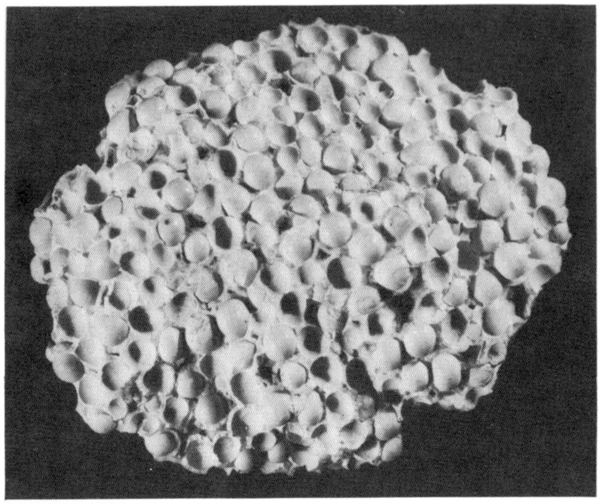

FIGURE 7–7
Oölitic limestone from Carlsbad, Bohemia.
(U.S. National Museum)

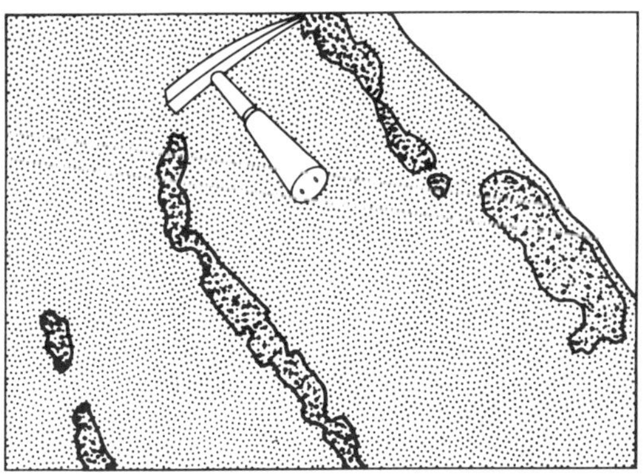

FIGURE 7–8
Nodules and stringers of chert in a massive limestone. Traced from a photograph of the Beekmantown formation.

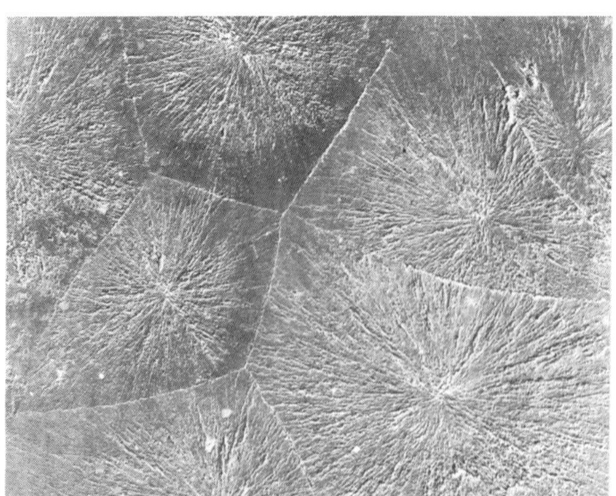

FIGURE 7–9
Intergrowths of spherulitic vaterite ($CaCO_3$) shown at a resolution of ×8,000.
(K. M. Towe, Smithsonian Institution)

Crystalline Sedimentary Textures

Crystalline textures form as a result of deposition through chemical reactions. Crystalline texture is characterized by interlocking aggregates of crystals. Crystals may be small, medium, or large, or they may even exhibit a mixture of sizes. Crystals may have certain definite shapes (i.e., equidimensional, fibrous, or scaly). It is often impossible to distinguish those formed by inorganic chemical reactions from those deposited by reactions caused by organisms.

Formation of Crystalline Textures

Most of the crystalline textures are formed by direct precipitation from a saturated solution. Crystals grow from many centers and spread until they form an interlocking framework or mosaic. Commonly, centers of crystals are small clastic particles of irregular shape and size. Crystalline textures are also formed through recrystallization. After recrystallization, new crystals may be larger crystals of the same mineral that was present originally, but elements from several minerals may combine to form new minerals that were not present originally. Usually part of the rock is slightly dissolved by solutions moving along boundaries of individual crystals. The dissolved substance is carried along, and it may eventually crystallize as part of another crystal.

When colloids coagulate, they may form a gelatinlike mass, a *gel*. If this mass loses water and hardens, it forms an *amorphous* solid. This amorphous substance is not stable, and it may be changed into a crystalline mass. When crystallization of a hardened mass of colloids does take place, the mass tends to have a fibrous structure first, but it may later become granular by recrystallization. Some of the textures of colloidal deposits are nodular, oölitic, and spherulitic (Figs. 7–7, 7–8, and 7–9). *Oölites* are small, nearly spherical bodies between ¼ and 2 millimeters in diameter. Oölites are made of concentric layers that have a fibrous radial or finely laminated structure. *Spherulitic* texture is characterized by small, spherical bodies with radial structures originating at the center and extending out to the margins. *Nodules* are rounded bodies that are generally harder than the enclosing rock. They may or may not have internal structure.

CLASSIFICATION OF SEDIMENTARY ROCKS

Because of their diversity in origin, texture, composition, and occurrence, sediments may be classified in many ways. Texture and composition are the criteria most frequently used, but these are not always easy to apply. One problem incurred in attempting to make exact classifications of sedimentary rocks is that many of them are mixtures of two or more types of material. The most common sedimentary rocks are shale (consolidated mud and clay); sandstone (made of sand); and limestone (composed of calcite in various forms). Completely pure limestones, sandstones, and shales are rare. Most limestones have a small amount of quartz or clay in them; most sandstones contain either calcite, clays, or feldspar impurities; and most shales contain some calcite or quartz, or both.

It is usually possible to represent either grain size or composition of sediment accurately on a ternary diagram (Fig. 4–10). Consider sediments that are composed almost entirely of small clastic particles of sand, silt, and clay size. These three components can occur in all proportions ranging from 100 percent sand, silt, or clay to a mixture composed of all three. One suggested subdivision of the field is shown in Fig. 7–10. The last name designates the type of rock, and the modifier indicates the second most abundant constituent. The percentage of a constituent can be estimated using Fig. 7–11.

Clay in conjunction with sand-size particles of quartz and feldspar is one of the common mixtures of clastic sediments. These are the main constituents of sandstone, arkose, and graywacke. Various suggestions have been made regarding the classification of these sediments. The nomenclature suggested by Pettijohn (1957) (Fig. 7–12) is based on the proportion of the three end members.

The names applied to sediments and sedimentary rocks are generally compound. The last name is the name of the main constituent and one or more modifiers are used to indicate major impurities of composition. These terms are used:

1. To indicate grain size:

   Rudaceous. Larger than sand size.
   Arenaceous. Sand size.
   Lutaceous. Mud.

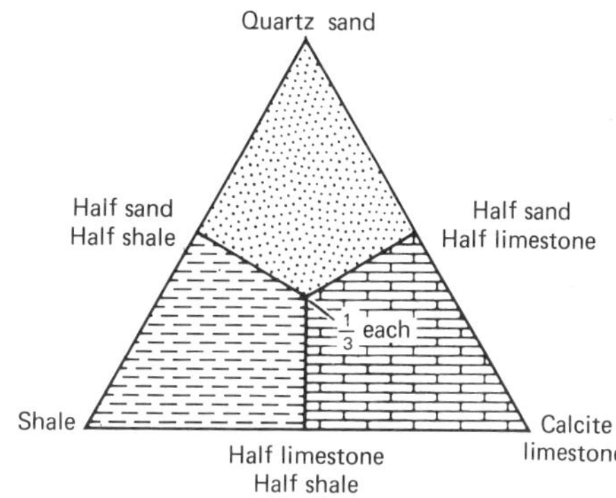

FIGURE 7–10
Most sedimentary rocks are mixtures of two or more constituents. The most common are mixtures of sand, lime, and clay. A triangular diagram such as this one may be used to classify such mixtures. Every point within the triangle corresponds to a particular rock. The points of the triangle (vertices) correspond to pure rocks. The outside boundaries of the triangle are mixtures of only two constituents. Points inside the triangle would correspond to mixtures of three constituents.

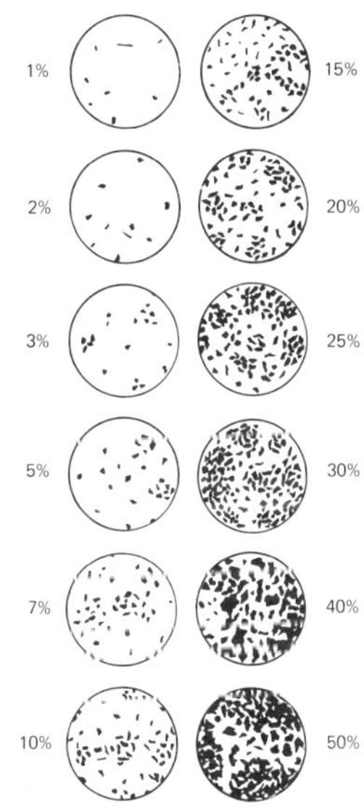

FIGURE 7–11
Chart for estimating the percentage composition of fragments in a sedimentary rock.
(after R. D. Terry and G. V. Chilingar, 1955)

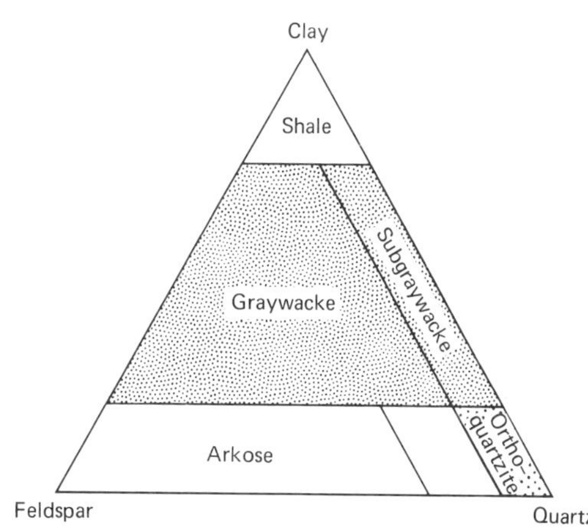

FIGURE 7–12
A classification of rocks composed of quartz sand, feldspar, and clay.

TABLE 7–1
Classification of sedimentary rocks..
(after Tyrrell, 1926)

PRODUCTS OF WEATHERING

RESIDUAL DEPOSITS
- Residual clays
- Bauxite

TRANSPORTED DEPOSITS

As solids and mechanically deposited

1. **Rudaceous**
   (size over 2 mm)
   a. Conglomerate
   b. Breccia
2. **Arenaceous**
   (sand size $\frac{1}{16}$ mm–2 mm)
   a. Sandstone
   b. Arkose
   c. Graywacke
3. **Silt**
   (silt size $\frac{1}{256}$ mm–$\frac{1}{16}$ mm)
   a. Siltstone
   b. Loess
4. **Argillaceous**
   (clay size $\frac{1}{256}$ or less)
   a. Mudstone
   b. Shale

As solution chemically deposited

1. **Siliceous**
   a. Sinter (hot springs)
   b. Chert-flint
2. **Carbonate**
   a. Tufa (springs)
   b. Travertine (caves)
   c. Oölitic limestone
   d. Dolostone
   e. Lithographic limestone
3. **Ferruginous**
   a. Ironstones
4. **Salts**
   a. Caliche
   b. Gypsum
   c. Anhydrite
   d. Halite

As solution organically deposited

1. **Siliceous**
   a. Radiolarian ooze
   b. Diatom ooze
2. **Calcareous**
   a. Fossiliferous limestone
   b. Chalk
   c. Marl
3. **Phosphatic**
   a. Rock phosphates
   b. Guano
4. **Ferruginous**
   a. Bog iron ore
5. **Carbonaceous**
   a. Peat
   b. Lignite
   c. Bituminous coal
   d. Anthracite

2. To indicate composition:

   Siliceous. Containing silica.
   Carbonaceous. Containing carbon.
   Ferruginous. Containing iron.
   Calcareous. Containing calcium carbonate.
   Argillaceous. Containing clay.

These may be used as follows: *argillaceous sandstone* (sandstone containing clay as a major impurity); *calcareous sandstone* (sandstone composed of or containing a large amount of calcium carbonate; and so on.

Most sediments are derived ultimately from weathering or decay of the earth's crust (Table 7–1). Decayed material can remain as residue where it originated or it can be transported to another locality by water, wind, ice, or mass movement. Deposition is eventually brought about through some mechanical process, such as settling from suspension, if it is transported as a solid. Those materials transported in solution as ions are deposited either chemically—by precipitation, for instance—or organically. Organisms accomplish deposition in two ways, (1) through direct deposits of waste products, and (2) through incorporation of materials into their own structures, in which case they are deposited as a part of the organism when it dies. It is most satisfactory to classify sediments transported as solids according to their size (their most obvious and easily measured characteristic). Sediments that originate from chemical and from organic deposition are conveniently classified first according to their chemical compositions, and second according to special features of either textural or genetic significance.

DESCRIPTIONS OF SEDIMENTARY ROCKS

Mechanically Deposited Sedimentary Rocks

Rudaceous rocks (coarse-grained fragmental or clastic rocks).

Conglomerate. The name is applied to a consolidated gravel. Fragments that make up conglomerates may be composed of any other rock. An appropriate prefix may be added to indicate the size of the major constituents of the rock; for example, pebble conglomerate or boulder conglomerate. Fragments in a conglomerate are rounded (Fig. 7–13), usually as a result of rolling in a stream or tossing and repeated impact in waves.

Breccia. The fragments are angular (Fig. 7–15) instead of rounded. In other respects it is like conglomerate except for its origin. A breccia may form where a rock is broken up but not rounded. Breccias occur in volcanic vents where sides of the vent or pipe are broken by explosions. They may form from fragments produced by the collapse of the roof of a cave, by fracture and deformation of brittle rocks that are folded, and by faulting.

Arenaceous rocks (sand-dominated rocks).

Sandstone. A sandstone (Fig. 7–16) is any rock composed of fragments of the size range $\frac{1}{16}$ millimeter to 2 millimeters, regardless of composition. Most sandstone is composed of quartz grains, but many other minerals may be present or even predominate. Sand dunes composed of fragments of ice are found in the Antarctic; some beaches in Italy are made of olivine sand derived from volcanic rocks; and sand dunes are composed of gypsum at White Sands National Monument, N. Mex.

Arkose. This arenaceous rock is made largely of a mixture of quartz and feldspar fragments. It may contain small angular rock and other mineral fragments and was originally thought to be derived from the breakdown of granite. Frequently arkose is red or pink, colors derived from oxidation of iron-bearing minerals and pink orthoclase feldspar.

Graywacke. An impure sandstone consisting of quartz and feldspar fragments and small fragments of igneous, metamorphic, and sedimentary rocks is called graywacke. However, graywacke is so varied in composition it is difficult to generalize. One common association is ash and volcanic dust with quartz and feldspar fragments. The color most commonly

FIGURE 7–13
Conglomerate.
(U.S. National Museum)

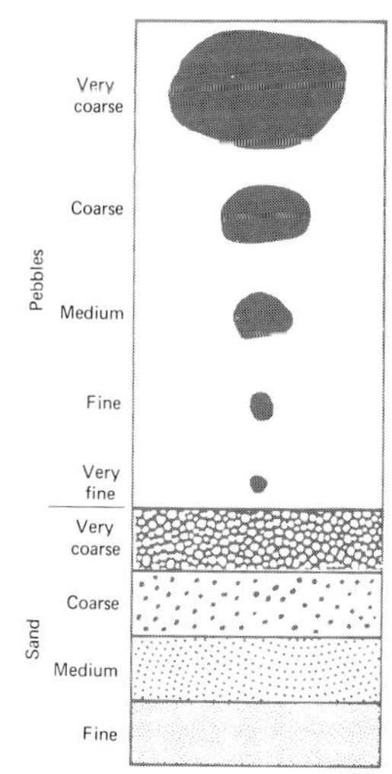

FIGURE 7–14
Gradations in relative size of pebbles and sand (see page 163 for actual range of measurements). This diagram is drawn to about one-half scale.

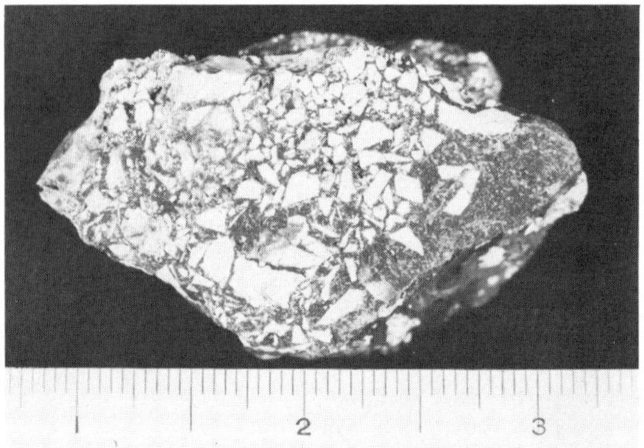

FIGURE 7–15
Chert breccia. Scale is in inches.

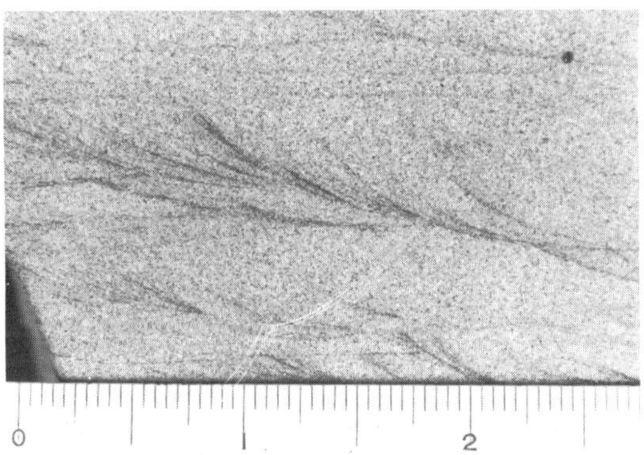

FIGURE 7–16
Cross-beds in a sandstone. Scale is in inches.

FIGURE 7–17
Shale outcrop in the Athens (Edinburg) formation of Virginia.

is gray, which comes from a matrix that is composed of a mixture of mica, chlorite, and quartz. The importance of graywackes lies in their widespread occurrence in mobile belts. They comprise a high percentage of the total volume of sediments.

Silty rocks.

Siltstone. This is a consolidated rock composed of clastic particles in the size range of $\frac{1}{16}$ millimeter to $\frac{1}{256}$ millimeter in diameter. Rocks composed of a large amount of silt are much less common than are sandstones or shales. Much of the silt is probably mixed in shales, which may contain up to 50 percent silt. Layers of siltstone are generally thin and hard. Unlike sand, a silt particle is usually angular, and most siltstones contain a large amount of platy minerals such as micas, although a microscope is needed to see most of the constituent minerals.

Loess. Loess is a light-buff, unconsolidated silt that is usually homogeneous and poorly stratified or even unstratified. Angularity of the particles in loess accounts for its ability to stand in vertical road cuts (Fig. 16–10), often for many years. The particles of loess are very well sorted; they are almost all the same size, averaging from 0.01 millimeter to 0.05 millimeter. It is fine enough to be a wind-blown deposit and is generally interpreted as being of eolian origin. Loess was probably derived from silts deposited on flood plains, outwash plains, and from dried-up lake beds along the margins of the continental ice sheets that covered parts of North America in the last 100,000 years.

Argillaceous rocks (rocks composed of clay or having a high percentage of clay in them).

Shale. Shale (Fig. 7–17) is consolidated mud and clay. It often contains sand (arenaceous shale) or lime (calcareous shale) in large amounts and usually has an earthy odor. Most shales are fissile; that is, they break or split along nearly parallel planes. Other important properties of shale are its low permeability, generally low porosity, particle size (less than $\frac{1}{256}$ millimeter in diameter), and composition. Shales are composed of clay minerals, quartz, sericite, chlorite, feldspar, calcite, and small quantities of many other minerals. Unlike most other sedimentary rocks, shale owes its rocklike character primarily to compaction of the particles rather than cementation.

Chemically Deposited Sedimentary Rocks

Siliceous deposits (containing silica). Many siliceous sediments are formed through chemical processes. Some are deposited directly from water; others are formed through processes of recrystallization in sediments during consolidation. There are few reliable criteria by which these two modes of origin can be positively identified. The most common occurrences of chemically deposited siliceous sediments are nodules, or concretions, in layers of limestone. Many of these nodules of siliceous material contain fossils at their centers, showing that they are of a secondary origin. Bedded cherts thousands of meters thick are found in the Coast Ranges of California, but many of these are biogenic.

Chert (flint). This is a dense, hard, siliceous rock in which color ranges from white through gray to black. Red chert is found in the iron deposits of the Great Lakes region. Chert is a form of quartz, SiO_2. It has the same hardness (7), a conchoidal fracture, and semivitreous luster. Chert occurs in other sediments as pebbles.

Siliceous sinter. This is a chemical sediment formed at mineral springs as well as by algae. It consists of silica, is white or light colored, and is porous. When it is formed around the vents of geysers, it is known as *geyserite*.

Limestone and other calcareous deposits (largely composed of calcium carbonate, calcite). Many of the calcareous sedimentary rocks may be classed as organic rocks. The calcareous rocks include all limestones, which are of similar composition, containing calcite as the dominant mineral constituent. Siliceous limestones, arenaceous limestones, and argillaceous limestones are all known and are common. Limestones tend to have many of the physical properties of their main constituent, calcite. They have a hardness of about 3, and they effervesce in hydrochloric acid.

Travertine and tufa. These limestone varieties are formed by evaporation of spring, stream, and ground waters containing calcium carbonate in solution. The name *tufa* is applied to spongy, porous, fragile deposits with an earthy texture. The deposits frequently contain branches, twigs, and other debris that fall into the water. Dense, banded, and compact deposits are called *travertine*. Stalactites and stalagmites are usually of this type.

Caliche. The deposits formed in the soils of semiarid regions underlain by soluble rocks are called *caliche* or *duricrust*. Capillary action draws ground waters in which ions are in solution to the surface where they evaporate, leaving the deposit as a cement in the soil or as thin layers. Caliche is often composed of calcium carbonate, but the name is also applied to nitrates of soda, and aluminous, ferruginous, and siliceous materials.

Dolomite (dolostone). A sugary-textured, dense rock that does not effervesce in dilute acid unless it is powdered, dolomite is a calcium magnesium carbonate $(Ca,Mg)CO_3$. Dolomites may be formed during recrystallization of calcareous sediments. Calcium and magnesium ions are nearly the same size, so that it is possible for magnesium to replace calcium in the calcite structure. Dolomitized fossils have been found, but no animal is known to construct its shell of dolomite. Dolomite occurs irregularly as patches through limestones, and dolomitized oölites have been located. Dolomite has been found in present day sediments near the Bahamas, in the southwestern Pacific, and in some lakes.

Ferruginous deposits (iron-bearing sediments).

Bedded siderites. These rocks, which are mined as iron ore in the Gogebic Range in Michigan, also commonly contain Mg and Mn as well as an association of siderite ($FeCO_3$) and chert. In this area both siderite and chert are thought to be direct chemical precipitates.

Iron-silicate sediments. Chamosite and glauconite are two silicate minerals that contain iron. Chamosite occurs in mudstones and limestones. Its origin is still in question. Glauconite (iron-potassium silicate) is forming on the continental shelves today. It is commonly known as *greensand* and looks very much like other sands except for its color.

Glauconite is closely related to the micas and is essentially an aggregate of colloids of hydrous potassium iron silicate formed through weathering of iron minerals. Sea water is an electrolyte and neutralizes charges on the colloids, making it possible for them to settle through the sea water to the bottom, where they form aggregates or granular masses. The iron-bearing silicates in the sedimentary rock taconite are probably of this origin. Taconite is the rock in which much of the Great Lakes iron ore formed. It is now hard, compact, often interbedded with jasper (red chert), and banded.

Sedimentary hematites. Many of these are oölitic hematites, and some provide important iron ores, such as the Clinton iron ores mined and used at Birmingham, Ala. Some of these are composed of fossil fragments that have been replaced by hematite; in others, the oölitic ores have cores of quartz grains around which layers of hematite were deposited.

Salts (deposits formed from concentrated solutions of brines). Three salts are of particular importance. These are halite (common table salt, sodium chloride), and gypsum and anhydrite, which are calcium sulfates. These, along with less common salts, are precipitated in a sequence from waters that become saturated. Such saturation of sea water with salts may occur when a part of the sea becomes cut off, leaving an isolated body of water that slowly evaporates. Similar conditions hold for inland bodies of water such as Great Salt Lake.

Experiments in the evaporation of sea water that contains about 3.5 percent by weight of dissolved solids give the following sequence:

1. Iron oxide and $CaCO_3$ are precipitated when half the water is evaporated.
2. Gypsum forms when one-fifth of the original volume is left.
3. Halite, NaCl, begins to crystallize next.
4. Finally, sulfates and chlorides of magnesium and potassium chloride form.

The most impressive thing about this sequence is the amount of water that must be evaporated before most of the salts begin to precipitate. Sedimentary formations of halite and gypsum are often several hundred meters thick. Such thicknesses would require evaporation of tremendous quantities of water. It is generally supposed that these bodies of water were inland seas, which either became restricted and cut off from the oceans or were only occasionally connected with them, at which times new supplies of water were introduced.

Organically Formed Deposits

Siliceous deposits (those formed from large amounts of silica).

Radiolarian oozes. Radiolarians (Fig. 11–4) are a group of single-celled animals that construct skeletons of silica. They float in upper parts of the ocean but when they die, their remains sink to the bottom. These accumulate in great quantities and may form a large part of the sediment in areas where rates of sedimentation are slow, and particularly where water is deep, for example, at depths between 4,000–8,000 meters. Silica is stable at the cool temperatures and high hydrostatic pressures found in the deep seas, but calcareous shells tend to dissolve under these conditions.

Diatoms. Diatoms are siliceous plants, also of microscopic size, that have many different shapes; they may be rodlike, spherical, or look like a circular disc (Fig. 7–3). These plants inhabit the oceans, particularly the surface waters, in tremendous numbers. Diatoms contain droplets of oil, and these may be the principal source of petroleum. The rock formed from remains of these plants has an earthy appearance and texture. It is a loose, fine, white, powdery rock, resembling chalk.

Calcareous deposits (containing calcite or calcium carbonate).

Globigerina ooze. Vast areas of the modern-day sea floors are covered by calcareous deposits that are composed largely of the shells of protozoans, particularly one group characterized by globular-shaped shells and known as *globigerina* (see Chapter 11).

Fossiliferous limestone. Most limestones are of organic origin. We include here primarily those limestones formed from the shells of marine animals. These shells are composed of the mineral calcite, which the animals are able to take from sea water and build into structures that house their soft parts. Cleavages of calcite are visible in the broken fragments of many shells. Shells accumulate in quantity in shallow seas and eventually become cemented by calcite, silica, or some such material that is precipitated in the water.

Chalk. This is a fossiliferous limestone composed of the shells of protozoans and particularly fossil *globigerina.* Chalk is white and has the property of being so soft that it will easily mark most things.

Marl. This name applies to mixtures of shells and shell fragments with muds, clay, particles of calcite or dolomite, or sand. It is an impure limestone usually found in a semiconsolidated state, held together loosely.

Coquina. This name is given to sedimentary rocks composed of a loose aggregation of shell fragments cemented together as rock (Fig. 7–18).

FIGURE 7–18
Cemented accumulation of shells (coquina).
(U.S. National Museum)

Phosphatic deposits. Bones and bird excrement are two organic sources for phosphate. Neither is common in rocks. Economically important deposits of bird excrement, *guano,* are confined to a few islands where birds have lived in large numbers for long periods of time. Accumulations of bones are rarely found in large enough quantities to be called rocks.

Ferruginous deposits. Certain bacteria and algae may cause deposition of ferric oxide and ferrous sulfide. The iron bacteria can extract iron from solution and deposit it around their cells. Others may perform the function of gathering the materials, which are then directly precipitated. An accumulation of bacteria casts and precipitated granules forms a rock called *bog iron.* As the name suggests, it is deposited in bogs, swamps, or marshes.

Carbonaceous deposits (containing large amounts of carbon). *Coal* is the most important naturally occurring material of carbonaceous composition. It is derived primarily from organic matter and does not have a definite mineral composition. The term "rock" is sometimes confined to naturally occurring materials composed of mineral matter. Coal is not a rock according to this definition, but it does make up large masses of crustal material, it is interbedded with rock, and its importance, both geologically and economically, certainly makes it worthy of our attention.

Coal is vegetable matter that has been altered both physically and chemically through geological processes to a black, rocklike substance. The chemical changes involve a loss of moisture and volatile constituents such as oxygen and hydrogen, some rearrangements of molecules in remaining matter, and an increase in the proportions of fixed carbon and ash. Physical changes include a darkening of color, increased hardness, and increased compactness.

Not all coal is alike; in fact, organic materials can be found in varying stages of alteration ranging from decaying mats of organic matter in swamps to the very hard, brittle anthracite coal. If we take the progression one step further, the material formed is almost pure carbon, graphite. There is no single place where all the varieties can be found together so that one can be traced into the other, but commonly two varieties can be found to grade into one another, and the evidence of progressive alteration is so convincing that it has been widely accepted. Varieties or ranks of coal are listed in order of progressive alteration.

Peat. An accumulation of vegetal matter that is in the initial stages of disintegration and decomposition. There is a high water and oxygen content, and the material ranges from a distinctly fibrous and woody light-brown material to a dark-brown and black jellylike substance. Peat grades into mixtures of peat and sediment in the swamps where it is most

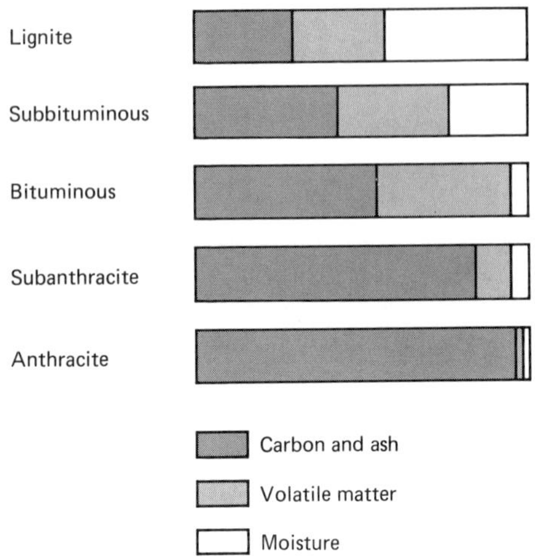

FIGURE 7–19
The proportions of carbon and ash, black, volatile matter, and moisture in various ranks of coal.
(after Campbell, U.S. Geological Survey)

commonly found. It may be cut out in blocks and used for fuel, but such use is infrequent today.

Lignite and **brown coals.** More compact than peat, lignite and brown coals may be woody or amorphous in texture, but twigs and branches can usually be found and identified in them. When dried they tend to split into slabs.

Subbituminous coal. This variety ranges from a glossy black coal to lignite. The coal is usually black with a pitchy luster, and it parts along a surface nearly parallel to the bedding, forming slabs.

Bituminous coal. Bituminous coal is pitch black to dark gray. It burns with a long, yellowish flame and gives off a strong odor. It is more or less laminated, and it breaks into small blocky fragments. It may be resinous, silky, pitchy, or dull and earthy in luster, and it soils the hands.

Semianthracite (86–92 percent fixed carbon) and **semibituminous** (averaging 15–20 percent volatile matter) **coals.** These are intermediate in properties between bituminous and anthracite. They are defined on the basis of the volatile matter. Semianthracite and anthracite are formed by alteration of lower ranks of coal and are often classified as metamorphic materials. These coals may also be classified as *subanthracite coal.*

Anthracite coal. Dark black in color, dull to brilliant and even submetallic in luster; it will not soil the fingers because it is too hard; it burns with a short, blue flame and emits little odor. It is the hardest coal, having a hardness of about 3 on the Mohs scale. In places, anthracite has some graphite mixed in with it (e.g., Rhode Island). It always has a high fixed-carbon ratio and a low volatile-matter percentage.

Coal Analysis

Analyses of coals are made to determine their suitability for use as fuels as well as for the special purposes to which they are put in the chemical and steel industries. An analysis may be a complete one, an "ultimate analysis," in which the amounts of moisture, volatile matter, fixed carbon, ash, sulfur, hydrogen, carbon, nitrogen, oxygen, as well as the calorific value of the coal are determined. But frequently a "proximate analysis" is made, in which only the amounts of moisture, ash, volatile matter, and fixed carbon are determined. These last components vary in a continuous way from peat to anthracite (Fig. 7–19).

The constituents of peat and lignite are easily recognized as plant matter, and parts of plant matter are readily identified in some bituminous coals. When constituents derived from woody tissues of plants, stems, branches, twigs, or roots are altered, they tend to form a material both homogeneous and black, which may appear as a smooth black or highly lustrous material called *anthraxylon*. The dull gray to nearly black, striped part of the coal is composed of *attritus*, material that is probably derived from parts of plants that are more resistant to decomposition, such as resinous, oily, waxy, and mineral matter.

Origin of Coal

Coal is regarded as a product of alteration of plant matter, according to most modern theories. Large accumulations of vegetal matter are forming in swamps, marshes, and bogs today. Tree trunks with roots firmly embedded in the underlying sediment occur in coal seams, and the trunks are coal. The old soils on which the plant matter grew is sometimes found under coal seams. Such coal must have been formed *in situ;* other coal beds have almost certainly formed from plant matter that was transported before it accumulated and was buried. For coal to be derived from peat, not only must the peat be buried and preserved; it must also accumulate in a very thick layer. An estimated 9 meters of peat are required to form 0.3 meter (or one foot) of bituminous coal. Modern peat bogs have been found in which there are tens of meters of peat. Thick accumulations might form in regions where there is regional subsidence or perhaps an advance of the sea so that the peat is buried under new peat. Peat is oxidized and bacteria tend to break down the woody matter; thus where peat is not protected by burial, it will ultimately decay to yield only ash. But if it is buried, the weight of the overburden will begin to compact and squeeze water out of the partially oxidized and decomposed plant matter. Biochemical processes probably play an important role in converting plant matter to coal, with bacteria and fungi being most important, though their action is confined to the upper portions of any peat accumulation because acids form in the lower portions in which they cannot survive.

Compaction, depth of burial, heat, and escape of volatile constituents are probably important in forming coal, and anthracite occurs only in areas where the sedimentary sequences in which it is found have been subjected to very high pressure (usually the strata are tightly folded and faulted). This suggests that dynamic deformation may play a role in its development.

Coal is of economic importance because it is necessary in the manufacture of steel. Coal is used not only as a fuel in heating the furnace, but special types of coal, *coking coal,* are used to produce a porous material, coke, that is used in the furnace with the iron ore to allow circulation of heat through the ore and to provide the carbon needed in the steel. Coal is also important as an industrial and domestic fuel, and it is used in the manufacture of a wide range of synthetic materials. Some of the finest coal deposits in the world are located in the United States (notably the bituminous coal of Pennsylvania, West Virginia, and Ohio). Major coal fields occur in the Rhine Valley, Silesia, central Russia, and China.

Conditions favorable for the formation of coal were very widespread in the Late Paleozoic in many parts of the world; that period of time is named *Carboniferous* for the coal deposits. Most of the above deposits are of that age.

PRIMARY STRUCTURAL FEATURES

Primary structural features are those features that form contemporaneously with sedimentation. Conditions of deposition determine what features form.

Stratification is the most nearly universal primary feature of sediments. Even the most uniform sediments usually possess some variations in texture, color, or composition that can be used to determine the bedding surface. Such surfaces usually indicate the interface between the sediment and the depositing medium. Beds may have been deformed subsequently by such processes as sliding, slump, plastic yielding during compaction, or even by folding and faulting in response to externally applied stresses. But bedding is important because it provides a system of reference planes. These serve to determine the extent and nature of any postdepositional deformation and make it possible to restore a sedimentary rock body to its original position during deposition. The type of stratification and its orientation may also provide valuable insight to the degree of agitation of the sediment by wave action, the existence, strength, and direction of currents, and even the water depth. Primary structural features may consist of internal structure within the layers of the sediment or they may be markings confined to the bedding surfaces. Many of these features (see ripple marks, swash marks, cross-bedding, and flow cast below) are directional and can be used to identify the top and bottom of the layers in which they occur, and all of them are valuable indicators of sedimentary environment. The determination of bed tops and bottoms is important because after deposition beds may be folded and faulted so that bedding is vertical, overturned, or even completely inverted.

Textural Variations

One type of primary structure may be attributed to variations in mineral content or texture that arise from the deposition of different materials in a particular sequence. Examples of features of this type are salt sequences, graded beds, varves, and flat-pebble or sharp-stone conglomerates.

Salt sequences. When salt water is evaporated,

FIGURE 7–20
Varved silt and clay. Light-colored bands are silty; dark-colored bands are clayey. Smooth surfaces of the exposure are vertical joints.
(F. O. Jones, U.S. Geological Survey)

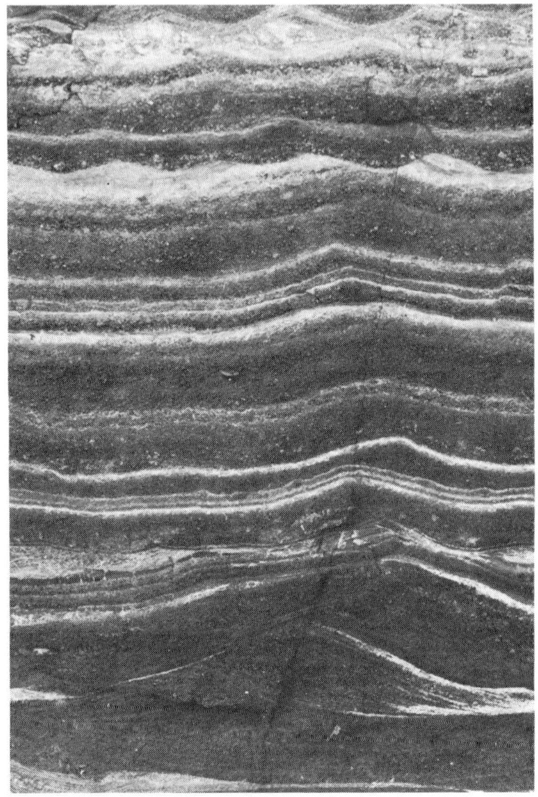

FIGURE 7–21
Graded bedding in layer of basaltic pyroclastic debris. These beds were deposited in lake water agitated by small-amplitude waves.
(H. E. Malde, U.S. Geological Survey)

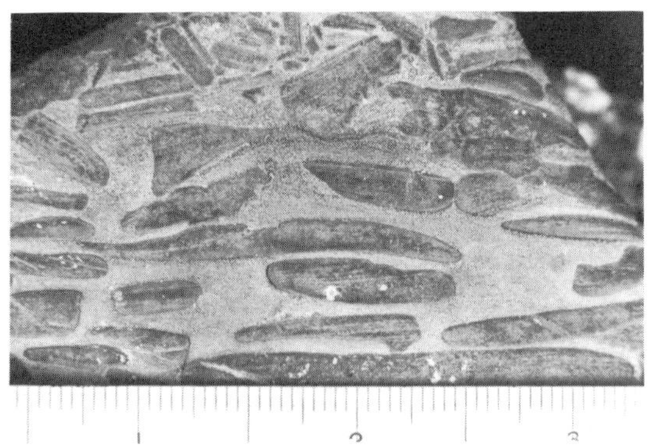

FIGURE 7–22
Flat-pebble conglomerate composed of elongate rounded limestone pebbles in a limestone matrix. From the Conococheague formation of central Virginia. Scale is in inches.

several salts are precipitated and they start coming out of solution in sequence as follows:

1. Calcium carbonate and iron oxide.
2. Calcium sulfate (gypsum and anhydrite).
3. Sodium chloride (salt).
4. Bittern [magnesium] salts.

Thus a particular sequence of beds of different composition is possible.

Varves. Varves are rhythmically banded sequences of sediment in which a particular sequence of beds is repeatedly deposited. The most common seasonal variations are found in glacial-lake varves (Fig. 7–20) where the lake is frozen over in winter, cutting off the supply of sediment, with the consequent development of a thin, often dark-colored layer. When the lake thaws, a thicker layer of sediment, newly brought into the lake, settles out during the summer. Varves can arise from seasonal variations in supply of sediment, type of sediment, or biological processes that affect sediment deposition.

Graded bedding. This is the name given to sequences in which there is a systematic vertical change in sediment grain size. Usually a graded sequence consists of lower layers of larger grain sizes with progressive diminution in size upward (Fig. 7–21). Graded bedding will form when a range of grain sizes of sediment are suspended and allowed to settle either in still water or in water with low-velocity current action. The currents will tend to remove the finest sizes, which can be maintained in suspension or transported. Graded beds are commonly associated with the relatively sudden introduction of sediment into a particular environment.

Flat-pebble conglomerates and sharp-stone conglomerates. These are conglomerates or breccia layers (Fig. 7–22) formed where a newly deposited sediment layer is broken by wave action and moved a short distance from its point of origin. If the pebbles in the conglomerate are standing on end, the rock is called an *edgewise conglomerate.* These form when the fragments from the disrupted layer are trapped as sediment is deposited all around them.

FIGURE 7–23
Current ripples formed in sand on a modern beach.

Surface Features

A great many features are found on bedding surfaces. Among these are ripple marks, mud cracks, swash and rill marks, impressions, pits and mounds, and scour and fill structures.

Ripple marks. These form as a result of movement of a granular sediment in the medium of deposition. Ripples are common features in sand both in sand dunes, where the sand is moved by air, and along a beach or in a stream, where the sand is moved by water. Many different types of ripples are found, but the most common ones are symmetric oscillation ripples and asymmetric current ripples. Oscillation ripples form as a result of wave action that induces movement in water to a depth equal to about half the wave length of

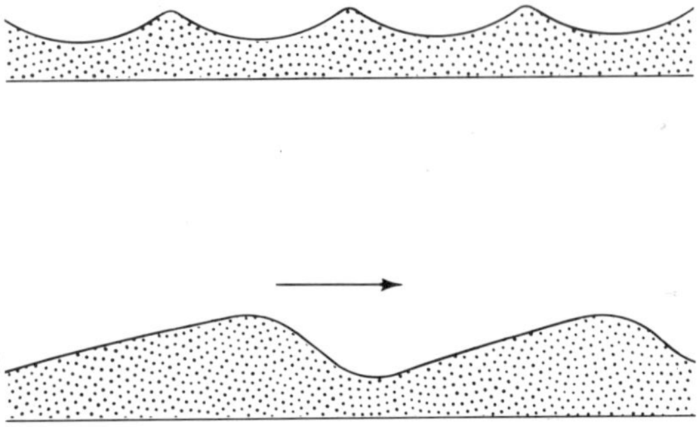

FIGURE 7–24
Ripple marks, asymmetric (below) and symmetric (above).

FIGURE 7–25
Swash marks.

FIGURE 7–26 *(below)*
Load casts developed in sandstone.

the waves. When the water depth is less than that distance, the motion touches the bottom sediment and moves the sand and silt-size particles back and forth along the bottom under the wave. The resulting ripples are symmetrical and have upward-pointed crests (Fig. 7–24). Current ripples form where there is a flow of water or air directed across the sediment surface. Sand is moved along by rolling and bouncing. As soon as irregularities develop in the surface, the character of the water or air movement on the upcurrent and downcurrent side of a crest is different. An eddying effect is created on the downcurrent side, which is protected from movement of the type seen on the upcurrent side. Grains slip and roll down the downcurrent side and the ripple develops an asymmetric profile (Fig. 7–24). The shape of such ripples cannot be used to determine bed tops, but sometimes the trough will accumulate coarser or heavier grains, indicating the top side of a layer. Current ripples formed in water can be distinguished from those formed in air by the ratio of wave length to wave height. Those formed in water have smaller ratios (e.g., 15 or less) than those formed in air (e.g., 15 or more). When currents move alternatively from two directions, an interference pattern may develop; other ripples are tongue-shaped.

Swash marks. Those patterns left in beach sand as a result of the forward rush of water after an incoming wave breaks are called *swash marks.* They are often concentric arcuate patterns (Fig. 7–25). As the waters return down the slope of the beach, they join to form a system of *rills* that feed into larger and larger flow channels. Other characteristic patterns are formed where the return flow of water is interrupted by rocks or shells.

Impressions. Many different types of processes and events leave marks or impressions on the sand or soft sediment on the sea floor. Animals leave *tracks, trails,* and sometimes *burrows* in the sediment. *Raindrop* and *hail imprints* with characteristic depressed center and raised edges are found. Drifting plants or other objects may *drag* objects across the sediment, leaving marks. *Bubbles* of gas may be held on the bottom in shallow water long enough for clay-size particles to accumulate on the bubble. When the bubble breaks or is compressed, it leaves a characteristic circular impression on the sediment. Small *pit* and *mound* features may develop from the escape of gas from sediment. *Ice crystals* may leave crystalline patterns when they form on the sediment surface.

Flow casts or load casts (Fig. 7–26). Rolls, lobate ridges, or other raised features may be produced in a sandstone bed as a result of the flowage of sediment in an underlying bed, usually composed of some soft sediment. Load casts may be bulbous, baglike, downward protrusions of sand produced by plastic yielding of an underlying, unevenly loaded soft sediment.

Mud cracks. These are cracks that form most commonly in mixtures of clay, sand, and silt when the sediment is dried. The cracks usually form polygonal-shaped patterns, and the mud layers frequently bend or curl upward (Fig. 7–27); the cracks are also wider at the top than at the bottom. Sand or silt sometimes fills the cracks when the cracks form in a mud layer that lies over a water-saturated sand layer.

Cross-Bedding

The third category of primary structural features includes those that are internal; that is, features found within the limits of the bedding planes. The most common of these are *cross laminations* or *cross-bedding,* bedding that consists of successive, systematic, internal bedding (called *fore-set bedding*) that is inclined to the principal surface of accumulation. Cross-bedding occurs under many different natural circumstances as a result of transport and deposition of a granular sediment in a current under conditions that give rise to inclined beds.

Deltaic cross-bedding. This bedding forms where fore-set beds are deposited on the front of a delta. It usually consists of three sets: (1) bottom-set beds that form from the deposition of fine sediment on the flat bottom beyond the edge of the delta; (2) fore-set beds that are thicker and coarser than the bottom-set beds and inclined; and (3) top-set beds that are laid on top of the delta (Fig. 7–28), more or less flat-lying.

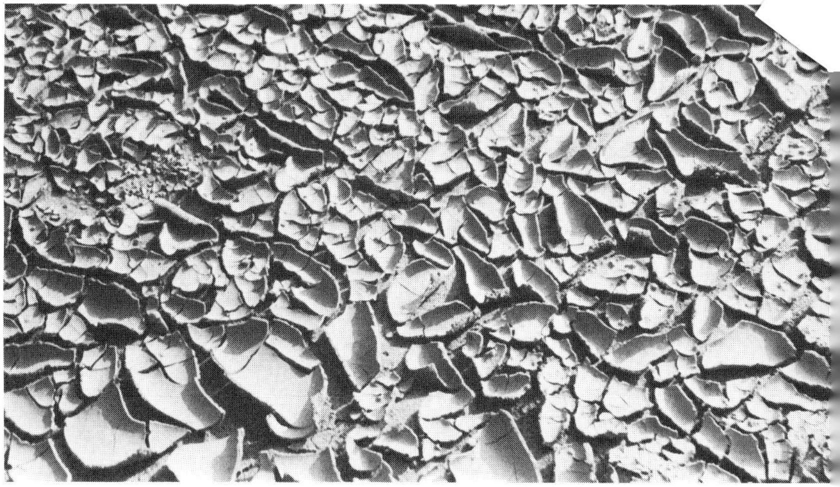

FIGURE 7–27
Mud cracks. This is a view of a lake bottom which has dried out. The mud layers on the lake floor have cracked open and curled. Such features are sometimes found in ancient consolidated rocks. Their presence is an indication that the place where they were deposited was once subjected to alternate wetting and drying.

FIGURE 7–28
Current-formed cross-bedding in glacial outwash deposits.

FIGURE 7–29
Festoon cross-bedding in layers of sand.
(H. E. Maide, U.S. Geological Survey)

Scour and fill. This term applies to small-scale bottom scour or channels that are subsequently filled. When filling takes place, cross-bedding of a type known as *festoon cross-bedding* is likely to form. This type (Fig. 7–29) consists of elongate, cross-cutting, erosional troughs filled in with conformably laminated strata. It is cross-bedding deposited on a concave surface.

CHAPTER OUTLINE

CHAPTER OUTLINE (cont.)

TERMS

abyssal
amorphous
anthracite coal
anthraxylon
arenaceous
argillaceous
arkose
attritus
bathyal
bedded siderites
bog iron
bottom-set beds
breccia
calcareous
caliche
carbonaceous
chalk
chemical precipitation
chert
clastic particles
coal
coking coal
colloids
compaction
conglomerate
coquina
cross-bedding or cross laminations
deltaic cross-bedding
diatoms
dolomite
duricrust
edgewise conglomerate
ferruginous
festoon cross-bedding
flat-pebble conglomerates
flow casts or load casts
fore-set beds
fossiliferous limestone
gel
geyserite
globigerina ooze
graded bedding
graywacke
impressions
iron-silicate sediment
laminae
lignite
lithification
loess
lutaceous
marl
mud cracks
neritic
nodules
oölites
peat
pit and mound features
radiolarian oozes
raindrop and hail imprints
rills
ripple marks
rudaceous
salt sequences
salts
sandstone
scour and fill
sediment
sedimentary hematites
sedimentary rocks
semianthracite coal
semibituminous coal
shale
sharp-stone conglomerates
siliceous sinter
siliceous
siltstone
solubility
spherulitic
strata or beds
stratification
subbituminous coal
swash marks
tracks, trails, and burrows
travertine
tufa
varves
Wentworth classification

REVIEW QUESTIONS

1. Which primary sedimentary structural features can be used to determine the top of a bed?

2. What conditions are necessary for the formation of asymmetrical ripple marks?

3. Contrast a pure sandstone composed of well-rounded, highly sorted quartz grains with a poorly sorted, coarse-grained arkose with regard to the sedimentary environment in which they formed.

4. What criteria might you use to distinguish a schist or gneiss formed from a sedimentary rock from a schist or gneiss formed from an igneous rock?

5. What criteria can you use to distinguish sedimentary rocks from igneous rocks?

6. Outline the principal factors that cause chemical precipitation of sedimentary materials.

7. How would you expect to be able to distinguish a deep-water sediment from a sediment deposited in shallow water?

8. Why is subsidence of peat bogs a necessary condition for the formation of thick coal seams?

9. What conditions favor the formation of poorly stratified sediments?

10. How would you recognize sediment formed in stagnant water with little oxygen?

SUGGESTED READINGS

Garrels, R. M., and **MacKenzie, F. T.,** 1971, Evolution of Sedimentary Rocks: New York, W. W. Norton, 397 p.

Pettijohn, F. J. and **Potter, P. E.,** 1964, Atlas and Glossary of Primary Sedimentary Structures: New York., Springer-Verlag, 370 p.

Pettijohn, F. J., 1957, Sedimentary Rocks, 2nd ed.: New York, Harper and Row, 718 p.

Shrock, R. R., 1968, Sequence in Layered Rocks: New York, McGraw-Hill, 507 p.

II
THE OCEANS

8
THE OCEAN BASINS

8
THE OCEAN BASINS

FIGURE (*previous page*) Air view of sediment being carried into the ocean. The sediment is being carried to the left by longshore currents. (See footnote on p. 260.) (U.S. Dept. of Agriculture)

OCEANOGRAPHY

Study of the oceans is one of the most rapidly expanding fields of science. The oceans are important as a source of food; they hold a vast potential wealth of mineral resources; knowledge of the oceans, the deep basins, their margins, and the sea floor is critical to our understanding of the structure and history of the earth.

The study of the oceans may be subdivided into four major fields: (1) *biological oceanography*, which involves the flora and fauna of the oceans and their relationship to marine environments; (2) *physical oceanography*, which deals with the physical characteristics of water masses and their movements—wave action, currents, and tides; (3) *chemical oceanography*, which treats the sea salts and other dissolved substances in sea water; and (4) *marine geology*, which is concerned with the nature and origin of marine sediments, the origin of submarine and coastal land forms, the structure of the ocean basins, and the history of the oceans. Each of these fields is highly specialized, but they are intimately interconnected. For example, the fauna and flora are dependent on environmental conditions, which are related to circulation of water and the supply of nutrients; marine sediments are largely composed of marine organisms and their distribution is thus influenced by the distribution of the fauna. The extent of these interrelations will become clearer as modern sedimentary environments are investigated in the following chapters.

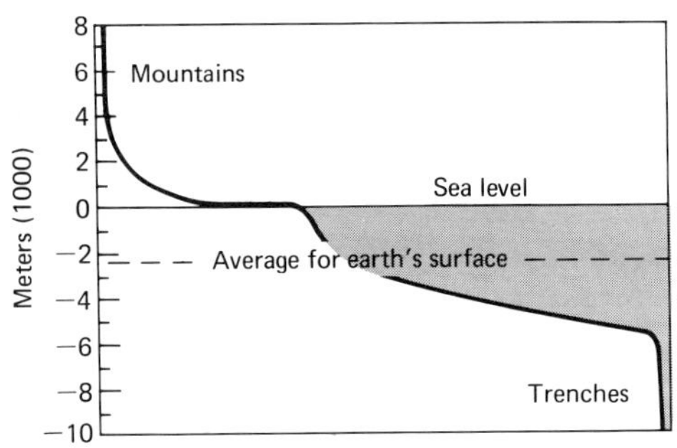

FIGURE 8–1
The distribution of elevations in various parts of the earth. The vertical scale is elevation. The horizontal scale is surface area of the earth.

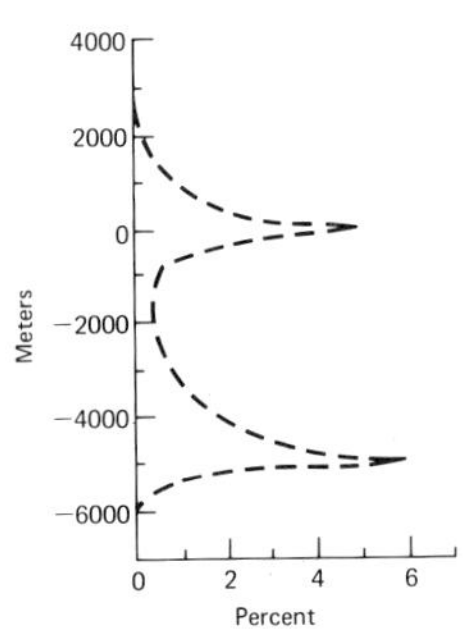

FIGURE 8–2
A plot of the percentage of the earth's surface at each elevation.
(after Bucher, 1933)

GEOGRAPHY OF THE OCEANS

Slightly over 70 percent of the surface area of the earth, or 361 million square kilometers, is covered by sea water to an average depth of 3,730 meters. The great difference in the average elevation of continents and ocean basins is one of the fundamental differences between these two parts of the earth's crust (Figs. 8–1 and 8–2). The oceans contain about 86.5 percent of the water on earth; the remainder is held in sediments and sedimentary rocks, in the atmosphere, in lakes and rivers, and in ice on land. The Pacific Ocean covers 180 million square kilometers to an average depth of 3,940 meters; the Atlantic Ocean, together with the Arctic, Mediterranean, and Black Sea, covers 107 million square kilometers to an average depth of 3,310 meters; and the Indian Ocean covers 74 million square kilometers to an average depth of 3,840 meters.

Land and water on the earth are distributed in an irregular pattern for which no satisfactory explanation has been advanced. Nearly 67 percent of the land area is located in the Northern Hemisphere. Almost no land occurs between latitudes 40° and 65° South, but the same latitudes in the Northern Hemisphere contain more land than any other latitude. The Pacific Ocean occupies over a third (180/510) of the surface area of the earth. Thus it is possible to divide the earth into hemispheres such that one hemisphere is almost entirely covered by water.

Even when considered superficially, the various ocean basins are far from identical. Considerable differences are found in shape, area, and depth of the Pacific, Atlantic, Indian, and Arctic oceans,

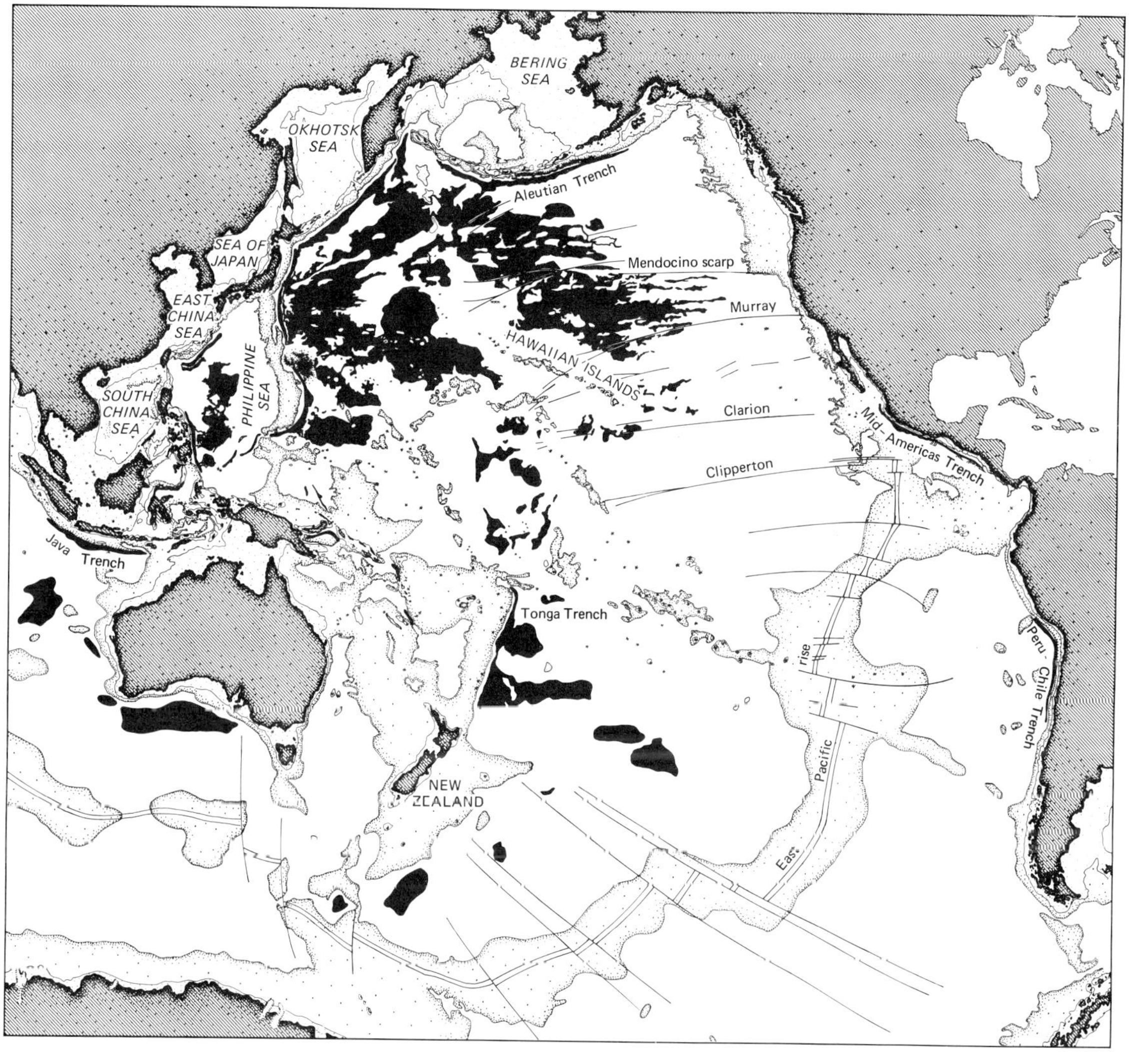

FIGURE 8–3
Map of the Pacific Ocean basin. Abyssal plains and trenches are shown in black. Small stars indicate seamounts. The crest of the East Pacific Rise and some of the major faults in the sea floor are shown.

and fundamental differences are indicated by the character of their margins and the various distributions of geomorphic provinces within each ocean (Figs. 8–3, 8–4, 8–5, and 8–6). The Arctic is almost completely landlocked; the Atlantic is a long, relatively narrow basin; the Pacific and Indian oceans are more nearly oval. The oceans differ greatly in the character and stability of their margins. The margins of the Arctic and Atlantic are stable. There is relatively little seismic activity, almost no volcanism, and no pronounced modern faulting or other diastrophic activity. Broad, flat continental shelves typify such margins. Exceptions to this are found in the Caribbean and in the Scotia Sea (between South America and Antarctica). These resemble the western Pacific margins where arcuate chains of volcanic islands lie east of the continents. These island festoons are typically separated from the continents by a shallow marginal sea. The islands are sites of vol-

FIGURE 8–4
Physiographic features of the Atlantic Ocean basin. The abyssal plains, crest of the Mid-Atlantic Ridge, and fault and fracture zones across the ridge are shown in black. (redrawn and modified after maps by Heezen and Tharp and charts of the U.S. Navy Hydrographic Office)

canism, earthquakes, and crustal movements. Some island arc systems are located between continents (e.g., in the Caribbean between North and South America; in the Scotia Sea between South America and Antarctica; in the Bering Sea between North America and Asia; in Indonesia between Australia and southeast Asia), but others are marginal to the Asian mainland, and still others lie far from continental margins, as does the Tonga Arc north of New Zealand (Fig. 8–15).

Portions of each ocean are cut off or partly separated from the main body of the ocean by land or shallow water to form seas. The Arctic Ocean is such a sea in the sense that it is a northern extension of the Atlantic, but the term *sea* is also used in a general way to denote a large body of salt water that is less than an ocean. For example, the marine waters between the Arctic and North Atlantic oceans are subdivided into the Norwegian, Greenland, and Barents seas.

Three seas, the Mediterranean, the Black, and the Red, are almost completely landlocked, and the Caspian Sea, which was once connected to the ocean, has been cut off by diastrophic movements of the land. The Black Sea is connected only through the narrow straits known as the Bosphorus to the Mediterranean, which is in turn connected to the Atlantic only at Gibraltar.

Many of the seas occupy marginal positions to continents where island arc systems partly cut the sea off from the ocean. This is true of the Caribbean, Scotia, Bering, South China, Philippine, Yellow, Coral, and East China seas, and the seas of Japan and Okhotsk (Fig. 8–14). Large parts of all of these seas are shallow, but most contain deep basins as well, and in the case of the Philippine Sea island arcs bound it on all sides, enclosing a basin that is as deep as most of the Pacific.

The Gulf of Mexico (Fig. 8–7) is like a small ocean basin in many ways. It has a deep-sea floor, the Sigsbee Deep, and its margins are geomorphically similar to those along the eastern coast of North America.

LAND FORMS OF THE OCEAN BASINS

Only within the last few decades have we had more than a vague notion of the deeper portions of the ocean basins, and even now observations are widely scattered in some parts of the world. Early knowledge of sea-floor configuration was based on highly unreliable soundings made with ropes and wires. The development of precision depth-recording instruments revolutionized the study of submarine topography. These instru-

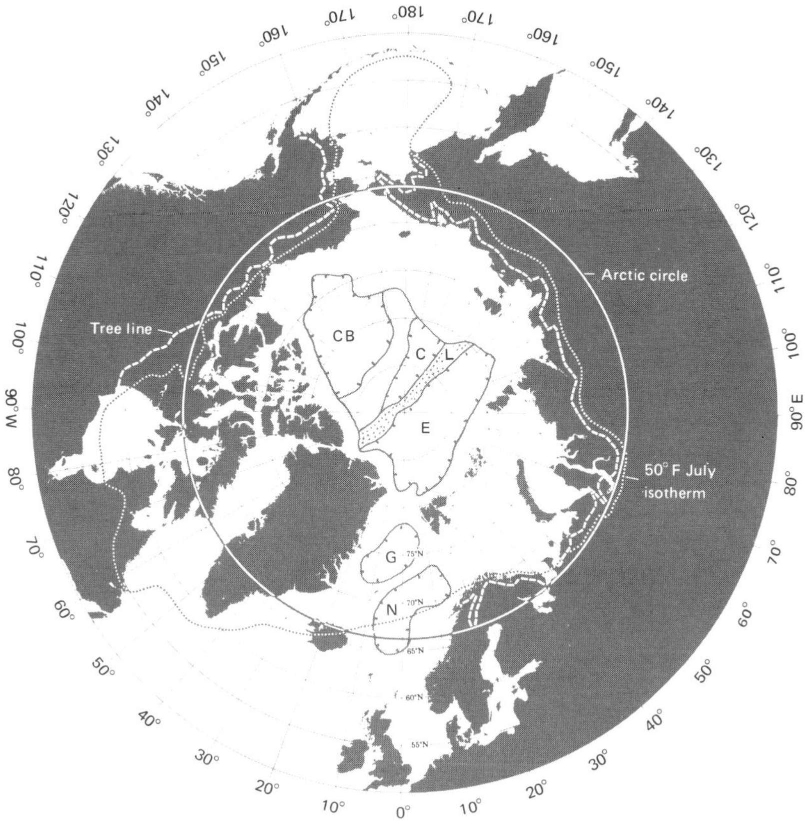

FIGURE 8–5
Outline map of the Arctic Ocean showing the Eurasian basin (E), Greenland basin (G), Norwegian basin (N), Central Arctic basin (C), Canadian basin (CB), and Lomonosov Ridge (L).
(base from Malloy, 1969; topography after King and others, 1966)

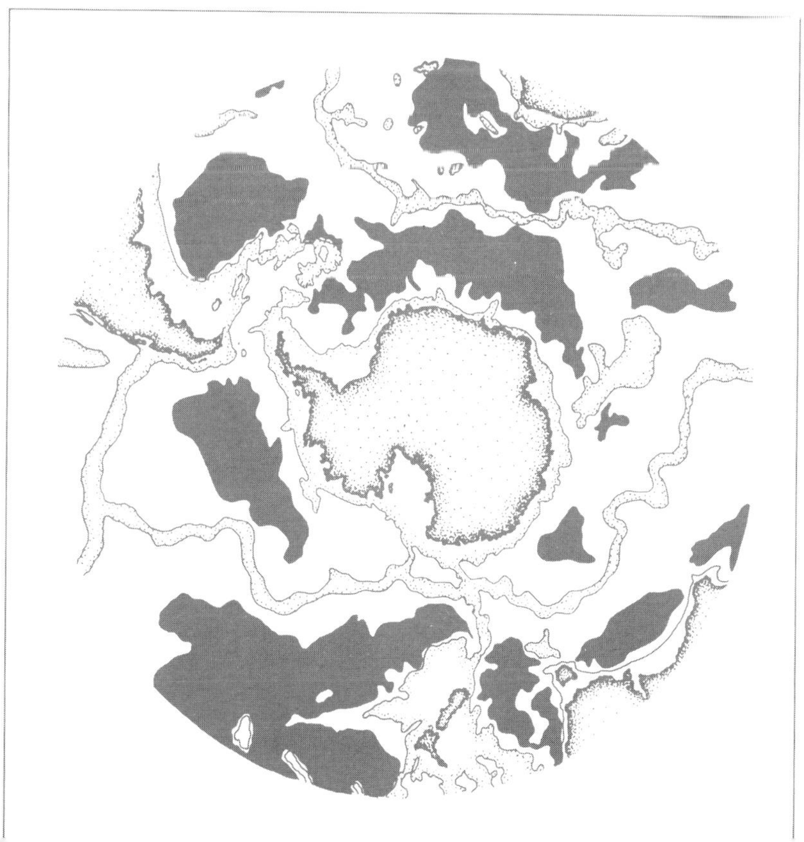

FIGURE 8–6
Major topographic features of Antarctic seas; the continental margins and the crests of the midoceanic ridges, areas of less than 2,700 meters, are shown by light stipling; the major basins, areas deeper than 4,500 meters, are shown black.
(after Ewing and Heezen, 1956)

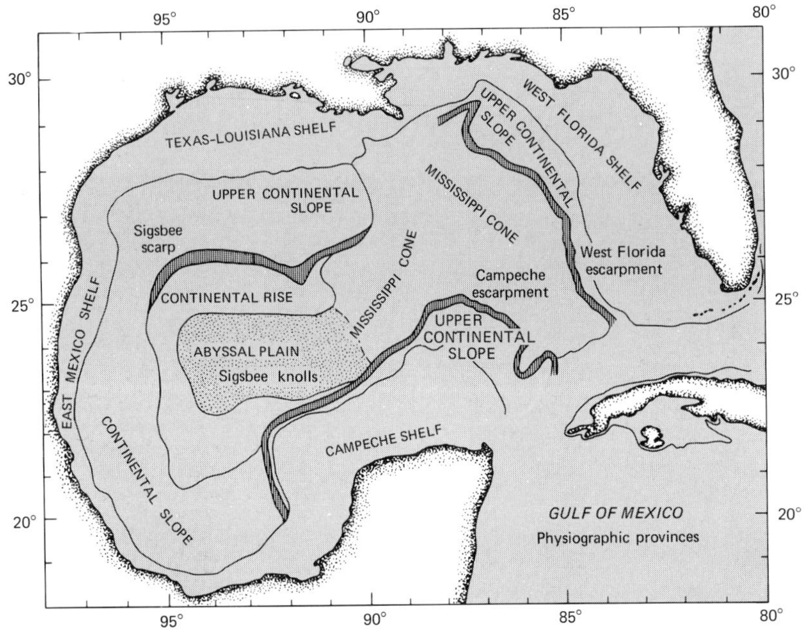

FIGURE 8–7
Physiographic chart of the Gulf of Mexico.
(after M. Ewing, D. Ericson, and B. Heezen, 1958)

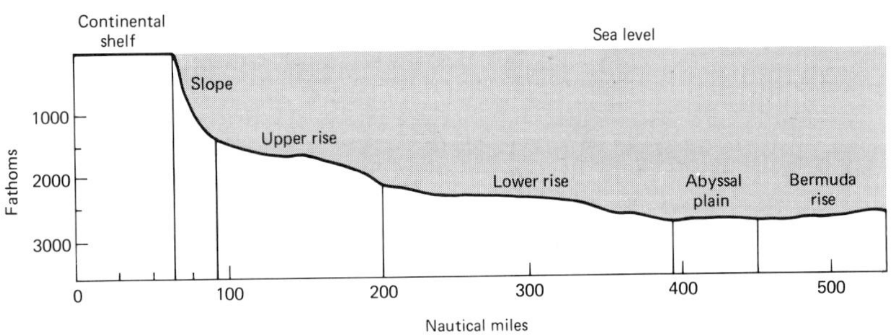

FIGURE 8–8
A typical profile of the sea floor off the northeastern United States.
(after B. C. Heezen, 1959)

ments use the time it takes a sound wave to travel from the ship to the sea floor and back; the travel time is a function of depth and velocity. The precision depth-recording method permits a continuous recording of water depth along the track of a ship.

Accurate depth determinations are of little value unless their geographic location is also known. Thus systems using radio signals (loran) from stations distributed throughout the world have been important in determining location. Now satellite navigational systems provide a means of location determination superior to the older methods.

Much time is still required to make the closely spaced network of tracks needed to obtain an accurate three-dimensional model of sea-floor topography, even in small areas. When more measurements are made, our present conceptions of sea-floor topography will become refined and we will gain a more complete understanding of the origin of sea-floor features. But already many interesting and varied types of topography are known. Among the major types of submarine land forms are the following:[1]

1. Continental margins of several types.
2. Abyssal plains—the deep-sea floor.
3. Island arc systems and related trenches.
4. Oceanic ridges and rises, such as the Mid-Atlantic Ridge.
5. Volcanic seamounts, ridges, and related features.

Continental Margins

An impressive body of observation (see Chapters 9 and 10) indicates that sea level has risen at least several hundred meters during the last few thousand years. Thus most of the edges of the continental plates are flooded by the seas. Some of these margins have been near sea level over the last few million years and are relatively wide and flat (e.g., Atlantic and Gulf Coastal Plain of the United States); others are sites of modern crustal deformation, faulting, and mountain building (e.g., the west coast of North America).

One of the most typical continental margins is like that found off the central Atlantic states and consists of a continental shelf, continental slope, and continental rise[2] (Fig. 8–8).

[1] Discussion of geophysical study of these provinces is in Chapter 25.

[2] Definitions of shelf, slope, rise, marginal plateau, and marginal escarpment are after Heezen and others, 1959.

Continental shelf. The continental shelf is a shallow (less than 180 meters), gently sloping (less than 1:1,000) surface of low local relief that extends from the shore line to the shelf break where the seaward gradient sharply increases to greater than 1:40. Its width ranges from a few kilometers to more than 300 kilometers. Continental shelves border all land areas.

Continental slope. The continental slope is the relatively steep (3–6°) portion of the sea floor that occurs at the seaward border of the continental shelf. It typically drops from depths of 90–180 meters to depths of 1,500–3,500 meters. The top of the slope is usually marked by a relatively sharp shelf break, and the base of the slope, although less definite, is generally abrupt.

Continental rise. The continental rise lies at the base of the continental slope. In general, gradients on the rise range from 1:100 to 1:700, and the width ranges from tens to hundreds of kilometers. The seaward limit of the rise is usually abrupt, and at this point the regional gradients decrease to less than 1:1,000. The continental rises generally occur between 1,500–5,600 meters with local relief low except for a few seamounts and submarine canyons.

The North American Atlantic margin (Fig. 8–9) has several portions that differ from those described above. The shelf region between New foundland and Nova Scotia was lowered as a result of erosion by glaciers during the glacial advances of 2 million years. The gulfs of Maine and St. Lawrence are examples of epicontinental seas.

Epicontinental marginal seas. These shallow seas (less than 3,000 meters) lie on the continental block and can be distinguished from the continental shelves by their greater depth (usually more than 200 meters) and possibly greater topographic irregularity. Most of them are enclosed by shallow banks of the continental shelf.

Marginal plateaus. A marginal plateau is a shelf-like feature at greater depths than the continental shelf and is separated from the continental shelf by an incipient continental slope. These plateaus generally occur at depths greater than 200 meters and less than 2,400 meters. Examples are found in the Blake Plateau east of Florida (Fig. 8–10), off the coast of Argentina, and off the coast of New Zealand.

Marginal escarpment. A marginal escarpment is a precipitous slope that forms the seaward face of a marginal plateau. Such escarpments begin in

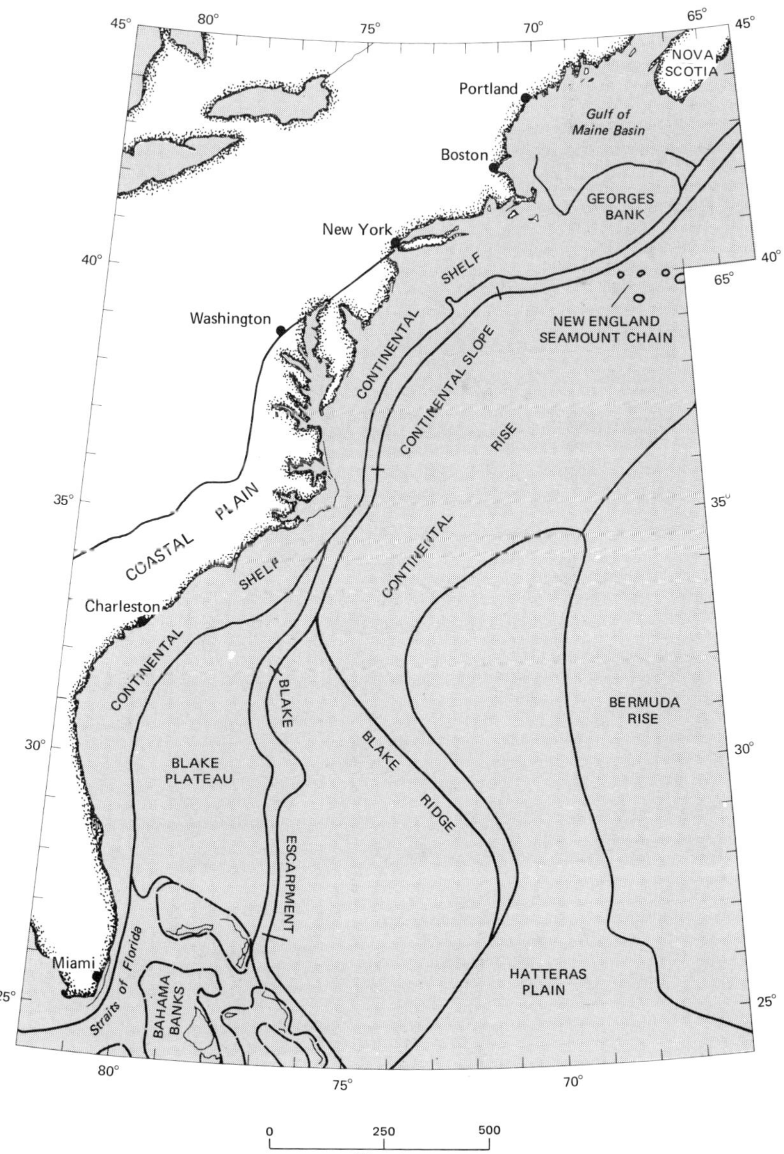

FIGURE 8–9
Major geomorphic divisions of the continental margin and ocean basin east of the United States.
(from Emery, 1966)

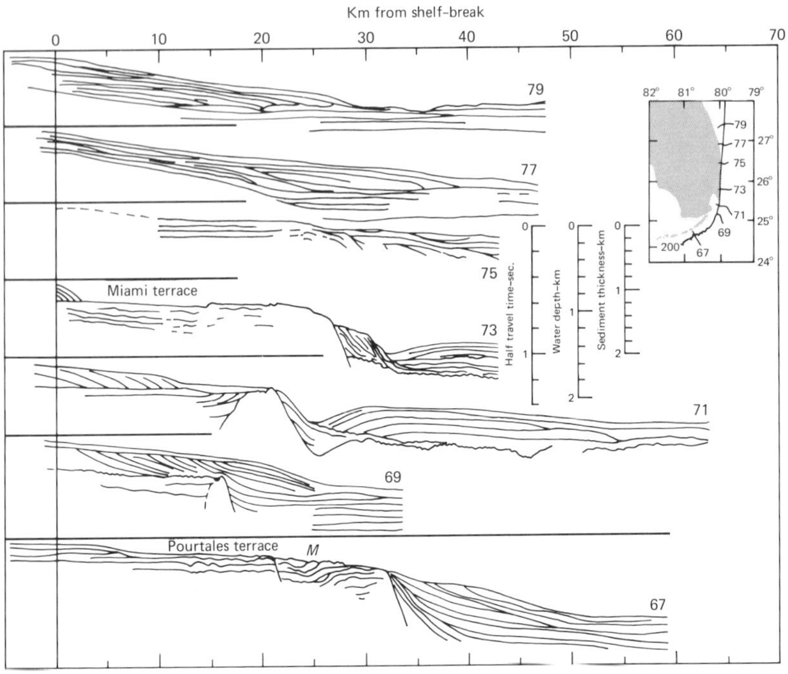

FIGURE 8–10
Profiles between Cape Kennedy and Key West. Miocene (M) rock.
(from Uchupi and Emery, 1968)

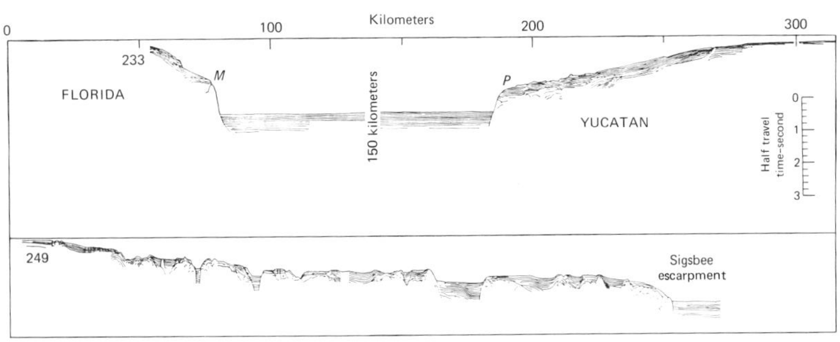

FIGURE 8–11
Continuous seismic profiles across the Florida and Campeche escarpments and across the continental margin east of Port Arthur, Tex.
(from Uchupi and Emery, 1968)

depths of 1,000–3,000 meters and are 2,000–4,000 meters high. Bases of escarpments are marked by an abrupt change in slope and gradients exceed 1:10. Escarpments are found in the Gulf of Mexico (Fig. 8–11) and off Argentina. The seaward edge of the Blake Plateau is marked by a prominent escarpment (Fig. 11–12).

Canyons on the Sea Floor

Submarine canyons on the continental shelves would be quite impressive features if they were exposed where we could see them. Several are deeper than the canyon of the Yellowstone River, 335 meters. One of the most prominent and best known submarine canyons is the canyon of the Hudson River (Fig. 8–12), which has been traced from the mouth of the Hudson River across the continental shelf, far out into the abyssal plain. Others, including the La Jolla, Cape San Lucas, and Carmel-Monterey Bay canyons, have been mapped and studied along the coast of southern California. Canyons have also been found on the continental shelf off the mouth of the Congo, the Amazon, and many other rivers.

Some of these canyons undoubtedly were eroded during the ice ages by streams that flowed on the continental shelves. But then, as now, cutting of the canyons beyond the shelves was most likely the result of submarine debris-charged currents. It may be that streams carry their loads farther out into the oceans than we think, but it seems likely that much of the erosion of these canyons is accomplished by streams that deposit their loads there, building up steep conelike piles of debris. When these become too steep, they are unstable. Slight earthquakes or the addition of more load may bring about a submarine landslide, which causes the debris to flow down very rapidly, eroding the canyon walls and floor.

Cones of Sediment

The term *cone* is applied to the large cone-shaped accumulations of sediment in the ocean basins; these reach an apex on the continental shelf near the mouth of a major river, where sands, silts, and clays are carried into the ocean. Cones form when sediment from the river mouth is dispersed into the ocean basin by submarine slumping and downslope movement as a sediment-laden, high-density current. Cones originating in this manner are well developed in the Atlantic Ocean north of the Amazon River, in the Arabian Sea from the Indus River, and in the Bay of Bengal from the Ganges River (Fig. 8–13).

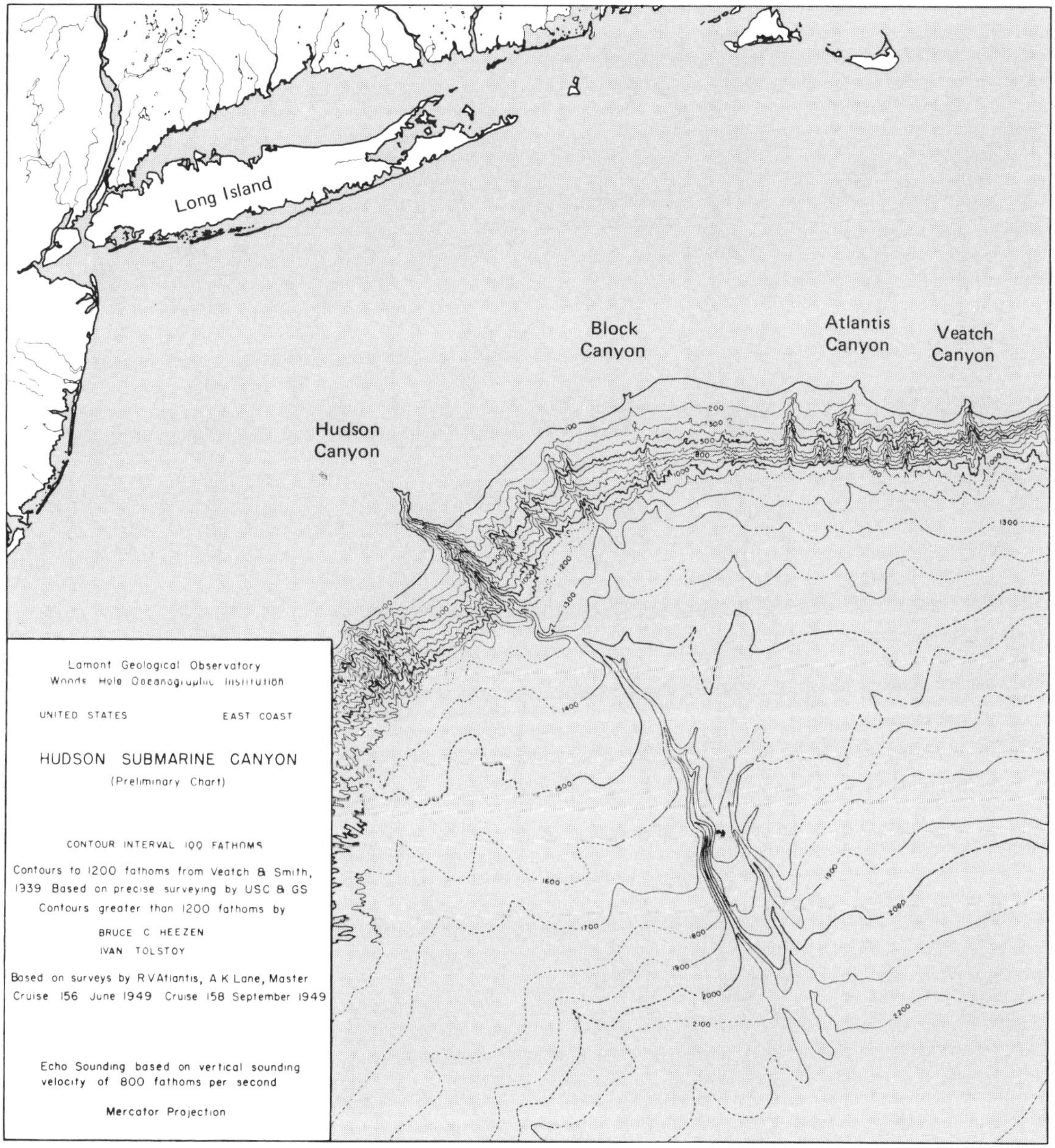

FIGURE 8–12
Chart of the Hudson submarine canyon. A contour interval of 100 fathoms (183 meters) is used. This chart was prepared by the Lamont Geological Observatory and the Woods Hole Oceanographic Institution. (from B. C. Heezen, 1959)

Island Arcs and Deep-Sea Trenches

Scientists have been intrigued by the arcuate character of many of the volcanic islands of the Pacific since the peculiar distribution and orientation of these islands were first recognized. It was early noted that these belts are highly active—sites of many earthquakes, volcanic eruptions, and modern folding and faulting. Attention of geologists has been increasingly turned to these, the most active of the crustal segments. Here more

FIGURE 8–13
Cone of sediment deposited in the ocean near the mouth of the Indus River.
(after Heezen and Tharp, 1964)

than any other place diastrophic processes involving the crust appear to be acting at the present time. The advent of geophysical techniques of studying the crust and emphasis on marine geology have added greatly to our knowledge of the island arcs and their relation to the continents and ocean basins. Some of the island arcs are almost circular arcs. Such near-perfect arcs include the following (Fig. 8–14): (1) Aleutian Islands; (2) Kurile Islands, which extend from Kamchatka to Japan; (3) Mariana Islands; (4) Indonesia, extending from Burma to New Guinea; (5) Ryukyu Islands, extending from Japan to Taiwan; (6) West Indies (Antilles of the Caribbean); and (7) Scotia Islands, between South America and the Palmer Peninsula.

Perhaps the name island arc is a misnomer, because several segments of the crust have most

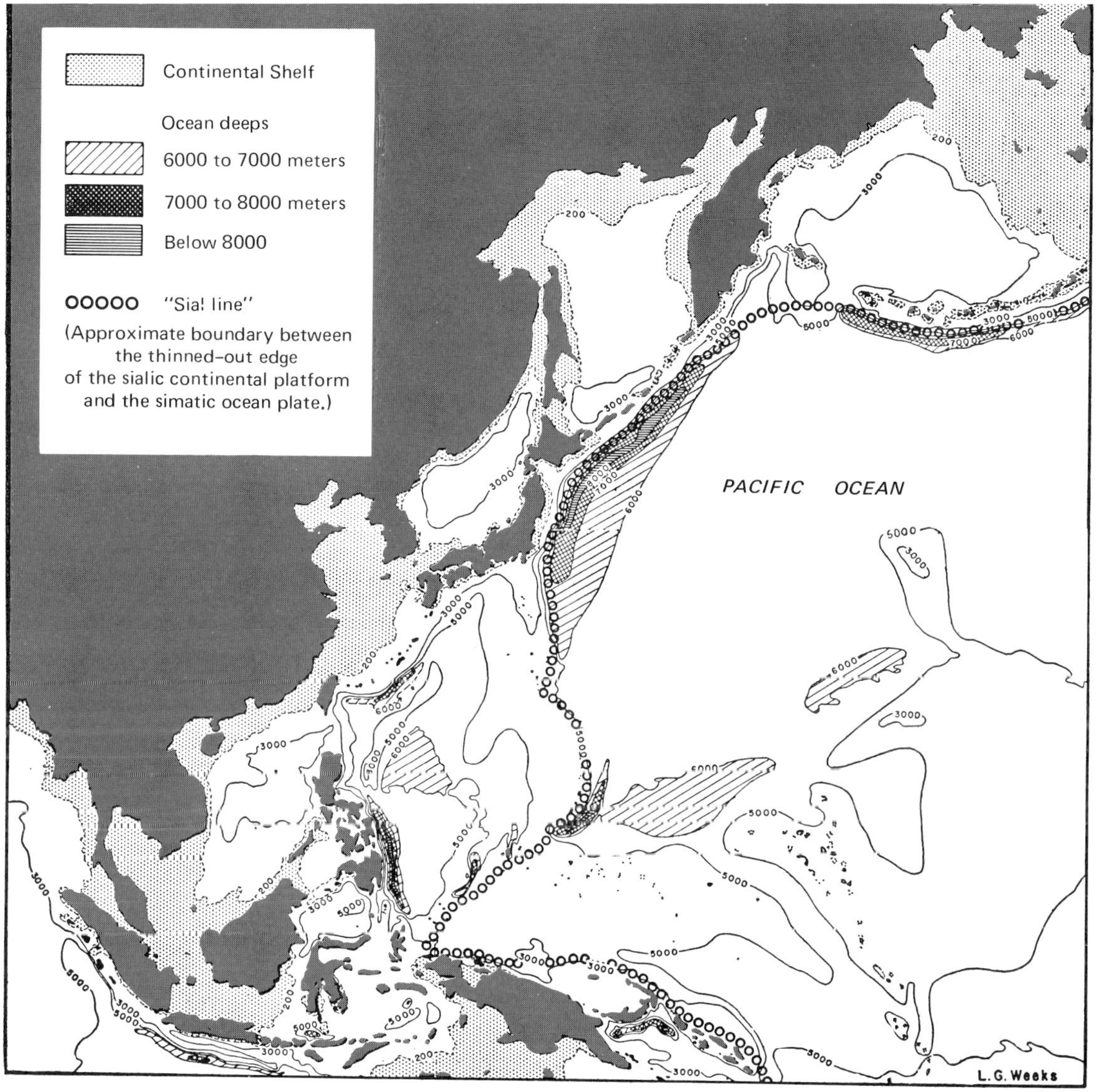

FIGURE 8–14
Mobile belts and island arc systems of the western Pacific
(from Weeks, 1953)

of the characteristics of the island arcs but either are not islands or are not distributed in an arcuate pattern.

Other less arcuate island groups that are usually classified as island arcs include (1) Japan, (2) Philippine Islands, (3) Solomon Islands, (4) New Hebrides Islands, and (5) Tonga-Kermadec Islands.

The Aleutian, Kurile, Japan, Ryukyu, Philippine, and Indonesian arcs are marginal to continents, being separated from the continents by shallow seas. The island arcs are convex toward the ocean basin and a deep-sea trench lies on the oceanward side of each of these arcs.

The composition of lavas from volcanoes in the Pacific seems to indicate that the crust from which the magma originates is basically different in all the circum-Pacific arcs from that in the ocean basin. Andesitic magmas are erupted from volcanoes in all arcs, but volcanoes in the Pacific basin itself erupt basalt. A line separating these two provinces, called the *andesite line* (*sial line* in Fig. 8–14), is often used to demark the boundaries of the Pacific basin. However, the crustal structure in the Philippine Sea, Coral Sea, Fiji basin, and Tasman Sea is more like that of the Pacific basin than that under the intervening island belts. The Tonga-Kermadec Islands form an almost straight line, arcuate only at the northern end, which is hooked; otherwise these seem very much like the more perfect island arcs. The Solomon-New Hebrides island groups are perhaps the most anomalous islands included (Fig. 8–15). They have been referred to as "strewn" islands—no arcuate pattern is evident, nor is there a distinct elongate trench on the Pacific basin side. Deep trenches are present, but they are on the Australian side.

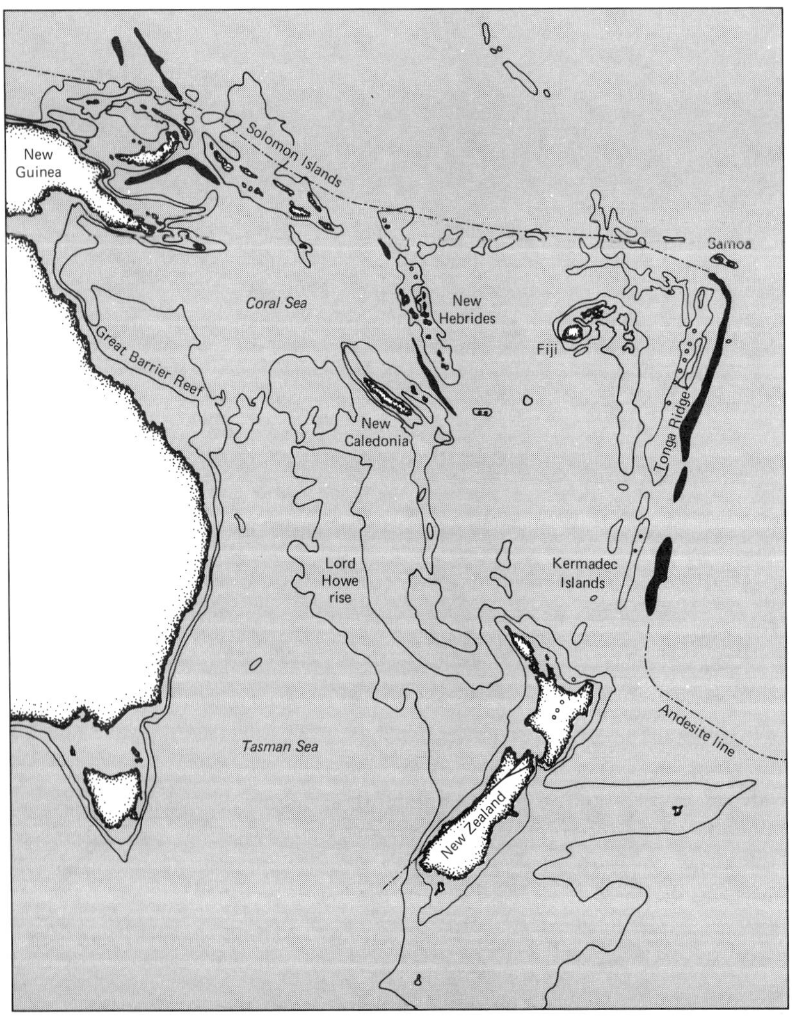

FIGURE 8–15 Major features in part of the southwest Pacific. Black areas are trenches; circles indicate major volcanic centers; basaltic lavas occur to east and north of the andesite line.

The two island arcs in the Atlantic, the West Indies and the Scotia Arc, are similar in that both are convex toward the Atlantic, both have trenches on their Atlantic side, and both are situated between, not marginal to, continental blocks.

Deep-sea trenches are characteristic features of island arc systems (Figs. 8–16 and 8–17). These are deep, elongate furrows in the sea floor. Many of the trenches have nearly flat floors similar to abyssal plains. With very few exceptions, the trenches are located on the convex side of the island arcs. The deepest trenches are over 9,000 meters. Seismic studies of the trenches have shown that parts of some trenches are filled by sediment. The eastern end of the Aleutian Trench becomes shallower as it extends into the Gulf of Alaska.

Island arcs are the most seismically active parts of the earth's crust (Fig. 20–16). Shallow-focus earthquakes occur under the trenches and under the volcanic islands. The foci seem to define a zone that dips under the arcs and toward adjacent continents. Some of these zones are inclined at angles of 30–50°. Almost all of the deep-focus earthquakes occur within these zones. The deepest are 700 kilometers—far below the level of the crust. A similar seismic zone occurs under the Mid-Americas and Chile trenches and the Andes. In this respect the Andes Mountains are similar to island arcs; modern active volcanoes also occur along the western side of the Andes. These seismic zones provide evidence that deep-sea trenches and island arcs are related to processes in the upper mantle. The upper level of this seismic zone is probably a fault in relatively brittle rock, but at depth the shearing and displacement occur in ductile material of the mantle.

The trench forms an outer convex feature that separates the arc from the typical deep-sea floor. Two concentric arcs of islands or arches in the crust are located behind the trench. The first row of islands is volcanically active. Large faults are often found within this volcanic belt. A second, nonvolcanic, belt is located behind the volcanic belt. Many large plutons are found in island arcs and very thick accumulations of sediment occur in some arcs where exposures and depth of erosion permit study. Large fold and fault features comparable to those in the Swiss Alps are found in Java and Sumatra.

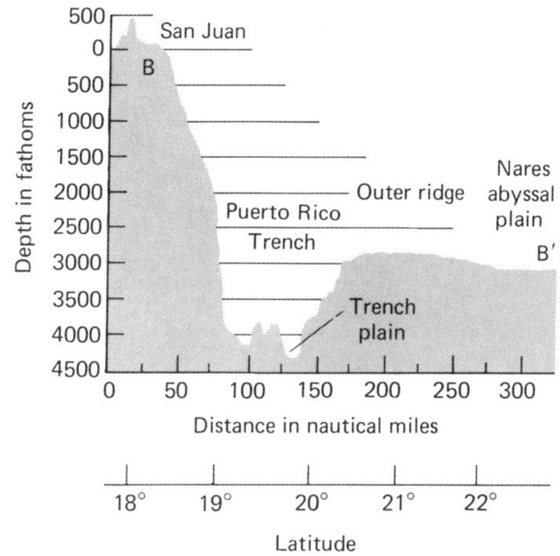

FIGURE 8–16 (*At left*) A profile of the Puerto Rico Trench. (from Ewing and Heezen, 1955)

Oceanic Ridges and Rises

The oceanic ridge system is one of the longest and most continuous features of the earth's crust. The existence of a mid-Atlantic ridge has been known for many years, but its true connection to other ridges was discovered only in 1958. This ridge runs the length of the North Atlantic, South Atlantic, and Indian oceans, and across the South Pacific Ocean. It is more than 64,000 kilometers long (Fig. 25–1). The topography associated with this system varies from place to place. It extends through the Greenland Sea, through Iceland, and south across the relatively shallow ocean in the northern Atlantic until it emerges as a prominent and distinct feature in the Mid-Atlantic Ridge. The Mid-Atlantic Ridge occupies a central position in the Atlantic, following a pattern that reflects the shape of the African and South American coasts (Fig. 8–18). The ridge system can be traced south of Africa into the Indian Ocean where long, linear ridges extend northward into the Gulf of Aden. Another branch of the ridge, also characterized by very rugged topography, splits off toward the southeast, but the relief decreases along this branch of the system and it continues across the southern Pacific as a broader topographic rise. This extends as the East Pacific rise across the eastern part of the South Pacific (Fig. 8–3). This rise appears to branch at the equator into two sections, one on which the Galapagos Islands are located, and one that extends northward into the Gulf of California.

FIGURE 8–17 (*Below*) Seismic reflection profiles across the Japan Trench. (from Ludwig and others, 1966)

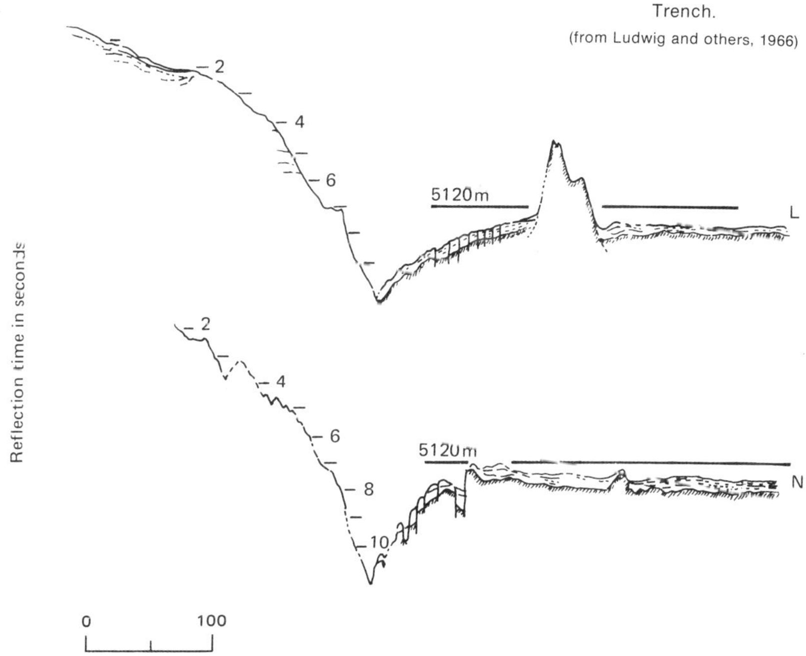

Vertical exaggeration 25:1

An elongate depression or trough runs along the crest of the ridge throughout most of its extent. This shows as a notch or cleft on most profiles. The trough is as much as 2,000 meters lower than peaks on either side, and the two sides are 30–45 kilometers across. This trough appears to be a down-dropped section of the crust, bounded on either side by faults. This feature exists not only on the crest of the Mid-Atlantic Ridge; there is positive evidence of its existence at many other places along the midoceanic ridges, and it is possible that this rift zone follows the midoceanic ridges around the world. If it does, it may be of profound significance, since a rift of this sort could be created by the crust of the earth being pulled apart as a result of tension within the crust.

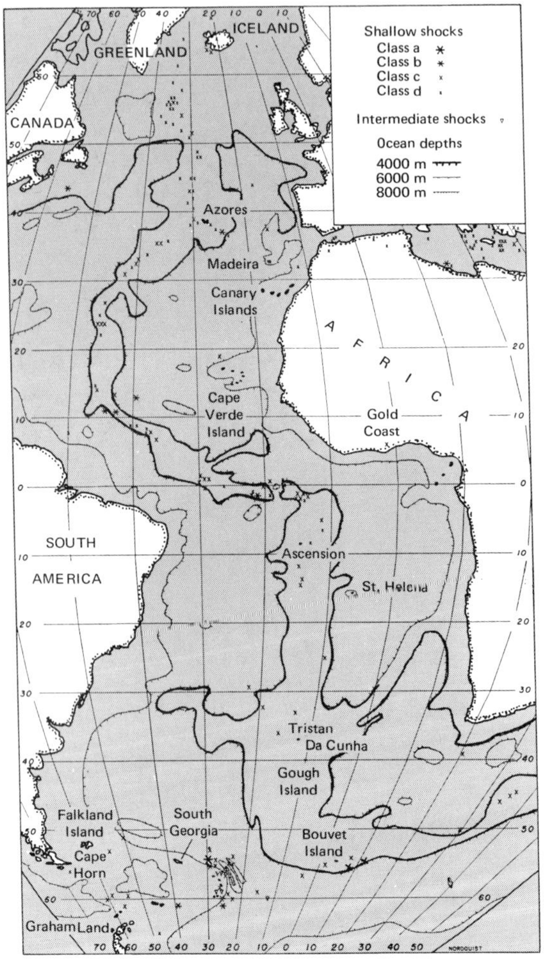

FIGURE 8–18
Map of epicenters in the Atlantic showing concentration of earthquakes along Mid-Atlantic Ridge.
(from Gutenberg and Richter, 1941)

Tolstoy and Ewing (1949) subdivided the Mid-Atlantic Ridge (Figs. 8–19 and 8–20) into two quite different physiographic provinces: (1) a central ridge or main range that is shallower than 1,600 fathoms (2,800 meters), also called the crest province, and (2) a flank or terraced-zone province (Fig. 8–21). The crest province includes the fault valley. Sometimes a single depression is found, on other crossings several deep valleys are found running subparallel to one another, and sometimes the depression is absent in profiles. The mountains in the crest province have been described as resembling tilted blocks. The mountains compose the rim of the block and the flanks form what is called a *high-fractured plateau.* Depressions in this region are deep and narrow.

The flank provinces of the Mid-Atlantic Ridge have the form of a series of steps. It is uncertain if these steps might be tilted blocks in the sense of fault-block mountains, but they are broken in a few places by smooth-floored intermontane basins that sometimes are tens of kilometers wide.

The continuity of the linear topographic features that parallel the Mid-Atlantic Ridge is broken by a series of relatively narrow, steep-sided troughs that trend east-west and cut across the ridge. These are sometimes called trenches, although their geologic setting is very different from that of the deep-sea trenches associated with island arcs. The Vema Trench is one such feature in the Indian Ocean (Fig. 8–22). Many similar features have now been found in the Atlantic and

FIGURE 8–19
(*Below*) Profiles of the Atlantic Ocean sea floor. Note the great vertical exaggeration of the scale, 40:1.
(from Heezen, 1959)

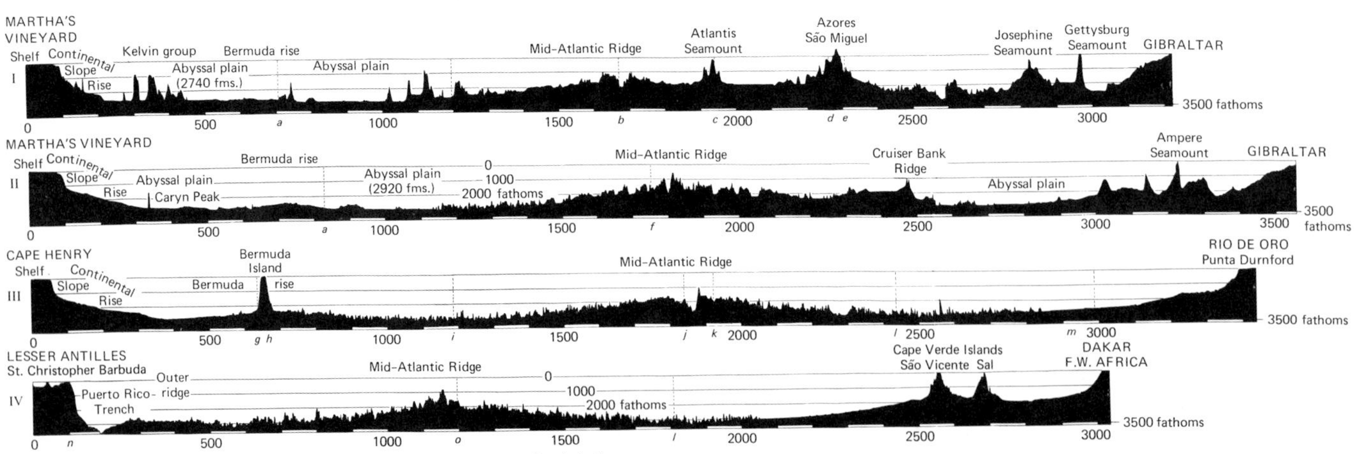

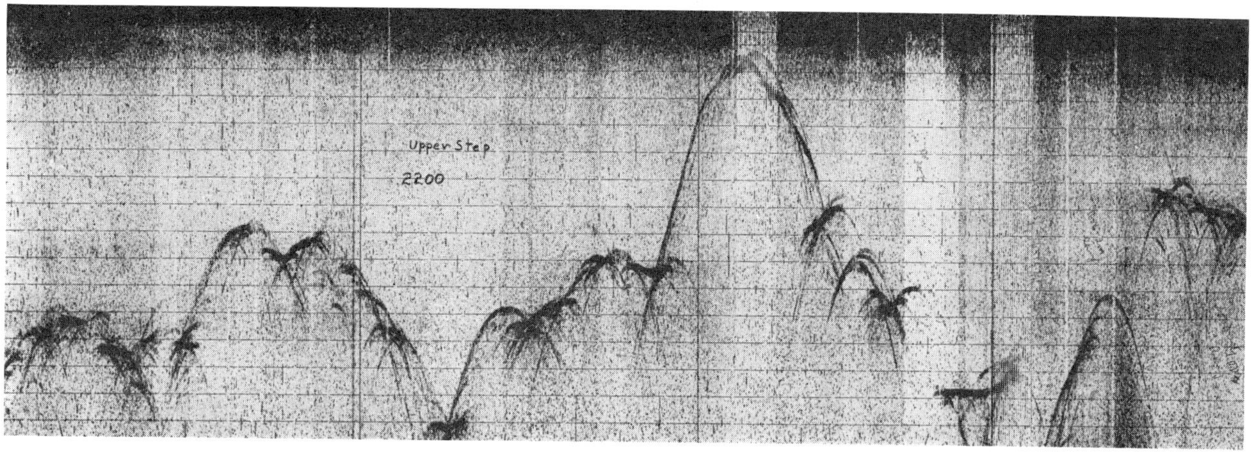

FIGURE 8–20
Profile of the ocean floor. The record was made during a crossing of the Mid-Atlantic Ridge. The 2,200-fathom (3,960 meters) line is indicated.
(from Heezen, 1959)

FIGURE 8–21
Topographic profile with 12x vertical exaggeration across the Mid-Atlantic Ridge at 23°N latitude. MV indicates median depression.
(after van Andel and Bowin, 1968)

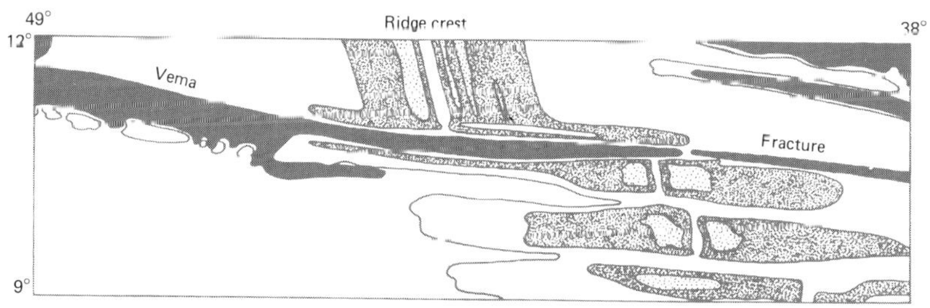

FIGURE 8–22
Bathymetric sketch of the Vema fracture zone. Depths below 2,000 fathoms (4,680 meters) are black; dark stipling covers areas between 2,000–1,600 fathoms (3,600–2,880 meters), and light stipling 1,600 fathoms (2,880 meters) or less. Ridge crest trends almost north-south.
(after Heezen, Gerard, and Tharp, 1964)

Indian oceans. These are thought to be deep-seated fractures or faults in the oceanic crust. A large number of such fracture zones of generally north-south trend are now identified in the Indian Ocean, whereas the longest and best documented fracture zones in the Pacific trend east-west (Figs. 8–3 and 25–1).

Unlike the continental margins and the deep oceanic basins, the oceanic ridge and rise system is the center of much seismic activity. When the epicenters of earthquakes occurring in the oceans are plotted, they form two striking patterns. One is associated with the island arcs, and the second is a very narrow zone along the crest of the midoceanic ridges. Many of the epicenters fall in or on the edges of the fault valley along the crest of the ridge. Other epicenters are closely associated with linear breaks that run across the topography of the ridges and probably are faults that cause lateral displacement of the ridge crest (Fig. 8–22). Some of the first recognized and most prominent of these faults lie in the equatorial belt where they cut across the Mid-Atlantic Ridge. These are represented by long notches thousands of meters deep. The sense of displacement of the ridge crest across them is such as to indicate that the northern section of the ridge has been displaced westward relative to the southern section. It is interesting to note that North America is located north of this shear zone and that this continent is west of the main mass of South America.

A number of islands are located along the oceanic ridges. These include the Azores in the northern Atlantic and São Paulo, Ascension, Tristan da Cunha, Gough, and Bonvetya in the South Atlantic. These and others are small peaks or groups of volcanic origin. Some of them, like the Azores, have been sites of eruption within recorded history—some have erupted in recent years. The eruptives are predominantly of basaltic composition. Basic igneous and volcanic rocks have been dredged from many places on the ridge system, and when coring is attempted, volcanic rocks are often encountered at shallow depth. Sediment is usually very thin over the ridge and much bare volcanic rock is exposed on the sea floor, particularly on the Mid-Atlantic Ridge. Although basalt predominates among these eruptives, many other rock types are found, including dunite, serpentine, gabbro, and diabase. Significantly, these are the rock types we associate with the earth's mantle. Granitic rocks and consolidated sedimentary rocks like the Paleozoic sequences found on continents are absent.

The Deep-Sea Floor

The continental rise slopes gradually to depths of about 3,600 meters, and here the floor of the ocean basin is found. On the basis of topography, at least three major types of ocean floor can be recognized: (1) the *abyssal floor,* (2) *oceanic rises,* and (3) *seamounts.*

Abyssal floors. The abyssal floors or *plains* are flat—much flatter, in fact, than most of the Great Plains of North America (Fig. 8–23). Abyssal plains have low relief, are irregular in shape, and are found in a great range of sizes up to widths of several hundred kilometers. Margins of abyssal plains frequently have greater relief, consisting of isolated hills that project above the extremely flat surface.

Abyssal plains have been found in all oceans. However, they do not occupy central parts of the oceans. In the North Atlantic, abyssal plains are located on either side of the Mid-Atlantic Ridge (Fig. 8–4). A similar, almost symmetrical arrangement is found in the South Atlantic. Deep basins also occur on either side of the Lomonosov Ridge in the Arctic Ocean. The abyssal plains of the Pacific occupy a vast region in the North Pacific broken primarily by volcanic seamounts and the Hawaiian Island Ridge.

Oceanic rises. Oceanic rises are large areas that are neither isolated seamounts nor parts of mid-ocean ridges. They rise a few hundred meters above the level of the surrounding abyssal floor. Such rises are found around Bermuda and in the South Atlantic, Indian, and Pacific oceans. Abyssal plains and oceanic rises are seismically inactive.

Oceanic Crust

The deep-sea floor, a vast region of the earth's surface, is characterized by great water depth and

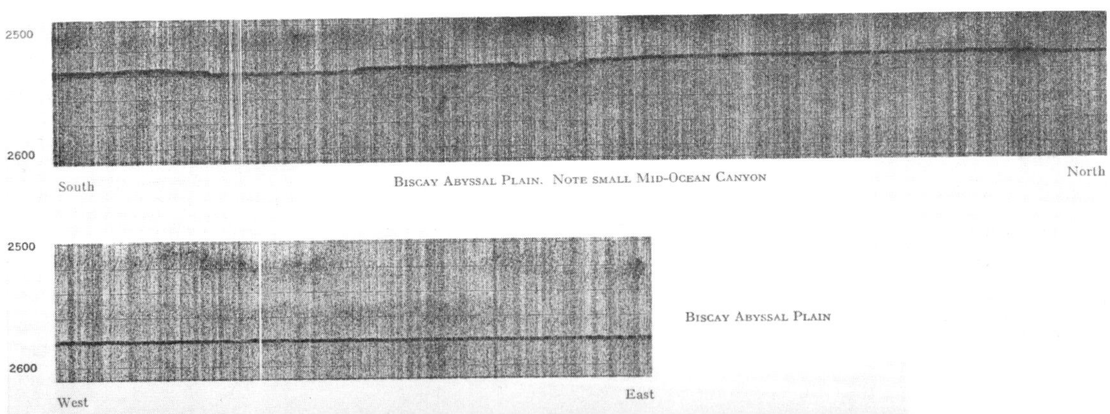

FIGURE 8–23 Echo-sounding profiles of the abyssal plains off the western coast of France. Depths are given in fathoms. (courtesy of Lamont-Doherty Geological Observatory)

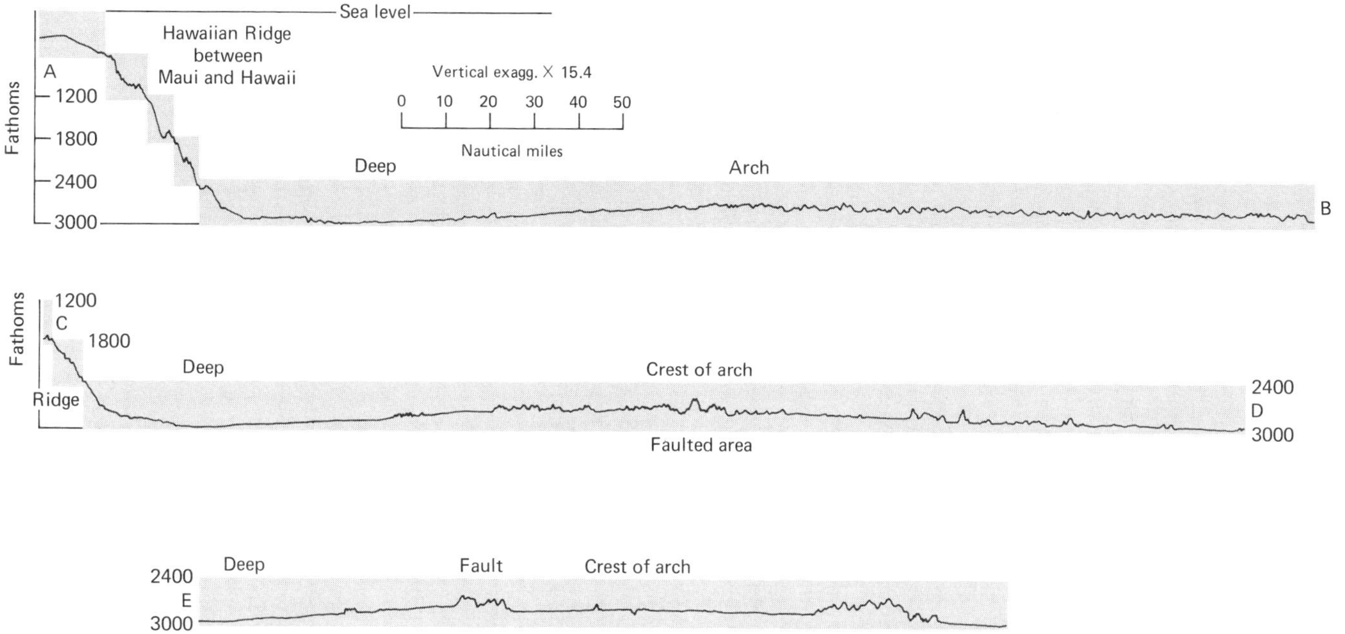

FIGURE 8–24
Echo-sounder records across the Hawaiian Deep and Arch.
(from E. L. Hamilton, 1957)

a thin crust approximately 3,300 meters of water, 0.5–1.0 kilometer of unconsolidated sediment and sedimentary rock, 3–4 kilometers of oceanic crustal rock (probably basalt)—above the mantle. This places the bottom of the oceanic crust at a depth of about 8–10 kilometers below sea level. Typical deep-sea-floor crust underlies the abyssal plains. These plains owe their flatness in large part to burial of rougher topography by sediments transported by turbidity currents. Deep-sea-floor crust extends close to the edges of most deep-sea trenches. Oceanic crust also underlies large portions of the continental rises. All of these parts of the deep sea are seismically inactive.

The Pacific Ocean is by far the largest and the deepest of the oceans, and the crust of the Pacific basin is correspondingly thinner. Vast regions in the central parts of the ocean appear to have unusually uniform crustal thickness. The anomalous areas in the western and southern Pacific are related to the island arcs. The areas of deeper crust in the central Pacific lie under the Hawaiian, Marshall, and Tuamotu island groups, all of which are volcanic in origin. Many other seamounts of volcanic origin are found in the Gulf of Alaska and in the southwestern Pacific.

Faults in the Deep-Sea Floor

Long escarpments that mark the position of faults in the sea floor have been traced hundreds and even thousands of kilometers into the basin from the eastern margin of the Pacific. These zones are roughly aligned (Fig. 25–1) so that they trend almost east-west. They were first recognized because of the high relief across them and the presence of enclosed basins along them. Concentration of shallow-focus earthquakes along them indicates modern activity. The most striking evidence for the faults is found in the offset patterns seen in the magnetic anomalies that parallel the crest of the East Pacific rise (Fig. 25–1).

Many other lineaments are suggested in the Pacific sea floor by the strong alignment of volcanic islands and submarine ridges, but most of these have not yet been studied.

Seamounts. Seamounts are isolated mountain-shaped rises that usually protrude more than 900 meters above the ocean floor. Many extend to the surface, where they stand as islands. Seamounts are found on oceanic rises and on mid-ocean ridges as well as on abyssal floors. Most seamounts appear to be remnants of volcanoes,

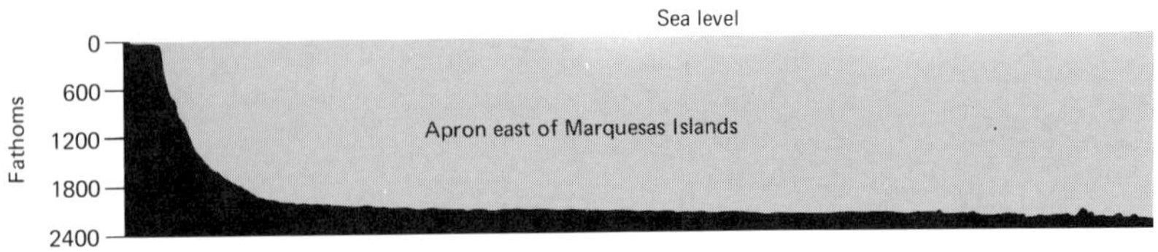

FIGURE 8–25
Apron east of the Marquesas Islands. Note the hills at the far right.
(from Menard, 1956)

having the general shape of volcanoes and being covered by ash and volcanic debris; indeed, some are still active. Some submerged seamounts, called *guyots,* have flat tops formed by wave erosion at a time when the wave-cut platform was at sea level. The tops of other seamounts are capped by coral reefs to form atolls.

Volcanic ridges like that of which the Hawaiian Islands are a part are prominent features of the Pacific sea floor. The weight of the pile of lava flows has caused the crust to be depressed around the ridge (Fig. 8–24).

Archipelagic Aprons

The coalescence of submarine lava flows, usually around groups of volcanic islands and seamounts, may lead to the production of broad, smoothed topography called *archipelagic aprons.* The topography may owe its exceptional smoothness in part to the deposition of sediment between flows and volcanoes. A profile of the archipelagic apron near the Marquesas Islands of the South Pacific is shown in Fig. 8–25. The smoothness of the archipelagic aprons is very similar to that of the abyssal plains, which are located off the coasts of continents. The upper slopes of some aprons slope one or two degrees and they may be cut by deep channels, but at greater distance the slopes diminish. Abyssal hills form the most characteristic type of topography beyond the edges of the aprons. Such aprons are widespread in the Pacific Ocean basin where most seamounts and modern volcanoes are located (Fig. 8–3).

CHAPTER OUTLINE

TERMS

abyssal floors or plains
andesite line
archipelagic aprons
cone
continental shelf
continental slope
continental rise
epicontinental marginal seas
high-fractured plateau
guyots
marginal escarpment
marginal plateaus
oceanic rises
seamounts
strewn islands

REVIEW QUESTIONS

1. Explain why the average elevation of the earth's surface is not a very useful description.
2. Compare the maps of the Atlantic (Fig. 8–4) and the Pacific (Fig. 8–3) outlining similarities and differences in the two ocean basins.
3. What major changes in shape would occur along the eastern United States coast if sea level dropped to the edge of the continental shelf? Refer to Fig. 8–9.
4. Describe the shape of the layers shown in cross section in the subbottom profiles across the Florida shelf (Fig. 8–10).
5. Summarize the distinctive features of the island arcs.
6. What types of evidence indicate the presence of major faults in the sea floor?
7. Classify each of the major submarine land forms as to mode of origin (depositional, erosional, structural, volcanic).
8. Summarize the distinctive features of the oceanic ridges.
9. Which features of the ocean basins can be traced into related features in the continental margins?
10. In what ways is the Gulf of Mexico (1) like and (2) unlike the Atlantic Ocean basin?

SUGGESTED READINGS

Fairbridge, R. W., ed., 1966, The Encyclopedia of Oceanography: New York, Reinhold, 102 p.

Hill, M. N., 1963, The Sea: New York, John Wiley, 568 p.

Keen, M. J., 1968, An Introduction to Marine Geology: Edinburgh, Pergamon Press, 218 p.

Kuenen, Ph. H., 1950, Marine Geology: New York, John Wiley, 568 p.

Mero, J. L., 1965, The Mineral Resources of the Sea: New York, Elsevier, 312 p.

Turekian, K. K., 1968, Oceans: Englewood Cliffs, N.J., Prentice-Hall, 120 p.

9
PHYSICAL OCEANOGRAPHY

9 PHYSICAL OCEANOGRAPHY

WATER OF THE OCEANS

The oceans play a far more important role in influencing man's life on this planet than generally has been recognized. The oceans influence the weather and climate of continental areas through interactions and exchange of materials between the surface layers of the ocean and the air masses in contact with them. Large air masses often obtain their characteristics over ocean waters; thus movements within the ocean as well as those within the atmosphere are important factors in determining climate. The vast expanses of ocean water and the capacity of water to absorb and store heat energy make the oceans of critical importance in the heat balance of the earth. The oceans exert a moderating influence by absorbing heat energy in the daytime and in summer and by liberating heat energy at night and in winter.

Variations in chemical composition and physical properties are slight and usually the changes are very gradual and over vast areas. But do not let the scale of these variations cast doubt on the significance (particularly in density) of those that are important in determining the movements of water in the oceans. Such water movements influence climate and determine where nutrients that support marine populations are abundant.

Composition of Ocean Water

Ocean water contains a variety of salts and gases in solution, but the primary constituent is water —H_2O. Most known elements are found in ocean water in at least trace quantity, but only a few elements make up most of the salts. An average sample of sea water weighing 1,000 grams would contain approximately the following:

Sodium	10.47 grams
Magnesium	1.28 grams
Calcium	0.41 grams
Potassium	0.38 grams
Strontium	0.01 grams
Chlorine	18.97 grams
Sulfate	2.65 grams
Bicarbonate	0.14 grams
Bromide	0.06 grams
Borate	0.02 grams

The composition of sea water is close to the composition of man's body fluids, an interesting note when considered in terms of the probable evolution of man from marine vertebrates. The relative abundance of the salts in sea water is also similar to that of the same constituents in crustal rocks (with the notable exception of chlorine). This might be expected since the salt content of the oceans has been derived at least in part from weathering products carried into the oceans by streams. Possible sources for chlorine are volcanoes (see composition of volcanic gases, Table 3–2) and the primitive atmosphere of the earth soon after the earth formed.

The relative abundance of the various salts in sea water is nearly constant because the waters of the ocean basins are so thoroughly mixed. Thus the main cause of variation of the composition of sea water is variation in the amount of pure water present. This greatly simplifies the task of determining the composition and interpreting the influence of composition on physical properties. Since the relative abundance of the salts is constant, the composition of sea water can be expressed simply in terms of total salt content per unit volume of sea water, *salinity*, or in terms of the abundance of any single constituent. Chlorine is often selected for this because it is relatively easy to measure and it is the most abundant element. Salinity is expressed in grams of salt per kilogram of water, or in terms of parts of salt per thousand parts of water (‰). If chlorinity is measured, it can be converted into salinity as follows:

$$\text{salinity (‰)} = 1.8 \text{ chlorinity (‰)}$$

Salinity determination can be made by methods of chemical analysis, by measurement of the electrical conductivity, or by measurement of the water's index of refraction.

Salinity ranges from about 33 ‰ to 37 ‰ in the open ocean with average values of 34.9 ‰ in the Atlantic and 34.62 ‰ in the Pacific. The values are lower in coastal waters but go up very high in partially enclosed water, for instance, in the Mediterranean Sea (38.5 ‰) or the Red Sea (40 ‰). Variations in salinity are caused by rainfall, addition of fresh water where streams enter the sea, evaporation, and melting or freezing of sea water.

FIGURE 9–1
Hurricane winds pile up waters of Biscayne Bay on Miami's North Bayshore retaining wall, September 21, 1948
(*Miami Daily News* photo, courtesy of U.S. Weather Bureau)

Dissolved Gases in Sea Water

The surface of the sea may be thought of as an air-water interface, a contact zone between a complex of gases and a salt-water solution. Even if the surface of the contact were perfectly still, there would be an interchange of gaseous constituents (e.g., water can be transformed into water vapor). The interchange of gases will go on until an equilibrium condition exists between gases in the upper layers of water and those in the air. Temperature and pressure are important determinants of equilibrium at a gas-liquid interface, but because pressure at sea level is so nearly uniform, temperature is more important as a cause of variations on earth. Interchange of materials is faster and mixing of atmospheric gases in sea water is more complex than in a still liquid because of the sinking and mixing due to such processes as wave action.

Oxygen and carbon dioxide are particularly important among the dissolved gases in sea water because they are involved in chemical and biological processes in the upper layers of the sea; however, nitrogen makes up nearly half of the dissolved gases. With respect to their involvement in chemical and biological processes, some gases, such as nitrogen, argon, helium, neon, and krypton, which have relatively constant concentrations, are called *conservative constituents*. Oxygen, carbon dioxide, and nutrients such as nitrates, phosphates, and silicates do enter into biological processes and are thus nonconservative. The *nonconservative constituents* are more likely to occur in marine sediments ultimately.

Carbon dioxide is one of the most important nonconservative constituents; it enters into most biological functions through the processes of respiration and photosynthesis. Carbon dioxide is present in sea water in large quantities when the water is saturated in CO_2, but it does not remain as a CO_2 molecule because it will react with water. Carbon dioxide combines with water to form carbonic acid (H_2CO_3), which in turn breaks down to form hydrogen ions and bicarbonate ions, and bicarbonate forms more hydrogen ions and carbonate ions as follows:

$$CO_2 + H_2O \rightleftharpoons H_2CO_3 \text{ (carbonic acid)}$$
$$H_2CO_3 \rightleftharpoons H^+ = HCO_3^- \text{ (bicarbonate)}$$
$$HCO_3 \rightleftharpoons H^+ = CO_3^{-2} \text{ (carbonate)}$$

These three equations are shown at equilibrium conditions; the reactions are reversible. Removal of constituents from the sea water by chemical precipitation or by involvement in biological processes causes the reaction to proceed in one direction. Hydrogen ions in the above equations are important determinants of acidity of sea water.

FIGURE 9–2

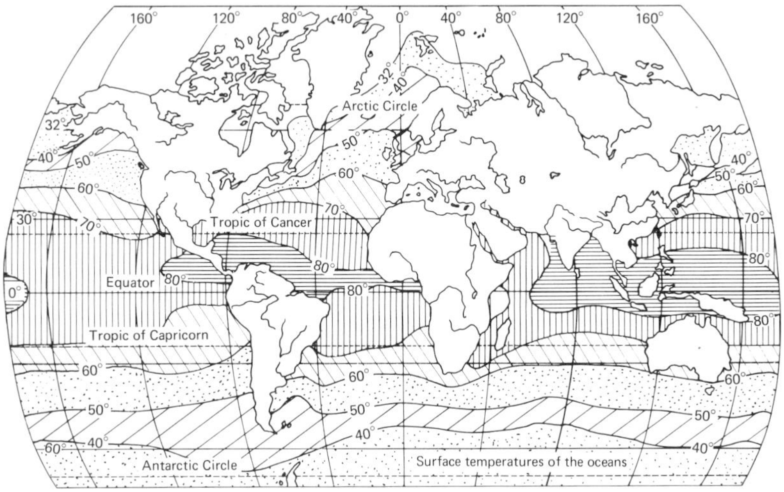

(*a*)
Mean annual temperature of the sea surface in degrees Fahrenheit.
(after Schott; from Miller and Langdon, 1964)

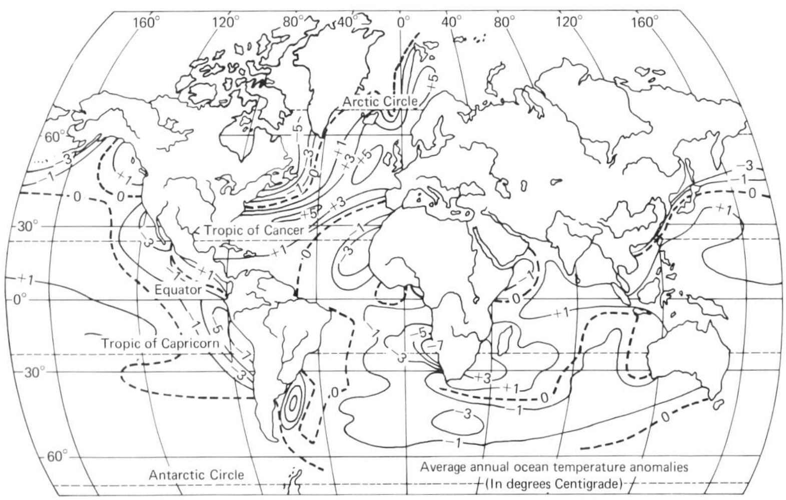

(*b*)
Average annual ocean-temperature anomalies. Positive anomalies indicate temperatures that are higher than the average for the particular latitude, and negative anomalies indicate the converse.
(after James, Markus, and Sokalsky; from Miller and Langdon, 1964)

Both oxygen and carbon dioxide are involved in the processes of photosynthesis in the upper layers, the *photic zone*, of the ocean, where sunlight penetrates. This process that takes place in plants involves the transformation of carbon dioxide and water plus sunlight into carbohydrate with the liberation of oxygen. Animals use oxygen in the process of respiration, releasing carbon dioxide into the water. Minute plants called *phytoplankton* that float within about 200 meters of the surface are the cause of most photosynthesis in the oceans. Thus dissolved oxygen is most abundant near the surface, and oxygen at depth is due primarily to deep-water circulation. Oxygen is consumed by the processes of decay of organic matter. This is probably responsible for the rapid diminution of oxygen at a depth of several hundred meters, and it is surely the cause of anerobic (oxygen-poor) conditions found in some deep basins within which there is little circulation and very slow renewal of oxygen.

Physical Properties of Sea Water

Density (mass per unit volume) is the most important physical property of sea water from the point of view of defining water masses and mass movements. Density of sea water is determined by temperature, salinity, and pressure; however, these cause only slight variations. For example, water at the surface of the ocean with a salinity of 30 ‰ and at 30°C has a density of 1.018 grams per cubic centimeter, whereas that at 8,000 meters depth, 35 ‰ salinity, and 0°C has a density of 1.063 grams per cubic centimeter. For this reason it is necessary to determine density, which is calculated from measurements of temperature and salinity (Figs. 9–2a, 9–2b, and 9–3) with great precision in oceanographic research.

Of the three factors (temperature, pressure, and salinity), pressure varies most regularly as a function of depth. Hydrostatic pressure increases with depth at a rate of $\frac{1}{10}$ atmosphere of pressure per meter. Figure 9–4 shows the relationship between freezing point, maximum density, temperature, and salinity, and Fig. 9–5 shows a series of curves indicating the temperature-salinity values for which density remains constant. The density of sea water increases with increase in

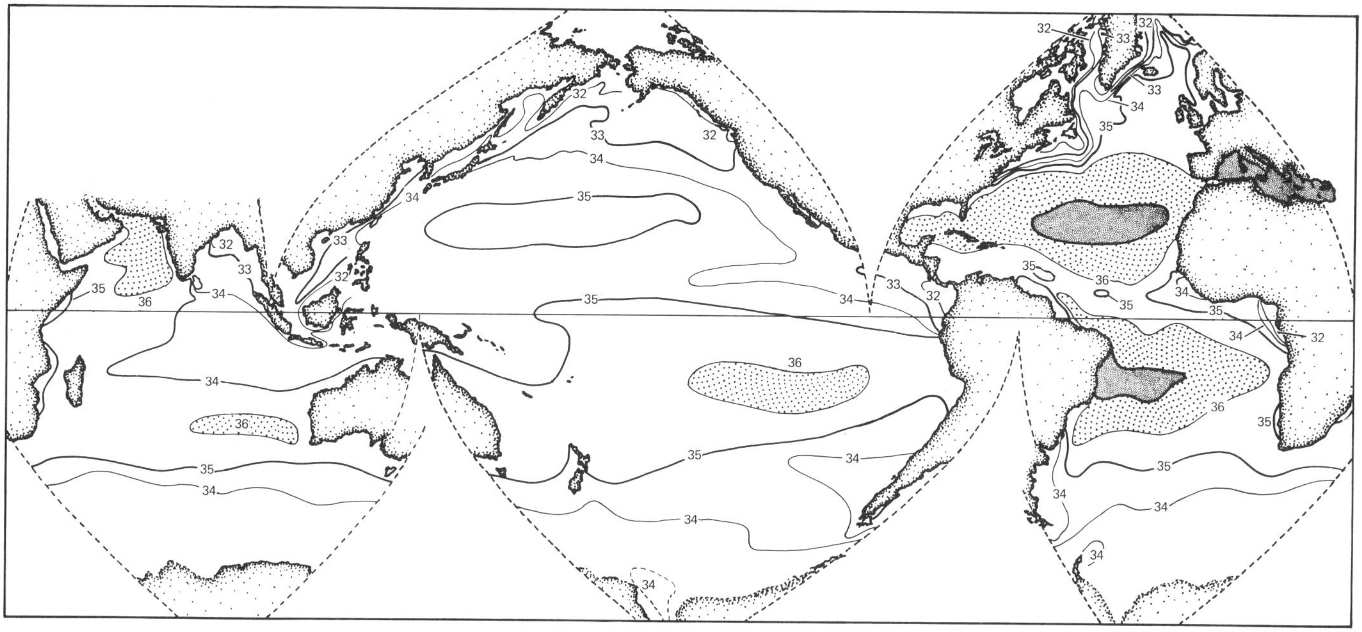

FIGURE 9–3 (*Above*) Average sea-surface salinity in parts per thousand. Areas with averages in excess of 37 ‰ are covered by fine dot pattern; 36–37 ‰ has open dot pattern.

salinity, increase in pressure, or decrease in temperature.

If no external forces acted on ocean waters and if no processes took place in the oceans, a stable density stratification would exist in the ocean. This condition fails to occur because (1) surface temperatures vary (more or less with latitude), reaching a minimum at the poles; (2) the earth is rotating; (3) surface rates of evaporation vary; (4) fresh water is added in most coastal areas; (5) distribution of plants and animals in the sea varies; and (6) some salts removed from sea water by biological and chemical processes go into formation of sediment.

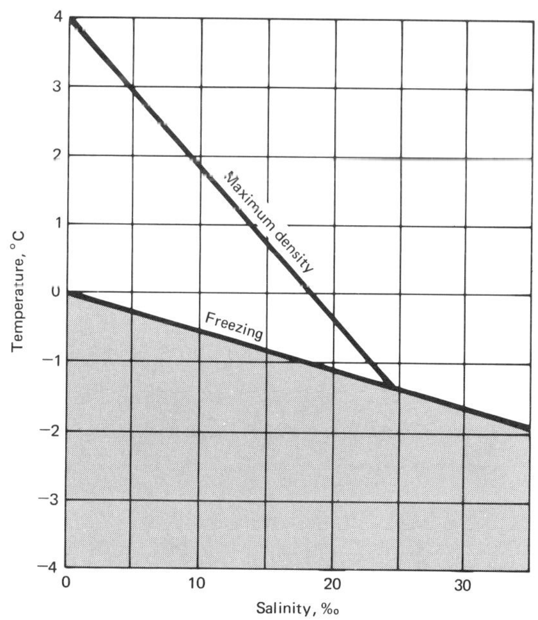

FIGURE 9–4 (*At left*) Relationship between maximum density and freezing temperature for sea water of various temperatures and salinities.

TEMPERATURE AND SALINITY DISTRIBUTION IN THE OCEAN

The range of near-surface temperatures is between 35°C in equatorial latitudes to freezing or −2°C in polar latitudes. Although temperature range is great, the bulk of sea water has a temperature less than 6°C and nearly half is under 2°C. There is usually a pronounced vertical *temperature gradient* (rate of change of temperature with depth) as well as one associated with latitude. Approximately the upper two hundred meters constitute a surface layer (Fig. 9–6) within which a strong seasonal variation in temperature is usually found. Wind action and warming in the spring and summer produce a shallow warm

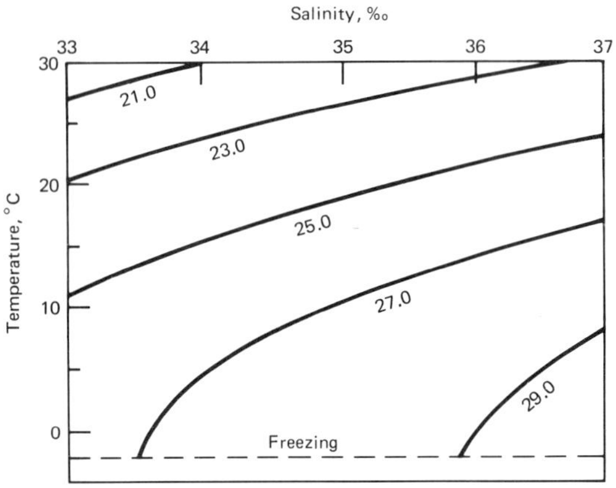

FIGURE 9–5 Density of sea water in grams per cubic centimeter as a function of temperature and salinity. (after U.S. Navy Publication 607)

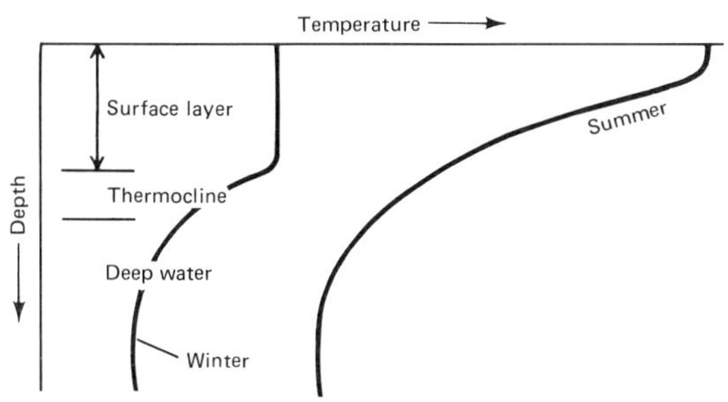

FIGURE 9–6 Temperature vs. depth plots for winter conditions (left) and summer conditions (right).

layer, but in the winter the temperature drops and mixing by strong wind and storms produces a less pronounced gradient. A zone called the *permanent thermocline,* in which the most rapid change of temperature with depth occurs in winter, is located below the surface layer. A deep-water zone in which temperature declines gradually with depth underlies the permanent thermocline.

The vertical distribution of salinity has a wider range of variation than the temperature distribution, but in general salinity reaches a maximum between 100 and 1,500 meters and then drops off to a value that remains relatively constant with depth, as is indicated by the sample distributions illustrated in Fig. 9–7.

CORIOLIS EFFECT

G. G. Coriolis, a French mathematician, first demonstrated that any object in motion on a rotating sphere is subject to a force that tends to deflect it. There is a certain amount of frictional drag between the earth and the atmosphere or water masses, so that they are not completely independent of the earth's surface, but the frictional drag is small enough so that the deflection becomes important in atmospheric and oceanic circulation. One of the easiest ways to visualize the *Coriolis effect* is to imagine a set of four cannon, one at each pole pointing toward the equator, and two at the equator pointing toward the poles. Because of the rotation of the earth, shells that are loaded in guns at the equator are moving in an easterly direction at the rate of movement of all objects on the earth at the equator. Shells in the guns at the poles are not moving except to turn around once each day. When the guns are fired, the shells start out in north and south paths along lines of longitude. The shell fired from the gun at the north pole travels south, but the earth continues to rotate from west to east beneath the path of the shell while it is in the air, so that it lands on a longitude line that is west, or to the right, of the line along which it was fired. The shell fired from the gun on the equator is deflected increasingly to the east as it travels north, because it is moving into latitudes where the velocity of the earth's surface is progressively decreasing. Thus both shells follow paths of deflection to the right. The same reasoning will show that deflections in the Southern Hemisphere are always to the left.

The Coriolis force is usually so small that it has little effect, but in wind and water movements the other forces acting on them are so slight that the Coriolis effect is noticeable. It is at a minimum along the equator and increases toward the poles. As a consequence of this force, surface water in the Northern Hemisphere moves along a path 45° to the right of the prevailing wind direction in the open sea.

One of the first evidences of the Coriolis effect in the oceans was observed by the Norwegian scientist, artist, and Arctic explorer, Fridtjof Nansen, who noticed that Arctic pack ice drifts some 20–40° to the right of the prevailing wind direction. Somewhat later Ekman showed theoretically that wind-driven currents at the surface of the ocean should be directed 45° to the right of the wind. This effect is transmitted to successively lower layers of water so that the angle of deflection increases and the velocity of the current decreases with depth, as shown in Fig. 9–8 (the length of the arrows corresponds to the velocity). The effects are largely confined to the upper 100 meters of water. Currents thus can be expected to flow in different directions at different depths and, at some depth, even counter to the surface currents. The mass transport of water near the surface is almost at right angles to the direction of the wind as a result of the Ekman effect.

If the wind blows from the north along a coast in the Northern Hemisphere, wind-driven currents directed out to sea will be set up along the coast. Thus surface water will be pulled from along the coast, allowing upwelling of deeper water. Where this occurs, the deeper water may be nutrient-rich and provide excellent conditions for a large fauna and flora near the surface. If the wind direction is reversed, water would be piled up along shore.

MOVEMENTS OF SURFACE WATERS OF THE OCEANS

Study of ocean currents began soon after the first attempts to navigate the seas, but the complexity of the movements of water even near the surface is such that many unanswered questions remain.

Several groups of currents can be recognized, although they often tend to be superimposed. These include the following:

1. Wind-driven currents.
2. Large-scale currents related to density distribution—*thermohaline circulation*.
3. Tidal currents.
4. Currents associated with internal waves.

Factors in Surface-Water Movement

Figure 9–9 illustrates the important surface currents of the oceans. Anyone who is familiar with the sea is well aware of the great variety of its moods. It is sometimes whipped into huge waves by storm winds, and at other times it is quiet and calm. Twice daily it responds to the gravitational pull of the sun and moon as tides come and go. Many factors play important roles in the move-

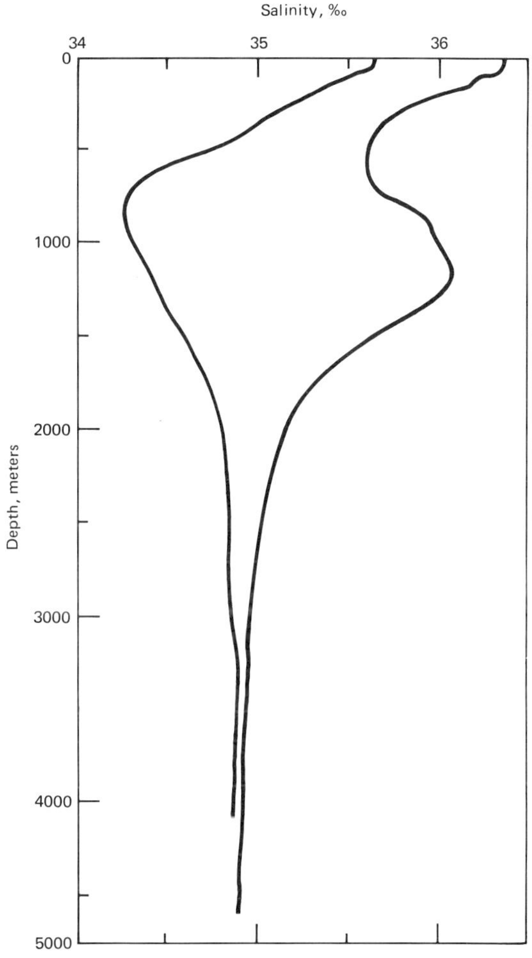

FIGURE 9–7
A plot of salinity (horizontal scale) vs. depth (vertical scale) in the northeastern Atlantic (right curve) and South Atlantic (left curve).
(after McLellan, 1965)

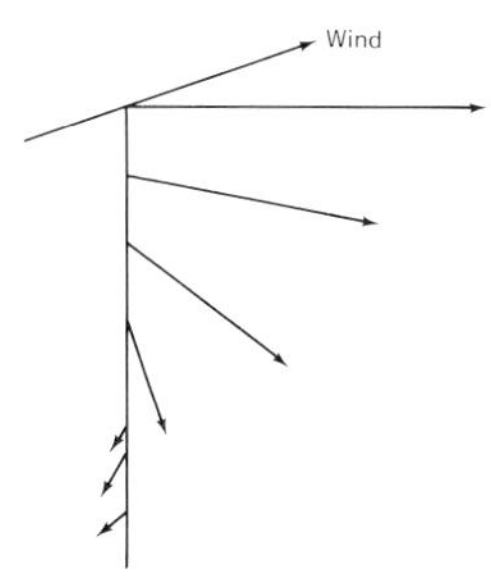

FIGURE 9–8
Current deflection due to Coriolis effect. Currents set up by wind currents are deflected to the right in the Northern Hemisphere. The amount of the deflection and the velocity of the resulting currents vary with depth as shown.
(after Ekman, 1905)

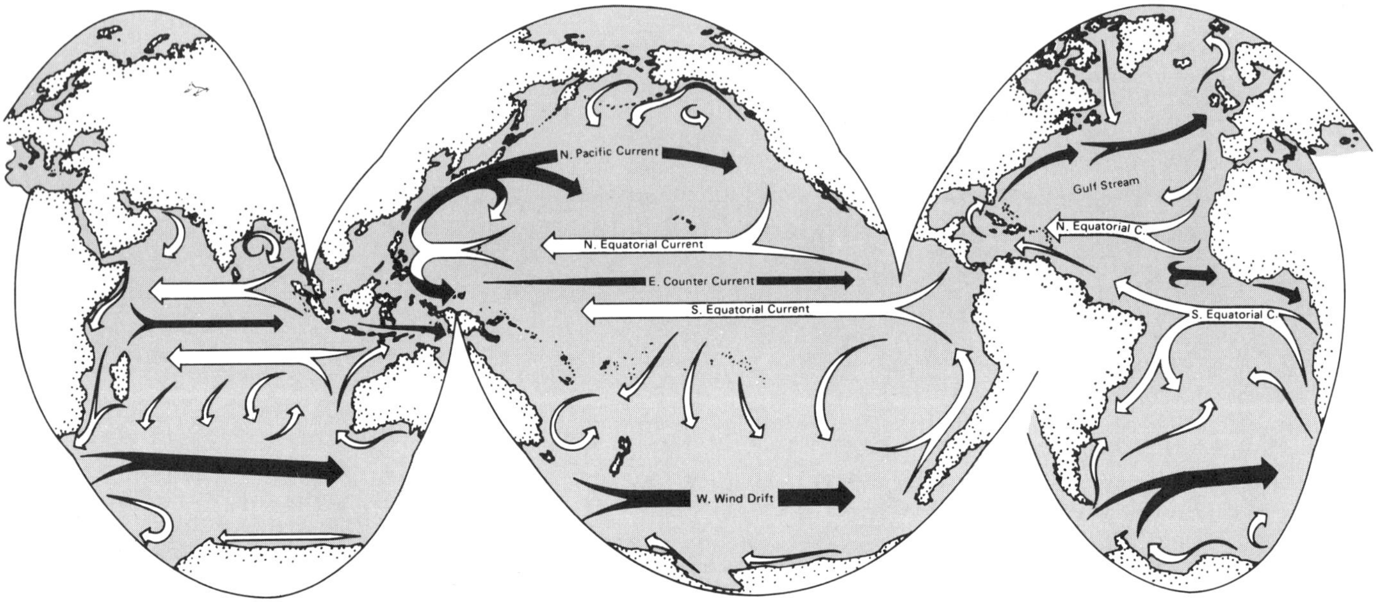

FIGURE 9–9
Generalized pattern of surface circulation in February and March. West to east currents are shown in black; east to west currents are white arrows.

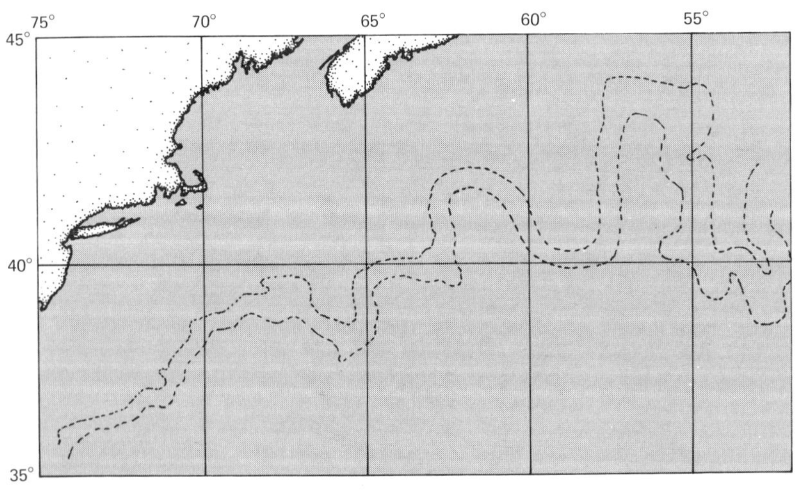

FIGURE 9–10
Schematic representation of the Gulf Stream, showing its general meandering pattern.
(after Stommel, 1966)

ments of the seas. Most important of these are the following:

1. Slope of the surface of the ocean.
2. Wind blowing over the water.
3. Density variations within the water.
4. Coriolis force.
5. Gravitational attraction of the moon and sun (see *tides*).
6. Configuration of the ocean floor, and the shape of the basins.

Ocean Currents

The complicated details of movement of surface water can be reduced to a relatively simple picture. Surface currents (Fig. 9–9) flow from west to east in what is called the *west-wind drift* in the middle latitudes of both the Northern and Southern Hemispheres in all oceans. Along the equator, currents flow from east to west in the equatorial drift of the Atlantic, Pacific, and Indian oceans. Between the prevailing westerly drifts and the equatorial drift, currents are deflected to the right in the Northern Hemisphere and to the left in the Southern Hemisphere. Where islands or irregular coast lines lie in the paths of these currents, they are deflected both right and left in streamlined patterns around the land.

The Northern Equatorial Current flows from east to west across the Atlantic. It strikes the northern coast of South America and is deflected to the north, where it flows through the Caribbean and into the Gulf of Mexico. Here the *Gulf Stream* flows from the Gulf through the strait between Florida and Cuba at the rate of almost 5 kilometers per hour. This segment, known as the *Florida Current,* flows northward along the southeastern coast of the United States, where it is joined by currents that have been deflected to the northern side of the Caribbean Islands. This stream of warm, northward-moving, equatorial water is as much as 80 kilometers wide and over 1,000 meters deep. It follows a meandering course at times and is occasionally broken up into several threads flowing in a braidlike pattern (Fig. 9–10). At about the latitude of Washington, D.C., the stream moves farther from the coast. Part of it diverges as the North Atlantic Current flows eastward almost all the way across the Atlantic. It is broken up off the western coast of Europe and split into two parts; one moves southward back toward the equator, and the other moves northward along the coast of Norway.

A second important current along the eastern coast of North America is the *Labrador Current,* which flows southward from the Arctic Ocean between Greenland and Labrador. It is largely broken up south of Newfoundland where it mixes with the Gulf Stream, which has continued to flow northeast after forming the North Atlantic Current.

Along the western coast of North America the principal current is the *North Pacific Current,* which flows from west to east and impinges on the coast at the state of Washington. Here it is split; one part is deflected northward to form the *Alaska Current* along the coast of Canada and southern Alaska. The other component is deflected to the south to become the *California Current,* which continues to move southward and eventually rejoins the equatorial drift.

The surface currents of the Indian Ocean (Figs. 9–11a and 9–11b) are particularly interesting because there is a seasonal reversal in the northern part of the Indian Ocean. The west-wind drift in the middle latitudes creates a persistent west to east movement in the southern Indian Ocean, and the counterclockwise movement between Australia and Africa persists throughout the year. But from the equator northward the currents reverse, following the direction of the prevailing winds. During the *monsoon season* (May–September), the winds blow from the southwest toward India and Asia, and at this time the currents move in a clockwise direction in the northern Indian Ocean, but when the monsoons cease and the winds switch direction in the winter, the currents also change. This pattern emphasizes the importance of wind direction in determination of surface currents.

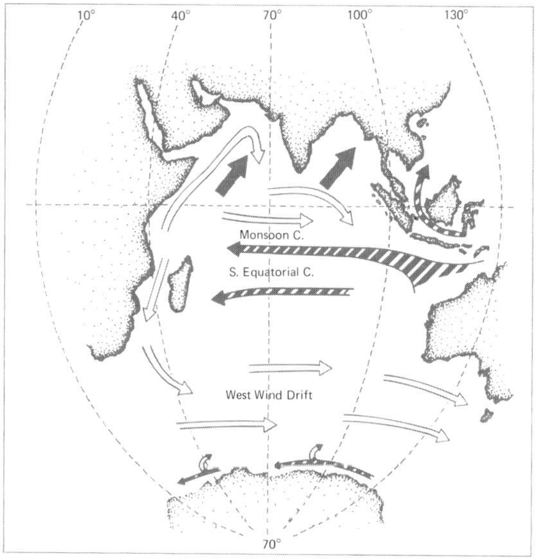

FIGURE 9–11

(*a*) Current pattern in the Indian Ocean during May to September when monsoon winds flow from the southwest, indicated by solid arrows.

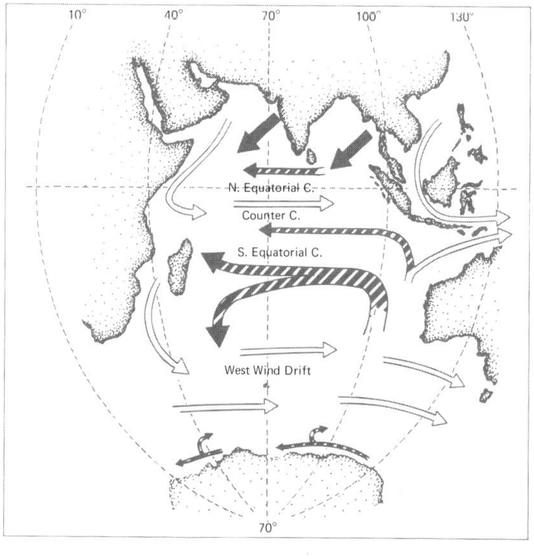

(*b*) Generalized current pattern in the Indian Ocean during November to March when winds blow southwest in the northern part of the ocean.

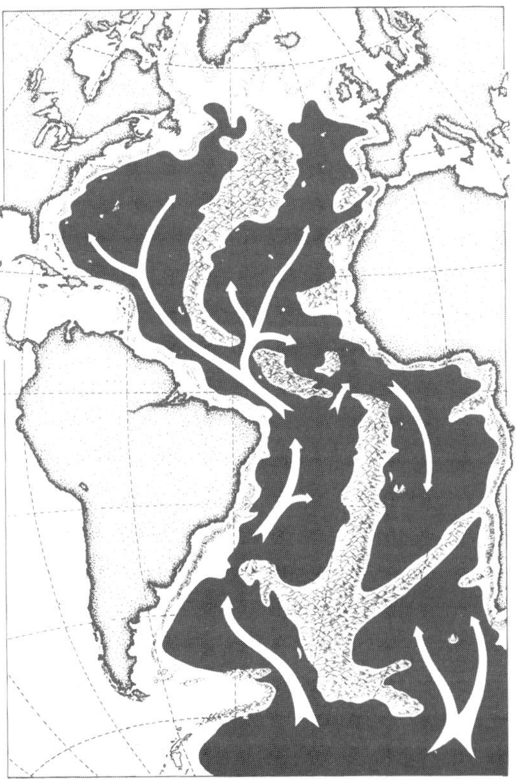

FIGURE 9–12
(*Above*) Deep-sea ripple marks. This picture was taken at a depth of 1,260 meters in the North Atlantic Ocean. The presence of ripple marks there is a positive indication of current action.
(Lamont Geological Observatory)

FIGURE 9–13
(*At right*) Deep-sea circulation in the Atlantic. Shown is a schematic view of circulation at a level approximately 3,600 meters deep. At this depth the Mid-Atlantic Ridge breaks up the flow of cold, dense water from the Antarctic.
(after Stommel, 1955)

Influence of the Shape of the Ocean Basins

The shape of the ocean bottoms and of the coasts of continents provides a framework within which the other factors operate. Surface movements are confined within this framework and deflected by it. For this reason, surface movements in the Southern Hemisphere, which has very little land area, are much simpler than those in the Northern Hemisphere, which is irregularly broken up by odd-shaped continental masses that make up four-fifths of all the land on earth.

The west-wind drift causes prevailing surface currents to follow a simple circular pattern, flowing from west to east around Antarctica. The only obstacle is the narrowness of the passage between South America and Antarctica, which constricts the currents. This circum-Antarctic current is important in bringing about mixing of waters in all oceans. In the Northern Hemisphere the equatorial currents follow the equatorial drift from east to west, but in the Atlantic these currents are broken up and deflected north and south by the eastward protrusion of South America.

Slope as a Cause of Oceanic Circulation

Differences in the level of the ocean's surface are an important cause of oceanic circulation. Variations in level are caused and maintained by mass transport of water due to wind-driven currents. Some of the differences in level are relatively constant, whereas others are transient. One such constant variation is that across the Florida Peninsula where the waters of the Gulf are nearly 18 centimeters higher than the Atlantic on the other side of the peninsula. This example is typical of the piling up of water that occurs on the western side of the oceans in the trade-wind belts.

The mass transport of water is at right angles to the prevailing wind directions in the Northern Hemisphere and the deflection of water is to the right due to the Coriolis effect. Thus in the North Atlantic, less dense water is piled up to the right of the trade winds and to the right of the westerly winds, producing a high area in the center of the North Atlantic Ocean. As a result of this high, water tends to flow downslope, but as soon as it begins to move it is deflected to the right and produces currents that move around the high, called *geostrophic currents*. The prevailing wind patterns, the Coriolis effect, and the distribution of land lead to the development of large *gyres* both north and south of the equator in the Atlantic and Pacific oceans (Fig. 9–9). In all these instances the level of the ocean is raised differentially, and compensating movements of the surface are initiated to restore the level. A perfectly level surface is never obtained. Instead, a balance is reached so that the rate of movement away from elevated areas is approximately equal to the rate of movement toward them.

DEEP-SEA CIRCULATION

Although scientists assumed until recently that no currents strong enough to transport sediments moved on the deep-ocean bottom, we know now that sediments on the floors of the oceans commonly contain current ripple marks (Fig. 9–12), which are produced by moderately rapid movement of water. Not only have the ripple marks been found in connection with turbidity currents, where they are readily explained by the rush of submarine landslide; they have also been found in places where no downslope movements could have contributed to their formation. Ripples have been photographed on almost all seamounts. This fact indicates that relatively strong deep-sea currents must exist. It has long been recognized that currents associated with the Gulf Stream extend down to depths of over 1,000 meters. Other forms of deep-sea circulation that may be strong enough to produce ripples are being studied.

Movement of water in the deep sea is controlled by the following:

1. Temperature of the water.
2. Salinity of the water.
3. Shape of the ocean bottom.
4. Surficial currents.
5. Internal tides.

Density variations are prime determinants of deep-water circulation. The density of sea water is increased by evaporation, by cooling, and by freezing (salts are excluded from the ice structure); density is decreased by heating or by dilution resulting from precipitation or melting of ice.

Water temperature, which affects relative density, is an important determinant of water movement. Water near the poles is cooled at the surface in this region where heat loss by radiation exceeds heat supplied by insulation. Because it is more dense than warm water, cold water sinks, sliding downslope into the deep sea and forcing bottom water toward the equator. This is just one example of the power of a density differential to set deep waters in motion. The shape of the sea floor also directs the path of sinking waters (Fig. 9–13).

Water masses are usually described in terms of their temperature and salinity, both of which can be readily measured. The density distribution and the movement of water masses can be determined from them. Oxygen distribution is also useful in determining movement patterns. Figure 9–14 shows both the distribution of salinity and temperature in a north-south section through the Atlantic Ocean and the movement pattern. The simple pattern of surface circulation in the Antarctic continues to great depth. This water is uniformly cold with salinity of about 34.7 ‰, and it is called the *Antarctic circumpolar water*. Very cold water, 0°C, is found to sink from the continental shelf around Antarctica, par-

FIGURE 9–14
North-south sections through the Atlantic Ocean to show temperature, salinity, and the movement of the major water masses.
(modified after maps by Sverdrup, Johnson, and Fleming, 1942)

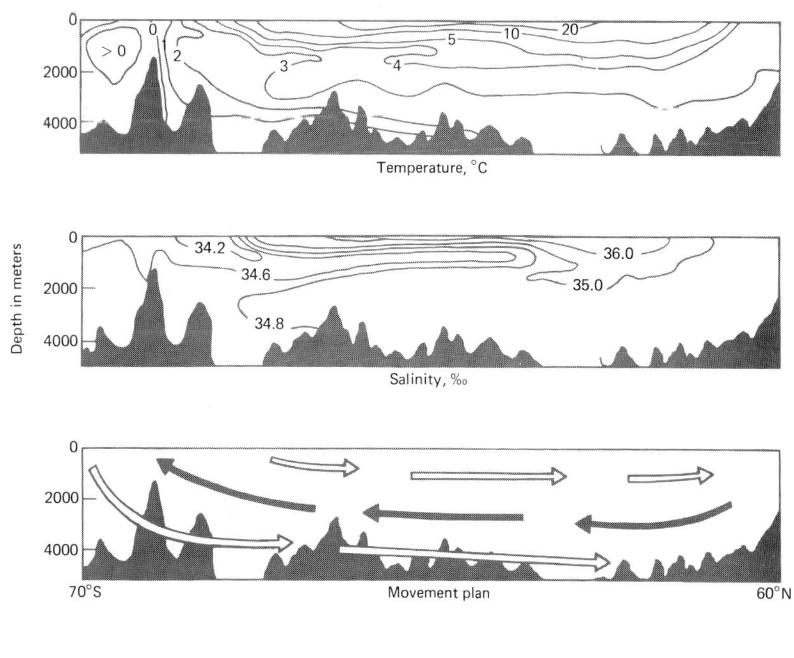

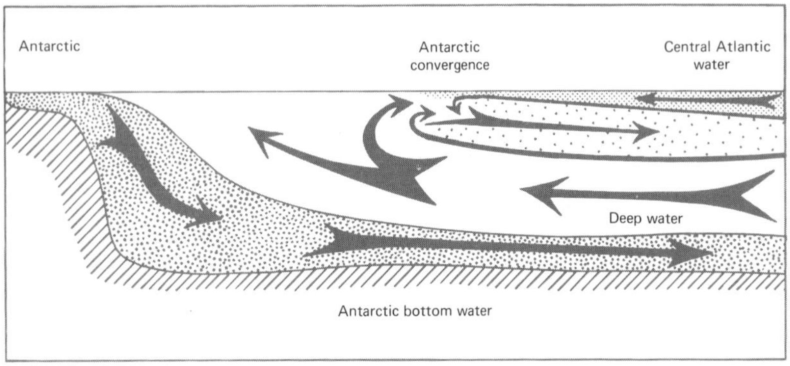

FIGURE 9–15
Simplified circulation pattern at depth in the southern part of the Atlantic. The lightly dotted pattern shows Antarctic intermediate water mass.
(after U.S. Hydrographic office)

ticularly in the Weddell Sea. This water, called the *Antarctic bottom water,* moves into the southern oceans, spreading northward across the sea floor (Fig. 9–15). Waters similarly cooled in the North Atlantic sink and spread toward the south, but these waters are 2–4°C and thus less dense than the Antarctic bottom water; therefore when they converge, the North Atlantic waters move over the Antarctic bottom water and eventually rise in the South Atlantic.

A second mass of water that receives its characteristics at about latitude 50–60° South, called *Antarctic intermediate water* (4°C and 34.2 ‰), moves northward at a depth of 500–1,000 meters.

EFFECTS OF WIND IN CONTACT WITH THE SEA SURFACE

When a strong wind blows across a body of water, waves are set in motion on the surface and a wave current in the direction of the wind is generated. Wave currents are often attributed to frictional drag of wind on the surface of the water. A very strong wind is required to start this motion if the wind is moving in a smooth and steady flow, but usually enough turbulence is involved in wind motion to create eddies, minor updrafts, and irregularities in the flow pattern of air over the water surface. These in turn initiate depressions and rises in the surface of the water. Once such motion is started, only a slight wind is needed to continue the action. If the wind is consistently from one direction, it exerts a more or less continuous pressure on one side of the wave, tending to push it along. As the wind passes the crest of the wave it exerts a pull, a form of suction, on the other side. In this manner wave currents in the oceans are established where wind directions are consistent. This is particularly important in trade-wind belts of the tropics and along the equator, where wind directions are predominantly from east to west (Fig. 9–9). In middle latitudes westerly winds set up a drift of currents from west to east.

Waves

The surface of the ocean is in an almost constant state of motion. These up and down movements are called waves. Most waves, or at least most of the ones we see, are elongated humps separated by irregular troughs; often waves of a great range of sizes are present and in this respect natural waves differ from the long, regular, sinusoidal waves produced in simple laboratory experiments. Many of the largest surface disturbances are not readily detected by the unaided eye.

Figure 9–16 shows a profile model of a theoretically perfect wave. This model is approximated in nature by some waves, but waves are not necessarily continuous over a long distance. In profile the ideal wave is a sinuous, curved surface. The top of the wave form is called the *crest;* the depression between two crests is the *trough.* The distance between two crests or two troughs is called the *wave length, L.* The vertical distance between the top of a crest and the bottom of a trough is the *wave height, H.* The time required for a crest to progress one wave length is called the *period, T,* of the wave motion, and the reciprocal of the period is called the *wave frequency.*

Natural waves are generated by a variety of phenomena and they have different forms. Several types are often superimposed on one another. Disturbance of the surface of the ocean is commonly classed according to period and nature of the generating disturbance. The periods of wave motion range from 0.01 second to 10^5 seconds.

Two distinctly different kinds of wave motion, *progressive* and *standing,* are found. *Progressive waves* are those in which the wave form moves laterally. Water does not move along with the wave form; if it did, seas would be impassable to most surface vessels (Figs. 9–17a and 9–17b). Instead, the water particles travel in a nearly circular path, like a cork bobbing, as waves pass. At the crest of a wave, motion is in the direction of the wave's movement, but in the trough, motion is in the opposite direction. Beneath the surface of the wave, the motion of the water is also almost circular, but the size of the orbits decreases with depth until at a depth equal to about half the wave length the motion is negligible. Below this depth, wave movements have no effect either in the water itself or on the ocean bottom. Usually a submarine submerged to this depth experiences no movement even if violent storm waves occur on the surface. However, wave lengths of some *swells* (wind-generated waves that have advanced into regions of weaker winds or calm) and storm waves are great—several thousand meters—and these can be felt on the bottom at depths of more than 300 meters.

The wave form does not move laterally in *standing waves,* but the crest and troughs do reverse. The water particles move up and down in straight lines except at points, called *nodes,* where no movement occurs.

Wind-Generated Waves

When the wind blows across a stretch of water for a period of time the surface of the water is drawn into progressive wave forms. Initially small ripples, called *capillary waves,* with periods of less than 0.1 second are generated. These are called capillary waves because their form is governed in large part by surface tension. Wind-generated waves will grow as the amount of energy imparted to the water from the wind increases. The wave motion is started by variations in atmospheric pressure due to turbulence in the air and frictional drag. Once the waves have been generated, they are propagated in the direction of the wind that formed them. They become aligned roughly perpendicular to the wind direction and tend to form long, subparallel rows if the wind is continuously from the same direction. Where the air is turbulent, waves are short and choppy. *Wind waves* are usually short-crested; their crests are more peaked than the ideal sine waves and they are skewed in the direction of propagation. The horizontal extent of water over which the wind blows is called the *fetch.* Once the wave motion is started, the progressive waves will travel out of the area in which they are generated as long, more regular waves with periods of 5–30 seconds, called *swell.* The swell is close to a sine function in form, the crest may be long and straight and the height is small relative to the wave length.

When the water depth is great relative to the wave length, L, the velocity of surface waves, C, is given approximately by the relation

$$C = \frac{L}{T}.$$

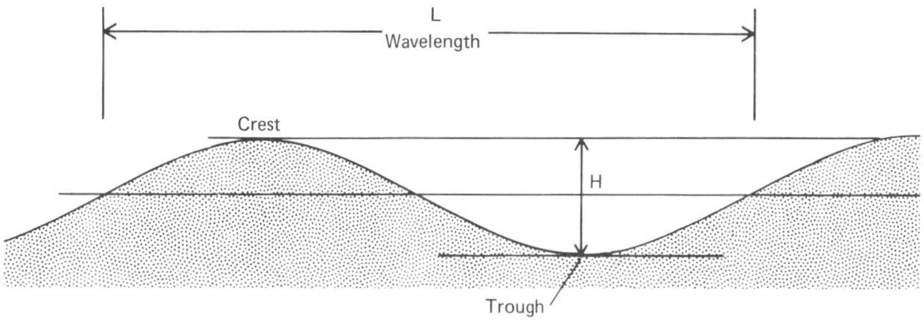

FIGURE 9–16
Profile of an idealized wave form showing wave length (L), crest, trough, and wave height (H).

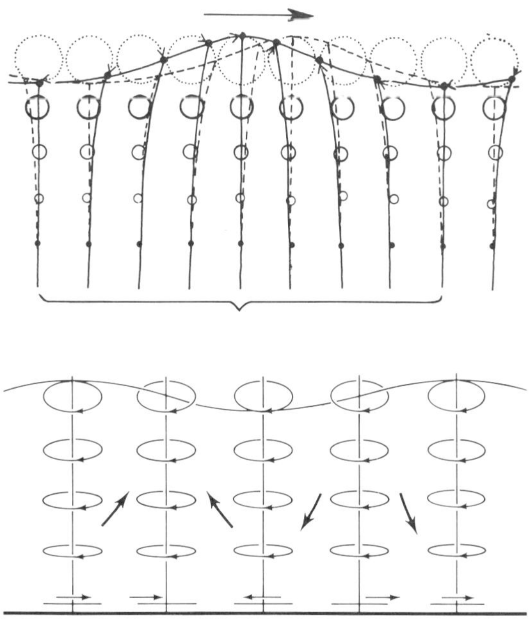

FIGURE 9–17

(a)
Theoretical wave form seen in cross section. The movement of water beneath the wave surface dies out at a depth equal to about one-half the wave length. Particles of water in the wave, shown here for a progressive wave in deep water, move in slightly elliptical paths.

(b)
Progressive wave showing trajectories of water movement when water is shallow relative to the wave length.

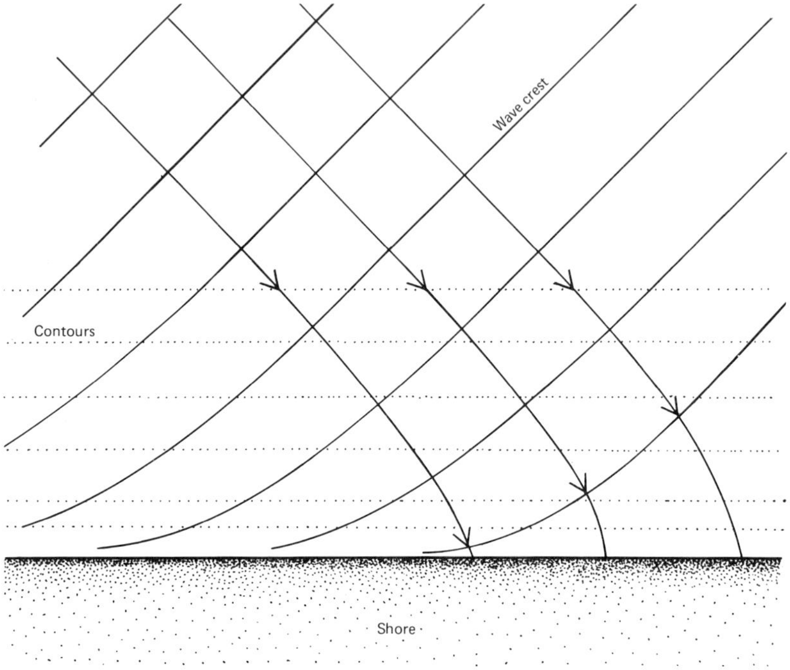

FIGURE 9–18
Schematic illustration showing wave refraction as waves approach land at an angle. Longshore currents would move left to right.

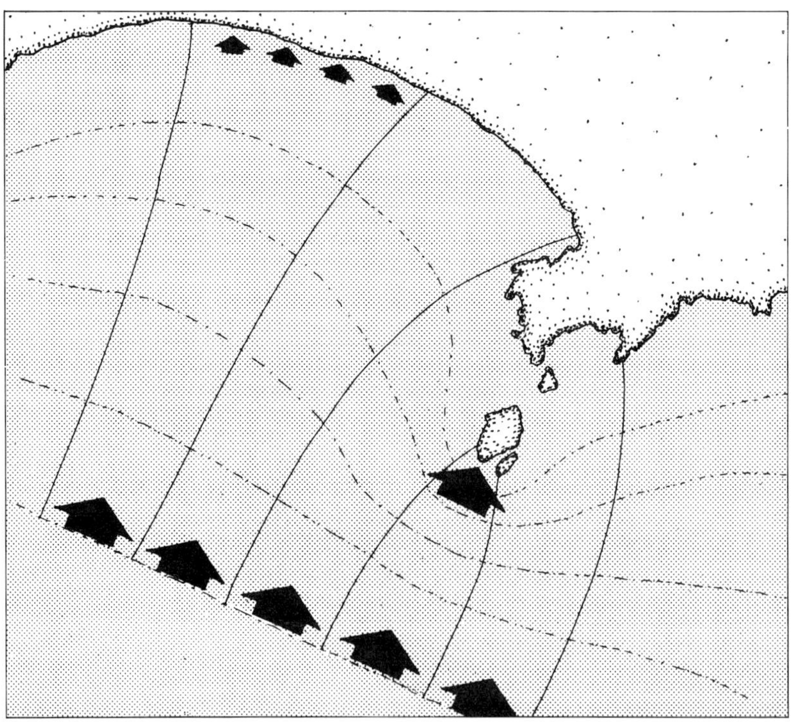

FIGURE 9–19
Refraction of incoming waves along a rugged shore. Contours of the sea floor are indicated by dashed lines. The energy possessed by the wave at the position first shown is equally distributed along the wave, but as the wave front is bent, the energy is spread over areas of varying sizes. Thus a much larger portion of the wave's energy is dissipated on the headlands than along an equal distance of bay shore.

where C = phase velocity, L = wave length, and T = period. When very long waves are considered or when a wave enters shallow water, the velocity is dependent on water depth. This change in velocity dependence begins to be noticeable when the water depth is less than $\frac{1}{20}$ of the wave length. The water particles begin to move in very flat ellipses and the velocity is given approximately by

$$C = \sqrt{gh}$$

where g is the acceleration due to gravity and h is the water depth.

When speaking of *wave velocity*, it is important to distinguish between the velocity of progression of the individual wave and the movement of the group of waves of which it is a part, called the *group velocity*. This distinction is necessary because any given individual wave tends to die out over a distance, but the group of waves representing the energy associated with the disturbance will be dissipated very slowly and may travel until it reaches land where it is dissipated in breakers. The group velocity is usually about one-half of the individual wave velocity (also called *phase velocity*).

Bending of Waves toward Shore

If the coast is oblique to the direction of wave movement, the waves are bent, or *refracted*, so that they tend to approach the shore perpendicularly (Fig. 9–18). The wave is slowed down in shallow water where subsurface water involved in the water motion is impeded by the bottom. That part of the wave that is moving in deeper water continues to travel at its original rate until it, too, reaches shallow water.

Because waves are not completely bent to be perpendicular to the shore, there is a component of motion in the water directed along the shore line, called *longshore current*. It is important in transporting sand and silt. Refraction is also of importance in that it focuses energy of the waves on the *headlands*, those parts of the shore that protrude into the ocean, rather than on bays or recessed parts of the shore. How this is accomplished is shown in Fig. 9–19. Energy of a wave is evenly distributed along the wave front while it is out at sea, but as it comes into shore and is refracted, that part of the wave's energy that enters a bay is spread out and distributed over a long shore line, and that part of its energy that is directed against the headlands is concentrated along a short portion of the shore (Fig. 9–19). For this reason headlands become eroded back by wave action much more rapidly than does the shore line as a whole.

Breakers

As a wave enters increasingly shallow water, water movement accompanying the wave motion is restricted. The circular paths of particle movement are forced to become flattened into flat ellipses and movement is reduced by drag on the bottom. The restriction of the wave form causes the wave height to increase and the velocity of the wave to decrease. This change takes place rapidly and at a specific depth for waves of any given characteristics. The velocity of the top part of the wave begins to exceed the forward velocity of the wave as a whole; the top outruns the rest of the wave and spills over in front of it, forming a *breaker* (Fig. 9–20). The wave form is lost; water is thrown onto shore, often with great force, as a rushing, turbulent mass. Where waves break there is a great deal of movement of sediment on the sea bottom. Out beyond the zone of breakers there is little more than a slight shifting of sediment back and forth by wave movement, but within the zone of breakers sediment is churned up into the water. A slight depression, or trough, may form just within this zone. Sediment from the depression is piled up on the landward side to build a *bar* and, eventually, in some instances, a *beach*.

Rip Currents

Once a wave breaks, the water rushes onto the beach, stops, reverses its direction, and begins its return to the ocean. The belief that water from a dissipated wave slips under the incoming wave and goes back out to sea as an *undertow* is a popular misconception. Recent studies have revealed that undertow is not common; neither is it responsible for large-scale movement of water. Instead, water from a wave finds its way back to sea by moving along close to shore and almost parallel to the shore line as part of the longshore

FIGURE 9–20
(*Above*) Breaking wave. The water in the foreground is returning to the ocean under the breaking wave as an undertow. The wave is just beginning to break at the right end.

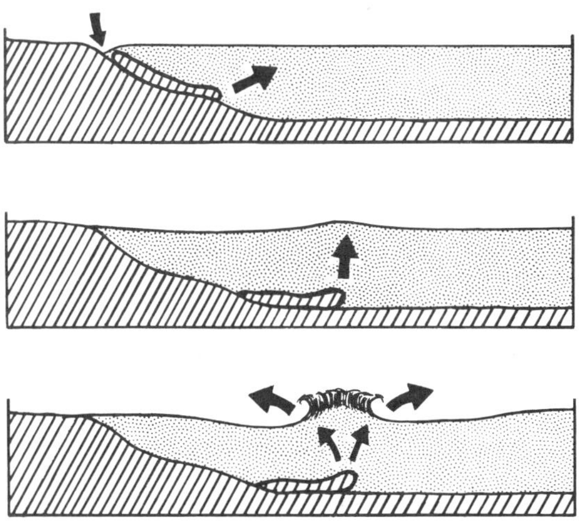

FIGURE 9–21
(*At right*) Schematic sections showing generation of waves associated with submarine landslides.
(after Kachadoorian, 1968)

current, until it reaches a current moving rapidly out to sea. This outgoing current is called a *rip current* (also a *rip tide* or a *sea puss*). Rip currents are probably responsible for most drownings attributed to undertow. They are rapid and forceful movements of water, certainly capable of knocking the unsuspecting wader off his feet. Rip currents are usually more or less confined to channels running out to sea perpendicular to the shore. The best way to escape one is to swim parallel to the shore across and out of the current, rather than to try to swim directly against it back to land. A rip current is fed almost continuously as each successive wave breaks, piling up more water along the shore.

TSUNAMI

Tsunami is a Japanese term that applies to any gravity wave system formed in the sea as a result of a large-scale, short-duration disturbance of the surface. Such waves are popularly known as *tidal waves,* but because they are not directly related to tidal effects the term is inappropriate. The effects of tsunami are such as to inspire great fear. The documented evidence from recent tsunami lend credence to the legends of ancient ones. They appear particularly mysterious because the waves travel across broad oceans from the place of origin, and the arrival of a wave is often preceded by no disturbance save the withdrawal of waters from the shore, followed by the arrival at and inundation of the shore by waves that reach anywhere from a few meters up to tens of meters above sea level.

Tsunami are generated by disturbances having a vertical component of movement that is dispersed under gravitational forces. Tsunami may be generated by shallow-focus earthquakes, by landslides into the ocean, by submarine slumping, and by volcanic eruptions. It was the tsunami formed during the eruption of Krakatoa that accounted for the large death toll following that gigantic explosion. Many of those deaths occurred hundreds, even thousands of kilometers away from the site of the eruption.

A suggested mechanism for the generation of tsunami by slumping is illustrated in Fig. 9–21. As the slump mass moves downslope, water is displaced, moving toward the site of the scar and away from the slump mass. A disturbance of the surface may be induced in this way or as a result of a sudden displacement in the bottom across a fault. When the disturbance is reasonably localized, the wave system that is set up resembles that produced when a rock is thrown into a pond; the front of the wave system advances out in all directions from the disturbance and the system consists of concentric rings of crests and troughs. The velocity of the wave front, C, is given by

$$C = \sqrt{gh}$$

where g is the acceleration due to gravity and h is the depth of the water. The waves may be refracted, but in the absence of boundaries to cause such breaks the individual waves retain their identity as they spread outward. The amplitude of the waves tends to decrease with time as the waves move farther away from the source in deep water. It is convenient to think of the energy of the wave as being spread more and more thinly over a larger area; therefore tsunami in midocean are such low-amplitude waves that they are not detected by ships, despite their velocity and size. But as the wave enters shallow water, the energy that was spread vertically through great depth becomes concentrated, resulting in increased amplitude. The local configuration of the shelf and shore line exerts great influence on the form the wave motion takes from place to place and on the height of the waves. Note, for example, the great range of heights of *runup* (the highest elevation reached by water along the shore) along the Hawaii shore during the 1960 tsunami following the Chilean earthquake. A high runup of 10.7 meters above mean lower low water was recorded at Hilo. Tidal gauges at Hilo recorded the oscillations of sea level that accompanied that disturbance (Fig. 9–22). Such records typically show a number of major oscillations followed by a gradual return to normal conditions over a period of a couple of days. Runup studies are important because they indicate the potential hazards along any given coast. In general, runups are higher along coasts with steep offshore slopes; runup is greatly reduced by broad, shallow continental shelves and fringing reefs.

Most tsunami have occurred in the Pacific Ocean and severe ones recur at a rate of about one every ten years. Because of the relative concentration of tsunami in the Pacific, a seismic sea-wave warning system has been set up with headquarters in Hawaii. Epicenters for earthquakes surrounding the Pacific can be determined within about 45 minutes. If the epicenter occurred in the sea, certain tide-measuring stations are alerted and any unusual wave action is reported. If a tsunami forms, an alert is then sent out to the nations surrounding the Pacific indicating arrival times. The arrival time for Hawaii from any point in the Pacific can be read from Fig. 9–23.

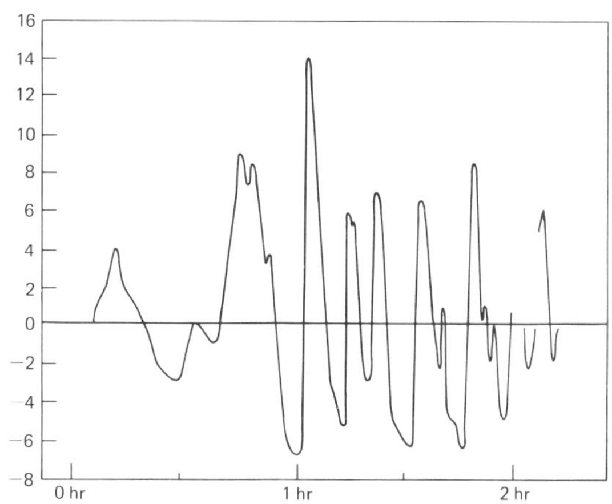

FIGURE 9–22 Water levels as observed at Hilo Bay, showing the initiation of the tsunami of May 23, 1960. (after Eaton, 1960)

TIDES

Tide is the name applied to the periodic rise and fall of sea level. Even in ancient times a relation between tides and the phases of the moon was recognized. As men learned to measure the *period* (time between successive stages), *range* (variation in level of the water surface), and variability of tides from time to time and from place to place, they came to recognize the complexity of the causes of the earth's tides. The two dominant factors are gravitational forces, due primarily to the attraction between the moon and sun and the oceans, and the hydrodynamic behavior of ocean waters confined by the shape of the ocean basin and subjected to disturbing forces that set the water in motion. Yet even today tidal observations cannot be completely explained by tidal theory. An appreciation of the variability of tides will be gained by study of the charts (Fig. 9–24) that show fluctuations of sea level at various coastal points over a period of time. Note that there are variations in period, range, and number of tides per day, as well as over longer spans of time.

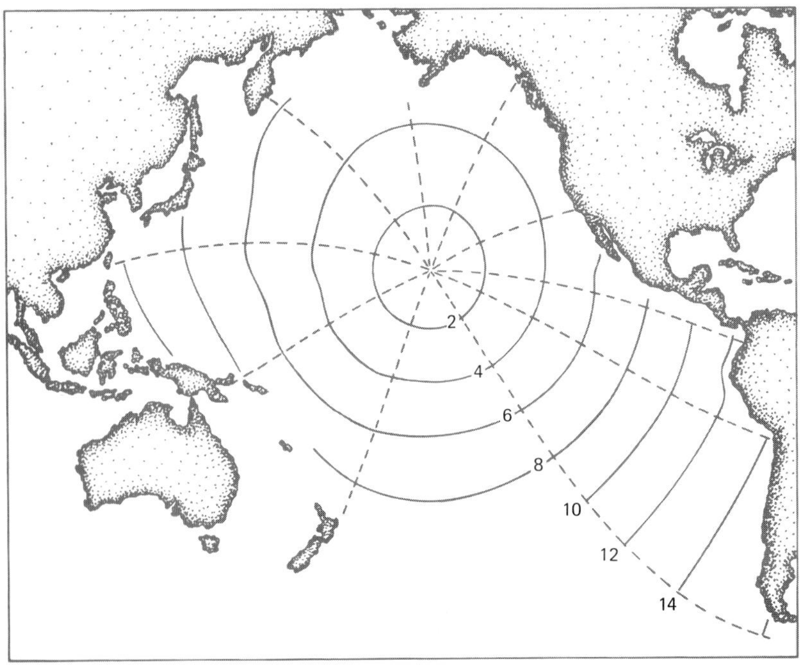

FIGURE 9–23
Travel time for a tsunami to reach Hawaii is shown by solid lines. Time is given in hours.

Effects of Gravitational Attraction

Pliny the Elder (A.D. 23–29) documented the relation between the tidal ranges and the phase and position of the moon. Nearly 1,600 years later, Sir Isaac Newton developed a theory that explained this relationship in terms of gravity.

The sun and moon are the two bodies of the solar system that exert great attraction on the surface water of the earth. Although it is 148.8 million kilometers away, the sun is such a massive body that it exerts nearly 167 times as much attractive force on earth as does the moon. Therefore it seems unlikely that the moon should be more closely linked to tidal phenomena. Yet this is the case, and the reason is that the attraction of the moon for water on opposite sides of the earth differs to a much greater extent than does that of the sun. Water on the side toward the sun or moon is 2,440 kilometers closer to it than water on the opposite side. The 2,440 kilometers makes a significant difference in the distance to the moon (382,200 kilometers) but only a very little difference in the much greater distance to the sun.

To understand how the earth's tides are affected by the gravitational attraction of the moon and sun, assume that the earth is a rotating and revolving sphere covered to a uniform depth with water. The moon and the earth revolve about a common center of gravity. If it were not for this revolution, the moon would be pulled to the earth. Similarly, if the moon and earth were held in fixed positions in relation to each other, one large bulge of water would extend out toward the moon. But in the case of a stable system of two bodies in revolution about a central point, the forces of attraction between the two bodies must be counterbalanced by opposed *centrifugal force*. The amount of the centrifugal force is constant and is directed in the opposite direction from the gravitational forces (Fig. 9–25). The gravitational forces on the earth vary from place to place, depending on the distance from the moon. Thus two tidal bulges occur on our model earth with a uniform hydrosphere—one on the side toward the moon, where the force of attraction exceeds the centrifugal forces, and a second on the opposite side, where the centrifugal forces exceed the force of attraction. It happens that the two forces are close but not equal in strength. This gives

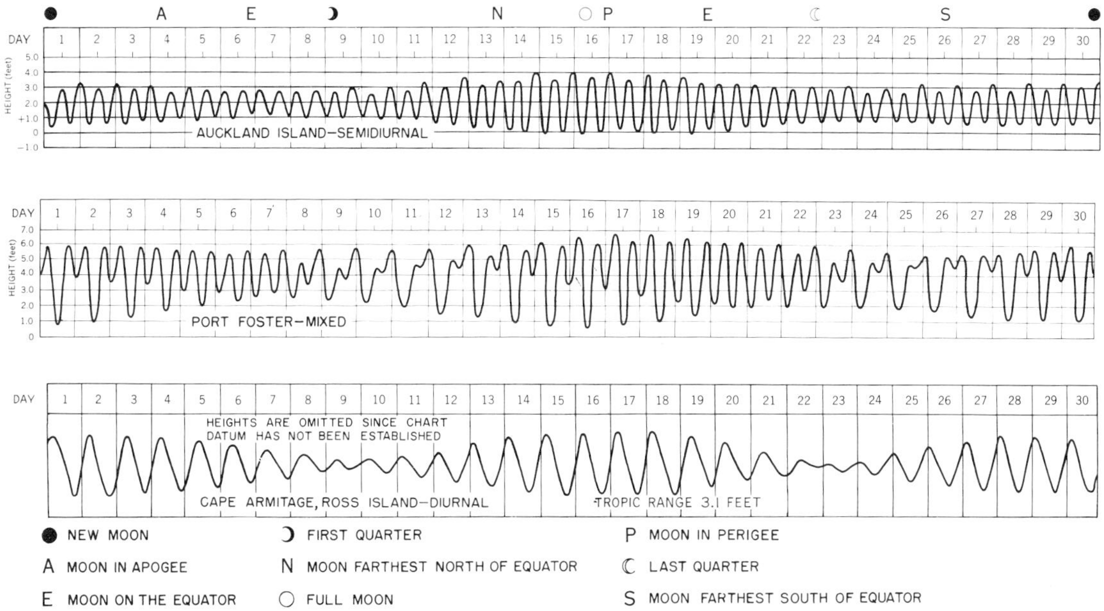

FIGURE 9–24
Examples of diurnal, semidiurnal, and mixed tides.
(from U.S. Navy Hydrographic Office, *Oceanographic Atlas of the Polar Seas*, 1958)

rise to a slightly higher bulge on the side toward the moon. The effects of the sun are similar in character, but slightly less than half as strong as those of the moon. At most times the sun and moon do not lie on a straight line with the earth, and the two effects are usually out of phase. Many components of observed tides can be predicted and explained in terms of the movements of the sun and moon.

Two high tides and two low tides should occur each day as a result of the earth's rotation about its axis. On most coasts the interval between high tides is 12 hours and 25 minutes, the slight difference being due to the revolution of the moon around the earth. Thus the high tides occur about 50 minutes later each day. These high tides are generated by the moon, but they do not normally coincide exactly with the passage of the moon overhead (that is, over a given meridian). Frictional drag usually causes the tide to arrive later than the passage of the moon; the interval is known as the *lunitidal interval*. It varies greatly from place to place, but for a given locality it is relatively constant.

When the sun and moon lie in line with the earth, as at the times of new and full moons, the attraction of the two bodies is cumulative, and unusually high and low tides, called *spring tides*,

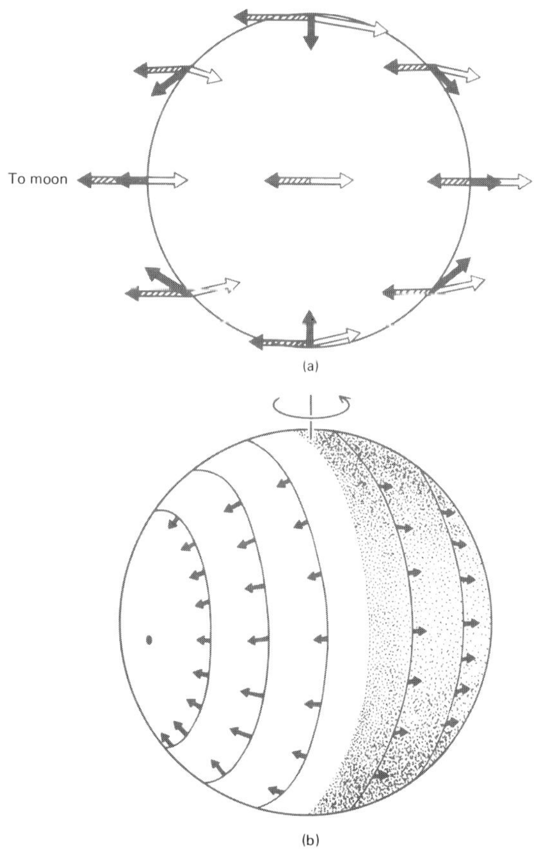

FIGURE 9–25
Tide-producing forces. (a) The direction and magnitude (indicated by length of black arrows) of the tide-generating force arise from a combination of the attraction of the moon (open arrows) and centrifugal forces (diagonal stripe arrows).
(after Defant, 1961)
(b) The horizontal tide-generating force as envisioned by George Darwin in 1898, after the upward and downward components of force shown in (a) are removed.

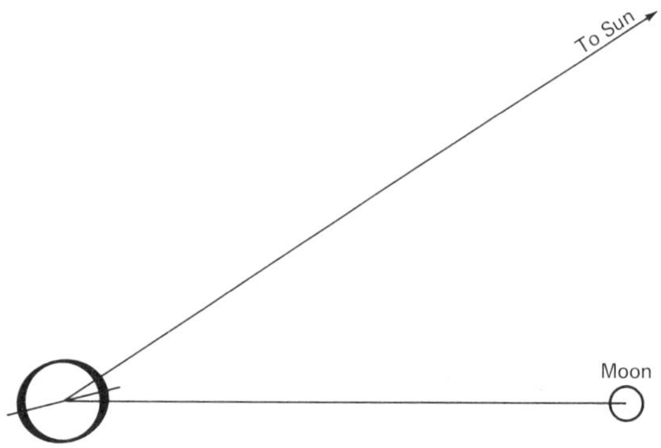

FIGURE 9–26
Conditions causing lagging tides (the high tide arrives after the meridian passage of the moon) are shown. Lines pointing directions to the sun and moon and high tides are indicated. Note that the line connecting the high tides lags behind and does not coincide with the line to the moon.

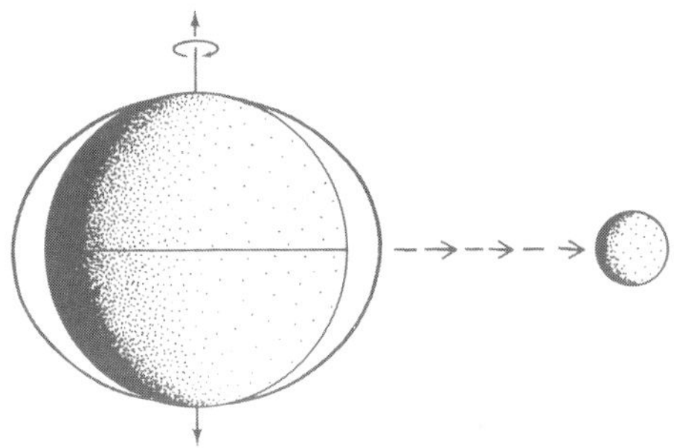

FIGURE 9–27
The moon is shown directly over the equator. Under this condition the height of the tide is essentially a function of latitude.

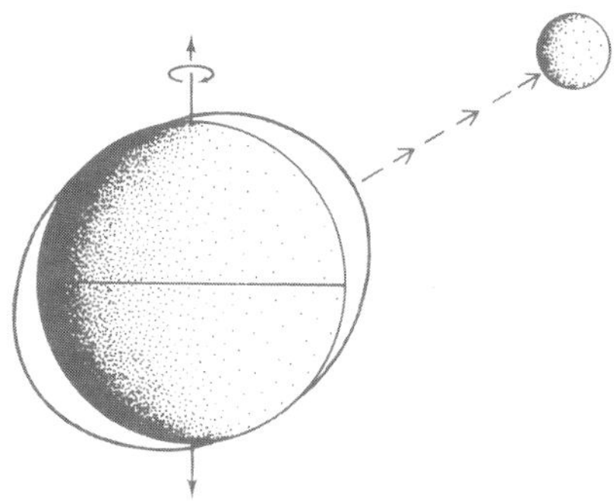

FIGURE 9–28
Here the moon is shown out of the plane of the equator. Under this condition there will be unequal tides at any given latitude. One of the tides will be much higher than the other.

result. The name has no relation to the season. When the sun and moon are located at right angles to the earth (*quadrature*), the tide-producing forces act at right angles, tending to cancel each other out and giving rise to unusually low tidal ranges, called *neap tides*. Both these conditions occur twice each month. When the sun and moon approach alignment with the earth, the resultant tidal bulge may arrive a little before the passage of the moon overhead, and the tide is said to be *priming*. After the passage of the moon out of line with the sun, the bulge will tend to be retarded, and the tide is said to be *lagging* (Fig. 9–26).

The orbit of the moon is inclined to the ecliptic (the path of the earth around the sun) by about 5°. The earth's orbit is inclined by $23\frac{1}{2}°$ to the plane of the equator. Thus the moon is over the earth's equator twice a month, but its path takes it as far as $28\frac{1}{2}°$ above and below the equator at times. When the moon is directly over the equator, the two daily tides are about equal in range (*equatorial tides*) (Fig. 9–27), but when the moon is far out of the plane of the equator great differences in the range of the two daily tides occur, as is indicated in Fig. 9–28. Such tides are called *tropical tides*.

The distance to the moon varies because the moon's path around the earth is elliptical. Gravitational attraction varies inversely with the square of the distance. Thus tides vary with the position of the moon in its orbit. The amount of variation due to this cause reaches about 20 percent. Similarly, a slight amount of tidal variation may be related to the varying distance to the sun as the earth moves in its elliptical orbit around it.

All five of these factors bring about variations in tides; in addition, each of these varies with a given period. At times, the effects of two or more of these factors are cumulative; at other times, they tend to cancel each other out.

Hydrodynamics of the Oceans

Many tidal observations cannot be reconciled with the gravitational theory. On the Pacific coast of North America, for example, where the direction of the coast line is essentially north-south, high tides sweep from south to north. High tides normally come to Cape Flattery in northern Washington three hours later than they do to San Diego in southern California. Similar wavelike motions, as if the tide were a long wave striking the shore at an angle, are found elsewhere and cannot be directly related to the rotations and revolutions of the earth, sun, and moon. In the Society Islands of the South Pacific high tides come at high noon and midnight, whereas low tides come at 6:00 A.M. and 6:00 P.M. Tidal behavior here is apparently related only to the attraction of the sun, the area being apparently totally immune to the attraction of the moon. Both of these effects can be explained in terms of *hydrodynamic* principles.

If a container of water is periodically tilted, *standing waves* will be set up. The size, period, and number of these waves depend on the size and shape of the container and the depth of the water. In a standing wave, certain points along the wave remain static while the other points oscillate. In a three-dimensional container the static points, called *nodes,* may be a line, or under certain circumstances a single point, about which the wave moves. As early as 1900 the oceans were divided into oscillating areas. This division was made in such a way that the observed differences in time of arrival of high and low water at opposite ends of an area would be about 6 or 12 hours. Since then the view of the oceans as complex systems of standing-wave oscillation has become well established.

The tidal phenomena along the Pacific coast and in the Society Islands can be explained by the concept of standing waves. The islands must lie close to a node of a standing wave whose period is about equal to that of the semidaily tides produced by the moon. Oscillating standing waves of the sort envisioned could not be simple standing waves because they are subjected to gravitational influences and exist in water on a rotating earth. The oscillating standing waves move in a more or less circular motion, called *amphidromic,* around the nodes. The motion can be likened to that of a liquid in a cup that is first tilted back and forth, creating a standing wave, and then swirled around so that the standing wave moves in a radial motion about the center of the cup. Radial lines, called *cotidal lines,* may be drawn from a node to connect points that experience high tides at the same time. These tides are not equal in range because they are at different distances from the node. The cotidal lines sweep counterclockwise in the Northern Hemisphere and clockwise in the Southern Hemisphere. One of the best documented instances is in the North Sea (Fig. 9–30).

Effects of Tides

Most points on the world's sea coasts are affected by tidal movements. Some typical tidal ranges are

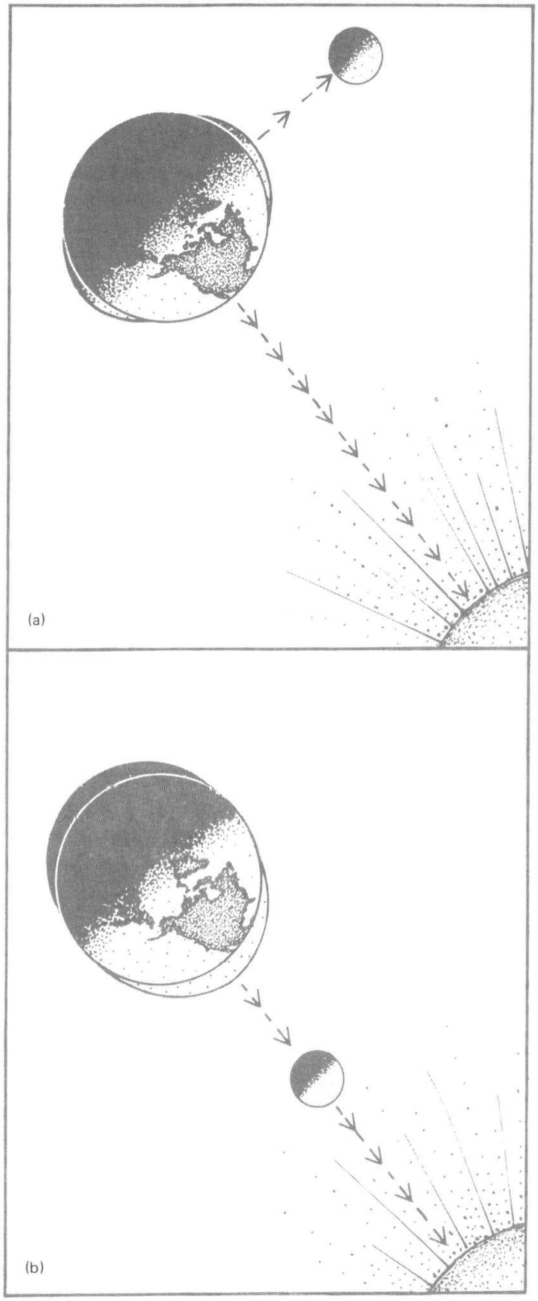

FIGURE 9–29

(*a*) Conditions depicted are those characteristic of quadrature when the moon is at right angles to the line connecting the sun and earth. At this time tides tend to have the least range.

(*b*) At conjunction the effects of the sun and moon on the tides are additive. Unusually high tides occur under these conditions.

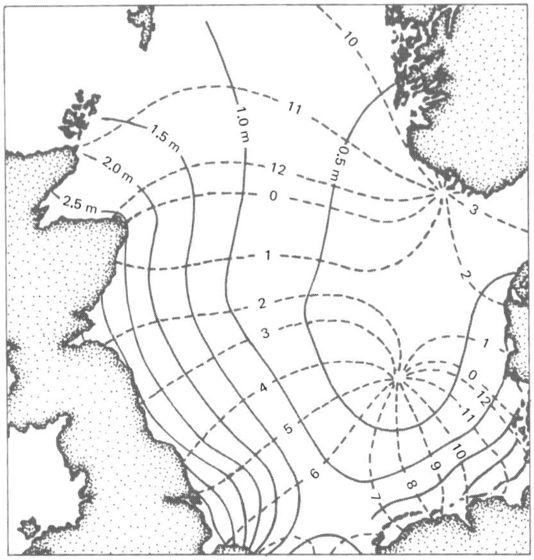

FIGURE 9–30
North Sea cotidal lines. The solid lines connect points that have equal tidal range. The dashed lines show successive times of arrival of high tides.
(after *Admiralty Manual of Tides*)

2.7 meters at Boston, 2 meters at New York, 0.75 meter at Key West and Galveston, and 7 meters at London. At certain locations, as in the Bay of Fundy, even higher ranges are known. The effects of tides in restricted portions of the coasts and in rivers along a coast having high tides may be pronounced. As the tide approaches the coast it creates a horizontal current, called the *flood,* which moves into bays and rivers and continues upstream great distances in some instances. The Hudson River, for example, is affected by tides as far as 208 kilometers upstream at Troy, where the tidal range of the river is 1.5 meters. In a number of rivers where tidal ranges are particularly high, the movement of the water takes the form of a turbulent wave, called a *tidal bore,* which moves with high velocity. The tidal bore formed on the Amazon River is 4.8 meters high and moves at a velocity of 22 kilometers per hour. Tidal currents are effective agents of marine erosion, keeping fine sediment in suspension and preventing its deposition. For this reason tidal bores help scour and shape the channels of the streams in which they are found.

The force of tidal attraction on the hydrosphere is not confined to the surface of the oceans. Water at all depths experiences the pull, and it is likely that deep tidal currents result. The principal effects of these currents are thought to occur where the deep waters off the continental slopes are pulled landward onto the continental shelf. These tidal currents may be responsible for the shifting and grading of the sea-floor sediments near margins of the shelves. The incoming tidal currents would bring with them little or no sediment at all, but as they reversed direction they could carry away the finer sediments. This may explain why sediments at the outer edge of the continental shelf are often coarser than those closer to the coast.

SEA LEVEL

The level of the ocean's surface varies from one place to another at any given time and it fluctuates with time at any given place. The surface is a free boundary that can assume different forms in response to external and internal forces applied to it. Only rarely is the sea surface quiet; therefore determination of sea level is not as simple a task as it might appear. In practice, sea levels are determined at any given point by use of a tidal gauge. One type of gauge is essentially a long pipe with openings in it mounted vertically. As water levels in the pipe change, they can be read from calibrations on the pipe or they can be recorded mechanically. Other gauges are pressure-sensitive devices mounted on the sea floor. Pressure varies as a function of water depth.

Sea level at a point undergoes a number of periodic and short-term fluctuations. Among these are (1) waves and swell caused by wind blowing across the water (2) tidal variations that will vary systematically through the day and through the year and (3) variations due to atmospheric pressure. Sea level is lowered when a high-pressure system in the atmosphere is located over a station.

These short-term and periodic variations may be thought of as oscillations on either side of *mean sea level.* Most wave effects can be evened out by taking the daily average of observations, and tidal effects can be removed by averaging the level over a year's time.

Even when effects of wave action, tides, and atmospheric pressure variations are taken into account, mean sea level is not a level surface.

Ocean water is held to the earth by gravitational attraction of the water to the earth. The force of this attraction is not uniform throughout the earth. It varies as a result of differences in density of the earth's crustal rocks and because the level of the crust varies. Water in the ocean basin is actually pulled toward the continents. If the irregular distribution of mass in the earth's crust were disregarded, if the earth were uniformly covered by sea water, and if there were no tides, atmosphere, or waves, then the surface of the sea would have the shape of a rotational ellipsoid. When gravitational effects due to the crust are considered, the irregular sea-level surface resulting from these is called the *geoid*. The level of the geoid is below that of the rotational ellipsoid in the ocean basins and would be above it over continental areas if a system of sea-level canals were available in which it could be measured.

Sea level is important as a reference level at the present time because most topography, both submarine and continental, is described in terms of elevations relative to the mean sea level. The system is set up with reference to mean sea level at certain selected points along the coast, and these are related to one another and to other points along the coast by means of very careful surveying (leveling) over land.

Many ancient sedimentary rocks were formed along coasts near sea level, and the type of environment in which they were formed can be recognized easily in terms of its position relative to sea level (e.g., continental shelf, marine swamp, etc.). The position of sea level relative to the land has undergone long-term variations due to several causes:

1. The crust of the earth is unstable and has been deformed, causing uplifts in some places, subsidence in others. Uplift along a coast would obviously change sea level, at least locally. Also variations in the shape of the ocean basin caused by deformations would result in changes in sea level.
2. Changes have occurred in the volume of water in the ocean as a result of the gradual addition of water to the earth's surface through volcanic activity (estimated by Penck at a rate to raise sea level about 3 millimeters per thousand years) and through changes in the balance between ice on land and water. Changes in sea level due to ice-water balance have been particularly important during the last million years of the earth's history. During times of maximum glacial advance it is estimated that an additional 40 million cubic kilometers of ice was stored on land, enough to lower sea level throughout the world more than 100 meters. The maximum range of sea level fluctuation during the Pleistocene ice ages is estimated at about 150 meters (Fig. 9–31), and much of this range has been realized in the last 15,000 years; therefore variations in sea level have very much influenced the development of modern coastal features and the submarine topography of the continental shelves.
3. Sediment is continually carried into the ocean basins from land, and this would have some effect in raising the level of the sea at a rate estimated by Penck at about 30 millimeters per thousand years.

In summary, sea level is influenced by many factors. Some are of such short duration that we can correct easily for them; others can be determined only by inference from geological data. Sea level changes with time because of the following factors:

1. Short-term variations:
   a. Wave action.
   b. Tides.
   c. Atmospheric pressure.
   d. Long-duration winds toward or away from coasts.
2. Long-term variations:
   a. Increase of new water.

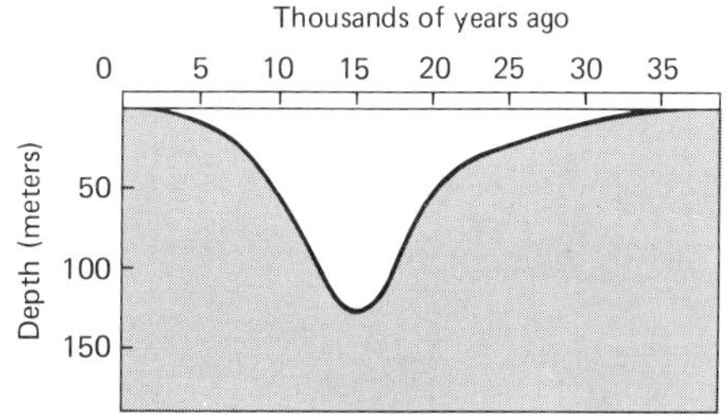

FIGURE 9–31
Sea-level variation with time for the Atlantic continental shelf.
(simplified after Milliman and Emery, 1968)

b. Changes in size or shape of the ocean basin.

c. Uplift or downwarping of crust.

d. Deposition of sediment in the ocean.

e. Water-ice balance.

RESOURCES FROM THE SEA

Food supplies and ocean water itself are undoubtedly the most important potential resources of the sea: These resources will have to be developed and used unless the present rate of population growth is either leveled off or actually reversed. The present rates of food production are destined to become vastly expanded and probably accelerated. The most abundant potential food resources are the free-floating and swimming microorganisms such as plankton and nekton of the upper layers of the sea. It has been estimated that these could be collected and dried to supply the annual animal protein needs of the world. Unfortunately, the product is not very tasty. Kelp is already used in large quantities as a feed for cattle.

In some areas it is already possible to produce fresh water from sea water as cheaply as it is to bring in surface fresh water or ground water. Processes are available by which large quantities of sea water can be evaporated to produce fresh water. The problem is to supply the necessary energy cheaply enough. Sea water will probably become an important source of water in many coastal areas, as it is already in southern California.

When sea water is evaporated, the salts in the water are concentrated and eventually precipitated. Sodium chloride has long been produced in evaporating basins where sea water is introduced and allowed to evaporate by solar radiation. Not all the salts come out of solution at the same time; thus it is possible to avoid the undesirable result of having epsom salt and calcium carbonate precipitated at the same time sodium chloride is. At the present time, bromine and magnesium are being produced commercially from sea water. Many other salts are present and these would probably become important by-products of the production of large quantities of fresh water from sea water.

Beach materials have long been mined in many parts of the world. Insoluble minerals that do not readily break down mechanically are most likely to occur in significant concentrations. Among the most important of such resources are gold, cassiterite (tin ore), magnetite (iron), ilmenite (titanium), scheelite (tungsten), and platinum. Diamonds have been found offshore in South West Africa.

Shells are used widely as road metal and they can be used as aggregate for cement in place of crushed stone; phosphorus occurs in nodules formed in shallow water; and barium sulfate concretions are present in significant quantities. Beach deposits may be expected to depths of several hundred meters on the continental shelves, and some of these surficial deposits may have moved into much deeper water.

Manganese nodules are known in many places on the sea floor in water of great depth. They were discovered by the Challenger Expedition in 1873, but only recently have plans been made to mine them in quantity. The manganese appears to have been removed from the sea water and deposited on a nucleus composed of various rock and mineral types in such a way as to build up layer on layer of the mineral. Both biological and chemical processes have been suggested as a probable cause of the removal of manganese from solution.

Additional important mineral wealth lies buried within marine sediments. The continental shelves are already known to be the sites of large quantities of oil and gas. Sixteen percent of all oil and gas production in 1966 was from the shelves, yet subsurface exploration of the shelves is only starting.

CHAPTER OUTLINE

TERMS

amphidromic
bar
beach
breaker
capillary waves
conservative constituents
Coriolis effect
cotidal lines
crest
equatorial tides
fetch
flood (tide)
geoid
geostrophic currents
group velocity
gyres
headlands
lagging tide
longshore current
lunitidal interval
monsoon season
neap tides
nodes (tidal)
nonconservative constituents
period
phase velocity
photic zone
phytoplankton
priming tides
progressive waves
quadrature
range (tide)
rip current (rip tide, or sea puss)
runup
salinity
spring tides
standing waves
swells
thermohaline circulation
tidal bore
tidal waves
tide
tropical tides
trough
tsunami
undertow
wave frequency
wave height
wave length
wave velocity
waves
west-wind drift
wind waves

REVIEW QUESTIONS

1. For what reasons would you expect that the composition of sea water in early Paleozoic seas was different from that in modern seas?
2. Compare the composition of normal sea water with the salts precipitated from sea water as sediments described in Chapter 7. Describe the conditions that favor accumulation of thick deposits of salt-sequence sediments.
3. Why is the photic zone so important in determining the gaseous composition of sea water and the lower atmosphere?
4. What is the approximate density of sea water at 15°C and 34 ‰ salinity? Refer to Fig. 9–5.
5. Describe the effects on water movement if a wind blows toward the northwest along a coast line that trends north-south in the Southern Hemisphere.
6. Examine the map of surface currents (Fig. 9–9). Which currents are most consistent with the idea that wind causes them?
7. Compare the Indian Ocean with the Gulf of Mexico. Can you suggest reasons why a monsoon climate is not found in the southeastern United States.
8. What factors contribute to causing the calm area known as the Sargasso Sea in the Atlantic?
9. How could you identify and trace the movement of a water mass formed in the Weddell Sea?
10. What zoning regulations would you suggest for any coastal arc subject to tsunami in order to protect lives and property?

SUGGESTED READINGS

Bascom, Willard, 1964, Waves and Beaches (Anchor Books): New York, Doubleday, 267 p.

Defant, A., 1958, Ebb and Flow, the Tides of Earth, Air, and Water: Ann Arbor, Mich., The University of Michigan Press, 121 p.

Fox, W. T., 1969, Computer-oriented laboratory exercises for geology and oceanography: Jour. Geol. Educ., v. 17, no. 4, p. 125–134.

Gross, M. G., 1967, Oceanography, 2nd ed.: Columbus, Ohio, Charles E. Merrill, 135 p.

Milliman, J. D., and **Emery, K. O.,** 1968, Sea levels during the past 35,000 years: Science, v. 162, p. 1121–1123.

Pickard, G. L., 1968, Descriptive Physical Oceanography: New York, Pergamon Press, 200 p.

Weyl, P. K., 1970, Oceanography, an Introduction to the Marine Environment: New York, John Wiley, 535 p.

10 ENVIRONMENTS AT THE MARGINS OF THE OCEANS

10 ENVIRONMENTS AT THE MARGINS OF THE OCEANS

Marine environments are important not only because they are the principal sites of formation of modern sediments, but also because they have unique potential to reveal to us the geologic past. Many sedimentary rocks now exposed on continents were once deposited in oceans or in the zone of transition between oceans and continents. These sediments and sedimentary rocks have individual characteristics that reflect the influence of the environment in which they were deposited. Recognition that certain characteristics of a rock reflect some particular environment is one of the important keys to unraveling the history represented by the stratigraphic successions—the sequences of layers of rock deposited in ancient seas. Just as sediments deposited from air, water, or a glacier differ from one another in textural, compositional, and structural characteristics, so do other, more subtle environmental conditions leave their imprint on sediment.

Fossil faunas found in a sediment provide valuable clues to such facts about the water in which they were deposited as temperature, depth, oxygen content, and penetration by light. Many sediments are composed largely of fossils or calcium carbonate secreted by animals. Fossils are almost entirely absent in other sediments. The type of water movement is indicated by the presence or absence of certain structures. Deposition in a quiet lake is different from the turbulent flow of sediment set off by a submarine landslide; likewise, it is different from deposition on a rocky coast where wave action breaks up the sea cliff, or on a low shore where currents sweep in sand and silt to be deposited on offshore bars. Other environmental factors governing deposition include the extent and configuration of the environment, the climate of the region, and the type of sediment supplied to the environment.

THE SHORE—PROCESSES AND ENVIRONMENTS

A great variety of environments are found where the land and the sea meet, and these margins of the sea are very important and interesting to us because they are places where land forms are being modified, where we can study the processes of modification in progress. Variations in the relative position of the land and sea level are responsible for much of the variety of the coasts. Sea level has fluctuated as the balance between ice and water has shifted during the ice ages of the last million years; in addition, the land itself has been uplifted in some areas and has subsided in others. These movements have resulted in the submerging of some land forms that were initially formed under subaerial conditions and the emerging of some portions of what was the sea floor.

The following terms are applied to the margin of the sea:

Shore line. The line along which the sea and the land meet.

Shore. The zone between mean low tide and the landward edge of wave-transported sand.

Coast. The broad zone directly landward from the shore.

Although our definition allows only a vague inner limit for the coasts, it would certainly include at least that region near the shore in which land forms directly attributable to marine erosion or deposition are found.

It has long been recognized that some coasts are rising or have recently risen relative to sea level, and these are called *emergent coasts;* others are *submerged.* Early classifications of coasts attempted to use submergence and emergence as a basis for classification, but it is often difficult to establish criteria by which these two can be distinguished. Thus it is advisable to use a classification that is based on characteristics that can be recognized readily.

When land near the continental margin is inundated, the land forms along the coast are, at least initially, those formed under subaerial conditions, and they have shapes determined by the nature of the processes by which they developed. Each of the primary agents of gradation produces certain characteristic land forms. For example, running water produces stream systems with various types of patterns (Fig. 14–29); valley glaciers tend to streamline and deepen old stream valleys and give them a U-shaped profile; deposits comprised of debris laid down at the edge of glaciers are characteristic land forms in glaciated regions; topography with enclosed depressions, sinks, and caverns form where groundwater solution has been effective; and sand dunes reflect the power of the wind to shape the land.

TABLE 10–1
F. P. Shepard's (1963) classification of shore lines and coasts.

I. Primary shore lines and coasts. Configuration due to the sea coming to rest against a land form shaped by terrestrial agencies rather than marine agencies.

- A. Land Erosion Coasts. Coasts along which the land forms were developed under subaerial erosion followed by partial inundation by the sea.
  1. Ria coasts (drowned river valleys).
  2. Drowned glacial erosion coasts (e.g., fiords, glacial troughs).
  3. Drowned karst topography.
- B. Subaerial Deposition Coasts.
  1. River deposition coasts. Where rivers have extended the shore line.
     - a. deltaic coasts
     - b. alluvial-plain coasts.
  2. Glacial deposition coasts.
  3. Wind deposition coasts.
  4. Landslide coasts.
- C. Volcanic Coasts
  1. Lava-flow coasts.
  2. Tephra coasts. Where the volcanic products are fragmental.
  3. Volcanic collapse or explosion coasts.
- D. Shaped by Diastrophic Movements.
  1. Fault coasts.
  2. Fold coasts.
  3. Sedimentary extrusions such as salt domes and mud lumps.

II. Secondary coasts. Those coasts shaped primarily by marine agencies or by marine organisms.

- A. Wave Erosion Coasts.
  1. Wave-straightened cliffs. Bordered by a gently inclined sea floor.
  2. Made irregular by wave erosion.
- B. Marine Deposition Coasts.
  1. Barrier coasts (e.g., barrier beaches, barrier islands, barrier spits, and bays created by spits that have enclosed bays).
  2. Cuspate forelands. Coasts with large projecting points with cusp shape.
  3. Beach plains. Sand plains differing from barriers by having no lagoon inside.
  4. Mud flats or salt marshes.
- C. Coasts Built by Organisms.
  1. Coral-reef coasts.
  2. Serpulid-reef coasts. Where cementing of worm tubes onto rocks or beaches has built out the coasts.
  3. Oyster-reef coasts.
  4. Mangrove coasts.
  5. Marsh-grass coasts.

Regions where any one of these types of subaerial gradation has been important in developing land forms may be submerged; thus the coastal land forms become a composite of those due to the sea's presence superimposed on those formed under other conditions. It is usually easy to differentiate land forms produced by subaerial erosion from those produced by subaerial deposition for each of the geomorphic agents (wind, streams, and the like).

Igneous activity and crustal deformation are operative along many coasts. Volcanism is concentrated in the ocean basins and around their perimeters. Many of the islands of the Pacific are volcanoes, as are some islands in the Mid-Atlantic. Lava flows, pyroclastic cones, and collapse and explosion calderas (see Chapter 3) are land forms associated with volcanic activity and volcanically active coasts.

Much of the emergence and submergence along coasts is due to crustal deformation or *diastrophism.* Whether land forms directly attributable to diastrophism are found depends on whether structural features were produced during the diastrophism that would influence land-form development. Fault scarps, folds, tilted strata, and fractures all are found along coasts (Fig. 10–1). The expression of such features depends in great degree on the type of rock, degree of consolidation, and degree of homogeneity of the rocks in which the structural features are developed. For example, folds in a sequence of alternating resistant and weak sedimentary layers will probably result in coastal forms quite different from folds in a metamorphic or more homogeneous rock mass.

Processes of marine gradation, erosion, and deposition begin to modify subaerially produced

FIGURE 10–1
Shore lines along which structure of the rocks influences the shape of the shore line. Broad, open folds control this Canadian shore line (top, left).
(courtesy of the Royal Canadian Air Force)

Strata are nearly horizontal at the chalk cliffs (White Cliffs) of Dover (top, right). The wave-cut cliff and a broad terrace are prominent (middle, left). Strata are almost vertical, but they have been eroded to a nearly flat wave-cut terrace that is now high above sea level (middle right). Wave action has cut a nearly flat terrace across steeply dipping strata (bottom, right).
(Crown Copyright)

land forms and volcanic or diastrophic features as soon as any of these come into contact with the sea. Submergence is usually a very slow process, and modification is usually of a slow, progressive character. The imprint of the processes of wave erosion and marine deposition is imparted to the land forms, and along many coasts shaping by marine agencies dominates. Such coasts may be subdivided into three groups, depending on which marine process dominates: (1) marine erosion; (2) marine deposition; or (3) work of marine organisms or plants.

MECHANICS OF WAVE EROSION

Rock and sediment exposed along coasts are undergoing an almost constant attack by waves where the water is deep enough for the waves to reach shore before they break. The effectiveness of wave action as a process of erosion is particularly evident when we can follow the destruction of land features. Several new volcanoes have risen from the sea in recent years. A new cone was formed at the edge of the volcano Fayal in the Azores in 1957. Cinder cones several hundred meters high have been destroyed or greatly modified by wave action in only a few weeks' time after the volcanic eruption subsided. Wave erosion is rapid in such cases because the volcanic debris is loose, granular, and readily eroded and removed.

As the swell or wind-driven waves approach the shore line, a number of changes in the waves occur. The direction of the wave front is refracted as the wave reaches shallow water, the wave length decreases, the speed of the wave decreases, the orbital paths of water particles are flattened into ellipses, and the wave becomes asymmetrical and steeper until it breaks. Breaking of waves is due to both their increased steepness and the drag of the bottom. Many breakers form when the water depth is about equal to the wave height. The breaker may plunge or spill forward in a froth of foam. After the break occurs, the wave becomes a wave of *translation*, called *swash*, moving almost straight up the beach. Erosion results where the breaker forms. The water digs several centimeters into the bottom sediment where the water plunges forward, material is rolled and sorted in the swash, and some particles are brought into suspension. Where the wave breaks against a cliff, erosion is a result of pressure and impact. Where erosion is active, waves are able to break down even the most solid rocks. Carrying the sand, pebbles, or even blocks that form rocky shores, the waves rapidly break down loosely consolidated or weathered rocks. Weak rock units are eroded first, sometimes undermining the more resistant units, but the impact of large boulders that may be moved or even thrown by large waves is capable of breaking up even the most resistant granites and massive sedimentary rock units.

Erosion is most rapid when a supply of "tools" for the water to use in cutting into the land exists, but the force of the water alone is enough to be effective in dislodging fractured blocks. Water is forced into the fractures and becomes highly compressed by the thrust of the water behind it. The amount of this pressure is greater than is generally recognized. A moderate-sized wave 1.5 to 3 meters high can exert pressures from 24,000 to 48,000 kilograms per square meter against the rocks exposed where it breaks. This is enough to extend or enlarge pre-existing fractures and to dislodge any loose blocks.

Other processes of erosion act in combination with impact of "tools" and water. When rock fragments or even sand is present, abrasion occurs. Solution of some rock types and cementing materials, chemical and biological weathering, and the alternate wetting and drying caused by sea spray are weathering processes indirectly due to wave action, and they serve to loosen the rock and make it more open to erosion.

The effectiveness of wave action as an erosion process is closely connected with the size of the waves, where they break, and the type of rock or sediment present along the coast. Larger waves possess more energy, they are capable of picking up larger "tools," and they are capable of causing more erosion. Where a wave breaks depends on the water depth and the wave length of the wave. Thus erosion of solid rocks along a coast will be most pronounced where deep water comes in close enough to the shore so that the largest waves actually break against or near the exposed rock.

Waves are also effective in stirring up the sediment on the bottom in areas where unconsol-

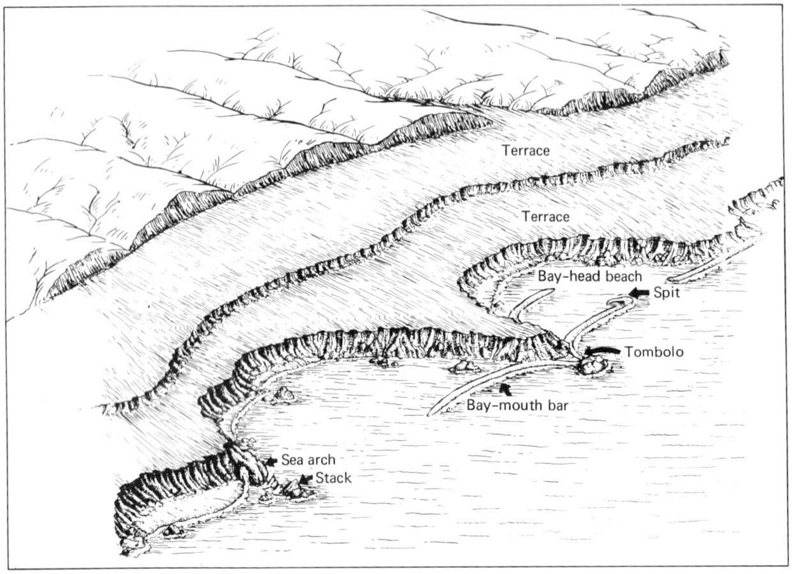

FIGURE 10–2
Characteristic physiographic features found along a rugged coast.

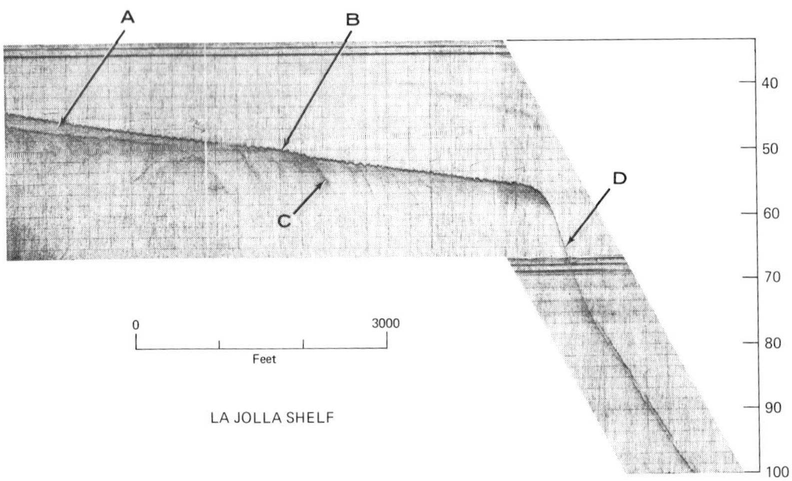

FIGURE 10–3
Sonoprobe record shows a lens of shelf sediment (A) overlying a bedrock surface (B) that appears to crop out and remain at or near the surface on the outer shelf; dipping bedrock has been truncated by wave erosion during lowered Pleistocene sea level. Also shown are west-dipping beds on the seaward flank of the anticline (C) and a steep upper slope apparently having little or no sediment cover (D).

idated sediment rather than solid rock lies along the shore. Waves in such cases may be effective in breaking up loosely consolidated layers of mud and silt on the bottom, and they entrain sand and finer sediment, bringing it into suspension so that it can be moved by currents present along the shore.

FEATURES FORMED BY WAVE EROSION

Erosion is the dominant process along the coast of New England and the western United States. Wave action is gradually gnawing away at the land, reducing it to the level of the seas. The rate is not always slow. On the south shore of Nantucket Island the shore retreats as much as 1.5 or 2 meters per year. Wave action has undercut the sea cliffs in California and the chalk cliffs of Dover in England (Fig. 10–1), bringing about large landslides and slumps, the products of which are broken up and graded by further wave action.

Cliffs and Terraces

Wave-cut cliffs and *terraces* are forms usually found where erosion is the dominant shore process. Almost all sea cliffs are steep, and many are hundreds of meters high. A sudden break in slope usually occurs at the foot of the cliff. The sea floor is nearly flat or gently sloping seaward (Figs. 10–2 and 10–3). This platform or terrace may be partially covered by the rocks that have been dislodged from the cliff. It is this material the sea uses for tools to further undercut the cliff. Sand can always be moved by the incoming waves, and in moderate waves cobbles and pebbles can be picked up and hurled. During storms even the largest boulders may be smashed against the base of the cliff. The effect of the breaking waves is negligible below the surface of the water, so the cliff stops abruptly just below water level. The width of the wave-cut terrace is limited by the fact that the water is shallow. As the terrace becomes wider, more and more of the wave energy is dissipated before the wave reaches the base of the cliff, so its ability to erode the cliff is reduced. Of course, if sea level is rising even slightly in relation to the land over a long period of time, the terrace will become wide because the part of the terrace formed early is increasingly submerged below the level of the wave action, and waves continue to break near the shore line.

Stacks, Arches, Sea Caves, Notches

Irregularities in the coast develop as a sea cliff is eroded back. These may be due to differences in the concentration of wave energy, to differences in the hardness of the rock, or to the structures in the units. Retreat of the cliff may leave some columnar masses of rock standing isolated as islands just off the shore. Because these frequently resemble smoke stacks sticking out of the water, they are called *stacks* (Fig. 10–4), although they are not necessarily small or round. *Sea caves* and *arches* (Fig. 10–5) are likely to develop when the rocks exposed in the cliff are stratified sediments of varying hardness. The weaker units are undercut rapidly, sometimes leaving the resistant capping layers as a natural bridge or arch over them. Such features, like the stacks, are transient. They will disappear and perhaps be replaced by similar features farther inland at a later date. Even if the cliff is composed of some massive, resistant rock type, the wave action may produce a *notch* near the base (Figs. 10–6 and 10–7). Such a feature is the initial stage of undercutting; the rock above the notch may eventually break away and fall into the sea.

Beaches of sand or larger size particles are often found along coasts that are being eroded by wave action. Such beaches are usually transient features, being well developed during summer but often completely absent in winter when larger waves cause greater transportation. It is during storms and in the winter months when the beach has been removed that erosion on the cliff is most active.

FIGURE 10–4 Sea Stacks. (Crown copyright)

FIGURE 10–5
Sea arch formed by wave erosion along the West Coast of the United States. Note the smooth surface above the cave. This is a remnant of an elevated sea terrace.

FIGURE 10–6
(*At left*) Marine terrace formed just above mean sea level. Fractures in the rock have also acted to control erosional form. The coast is just north of Auckland, New Zealand.

FIGURE 10–7
(*Below*) Tidal terrace showing Lithothamnium encrustation and nip (notch) above and below terrace.
(from Kaye, 1959)

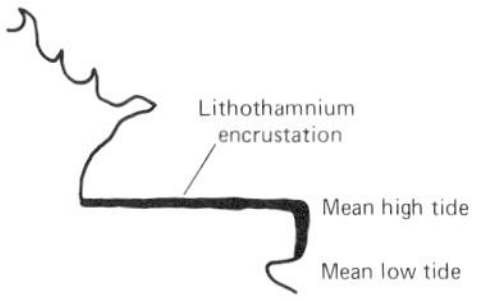

MOVEMENT OF SEDIMENT IN SHALLOW WATER

An incoming wave is capable of moving sediment as soon as the water depth is about half its wave length, but until the wave begins to break this capacity does not account for much erosion. The agitation and shifting associated with waves offshore does, however, separate sediment according to sizes. Long-period sea swells can affect the bottom at depths of hundreds of meters in this manner.

Most movement of sediment is accomplished by breaking waves and longshore currents. Even

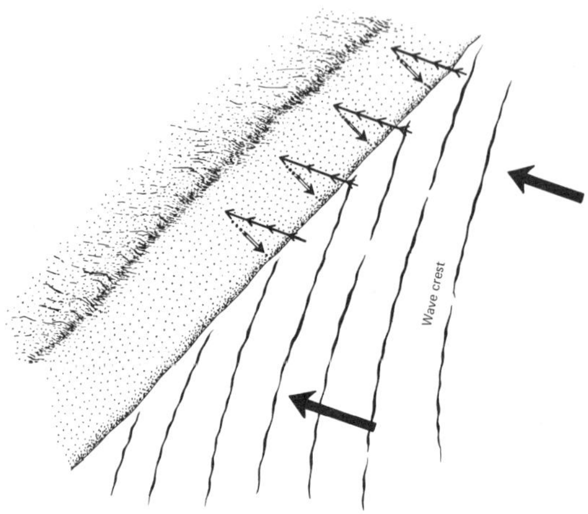

FIGURE 10–8
Beach drifting. The path of sand is shown; solid arrows indicate the path of sand carried onto the beach by incoming waves; dotted line shows the path of sand back down the beach.

after refraction, incoming waves hit the shore at a slight angle. Pebbles or grains of sand are rolled or tossed along in this water that is thrown on the beach obliquely. When the water returns, they are moved directly down the slope of the beach, perpendicular to the shore. Thus with each new wave the sediment moves along the beach (Fig. 10–8); this movement is called *beach drifting*. As a result of beach drifting beaches can be extended, as is shown by the photos taken at Little Egg Bay, N.J. (Fig. 10–9); inlets may be closed; and beaches may be built completely across bay mouths.

Waves are capable of moving the finer sizes of sediment in suspension, and they may carry these in the longshore current until the velocity of the current diminishes.

MARINE DEPOSITS

Beaches

Beaches are transient features; the sandy beach that appears so permanent in the summertime may be reduced to a narrow strip covered by coarse pebbles in the winter, or perhaps bedrock may be exposed. Most beaches are made up of sand-sized materials that are easily moved by waves, and they can be transported in moderate currents. Thus when storm waves reach great heights beaches may be completely washed away or covered with coarse debris. But storms are not the only mechanism for the movement and destruction of beaches. Any upset in the supply of sand to the beach or in the currents moving along the beach will influence its growth or depletion. Coastal inhabitants are well aware of some of the factors involved, and attempts are often made to control erosion and deposition along the shore by building jetties and piers or breakwaters. Care should be taken in planning such obstructions, however, because jetties occasionally introduce side effects that cause more damage than they prevent. For example, building a jetty may protect one segment of the shore line from erosion, but other parts may as a result become subjected to increased erosion. Or breakwaters placed along the coast to reduce wave action and to offer protection to a harbor may bring about deposition of sand in the harbor and clog the channels.

Beach Profiles

Most beaches (Fig. 10–10) may be divided into the following three parts:

Offshore. From the position of low tide seaward.

Foreshore. The zone between low tide and the point where the beach becomes either horizontal or slopes landward.

Backshore (berm). That part of the shore that is level or sloping landward.

The profile of a beach changes from hour to hour and particularly from one season to another season. The short-term variations are largely due to changes in sea level related to tidal conditions and to constant fluctuations in strength and direction of the wind that are rapidly reflected in wave size and beach drift. Seasonal variations arise from protracted periods of storm activity, storm waves, and the balance between sediment supplied to the shore and rates of beach drift. Common features of a beach profile are illustrated in Fig. 10–11. This profile is more typical of a coast like that in the western United States than of the Gulf or Atlantic coast south of Long Island, where the sea cliff is absent and a backshore consisting

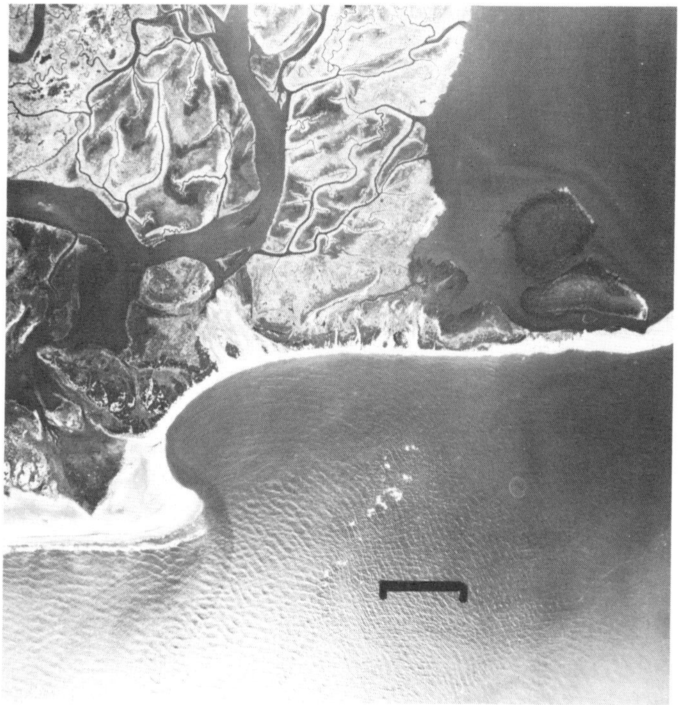

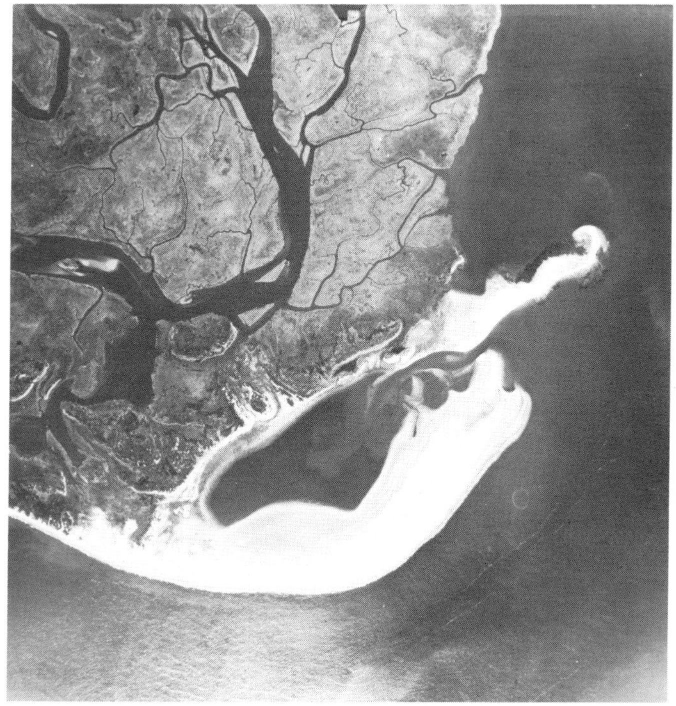

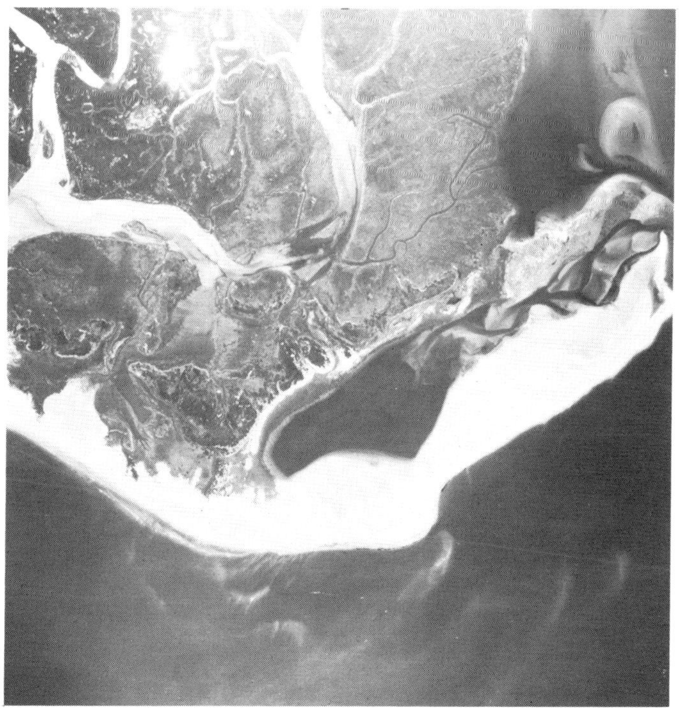

FIGURE 10–9
These three vertical photos were taken of Little Egg Harbor, N.J., 1940 top left, 1957 top right, and 1963 left. They show the changes in shape of the island beach spit due to long-continued beach drifting. The black line right of the beach in the top left photograph is 1.6 kilometers.
(U.S. Geological Survey)

FIGURE 10–10
Cape Hatteras beach. The berm ridge is very prominent in the foreground (left) where it has been protected by grass plantings and use of sand bags. A ground-level view is shown below.

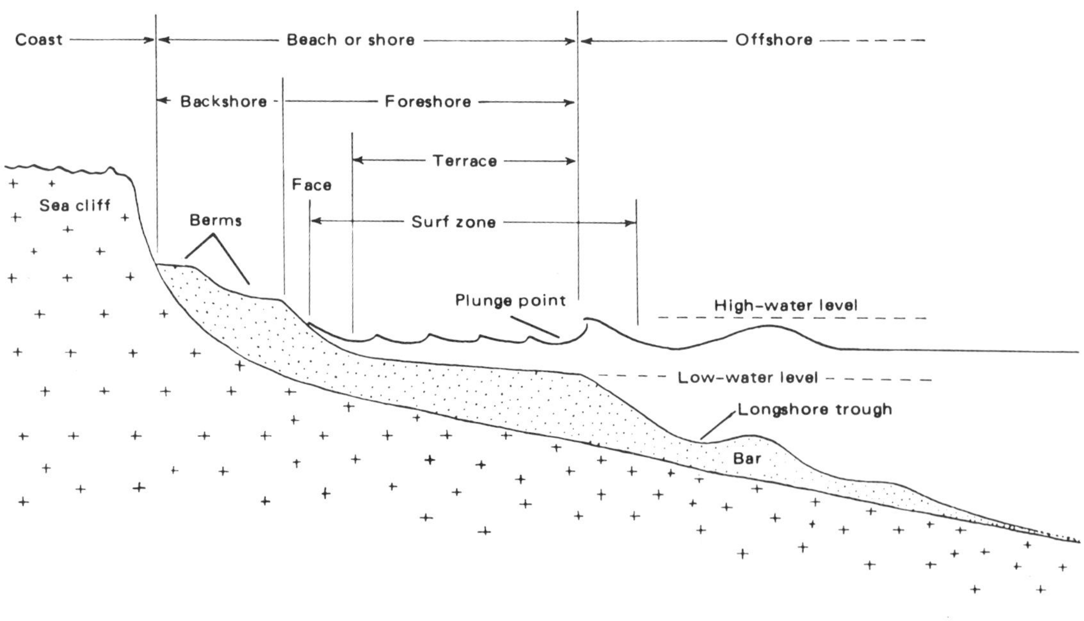

FIGURE 10–11
Profile of the beach and near-shore region modified after Inman (1962). The principal effect of waves is to push sand against the sea cliff rather than to construct a wave-built terrace.
(from Dietz, 1963)

of gentle slopes leading into a marsh or lagoon is found. This profile shows the sand of the beach as a lens-shaped body that thins out toward the deeper water and emphasizes the effects of wave action in holding the sand against the shore.

Seasonal variations in beach profiles are seen in Fig. 10–12, which depicts the different profiles of a beach at Carmel, Calif., from April of one year to February of the next. The beach was much wider and the offshore bars and troughs were smoothed out in the summer months. The larger waves increased turbulence, and greater transporting power narrowed the beach and accentuated offshore bars and troughs in the winter.

Conditions Favorable for Beach Formation

Beaches are characteristic of shore lines that are dominated by deposition, but they occur as patches along even the most rugged shores. Of prime importance is a source of sediment supply. Most of the sediment comes from streams entering the ocean from the continent or from wave erosion and breaking down of rock units exposed along the shore. During storms some sediment from the sea floor may be brought in from deeper waters offshore, but ordinarily the movement of this material is minor. Locally the sediment may come from volcanoes, melting glaciers, or deposition by the wind.

If beaches are to persist, the rate of supply of sediment to the shore line must equal or exceed the rate at which it is being removed either by storm waves or by longshore currents. If sediment is abundant the shore line may have a beach that extends for hundreds or even thousands of kilometers. If the amount of sediment available for transportation and deposition is limited, or if longshore drift is rapid, beaches develop only along protected stretches of the shore.

Along irregular coast lines the most common site for a beach is at the head of bays or inlets. Here the beach is relatively protected from wave action and at least partially protected from the longshore current. Headlands projecting out on either side of the bay prevent the rapid removal of beach materials by beach drifting and the longshore current. Protection from wave action in the bays comes from the fact that the wave energy is spread out over a long stretch of beach due to

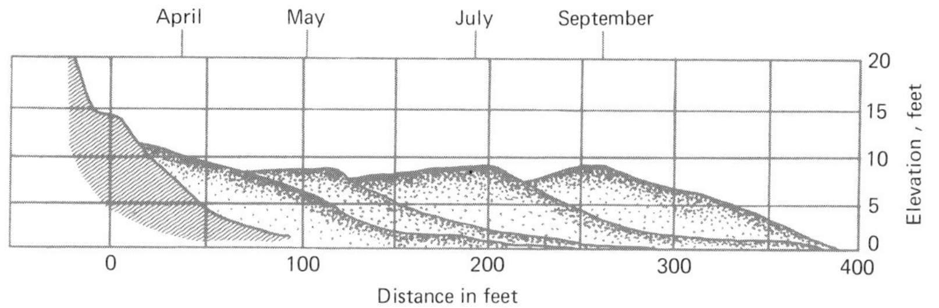

FIGURE 10–12
Growth of the berm at Carmel, Calif., during the spring and summer months. The deposits are cut back during the winter.
(after Bascom, 1964)

wave refraction along the sides of the bay. The headlands are the sites of most intense wave erosion, but when the supply of sediment is excessive, beaches may develop even at the foot of the headland sea cliffs. These beaches are likely to be destroyed during storms, and they will show the marked influence of changes in the supply of sand. Beach sands drift in the direction of the longshore current, and sand beaches may even be built out across the mouth of the bay. At first a submerged bar is built out from the tip of the headland beach. With the addition of more sand, this may emerge and form a projection from the land mass into the bay. Such a feature is called a *spit* (Fig. 10–13). The end of the spit is generally curved back into the bay, but with sufficient sediment supply it may continue to be built out and even extend itself completely across the bay to form a *bay-mouth beach* (Fig. 10–15). Such beaches are not likely to cross large bodies of water because rivers empty into most of them, and this water must find an outlet into the sea even though the outlet may be no more than a narrow channel.

Tombolos are bars or beaches of sediment built between islands and the mainland. When the rate of wave erosion is reduced or when the supply of sand is increased, this small strait between the island and the land interferes with the longshore drift of sand, and the strait becomes silted in. As more sand is brought into the zone, the submerged bar rises above sea level and is built up more during high tides and high waves.

Offshore Bars and Beaches

Long bars and beaches are located offshore along most of the coast of the United States from Long Island to Florida and in the Gulf Coast (Fig. 10–

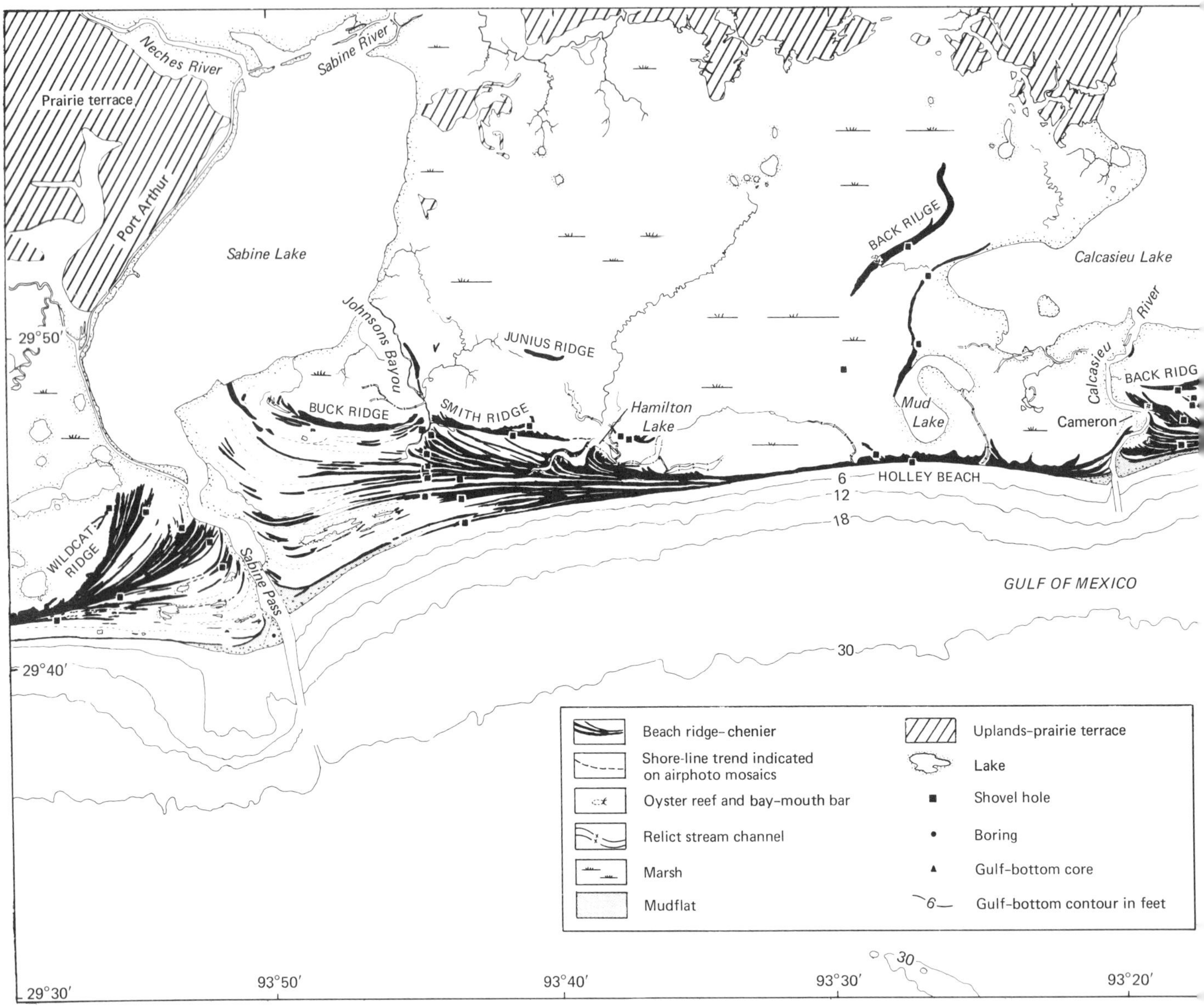

FIGURE 10–13
Physiography—chenier plain of southwestern Louisiana.
(from Byrne and others, 1959)

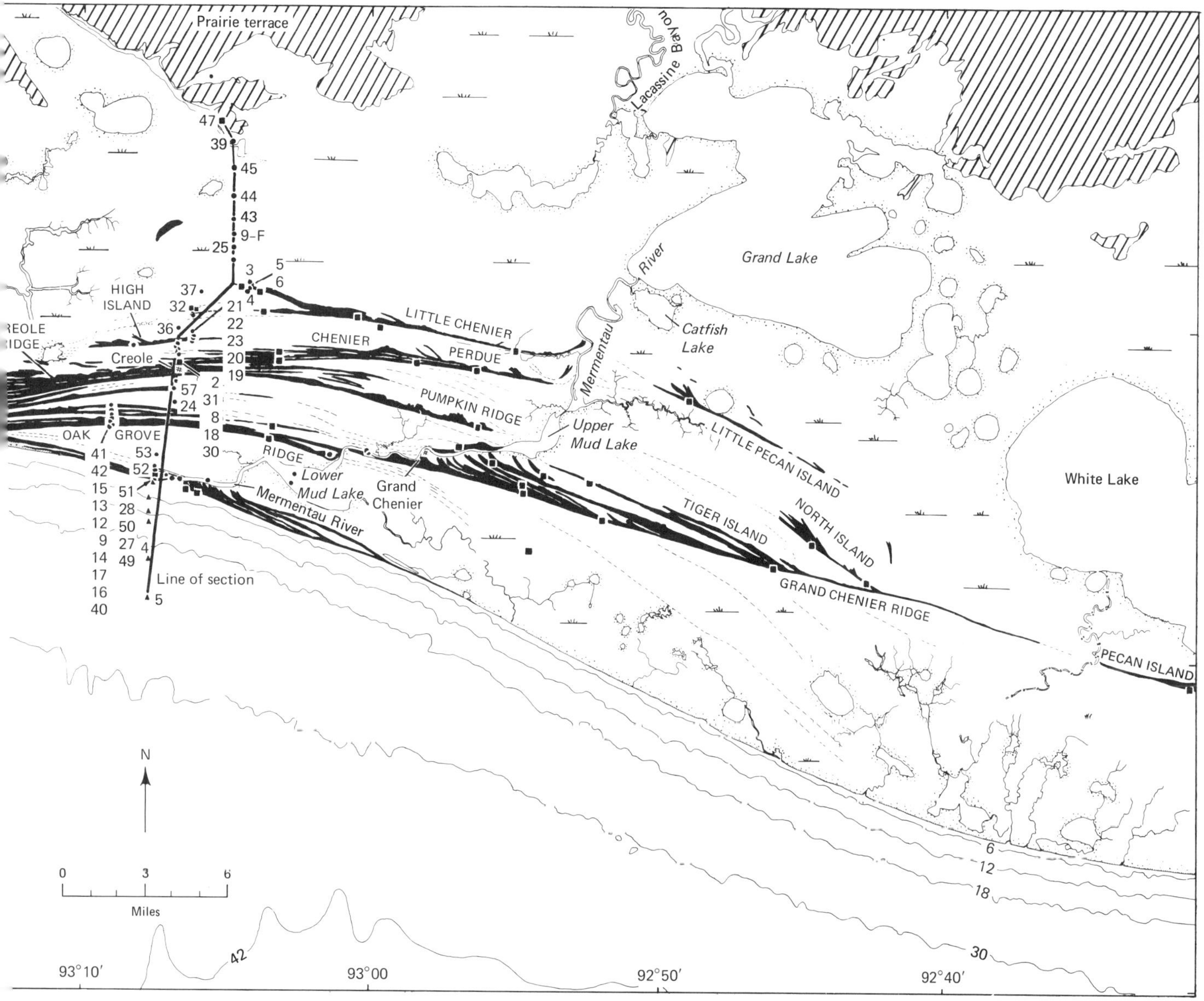
Prairie terrace
Lacassine Bayou
Grand Lake
River
Mermentau
Catfish Lake
HIGH ISLAND
CREOLE RIDGE
Creole
CHENIER
LITTLE CHENIER
PERDUE
PUMPKIN RIDGE
OAK GROVE
RIDGE
Upper Mud Lake
Lower Mud Lake
Grand Chenier
Mermentau River
LITTLE PECAN ISLAND
TIGER ISLAND
NORTH ISLAND
GRAND CHENIER RIDGE
White Lake
PECAN ISLAND
Line of section
N
0 3 6
Miles
93°10′
93°00
92°50′
92°40′
6
12
18
30
42

FIGURE 10–14
Headlands separated by sandy beaches. Note that former marine terraces cap headlands. Near Westport, New Zealand.
(National Publicity Studios, New Zealand)

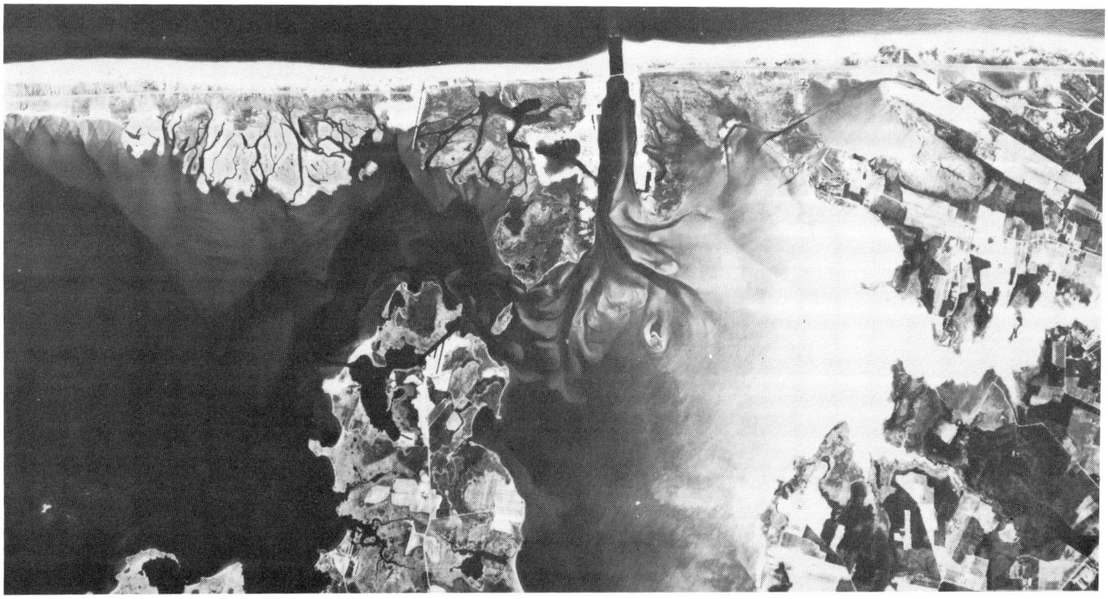

FIGURE 10–15
Rehoboth Bay, showing the barrier beach separating the Atlantic from the bay. The inlet (Indian River Inlet) is artificial. A tidal delta is seen at Indian River Inlet and sites of two former natural inlets (left of the present inlet) can be recognized by remnants of former deltas.
(U.S. Geological Survey)

13). Generally a lagoon (Fig. 10–15) separates the offshore beach from the mainland, and there are occasional breaks through the bars or beaches where streams enter the ocean or gulf. A submerged offshore bar is often found along coasts that have offshore beaches.

The origin of these beaches is still being debated. Some may be formed by breaking waves where the water is extremely turbulent and loaded by sand picked up from the sea bottom. This sand is rapidly dropped out as soon as the turbulence dies down. Since the initial direction of movement after the breaker forms is toward the shore, most of the sand moves in that direction before it settles. This gives rise to a shallow depression within the zone of the breakers and a bar on the landward side of it. This bar may be built up by the addition of sediment carried along shore inside the bar, and it is raised during periods of high waves when sediment is brought in from great depths offshore. This bar eventually becomes emergent to form the *offshore barrier beach.*

The above theory of formation of barrier beaches has come under attack because there are few instances where the conversion of an offshore bar into a beach can be demonstrated. One alternative hypothesis suggests that barrier beaches and islands form as a result of the submergence of dune and beach ridges adjacent to the shore line, the lagoons being flooded areas behind the earlier shore line.

Offshore beaches also may form as a result of the growth of spits along shore. The progressive extension of a spit is seen in the photographs of Little Egg Harbor, N.J. This mode of origin is accepted for a number of the beaches along the Gulf Coast (Fig. 10–13). A block diagram showing the subsurface structure inferred for this hypothesis is shown in Fig. 10–16.

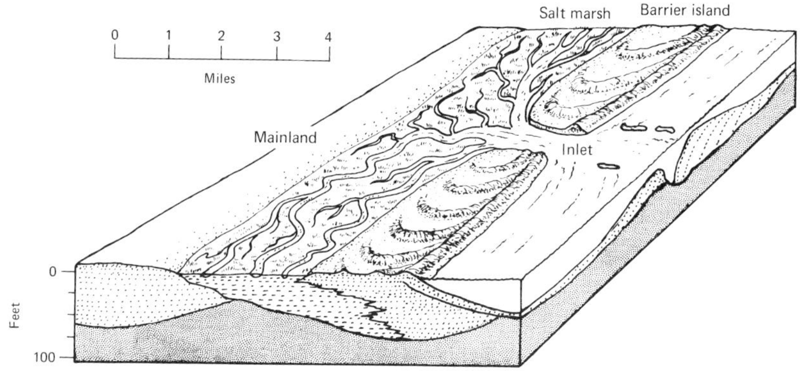

FIGURE 10–16
(*Above*) Idealized diagram of a migrating barrier island.
(modified after Hogt and Henry, 1967)

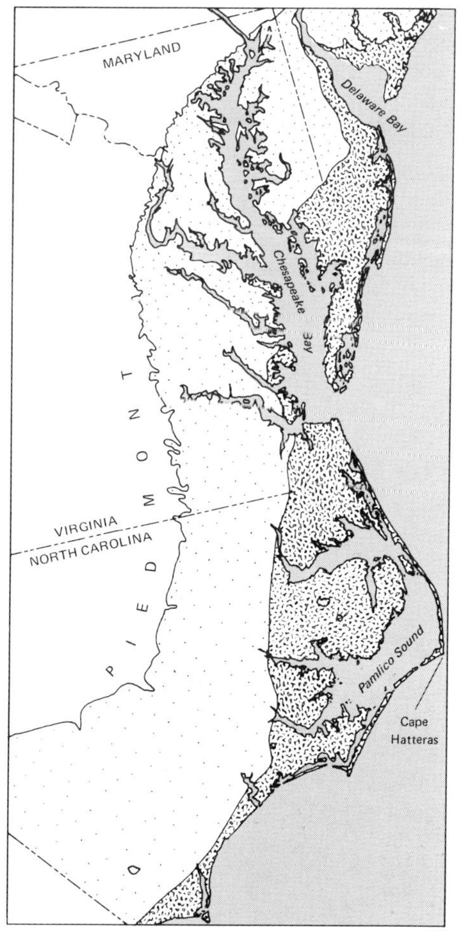

FIGURE 10–17
(*At left*) A portion of the East Coast of the United States. Recent deposits of sand and swamps are shown dark. Dots cover the remainder of the Coastal Plain.
(after U.S. Geologic Map, U.S. Geological Survey)

EVOLUTION OF COASTS

If a hilly or mountainous area is rapidly lowered below sea level, the river valleys will be drowned; the coast line will be irregular as valleys are filled with water (Figs. 10–18 and 10–19); islands will protrude offshore; the sides of hills will drop off

FIGURE 10–18
Evolution of shores, according to the idealized cycle of W. M. Davis (1909). The early stage is at the top.

sharply into the water; and, initially at least, no beaches, bars, or other marine deposits will occur along the coasts. Because water drops off suddenly to a great depth, waves will not break until they are almost against the land. They will break with great force, and the maximum amount of their energy will be dissipated in cutting into the land. Cliffs will form rapidly (Fig. 10–20). The debris from the cliffs will consist in large part of blocks too large for the longshore currents to move. As this material is broken down the fine particles are moved along the beach by beach drifting or carried into water by rip currents or turbidity flows. The sediment may become deposited on the shelf or on the continental rise. Continued wave action and mass wasting cut the cliff back, giving rise to a wave-cut terrace at the foot of the sea cliff.

More wave energy is concentrated at the islands and projecting headlands than along the more protected parts of the shore line, so cliffs and terraces will be built at these points first. The effect of the concentration of wave erosion on the headlands is to produce a shore line with fewer irregularities. This is aided by the deposition of bars along the shore. Bars are built out from the headlands into the bays. Others form at the heads of the bays and along their sides. Eventually bars may entirely close off the bay, and the coast line becomes relatively smooth. Beaches that may form at the headlands become essentially continuous with the bay-mouth bars. The bays and lagoons left behind these bars are gradually filled in by sediments furnished from the land.

A great deal of time is involved in the evolution of all land forms. Once the coast has attained the characteristics described above, it changes even more slowly. But it does not cease to change. Waves continue to break up any rocks exposed in the sea cliff, and longshore currents continue to move the material along the shore. The shore line gradually retreats. As it moves back, bars and beaches move with it. Eventually the shore line may lie near the heads of the drowned valleys. Evolution of most modern shore lines has been dominated by fluctuations in sea level. These fluctuations have involved vertical changes of sea level on the order of 90–120 meters. At times of greatest ice advances most of the continental shelves were exposed to subaerial erosion, but

FIGURE 10–19
Topography of a drowned land area, Marlborough Sound, New Zealand.
(National Publicity Studios, New Zealand)

FIGURE 10–20
Beaches showing growth lines of many new berms connect with an extinct volcanic region, Banks Peninsula, New Zealand.
(photo copyright V. C. Browne)

sea level has risen in most places over the past 11,000 years. The continental rise in sea level allows wave action to remain concentrated on cliffs and headlands until broad wave-cut terraces are formed.

GENERAL FEATURES OF THE CONTINENTAL SHELVES

Continental shelves are far from uniform in nature. Off the eastern and southern coasts of the United States a lowering of sea level to the edge of the shelf would reveal a wide, gently sloping land surface. Florida would be roughly twice its present width. Off the coast of southern California the shelf area is very irregular and includes deep basins and high mountain blocks situated between the coast line and the steep escarpment that borders the deep Pacific basin. The outer margin of the shelf off the Mississippi Delta is also irregular. These irregularities are probably the results of subsidence and slumping of the edge of the shelf.

The shape and extent of the shelf is determined by the nature of the continental margins. Along coasts characterized by young folded and faulted mountains the shelf is narrow. Wide shelves appear most often off coasts characterized by broad lowlands, off glaciated lands, and off many of the world's largest rivers. The largest shelf area is in the Arctic, where the width reaches 1,300 kilometers. Shelves are commonly between 16 and 32 kilometers wide.

Sunlight tends to make shelf environment warmer and more suitable than the deeper areas for a larger number of marine organisms. Almost all the shelf is disturbed periodically by wave motion. Even the outer margins are affected by great waves associated with storms and tides, and longshore currents move continuously over the shelves. Temperature is an important environmental factor. It depends primarily on latitude and the temperatures of the water brought onto the shelves by oceanic currents.

Sediments on the shelves are dominated by terrestrial debris near shore, but at greater distances offshore biological factors become increasingly important. Several generalizations may be drawn regarding distribution of sediment.

1. Sediment textures do not vary systematically either along or perpendicular to shore lines. Many areas show a progressive decrease in size of sands or silts out from beaches for several thousand meters. But this often grades back into coarser sediments again. Several explanations have been offered to account for coarsening of sediment toward the continental shelf margins. One is that storm waves have carried both coarse and fine material out to that position, and that later the fine sediment has been sorted out and removed by weaker current and wave action. Along the edge of the shelf off the eastern coast of the United States it may be that wave action has sorted the sediments and that the Gulf Stream has removed the finer sizes and carried them into deeper waters. Not all shelves have coarse sediments at their outer margins. Some show a progressive decrease in the size of sediment from the shore line out to the margin of the shelf.

2. Bottoms formed of bare rock are found on some shelves. They are more common on narrow shelves than on broader ones. Most are covered by a thin veneer of sediment that is in the process of being transported to deeper water. For many years it has been assumed that these were wave-cut and wave-built terraces. Photographs taken underwater now show that the outer margins are not all wave-built terraces composed of deposits. Instead, rocky outcrops appear right at the edge of the shelf. The sediment that is deposited near the edge of these solid shelves periodically slumps off or slides downslope.

3. The character of sediments on shelves is closely related to the types of land that are located near them. Muddy sediments are generally found on shelves beyond the mouths of large rivers such as the Amazon, Mississippi, Elbe, Yangtze Kiang, and Indus. On shelves off regions recently covered by continental glaciers sediment is a mixture of rock fragments, sands, and mud like that produced by glacial erosion. This is typical of the Barents Sea north of Norway and Sweden, Baffin Bay between Greenland and Canada, Ross Sea in the Antarctic, and the Gulf of Maine. Where high mountains border the coast, an abundance of gravels, coarse sands, and other

coarse sediments flood out onto the shelves. If lowlands or plains border the sea, sediments are usually fine sands and contain a large proportion of animal and plant remains.

FIGURE 10–21
(*Above*) Tidal flats near Mont St. Michel.
(French Cultural Services)

Tidal Flats

Tidal flats develop on low-lying land that is near sea level yet protected from wave erosion and strong currents (Figs. 10–21 and 10–22). They are generally located near an abundant source of sediment, such as the mouth of a large river. Sediments on tidal flats consist of soft, water-soaked mud. Fine silt, clay, and sand are mixed with varying amounts of shell fragments, sea-urchin spines, fine plant detritus, and fragments of the shells of diatoms, foraminifers, and ostracods. In some areas excrement of worms and other mud-ingesting invertebrates makes up a significant part of the sediment. Many invertebrates inhabit the tidal-flat environment. These include gastropods, worms, and pelecypods.

Deposits on tidal flats are stratified and cross-

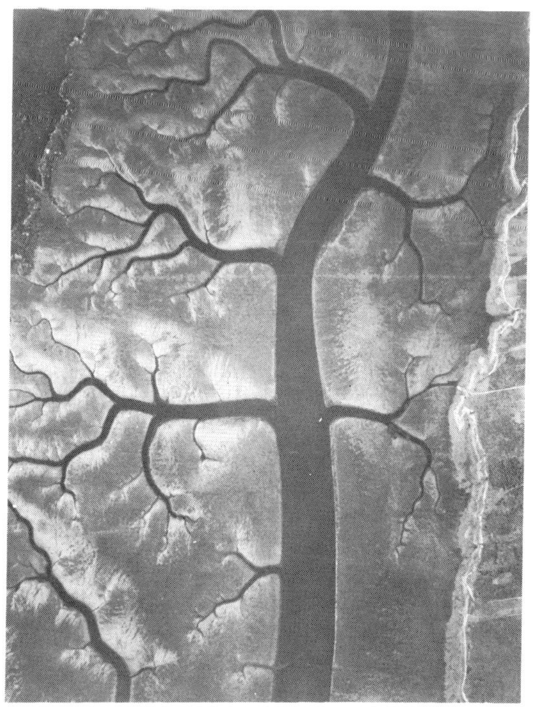

FIGURE 10–22
(*At left*) Drainage pattern in the tidal mud flats at low tide, near Yarmouth, Nova Scotia.
(Royal Canadian Air Force)

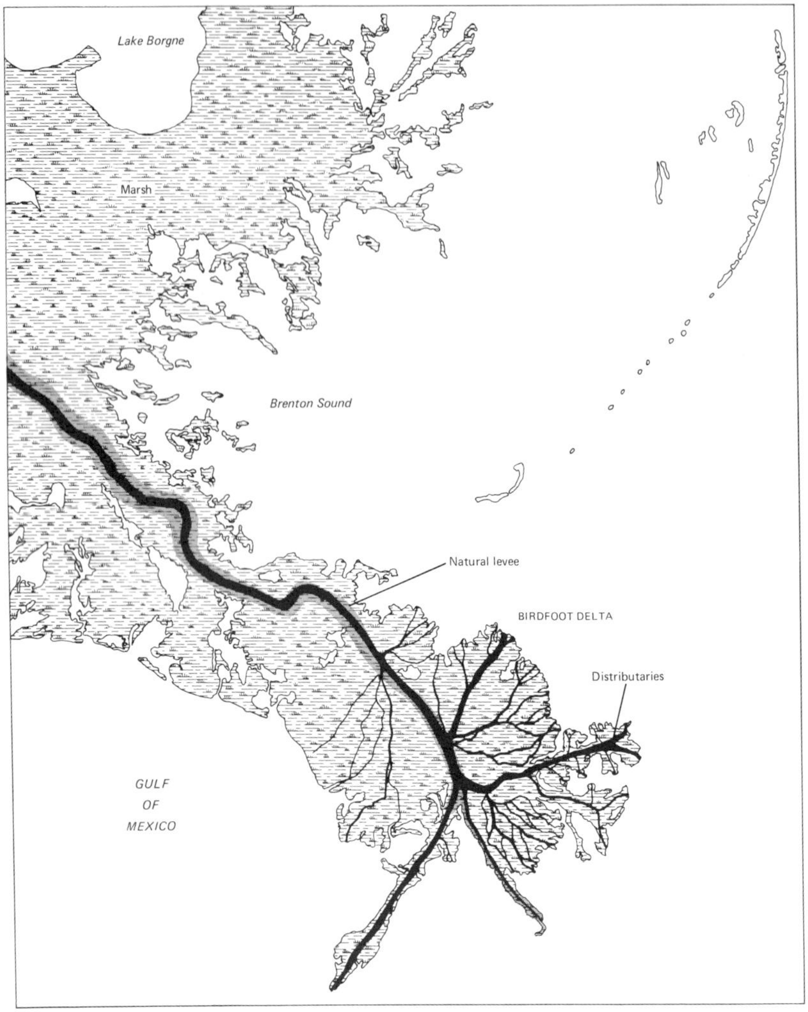

FIGURE 10–23
(*At left*) Map view of the delta of the Mississippi River, the Birdfoot Delta. The various environments depicted are described in the text.

FIGURE 10–24
(*Below*) Physiographic features of the Mississippi River deltaic plain and outlines of the delta complexes.
(from Frazier, 1967)

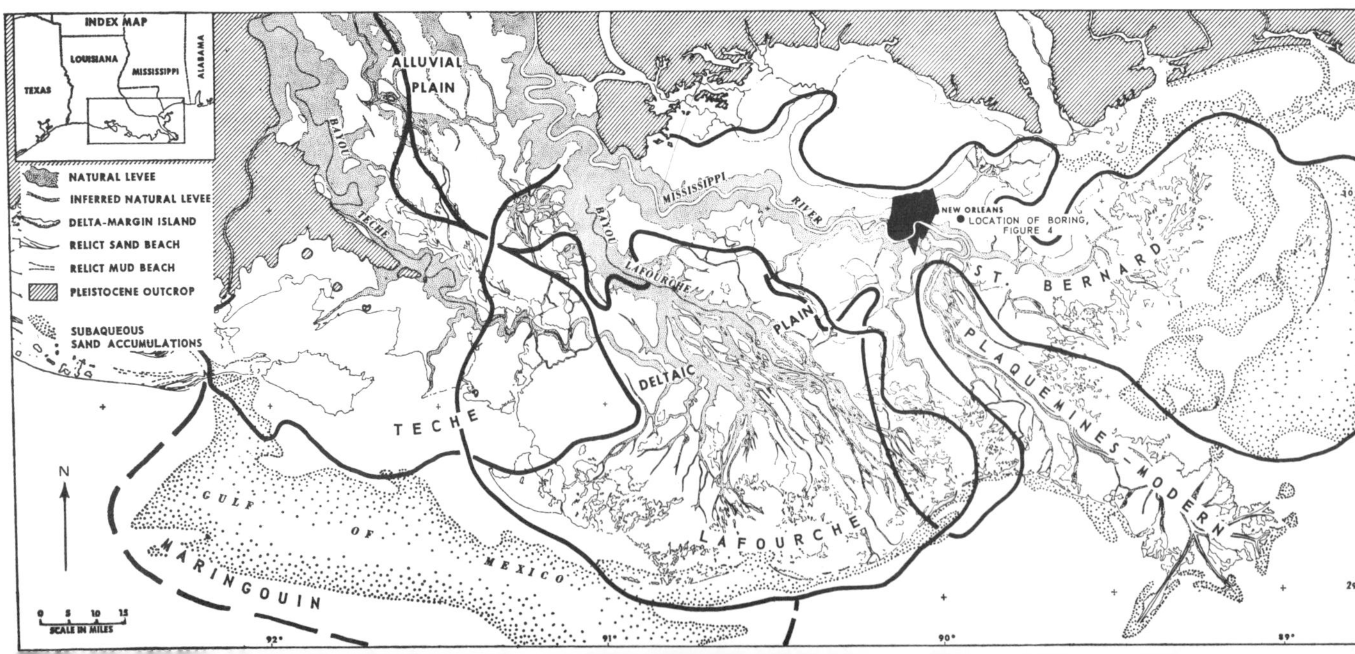

bedded, reflecting reworking of the sediment by shifting currents. The amount of sediment and rate of sedimentation on certain present-day tidal flats are surprising. As much as 7.3 meters of sediments were deposited within 3 years in the harbor entrance to Wilhelmshaven, Germany. The rate is sufficient to make continuous dredging necessary to keep the channel open for shipping.

Sources of muds vary from place to place. Some tidal flats are thought to originate almost entirely from deposition of suspended matter in streams, but others appear to be swamp or glacial deposits reworked by the forces acting in the tide zone. Sediments are often rich in organic matter derived primarily from plants. In Holland, when a tidal flat is built up close to sea level or slightly above it by storm waves, the Dutch construct dikes around a section of the flat, then drain it and use it for agricultural purposes.

Deltas

Deltaic environments are composites of a number of different types of sedimentary conditions. No other type of environment contains so many different physical, biological, and chemical conditions in such intimate association.

A delta forms when the amount of sediment being deposited by a stream at the margin of the sea exceeds the amount of erosion and removal accomplished by wave action and currents. If the stream supplies more than can be distributed, the delta grows. If a balance is reached, the delta becomes stabilized in size. And if the capacity of wave and current action exceeds deposition, the delta will eventually disappear. Deltas are not confined to oceans and transitional environments. They form in lakes, and in streams where tributaries supply more debris than the main stream can transport.

Usually the stream enters and flows across the delta on a low slope. Fresh water and sediment are carried out to the edge of the delta, where the stream enters a larger body of water. At this point the stream loses its transporting power. Coarse sand, silt, and other sediments are deposited. Clay materials, which are in suspension in fresh water, are deposited when the fresh water mixes with salt water and form an extensive layer of clay. Heavier substances are deposited on the slope of the delta front, and finer matter is carried farther out in suspension before it settles to the bottom. Three sets of beds are formed almost simultaneously. The thickest are the layers of inclined sediment on the front slope of the delta, called the *fore-set beds*. Beyond them finer sediments settle on a nearly flat bottom, creating the *bottom-set beds*. The third group, the *top-set beds*, is composed of flat-lying sediments deposited on top of the delta by streams flowing across the delta, marshes, or low areas on either side of the streams. These top-set beds are laid on truncated fore-set beds, forming cross-bedding, as the delta is built seaward (Fig. 7–28).

Mississippi Delta

During the last 10,000 years the Mississippi River has carried more than 32,000 cubic kilometers of sediments into the Gulf of Mexico. Most of this has been deposited in the delta (Figs. 10–23 and 10–24) and near the coast. The weight of this mass has caused the continental shelf to be depressed locally by an estimated 91 to 150 meters.

The Mississippi River flows over an extremely low slope along its lower reaches. As it approaches the delta the river begins to break up into a number of smaller streams called *distributaries*. The pattern formed by these resembles that of a braided stream. These distributaries, carrying a heavy load of silt and dissolved material, flow outward across the delta into the Gulf. Seen from far above they resemble the veins of a leaf or the bones of a bird's foot. Between distributaries there are extensive marshes slightly above sea level. Streams are actually above the level of these marshes over a large part of the area, since seasonal floods build natural levees that confine the stream channels. As the distributaries reach the end of their channels, the stream loads are dropped as lobate deposits on the delta front. The streams have built these lobes out great distances, leaving bays and troughlike depressions between them. Farther out, the slope of the delta diminishes and grades onto the nearly flat bottom of the continental shelf.

Marine Swamp Environment

The names swamp, bog, and morass are applied to low, spongy land generally saturated with

FIGURE 10–25
A restored Pennsylvanian coal swamp. It was from such forests as these that the coal fields of the continental interior of the United States originated. The ferns are similar to many living plants, but the lycopods, *Sigillaria* and *Lepidodendron*, are unlike any living trees. Notice also the small amphibian at the base of the largest trees. A large cockroach is in the foreground on a fallen *Lepidodendron*.

(Chicago Natural History Museum)

moisture. This type of land is not confined to areas near sea level or even to flat land, although it is much more likely to form and less likely to be drained in one of these locations. Some estimates put the amount of land covered by swamps of one sort or another at more than 2.6 million square kilometers. Not only are they important environments today, but there is a rich and extensive record of this type of environment in the geologic past. At various times swamps have covered much larger areas than they do today. Late in the Paleozoic era swamps became so extensive throughout the world that the periods are named for the carbonaceous sediments formed from the swamp deposits (Fig. 10–25).

Two general types of swamps are recognized—marine swamps, located along coasts where both brackish and fresh water are present, and fresh-water swamps. Most fresh-water swamps are located in basins or on flat or gently sloping land that may be far removed from the sea. But the most favorable position for the location of a swamp is near sea level, thus explaining why this particular environment is so commonly preserved in the rock record; relatively slight elevations of sea level will cover them so that they are preserved under marine deposits. For this reason swamp deposits usually grade into marine sediments in stratigraphic sequences.

An impermeable substratum, abundant rainfall, or some other water supply is necessary for the continuance of a swamp, which is generally characterized by slow circulation of water or by near stagnation. Biological factors dominate this environment. Swamps are especially favorable for the growth of plants, and remains of plants make up a large part of the sediment. The energy sources in this lush environment are largely chemical reactions and thermal energy supplied from the sun.

Marine swamps are common along coasts of the eastern and southern United States. Many of them have formed behind offshore bars. These were initially lagoons that have since been filled by sediment from the continent and by remains of plants. Plant life is most abundant where there is little or no agitation of the water; thus the most favorable areas for swamp conditions are regions with a relatively small range between high and low tides. Quite a few swamps, however, are at least partially, and some completely, covered by sea water during high tide. Within a marine swamp different plants are favored by certain special conditions; in tropical climates, for example, marine swamps contain mangrove or cypress forests (Fig. 10–26). However, a large percentage of marine swamps are partially covered by moss, and most contain grasses and reeds in abundance. As plants die they fall into the water and begin to decay. If they can be buried rapidly enough by other plant remains or by marine sediments, there is a good chance that they will be preserved and may eventually be transformed into coal. If burial is not rapid, oxidation and bacterial action soon break them down. For this reason only thin accumulations of swamp deposits are found if the level of the swamp remains stable. Probably no more than a few yards of plant matter can accumulate, since water depth is seldom great. The thick sequences of swamp sediments so common in stratigraphic sequences of the Carboniferous period must indicate changes in the position of sea level in relation to the swamp. A slowly subsiding area would obviously be the most favorable site for preservation of great thicknesses of such sediment.

Barrier Reefs

The name *barrier reef* is given to all coral reefs that run parallel to the shores of islands or continents and are separated from them by lagoons. Frequently they are found around volcanic islands (Fig. 10–27), which are abundant in the latitudes where reef forming organisms abound. They vary in width and distance offshore, and many are 100 meters or more wide. Debris broken off by storm waves on the oceanward side of the reef lies at the base of the living reef. The reef surface is close to mean sea level; it is submerged during high tides and exposed at low tides. Fine sands and algal remains cover the bottom of the lagoon, which is relatively quiet water because waves break at the reef rim. The shore may be fringed by reefs similar to the barrier reef.

Organisms that make up a reef are types that flourish in near-surface waters. Many, like corals, favor warm water that is relatively free of suspended dirt. Corals are attached to the sea floor; they are not free to move about to gather food, so they live where food is brought to them. Clean, agitated waters near the surface are most favorable, and breaks in these reefs are found where muddy waters enter the ocean from streams flowing off the land.

The origin of barrier reefs was explained by Charles Darwin in 1839, and his explanation is still considered valid by most geologists. He held that for islands to have fringing reefs they must be situated in climatic belts favorable for corals and other shallow-water animals and plants and that the position of sea level in relation to the islands must have fluctuated. During the ice ages up to 100 meters or more of water was removed from the sea and stored on land as ice. Warping of the sea floor can also cause sea-level fluctu-

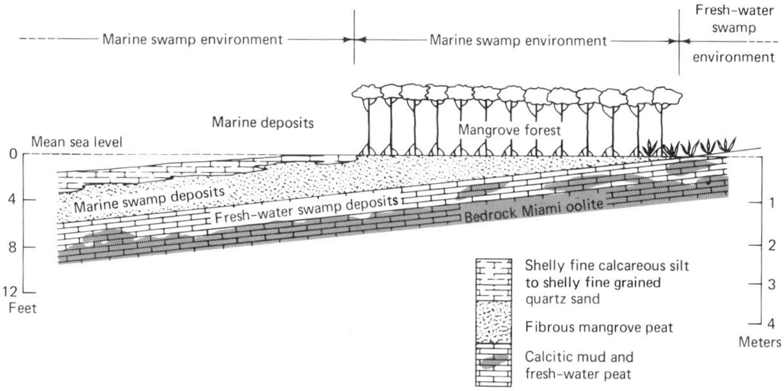

FIGURE 10–26
Transgressive stratigraphic section under edges of bays, swamps, and marshes of southern Florida
(from Scholl and Stuvier, 1967)

FIGURE 10–27
Oblique aerial photograph of the barrier coral reef around Boro Boro Island in the Society Islands.
(U.S. Naval Oceanographic Office)

ations. When an island with a fringing reef begins to be submerged the reef is built upward and will maintain a position near sea level. Thus over a long period of time the original fringing reef becomes a barrier reef, and the island is separated from it by a lagoon.

The Great Barrier Reef

The Great Barrier Reef of Australia covers nearly 200,000 square kilometers and extends 1,900 kilometers along the east coast of Australia from Papua nearly to the Tropic of Capricorn. The shelf here is a complex of scattered reefs separated by channels. This complex is not a single, continuous reef mass, although one reef in it is 960 kilometers long. Long reefs aligned parallel to the coast form an outer barrier near the edge of the continental shelf. A relatively open body of water called the Barrier Reef Channel (although it is 240 kilometers wide near its southern end) lies between the outer barrier reefs and the mainland. Some six hundred islands are found in this channel. A number of these are small islands composed of coral sand and fragments. Other islands are composed of rock masses like those on the mainland. Most of these islands are surrounded by fringing reefs.

FIGURE 10–28
View across the Great Barrier Reef, northeastern shore of Australia.
(Qantas Empire Airways Ltd.)

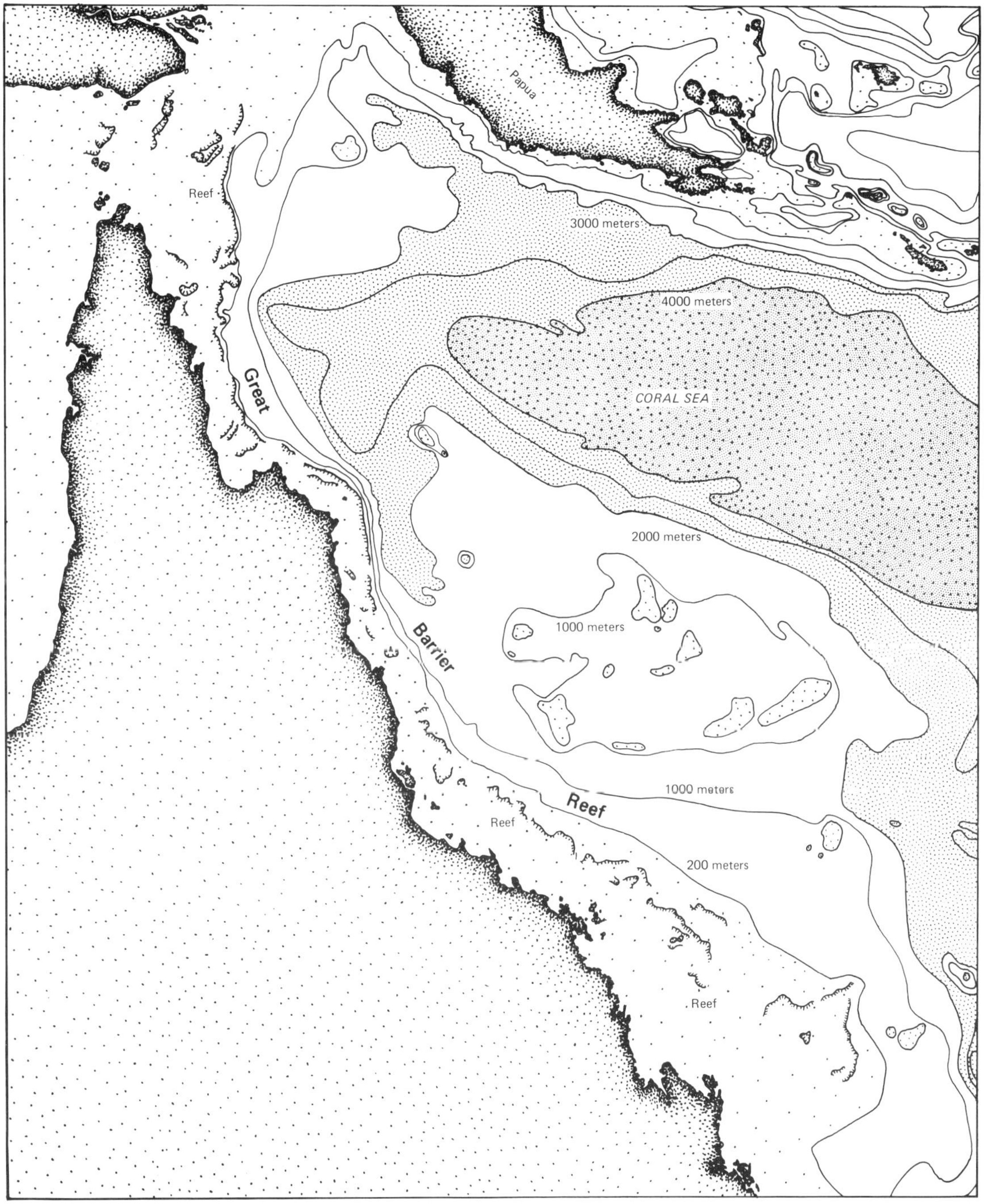

FIGURE 10–29
Location of the Great Barrier Reef. Contour values are in meters.

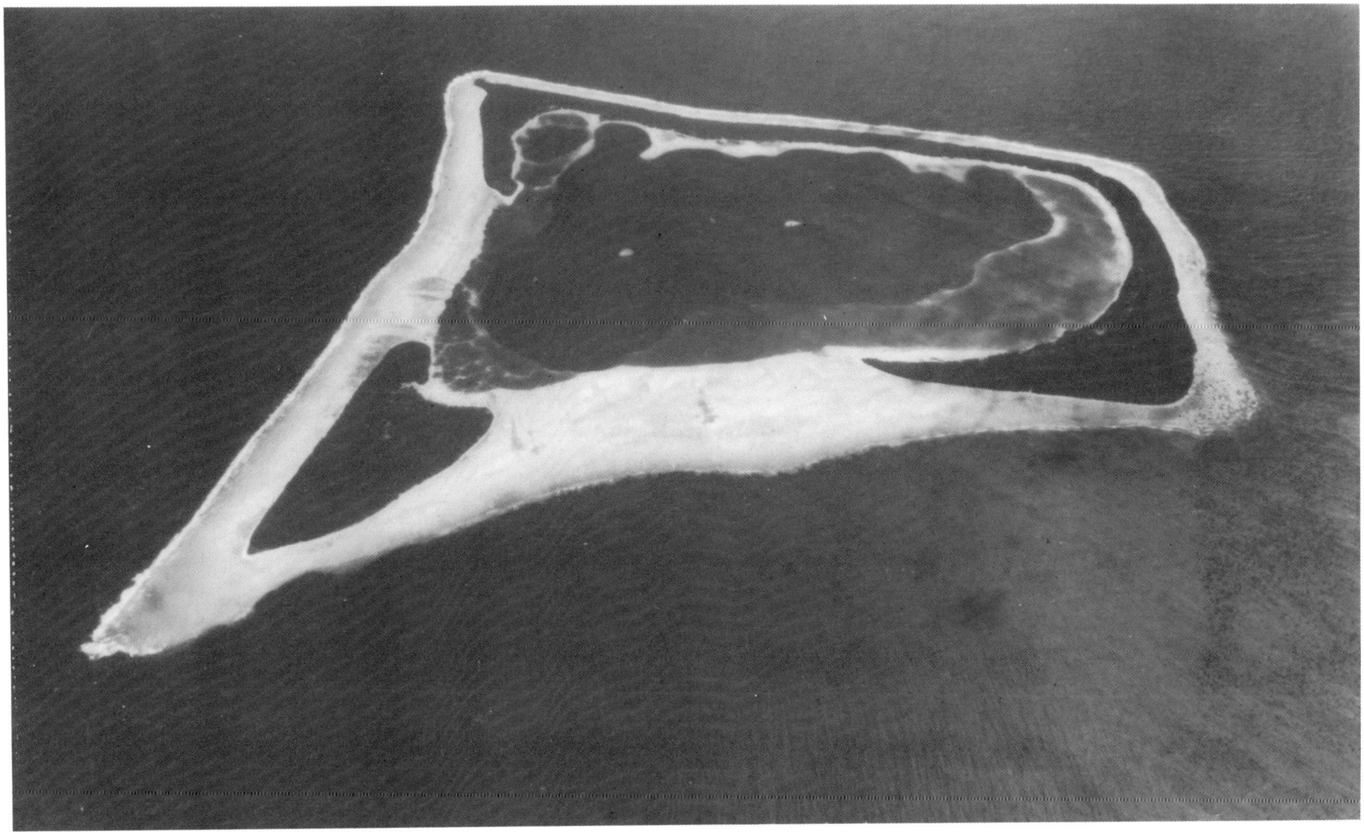

FIGURE 10–30
(*Above*) Oblique aerial photograph of a small coral atoll, Namorik Atoll in the Marshall Islands.
(U.S. Naval Oceanographic Office)

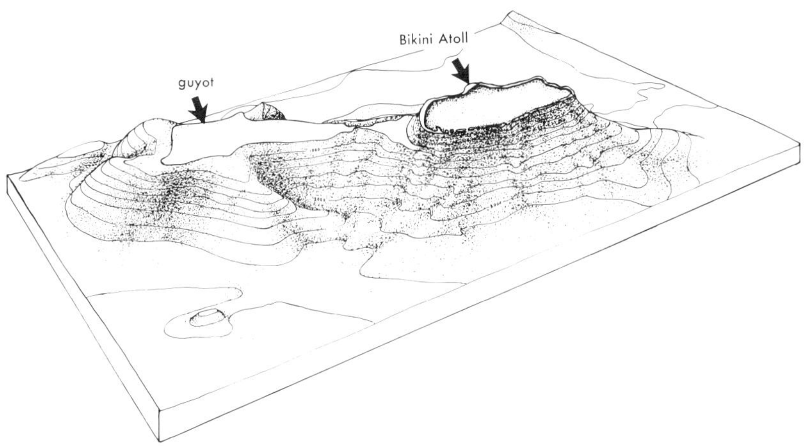

FIGURE 10–31
(*At right*) Configuration of Bikini Atoll. The atoll is at the right. A similar flat-topped submerged mountain lies to the left of the atoll.
(after Emery, Tracey, and Ladd, 1954)

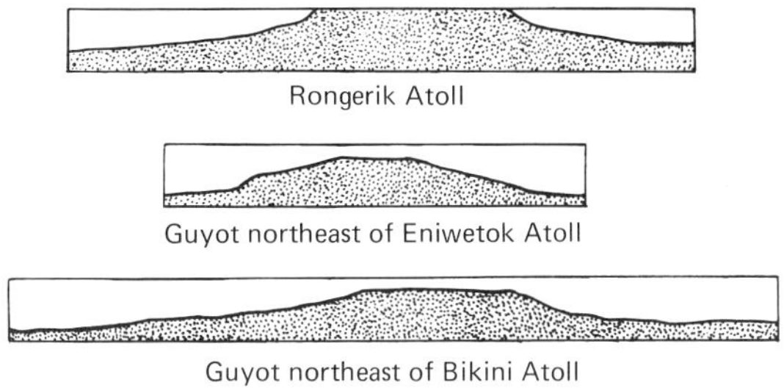

FIGURE 10–32
Profile of a guyot, and comparison with a modern atoll (top).

Atolls

Atolls (Fig. 10–30) are ring-shaped coral reefs that enclose lagoons. Darwin suggested that they are formed when an island with a fringing or barrier reef is submerged. The reef continues to grow upward to maintain the living organisms at the necessary water depth. When submergence has gone far enough the island is covered by water, leaving the atoll.

An alternate hypothesis was advanced by R. A. Daly in 1910. He reasoned that during the Pleistocene, when sea level was over 100 meters lower than it is now, the ocean was cooler than at present and thus less favorable for the growth of corals. As sea level dropped, islands and their fringing reefs were exposed above sea level. Corals could not grow around the margins of the islands fast enough to protect the islands from wave erosion. Thus the islands were beveled off to form platforms. When sea level began to rise from the melting of glaciers and oceans began to warm up, new reefs formed near the margins of the platforms and maintained their positions during further rises in sea level, thus creating the atolls with their flat-bottomed lagoons.

Bikini Atoll

Bikini Atoll is an oval-shaped group of islands about 42 kilometers long and 24 kilometers wide (Fig. 10–31). Marginal reefs are almost continuous around the atoll. In places they are 1.6 kilometers wide. The central lagoon, which has an average depth of 48 meters, covers approximately 648 square kilometers. Twenty six islands surround the lagoon and stand at a maximum height of 7.6 meters above sea level. A number of passes 1.6 kilometers wide cut through the reef.

Reefs on the seaward side of the islands are abundant with life. The open sea brings in a constant supply of water rich in food and nutrient salts. The character of the seaward margin of the reef and the assemblages of organisms that live there is determined by the slope on which the reef is located, the prevailing waves, the currents, and other ecological conditions. Reef rocks fall into two groups, *algal* or *coral-algal-reef limestone*, and *detrital-reef limestone*. Coral-algal limestone is formed of organisms mostly in their original position of growth. Detrital limestone includes reef fragments mixed with finer fragments. The two types are mixed and grade into one another.

Where the reef is strongly grooved on the seaward side, ridges between grooves are flattened and covered by massive coralline algae, which appear to be well adjusted to the strong, steady surf. Where marginal reefs are low, there is a rich growth of corals on the flat, inside portion of the reef. Where marginal reefs are high, a meter or so above the main reef flat, water circulation is restricted, and few living corals are found on the flat.

When grooves are weak or absent, no massive coralline algae are present. These smooth reefs are usually on the leeward side of the island. The seaward slope of the reef is steep, plunging into water about 61 meters deep. The marginal zone here contains coral and algae, but they rise only a few inches above the reef flat. On these smooth reefs, growth of the reefs exceeds the rate of erosion. Reefs of a slightly different nature are found where the most severe storms occur and erosion is more active. These storm waves break up the reef into large blocks that slump into deeper water but continue to be covered by a growth of new coral.

Bahamian Platforms

The Bahama Banks, located south and east of Florida, cover more than 250,000 square kilometers of the Caribbean Sea (Fig. 10–35). They are most notable as an extraordinary example of a modern shallow-water area in which almost pure limestone is being formed. This condition is not new for the area. Calcium carbonate sediments have been forming there for approximately 130 million years, since early in the Cretaceous period. Since that time virtually no sediment other than

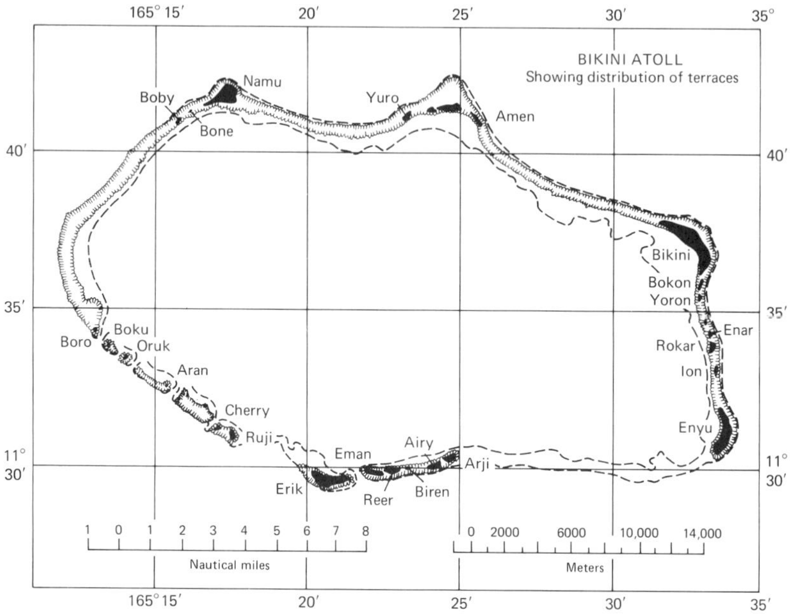

FIGURE 10–33
(*At right*) Sketch map showing the distribution of terrace inside lagoon and outside reef. Seaward terrace generalized to include all areas with a gentle offshore slope. Terrace shown by heavy dashed line.
(from Tracey, Ladd, and Hoffmeister, 1948)

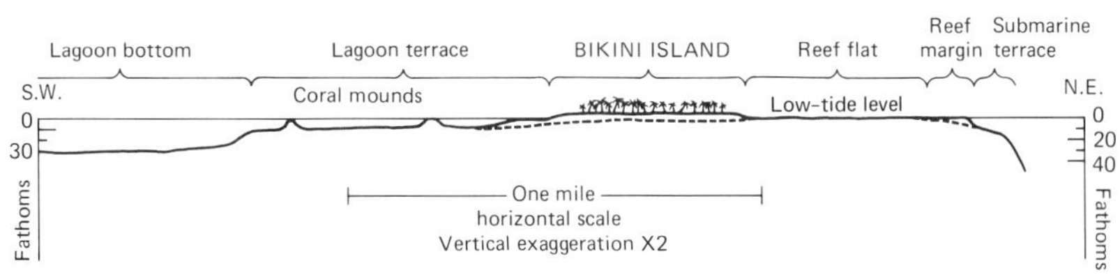

FIGURE 10–34
(*Below, at left*) Generalized cross section through the windward reef of Bikini. Figure shows relation of present reef, island, and lagoon bottom surface (solid line) to inferred older surface (dashed line) and terraces.
(from Tracey, Ladd, and Hoffmeister, 1948)

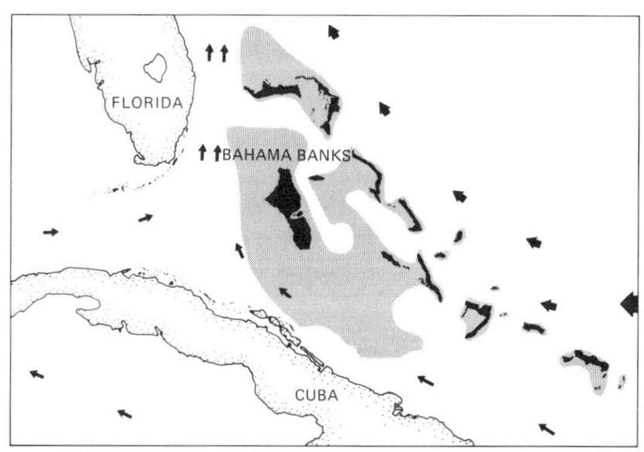

FIGURE 10–35
(*At right*) Bahama Banks. The islands are shown in black; the shallow-water banks surrounding them are shaded. The shallow water surrounding Florida and Cuba is not shaded. The arrows indicate the direction of prevailing ocean-surface currents. They bring warm waters rich in calcium carbonate up from the equator.

limestone has been deposited. These beds have an accumulated thickness of more than 4,300 meters, indicating a probable slow subsidence of the platforms as limestone was deposited near the surface.

The banks are low-lying platforms covered for the most part by approximately 9 meters of water and separated by broad, flat-bottomed, deep channels. Coral reefs form rims particularly along the eastern side of the banks. Islands made up largely of beaches and dunes of calcium carbonate sand have formed on sites of ancient reefs. The islands are separated by broad lagoons. Only a small part of the exposed limestone is made up of reef debris. Most of it has been precipitated from solution to form oölites that make up shoals on the banks and cover large areas of sea floor. According to Newell, the platforms developed early in the Cretaceous from oceanic coral atolls that gradually became incorporated into the continent by the spread and coalescence of calcareous deposits.

The region is free of terrigenous sediment and is usually covered by warm, dry air. This favors precipitation of $CaCO_3$, since the rate of evaporation of water is accelerated, leading to supersaturation. A solution is *saturated* when it contains all *the dissolved chemicals it can hold at that temperature without precipitation;* if the temperature is subsequently reduced or the solution is concentrated through evaporation, the solution will remain above saturation until conditions cause the dissolved material to be precipitated (to become a solid and settle). Average temperature in the Bahamas in winter is 21°C., and rainfall averages approximately 127 centimeters per year. The banks lie in the path of the easterly trade winds, which also accelerate evaporation. Surface waters of the Caribbean Sea are almost 40 percent above saturation in calcium carbonate. As these waters are carried across the shallow banks there is naturally more surface evaporation, which further concentrates the calcium carbonate solution, and agitation and heating cause a loss of the carbon dioxide essential in keeping $CaCO_3$ in solution. These conditions accelerate precipitation and a resulting accumulation of $CaCO_3$ over the shallow banks.

The platforms may be subdivided into several sedimentary regions. Sand composed of oölite grains and skeletal fragments of the reef faunas are found on the barrier rim. Sand is produced by wave action breaking up the shell and reef remains. Just behind the reef there are many lime-secreting organisms, which also account for a large part of the rim sediment. Over the platform as a whole much of the $CaCO_3$ is chemically precipitated ooze, oölite sand, and fecal pellets. Oölite sand is originating near the edge of the banks. Oölites, round balls named for their resemblance to fish eggs, tend to form in agitated supersaturated water where it first breaks and suddenly loses carbon dioxide. Oölites are made of concentric layers of calcium carbonate built around a central nucleus of aggregate grains of lime or fecal pellets (Fig. 7–7). The shelf lagoon is the site of accumulation of calcium carbonate ooze, a mud that consists of microscopic crystals formed by direct precipitation.

CHAPTER OUTLINE

TERMS

abyssal
arch
atolls
backshore (berm)
barrier reef
bathyal
bay-mouth beach
beach drifting
bog
bottom-set beds
coast
coral-algal-reef limestone
cuspate
detrital-reef limestone
diastrophism
distributaries
emergent coast
fore-set beds
foreshore
karst
Lithothamnium
morass
notch
offshore
offshore barrier beach
saturation
sea cave
serpulid
shore
shore line
spit
stacks
submerged coast
swash
swamp
terraces
tombolos
top-set beds
wave-cut cliffs

REVIEW QUESTIONS

1. Sedimentary environments can be classified according to the amount and type of energy expended in each. Which environments are characterized by high mechanical energy? By high biochemical energy reactions?
2. What evidence proves that sea level has had much higher stands than it has now?
3. Describe the causes of the changes shown on the three photographs of Little Egg Harbor (Fig. 10–9).
4. In what ways does the beach at Hatteras (Fig. 10–10) differ from the idealized beach profile. Explain why.
5. Explain the origin of the tidal delta in Rehoboth Bay (Fig. 10–15). Why is not the delta on the oceanward side?
6. Why do not all volcanic islands in the Pacific have barrier reefs? What differences cause some volcanoes to become atolls, some to become seamounts, some to become guyots, and still others to have barrier rims?
7. Briefly summarize sedimentary rock types and forms that would enable you to recognize ancient (many millions of years old) deltas, barrier reefs, beaches, and marine swamps if the evidence is in cores taken from deep wells.
8. What conditions favor development of sea cliffs?
9. Why does the type of sediment being deposited on continental shelves differ from one place to another?
10. What features in sedimentary rocks give an accurate indication of the depth of water in which they formed? What features indicate *relative* accuracy? Refer to Chapter 7, also.

SUGGESTED READINGS

Bird, E. C. F., 1969, Coasts, an Introduction to Systematic Geomorphology: Cambridge, M.I.T. Press, 246 p.

Emery, K. O., 1966, Atlantic continental shelf and slope of the United States: U.S. Geol. Survey, Prof. Paper 529–A, 23 p.

Guilcher, Andre, 1958, Coastal and Submarine Morphology: London, Methuen, 274 p.

King, C. A. M., 1959, Beaches and Coasts: London, Edward Arnold, 403 p.

Maxwell, W. G. H., 1968, Atlas of the Great Barrier Reef: Amsterdam, Elsevier, 258 p.

Russell, R. J., 1967, River Plains and Sea Coasts: Berkeley, Calif., University of California Press, 173 p.

Wiens, H. J., 1962, Atoll Environment and Ecology: New Haven, Conn., Yale University Press, 532 p.

11
DEEP-SEA ENVIRONMENTS

11 DEEP-SEA ENVIRONMENTS

FIGURE (*previous page*)* "Sandfall," Cape San Lucas (California) submarine canyon. The "fall" is about 10 meters high. Currents feed sand from the nearby beaches into the canyon.
(photo by Conrad Limbaugh, University of California, La Jolla, Scripps Institution of Oceanography)

* Photographs appearing at beginnings of chapters but not also in the chapters are identified separately, as here.

INTRODUCTION

About 65 percent of the surface of the earth is covered by deep marine waters. Included are the extensive abyssal plains, deep-sea trenches—some reaching 11,600 meters below sea level, and the continental slopes and rises that are covered by more than 183 meters of water. With few exceptions these areas are many kilometers distant from land. The exploration of the deep sea has until recently posed problems that were largely insurmountable. Even today we know far less about the oceans than about any other surficial part of the earth. This is because of the very limited access we have to the deep parts of the oceans. Skin divers are limited to the upper 100 meters of the oceans, little over half of the way down to the shallow edges of the continental shelves. The diver wearing a suit may be able to work at depths of 150 meters, but even this depth does not reach all continental-shelf areas. The bathyscaphs of Auguste Piccard, the French Navy, and the United States Navy have made it possible for man to descend into the deep-sea trenches, but so far these vehicles have had only limited use. Most data about deep-sea areas have been obtained through indirect observations by means of cameras, core samplers, dredges, and echo-sounding devices.

Before the twentieth century, man's view of the deep seas was based almost entirely on speculation. There was little information on which to base any concrete understanding. The early geomorphologists reasoned that running water, glaciation, wind, and ground water were effective agents of erosion only above base level, and that ultimate base level was sea level. At sea level glacial ice floats, and it appeared that running water ceases to transport and erode. Samples of sea water collected throughout the oceans were always free of suspended sediment when collected far from shore. It seemed logical to assume that the deep-sea basins were untroubled by currents and erosion and therefore subject to very little change. Such a view was expressed by Admiral M. F. Maury in 1855:

> The geological clock may strike new periods; its hands may point to era after era; but, so long as the ocean remains in its basin—so long as its bottom is covered with blue water—so long must the deep furrows and strong contrasts in the solid crust below stand out bold, ragged, and grand. Nothing can fill up the hollows there; no agent now at work, that we know of, can descend into its depths, and level off the floors of the sea. . . .

Later, when it was recognized that surface waters of oceans contain vast quantities of small free-floating and swimming plants and animals, this view of the deep sea changed. Instead of bare exposed rock, the ocean floor was visualized as covered with a thick homogeneous blanket of sediment composed of remains of microscopic-sized marine plants and animals. Rachel Carson depicts this concept in her book *The Sea Around Us:*

> When I think of the floor of the deep sea, the single overwhelming fact that possesses my imagination is the accumulation of sediments. I see the steady, unremitting, downward drift of material from above, flake upon flake, layer upon layer—a drift that has continued for hundreds of millions of years, that will go on as long as there are seas and continents.

Sampling Sediment

A number of successful attempts were made in the early 1900's to obtain samples of the sediments on the continental slopes. Cores of the top layers of the sediments in the oceans were obtained, and it came as a surprise that some of these cores were stratified. This meant that changes in sedimentation do occur in the deep sea, contrary to the prevailing view that the deep seas were not subject to change. Processes of gradation that might cause stratification were not thought to be effective in the deep seas. That scientists had been laboring under a misconception became even more apparent when it was discovered that layers on the bottom were composed not only of layers of shell fragments of microscopic marine animals, called *oozes,* but that these were sometimes found interbedded with layers of silt and sand, sediments thought to be formed only at shallow depths on the continental shelves.

Cores are obtained with various types of coring devices. One of the most effective coring devices, the *piston corer,* consists of a long section of heavy steel pipe with a sharp rim at the bottom and a large weight attached to the top. The corer is lowered to the bottom on a long steel cable. Just above the top of the device the cable is looped and locked by a clamp, which is released when a small weight hung from it hits the bottom. This triggers the clamp and releases the pipe and weight, which fall freely into the sediments of the bottom. Inside the pipe a small piston moves up, easing the movement of the core into the pipe. Cores 17 to 20 meters long have been obtained with this instrument.

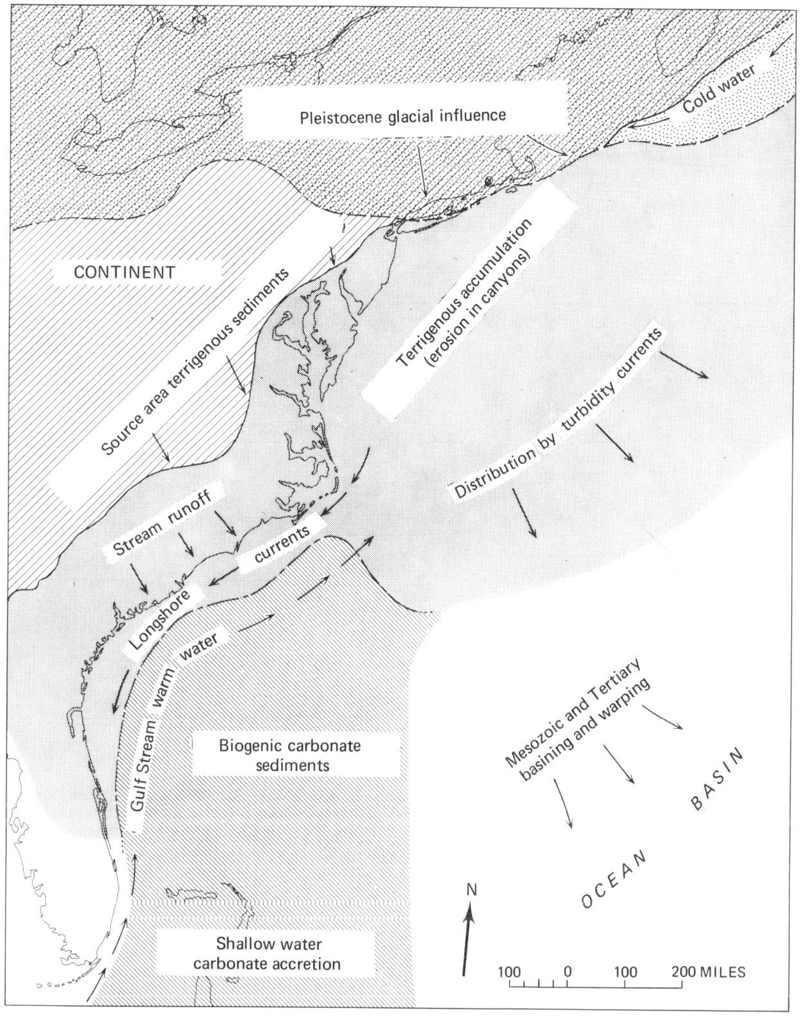

FIGURE 11–1
Diagrammatic summary of factors influencing sedimentation off the East Coast of the United States.
(from Pratt, 1968)

SOURCES OF DEEP-SEA SEDIMENT

The edge of the continental shelf, which is generally found at a depth of about 180 meters, is the boundary between deep-sea and shallow water. Water depths between 180 and 1,800 meters are called *bathyal*, and below 1,800 meters, *abyssal*. Both bathyal and abyssal environments receive most of their sediment from shallow water where it originates as either inorganic or organic matter.

Inorganic Matter

Types of inorganic matter include the following:

1. Finely suspended inorganic matter, largely composed of clay derived from continents and carried into the sea by streams. Dust derived from beaches, coastal areas, deserts, and volcanic eruptions may be blown far out to sea to become a component of the suspended or slowly settling debris in the oceans. Some of the sediment produced by glacial erosion remains suspended in sea water long enough to be transported into deep water. Sands derived from glacial erosion have been carried into the Atlantic through submarine canyons and dispersed as shown in Fig. 11–1.
2. Ice from continental ice sheets and valley glaciers that breaks free in the ocean can transport even large boulders into deep water.
3. Some bottom currents appear strong enough to move sediment and may be important in shifting coarse silt and sand into deep-water environments.
4. Shallow-water deposits of sand, silt, and mud initially deposited on the shelf, particularly near the edge, may slump and slide into deep water as turbid, dense flows of suspended sediment.
5. Meteorites that often are very small yet numerous contribute to deep-sea sediment.

Organic Matter

Included in organic matter are the following:

1. Plankton, the free-floating and drifting plants and animals largely composed of single-celled animals, protozoans, such as

the foraminifers (Figs. 11–2 and 11–3) and radiolarians (Fig. 11–4), and microscopic-sized plants such as diatoms and flagellates. The skeletal remains and hard parts of these organisms are a major source of sediment.

2. Nekton, the swimming organisms, dominated by the fish, of which there are some 20,000 species. The remains of such organisms become part of the sediment as do some of the waste products of their life processes.

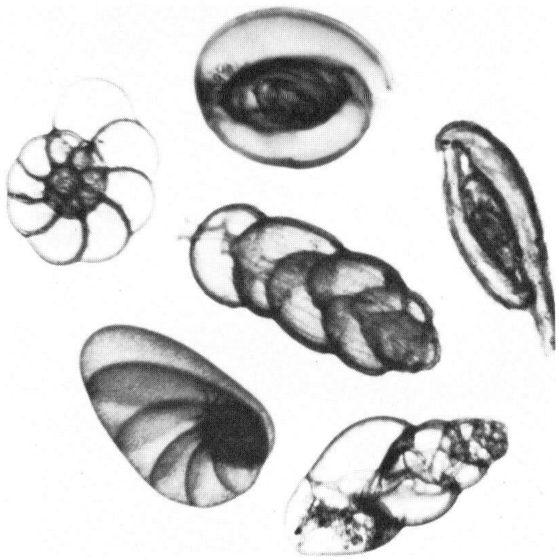

FIGURE 11–2
Foraminifer shells such as those found on the ocean floors. The protozoans shown here are about the size of a grain of wheat, but many are much smaller.

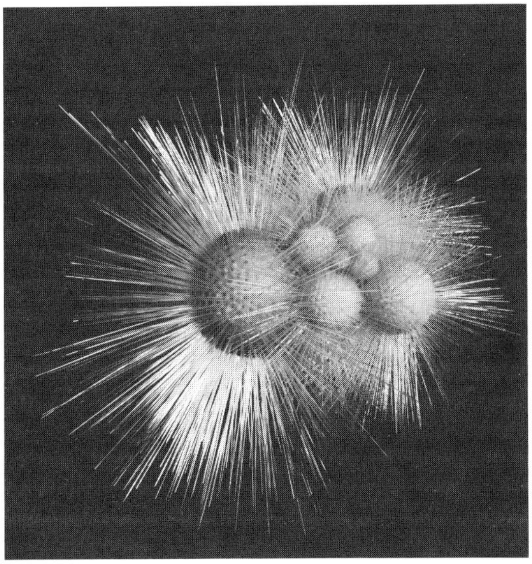

FIGURE 11–3
Globigerina, a type of foraminifer. Their remains make up a large part of the sediment on the ocean floors in middle latitudes.
(American Museum of Natural History)

Both bathyal and abyssal zones are free of surface-wave disturbance and relatively free of current action over vast areas. Internal tidal currents may affect them, but little is known about these currents. Neither zone receives light, nor do they contain nearly as abundant a *benthonic* (bottom-dwelling) population as shelf waters. Hydrostatic pressure is high; at 5,000 meters the pressure is about 500 bars (1 bar = 1 atmosphere), and temperatures are low, close to freezing in the abyssal plains.

DEEP-SEA SEDIMENT TYPES

Red clay. This clay, actually brown to reddish in color, is found at the greatest depths and is the second most extensive deep-sea deposit. It contains films of manganese and manganese nodules. The origin of red clay is subject to considerable debate; it may be derived from the finest clay fractions taken into the oceans from land areas, but it has also been attributed to submarine weathering of basic volcanic material as well as to accumulation of meteoric dust.

Red clays cover large areas in many parts of the ocean. They appear to be abundant where other types of sediment are lacking. Possibly these clays are deposited uniformly everywhere but are inconspicuous when mixed with large amounts of other sediment. The lowest measured rate of accumulation of deep-sea sediment is about 0.1 centimeter per 1,000 years. If it is assumed that this is the average rate for accumulation of cosmic dust, it amounts to about 10 million tons per day.

Coral sand and mud. These are locally abundant in deep-sea areas around coral reefs, having been transported and dropped by surface currents.

Muds of various colors and *volcanic sand* are among the less widely distributed sediments. Red mud is most prominent in the coastal areas close to the mouths of the Amazon and Yangtze Kiang rivers. Blue and gray muds are of continental origin and occur in both shallow and deep waters near land masses. Green muds probably derive their color from the mineral glauconite or chlorite. Volcanic muds and sands are frequently mixed with other deep-sea sediments in volcanic belts throughout the world.

Deposits carried into the abyssal plains as turbulent high-density flows exist in an abundance that is as yet undetermined. These may account for a large part of the total volume of ocean-basin sediment. Activated through slumping along the edge of the continental shelf, sediments of the neritic zone move downslope into deeper water. That these sediments can move down low slopes for hundreds of kilometers has been definitely established. They account for the presence of neritic fossil fauna, ripple marks, and coarse sands in the abyssal environment.

Sedimentary bodies deposited by density currents are often lobate in form. The lobes originate from areas on the shelf where slumps have occurred. Most of these areas have slopes that have become oversteep as a result of long-continued or rapid sedimentation. For this reason, density-current deposits are most likely to occur off steep continental shelves, where sediment is transported into the ocean by large rivers. Modern deep-sea trenches act as traps for sediment from adjacent land areas and prevent movement of turbidity currents into abyssal plains farther off the coast. Some trenches are filled or partially filled by such deposits.

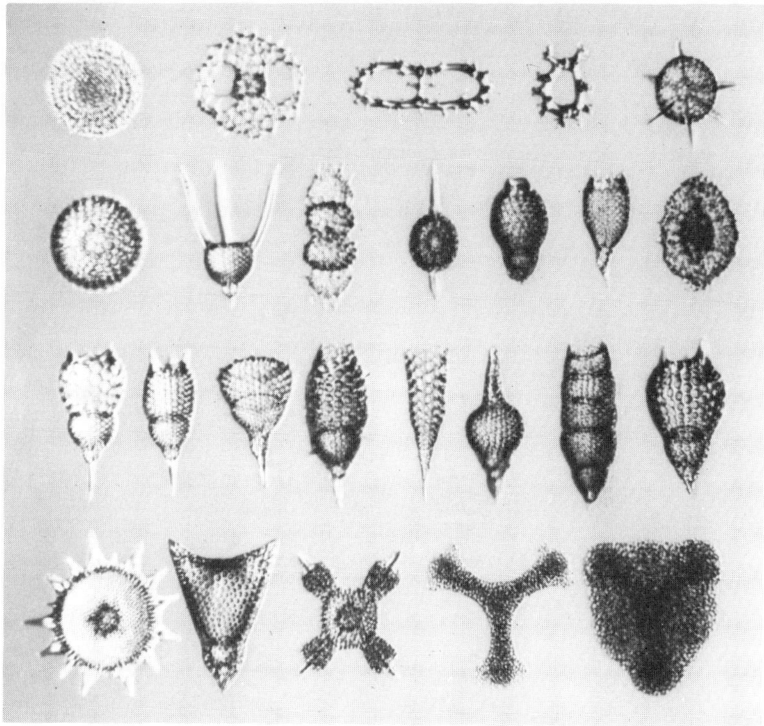

FIGURE 11–4
(*Above*) Radiolarians. These are important constituents of marine sediments, particularly in very deep water and in high latitudes.
(photo copyright by General Biological Supply House, Chicago)

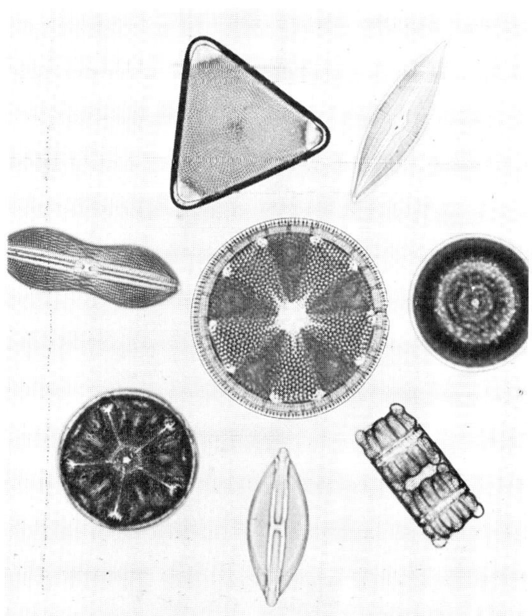

FIGURE 11–5
Diatoms. The siliceous remains of these microscopic-sized plants are important constituents of marine sediments.
(photo by courtesy of the General Biological Supply House, Chicago)

Plankton

The upper layer of sea water, the zone penetrated by sunlight, is far more densely populated than most of us would imagine. This is the domain of the plankton, the drifting plants and animals, most of which are no more than 0.05 centimeter in diameter. Marine plants constitute the base of the food chain of marine life; of these plants two groups, diatoms and flagellates, are most important.

Diatoms. Between 15,000 and 20,000 species of diatoms (Fig. 11–5) inhabit the widely varying marine environment. They are simple plants, algae, consisting of a blob of protoplasm with *chloroplasts* that contain the chlorophyll necessary for photosynthesis. They reproduce by cell division, and under favorable conditions the single cell may divide several times a day. Upwelling of nutrients and plenty of sunlight may result in growth of a prodigious number of diatoms, discoloring the water to produce what is sometimes called the *red tide.* As many as several million diatoms may be contained in a single liter of water. Diatoms have many different shapes (Fig. 11–5), but each contains a framework made of silicon dioxide.

Diatom ooze. This is a deposit consisting of siliceous remains of diatoms. The most extensive diatomaceous deposits are located between 40°S and the Antarctic Circle. These beautiful minute plants contain droplets of oil and may be the source of petroleum.

Flagellates. The flagellates are a second important group of microscopic-sized plants; they vary in shape, some resembling a balloon with a string attached, others a pot or fancy mask; all have whiplike projections that move. Because they can move, the flagellates are sometimes classed as animals but, like plants, they contain chlorophyll and manufacture food by photosynthesis. They may be even more abundant than diatoms; 60 million were counted in one liter of sea water collected off Florida, but most of them do not possess hard parts that are preserved in sediment. One exception, *coccoliths,* possess minute calcareous shieldlike plates.

Protozoans

Diatoms and flagellates are the food source of the single-celled animals of the sea, notably the *foraminifers* and the *radiolarians.* Foraminifers possess small shells composed of calcite. One of the most abundant of this group, the *globigerinas,* have globular-shaped shells that resemble snail shells (Fig. 11–2). These shells are a main constituent of chalk deposits such as the famous White Cliffs of Dover where some 50,000 shells may be counted in a single spoonful of chalk.

Radiolarians (Fig. 11–4) have an internal skeleton with radiating projections composed of silica; their skeletal remains are important constituents of sediment in some areas of the ocean.

Globigerina ooze. The most extensive deposit in the deep sea, globigerina ooze covers about 48 percent of the areas of deep-sea sediment. It is a calcareous mud formed from the shells of globigerinas. Almost no globigerina ooze is found below depths of 4,500 meters because calcite is soluble under high pressure and low temperature. Clays are usually mixed with globigerina shells in these sediments and may constitute more than 50 percent of the deposit.

Distribution of Planktonic Sediment

Diatoms are most abundant near the surface in high latitudes where the ocean waters are upwelling and fertile. Radiolarians and foraminifers are most abundant toward the equator, favoring the warmer water; they are also more abundant near shore than farther out because their food supply is greatest in shallow waters. Although the skeletons of these tiny creatures are the main constituents of deep-water sediment, they are of little importance in shallow water where their remains make up a small fraction of sediment dominated by sand and silt.

The relative solubility of calcareous and siliceous material is an important factor determining the dominance of siliceous sediment over calcareous sediment in very deep water. Calcium carbonate is more soluble than silicon dioxide in deep, cold waters. Thus siliceous remains of diatoms and radiolarians dominate these bottom sediments in trenches and other deeps.

DEEP-SEA SEDIMENTS OF THE PACIFIC OCEAN

The Pacific is different from other oceans in its great size and in the small number of major rivers that flow into it. The margins and southwestern part of the Pacific are highly active seismic and volcanic zones and deep-sea trenches located along the margins act as sediment traps. The deep water is more alkaline than that of most oceans and contains higher concentrations of silicates and phosphates. Finally, the equatorial current system is unusually well developed in the Pacific, and varieties of plankton are extremely abundant in these waters.

Diatom ooze forms bands across the northern part of the Pacific floor and across the Antarctic

Ocean (Fig. 11–6). A belt of calcareous oozes is found at the equatorial belt and particularly in the southern Pacific. Radiolarian ooze borders this calcareous belt on the north, but red clay dominates most of the bottom sediment in the northern Pacific and in large, scattered areas of the southern Pacific. This clay covers almost twice as much area in the Pacific as in the Atlantic, probably because of the greater areas of deep water in the Pacific. Turbidity-current deposits are known to be present in great abundance off the west coast of North America. The high coastal mountains that erode rapidly furnish a steady supply of terrigenous sediment to the narrow continental shelf on the west coast, thus favoring oversteep deposits and consequent slumping.

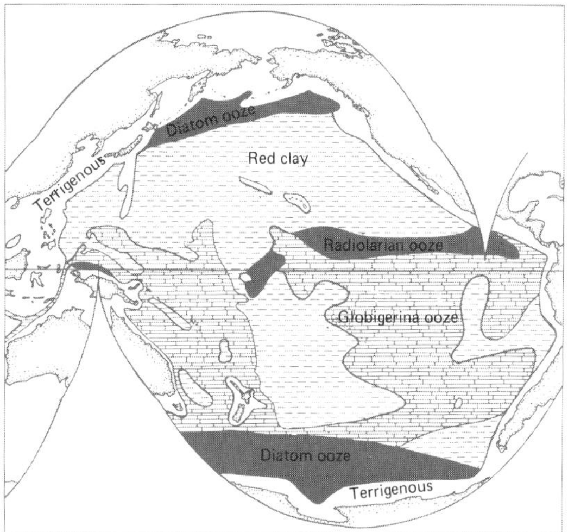

FIGURE 11–6
Distribution of principal marine sediments in the Pacific.
(after Sverdrup, Johnson, and Fleming, 1942)

Sediments of the Indian Ocean (Fig. 11–7) include almost all known types of recent deep-sea sediments. Globigerina ooze, red clay, and radiolarian ooze are found in the equatorial and temperate latitudes. Diatom ooze and glacial-marine sediments are prominent in subpolar and polar regions.

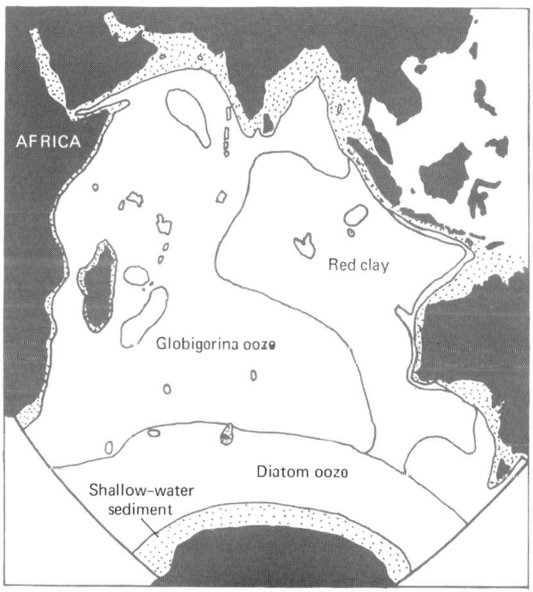

FIGURE 11–7
Distribution of sediments in the Indian Ocean.
(after Wolfgang Schott, 1939)

TURBIDITY CURRENTS AS AGENTS OF GRADATION

Early sounding methods revealed submarine canyons located near the mouths of rivers and traceable across the continental shelves. They were thought to be seaward extensions of the rivers during the ice ages when the water levels of the oceans were lower. But with improved sounding devices it was found that many of these canyons do not begin at the mouths of rivers and that most of them extend beyond the continental shelves. In fact, they extend down the slopes and across the rises, and a few can be traced into the abyssal plains. How can such features be explained? Could sea level have been lowered far enough to account for channels 5,500 meters below sea level? No other evidence supports such a hypothesis.

The failure of transatlantic cables laid across the deep-sea floor led to investigations that have yielded important insights to the processes of gradation in the deep sea. In some instances these breaks were accompanied by earthquakes, and it was assumed that faulting was the cause of the breakage, but when the cables were brought up for repair it was usually discovered that the cables were irregularly abraded, as might be expected if they were laid across the channel of a river. No evidence indicated that the cables had been stretched and pulled apart. It was later noted that along certain coasts the cables broke in the spring year after year.

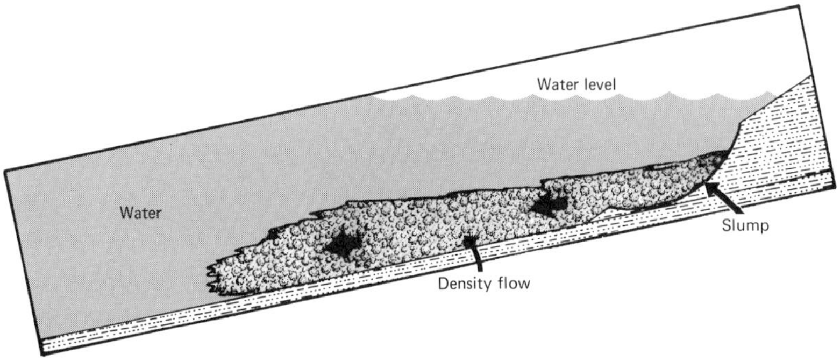

FIGURE 11–8
Formation of a density or turbidity current by experiment. A layer of fine sediment is first placed in the tank. Then a pile of sediment is laid down at the right end of the tank to approximate the continental slope. As the tank is tilted, the oversteepened slope slumps and a turbidity current is started.

The Concept of the Turbidity Current

The concept of turbidity or density currents has proved to be a unifying explanation for what were previously thought to be the unrelated problems of cable breaks, submarine canyons in the deep sea, and the presence of clastic sediment in the abyssal plains. The idea of turbidity currents is old, but its application to the gradation of the deep seas is new. A *turbidity current* (or density current) is a mass of water highly charged with material in solution or suspension that flows with turbulent motion down slopes through normal marine waters. The turbidity current flows beneath water by virtue of its greater density. It sinks and flows downslope as a mixture of mass movement and stream flowage in a manner analogous to a mudflow.

The nature of turbidity currents was first studied experimentally by a prominent Dutch geologist, Ph. H. Kuenen (1936), and later by many others. Kuenen induced turbidity currents in two ways: (1) by introducing mixtures of water with colloids and fine sediment suspended in it at the top of a *flume* (artificial channel) or (2) by laying a thick layer of sediment at the upper end of the flume and increasing the inclination of the flume until a slump or slide occurred (Fig. 11–8). When small amounts of muddy water were introduced, a slow current was produced that flowed downslope, but very little erosion or deposition accompanied the flow. With an increase in the amount of load in the current, however, the velocity was greatly accelerated, and the stream of turbid flow eroded the sediment on the bottom and carried it rapidly down into the lower end of the flume.

Evidence of Turbidity Currents

A series of breaks occurred in cables on the sea floor off the Grand Banks of Newfoundland in November of 1929. The exact time of each break was recorded, since the communication along that cable was disrupted at that moment.

A full-scale investigation of the breaks was made in the early 1950's by the Lamont Geological Observatory. Bruce Heezen and Maurice Ewing advanced the hypothesis that the breaks had been caused by a turbidity current. Presumably, the turbidity current was initiated by a slump or landslide from the steep edge of the Grand Banks (Figs. 11–9 and 11–10). This might have been caused by the earthquake, or the earthquake might have been simply a shock produced by the slump. The movement of the unconsolidated sediments from the edge of the Grand Banks set much of the sediment into suspension, and it moved down across the slopes as a density flow traveling at very high rates. The velocity could be calculated from the times of the cable breaks. It required a movement of 25 meters per second. The turbid flow extended all the way down the slopes a distance of 700 kilometers into the abyssal plains.

It is now generally accepted that the hypothesis is an accurate account of what happened. Proof lay in the following discoveries made through extensive studies of the area. Deep-sea cores were taken throughout the area. It was found that a layer of graded silt almost 1.2 meters thick covered the normal deep-sea sediments and that this layer contained shallow-water microfossils. This layer covered a normal deep-sea sedimentary unit, such as red clay. It was possible to outline the lobate form of the turbidity current from the core samples. No sign of the turbidity-current deposits could be found on top of the hills near the foot of the continental slope. Because of its high density, the flow moved along the low places in the submarine topography and did not go over the small hills. Photographs obtained by use of a specially designed underwater camera showed that the silty sediment was covered by ripple marks—a definite sign of relatively strong currents.

Further evidence has been obtained from a number of other sites. Cable breaks off the mouth of the Magdalena River of Colombia occurred year after year at just about the same season. Invariably the breaks took place during the highest flood stages of the river. At these times the river carried large quantities of debris out from the continent, deposited it in the waters off the mouth of the river, and usually built up an alluvial cone of debris in the submarine canyon of the river. During the floods it was noted that the top of these deposits disappeared just before the cable broke. In 1935 when the cable was brought up from its position 19 kilometers offshore in nearly 1,500 meters of water, it had green grass wrapped around it—dramatic proof that the river debris had been carried out to the cable and almost certainly played a major role in its failure.

Soundings of the deep-sea trenches show that alluvial conelike deposits are accumulating on the sides of the flat bottoms of some of these trenches. Thus, the trenches that almost completely encircle the Pacific Ocean are collecting much of the sediment that otherwise would be transported out into the ocean by turbidity currents. This seems to be verified by the absence of turbidity-type deposits in the central Pacific basin. Large areas of the Pacific Ocean bottom are covered by scattered nodules of manganese. The nodules are estimated to form at the rate of 1 millimeter in 1,000 years. These nodules often contain teeth of Tertiary sharks as nuclei. Therefore it is possible that either little deposition or much sediment movement has occurred since the teeth were first deposited. If the trenches continue to receive the turbidity-current deposits, one may expect they will eventually become filled. Parts of the Aleutian and Peru trenches are filled by sediment and a former trench, now filled, has been discovered west of the Philippines.

Once the trenches are filled, turbidity currents transport their loads into the abyssal plains. Modern trenches prevent this in the Pacific, but few trenches are present in the Atlantic to intercept the sediment. Sediment transported into the ocean basins by turbidity currents may be the main factor in the formation of the abyssal plains. They may be plains of deposition covered by the combined processes of the slow accumulation of shell fragments and extraterrestrial dust and the rapid deposition of mud and sand transported by turbidity currents.

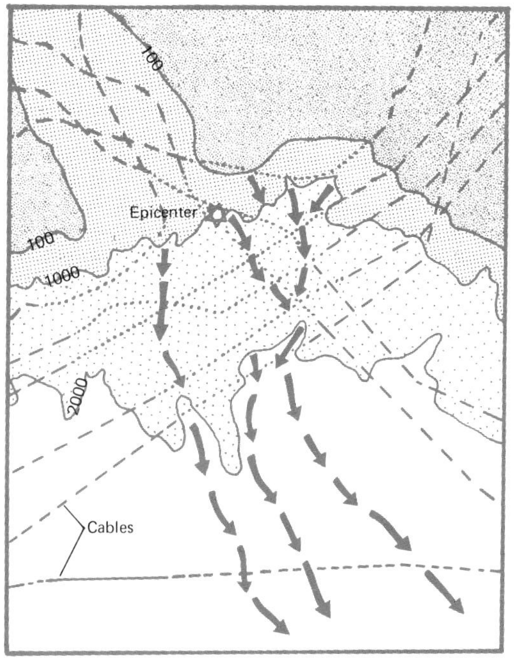

FIGURE 11–9
Map of the Grand Banks slump (depth is in fathoms). All cables within about 96 kilometers south of the epicenter of the Grand Banks earthquake broke at the instant of the earthquake. A sequence of later cable failures occurred to the south, the first one 59 minutes after the earthquake. The star indicates the position of the epicenter. Arrows show flow of density currents.
(after Heezen and Drake, 1964)

FIGURE 11–10
A tracing of the seismic reflection records made along a profile across the continental slope and continental rise south of the Grand Banks showing the Grand Banks slump. All cables crossing the area of the slump were broken at the instant of the earthquake.
(after Heezen and Drake, 1964)

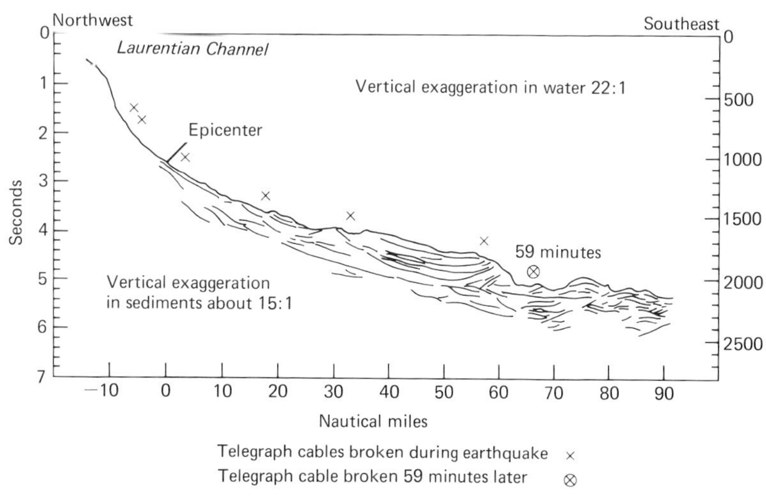

This view helps explain why the midocean canyon in the Atlantic has sides slightly raised above the level of the ocean floor. The geomorphic form is much like a river levee and probably formed in the same manner. The turbidity currents flowed down the canyon, overflowed the sides of the canyon, and the debris-laden water deposited sediment near the sides of the canyon in a wedge that thinned out away from the main path of flow.

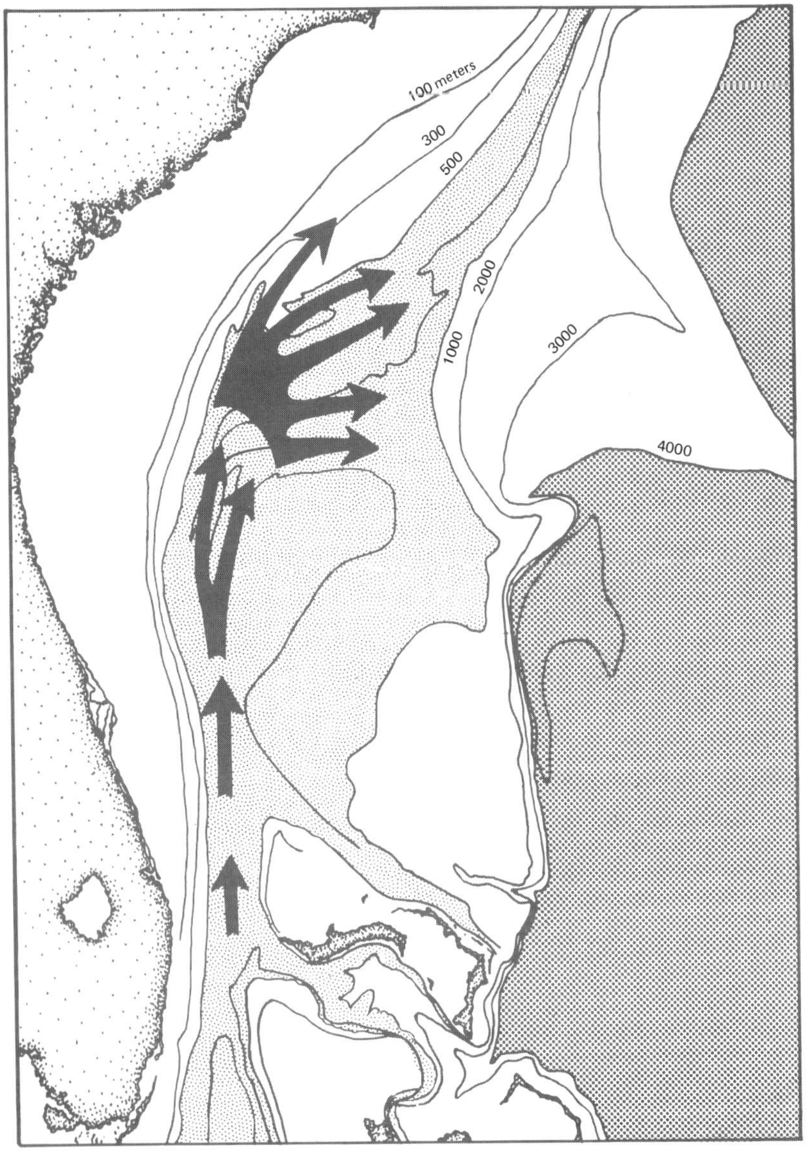

FIGURE 11–11
Arrows indicate position of channels through which sediment has been transported. The broad relatively flat area covered by a light dot pattern is part of the Blake Plateau.
(after Uchupi and Emery, 1968)

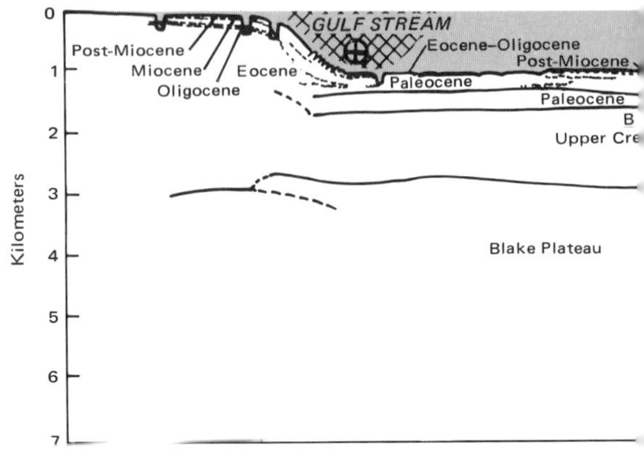

DEEP CURRENTS AS AGENTS OF GRADATION

Bottom currents were little known until recently, and it was assumed that movement of water along the ocean floor was of such a low velocity that it would not be capable of transporting sand or silt. Thus erosion and deposition of sediment by current action in deep water was considered negligible. We are still in the early stages of developing a knowledge of bottom currents, but recent evidence indicates these currents may be far more important than was previously supposed. One such investigation was made along the continental rise and the Blake Plateau off the southeastern United States (Fig. 11–11) (Heezen, Hollister, and Ruddiman, 1966; and Uchupi and Emery, 1968). The complex topography and current systems in this area make it a particularly interesting situation. The continental shelf east of Florida leads onto a slope that levels out at a depth of about 1 kilometer to form the Blake Plateau (Fig. 11–12). A sharp escarpment defines the outer edge of the plateau and separates it from a sea floor 5 kilometers deep. The Blake-Bahama outer ridge lies between the escarpment and the Hatteras abyssal plain.

The Gulf Stream moves along the coast with its northerly flow concentrated on the inner side of the Blake Plateau (Fig. 11–12) at rates in the range of 100 to 300 centimeters per second, a velocity easily capable of moving the terrigenous sediment that crosses the shelf. The position and strength of the stream prevent the movement of such sediment toward the eastern part of the Blake Plateau and the Blake-Bahama Basin. Yet terrigenous sed-

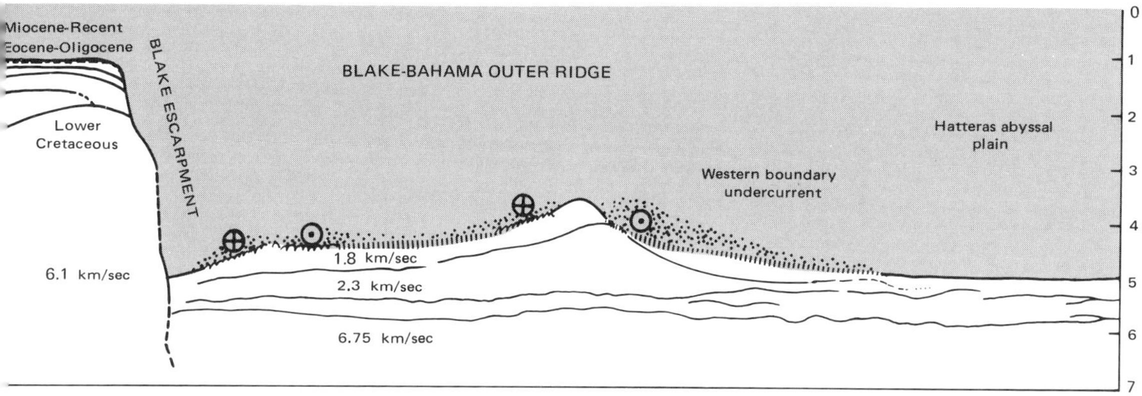

FIGURE 11–12
Profile across Blake Plateau and Blake-Bahama outer bank showing current directions and seismic cross section. The circular symbol with crosshairs indicates a current flowing into the figure; the circular symbol with a dot indicates a current flowing outward.
(from Heezen and others, 1966)

iments like those found on the continental slope and rise are present east of the Blake escarpment. These terrigenous sediments include such materials as fine sands, silts, and clays that must have been transported into their present position by current action. Since the sediment could not have been moved across the Gulf Stream, it probably was brought in by the currents that flow along the bottom (Fig. 11–12). Photographs of the bottom have revealed ripple and scour marks on the Blake-Bahama outer ridge, particularly concentrated on the eastern side of the ridge, confirming current action there.

Coring in the region reveals the following general pattern of sediment thickness and distribution:

1. Quartz sands with a high percentage of shell fragments are prominent on the continental shelf.
2. Shelly green sand and phosphorite occur on the western side of the Gulf Stream, and coral sand and manganese nodules occur beneath the Gulf Stream. Thickness is variable.
3. Globigerina oozes are dominant on the Blake Plateau east of the Gulf Stream except near the outer edge where the Recent and Pleistocene (ice age) sediment is absent and Tertiary sediments are exposed.
4. The outer ridge area is covered by a relatively thick sequence of Pleistocene sediment. This thickens from the Hatteras abyssal plain toward the outer ridge. Pleistocene sediment was not penetrated in some cores of 10 to 20 meters length. The thickness variations appear to follow a pattern of thinning out downslope. This has been detected in seismic profiles along the continental rise in other places and has led to the suggestion that these deposits on the continental rise owe their origin to deep-sea currents moving along and more or less parallel to the rise (Fig. 11–13).

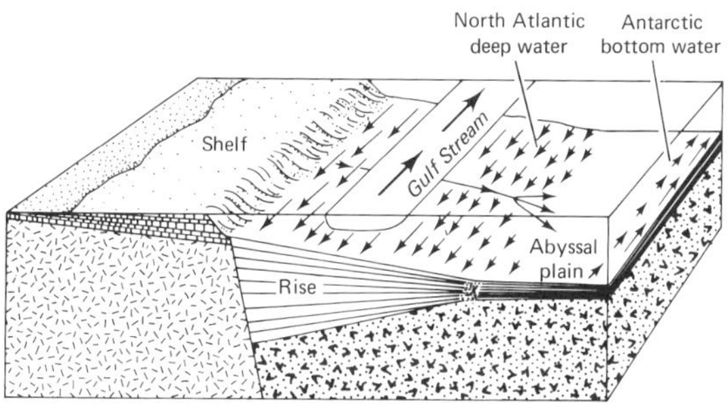

FIGURE 11–13
Shaping of the continental rise by contour currents. Arrows indicate prevailing bottom currents. Continental and oceanic crusts are shown by patterns. Sedimentary rock is shown by conventional symbols; turbidity deposits by horizontal ruling, and rise deposits by open wedges. Although it does not take into consideration the shifting position of the Gulf Stream, this schematic diagram, drawn on the basis of an average section off eastern North America, is intended to illustrate the principal processes shaping a "normal continental rise" in any part of the world.
(Heezen and others, 1966)

The bottom-water movement along the outer ridge system is from north to south in what is known as the *western boundary undercurrent.* This current is strong enough to move clay and fine silt, but not strong enough to erode it; therefore, it is probable that the current has moved sediment down along the rise. The Coriolis force has deflected the mass transport to the right against the rise where deposition has occurred with increasing thicknesses up the slope. Heezen, Hollister, and Ruddiman (1966) suggest that the continental rises may have been built through deposition of sediment transported by bottom currents moving along the slope (not downslope). Where faulted continental margins are found, the rises could have been maintained by this process, eventually producing a succession of wedge-shaped sedimentary units, all of which thin out toward the ocean basin.

CHAPTER OUTLINE

Introduction
- Sampling Sediment

Sources of Deep-Sea Sediment
- Inorganic Matter
- Organic Matter

Deep-Sea Sediment Types
- *Red clay*
- *Coral sand and mud*
- Plankton
  - *Diatoms*
  - *Diatom ooze*
  - *Flagellates*
- Protozoans
  - *Globigerina ooze*
- Distribution of Planktonic Sediment

Deep-Sea Sediments of the Pacific Ocean

Turbidity Currents as Agents of Gradation
- The Concept of the Turbidity Current
- Evidence of Turbidity Currents

Deep Currents as Agents of Gradation

TERMS

benthonic	flume	plankton
chloroplasts	foraminifers	radiolarians
coccoliths	globigerina ooze	red clay
diatom ooze	globigerinas	red tide
diatoms	oozes	turbidity current (density current)
flagellates	piston corer	

REVIEW QUESTIONS

1. After studying Figs. 11–1 and 11–11, give several reasons why terrigenous sediments do not dominate sediment accumulation along the shelf of southern Florida, and why sediment types differ on the slope and rise north of Cape Hatteras from those found southwest of Hatteras.

2. In what ways is the sediment accumulation in the Pacific Ocean similar to that in the Indian Ocean? Compare Figs. 11–6 and 11–7.

3. Obtain a physiographic map of the areas surrounding the Pacific and one for the Atlantic and compare the number of major streams entering each ocean. How does this help you understand the distribution of sediment in the oceans?

4. What physical conditions at the surface (e.g., climate, currents, location of land, etc.) are indicated by deep-sea sediment? How could you use these as evidence to describe conditions at a relatively specific time in the geologic past?

5. Explain possible origins for ripple marks and graded beds in deep-sea sedimentary deposits.

6. What conditions cause turbidity currents?

7. Describe the techniques you would use if you were planning to undertake a study of an area of the sea floor in an effort to determine the types of modern sediment present and the processes by which the sediment is formed.

8. What conditions are most favorable for the formation, preservation, and exposure of manganese nodules on the sea floor?
9. Suggest some ways deep oceanic currents might be detected.
10. There were few marine organisms and no known terrestrial plants or animals on earth during the Cambrian period. How would this have affected sedimentation in the deep sea?

SUGGESTED READINGS

Ericson, D. B., and **Wollin, G.,** 1964, The Deep and the Past: New York, Alfred A. Knopf, 292 p.

Idyll, C. P., 1964, Abyss, The Deep Sea and the Creatures That Live in It: New York, Thomas Y. Crowell, 396 p.

Pettersson, Hans, 1969, The Ocean Floor: New York, Hafner, 179 p.

Shepard, F. P., 1964, The Earth beneath the Sea: New York, Atheneum, 275 p.

Shepard, F. P., and **Dill, R. F.,** 1966, Submarine Canyons and Other Sea Valleys: Chicago, Rand McNally, 381 p.

III
THE CHANGING LANDSCAPE

12 GRADATION OF THE EARTH'S CRUST

12 GRADATION OF THE EARTH'S CRUST

GRADATION

Man is one of the main agents causing changes in the surface of the earth today. Yet all around us there is evidence of changes in the surface of the earth of a sort man could never hope to duplicate. In August of 1959 an earthquake in central Montana set off seismic waves that triggered a gigantic landslide. In less than five minutes 27 million cubic meters of rock moved half a mile. In 1960 violent earthquakes in Chile caused landslides that led to the loss of many lives and extensive property damage over thousands of square kilometers. Tidal waves caused by these earthquakes crashed on shores surrounding the Pacific Ocean, resulting in millions of dollars in damage as far away as Japan. Hurricanes annually batter the southeastern United States and the islands of the Caribbean, creating severe wave and current action, and sometimes accompanying rains cause vast flooding inland, as happened along the Rio Grande Valley in 1967 and in Virginia in 1969. Rapid changes of this sort are certainly catastrophic in character, and there are hundreds of earthquakes, floods, and other natural disasters each year. These catastrophes are important in modifying land forms, but most sculpture of the earth's crust takes place through processes acting so gradually that man rarely even thinks of them as significant. It is the slow, downhill movement of soil, the bits of dislodged rock that fall downslope, the gradual solution of limestone by rainwater percolating through it, the movement of grains of sand by the wind, and the constant transport of suspended loads, dissolved solids, and channel fill in streams that over long periods of time account for the removal of most material. We are conscious of the muddy water in streams after heavy rainfalls. We take note of the shifts in sand bars along the coast. We see glaciers, covered by rock fragments, slowly pushing their way down mountain valleys. But we do not always perceive that this is part of a large-scale pattern by which the land forms so familiar to us were brought into existence and by which they will one day vanish from the face of the earth.

The pattern is one of lowering of the earth's surface, shifting of sediments, and redeposition elsewhere. This lowering of the surface, or *degradation,* is accomplished through processes known as *weathering, denudation,* and *erosion. Weathering* includes all those physical, chemical, and biological processes by which rocks and minerals exposed at or near the ground surface are caused to disintegrate or decompose. The stripping away of vegetation, soil, or unconsolidated sediment to expose the underlying bedrock is known as *denudation. Erosion* includes all processes by which earth materials are loosened and moved from place to place. It is often used to include both weathering and transportation, but many geologists prefer to restrict its use to processes involving movement of materials. *Degradation,* then, is the lowering of the surface of the land by erosive processes (as defined by La Forge).

Movement of materials is accomplished by water, glacier ice, downslope movement of surficial materials, and wind. It is almost always directed from higher elevations to lower elevations on land, and it is driven primarily by the force of gravity. Thus the long-range tendency of these processes is toward a leveling of the irregularities of the earth—a cutting down of the high places and a filling in of the lower ones. It is estimated that the rate of removal of materials (primarily by stream action) from the continents today is such that between 20 and 40 million years would be required to level the continents. The earth is estimated to be at least 4.5 billion years old; thus enough time has elapsed for the surface to have been eroded flat many times over. It is obvious, therefore, that there are some processes that more than balance degradation. These are constructive forces; they are diastrophism (mountain-building processes), the upheaval of the earth accompanying rock deformation, and volcanism. By these phenomena material is moved from lower to higher positions. Thus the particular configuration of the surface of the crust at any given time, the present, for instance, represents a sort of balance between the forces of upheaval in the crust and the forces of erosion, denudation, and degradation. Throughout most of the geologic time for which a rock record is available, continents have persisted as land areas, although their shapes and sizes have varied.

Gradation—a threefold process. First, the surface is decayed and eroded; second, the products of this decay and erosion are transported; and third,

FIGURE 12–1
Aerial view of the Esk and Waimakariri rivers in New Zealand showing the effects of long-continued erosion by streams. On both sides of the rivers, streams have cut down through the smooth, terraced surfaces that were once continuous across the valley.
(copyrighted by V. C. Browne)

they are deposited, usually at lower levels. Erosion is facilitated by chemical decomposition and mechanical disintegration of rocks, called weathering, where the rocks are brought into contact with the atmosphere, waters of the surface, and plant and animal life. Both types of weathering and processes of erosion are determined in large part by the climate. Climate is often expressed in terms of temperature and amount of precipitation; high temperatures and precipitation favor chemical decomposition and stream action, whereas low temperatures and high precipitation favor glaciation and mechanical disintegration. Once the rocks are decomposed or have partially disintegrated, they are easily moved. If they are on sloping surfaces they tend to move down those slopes under the constant pull of gravity. Eventually, they may come in contact with running water or glacier ice, and they are carried along until either the velocity of the water becomes too low to move them or the ice melts. Then they are deposited and may eventually become consolidated, compacted, and cemented into sedimentary rocks. The sediments may be deposited on land, but the transitional environments, the continental shelves and slopes, receive much larger quantities of sediment. Thus the processes of gradation are divisible into two major categories:

Degradation. Weathering, mass movement, and erosion, including removal and transportation.

Aggradation. Deposition of sediments.

GEOMORPHIC AGENTS

Certain agents function to bring about gradation and the resulting changes in land forms. These may be called geomorphic agents. The most important of these are the following:

ON LAND:

Mass movement: driven by the pull of gravity.

Running water: driven by the pull of gravity.

Ground water: driven by the pull of gravity.

Glaciers: driven by the pull of gravity.

Wind: driven by the rotation of the earth and thermal activity within the atmosphere.

IN OCEANS:

Mass movement: driven by the pull of gravity.

Waves: driven by the rotation of the earth and wind.

Currents: driven by the rotation of the earth, wind, and by the temperature and density distribution of water masses in the ocean.

Tides: driven by gravitational attraction of the moon and sun.

Tsunami (tidal waves): caused by earthquakes and volcanic explosions underwater.

These agents should not be thought of solely as agents of erosion because each serves three functions: (1) they make erosion possible; (2) they provide a means of transportation; and (3) under certain circumstances they promote deposition.

Mass movement is the most pervasive of these agents. It functions almost everywhere to some extent, and while it usually plays an important part in the effect of all the other agents, it may be important by itself in the absence of other agents, particularly where the ground slopes are steep. Slump, slides, dislodgment, rolling, and falling of masses of soil or rock are all expressions of the phenomenon of mass movement. As the amount of water present increases, the processes take on the character of those associated with running water and stream action. As the amount of snow and ice present increases, the processes resemble those associated with glaciers. Mass movement is effective both on land and underwater, and it is an efficient agent of gradation in both environments.

Streams, glaciers, ground water, and wind are more effective as agents of transportation over long distances than is mass movement; however, wind can move material of only relatively small size and ground water transports mainly dissolved constituents, so these two are most effective under special conditions. Ground water is most efficient where soluble rock is exposed at or near the ground surface; wind is most effective where there is fine material on the ground and little moisture. Streams and downslope mass movement account for most transportation of materials on the land surface, just as a combination of slump and high-velocity turbidity currents charged with a load of suspended sediment accounts for most movement of solids in the ocean basin. The processes of gradation that take place on the continents and on the sea floor are similar in some respects but quite different in others. The constant pull of gravity downward on all materials is pervasive in both environments, and mass movement is also effective in both. The physical properties of air and water are quite different, and these differences have a marked influence on the processes of gradation. All materials underwater are buoyed by the water. Almost all mass movement in the sea involves saturated material that has lower internal friction than similar dry material. Finer sizes can become suspended and move as flows of high-density fluids. Glaciers and streams become relatively ineffective agents of erosion when they enter the sea. Glacier ice is not only subjected to melting because most sea water is above freezing, but also the ice floats and the effectiveness of the glacier's scouring and gouging action is reduced in shallow water and eliminated in deeper water. Similarly, the efficiency of streams as agents of gradation is greatly reduced or eliminated when they enter the sea. The effects of wind are completely different in the sea from on land. At sea, the wind drives surface currents and generates waves; on land, however, materials can be rolled and even taken into suspension by the wind.

Chemical and biological factors play a major role in gradation in marine and transitional environments. Salts dissolved on continents are carried by streams into the ocean where the ions are added to the salt content of ocean waters. Chemical precipitation and biological processes in the oceans lead to the eventual return of the salts to a solid sediment. For example, halite may be directly precipitated; calcium carbonate is precipitated and removed by organisms for use in construction of shells, which in turn sink to the sea floor; silica is used by some marine plants and animals to construct hard siliceous parts; and magnesium is precipitated in the form of nodules now known to be on many parts of the ocean floor.

CONCEPT OF BASE LEVEL

If the surface of the earth were not affected by diastrophism and volcanism, it would be reduced slowly to a uniform level. That level is called *ultimate base level*. Most of the processes tending to lower the continents cease to be effective when the area approaches the level of the oceans, so sea level may be taken as the ultimate base level. It should, however, be remembered that if reduction of the entire earth to sea level did occur, sea level would be substantially above its present position. All the volume now above water would be under the sea just before attainment of ultimate base level.

The term *temporary base level* is used to describe a transient condition for a region or small area. It is a level below which the dominant forces of erosion will not reduce the area. If a region is being eroded mainly by running water, it will not reduce the level of the region below the level of a lake into which the river drains, so long as the lake remains. Why? Because the lake stands in relation to the local area just as the oceans do to the continents.

FIGURE 12–2
The action of geomorphic agents gives the land its characteristic shape. Streams have been effective here in Lindis Pass, New Zealand.
(National Publicity Studios, New Zealand)

FIGURE 12–3
The sedimentary layers of the earth's crust have been folded and subsequently eroded through to produce unusual patterns as shown in this oblique photograph in Africa.
(U.S. Air Force)

RELATION OF TOPOGRAPHY TO GRADATION PROCESSES

Geomorphic processes leave their imprint upon land forms. Each of the geomorphic agents functions in ways particular to it and as a result produces erosional and depositional land forms that are characteristic of that agent. It is possible to recognize the origin of land forms even though the processes by which they were developed may have ceased to operate. For example, valley glaciers erode the floor and valley sides to produce a typical U-shaped profile; high, sharp peaks like the Matterhorn are produced near the heads of glacial valleys, and deposits of the debris eroded from the valley are laid down in characteristic forms along the sides and near the ends of glaciers. Thus glaciated valleys are recognizable long after the last traces of ice are melted. Similarly, wave-eroded platforms and beaches now elevated high above sea level can be recognized by their shape and the nature of the deposits associated with them.

The structure of the crust acts as a control in the creation of land forms. Such representative features as folds, faults, unconformities, differences in the hardness and susceptibility of rock units to decomposition, and disintegration provide a framework within which the geomorphic agents function. The resulting forms are always influenced by the structure of the underlying rock. Thus a particular structure being eroded primarily by one geomorphic agent may often result in a predictable type of land form.

If a particular structure is subjected to the effects of certain geomorphic agents acting over a long period of time, a particular sequence of land forms will develop. These slowly will undergo systematic change or evolution unless some external conditions bring about a change in either the dominant structure or the geomorphic agents. Several quite different views have been proposed to explain the evolution of land forms. The essential difference is that some geologists think there are progressive changes involving lowering of slope, whereas others think that land forms are modified until a state of equilibrium is reached when there is a gradual lowering of the region toward its base level, during which the shape of the land and slopes are not progressively changed.

CHAPTER OUTLINE

Gradation
Gradation—a threefold process
Geomorphic Agents
Concept of Base Level
Relation of Topography to Gradation Processes

TERMS

aggradation
degradation
denudation
erosion
temporary base level
ultimate base level
weathering

REVIEW QUESTIONS

1. Distinguish between degradation, weathering, erosion, and denudation.
2. Give examples of various types of temporary base levels.
3. Why is modern sea level not the elevation of ultimate base level?
4. Which geomorphic processes acting on land are not effective in the oceans and which processes are effective in the ocean but not on land?

SUGGESTED READINGS

Bloom, A. L., 1969, The Surface of the Earth: Englewood Cliffs, N.J., Prentice-Hall, 152 p.

Chorley, R. J., Dunn, A. J., and **Beckinsale, R. P.,** 1964, The History of the Study of Landforms or the Development of Geomorphology: London, Methuen, 678 p.

Fairbridge, R. W., 1968, The Encyclopedia of Geomorphology: New York, Reinhold, 1295 p.

Shimer, J. A., 1959, This Sculptured Earth, the Landscape of America: New York, Columbia University Press, 255 p.

13 MOVEMENT OF SURFICIAL MATERIALS BY GRAVITY

13 MOVEMENT OF SURFICIAL MATERIALS BY GRAVITY

INTRODUCTION

Downhill movement of rock and soil caused by the force of gravity is the most universal of all processes of erosion. Such mass movement mechanisms as landslide, slump, earth flow, sheet wash, soil creep, and subsidence in combination with transportation by running water, glaciers, wave action, wind, ground water, and sea currents are responsible for most erosion. No matter where you are, you do not have to look far to find evidence of mass movement. Principles governing it are simple, but the variety of combinations of types of movement, of materials moved, and of geomorphic forms assumed by these masses is great.

The driving force behind all mass movement is the force of gravity. This force is directed toward the center of the earth, but components of it act along any inclined plane. The steeper the inclination, the greater will be the component of force acting down the slope. This force is most effective in moving materials that are unstable in their existing position, such as loose or water-saturated soil, or masses of rock and soil resting on surfaces over which they might slide, for example, fractures or bedding planes.

Downslope movements are of major importance locally and in combination with streams that are responsible for much long-distance transportation. Downslope movements occur under all climatic conditions—in the air and under the waters of the oceans. Mass movement is even significant on the moon where there is neither atmosphere nor water.

Not only natural mass movements and resultant dangers are important to man; instability may be induced in rock and earth masses by our own modifications of the natural environment. As use of land is intensified, it becomes increasingly difficult to avoid areas where potential natural mass instability exists—areas where floods, wave action, or earthquakes may trigger disastrous mass movements. Engineers must take great care to circumvent costly mass movements in planning and constructing building foundations, dams, reservoirs, bridge abutments, tunnels, and in designing cuts and fills along highways and canals. Costly and sometimes even disastrous results have followed where the dangers of potential mass movement have not been fully recognized or efforts to meet the danger have been too limited.

CLASSIFICATIONS OF MASS MOVEMENT

Numerous classifications of mass movements have been proposed. These are based on the type of material moved, the rate of movement, the presence or absence of water, or the mechanisms of movement.

In referring to materials moved, the terms rock, earth, soil, debris, and mud are commonly used.

Rock. Applied to solid rocks and consolidated fragmental rocks.

Earth. General term used to describe disintegrated rocks and loosely consolidated sediments.

Soil. Product of rock disintegration and decomposition by weathering modified by biological agents; capable of supporting plant life.

Debris. General term applied to mixtures of rock, soil, plant matter, or mud.

Mud. Mixture of water and the finer particles of earth and soil.

Movements may take place through sliding along some surface or through an internal rotation of constituent particles, resulting in general flowage of the entire mass. Sliding is promoted by the existence of faults, fractures, bedding planes, and other planes of weakness. When these are inclined downslope, the material above can move along them. Flowage is a much more complicated process, involving rotation, slippage, or sliding of the materials inside the moving mass. Flowage of masses may take place slowly in nearly imperceptible movements called *creep* or at velocities of up to 96 kilometers an hour, as in debris avalanches. Flowage is possible in dry masses, but high water or ice contents have pronounced effects on the nature and velocity of the flow.

The following classification, devised by Sharpe (1938), is widely used. It separates the major categories on the basis of rate and type of movement, and subdivisions are made on the basis of the type of material moved.

Slow flowage
- Rock creep
- Talus creep
- Soil creep
- Rock-glacier creep
- Solifluction

Rapid flowage
- Earth flow
- Mudflow
- Debris avalanche

Sliding
- Slump
- Debris slide
- Debris fall
- Rock slide
- Rock fall

Subsidence

FIGURE 13–1
Soil creep causes these wrinkles on the ground surface.

Creep

Evidence of the slow downhill movement of soil or unconsolidated sediment is apparent on close inspection of almost every hillside. Weathered remnants of rocks and boulders may be drawn out into long, lens-shaped masses by creep. The small, parallel rows (Fig. 13–1), originally paths used by grazing animals, that circle many hillsides are signs of this creeping downslope movement. Fences and telephone poles set on slopes give indication of these surface movements as they slowly become inclined downhill. Observations of creep phenomena like the ones cited above indicate that creep takes place as a result of combinations of rotation of particles in the soil, drawing out of plastic materials, and probably other types of internal readjustments in surficial materials.

Rock Fall and Talus Accumulation

Cliffs are formed along recent fault scarps, where resistant beds are eroded by stream action, and many of the most spectacular bare-rock cliffs, some thousands of meters high, have been formed where valley glaciers have eroded and deepened stream valleys in high mountains. Freezing and thawing are very effective in loosening rock fragments from cliffs, and other weathering processes are also responsible for inducing rock falls. Rocks usually dislodge other rocks as they fall, sometimes starting landslides. As this rock debris, called *talus,* reaches a lower slope, it piles up to form a cone-shaped feature called a *talus cone* (Fig. 13–2). The cones may coalesce or, if rock fragments are not channeled into cones, a nearly continuous sheet of talus may form at the base of the cliff. Talus cones or sheets are composed of whatever rocks make up the face of the source cliffs. Mechanical weathering on such steep slopes, particularly freezing and thawing, is so much more rapid than chemical decay that the talus rocks on top of the piles usually appear freshly broken. These broken blocks are of all sizes, but seldom have dimensions smaller than several centimeters. Talus may form anywhere rock outcrops are subjected to weathering and erosion over long periods of time, where there is a cliff above a lower slope from which removal of materials lags behind the rate of accumulation.

Rock Glaciers

On the floors of mountain valleys, talus may lie where it falls, moving at very slow rates and then only because it is disturbed by freezing and thawing, by animals, or by impact of new rocks falling

FIGURE 13–2
Talus cones. These cones line the side of this high mountain cliff in the Madison Mountains of Montana.

FIGURE 13–3
A rock glacier seen from the trail pass near the head of Horseshoe Basin, Silverton quadrangle, Colorado.
(U.S. Geological Survey)

on it. But if the valley floor is inclined, converging piles of talus may form a tongue-shaped projection down the valley. These lobes of talus move so slowly that their motion can be detected only by checking the position of the end of the lobe over a period of years. Such is the movement of the rock glacier (Fig. 13–3). As a rock glacier advances it assumes garland-shaped loops, giving the appearance of a very viscous liquid. Some of these sheets have considerable amounts of ice mixed with rock, making the movement of the rock glacier more rapid than glaciers at lower elevations and lower latitudes.

Solifluction

This term, which literally means soil flowage, is applied to downslope movement of soils, rock debris, and other fragments in climates where the ground is solidly frozen in the winter and only partially thaws in the summer months. When thawing occurs, upper layers of the soil, which have been forced up and deranged by frost heaving, are bathed in melt-water. Lower layers of the soil remain frozen. Water above the frozen layer lubricates the interface. This facilitates flowage and slow movement of the upper layers over the lower frozen layers, even on very low slopes. The moving mass of soil and debris takes sheet, lobate, or tonguelike forms as it moves. Where valleys are present, it will move into and down them. The nature of the debris depends on the composition of the soil. On Bear Island in the North Atlantic the surface is covered with a thick flowing mud in the warm months, whereas in places on the Falkland Islands the debris is made up of quartzite fragments.

Mudflow

Mudflows occur most frequently in deserts where fine weathering products are dry most of the year. When rain does fall it often comes in large quantities, perhaps with a large part or all of the annual rainfall coming in a single rain. Water seeps into the weathering products, and the mixture gathers momentum as it moves rapidly down steep slopes. These flows follow existing channels and gullies. Although a mudflow is not dry, the mass does not necessarily contain much free water. Sometimes even large boulders will float in the mud. Mudflows are favored by intermittent water supply, lack of vegetation, and abundant unconsolidated rock debris.

Some of the most disastrous mudflows have occurred on the steep slopes of the San Gabriel Mountains of southern California. Mudflows there are often preceded by extensive forest fires that destroy the vegetative cover. One such flow started on a warm day when the heat caused melting of the snow cover in the high mountains. Melt-waters percolated into the thick, weathered cover of decaying schists. When the weathered material became saturated, it broke loose, leaving a scar extending 300 meters up the mountainside with cliffs 46 meters high around the edge. The mudflow continued for five days and at its height splashed mud 6 meters into the air, leaving a mud coating on treetops. The mixture was about 25 percent water.

TYPE OF MOVEMENT		TYPE OF MATERIAL			
		BEDROCK		SOILS	
FALLS		Rockfall		Soilfall	
SLIDES	Few Units	Rotational Slump	Planar Block glide	Planar Block glide	Rotational Block slump
	Many Units		Rockslide	Debris slide	Failure by lateral spreading
FLOWS		ALL UNCONSOLIDATED			
		Rock Fragments	Sand or Silt	Mixed	Mostly Plastic
	Dry	Rock fragment flow	Sand run; Loess flow		
			Rapid earthflow	Debris avalanche	Slow earthflow
	Wet		Sand or silt flow	Debris flow	Mudflow
COMPLEX		Combinations of materials or type of movement			

TABLE 13–1
Classification of landslides, abbreviated version.
(Highway Research Board)

LANDSLIDE CLASSIFICATION

The Highway Research Board landslide committee has proposed a classification of landslides (Table 13–1) that includes most types of mass movement (Eckel, 1958). This classification is structured to bring out the interrelationships of the various factors involved in mass movements. It does not, however, differentiate such forms of movement as solifluction, creep, rock glaciers, and talus accumulations, which are described separately.

Landslides are defined as "the downward and outward movement of slope-forming materials composed of natural rock, soils, artificial fills, and combinations thereof" (Eckel, 1958). (Sharpe [1938] restricted the term landslide to relatively dry masses.) The type of material involved is classified according to its state prior to initial movement or, if the type of movement changes, according to its state at the time of the change of movement.

The type of movement and type of material provide the primary basis for the classification, with the amount of moisture, the nature of the surface on which movement occurs, and the speed of movement allowing finer distinctions within the broader framework of the classification.

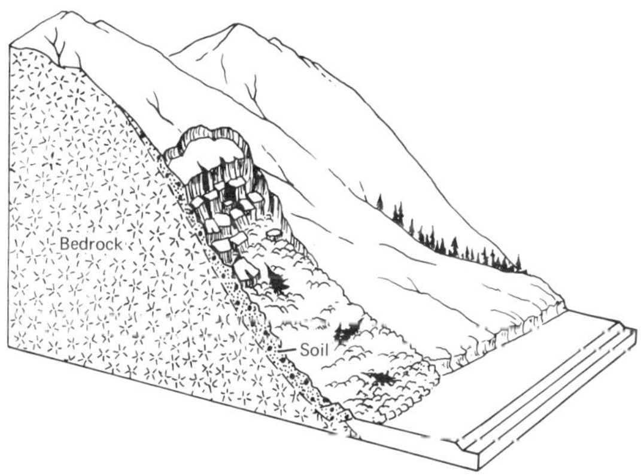

FIGURE 13–4
Debris slide. Sliding of rock, soil, clay mixtures over solid bedrock. Movement is slow to very rapid.
(after Varnes, 1958)

Falls. Falls occur when the mass in motion travels most of the distance through the air. Included are free falls, bounces, and rolling of rock and debris fragments without much interaction of one fragment with another.

Slides. Slides include movements caused by finite shear failure along one or several surfaces (Figs. 13–4, 13–5, and 13–6). The mass may divide into only a *few units* or it may be broken into *many units.* When there are only a few units the material in motion is not greatly deformed.

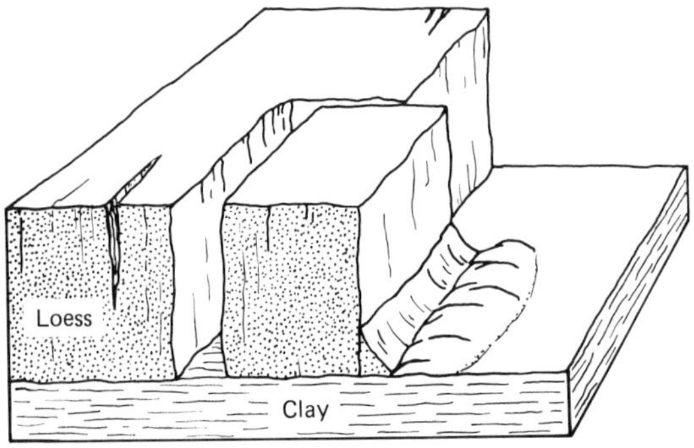

FIGURE 13–5
Block glide. Slow movement of blocks over a base with a low coefficient of friction—loess over clay in this case.
(after Varnes, 1958)

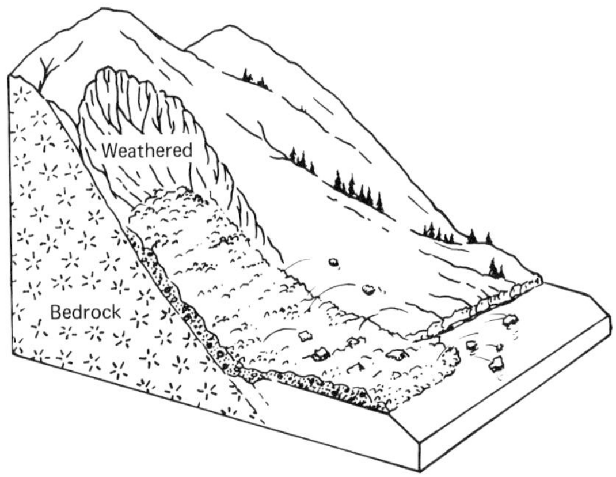

FIGURE 13–6
Debris avalanche. Dry to moderately wet debris moving rapidly over bedrock. Movement is very rapid to extremely rapid.
(after Varnes, 1958)

FIGURE 13–7 (*below*)
Debris flow. A very rapid flow in which there is a high water content in materials usually consisting of a mixture of rock, soil, and clay.
(after Varnes, 1958)

The movements may be controlled by surfaces of weakness such as fractures, faults, or bedding planes. The masses may move as a *slump* with movement along internal slip surfaces, which are usually concave upward, or as a *block-glide movement* (Fig. 13–5) in which the units move out and down along a more or less planar surface of weakness, for example, bedding, fractures (Fig. 13–15), faults, or the original ground surface. The mass in movement may break down into many smaller units when it is intensively fractured or when thin beds are separated by zones of weakness. The size of the individual unit is comparable to or less than the displacements between units. Movements of such masses are likely to go beyond the original slip surface so that parts of the mass slide over the ground surface.

Flows. This designation is used when the form taken by the moving material or the apparent distribution of velocities and displacements resembles that of viscous flow (Figs. 13–7 to 13–11). Slip surfaces usually are either not visible or are temporary. The size of the individual particles in the moving mass is generally very small compared with the amount of movement.

When the movement is a combination of the three principal types of movement, the landslide may be designated *complex.* The following scale of rate of movement has been used by the Highway Research Board:

Extremely rapid	10 ft/sec or more
Very rapid	0.01–10 ft/sec
Rapid	5 ft/day–0.01 ft/sec
Moderate	5 ft/month–5 ft/day
Slow	5 ft/year–5 ft/month
Very slow	1 ft/5 years–5 ft/month
Extremely slow	less than 1 foot in 5 years

PRINCIPLES GOVERNING MASS MOVEMENTS

The primary force involved in mass movement is the force of gravitational attraction, a body force that acts on every particle throughout the materials of the earth's crust. This force is directed

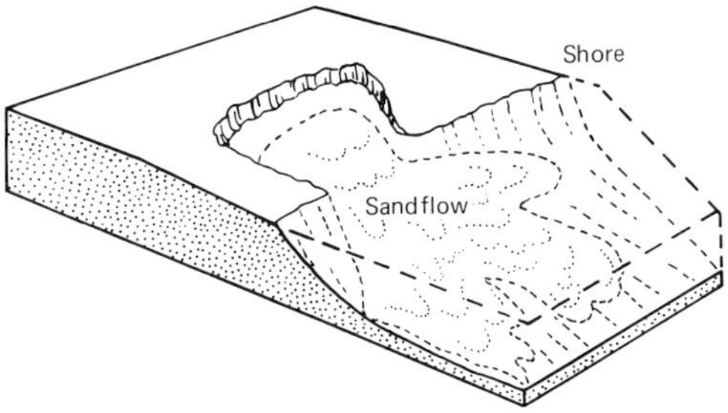

FIGURE 13–8
Sand or silt flow underwater in uniform, saturated sand or silt. Movement is rapid.
(after Varnes, 1958)

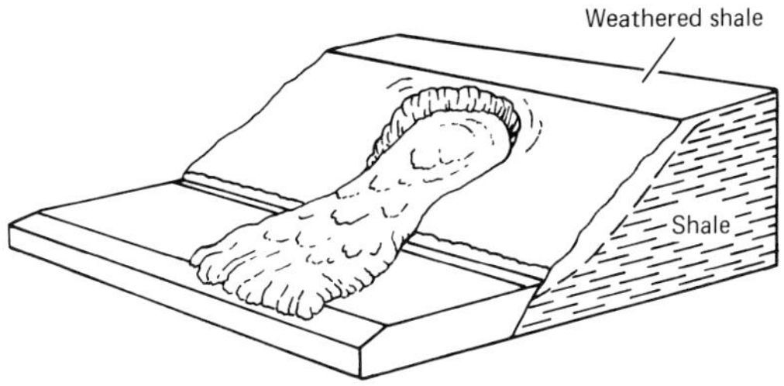

FIGURE 13–9
Earth flow. Mostly plastic flow here of weathered shale over unweathered shale.
(after Varnes, 1958)

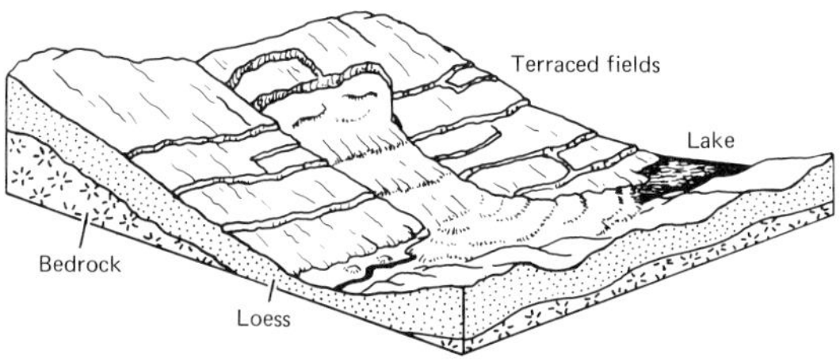

FIGURE 13–10
Loess flow (dry) as it occurred in Kansu Province, China (1920). Movement was extremely rapid.
(after Varnes, 1958)

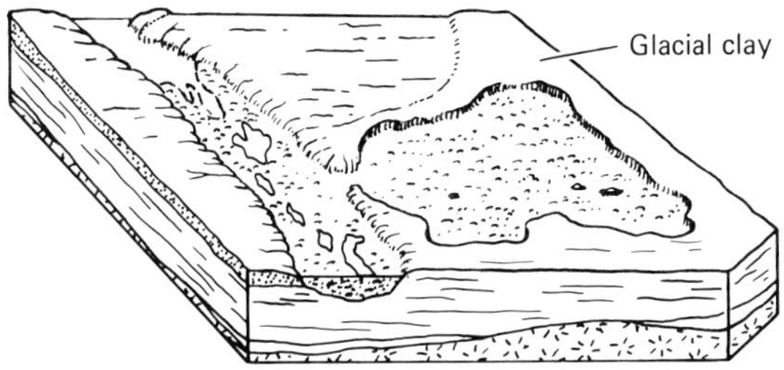

FIGURE 13–11
Rapid earth flow of wet soil. Flow has occurred in glacial clay and silt in this case.
(after Sharpe, 1938)

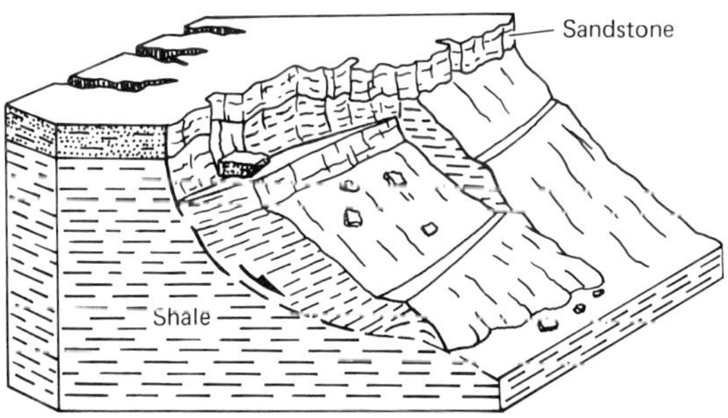

FIGURE 13–12
Slump. A rotational slump in bedrock is shown. Movement is at slow to moderate rates.
(after Varnes, 1958)

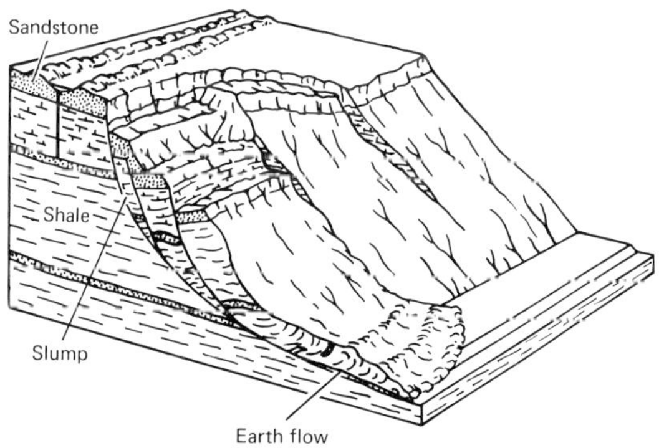

FIGURE 13–13
Slump and earth flow combined, involving rotational slump of bedrock and soil.
(after Varnes, 1958)

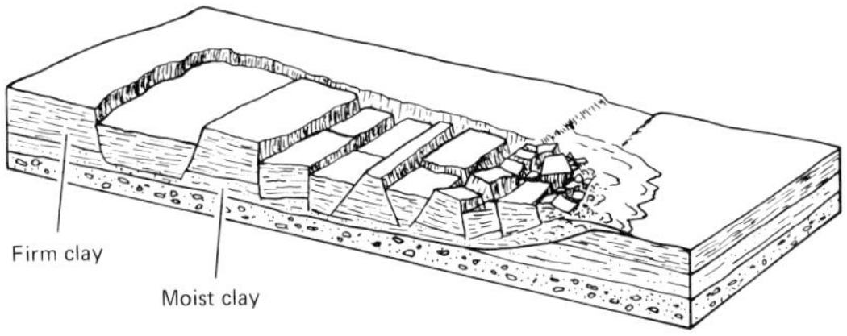

FIGURE 13–14
Failure by lateral spreading in clay over a soft clay. Movement is usually very rapid. See section on Alaskan earthquake. (pp. 300–01).
(after Varnes, 1958)

FIGURE 13–15
Close-up detail of some of the large rocks dislodged in the famous Frank slide, Alberta. These massive blocks of limestone slid along fractures during the landslide.
(Canadian Pacific Railway)

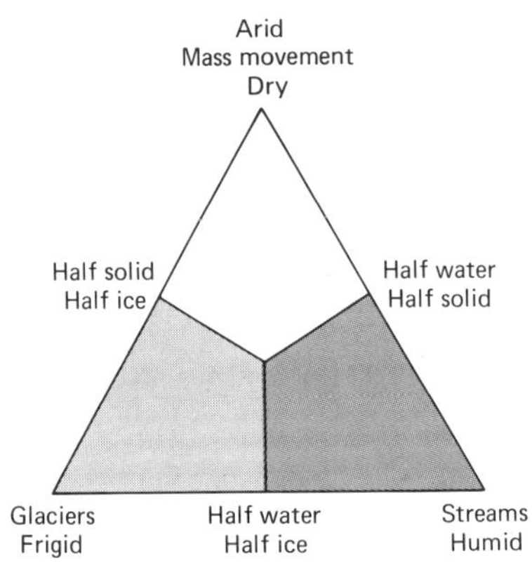

FIGURE 13–16
Process of degradation. Degradation is accomplished through breakdown and transportation of the crust. Under the pull of gravity these decay products move downhill, transported dry, in water, in ice, and in combinations of these three.

downward toward the center of the earth and it acts with nearly uniform magnitude on surficial materials, yet some of these materials are so situated in their environment that they become unstable and move downslope while others remain stable. Chief among the factors that influence stability are the following:

1. The nature of the material.
2. The structure of the material and its surroundings.
3. The slope of the ground surface.
4. The presence of water, ice, compressed air, or steam in the material.
5. The presence or absence of stabilizing ground cover such as forests.

The factors combine in a variety of ways to determine the degree of stability.

Streams, glaciers, and dry mass movements are all driven by gravity. One might reasonably expect to have little difficulty in distinguishing a stream from a glacier, and either of them from a dry mass movement, but the distinction is not so clear-cut. Glaciers carry heavy loads of rock, soil, and debris; so do streams; and mass movements usually contain some proportion of water or ice. The three types of movement are represented in Fig. 13–16 by a triangular diagram that is indicative of the gradational character of each.

Nature of the Material

Some earth materials are inherently strong, for example, massive granite, basalt, and quartzite, where the mineral components of the rock form an interlocking network of crystals or where the fragments are tightly cemented together. The bonds between the particles of such rocks are more uniformly strong in all directions through the rock mass than is found in most sedimentary and metamorphic rocks. Weakness in such rock masses usually is caused by structures (e.g., fractures) that may divide the massive rock into blocks. Among metamorphic rocks, planes of inherent weakness may arise as a result of compositional layering. Planes of foliation, schistosity, slaty cleavage, as well as fractures, are planes of weakness along which the rock may break.

The physical properties of sections of sedimentary rocks vary from layer to layer so that the layering generally imparts a plane of inherent weakness although individual strata may be uniformly strong in all directions. Some layers may be soluble (e.g., limestone, gypsum); others may have open interconnected pore spaces (e.g., sandstone, conglomerate); some, like shale, may be cleaved or subject to expansion on wetting.

Unconsolidated sediment also has a wide range of physical properties. The stability of fragmental materials such as sand, silt, and pebbles is greatly influenced by the degree of sorting, the angularity of the particles, the degree of cementation, the porosity and presence of fluids in the pore spaces, and even the texture of the surfaces of the individual fragments. The stability of clay is closely related to the degree of saturation with water and to the weight of the overlying material.

The principles governing mass movement differ with the types of material involved. Some materials owe their weakness to the existence of planes of weakness along which parting or sliding is possible. Instability in such materials is increased by anything that tends to induce slip or parting along those planes. Each fragment is capable of movement in granular materials so processes or conditions that favor rotation, rolling, or mass slip may induce instability.

The Role of Water

Water is usually present in the ground and it may play a crucial role in causing movement by one of the following mechanisms:

1. When clay colloids absorb water their plasticity is increased; the shearing resistance of wet clay is less than that of dry clay.
2. The chemical weathering of a number of common minerals is facilitated or caused by water, and clay is often a by-product of this weathering.
3. Water, especially rainwater, can dissolve rocks such as limestone and thus lead to reduction of strength of the rock and to development of cavities and the opening of fractures.
4. Water may act as a lubricant, reducing the frictional resistance between surfaces or between points of contact between fragments.
5. When a granular mass is saturated with water, the water has a buoyant effect, reducing the effective weight of each submerged grain by the weight of the displaced volume of water.
6. Under some conditions (treated later under *ground water*) water rising to the surface of the ground may be under pressure and will exert upward force on the material through which it moves.

PROCESSES OF MOVEMENT

Fall

Free falls occur only on the steep slopes that are characteristic of massive rocks, such as igneous and metamorphic, or tightly cemented sedimentary units, although a few unconsolidated sediments may also stand in near-vertical cliffs. Rocks from the cliff may be wedged off by frost action, causing a face bounded by a fracture to fall away, or undercutting may simply proceed until the unsupported weight of the overhanging rock exceeds the breaking strength of the rock and it fails.

Subsidence and Collapse

Subsidence and collapse are significant forms of mass movement in areas where subsurface water is actively dissolving and removing rock, effectively undermining the surficial materials. Limestone, gypsum, salt, and anhydrite are the most common rocks that are likely to dissolve in sufficient quantity to produce large-scale subsidence or collapse; and of course the melting of ice frequently produces this result. Subsidence may also occur following release of gas or removal of large quantities of ground water or oil from reservoirs in unconsolidated or semiconsolidated sediment.

Subsidence is the gradual lowering of surface materials into space opened as a result of removal of rock below. If the rock being removed is strong or if it is overlain by rock that is strong, a large cavity may be developed before failure takes place. Failure in such cases is likely to be sudden, followed by the collapse of the roof into the cave. The broken rock may form a coarse breccia when collapse occurs, but a crude layering may be preserved when subsidence takes place.

Sliding on a Surface

Sliding on a surface may be examined in terms of

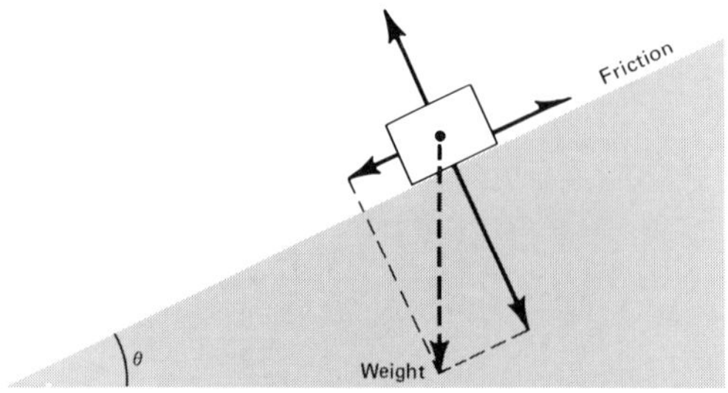

FIGURE 13–17
Forces acting on a sliding block on an inclined plane. The weight is shown resolved into two components—one perpendicular to the inclined plane and the second acting down the plane.

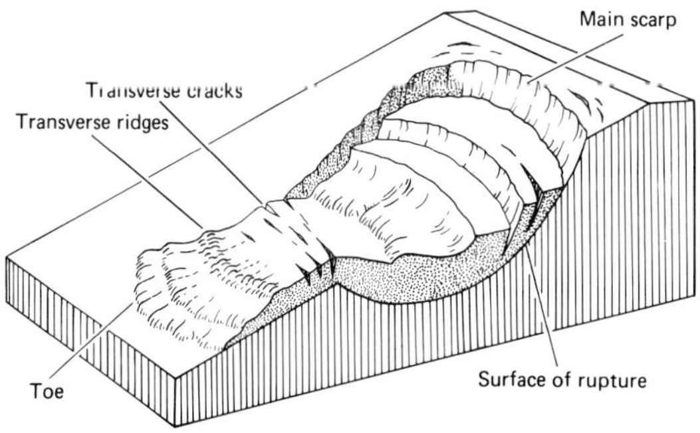

FIGURE 13–18
Terms applied to parts of a landslide.
(after Varnes, 1958)

a block on an inclined plane (Fig. 13–17). The weight of the block, which is directed down, may be resolved into two components—one acting down the inclined plane and the other perpendicular to the plane, which is counterbalanced by the push of the plane against the rock. If the block is not moving, the component of force directed down the plane is counterbalanced by frictional resistance to such movement along the contact between the plane and the block. Instability can be induced by either increasing the inclination of the plane, steepening the slope, or reducing the frictional forces between the plane and the block.

The slope can be increased in nature as a result of either erosion or uplift; or man may cause the increase by removal of near-surface materials. The friction between the block and the plane is due to the nature of the surface, the presence of lubricants, and the weight of the block. The slope required for sliding is steep when the friction is great, but movements on very low slopes are possible when frictional resistance is low.

The plane may be a fracture surface, a fault, a bedding plane, or the surface of the ground. The block may be a large or small block of rock, massive or not, bounded by joints or masses separated by erosion from their natural continuations. One of the most common and potentially dangerous situations in which this type of mass movement occurs is that found where steeply inclined strata or fractures dip in the same direction as the slope of the ground surface and are undercut in the valley by streams, glaciers, or highway construction. When undercutting penetrates massive rocks and cuts into shales or other rocks with low shearing resistance, failure is likely because most strata are fractured and these permit separation as part of the mass slides downslope. Water may enter the potential zone of movement through fractures, and if the potential slide zone contains clay, resistance to shear in the zone will be significantly reduced.

Slump

The term *slump* is applied to movements that involve a rotational shear type of sliding motion (Fig. 13–18). These often combine with other types of mass movement, earth flow or mudflow, slides, falls, and so on, and are one of the most common types of mass movement. Movement in a slump is more or less rotational about an axis that is parallel to the slope. Movements take place on internal slip surfaces; these surfaces appear as cracks in the ground at the head of the slump mass and usually have arcuate form, being concave toward the slumping mass. The slump may assume a spoonlike shape, as is common when slumps occur in uniform, granular material, or they may have a cylindrical form where the slip surface runs parallel to the slope for some distance. Several varieties of slump are illustrated in Figs. 13–12, 13–13, and 13–18; these differ in the nature of the internal slip surface and its relationship to the slope and to the structure of the slump mass. The slip surfaces tend to follow planes or zones of weakness, such as the old soil surface where land fills have been placed, or jointed surfaces, faults, clay or shale zones, and highly cleaved zones; all of these are marked by low frictional resistance to sliding. Frictional resistance may be reduced further by wetting. Water flows into these zones of movement through cracks at the head of the slump, which pull open. As the slump mass rotates, the surface is tilted and water is directed toward the open cracks.

Plastic Flow

Clay and clay mixtures are plastic; that is, they behave as elastic material if they are stressed below some limiting value, but if loaded beyond that limit they flow and deform continuously. Plastic behavior can be induced either by adding load on the material or by changing the plasticity of the material, notably by adding water to it. Water seeping down along fractures or through pore spaces into a claystone or mudstone may make the clay more plastic and thereby induce plastic flow or sliding. Shale can be altered to clay by water; limestone can be dissolved by ground water, leaving clay residue; and fault breccias are subject to similar weathering effects. Plastic flow can be initiated by adding weight artificially, as has happened during road construction over salt beds in the salt basins of Utah, and over clay beds in the Coastal Plain. Failure of the Fort Peck Dam in Montana was attributed to swelling and plastic flow of clays under the dam when water accumulated in the reservoir and seeped down into the clay layers. Plastic flow in the clays under the Panama Canal was caused by excessive loading when dredged material was piled along the sides of the canal.

Failure of Granular Materials

Uncemented granular materials assume slopes of a certain steepness if they are free to move. If the granular material is uniform, the slope of that material does not exceed a particular value, called the *angle of repose,* usually 30 to 40 degrees. That angle depends on the shape of the fragments: rounded fragments tend to have lower angles of repose than angular fragments; smooth-surfaced fragments tend to have lower angles than rough-surfaced fragments; and the angle of repose is a function of the size of particles, degree of sorting, density of materials, and the amount of anhydrous constituents. In short, the angle of repose is essentially a reflection of the cohesiveness of the fragments.

Fluid Flow

We find in nature a completely gradational transition from forms of entirely dry mass movement to stream flow. The behavior of water in stream flow is true fluid flow of a viscous liquid. Liquids cannot support shear deformation; instead they move continuously under any applied shearing stress. The viscosity of a fluid is the internal friction or internal resistance the fluid offers to applied stresses. Because the viscosity of water is low, it cannot stand at different levels in a container or in a natural stream system. Fluid flow may be nearly *laminar,* in which one layer slips over another, or *turbulent,* in which the flow pattern of a particle of water follows an irregular path of swirls and loops.

As the water content of soil or debris on a slope increases, the clay fraction absorbs water and the plasticity of the mass increases; the contacts between solid fragments are lubricated; and, eventually, if water content becomes very high the fragments are buoyed by the water. Under these conditions the mass flows in a manner analogous to a viscous liquid as the internal frictional resistance of the mass is reduced. Mudflows and solifluction are examples of this general type of mass movement.

CASE HISTORIES OF SOME LARGE-SCALE MASS MOVEMENTS

Madison Canyon Landslide

About midnight on August 17, 1959, a severe earthquake shook the region west of Yellowstone Park. Surface waves rocked water in nearby lakes, threw the soil up in waves, and cracked the highways and ground surface over a large area. The movement was caused by a displacement along a fault. This fault extends near a narrow and very steep canyon, 400 meters (1,300 feet) deep.

The earthquake triggered a movement in decayed soils and rock debris on the side of the canyon, and in the chaos that followed about 27

FIGURE 13–19 (*Above*) Air view of Madison Canyon slide from over Earthquake Lake somewhat east of drowned toe of slide. Madison County, Mont. August, 1959.
(J. R. Stacy, U.S. Geological Survey)

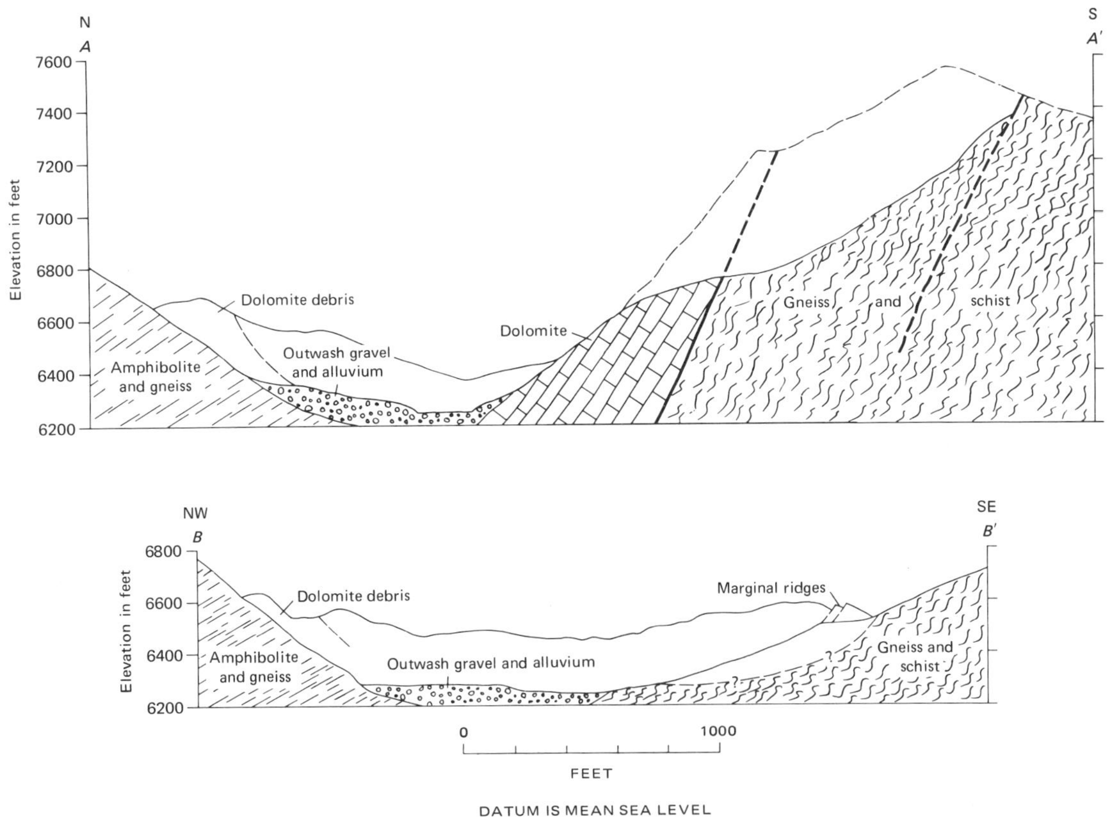

FIGURE 13–20 Sections of the Madison slide.
(from J. B. Hadley, 1964)

FIGURE 13–21
The Gros Ventre slide, Wyoming.

million cubic meters of rock, soil, and trees slid off into the canyon, leaving a scar from the top of the mountains almost to the river. The debris engulfed the valley, a campground, and the highway, burying about 135 acres of land with an average thickness of 45 meters of rock. The movement was extremely fast—perhaps as much as 96 kilometers per hour. The mass forced air in the valley aside, setting up a wind of gale force. When the slide hit the river, water was splashed out, leaving mud splashes in trees high up on the sides of the valley. Material broken loose moved with such impetus that it continued up the opposite valley wall 122 meters. As the slide filled the valley, a wave of water was forced up the river for nearly 1.6 kilometers. Material in the slide ranged in size from blocks 9 meters across to fine soil particles.

The potential for mass movement had existed for many years. Slopes had been oversteepened to angles of 40 to 60 degrees, and deeply weathered Precambrian gneisses and schists were exposed on them. These deep soils were held on the slope by the presence near the bottom of the valley of a massive layer of dolostone, which is resistant in such semiarid climates. The gneiss, schist, and dolostone are inclined toward the river on the south side of the valley and are fractured. Thus there were many potential planes of slippage. When earthquake movements began, blocks of the dolostone dislodged and released the main mass movement.

Gros Ventre Slide (Earth Flow)

Where the Gros Ventre River cuts through the mountains east of Kelly, Wyo. (just east of the Teton Mountains), earth flowed slowly from one side of the river valley into the valley during 1909–11. Movements there were facilitated by the structure of the sedimentary strata exposed on the mountainside. Clays and shales interbedded with thin sandstones dip at about 20 degrees toward the river. When the stream cut through the sandstones overlying the muds and shales, slow flowage and slipping began in layers of plastic clay. Flowage was most rapid after spring thaws, when melt-water from ice and snow seeped into the top of the flow scars and saturated underlying clays. Sliding occurred on this slippery clay base. The mass left a large scar extending up the mountain, and a lobelike mass of earth filled the river valley and moved up the opposite side of the valley during the most rapid movements. It formed a dam across the river and backed up a large lake (Fig. 13–21). The river topped the dam, and in 1925 it rapidly cut through the weak clays and sand of the dam, causing a flood that destroyed most of the town of Kelly.

Panama Canal—Slides, Slumps, and Plastic Flowage

Construction of the Panama Canal was plagued by many costly earth movements. At times they were so extensive and costly to remove that some question arose as to the feasibility of completing the canal. The company that originally contracted to build the canal went broke because it was unable to cope with the movements. Slumping of the sides of the canal became worse where the canal was cut through the Cucaracha formation. This unit is partially composed of a highly plastic clay. At first slides were a result of cutting the sides of the canal too steeply—material slid because the slope exceeded the angle of repose for the material. Conditions were aggravated when the sides were cut down to lower slopes because the material removed was dumped just above the top of the cut. The weight of debris piled up in that position led to the development of slumps.

Other mass movements plagued construction. In places plastic clays underlie the bottom of the canal. When the canals were dredged to deepen the channel, the mixture of water and clay was pumped up on the sides. The weight of this mass further compressed the layers of clay, and the plastic clay was squeezed along within the clay zone until it came to a point where the pressure was less—this point, of course, was where the canal had been cut. So lumps of clay continued to rise in the canal. This process continued until the clay layer was squeezed off, and the canal became stabilized.

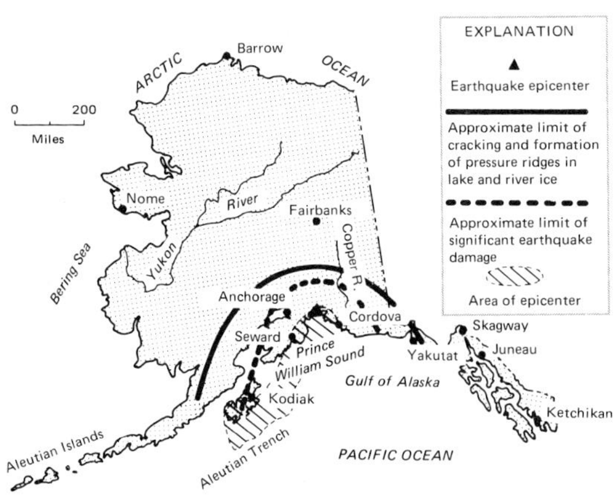

FIGURE 13–22
Index map of Alaska, showing areas of earthquake damage and surface cracking.
(after U.S. Geological Survey Circular 491)

1964 Alaskan Earthquake

As in the Madison Canyon slide, an earthquake of high intensity initiated extensive mass movements in south-central Alaska on Good Friday, March 27, 1964, which were responsible for the loss of over a hundred lives and property damage of hundreds of millions of dollars. The earthquake resulted from movement on a major fault that extends over 800 kilometers along a south-southwest trend from the point on the ground surface over the focus of the earthquake, located about halfway between Anchorage and Cordova, Alaska (Fig. 13–22). As a result of the faulting, blocks of the earth's crust were displaced. At Cordova the area was uplifted over 2 meters, at Anchorage the region was depressed relatively about 1 meter, and at Kodiak, nearly 800 kilometers south, the relative depression amounted to nearly 2 meters.

The secondary mass movements at Anchorage were described as follows by Oakeshott (1964):

> The city [Anchorage] lies on a glacial outwash plain which slopes from the base of the high and rugged Chugach Mountains down to the sea, where it drops off abruptly along a series of bluffs or seacliffs about 30 to 50 feet above sea level. Under the entire city is the flat-lying formation called by geologists the Bootlegger Cove clay. The Bootlegger Cove clay is gray clay, silt, and mud deposited from melting glaciers in lagoons during the Pleistocene Epoch; it ranges in thickness from about 100 to 250 feet and lies only a few feet below the land surface under most of Anchorage. When water-saturated, as it is in the Anchorage area, this material completely lacks structural strength. When the tremendous shaking of the earthquake occurred, the Bootlegger Cove clay failed, partly by sliding out toward the waters of Cook Inlet on slip-surfaces of especial weakness, and probably also partly by "liquefying" to flow at low angles toward the waterfront. . . . Great landslide cracks opened back of the bluff; some of the houses were carried toward the sea as far as 500 feet. About 200 acres of the terrace surface broke into a series of blocks each a few feet across, tilted at various weird angles to form a chaos of blocks of the bluish-gray Bootlegger Cove silt and clay, brownish peat and muskeg, black soil, and road pavement [Fig. 13–23].
>
> . . . In front of some of the advancing blocks

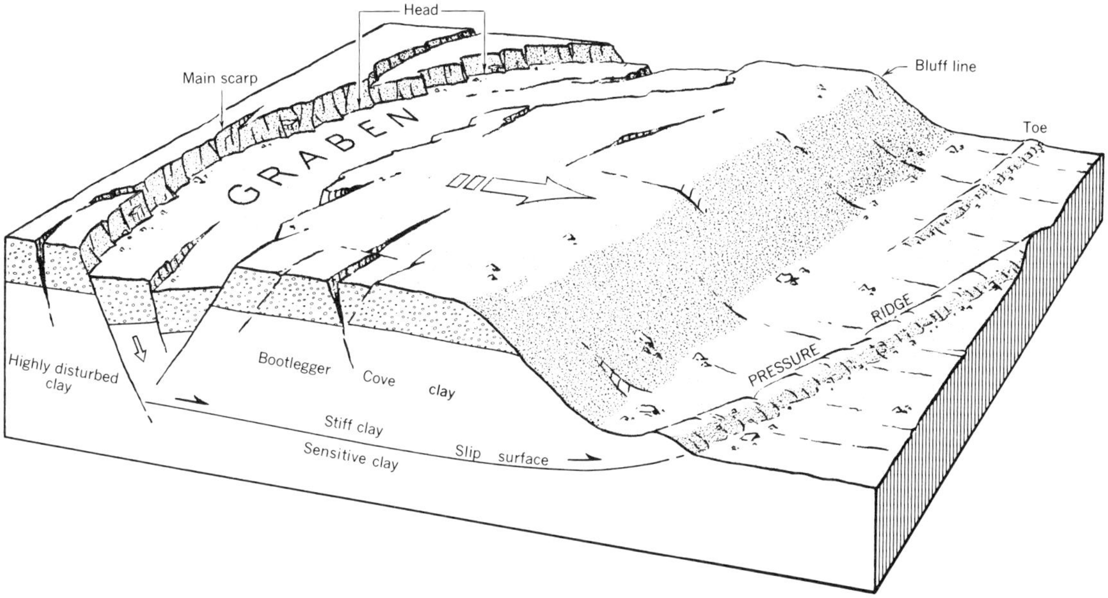

FIGURE 13–23 Block diagram of a translatory slide caused by Alaskan earthquake of March 1964.
(from W. R. Hansen, 1965)

> compressional "pressure ridges" were formed. Most of the damage in Anchorage was of this type; that is, due to direct landsliding and to landslide fractures [Fig. 13–24], cracks, and in some cases to the pressure ridges built up along the fronts of the slides. . . .
>
> Seismic sea waves, or "tidal waves," generated by the large displacement of the surface of the earth under Prince William Sound affected various coastal towns selectively, depending on the configuration of the coastline, the depth of water, the configuration of the sea bottom, proximity to the movement of the surface of the earth in the Sound below sea level, and probably other factors. The little town of Kodiak on Kodiak Island, besides being dropped about six feet in the western block, was hit by tidal waves up to 30 feet high.

The Vaiont Reservoir Landslide[1]

On October 9, 1963, 240 million cubic meters of rock and soil slipped from the side of Mount Toc near Longarone, Italy, into the reservoir of the Vaiont Dam, the world's second highest dam (266 meters high), filling the reservoir for a distance of almost 1.6 kilometers and setting up waves estimated to be 244 meters high. The mass of debris moved at rates of 25 to 30 meters per second, so fast that a blast of highly compressed air was forced into the valley; the water from the reservoir poured over the dam, destroying the town of Longarone where 2,600 lives were lost within minutes of the slide. The Vaiont Dam, a thin-arch reinforced concrete structure, withstood the force of the water and did not fail.

This disaster, the worst of its type, was caused by mass movement of an unstable mountainslope set off by progressive weakening of the rock mass with time and accelerated by excessive groundwater recharge resulting from heavy rains and the water level in the reservoir.

Instability of the slopes on the valley sides of the reservoir arises from a number of geologic conditions. This region lies within the Alpine Mountain system where intense deformation has folded, fractured, and faulted the sedimentary rock units. The mountain-building in this region was followed by mountain glaciation, which caused deep U-shaped valleys to be carved out of the rock masses (Fig. 13–26). Earlier formed fractures near the ground level opened and new fractures probably developed as the ice melted and

[1] Based on a report by Kiersch (1965).

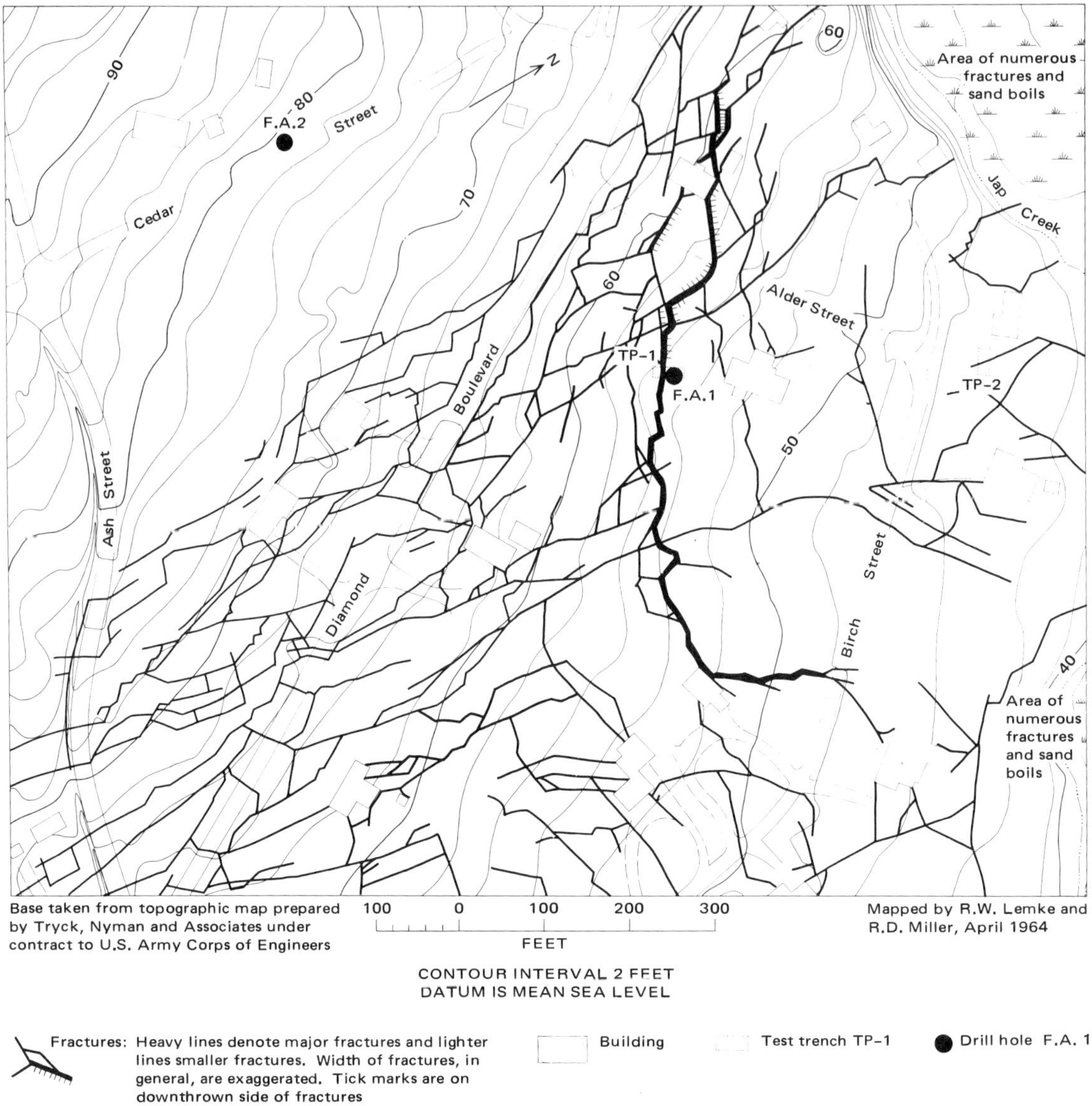

FIGURE 13–24 Map of the Forest Acres subdivision showing fractures produced by the earthquake of March 27, 1964, at Anchorage, Alaska. (from R. W. Lemke, 1967)

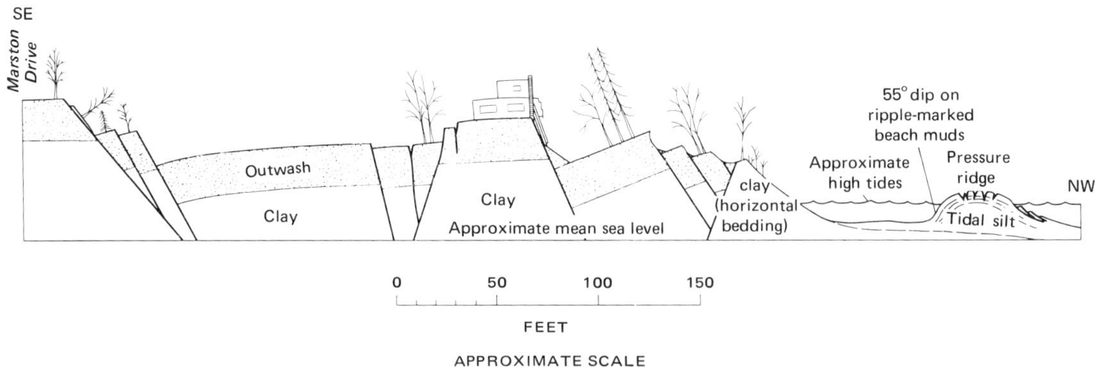

FIGURE 13–25 Sketch section through eastern part of Turnagain Heights slide (Anchorage, Alaska). Note pressure ridge, upright blocks at center and to left of pressure ridge, and tilted collapsed blocks between. (from W. R. Hansen, 1965)

the pressure due to the weight of the ice was released. The floors of the U-shaped valleys were subsequently cut down to form a narrow and deep inner V-shaped valley, and new release fractures formed along the sides of this valley. These events led to the development of the structural and topographic conditions depicted in cross section in Fig. 13–27. The slide occurred in an area where the layered sedimentary rocks dip, or are inclined, toward the center of the valley. The rock units involved are inherently weak and possess low shearing resistance due to seams of clay and thin beds of marl interbedded with limestone and claystone. The steep inner canyon no longer affords support for the units, which have been completely truncated. Thus the rock masses above the zones of low shearing resistance are free to move downslope. Furthermore, the whole mass is substantially weakened by the open fractures. The open fractures and the presence of limestone have led to the extensive development of solution cavities, caves, tubes, and so on, and a system of sink holes has formed on the upper slopes of the valley. These act as *catchment areas* and traps for surface water that is directed underground where it can act to lubricate the clay zones, weaken bonds, and increase hydrostatic uplift.

All of the above conditions existed before the construction of the dam. As the reservoir filled, water moved into the open fractures along the sides of the inner canyon, artificially raising the ground-water level. The level of the reservoir was raised about 20 meters in the months immediately preceding the slide, suggesting that water levels were important in upsetting the equilibrium. There were also heavy rains during August, September, and October that caused excessive infiltration of surface water.

Landslides have been common in the Vaiont Valley. A large slide of about a million cubic meters took place in 1960; this movement was accompanied by the development of creep over a large area and opening of ground cracks upslope from the slide. These cracks ultimately outlined the October 9th slide (Fig. 13–28). The slide area moved very slowly during the spring and summer of 1963; a rate of 1 centimeter per week was measured. About September 18th, some of the survey stations constructed after the 1960 slide started moving 1 centimeter per day. Heavy rains

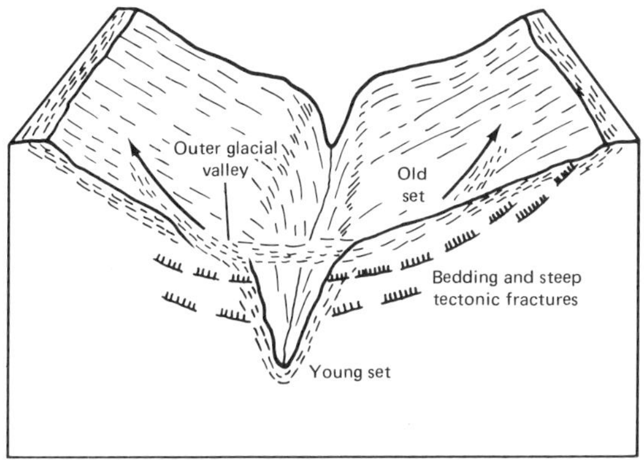

FIGURE 13–26
On this sketch of inner Vaiont Canyon and remnants of the outer glacial valley are shown fractures —old and young set— from stress relief within the walls of the valley to depths of 100 to 150 meters (330 to 500 feet).
(from Kiersch, 1965)

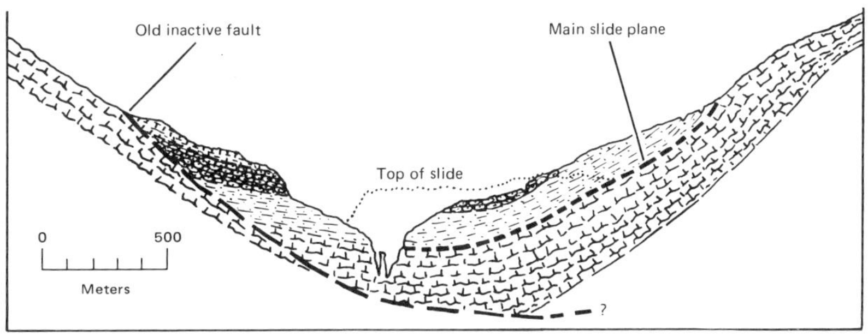

FIGURE 13–27
Geologic cross sections of slide and reservoir canyon, extending from north to south, showing principal features of the slide plane and rock units.
(from Kiersch, 1965)

FIGURE 13–28
Map of Vaiont Dam area and Piave River valley showing limits of the slide.
(after Kiersch, 1965)

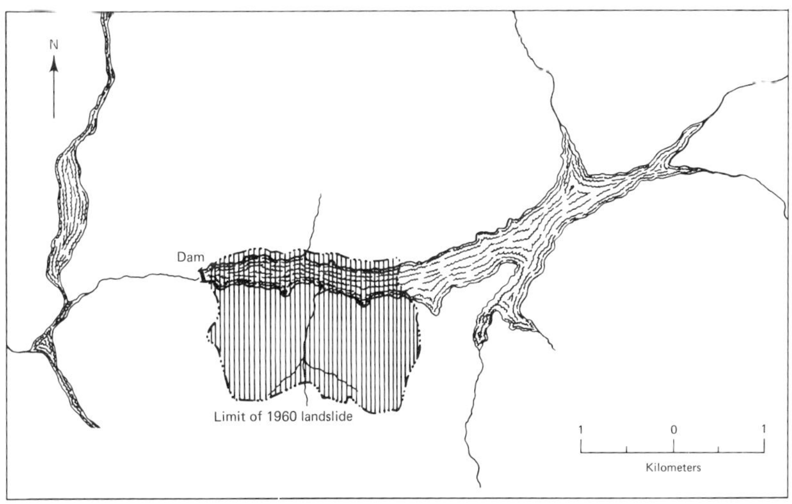

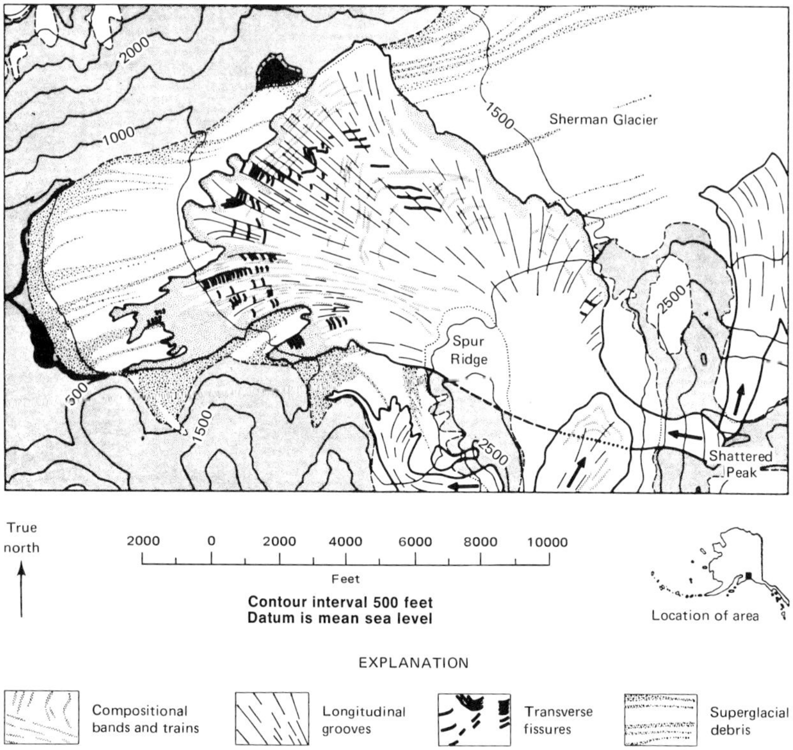

FIGURE 13–29
Sketch map of Sherman landslide showing schematically compositional bands, compositional trains, longitudinal grooves, and transverse fissures.
(from Shreve, 1966)

started September 28th and continued until after the landslide. About October 1st, animals grazing on the north slopes of Mount Toc moved away. At first it was thought that only isolated blocks were moving, but as the rate of movement over the large area accelerated from 1 centimeter per day to 80 centimeters per day on the day of failure, it became apparent that the stations were moving together and involved an area five times larger than initially estimated. On October 8th the outlet tunnels were opened in an attempt to lower the level of the reservoir, but the amount of rainfall was too great and overbalanced this outflow.

A witness to the slide, whose house stood 255 meters above the reservoir on the opposite side from the slide, reported that he was awakened at 10:15 P.M. by the sound of rolling rocks, not uncommon in this area, but the sound grew louder and at 10:40 P.M. his house was struck by a strong wind that broke the window panes. The house shook violently, and there was a loud rumbling noise; the roof of the house was lifted, rocks and rain fell in, and then the roof collapsed and all was quiet. A massive wall of water 69 meters high at the mouth of the Vaiont Canyon buried Longarone at 10:43 P.M., destroying everything in its path. By 10:55 P.M. the flood had receded and all was quiet in the valley.

The Sherman Landslide, Alaska[2]

The Sherman landslide was one of many landslides triggered by the Alaskan earthquake of March 27, 1964. This one is of special interest because the landslide debris moved across and covered a large area of the Sherman Glacier and because the features associated with it suggest that the debris might have traveled on a layer of trapped and compressed air, a variety of slide that is known as the "Blackhawk-slide type."

When the earthquake occurred, a massive slab of sandstone and argillite that stood as a steeply dipping, pervasively jointed mass on the side of a Matterhorn-like peak was shaken free along a bedding surface. The mass of loose rock 450 meters long, 300 meters wide, and 150 meters deep plunged down the 40° slopes of the mountainside and slid up to 5 kilometers across the snow-covered surface of the Sherman Glacier (Fig. 13–29). The debris spread out to form an irregular digitated lobe 1 to 3 kilometers wide with a remarkably uniform thickness of 3 to 6 meters. Despite the thinness of the layer, it was traveling with sufficient momentum to climb 25 meters up the opposite valley wall 5 kilometers from its source.

A number of the features of this slide throw light on its mode of origin. For example, the topography of the surface underlying the slide is faithfully revealed by the surface of the slide, and there are distinct compositional bands and trains. The outer edge of the slide is formed into a rim where the slide debris apparently plowed into the glacier surface; ice and snow make up much of this rim, which is up to 15 meters high, 15 to 150 meters wide, and characterized by a chaotically *hummocky*, or lumpy, topography. Where crevasses have opened through the debris, the lower boundary is sharp and brings the angular boulders of the slide material into contact with ice or evenly stratified snow.

[2] Based on a report by R. L. Shreve (1966).

The surface of the slide is marked by longitudinal grooves up to 8 meters wide, 2 meters deep, and 600 meters long. Some of these end at boulders, indicating that their origin could be due to movement of the boulders, but more commonly there are no such boulders. The origin of such grooves is uncertain, but they might be splits in the slide sheet caused by the lateral spreading of the sheet as it advanced. Transverse fissures, many concave toward the outer edge of the slide, mark the surface of the slide mass. Among the most unusual features are small cones of debris piled on top of boulders as though the debris were dropped onto the slide and accumulated only above the larger boulders.

> According to the hypothesis of air lubrication, huge rockfalls like the Sherman . . . acquire so much momentum in their initial fall that at a break in slope they leave the ground, overriding and trapping a cushion of compressed air upon which they slide at high speed with little friction (Shreve, 1966).

The possibilities of lubrication by water and mud are ruled out by the low temperatures in this area; however, dry snow or a mixture of snow and air are possibilities in the Sherman slide. If air were a primary lubricant, then the leakage of air from the leading edge would account for the pronounced terminal increase in friction with the consequent plowing up of snow into ridges. Lubrication of some sort is required to account for the high speeds (80 to 180 kilometers per hour) and thinness of the sheet.

CHAPTER OUTLINE

Introduction
Classifications of Mass Movement
- Creep
- Rock Fall and Talus Accumulation
- Rock Glaciers
- Solifluction
- Mudflow

Landslide Classification
- *Falls*
- *Slides*
- *Flows*

Principles Governing Mass Movements
- Nature of the Material
- The Role of Water

Processes of Movement
- Fall
- Subsidence and Collapse
- Sliding on a Surface
- Slump
- Plastic Flow
- Failure of Granular Materials
- Fluid Flow

Case Histories of Some Large-Scale Mass Movements
- Madison Canyon Landslide
- Gros Ventre Slide (Earth Flow)
- Panama Canal—Slides, Slumps, and Plastic Flowage
- 1964 Alaskan Earthquake
- The Vaiont Reservoir Landslide
- The Sherman Landslide, Alaska

TERMS

angle of repose	laminar flow	slump
block glide	landslides	solifluction
catchment area	mudflow	talus
creep	plastic flow	talus cone
falls	rock glaciers	turbulent flow
fluid flow	sheet wash	
hummocky	slides	

REVIEW QUESTIONS

1. What information would you need in order to determine the stability of a claystone to be used as a foundation for a heavy pier?
2. Why do laboratory tests on materials often fail to give a satisfactory understanding of the stability of the material as it occurs naturally?
3. Review the descriptions of igneous, metamorphic, and sedimentary rocks and prepare an outline of the characteristics of each major rock type that may have a pronounced effect on slope stability (e.g., slaty cleavage in slate).
4. What zoning regulations would have been desirable in Anchorage, Alaska, to reduce or prevent damage from the types of mass movement that occurred in the 1964 earthquake?
5. How might you recognize an area that is in danger of slumping?
6. If you had a topographic map and a map showing aerial distribution for rock and sediment type, how might you recognize an area of potential landslide?
7. What additional information would be most helpful in making the type of decision required in question 6?
8. What methods would be effective in controlling mudflows in an area subject to them (e.g., San Gabriel Mountains, Calif.)? In controlling solifluction?
9. Why do retaining walls prove ineffective in stopping most slumps?
10. What factors contributed to the mass movements that caused the failure of the slopes above the Vaiont Reservoir?
11. Describe the aspect of the Sherman landslide (Alaska) that is of special interest.

SUGGESTED READINGS

Eckel, E. B., ed., 1958, Landslides and engineering practice: Highway Research Board Spec. Rept. 29, NAS-NRC Publ. 544, Washington, D.C.

Thornbury, W. D., 1964, Principles of Geomorphology: New York, John Wiley, 618 p. Refer to Chapter 4.

Zaruba, Quido, and **Mencl, Vojtech,** 1969, Landslides and Their Control: Amsterdam, Elsevier, 205 p.

14
STREAMS AND THEIR ROLE IN LANDSCAPE DEVELOPMENT

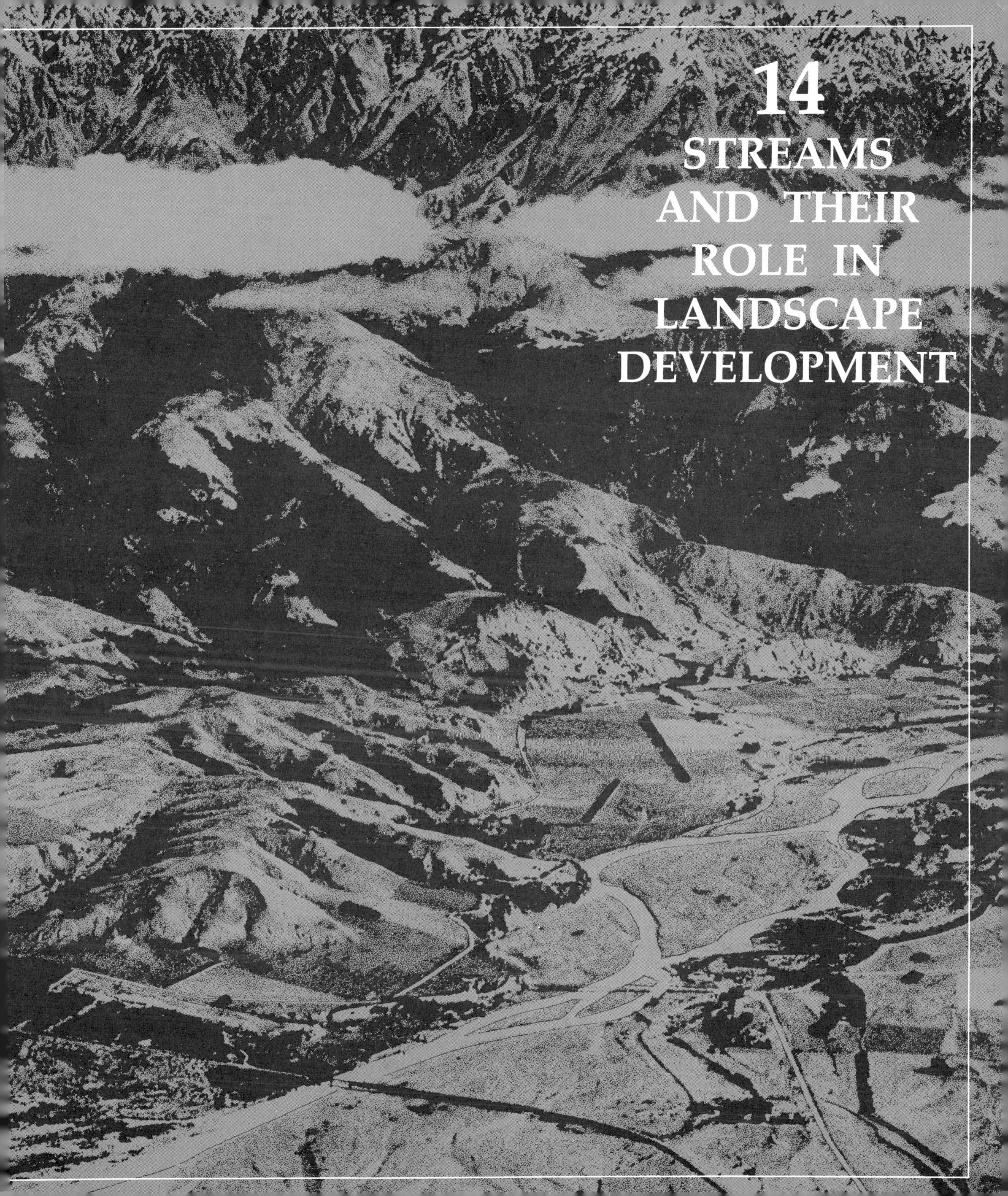

14 STREAMS AND THEIR ROLE IN LANDSCAPE DEVELOPMENT

THE IMPORTANCE OF STREAM ACTION

The role of streams in man's environment is an important one. Water for drinking, for agricultural purposes, and for means of transportation led to the early concentration of cities, farms, and population along rivers and stream valleys, particularly on the *flood plains.* All streams are subject to changes in discharge as a result of seasonal fluctuations in precipitation, and therefore the size of the channel required by a stream to carry its discharge varies from day to day and from one season to the next. However, unusually high discharge leads to high water and flooding. The area over which these flood waters spread is called the flood plain.

When man took advantage of the rich soil of the flood plains and the easy access to water to build his cities, a direct conflict between natural processes and artificial modification of the environment was started. This conflict and its ramifications have burgeoned as man's ability to modify the environment has increased. Too often the complex interrelationship between various parts of a natural drainage system has not been understood or, more recently, what is known has been ignored. The extensive development of agriculture and forestry has hazardous implications for drainage basins and flood potential, and even greater dangers exist where urbanization is taking place.

When land is put into cultivation, the potential for increased erosion exists because the natural plant cover is destroyed and the land is laid bare for at least part of the year. At that time heavy rains can move the loose soil easily and surface-sheet erosion and gullying where wash is concentrated can be greatly accelerated. As this soil gets into the local streams, their behavior is modified. Removal of natural vegetation may increase surface runoff, increasing discharge and possibly leading to floods downstream. The same problems of erosion and excessive runoff exist where forests are destroyed by fire or are cut down.

Some results of urbanization are the following (Parker and others, 1964):

1. Water use is sharply increased.
2. Small, scattered ground-water supplies may be replaced by a single public surface-water supply, or ground-water withdrawals may be multiplied by private or municipally owned wells.
3. Large areas are covered by roofs and pavements that increase runoff and reduce ground-water recharge.
4. Storm-drainage systems provide rapid runoff and decrease recharge.
5. The increased use of water causes lower flows in drought periods, and the more rapid runoff causes higher flood flows in streams.
6. Sewers develop leaks. If sewers are below the water table, ground water is drained away and the water table falls. If sewers are above the water table, leakage adds to recharge and pollutes affected aquifers.
7. Large numbers of septic tanks in some suburban areas may pollute shallow aquifers, and where these areas are served by public water supplies, the "imported" water discharged through the septic tanks recharges the local aquifers and raises water levels.
8. Municipal and industrial waste disposal tends to increase pollution of streams and aquifers.
9. Urban developments often encroach on flood plains, which are natural waterways and an integral part of the river's discharge system. The encroachment on the flood plain impedes flood flows and increases flood hazards to life and property.

The way man has changed the nature of a stream with conscious intent is amply illustrated by the Anacostia River, as described by Withington (1967):

> The effect of erosion on the economy of a community can be seen in the history of Bladensburg, Prince Georges County. Bladensburg, the first organized settlement in the Washington area, was founded in 1742 as a seaport on a deep and protected harbor at the head of navigation on the Anacostia River. At first, ocean-going vessels reached Bladensburg with little difficulty, and the town became a major center for tobacco export (Bryan, 1914, p. 430). As the land around Bladensburg was cleared for crops, unconsolidated Coastal

Plain sediments were exposed to erosion, and the channel of the Anacostia began filling with sediment. By 1800, the maximum depth at low tide had been reduced from the original 24 to 14 feet [7.3 to 4.3 meters], and the width of the channel so narrowed that navigation was difficult. By 1830, the river became unnavigable to all except the shallowest draft ships, and Bladensburg was abandoned as a port. Few floods had occurred at Bladensburg before the 1930's, but by then, urbanization upriver from the town had increased the silt load in the river. The silt clogged the channel so that floods occurred with increasing frequency. By the 1950's, extensive flood-control measures were constructed.

Modification of river systems through construction of towns, docks, levees, and so on, has proceeded too far for us to consider allowing most streams to revert to their natural state. The question now is what can be done to control the streams and what steps must be taken to insure the proper future development of drainage systems. It is with these thoughts in mind that much of the quantitative study of streams has been done.

Streams are important as geomorphic agents; their role in the gradation of the earth's surface is highly significant and effective in all but arctic climates. Mass movement and running water are intimately associated. Streams are an effective means of transporting material from one part of the earth's surface to another. The rate of transportation is relatively rapid and the quantity of material moved by water is enormous. For example, the suspended load of the Mississippi River near its mouth is about 300,000,000 tons per year. Averages indicate a rate sufficient to remove between 2 and 8 centimeters every thousand years from the entire surface area of the country.

HYDROLOGIC CYCLE

Streams constitute but one element in a complex and continuous system of movement and interchange of water on earth. This system is best described in terms of the *hydrologic cycle* (Fig. 14–1), a representation of the interchange of water among the atmosphere, oceans, and land. Solar

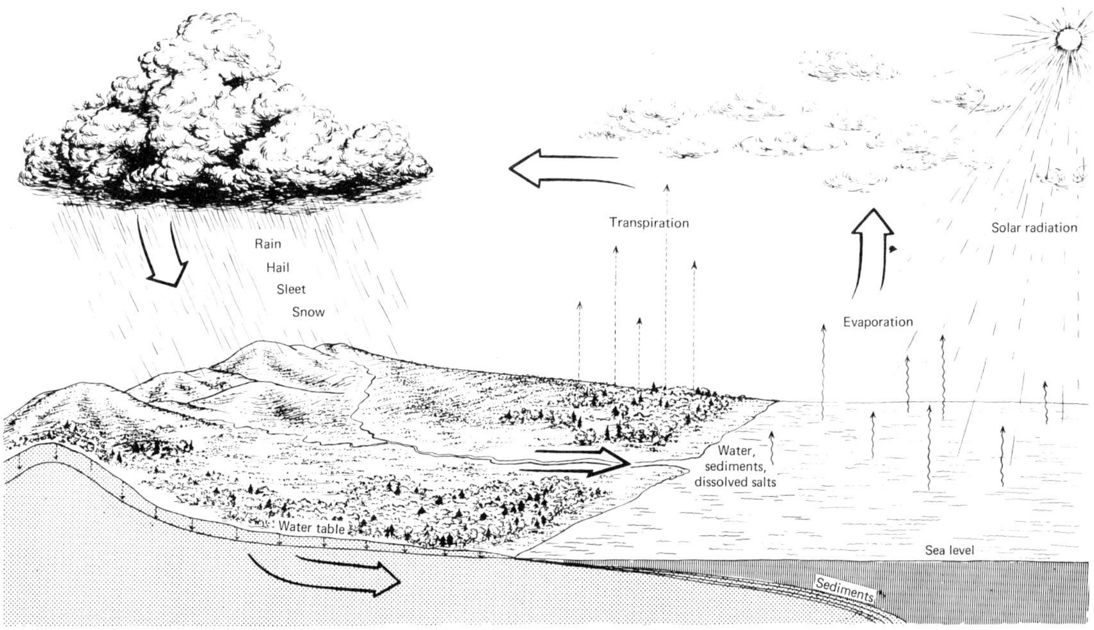

FIGURE 14–1
The hydrologic cycle. Water is an important agent in bringing about change near the surface of the earth. This is possible because it is recycled through the oceans, atmosphere, and land. Solar radiation causes evaporation of water from the oceans. Plants release water vapor directly into the air by transpiration. This water vapor then moves through the atmosphere until conditions favorable for its precipitation as rain, hail, sleet, or snow occur. Then the water falls to the earth. The runoff carries sediment in streams and erodes the land. The water that seeps into the earth moves much more slowly through the soil and bedrock, promoting solution and decomposition of rock. The broken sediments and dissolved solids are carried by the runoff and ground water into the ocean where they are deposited as sediments on the sea floor. Then the cycle is complete.

radiation causes evaporation of large quantities of water from the surface of the oceans, streams, lakes, and marshes, and significant amounts from the land surface, the soil itself. Water vapor is also transpired directly into the atmosphere from plants. Water vapor in the atmosphere is caught up in the circulation of air masses and eventually condenses or freezes and falls back to earth in the form of precipitation. This precipitation helps to bring about rock weathering and erosion through glaciation or stream action, it may dissolve soluble minerals in rocks or soil, or it may be absorbed by and become involved in chemical reactions with minerals of rock or soil. Water may be incorporated by plant roots and transpired back into the air; it may infiltrate through the soil and move through permeable rocks as part of the groundwater flow; or it may collect as runoff on the land surface and move in streams. Eventually much of the water, along with sediment and dissolved substances, finds its way into the ocean, and the cycle is complete.

Of course, the description of this interchange as a closed cycle is an oversimplification. Another way of viewing the hydrologic cycle is as a system of interchanges of water between natural reservoirs. Water is stored in the oceans, in the atmosphere, as ice in glaciers and pack ice at sea, in lakes and river systems, and underground in pore spaces and cavities. A continual interchange takes place through the processes of evaporation, transportation, precipitation, condensation, freezing, melting, percolation, and gravity drainage. Climate is an important determinant of the manner in which water occurs and the way interchanges between the various reservoirs take place.

The hydrologic cycle is but one example of the interplay of the air, oceans, and land. The earth's rotation, the shape of the ocean basins, and the density of sea water are among the factors that affect ocean-surface currents. The currents influence the temperature and moisture content of the overlying air masses, and these air masses are governing factors in climates. Climates are important determinants of rock weathering and soil development, which in turn influence the types of land forms developed. Such chains of interdependence are a common characteristic of the natural processes of the earth.

WATER BUDGET AND RUNOFF

As the need for increased water resources in the heavily populated parts of the world has grown in recent years, much more careful analysis has been made of the factors that affect the availability of water. Both the water in an area and that coming in and going out may be thought of conveniently in terms of the *water budget* of the area. At any given time, some water is held in storage in lakes, in the ground, in actual storage compounds, and in streams within the area. Water is continually coming into the area and going out via streams and through underground flow; it is moved by pipe lines and canals as well. Water is naturally supplied to the area through precipitation (rain, snow, etc.) and naturally lost from the ground by evaporation and transpiration. Some of it infiltrates through the soil to become ground water and the remainder runs off on the land surface.

In making a full-scale survey of water resources, all the above-mentioned sources and losses would have to be considered as well as the consumptive use of water and the artificial movements of water, such as pumping of ground water into streams or artificial recharge of ground water; but for the moment we will consider the water budget in terms of factors that are most important in natural streams. Those five factors are (1) *precipitation,* (2) *evaporation,* (3) *transpiration,* (4) *infiltration,* and (5) *runoff.* The amount of water available for surface runoff in an area is approximately equal to the amount of precipitation less the amounts lost by infiltration, evaporation, and transpiration.

Precipitation is measured by means of rain gauges and snow surveys. Runoff is calculated from stream gauges; averages taken over a period of years have been compiled for the United States as shown in Fig. 14–2. The precipitation for a single drainage system can be determined in the same manner. The amount of potential evaporation can be calculated by observing the amount of evaporation that takes place from open pans, but this value is far greater than actual water loss. The value of loss by evaporation from a pan may be used to calculate the loss from lakes and stream surfaces, but for other land areas it is usual to combine the losses through evaporation and transpiration. *Potential evapotrans-*

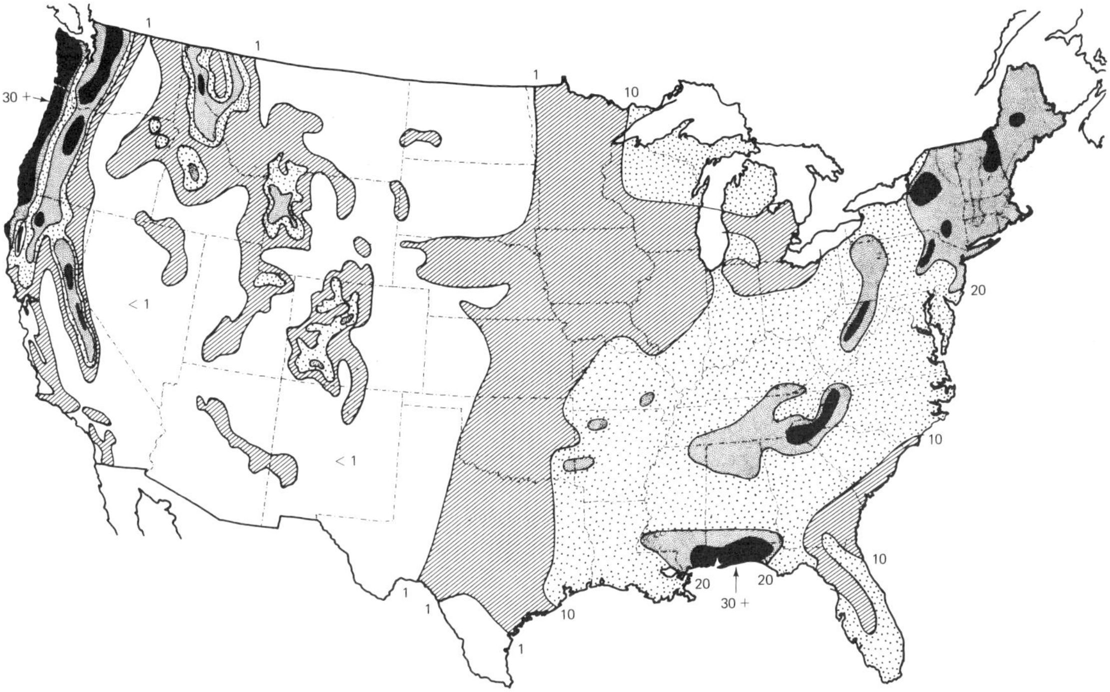

FIGURE 14–2
Average annual runoff in the United States, 1921–45, expressed in inches of runoff.
(after Langbein and others, 1949)

piration (Fig. 14–3) is the loss that would occur in a given area if the soil were supplied with all the water it could hold. *Actual evapotranspiration* is the amount of water loss actually observed. Accurate determination of this value can be made by careful measurements.

The amount of water that infiltrates the soil and the ground-water system is determined largely by the porosity and permeability of the soil and rock, the ground slope, and the rate of precipitation. Surface runoff is measured by gauging stations along streams. Discharge of a stream can be measured accurately if the cross-sectional area and stream velocity of flow can be measured accurately. The channel profile, water depth, and velocity are also periodically measured at many stations along streams in their natural channels. The U.S. Geological Survey has more than 8,000 gauging stations on rivers throughout the country.

The average precipitation over the United States is 76 centimeters per year. Nearly two-thirds of that is lost by evapotranspiration. The remaining third makes up runoff and ground water. The broad pattern of distribution of precipitation through evapotranspiration, infiltration, and runoff can be shown through long-range averages of these quantities, but stream action is also influenced by short-period variations. Precipitation is not uniformly distributed within drainage basins. For example, heavy rains fall in mountains where moisture-laden air is uplifted and cooled until condensation occurs. Even outside such areas a rainstorm will likely cross only part of a drainage basin. So the number, frequency, intensity, and duration of storms may be more important in determining the character of stream action in an area than the average annual precipitation.

Generalizations

So many variables influence surface runoff even over a short period of time that it is most meaningful to discuss the variables in terms of a specific place and specific conditions, but some

FIGURE 14–3
Average annual potential evapotranspiration (inches) in the United States.
(from C. W. Thornthwaite, 1948)

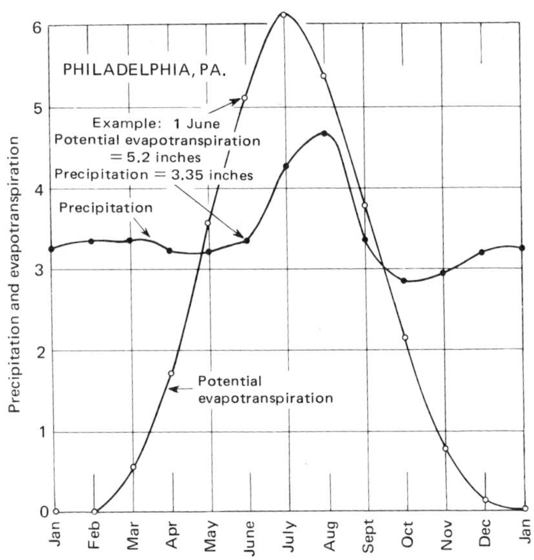

FIGURE 14–4
Average monthly potential evapotranspiration and precipitation at Philadelphia, Pa.
(from Parker and others, 1968)

generalizations can be made. Runoff is increased under the following conditions:

1. Where ground slope is steep.
2. Where the ground surface is impermeable or nonporous (e.g., where impervious rock is exposed; where clay is exposed; where the ground is covered by pavement, houses, etc.; where the ground is frozen).
3. When evaporation is low. Rates of evaporation and transpiration are seasonal (Fig. 14–4).
4. When plant cover is absent or low.
5. When the rate of precipitation is so high that it exceeds the rate of infiltration, as it often does during heavy storms.
6. Where natural water traps such as lakes, sink holes, and so on, are not extensively developed (usually areas of low slope).

PRINCIPLES GOVERNING STREAM ACTION

Stream Channels

Streams differ in such characteristics as velocity, amount of water in the channel, relationship of the channel to the larger valley in which it is located, width and depth of the stream, pattern of streams in the drainage system, shape of the channel, and pattern of flow within the channel. Some streams, like the lower Mississippi, flow in large meandering loops across a broad flood plain through a region of low, rolling to flat topography; others, like the Arkansas River in Colorado, follow an irregular but relatively straight path in deep narrow canyons where no flood plain exists; still others have inherited their valleys from glaciers that have retreated or melted away in the relatively recent geologic past; and others, like the Platte River of Nebraska or the Waitaki River in New Zealand, flow in broad shallow channels with the stream flow broken into a braided pattern (Fig. 14–5). These are but a few examples to illustrate the diversity of streams.

Most streams normally flow in well-defined channels, but great differences exist among the various channels and the type of channel is important. Some streams flow over consolidated rock with little loose sediment in the channel (Fig. 14–6). Usually the bedrock surface is covered, often to considerable depth, by *alluvium,* stream-transported material. When bedrock is exposed the profile of the channel is measured by taking a series of soundings along a line across the stream. When streams have a blanket of sediment in their channels, depth soundings can be used to prepare a profile of the sediment-water interface, but drill holes are needed to obtain the sediment-bedrock profile. The profile of the sediment-water interface changes as the amount of material being moved by the stream changes.

The profile of the sediment-water interface is important because it is used to determine the

FIGURE 14–5
Oblique aerial view of the Waitaki River, a braided stream in New Zealand.
(National Publicity Studios, Wellington, New Zealand)

FIGURE 14–6
(*Above*) Stream (Whistle Creek, Va.) flows in a channel cut in massive limestone (See Fig. 14–11 for potholes in channel.)

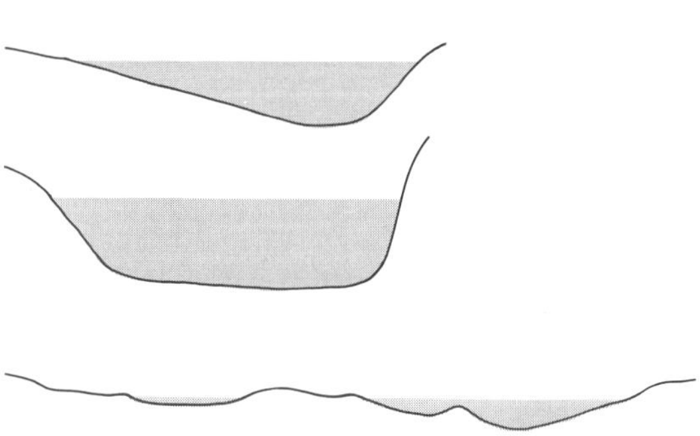

FIGURE 14–7
Schematic drawings of various types of stream-channel profiles.

cross-sectional area of the stream, *A*, a value that is used to calculate the quantity of water moving through any section of the stream. This profile also makes it possible to measure the length of the water-sediment contact, called the *wetted perimeter*. Flow in a stream is slowed by the frictional resistance and turbulence set up along this wetted perimeter where water is in contact with sediment and rock. The effect of channel roughness is much greater for any given volume of flow where the wetted perimeter (measured in cross section) is long. A broad, shallow stream offers more resistance to flow than one with a circular or parabolic profile. A few examples of channel profiles are illustrated in Fig. 14–7.

Stream Flow

Gravity causes surface water to run down slopes, forming streams, and to continue toward sea level. Although the pull of gravity is a constant, the nature of the flow undergoes many changes in the course of the journey. It may start in mountains where the water is confined to narrow channels filled with boulders and rock debris and literally jumps, boils, and foams as it rushes down steep slopes. At the foot of the mountains the stream is more likely to follow a sinuous course through rolling landscapes. The water is quieter, except for occasional floods and rapids, as it moves toward lower elevations. Finally, as the stream approaches sea level it may follow a broad meandering path across a flat countryside. The water flows gently with only mild surficial disturbances. Thus the amount of turbulence decreases downstream, although velocity may increase.

Along the course of almost any river obvious changes in the nature of the movement of the waters can be seen. These range from the extreme turbulence of water where it cascades down a rough stream channel or drops to the foot of a waterfall, to the smooth, even flow at the lip of a waterfall or wherever the slope of a smooth stream channel suddenly steepens. Turbulence is brought about by obstructions, which deflect the water from a straight path, disrupting smooth flow patterns and setting up eddy currents, whirlpools, and boiling movements. Flow in most streams on medium to low slopes is slightly turbulent and is called *streaming flow*. The surface of the water is smooth in some places, slightly undulating in others. A mild, slow eddying and boiling may be visible.

In *laminar flow*, water moves as though it is made up of fine layers that are free to slip over one another. The path of a particle in any one layer is smooth or streamlined. Laminar flow is most closely approached when water is moving slowly through a channel with smooth sides, or when velocity increases rapidly. A lip forms where the water, dropping over the suddenly steepened slope, runs ahead of that upstream. The surface of the water drops and becomes smooth.

Distribution of Velocity and Turbulence

Turbulence is greatest where irregularities are present at the water-sediment or water-rock contact. The net effective forward velocity of a stream is reduced by turbulence and frictional drag along the bottom. The greatest velocity in a straight channel is usually found in the deepest part of the stream channel, about one-third the way down from the surface (Fig. 14–8). If a stream channel is symmetrical, the greatest velocity is usually toward the middle, but if the channel is asymmetric, as in a curve, the greatest velocity is shifted toward deeper water (Fig. 14–9). The deepest part of the channel is generally on the outside of a curve. Maximum turbulence lies just beneath and to either side of the maximum velocity.

Sources of Stream Loads

The products of rock disintegration and decomposition produced by weathering processes are probably the most important ultimate source of stream load. These materials find their way into streams primarily as a result of sheet wash and various processes of mass movement. Sheet wash is effective in moving fine fragments. Urbanization, forest fires, and intensive agricultural development expose fine particles and often lead to silting and flooding in nearby streams.

The origin of the sediments being transported by a stream is the bedrock and soil that occur in the drainage basin. Ordinarily a stream channel is partially filled with alluvium left from the last flood or from former periods of high water when greater quantities of load could be moved. This potential load may again be entrained and moved when high water comes again, and at such times the bedrock or sediment below the channel may become a source as the channel is cut deeper.

Stream banks are important sources of load. When the stream removes alluvium from the channel along a bank, the bank may become oversteepened, causing slumping, sliding, or some other form of bank failure. This process is important along many streams, particularly where the streams flow through unconsolidated sediment. It is important along stretches of the Mississippi River where flood-plain deposits or wind-deposited silt form the banks of the river. Streams located in many high mountains and in areas formerly covered by glaciers (see Chapter 17) flow over unconsolidated glacial deposits that may be readily entrained, and near modern

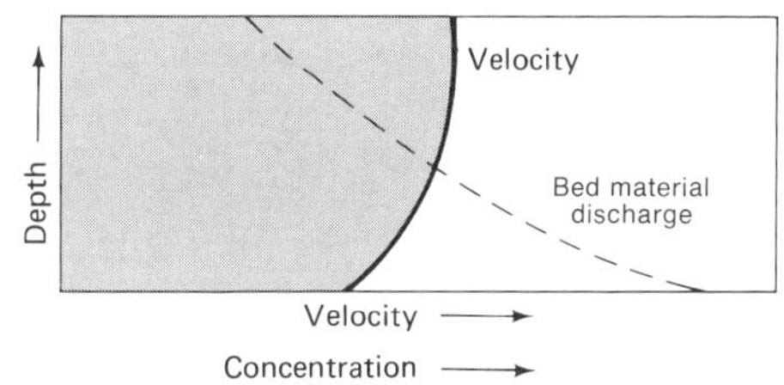

FIGURE 14–8 Schematic drawing showing vertical distribution of velocity (solid line) and bed-material discharge (dashed line).
(after Colby, 1964)

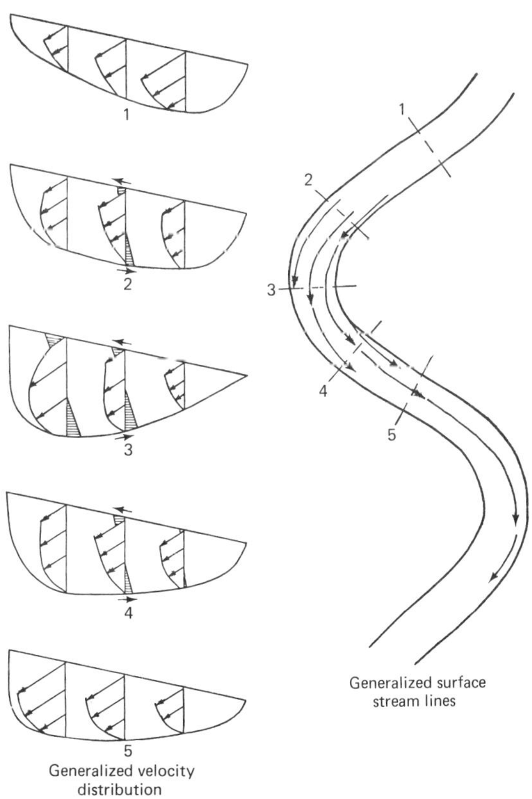

FIGURE 14–9 Generalized diagram of flow distribution in a meander, showing down stream (open parabolas with arrows) and lateral (closely lined areas) components of velocity as vectors, and surface stream lines.
(after Leopold and Wolman, 1960)

glaciers the melt-waters carry loads of rock powder, called *rock flour,* and other larger, suspended fragments into streams. Wind-blown sand and silt may account for large parts of the load of streams located in deserts, near lakes, or along coasts where strong prevailing winds and sources of loose sediment are available. Volcanic ash and dust are important sources of stream load in some localities.

Transportation by Running Water

Suspension. Muddy waters characterize streams of all sizes after a heavy rainfall, when increased surface runoff sweeps large quantities of soil and weathering products into them. Water appears muddy because part of its load is being transported in suspension. The lightest particles may float, as everything is buoyed up by a force equal to the weight of water displaced. Suspended load is usually made up largely of silt and clay sizes. Larger particles may be suspended if the water is turbulent.

Solution. Some weathering products are carried in solution in most streams. A large part of the soluble load in streams in humid climates comes from ground water that finds its way into surface streams. Streams flowing out of deserts where large quantities of soluble minerals may exist at the surface of the ground also have large loads in solution. In most streams the load in solution is a small part of the total load; slow, sluggish streams are an exception. Limestone, dolostone, and salts are notably soluble. Over 300 million tons are removed from the United States in solution each year.

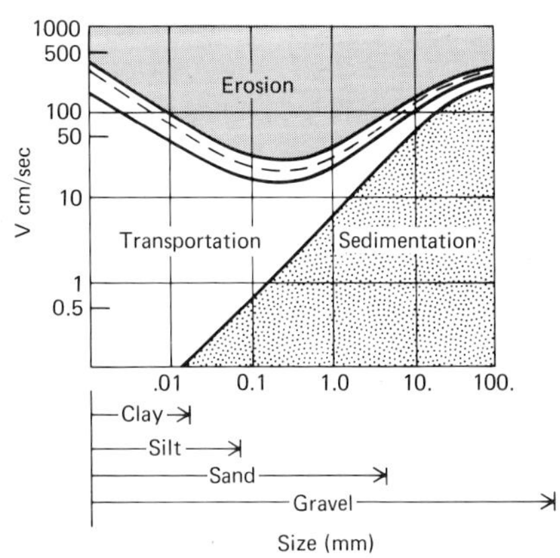

FIGURE 14–10 Relationship between stream velocity, sediment size, erosion, transportation, and deposition. (after Hjulström, 1935)

Traction. The remainder of the stream load is carried along the bottom in traction. Sand, pebbles, and small cobbles may bounce along the bottom, hitting irregularities and traveling as suspended material for a short distance until the pull of gravity brings them back to the bottom again. Larger cobbles and boulders are rolled along the bottom or slip and slide downstream. Experimental evidence indicates that the largest size that will roll along a gently sloping stream bottom varies as the sixth power of the velocity of the water (approximately). Thus a slight increase in velocity will bring about a great increase in the sizes that the stream can move in traction.

Part of the load of most streams is carried by each of the above means, but most of the load may be moved in any one of the three ways, depending on conditions at a particular time. Along the lower Mississippi River most of the load is carried in suspension and solution, and only a small part in traction. The load of most mountain streams is carried largely in traction. The amount of load in transport is also dependent in part on the availability of debris. Streams do not always have enough debris coming into them to fill them to *capacity,* which is the total amount that could be transported under the conditions existing at any given time. Similarly, a stream may not be filled to capacity because it cannot pick up the sizes available. This applies both to boulders that are not moved because of their mass, and to clay that is held together so tightly by cohesive bonds that it is not readily dislodged. In both cases the stream may have unused capacity in one of the three methods of transport. Fine sand and coarse silt are the most easily moved sizes.

Hjulström (1935) studied the relationship between stream velocity and ability to transport and erode. The results, depicted in Fig. 14–10, show the velocities of water movement at which various sediment sizes will be transported. An exponential relationship is found between velocity and size moved for materials of silt to gravel size. Clays will remain in suspension even when water is not moving, but a pronounced difference is seen when the relation between velocity and erodibility is considered (top curve). Because wet clay is cohesive, a much higher velocity is necessary to entrain clay than to entrain sand. In the case of particles larger than sand size there is little difference between the velocity needed to entrain initially and that required to maintain movement. The term *competence* is used to express transporting power. The competence of a stream is the largest size sediment it is capable of moving.

Stream Erosion

Streams accomplish erosion primarily because sheet wash and mass movement provide a constant supply of material. In addition, stream channels and banks are eroded through processes of abrasion, solution, hydraulic force of water, and impact of transported sediment.

Potholes (Fig. 14–11) are striking evidence of the work of solution and abrasion. These smooth, rounded holes found in the bedrock bottoms of many streams are formed by long-continued circular motion of eddy currents carrying sand, silt, and gravel. The surface of the hole is abraded by the grinding and mill-like action of sediment in the hole. Dissolved and suspended products are washed up and out of the hole. Most potholes are relatively small but they occasionally become 6 meters or more deep.

Debris transported in streams flowing rapidly on steep gradients is constantly moved by turbulence, but its motion is usually directed toward the bottom because of the downward pull of gravity. Streams of this sort tend to concentrate erosion at the bottom of the channel (Fig. 14–11). One part of the Bighorn River in Wyoming flows through a canyon that is more than 60 meters deep and not much wider at the top than at the bottom. At the Royal Gorge in Colorado the Arkansas River has cut a narrow chasm more than 300 meters deep. The Black Canyon of the Gunnison River in western Colorado is a similar feature. In these places lateral erosion is negligible, the rivers are cutting high above regional base level, and the streams flow rapidly with great turbulence down steep gradients, cutting valleys in resistant rock.

FIGURE 14–11
Potholes formed in massive limestone along Whistle Creek, Va.

Lateral cutting takes place in streams flowing in sinuous courses on low slopes. At each turn water is shifted toward the outside of the turn, where turbulence and erosion are concentrated. The channel is deepened and the bank steepened on the outside of the curve. This makes the outside bank unstable, and it slumps into the stream. Impact of larger transported materials against sides and bottom of the channel is also an effective means of stream erosion. Heavier materials dislodge loose debris or crack even the hardest and most solid rock. The hydraulic force of water itself can be an important cause of erosion and undercutting of banks, especially on curves. The hydraulic force is analogous to the erosion of soil by a stream of water from a hose.

Cavitation, the collapse of bubbles, in water is an important cause of erosion where water becomes charged with air bubbles, usually due to a velocity increase. This may be especially important in the development of potholes and erosion along rapids and at the base of waterfalls. Bubbles collapse when the water pressure increases, and a strong shock wave is generated by the implosion. These shock waves are capable of causing rapid erosion even in hard materials. Dam spillways must be constructed to prevent or minimize cavitation.

Deposition

The most important factor causing deposition is a decrease in velocity and a consequent loss of transporting power. Decreases in velocity occur where:

1. The slope of the stream channel is reduced.
2. Water is either lost or spreads out. Streams in arid climates lose water by evaporation and infiltration of water into the ground. When the water table is lowered below the level of a stream bed, water moves out of the stream toward the water table. Streams that lose water in this way are called *influent streams.* Spreading of the water may occur where the shape of the channel has altered or where the channel splits into several distributaries, as it does near the mouth of the Mississippi River.
3. Streams encounter obstructions in their paths. These obstructions include natural dams, timber rafts, or fans of debris piled into a stream by one of its tributaries.
4. Local stagnant pools form or the stream enters a standing body of water, such as a lake or the ocean. Flow may stop altogether. A similar condition exists where a stream enters the standing backwaters of a flooded area.

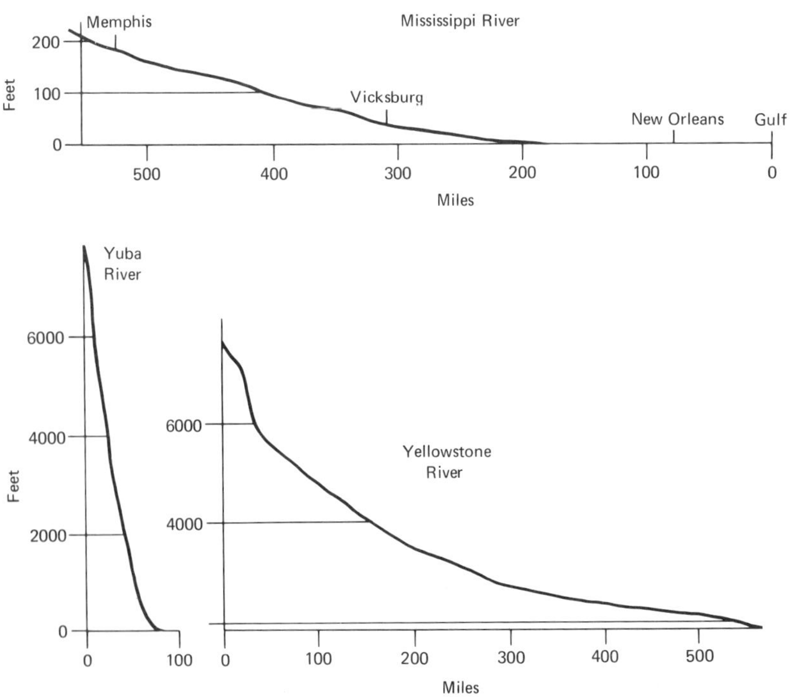

FIGURE 14–12 Gradients of the Yuba, Yellowstone, and Mississippi rivers. (after H. Gannett, 1901)

Streams also deposit part of their load when load exceeds stream capacity. Excessive loads may be produced by rapid weathering and surface wash. The contribution from surface wash may be enormously and suddenly increased following the loss of forest or other protective soil covers. Man is sometimes the major factor in the creation of these conditions. Through carelessness he causes forest fires and allows soil to be removed without planning for control of further erosion that may result. But he pays for his carelessness by loss of valuable topsoil and increased flood hazards. Estimates of the total amount of material moved into the oceans annually before the advent of man place it at 9.3×10^9 tons. The rate of removal since man has become active in modifying the land is 24×10^9 tons annually (Ritter, 1967).

A QUANTITATIVE APPROACH TO STUDY OF STREAMS—HYDRAULIC GEOMETRY

The cost of floods and the importance of water resources have led to demand for a precise understanding of stream behavior. Factors that affect streams have long been recognized, but now they are being evaluated quantitatively. The various elements that influence the behavior of fluids in motion are called *hydraulic factors.* These factors in a natural stream are the following:

1. **Discharge** (Q), which is the amount of water that passes through a given cross section of the stream at any given time. The quantity of the discharge may be calculated by multiplying the cross-sectional area (A) by the mean velocity (V) of the water.

Discharge is measured in volume per unit time (e.g., cubic feet per second, gallons per minute, etc.).

2. **Mean velocity** of the stream (*V*) measured at some particular cross section.
3. **Size and shape** of the channel, usually represented by measurements of mean *width* (*W*) and mean *depth* (*D*).
4. **Channel roughness** (*N*), an approximation of the resistance to flow caused by the wetted perimeter of the stream channel and the materials in the channel.
5. **Gradient,** the slope of the stream channel between two specified points, expressed as meters per kilometer, and so on. The longitudinal profile of a stream may be constructed by plotting channel elevation against distance along the stream. The gradient usually becomes progressively lower downstream, making the profile concave upward (Fig. 14–12).
6. **Load,** which is composed of three parts: a dissolved load, a suspended load, and a bed load or traction load. The first two components are easily measured; the bed load is much more difficult to measure although changes in stream cross-sectional profile show that this load does move. The suspended-load (*L*) fraction is usually the part considered in making quantitative analyses of hydraulic factors.

Discharge and load are determined by the climate, geology, and geomorphology of the drainage basin. Some of the factors that determine discharge are (1) the amount of rainfall, (2) the amount of surface runoff, (3) the size and shape of the drainage basin, and (4) the amount of water interchanged between the water table and the surface drainage. The zone of saturation in many arid climates lies below the level of the stream bottom so water is lost from the stream into the ground, creating intermittent streams. Similarly, the amount of load supplied is a function of the types of sources and how readily it can enter the stream. The width, depth, velocity, size of material in transport, channel roughness, and gradient are dependent on processes acting within the stream; they are dependent variables that are complexly interrelated and interdependent.

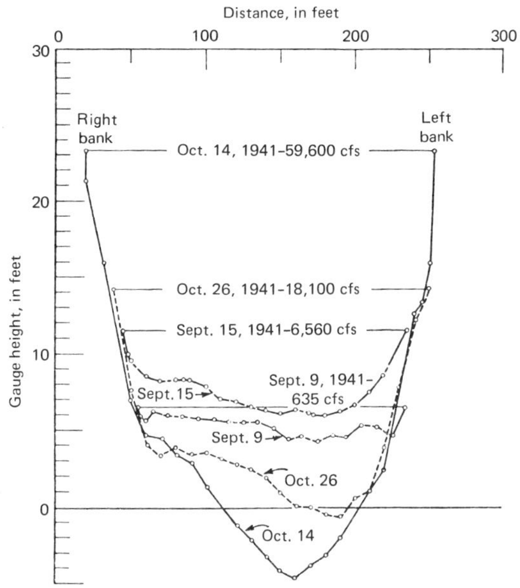

FIGURE 14–13
Channel cross sections during progress of flood, September–December 1941, San Juan River near Bluff, Utah.
(from Leopold and Maddock, 1953)

Gradient is dependent on many other factors. Adjustments of the slope of the bottom of a stream channel occur constantly. If the amount of load being brought into one part of the stream is more than the stream can carry, then some is left in the channel, and the downstream slope is increased. Processes acting within the stream tend to adjust the slope of the stream so it can carry its load.

The interplay of these factors in a stream has been studied primarily by making measurements at a gauging station over a period of time and determining how the factors that can be measured at a single station vary with one another over a period of time. Then comparisons are made among gauging stations down the stream to see how they are related along the longitudinal profile.

At a given station, average figures for velocity, depth, width, discharge, and suspended load can be measured over a short time span. Some observed changes in channel shape with discharge are illustrated in Fig. 14–13. This figure shows depth soundings and water-surface levels for several different dates. The date and discharge are indicated for the San Juan River. Records of these hydraulic factors have been kept over a long interval of time for many streams; so

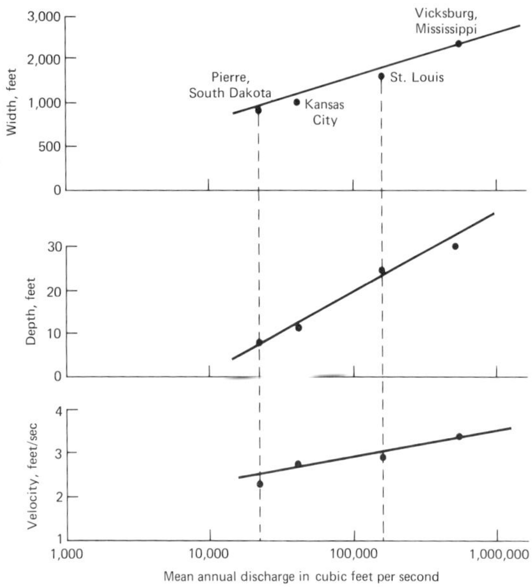

FIGURE 14–14
Variation of characteristics in a downstream direction on the Missouri and Mississippi rivers.
(from Leopold and Maddock, 1953)

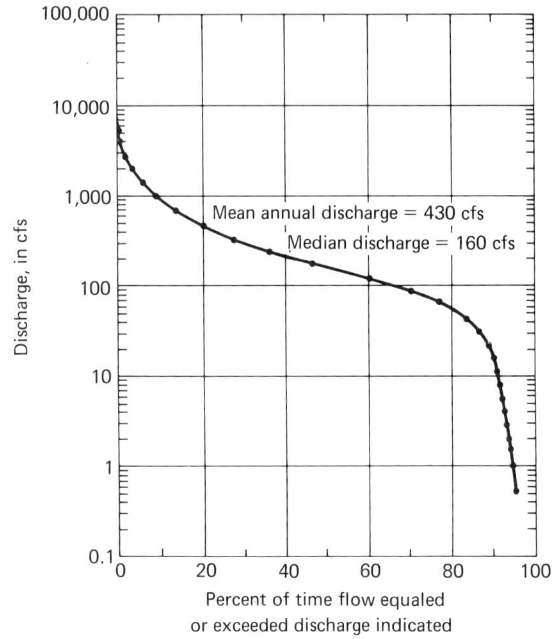

FIGURE 14–15
Cumulative-frequency, or flow-duration, curve for Powder River at Arvada, Wyo., 1917–50.
(from Leopold & Maddock, 1953)

FIGURE 14–16
Flood hydrograph on Trinity River above Coffee Creek and six-hour rainfall measurements at Coffee Creek Ranger Station.
(from Stewart and La Marche, 1967)

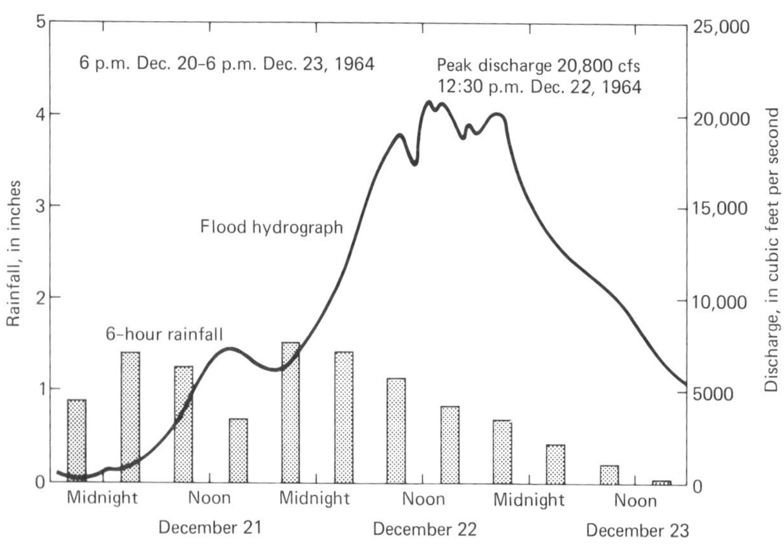

it is now possible to plot mean velocity, width, and depth for many different values of discharge. Width, depth, and velocity are plotted against mean annual discharge for five gauging stations on the Missouri and Mississippi rivers (Fig. 14–14). Plots of these variables usually approximate straight lines when plotted on log graph paper, so it is possible to express the relationships in terms of exponential equations:

$$W = aQ^b$$
$$D = cQ^f$$
$$V = kQ^m$$
$$L = pQ^j$$

where a, c, k, p, b, t, m, and j are constants. Because Q = area of stream cross section, multiplied by the mean velocity

$$Q = WDV$$

or

$$Q = aQ^b \cdot cQ^f \cdot kQ^m$$

or

$$Q = ack \cdot Q^{b+f+m}$$

Therefore $ack = 1$ and $b + f + m = 1$.

The values for the constants are determined at a single gauging station by making a plot of discharge against W, D, V from measurements made over a period of time. Examples of values determined for 20 streams in the Great Plains in this way are $b = 0.26$, $f = 0.40$, and $m = 0.34$. Note that these values vary from stream to stream and even from place to place on a stream.

Frequency of Discharge

The frequency of discharge at a station may be plotted as shown in Fig. 14–15. The percentage of the time the stream flow equaled or exceeded any given discharge may be read from this graph. More than 90 percent of the time the flow in this example exceeded 10 cubic feet per second (0.28 m^3/sec), and 4 percent of the time the flow exceeded 1,000 cubic feet per second (280 m^3/sec), as during floods. Another way of treating flow at a station is by plotting discharge against time (Fig. 14–16). The main factor causing variation in discharge is variation in precipitation in the drainage basin. The rainfall in late December in the Trinity River basin, California, caused a flood. Note the time lag between rainfall and the flood crest; the peak discharge came many hours after most of the rain had fallen. Normally a peak like this will travel downstream but it will tend to spread out and be lower as it moves downstream.

Downstream Variations in Hydraulic Factors

The hydraulic factors in streams vary from one cross section of the stream to another. In perennial streams, the discharge increases downstream as the size of the area drained increases and the number of tributaries contributing to the flow increases. Often this increased discharge is carried in a channel with relatively less wetted perimeter and less channel roughness. Increased discharge also generally means the width, depth, and velocity have increased.

One interesting way to represent the changes that take place in hydraulic factors is to plot the changes of velocity, depth, width, suspended sediment, channel roughness, and slope against discharge. This can be done for two (or more) stations as in Fig. 14–17. The dashed lines show changes that occur at a single station with increases of discharge, and the solid lines give a comparison between one cross section and another. The plots of average values for hydraulic geometry indicate the following:

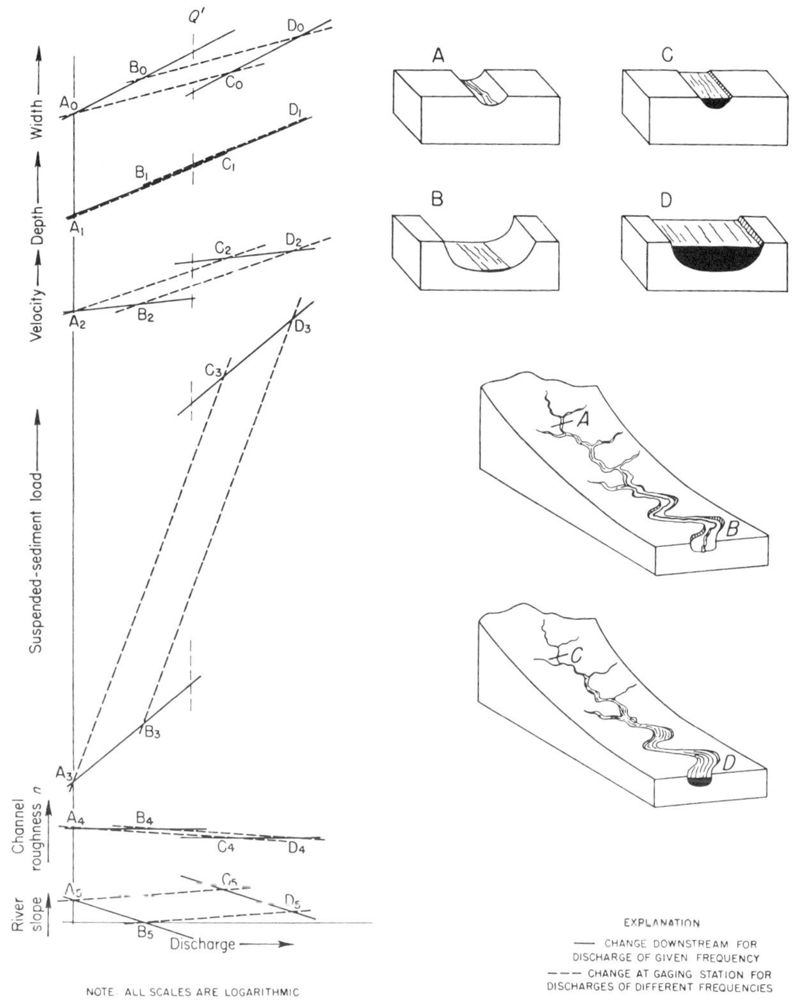

FIGURE 14–17
Average hydraulic geometry of river channels expressed by relations of width, depth, velocity, suspended-sediment load, roughness, and slope to discharge at a station and downstream.
(from Leopold and Maddock, 1953)

1. Channel roughness tends to be lower when discharge is increased, and channel roughness tends to decrease downstream.

2. Mean width, depth, and velocity all increase with discharge both at a station and downstream.
3. The most prominent changes occur in the suspended-sediment load. Small increases in velocity cause great increases in the load.

Longitudinal profiles of streams (Fig. 14–12) almost invariably show downstream decreases in slope, giving the profile a concave-upward appearance. Stream velocity is a function of slope, so we might expect streams to be faster in their headwaters and sometimes they are; but often the increase in discharge and reduction of channel roughness provide a much more efficient channel downstream and can lead to increased velocity despite the lower slopes.

Many streams have irregularities in slope where differences in erodibility of the bedrock exist. Sometimes these show up as waterfalls, cascades, or riffles. At such places the turbulence of the water and velocity are increased, which contributes toward removal of the harder rock; but once the obstruction has been removed a break in slope may remain, although it will likely migrate upstream. Such a break may be removed or smoothed out if the channel contains a thick alluvial fill above the obstruction.

Concept of the Graded Stream

One of the most useful concepts for understanding the behavior of streams in a qualitative way is the idea developed by Hoover Mackin that streams have a strong tendency to become graded.

> A graded stream is one in which, over a period of years, slope is delicately adjusted to provide, with available discharge and with prevailing channel characteristics, just the velocity required for the transportation of the load supplied from the drainage basin. It is a stream in equilibrium; its diagnostic characteristic is that any change in any of the controlling factors will cause a displacement of the equilibrium in a direction that will tend to absorb the effect of the change.

We should not consider any river to be perfectly graded. A graded stream is one in which the various hydraulic factors are constantly changing, shifting, and compensating to bring about a state of equilibrium. A stream is often considered to be at grade when it is just able to remove all the debris supplied to the stream from its drainage basin.

Quantitative methods of stream analysis provide a much more exact way of approaching stream action, and they are now replacing the older qualitative methods.

DENUDATION

The materials of the earth's crust have been subjected to the effects of erosion by geomorphic agents over such long periods of time that even very slow processes may be effective in changing the landscape. Man has witnessed many short-term effects of erosion, such as formation of gullies and even the natural diversion of streams, and it is apparent that materials eroded from one place may build up as deposits in another. However, when material is removed from a drainage basin a net loss occurs that cannot be replaced by processes of gradation. The long-term effect of erosion that leads eventually to a regional lowering of topography is called *denudation*.

Many efforts have been made to calculate the rates of denudation (see the article by Ritter, 1967). The rate at which the streams from a given drainage basin are removing sediment is calculated. The dissolved and suspended load can be estimated with accuracy, but the amount of bed load is difficult to measure. Once the total load is estimated, it is converted into an approximate volume of material removed. Then if the area of the basin is measured the denudation rate can be expressed as a uniform lowering of the basin per unit of time, usually in inches per 1,000 years. Some sample figures are given in Table 14.1. The present rate of denudation for the United States is between 1 and 3 inches per 1,000 years, and estimates of world-wide denudation are between 2 and 4 inches per 1,000 years.

The removal of weight from one part of the earth's crust usually induces a movement of subcrustal materials toward that region. Thus removal of rock and regional lowering due to erosion are partially compensated for (see Chapter 18). A distinction should be made between absolute lowering of the surface that is partially compensated by uplift and the amount of material removed from the surface.

Delaware River above Trenton, N.J.	0.8 in./1,000 yrs
Schuylkill above Philadelphia, Pa.	2.7 in./1,000 yrs
Rio Grande above San Acacia, N.Mex.	1.8 in./1,000 yrs
Mississippi River above Baton Rouge, La.	1.3 in./1,000 yrs
Colorado River above Grand Canyon, Colo.	5.6 in./1,000 yrs
Eel River above Scotia, Calif.	30.4 in./1,000 yrs
Columbia River above Palmer, Wash.	0.5 in./1,000 yrs

TABLE 14.1
Selected rates of denudation.
(Figures from Judson and Ritter, 1964.)

THE ROLE OF STREAMS IN LANDSCAPE DEVELOPMENT

Valleys, hill slopes, hills, and mountains owe their forms to geologic conditions and processes. Rock structure, the combination of active geomorphic agents (e.g., mass movement, streams, ground water, wind, waves, currents, and glaciers), the regional base level, and time are all important factors in land-form development. To these we might add the geologic history of the region. This shows up prominently in the relationship of the stream to the stream valley.

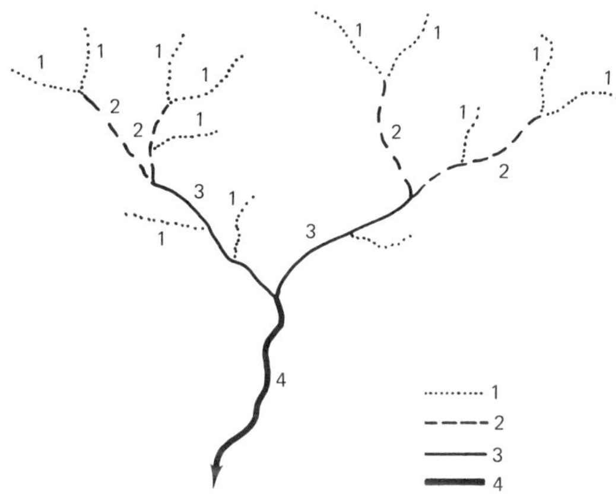

FIGURE 14–18
Method used to number streams
(after Strahler, 1956)

Quantitative Approach to Basin Analysis

The classical approach to study of a drainage system and its relationship to the topography has been essentially descriptive and qualitative. Recent studies employing a more quantitative approach are producing interesting results that ultimately may provide a more sound basis for our understanding of the processes of landscape formation. R. E. Horton (1945) and A. N. Strahler (1956) are among the pioneers in this field. Their approach is seen in the way they describe a drainage basin. The boundary of a basin is the divide that separates it from adjoining basins. A purely descriptive characterization of the basin might include something about its shape and the pattern of the streams. Quantitative descriptions might include measurements of its area, perimeter, and stream lengths, and ratios of basin area to basin length, basin length to basin width, and so on. *Drainage density* is the sum of all the stream lengths divided by the basin area. *Stream frequency* is the number of channels (stream segments) per unit of area. These two ratios provide an easy way to compare different drainage basins.

The drainage within a basin may be further described quantitatively by determining a hierarchy among the various streams. Strahler assigns a hierarchy as shown in Fig. 14–18, in which small streams at the head of the drainage are called first-order streams. When two of these join, a second-order stream is formed. When two second-order streams join, a third-order stream is formed, and so on.

R. E. Horton observed an orderly decrease in the number of streams of each higher order. He decided to test his hypothesis that drainage-system development follows certain laws by plotting (on semilog graph paper) stream order against the log of the number of streams. He found that the plots from most drainage systems yield straight lines, which indicate an exponential relationship (Fig. 14–19). From this conclusion Horton and others have derived several relationships that have been termed *Horton's Laws of Drainage Composition*. These relationships are described below.

FIGURE 14–19
Plot of stream order against number of streams of that order indicates an exponential relationship between order and drainage density.

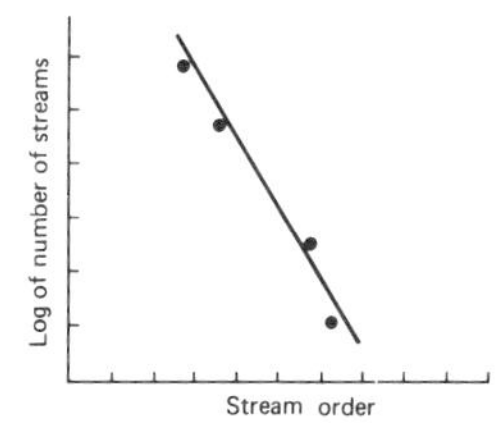

1. The number of streams of a given order (N_u) is equal to the ratio of the number of streams of a given order to the number of streams of the next higher order—called the bifurcation ratio (R_b)—raised to an exponent equal to the highest order stream in the basin (s) minus the order (u).

$$N_u = R_b^{s-u}$$

2. The average length of streams of order u (L_u) is equal to the average length of the first-order streams (L_1) times the stream-length ratio raised to the order ($u - 1$).

$$L_u = L_1 R_L^{u-1}$$

3. Average area of basins of order u (A_u) is equal to the average area of basins with area of order 1 (A_1) times area ratio (R_a) raised to the order ($u - 1$).

$$A_u = A_1 R_a^{u-1}$$

4. Average relief of basins of order u is equal to the average area of order 1 basins times basin-relief ratio (R_R) raised to the order ($u - 1$).

$$H_u = H_1 R_R^{u-1}$$

FIGURE 14–20
A V-shaped valley.
(Crown Copyright)

Here then are a set of quantitative relationships that have been found valid for a great many drainage basins. They provide a precise way to compare basins and discover differences.

R. L. Shreve (1966) studied Horton's Laws and concluded that "the statistical nature and remarkable generality of Horton's Laws suggest . . . that the law of stream numbers arises from the statistics of a large number of randomly merging stream channels."

Relationship of the Stream to Its Valley

A stream entrains material from its valley sides, from the channel, and from its banks; and it removes that material from the valley. Most valleys owe their origin to the streams flowing in them and the topographic configuration within most drainage basins is in large part the result of processes of running water imposed over a long period of time on rock of a particular structure. Compare the stream valleys shown in Figs. 14–20 and 14–21. Recent history of the earth has been marked by glaciation, changes in sea level, volcanic activity, and crustal deformation that have affected drainage systems in many areas. As a result the land forms in some areas where streams are the primary geomorphic agent today were shaped in large part by other agents. For this reason many streams are not as well adjusted to their valleys as might be expected, and some flow in valleys where most land forms were shaped by other agents. The most striking examples of this situation are found in high mountains where valley glaciation has been active. Glaciers still occupy the valleys in these mountains at the higher latitudes and small glaciers still exist even in middle and low latitudes. The valleys in such areas have glacial land forms that are relatively

unaltered by stream action. The ice ages have had many influences on streams and their valleys. The last large-scale advance of continental ice sheets ended about 11,000 years ago. The last of the large ice sheets in North America extended as far south as the Missouri and Ohio rivers. In fact, the present position of those rivers was largely determined by the edge of the ice. Pre-existing drainage that flowed north into the St. Lawrence was disrupted by the ice advance. Ice moved southward into Kansas during earlier glacial advances. The surface of the ground north of the terminus of the ice sheets is largely covered by glacial deposits, and the land forms there were shaped primarily by glacial deposition. South of the former edge of the ice sheet glacial outwash deposits are abundant and large areas are covered by dust (loess) blown from these deposits (Fig. 14–22). Streams are now cutting into these deposits, but the process is in an early stage of development.

The climate changed during the glacial ad-

FIGURE 14–21
Meandering stream patterns. Note the old meander scars, oxbow lakes, and typical flat terrain.
(U.S. Air Force)

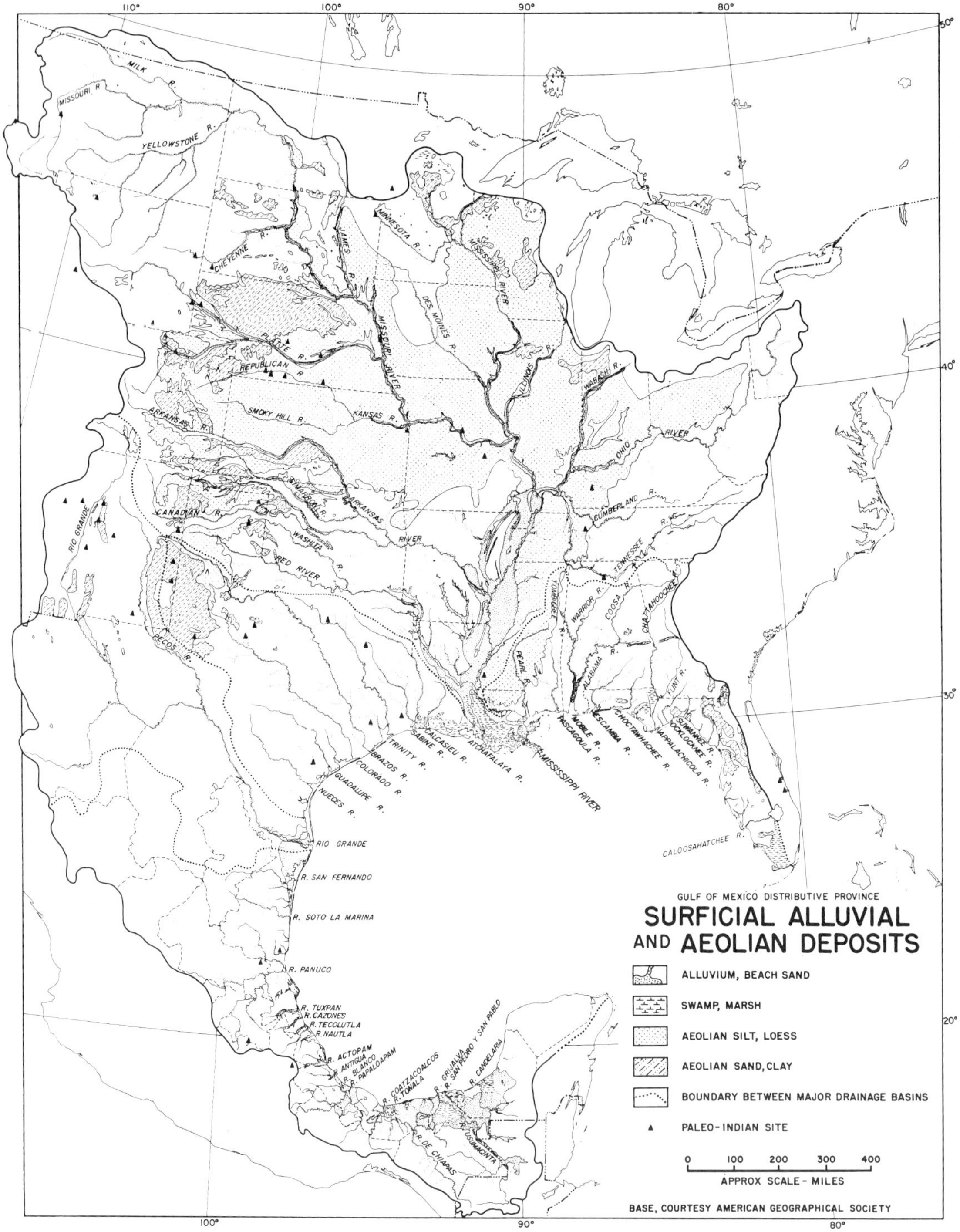

FIGURE 14–22
Surficial deposits in the drainage basin of the Mississippi River and its tributaries.
(from Moody, 1967)

vances and retreats. The patterns of circulation, temperature, and precipitation were all affected and this had marked influence on stream behavior. Valleys in areas where streams had much greater discharge than at present were shaped by streams of a different character from those that now flow. The balance between load, discharge, and other variables in streams has been subjected to many changes and reversals during and following the ice advances and retreats. Near the coasts this has been complicated by changes in the water-ice balance and consequent long-term fluctuations in sea level, the ultimate base level.

Volcanism has played a role in disrupting modern drainage systems, although to a much smaller extent than glaciation within the United States except at Hawaii, along the West Coast, and in Alaska.

Finally, we must consider the effects of earth movements. Regional tilting and warping have affected the area covered by the ice sheets; this deformation is continuing today. Slow regional warping and abrupt movements on faults affect large parts of the western United States. The history of these movements in the West extends far back in geologic time and has been an important influence on stream action and land-form development. A sudden movement causing an abrupt uplift on a fault may change a stream's gradient locally, create a waterfall, impound drainage, or even divert streams from one drainage basin into another.

Thus land forms in a particular area are a product not only of the action of geomorphic agents now active but also of those that have been active in the recent past as well as the stability or instability of the area, changing patterns of climate, and so on—in short, land forms are a product of an area's geologic history.

LAND FORMS RELATED TO STREAM ACTION

Flood Plains

A flood plain is a flat area near the level of a stream that is sometimes covered by water during flood stages on the stream. Usually they occur where stream channels on low and moderate slopes are not confined in location and change position by lateral erosion. When the channel shifts it leaves behind a filled channel. Shifts of this sort become increasingly common in regions of low relief and where unconsolidated sediments or easily eroded rocks underlie the channel or the stream banks. However, flood plains also occur on hard-rock surfaces. Eroded surfaces on either side of the stream are at approximately the same elevation as the main stream.

The flood plains on some streams are narrow; others cover vast areas, as in the case of the lower Mississippi. Many of the broad flood plains found in the Coastal Plain of the United States are related to the unconsolidated, easily eroded sediment that underlies the Coastal Plain and the gentle dipping structure of these beds. Lateral planation is easily accomplished in this sediment.

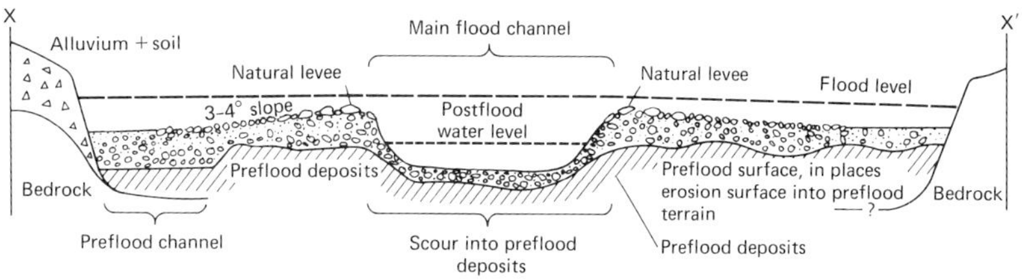

FIGURE 14–23 Generalized map and section showing flow patterns and natural levees formed during flood.
(from Stewart and La Marche, 1967)

Natural Levees

When the discharge in a stream becomes so great that it overflows the banks, water floods onto the flood plain, usually carrying with it a large load of suspended sediment. Flood-plain sediments are thickest near the stream channel (Fig. 14–23) where velocity is high and where most of the load is carried just before and after flooding starts. Once waters spread beyond the main channel they become shallow, frictional drag is great, and the heaviest material is deposited. As the water begins to recede, velocity decreases and more deposition occurs. A wedge-shaped layer of sediment is deposited. The thickest part of the wedge is on either side of the channel. Thus the river builds *natural levees* by the flooding process itself. Effects of this process are most interesting along the Mississippi south of Baton Rouge, where the river flows down a narrow ridge built up above the delta surface by flood-plain deposits. The surface of the stream actually flows at a level above that of the flood plain.

Natural levees contain the stream at normal

FIGURE 14–24
Gemini 4 spacecraft view of the Nile Delta, Egypt; Mediterranean Sea to left; Suez Canal and Red Sea in background. Dark areas are cultivated land.
(National Aeronautics and Space Administration)

FIGURE 14–25
An aerial view of Kaikoura Coast, Clarence River, and Inland Kaikoura Mountains, South Island, New Zealand, showing a stream and delta modified by current action.
(National Publicity Studios, Wellington, New Zealand)

discharge and may prevent tributaries from flowing into the main channel (the Yazoo River of Mississippi is an example). This causes sluggish drainage, facilitates deposition on the flood plain, and gives rise to bayous such as those found on the lower reaches of the Mississippi River flood plain.

Deltas

Alluvial deposits formed at the mouth of a stream are called *deltas* because some of them are nearly triangular in map view, resembling the Greek letter delta. A delta is built where the amount of load entering a body of water is in excess of that that can be moved. Counterparts of the famous deltas of the Nile (Fig. 14–24) and the Mississippi may be found in many small lakes and streams (Fig. 14–25). Such small deltas may be built where a heavily loaded tributary flows into a lake or stream. A huge area (Fig. 10–24) has been filled with debris eroded from the continental interior of the United States in the Mississippi River Delta. The debris is carried across marshes of the lower Mississippi Valley through a system of distributaries. During floods the sediment is spread over the marshes. At other times the sediment is carried to the edge of the delta and deposited on the slopes along the delta front. Fine sediments are carried in suspension into the Gulf of Mexico.

Whether or not a delta can be built in the ocean depends on the topography of the continental shelf, the crustal stability of the shelf, and the effectiveness of waves and currents in breaking up and moving the deposits.

Stream Meanders and Related Features

A meandering stream is one that has a looplike channel pattern (Fig. 14–26). Some meanders almost double back on themselves, and they usually occur as a train of meander loops of a nearly uniform size, expressed in terms of wave length, amplitude, or width of the train. They occur where the gradient of a stream becomes extremely low. The lower Mississippi, a classic example of a meandering stream, is flowing down a slope that is approaching the horizontal—0.4 meters per kilometer. Because velocity, turbulence, and consequently erosive power are greatest on the outside of a curve, it is possible for a stream to move laterally by undercutting banks or by bringing about collapse through sliding and slumping of the outside banks of bends. Since the process is largely one of slumping, it is slow when the river flows against resistant rock. This may cause the direction of lateral movement to be stopped or reversed. The presence of other loops up- and downstream limits the growth of a meander. If the loops become too large, they merge and a *cutoff* occurs, the flow is diverted, the river channel straightens out its path for a short distance, and a meander loop is left isolated from the main stream. These loops become detached from the stream when clay and silt are deposited by the main stream in the quiet waters at the entrance to the loop. Thus a loop-shaped lake, called an *oxbow lake,* is formed.

Some of the sediment removed from the outside of the curves migrates downstream and stops in the quiet waters inside the next meander. Thus curved bars, called *point bars,* are formed on the inside of the curves (Fig. 14–27). Sometimes part of the flow may be diverted and return more or less to an older path of flow inside these bar deposits. Such a course is called a *chute.* Chutes are formed during floods, but once formed they may remain after the flood waters recede.

Meanders undergo continual change as the positions of the meanders shift. Meanders may migrate back and forth across a flood plain and they may migrate downstream. Successive surface profiles across a meander over a six-year period show this lateral migration (Fig. 14–28). Most meandering streams flow through deposits laid down earlier when the stream was also meandering (Fig. 14–27). The bars deposited on the inside of curves form the pattern that shows so clearly in the photographs.

The lower Mississippi River is an example of a meandering stream that is flowing on a flood plain; such meanders are called *free meanders.* Other meandering streams flow through regions where there is either no flood plain or a very narrow one. Such meanders are called *incised meanders.* If they are cut into the bedrock either as a result of downward erosion, *entrenched* streams, or by lateral cutting, they are termed *ingrown.*

The relationship of the stream to its valley varies in other ways also. Some meandering streams flow in meandering valleys; others flow

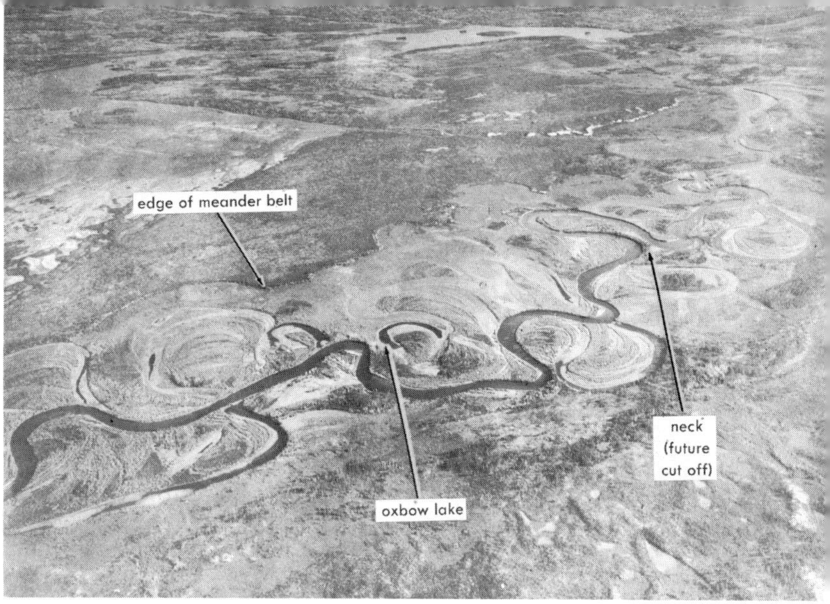

FIGURE 14–26
Features of the meandering stream and its valley. The edge of the flood plain is marked by a cliff. Oxbow lakes, meanders, meander scars, and a neck are clearly shown. The natural levee is not obvious.
(Royal Canadian Air Force)

FIGURE 14–27
Point-bar deposits laid on the inside of a meander loop. View is from a vertical aerial photograph.
(U. S. Geological Survey)

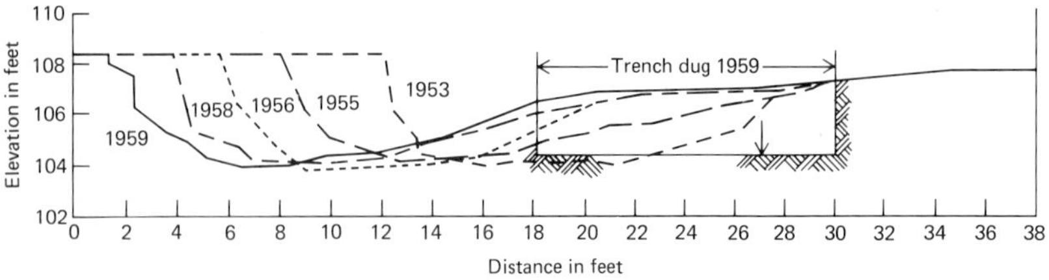

FIGURE 14–28
Successive surface profiles of meander channel and point bar, 1953–59.
(from Leopold and Wolman, 1960)

on flood plains within straight valleys; still others follow meandering paths along meandering valleys but with very different wave lengths. Where the valley in which a stream flows is larger than that needed by the stream, the stream is said to be *underfit,* a term not restricted to meandering streams. This would appear to be the case where a large meandering valley contains a stream that is meandering with much smaller wave length. Underfit streams occur in many parts of the United States, and their origin is likely related to changes in the amount of precipitation. It is suggested (Dury, 1965) that the condition that led to their development was increased precipitation and discharge that occurred in early postglacial times followed by generally decreasing precipitation.

Origin of Meanders

A number of ideas have been advanced to explain the origin of meanders. None is completely successful in explaining the characteristic geometric relations[1] found in most meandering streams. The initiation of meanders has been studied by means of flume experiments. Streams do not often follow straight courses in nature, nor do they long continue straight courses in flumes, even when the channel is initially straight, the slope relatively steep, and the sediment uniform. Any natural obstruction will cause the flow pattern to be broken and deflected, and once deflected the water particles tend to follow a sinuous path (more accurately, a helical path). This helical flow or sinuous pattern is formed in flume experiments even in the absence of obstructions. As soon as a curve is established the velocity maximum is shifted toward the outside of the curve and the water surface takes on a slight slope toward the inside of the curve, which directs the flow back across the channel and against the opposite bank. Soon deposits form on the inside of the curve and the initial conditions leading to meandering are started.

The relationship between gradient, discharge, size of sediment, and the way it is introduced into the system determines whether the stream in a flume develops meanders, continues to follow a sinuous course, or breaks up into a braided pattern. Meandering streams are usually found where the gradient is very low. One effect of the meandering pattern is to lengthen the stream; thus the development of meanders may be thought of as an internal adjustment along a stream that effectively lowers the gradient within the valley. Resistance to flow is greatly increased in the curves also.

Meanders with geometry similar to that of streams in alluvial valleys are found on the surface of glaciers where there is no load to deposit on point bars, and meanders of similar geometry are found in the Gulf Stream where neither load nor even confining banks exist. Some hydrologists have attempted to explain the oscillatory motion in meanders as a type of gravity wave. More recently meanders have been studied in terms of the rate of energy dissipation along the channel. These studies indicate that a meandering stream is closer to being in a state of equilibrium than one that is not meandering (Leopold, Wolman, and Miller, 1964).

[1] The following relationships have been established between wave length of meanders (λ), basin area (A), channel width (W), and radius curvature (R):

$$\lambda = cW^m = fR^t = aA^b$$

$$A_m = dW^n$$

where m, n, and t are constants approximately equal to 1, and a, c, d, and f are constants.

THE ROLE OF ROCK STRUCTURE IN LAND-FORM DEVELOPMENT

The structural configuration of the rock that underlies any region influences the land forms that will result from mass movement, stream erosion, or the action of other geomorphic agents. Structure in the broadest sense of the word applies to the internal structure of rock bodies and the variations in resistance to erosion as well as to the configuration of the rock masses. Variations in resistance of the underlying rock to erosion and the aerial distribution of these variations are important in determining land-form development.

Consequent Drainage

If a stream's path is governed primarily by the original slope of the ground surface, the stream is called a *consequent* stream. This condition is most likely to exist where the underlying rock is relatively uniform in its resistance to erosion, as where a gently tilted uniform sedimentary rock unit underlies a large region. Conditions favorable for consequent drainage may be envisioned in terms of the drainage that would exist on a newly formed volcanic cone or a recently emerged, nearly flat continental shelf. The most common patterns of consequent drainage systems are nearly straight streams flowing down steep slopes and radial patterns developed on volcanoes or newly uplifted structural domes (Fig. 14–29a).

In the initial stages of the formation of a drainage system on a new surface, slope is important. Sheet wash that covers a new surface begins to become concentrated in any natural depressions; small rills coalesce and these run together to form larger streams. As these concentrations of water reach a velocity and discharge needed to move sediment, a channel begins to develop. The density of the drainage is influenced by such variables as rock type and its erodibility, climate, steepness of land slopes, and stage of development. That a drainage system with deeply incised valleys can develop rapidly is seen where gullies form in the steep embankments of soil along many road cuts and other areas where the protective cover is removed and old drainage patterns are destroyed.

As a new drainage system becomes established the streams in it accomplish erosion *downward* by cutting the stream channel deeper, *laterally* by sheet wash and by undercutting banks and inducing mass wasting, and toward the stream

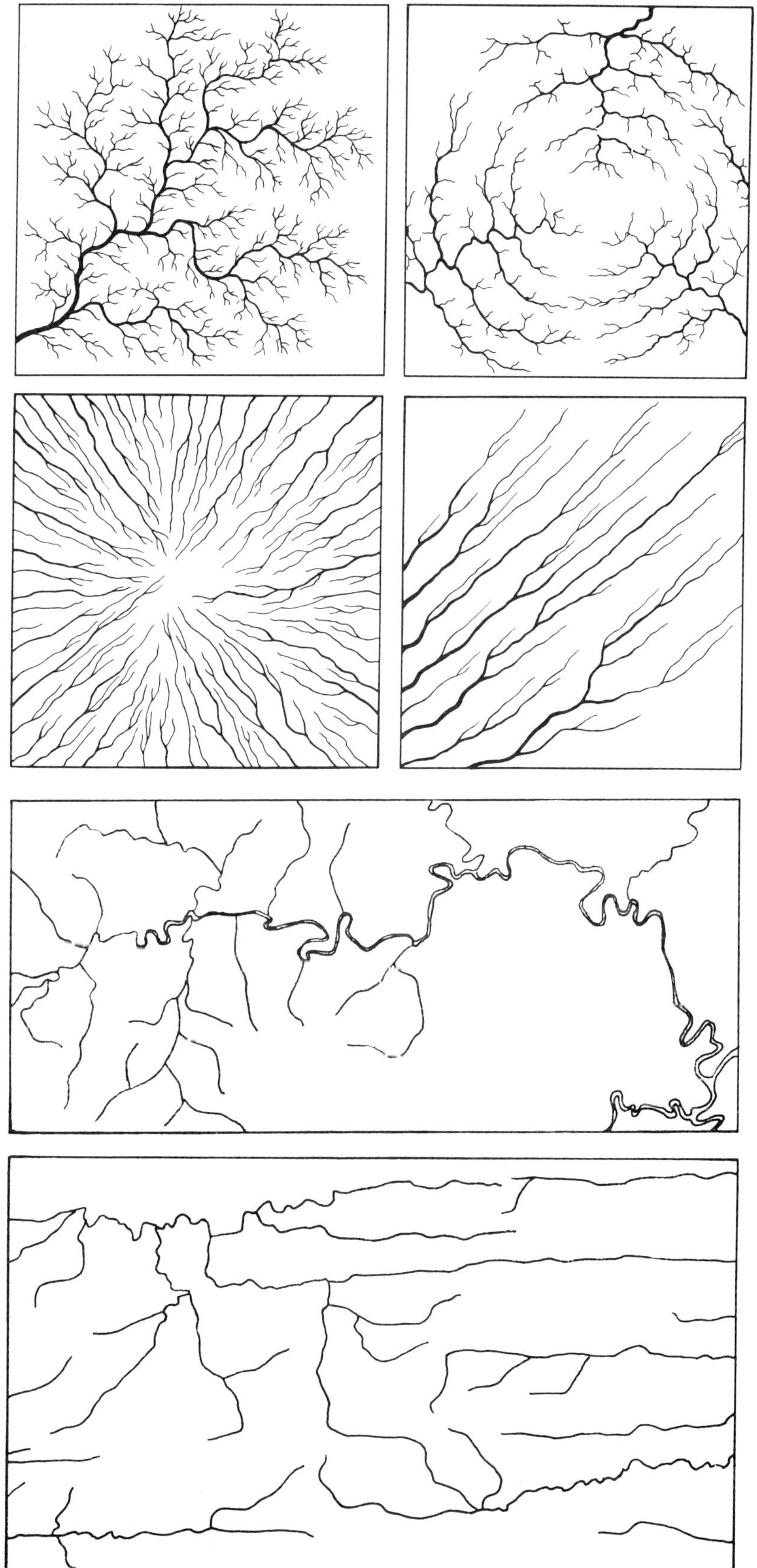

FIGURE 14–29(a)
Drainage patterns. The patterns shown are (top left) dendritic; (top right) annular; (middle left) radial; (middle right) linear or straight; (second from bottom) meandering; and (bottom) modified rectangular or trellis pattern.

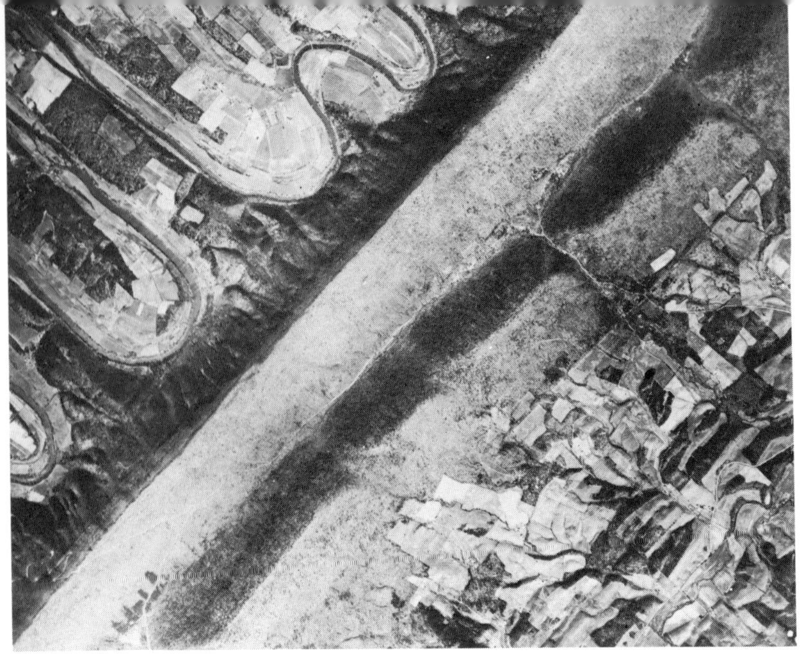

FIGURE 14–29(b)
Vertical aerial photograph of the North Fork Shenandoah River. The drainage here is well adjusted to the structure of the underlying bedrock. The river meanders in an outcrop belt of shale. The ridges that cross the photograph diagonally are underlain by quartzite and are limbs of a syncline. Note streams flowing downslope from the ridge and parallel to the synclinal valley.
(U.S. Geological Survey)

FIGURE 14–30
A planed-off dome. Note that the stream follows a pattern that reflects the structure.
(photo by courtesy of the U.S. Air Force)

head, *headward erosion,* by extending the channel at the head of the stream.

The development and subsequent evolution of hill slopes after the initial stages of drainage are a complex matter (reviewed by Schumm, 1966). Hill slopes may develop a straight, convex, or concave surface depending on relief, climate, rock type and structure, earth movements, rates of uplift or subsidence, rate of erosion in comparison to diastrophism, soil infiltration capacity and erodibility, and the processes of erosion and deposition active in the area. Some of the classic ideas concerning this evolution are reviewed later in this chapter.

Subsequent Drainage

The drainage pattern may reflect the structure and differences in erodibility of the underlying rock. Where the stream patterns show such an adjustment to the underlying rock structure the drainage is termed *subsequent.* In the Colorado Plateau drainage of many smaller streams is well established along fractures in sandstone layers. A different pattern is found in the Appalachian region where many of the tributaries of the major streams flow in valleys floored by shale and limestone and separated by ridges capped by sandstone and quartzite. The pattern (Fig. 14–29b) formed is a trellis or rectangle. Many of the tributaries in the trellis patterns in the Appalachians flow in the opposite direction from the dip of the underlying beds; these are called *obsequent streams.* Other subsequent patterns include the parallel streams formed where streams flow on weak zones in a uniformly dipping sequence, and the annular pattern found around some domes and basins (Fig. 14–30).

Headward erosion is particularly important in the development of subsequent drainage. Tributaries flowing on the most easily eroded material are extended and cut downward most rapidly.

Shifting of Drainage Divides

The divide between two drainages is the boundary that separates the direction of surface flow between the two. As denudation proceeds the position of the divide will remain the same only if the rates of erosion, both downcutting and headward, are the same on the two sides of the divide. If the rates are unequal the position of the divide will shift as the tributaries in one basin cut down or grow headward more rapidly than those in the other. Several conditions favor the shifting of the divide—increased discharge in one basin, the presence of more easily eroded rock in one basin, or differences in rock structure between the two.

FIGURE 14–31
Stream capture near Asheville, N.C. The stream flowing eastward has captured the upper tributaries of the westward-flowing river. (The lower edge of the large arrowhead—indicating north—covers a portion of the east stream at the divide.)

Stream Piracy

Stream piracy is the name given to the diversion of drainage from one stream into another. The headward part of one drainage can be diverted into another stream when the divide separating the two streams is eroded low enough to allow the connection of a tributary of one stream with the trunk of another (Fig. 14–31). Consider the case of two parallel valleys separated by a ridge composed of tilted rock strata. The channels of the main stream in each of these valleys are close to the same elevation. Tributaries flowing from the ridge between the two large valleys cut into the ridge. Those flowing down the steeper side of the ridge will cut more rapidly than those on the other side, because the steeper gradient will contribute to more rapid erosion. As this tributary stream cuts headward into the ridge, the crest of the ridge is breached and a notch is formed. As this notch is worn lower by headward erosion of the stream, the divide shifts toward the center of the second valley. Eventually, through continued headward erosion and lowering of the stream channel, the head of this tributary intersects the main channel of the stream in the upper valley. Now the path of movement of water from the head of the stream in the upper valley shifts, flows through the gap, called a *water gap* (Fig. 14–32), and into the main trunk of the lower stream valley. The process of piracy is complete. The tributary of the lower stream is called the *pirate stream.* The stream it has diverted or captured is said to be *beheaded.*

The process of stream piracy does not necessarily require that the slopes of the separating ridge be different. Essentially the same steps can take place if streams on opposite sides of the ridge have different gradients for any reason. The stream flowing on the steeper gradient will lower its channel more rapidly.

FIGURE 14–32
Vertical view of a water gap cut through a cuesta, a sloping plain developed on a dipping rock layer.
(U.S. Air Force)

FIGURE 14–33
Playa lake. This view is southeastward across Harney Lake playa, Oregon.
(A. M. Piper, U.S. Geological Survey)

LAND FORMS IN ARID AND SEMIARID CLIMATES

Arid and semiarid regions cover more surface area than do humid regions. Although the origin and evolution of land forms in different climates have long been a subject of controversy, many geomorphologists agree that differences between land forms of humid-temperate, semiarid, and arid environments are differences primarily in degree, not in kind.

DESERTS

Effects of wind erosion are most obvious in deserts, but most geomorphic features of deserts located in tropical and temperate climates are the result of mass movement and running water. Glaciation and freezing effects cause most features in deserts located in high-latitude frigid climates.

What exactly is a desert? Common fallacies are that deserts have no vegetation, that they consist mainly of drifting sand dunes, that they are always hot, and that wind is responsible for most desert land forms. Arid regions make up more than 30 percent of the land surface of the world. They are regions characterized by scarcity of rainfall and, consequently, relatively thin vegetative cover. Rainfall in deserts is typically torrential; that is, it comes down rapidly in a short period of time. The entire year's rainfall may be concentrated in no more than three or four rains, but several centimeters of rain may fall during each of these. Most plants found in humid regions do not grow in deserts because of the long, dry periods, but others adapted to the small supply of rain flourish. Tropical deserts may be hot or at least warm throughout the year, but those in the middle latitudes undergo seasonal temperature changes much like those of the humid regions, and are frequently more extreme. In high latitudes deserts may be cool or cold most of the year. Even in tropical deserts a high daily variation in temperature occurs. Because there are fewer leafy plants, less moisture, and few clouds, heat is lost rapidly from the desert at night.

Some deserts are characterized by *internal drainage*. That is, the amount of rainfall is so small and the rate of evaporation is so great that a well-developed drainage pattern is lacking and no water flows out of the desert on the land surface. Other deserts are drained by large streams such as the Nile and the Colorado rivers, but few tributaries are found within deserts even in these instances. In the southwestern United States many deserts occur in basins that are almost entirely enclosed by mountain ranges. Weathering processes in deserts produce more rock debris than the available water and wind are capable of moving.

Weathered debris on the ground surface in deserts is moved by wind, mass movement, rain wash, and stream action. Wind moves sand and finer sizes of material, mass movement is most effective where relief is great, and erosion by water is accomplished largely by sheet wash and stream action. During a typical desert cloudburst, rills and sheets of water containing a very heavy load form immediately. Most of the streams flow for a brief time following a rain but disappear soon afterward as water evaporates or soaks into the ground. Channels formed by these intermittent streams typically have steep walls, a sandy braided channel, and are called *washes, wadies,* or *arroyos.*

Desert Land Forms

Plains are the most prominent land form in deserts. These are formed in a variety of ways. Prominent structural plains, flat surfaces of structural origin, such as uplifted sea floors and plains produced by the erosion of flat-lying sedimentary layers, are found in the southwestern United States. Other structural configurations found in the same region include tilted and block-faulted mountains, folds, arches, and flexures. Plane or gently sloping surfaces have been superimposed on these structural features through erosion and deposition.

Playas. In deserts consisting of basins enclosed by mountain ranges, drainage is toward the center of the basin from all margins. Where these drainage lines converge we might expect to find a lake. In general, a nearly level plain has been built up by gradual influx of sediment from the basin sides. Sometimes this plain is covered by a broad shallow lake called a *playa* (Fig. 14–33). Playas usually retain water only a short time after a rainfall, then lose it through evaporation or percolation. The lake becomes the site of deposition of sediments from lake waters. As the lake water evaporates, the sediment and dissolved mineral matter are deposited to form the *alkali flat.* If the concentration of salts is high the flats are called *salinas.*

Alluvial fans. Alluvial fans form where streams flow out of an area of high relief into one of lower relief—where a stream passes from a steep to a low gradient. The channel becomes wider, several distributaries may form, water is lost into the sediment, velocity is reduced, and deposition takes place. The stream flowing from a mountainous region may carry great quantities of material that is too coarse to be transported across lower surfaces without large increases in velocity. Freezing and thawing, steep slopes and mass wasting, and glacial deposits provide load. Rock debris piles up on the slopes until a heavy rainfall washes it down into the stream valley and downstream to the fan. The result is formation of a broad, low, cone-shaped deposit called an *alluvial fan.* Such fans are composed of poorly sorted, stratified stream deposits (Fig. 14–34) and mudflows. Usually streams on them are not confined to particular channels very long. The stream shifts position back and forth across the fan, or it may form distributaries that spread the debris more evenly across the fan's surface. Fans are best developed in semiarid and arid climates where the process is facilitated by lack of abundant rainfall and high rates of evaporation, which create an even greater imbalance in the amount of water in relation to load. Water in arid climates may also percolate through the stream channels and be lost to lower water tables, and the relative absence of plant cover makes it easier for the stream to change position on the fan. The ground-water resources of alluvial fans are discussed in Chapter 16.

The apex of alluvial fans is often cut by stream channels that are entrenched into the top of the fan. Braided-stream channels and distributary channels several meters deep occur downslope, but their position is transient and they shift as the streams flood over the shallow banks. Several fans may coalesce near the foot of the mountains to form a broad, gently sloping plain.

Studies of fans in California and Nevada have shown that the slope of a fan's surface is closely related to the steepness of the longitudinal profile of the canyon above the fan (steep valleys have steep cones, low, sloping valleys have low cones). The size of the drainage area and the types of lithology also determine steepness. The radial profiles of some of these fans are not smooth

FIGURE 14–34
Section of material exposed in trench dug beneath pavement on Shadow Mountain fan.
(from Denny, 1965)

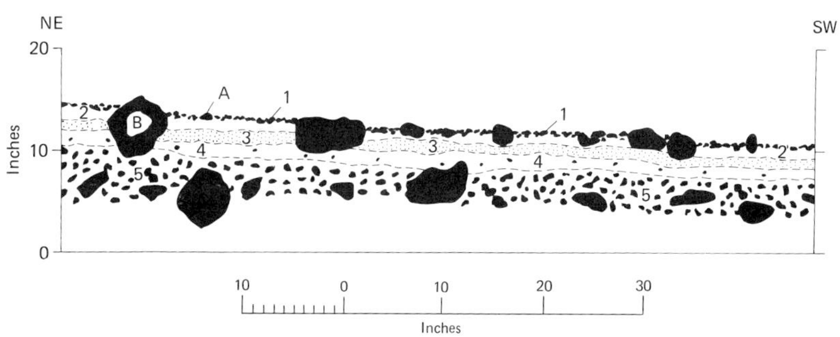

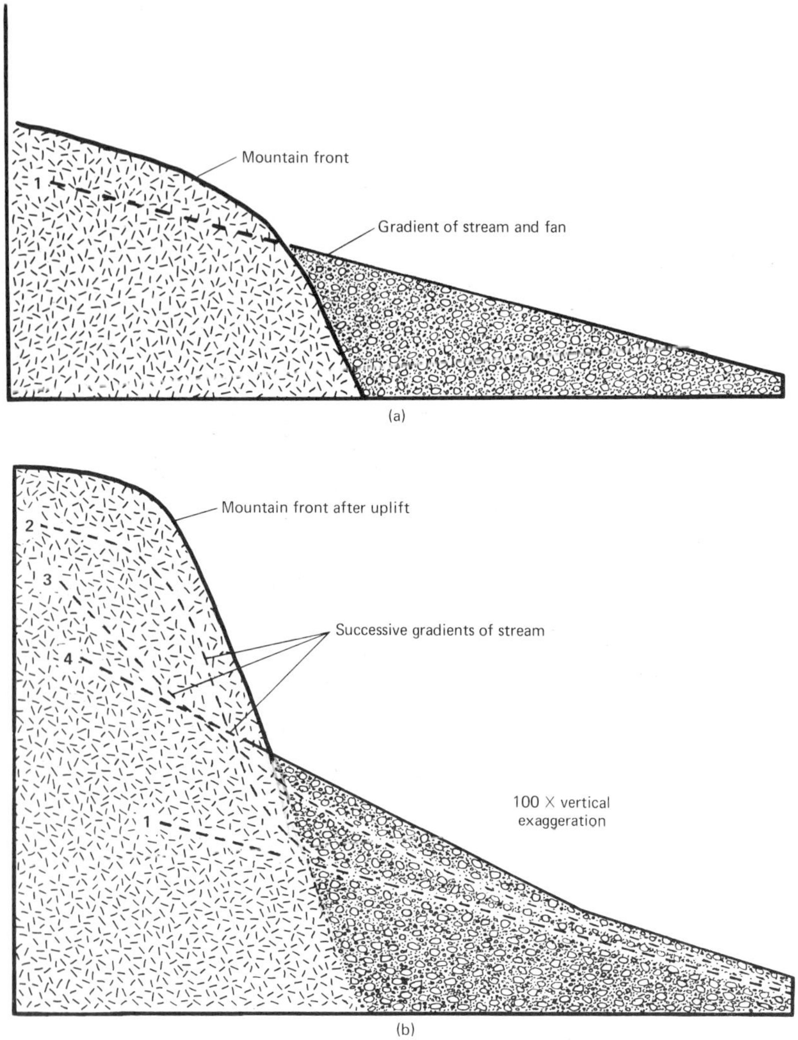

FIGURE 14–35 Diagrammatic sketch showing stages in the development of an alluvial fan. Initial condition (a) shows slope of fan equal to stream gradient in mountains. After rapid uplift (b) the stream becomes entrenched—profiles 2, 3, and 4. Equilibrium is again established in profile 4. The fan has two slope segments at this stage. (after Bull, 1964)

curves, but comprise three or four straight-line segments (Fig. 14–35). This segmentation is thought to be a result of intermittent uplift that has changed stream-channel slopes and depositional slopes (Bull, 1964). Each time the channel was steepened, the succeeding fan deposits formed a new fan-shaped segment of steeper slope than that deposited on the upper part of the pre-existing fan.

Bajadas. Playas occupy positions that have been formed by successive deposition of alluvial material washed into the basin by sheet flooding and channelized flow. These are often deposits of coalescing alluvial fans. Surfaces of these deposits slope upward from a flat plain at the center of the basin to the surrounding mountains. These low alluvial deposits at the base of the mountains and near the center of the basin are called *bajadas* (Fig. 14–36).

Pediments. Closer to the mountains the gently inclined plain that slopes up to their base may be primarily an erosional rather than a depositional feature. As the mountain front has retreated away from the center of the basin, solid rock that lies close to the surface has been eroded, forming a pediment (Fig. 14–37). The pediment surface may be covered by a thin veneer of debris, deeper in buried channels. The material is in the process of moving from the high slopes toward the center of the basin.

Evolution of land forms in deserts is primarily related to formation and extension of pediments. A number of different processes play a part. These include lateral erosion by streams, soil creep, and unconcentrated sheet wash. The net effect of these processes is to bring about retreat or back wearing of mountainslopes.

Fault scarps are prominent along edges of some mountains in the southwestern states. Sharp folds and downcutting of streams through rock units of varying resistance also give rise to cliffs and steep slopes. Many of these hillsides or mountainslopes may be divided into four elements. A broad, nearly flat, somewhat rounded surface occurs at the top. This passes into a cliff that has been washed clear of soil and debris. At the foot of this cliff talus slopes form. In hard, resistant rock a pronounced break occurs where the talus slope meets the low slope of the pediment, but in weak rocks there is a gradual lowering of the slope onto the pediment surface. Usually the most active of the hill slopes are the cliff and the talus slope. Mass movement is most effective on them, and surface wash is most capable of moving debris across steep surfaces. Material from these steep slopes is washed out across the pediment. In arid climates the most active movement of debris occurs from the beginning of rainfall to shortly after the rain stops. The amount of water moving on the pediment increases downslope.

FIGURE 14–36
Oblique aerial view of Death Valley showing a playa lake in the upper left with bajada and pediment surfaces leading into the mountains. Alluvial fan deposits spread out from the mountains.
(John S. Shelton)

FIGURE 14–37
Erosional remnants in an Australian desert. These isolated masses of rock are all that remains of a formerly higher and more rugged topography. The units of rock are stratified and are turned up on end, indicating that strong deformation once affected the area.
(Qantas Empire Airways Ltd.)

Water becomes deeper, turbulent, and highly erosive. Wadies form, and gullies are cut into the pediment. Moving water is the dominant process responsible for transportation of material across the pediment.

If the steep slope becomes covered with vegetation or stops actively retreating for any other reason, the slope above it becomes smoothed and gradually develops a convex form where it breaks away into the steep slope. This rounded form may become so pronounced that the cliff and the talus slope appear convexly rounded at the top.

Evolution of the topography in an arid region, once pediments are formed, consists essentially of back-wearing and retreat of the steep face above the pediment. The slope of the pediment is maintained. If pediments occur on both sides of a mountain range they gradually converge. Eventually the range is reduced to a gently sloping plain with slope equal to that of the pediment on either side of a crest, a narrow ridge of small, scattered, steep-sided residual hills known as *inselbergs* (island mountains) (Fig. 14–37). Once inselbergs are removed from the landscape, further reduction of the land is negligible. Compare the evolution of landscape according to the back-wearing concept with that according to slope reduction (Fig. 14–38).

TEMPORAL EVOLUTION OF LAND FORMS

The idea of land-form evolution was first conceived by Major James W. Powell, a Civil War veteran who headed the United States Geological Survey, studied the arid southwest, and made the first trip through the Grand Canyon of the Colorado River. William M. Davis developed the idea, and a student of his, Douglas Johnson, applied and elaborated the concept to the land forms of the world. Both Davis and Johnson concentrated their work on land forms in humid, temperate climates. Other regions and different land forms were treated as deviations from the cycle of evolution in humid regions, or as climatic accidents. They visualized landscapes as passing through stages in development somewhat comparable to the stages of life, that is, youth, maturity, and old age (Figs. 14–39a and 14–39b). Youth is a period following uplift, when erosion is active on land that has just experienced a lowering of base level. Maturity is reached as processes of erosion and deposition begin to become adjusted to the topography. Old age comes about when the topography, after millions of years of gradual reduction, begins to approach a level, nearly featureless surface. Davis and Johnson recognized that interruptions in the cycle would occur if uplift of the land lowered the regional base level at any time during the cycle. But if the cycle goes to completion—old age—then in the ideal case the cycle might be started once again by uplift, and the sequence of development would be essentially the same as in the previous cycle.

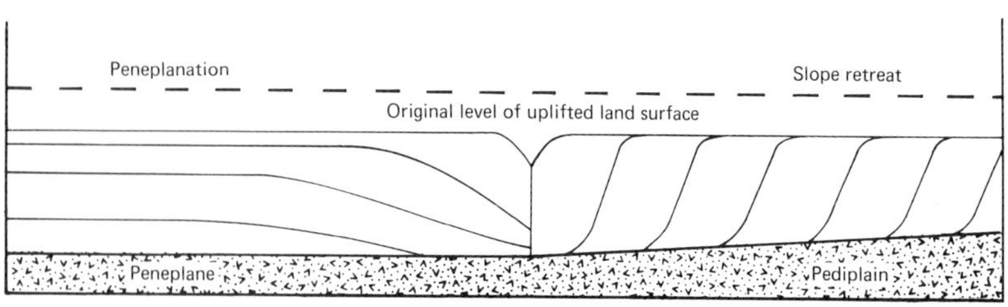

FIGURE 14–38
Schematic diagram showing the differences in the concepts of land-form evolution according to the back-wearing or slope-retreat concept (right side) and the down-wearing or peneplanation concept (left side).

DAVIS' IDEAL FLUVIAL CYCLE

The ideal erosion cycle envisioned by Davis commences with rather rapid uplift of sedimentary rocks from the sea floor. The landscape evolves in a temperate, humid climate in which erosion is accomplished mainly by running water and mass wasting.

A number of widely separated, deeply incised, and narrow valleys containing streams with many rapids, waterfalls, and no flood plains are formed at first. When the streams become graded, most irregularities, such as waterfalls and cascades, will have been removed from the river channel and lateral erosion begins to be more important. Valleys are broader, contain flood plains, the drainage pattern is better developed, broad upland areas near divides disappear, and streams follow sinuous courses. At this time the landscape is mature and streams cut their channels lower at a slower rate. Valley sides continue to be reduced

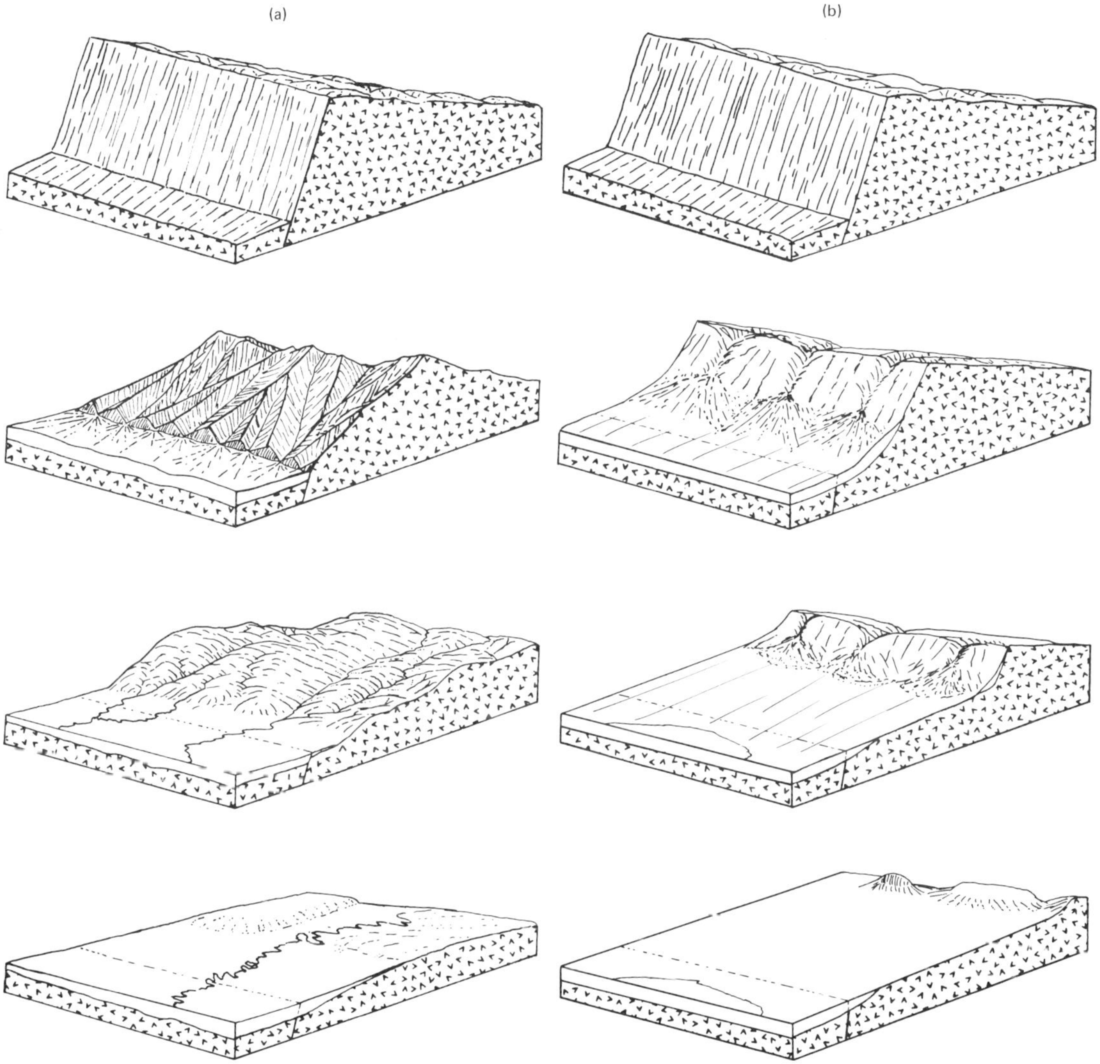

FIGURE 14–39
(a) Evolution of an uplifted block-faulted mountain according to the concept of slope reduction, as visualized by Douglas Johnson.
(b) Evolution of an uplifted block-faulted mountain according to the concept of slope retreat.
(after W. M. Davis, 1909)

by weathering, mass movements (particularly surface creep and wash), and lateral erosion. Gradually valley sides are reduced to lower and lower slopes. Divides between drainage basins are likewise lowered until they are almost completely flattened. According to Davis, a lowland of faint relief remains at this time. This land form, called a *peneplain* ("nearly a plain") or *peneplane*, is the ultimate in the erosion cycle (Fig. 14–40). Running water ceases to be an effective agent of erosion on such a surface.

Effect of Interruptions in the Cycle

Davis recognized that compound or multicyclic evolution is more common than single cycles. Evidence is found where topography generally characterized by features of old age or maturity

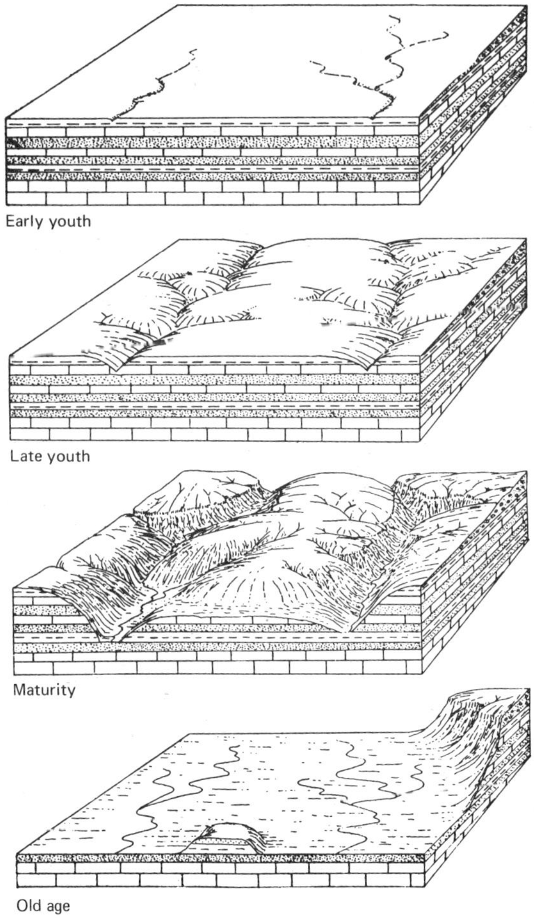

FIGURE 14–40
Four stages in the evolution of a plateau according to the idealized fluvial cycle of Davis and Johnson.
(after A. K. Lobeck, *Atlas of North American Geology*. By courtesy of Edwin Raisz)

contains elements of youthful landscape. The process by which the landscape begins to revert to youthful characteristics is called *rejuvenation*. It may occur in a number of ways:

1. *Uplift*. Old or mature topography may be uplifted as the crust is raised in relation to sea level.
2. *Lowering of sea level*. This changes the relative position of the base level just as effectively as lifting the land surface. Stream profiles are steepened, especially at the mouth, and channels are cut down toward the new level.
3. *Changes in hydraulic factors*. If load supplied to a stream decreases, or if discharge increases, the channel may be cut deeper.

Land Forms Resulting from Rejuvenation

Entrenched stream channels and river terraces are characteristic of rejuvenated landscapes. Entrenched meandering streams are found in the Colorado Plateau. The San Juan River follows a broad, meandering pattern that in some places almost forms cutoffs, but flood plains are absent (Fig. 14–13). Valley sides rise steeply from banks on one or both sides of the stream. It appears that these streams were once meandering on a broad, flat flood plain and that this was interrupted by a regional uplift with consequent downcutting of the stream. Once downcutting started, lateral erosion became minor, and the stream was unable to straighten out its path before it became entrenched.

Terraces

Elevated stream terraces are often clearly demarked in the topography and may appear at the same level on both sides of the river valley (Fig. 14–41). Such *paired terraces* are explained as resulting from entrenchment of a stream into its own flood plain. Multiple paired terraces signify repeated periods of entrenchment separated by periods of stability during which the stream eroded laterally. When terraces do not occur at the same level on opposite sides of the river, it appears most likely that entrenchment was a more or less continuous process. *Unpaired terraces* result if a river continues to cut downward as well as laterally as it migrates from one side of its flood plain to the other.

Terraces may result from regional uplift or changes of discharge and load that affect the ability of a stream to lower its channel. Change in discharge has been common in the ice ages and terraces in the valleys of many underfit streams are thought to have this origin.

Effects of Deformed Strata

If a sequence of folded and faulted sedimentary rocks of varying hardness buried beneath flat-lying sediments is uplifted, water initially flows from it as a consequence of the slopes. V-shaped

FIGURE 14–41 Stream terraces and braided stream, Waimakariri Gorge, New Zealand.
(Copyright by V. C. Browne)

FIGURE 14–42 (*Below*) Cross sections showing relationship between topography and structure in an area of folding at two stages in the history of the area.

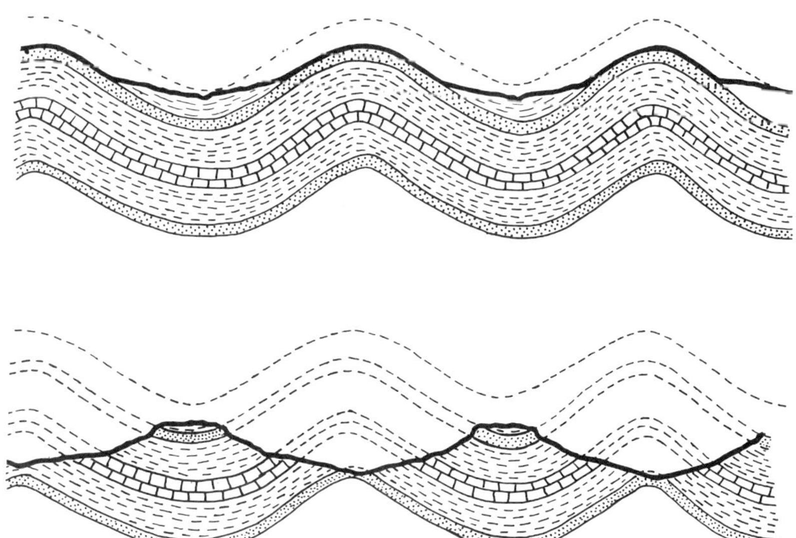

valleys are cut (Fig. 14–40). As the streams cut through the sediment they encounter the deformed rocks. The easily eroded sediment veneer is stripped from the surface. But positions of main stream channels remain fixed if the streams have already established their channels in consolidated rocks before the surface covering of sediment is removed. If this has happened, channels cut across the deformed rocks in spite of their varying resistance.

Tributaries to the main streams tend to assume positions on less resistant rocks, along fault zones, or on other lines of weakness. Weathering and erosion are most effective on less resistant units, which are lowered much more rapidly than harder units. Streams may be established in weaker units regardless of their position on the structures. If weak rock units are exposed along the crest of an anticline (Fig. 14–42), the crest is lowered and an *anticlinal valley* is formed. If the resistant units are exposed in the center of a syncline, a *synclinal mountain* may form. As a mountain range made up of folded sedimentary rocks is gradually denuded, the ridges may be anticlines and the valleys synclines at one stage. Later a reversal may occur, causing the ridges to be synclines and the valleys anticlines (Fig. 14–42). Further development depends largely on angles of slope of the ridges and valleys and on the gradient of the stream (see page 328).

Classical Fluvial Cycle Applied to the Appalachians

The fluvial cycle was formulated largely as a result of studies in humid regions and especially to account for land forms in the Appalachian Mountains.

Most of the main trunk streams in the Valley and Ridge province are subsequent and follow courses parallel to the ridges, which are usually flanks of folds and capped by sandstone and quartzite. From the James River north each major stream flows across the ridges through water gaps and into the Atlantic. Davis and Johnson postulated that the Appalachian Mountains were worn low to form a peneplain, called the fall-zone peneplain, after the Triassic. The sea advanced across this peneplain and a blanketlike deposit was laid down over the region. This cover was stripped away and the Schooley Peneplain formed following the first uplift (Fig. 14–43). The crests of the Blue Ridge and the ridges of the Allegheny Mountains in the central Appalachians (Pennsylvania to northern Virginia) are all that remain of this erosion surface.

To explain the present elevation and the formation of the valleys between the ridges it is necessary to postulate additional regional uplifts. The initial uplift following development of the Schooley surface resulted in formation of a second partial peneplain known as the Harrisburg Peneplain. A second uplift led to formation of a third partial peneplain, the Somerville Peneplain. Before this surface could be much enlarged another uplift occurred, causing present streams to be entrenched and downcutting. Thus repeated uplifts were needed to explain the various concordant levels recognized in the topography. To explain the fact that the Schooley surface is not flat over the region as a whole, unevenness in amount of uplift is postulated. Thus the region of the Smoky Mountains was uplifted higher than the region to the north. Some isolated peaks are described as being remnants that stood above the peneplain even when it was near base level.

OBJECTIONS TO THE HUMID-CYCLE EVOLUTIONARY CONCEPT

The concept of the peneplain is the most frequently challenged idea involved in the evolutionary concept. The Schooley Peneplain in the Appalachians bears no remnants of the sediment that was postulated as having covered that peneplain, although sediments of this age lie in the Coastal Plain. Thus we have no proof that this surface has ever been nearer to sea level than it is now since the mountains were initially uplifted.

A second major argument is based on estimates of the current rate of degradation of the continents, which indicate that only 40–50 million years would be required to level the continents. (Over 200 million years have elapsed since the Appalachian Mountains were folded into a high mountain system.) In view of this estimated time it is reasonable to ask why there are no modern peneplains—nearly flat surfaces near sea level where streams meander across rocks of complex structure. Many of the areas that have topographic features of old age are depositional surfaces that have never had high relief, as in the case of the Mississippi River.

A third argument involves the concept of the graded stream. According to Davis, streams become graded just at the start of maturity. At this point the streams cease to cut down and begin lateral planation. Opponents of the erosion cycle question whether streams begin lateral planation at any particular stage or as a result of reduction in slope of the channel. It may be that streams increase in capacity to remove material from their drainage basin as gradation proceeds. Still another argument is found in studies of the hy-

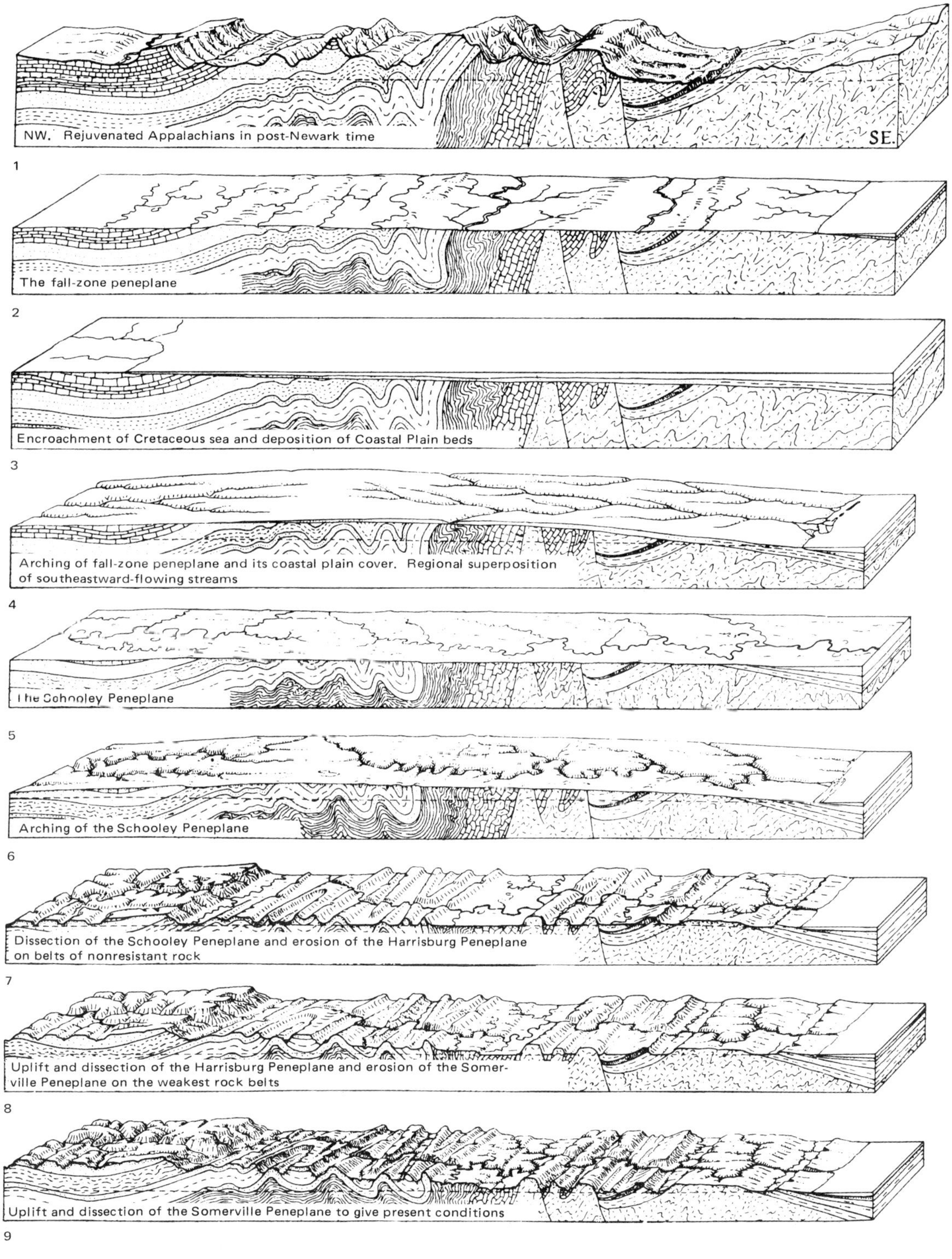

FIGURE 14–43
Stages in the development of the Appalachian region, as described by Douglas Johnson.
(from A. K. Lobeck, *Atlas of North American Geology*. By courtesy of Edwin Raisz)

draulic factors along modern streams. These studies reveal that the interrelationships between these factors and sediment load are just as consistent in graded as ungraded streams. It is also found that, contrary to Davis' theory, the capacity of a large stream on a low slope may be greater than the same stream's upper tributaries where they flow on steep slopes.

CONCEPT OF DYNAMIC EQUILIBRIUM IN LANDSCAPE DEVELOPMENT

> An alternative approach to landscape interpretation is through the application of the principle of dynamic equilibrium to spatial relations within the drainage system. It is assumed that within a single erosional system all elements of the topography are mutually adjusted so that they are downwasting at the same rate. The forms and processes are in a steady state of balance and may be considered as time independent (Hack, 1960).

According to this theory (Strahler, 1950; and Hack, 1960), streams achieve a graded condition almost immediately and maintain it throughout the period of degradation, or until some change in the drainage system occurs. The particular land forms found in a drainage system are related primarily to the geology of the basin; such features as structure and variable rock resistance to erosion are especially important. In the folded Appalachians, for example, the ridges that comprise the Schooley surface are similar in composition and structure. Many of the water gaps are located at structural zones of weakness in the ridges, places where headward erosion through the ridge would be facilitated, allowing stream capture of drainage in the Valley and Ridge province.

On the basis of the dynamic-equilibrium theory, waterfalls or rapids are features resulting from the high resistance of the underlying rock to erosion, whereas streams on weak rocks have low slopes. One of the particular advantages of this theory is that it offers an explanation for the unusual mixtures of topographic features that would belong to different stages of Davis' erosional cycle, thus avoiding postulation of the periodic uplifts necessary to explain the various levels or peneplains found in mountain systems.

The dynamic-equilibrium theory does not abandon the idea of evolution in landscape development entirely. Rapid evolution would follow certain types of changes, such as diastrophism or alterations due to nature and quantity of stream load. Changes of this sort might be brought about by variation in climate or by exposure of a weak rock unit as a result of removal of an overlying resistant unit. Thus where diastrophism is active an equilibrium condition may not be found, but in the long-stable areas such as the Appalachian Mountains the necessary adjustments have been completed.

As an example of an erosionally graded or equilibrium topography Hack uses several features of the folded Appalachian region as follows:

> . . . in a landscape like that of the [folded] Appalachian region in which large areas are mutually adjusted, the diversity of form is largely the result of differential erosion of rocks that yield to weathering in different ways. Such topography may be referred to as erosionally graded.
>
> Many of the erosionally graded landscapes in humid temperate regions belong to the almost ubiquitous type that is commonly known as the "maturely dissected peneplain." Preferring a term that has no genetic connotation, a more descriptive one such as "ridge and ravine topography" is suggested. This term refers to the monotonous network of branching valleys and intervening low ridges that make up the landscape of large areas. This topography may be concisely explained in terms of dynamic equilibrium. . . . The important elements of the topography could be divided into two domains. The first [is] a domain of stream sculpture represented by channels in which the slopes are concave upward, because the transporting power of a stream per unit of volume increases with the volume; the transporting power also increases with the slope; and a stream automatically adjusts slope to volume in such a way as to equalize its work of transportation in different parts. The other domain is that of creep, represented by the slopes between the channels. In this domain slopes are mostly convex.
>
> The forms of well-graded ridge and ravine landscapes vary within wide limits. The reg-

ularity of the landscape and the rather uniform height of the hills owe their origin to the regularity of the drainage pattern that has developed over long periods, by the erosion of rocks of uniform texture and structure.

Differences in form from one area to another, including the relief, form of the stream profile, valley cross sections, width of flood plain, shape of hilltops, and other form elements are explainable in terms of differences in the bedrock and the manner in which it breaks up into different components as it is handled on the slopes and in the streams.

TIME, SPACE, AND CAUSALITY IN LAND-FORM EVOLUTION

The concept of dynamic equilibrium and the idea that land forms at any given time represent a stage in a cycle of erosion are not mutually exclusive. S. A. Schumm and R. W. Lichty (1965) in analyzing the relationships of time, space, and causality in geomorphology show that cause-effect relationships in land-form development depend on the span of time and the size of the system studied. The variables that affect development of a drainage basin are listed in Table 14–2 in a hierarchy thought to indicate increasing degrees of dependence of the variables (variables low in the list depend on those above).

The relevance, independence, or dependence of each of these variables for three time spans is shown at the right. *Cyclic time* designates long intervals of geologic time measured in millions of years. It may best be thought of as extending back to the start of the last erosion cycle for any given system. *Graded time* is a short interval during which a graded condition or dynamic equilibrium exists. ". . . the progressive change during cyclic time is seen to be, during a shorter span of time, a series of fluctuations about or approaches to a steady state" (Schumm and Lichty, 1965, p. 114). *Steady time* is the time during which essentially a steady state exists in part of a system (i.e., amounts of water and sediment entering a section of a stream just equal the amounts leaving). Such a steady-state condition cannot be obtained throughout a system because discharge and load are not uniform throughout the system.

Variables	Cyclic	Graded	Steady
1. Time	Independent	Not relevant	Not relevant
2. Initial relief	Independent	Not relevant	Not relevant
3. Geology (lithology, structure)	Independent	Independent	Independent
4. Climate	Independent	Independent	Independent
5. Vegetation (type and density)	Dependent	Independent	Independent
6. Relief or volume of system above base level	Dependent	Independent	Independent
7. Hydrology (runoff and sediment yield per unit area within system)	Dependent	Independent	Independent
8. Drainage-network morphology	Dependent	Dependent	Independent
9. Hill-slope morphology	Dependent	Dependent	Independent
10. Hydrology (discharge of water & sediment from system)	Dependent	Dependent	Dependent

TABLE 14–2
Variables affecting development of a drainage basin.

Thus the steady-state condition is confined both to short intervals of time and to restricted segments of the system. Conditions of grade or dynamic equilibrium may be more general and apply to larger parts of the system as a stage in a cycle advances. Schumm and Lichty maintain that ". . . the entire system can not be graded because of the progressive reduction of relief or volume of the system above base level, which occurs through export of sediment from the system."

The dynamic equilibrium concept is concerned with "the relations between rocks and processes as they exist in space. The forms can change only as the energy applied to the system changes. . . . It is obvious, however, that erosional energy changes through time and hence forms must change" (Hack, 1960). Energy changes in the form of diastrophism or climatic change are important, as are changes in the potential energy of the system as relief and mass are reduced by erosion. Thus over an interval of geologic time, cyclic time, a drainage system progressively loses potential energy and mass. These long-term changes are probably not important in short time intervals when self-regulation or dynamic equilibrium exists.

Schumm and Lichty come to the following conclusion:

> The distinction of cause and effect among geomorphic variables varies with the size of a landscape and with time. Landscapes can be considered either as a whole or in terms of their components, or they can be considered either as a result of past events or as a result of modern erosive agents. Depending on one's viewpoint the landform is one stage in a cycle of erosion or a feature in dynamic equilibrium with the forces operative.

CHAPTER OUTLINE

The Importance of Stream Action
Hydrologic Cycle
Water Budget and Runoff
- Generalizations

Principles Governing Stream Action
- Stream Channels
- Stream Flow
- Distribution of Velocity and Turbulence
- Sources of Stream Loads
- Transportation by Running Water
  - *Suspension*
  - *Solution*
  - *Traction*
- Stream Erosion
- Deposition

A Quantitative Approach to Study of Streams—Hydraulic Geometry
- Frequency of Discharge
- Downstream Variations in Hydraulic Factors
- Concept of the Graded Stream

Denudation
The Role of Streams in Landscape Development
- Quantitative Approach to Basin Analysis
- Relationship of the Stream to Its Valley

Land Forms Related to Stream Action
- Flood Plains
- Natural Levees
- Deltas
- Stream Meanders and Related Features
- Origin of Meanders

The Role of Rock Structure in Land-Form Development
- Consequent Drainage
- Subsequent Drainage
- Shifting of Drainage Divides
- Stream Piracy

Land Forms in Arid and Semiarid Climates
Deserts
- Desert Land Forms
  - *Playas*
  - *Alluvial Fans*
  - *Bajadas*
  - *Pediments*

Temporal Evolution of Land Forms
Davis' Ideal Fluvial Cycle
- Effect of Interruptions in the Cycle
- Land Forms Resulting from Rejuvenation
- Terraces
- Effects of Deformed Strata
- Classical Fluvial Cycle Applied to the Appalachians

Objections to the Humid-Cycle Evolutionary Concept
Concept of Dynamic Equilibrium in Landscape Development
Time, Space, and Causality in Land-Form Evolution

TERMS

actual evapotranspiration
alkali flat
alluvium
alluvial fan
anticlinal valley
bajadas
base level
beheaded stream
capacity (streams)
cavitation
channel roughness
chute
competence (stream)
consequent stream
cuesta
cutoff
deltas
denudation
discharge
drainage density
entrenched (streams)
evaporation
flood plains
free meanders
gradient (stream)
headward erosion
hydraulic factors
hydrologic cycle
incised meanders
infiltration
influent streams
ingrown (streams)
inselbergs
internal drainage
laminar flow
mean velocity

meandering stream
natural levees
obsequent stream
oxbow lake
paired terraces
pediment
peneplain
pirate stream
playa
point bar
potential evapotranspiration
potholes
precipitation
rejuvenation
rock flour
runoff
salinas
stream frequency
stream piracy
streaming flow
subsequent stream
synclinal mountain
transpiration
underfit (streams)
unpaired terraces
washes, wadies, or arroyos
water budget
water gap
wetted perimeter

REVIEW QUESTIONS

1. Explain the reason for the shape of the curve between the fields of erosion and transportation in Fig. 14–10.
2. Under what conditions might a longitudinal stream profile be convex upward?
3. Which stream characteristics are changed most by an increase in discharge at a gauging station (refer to Fig. 14–17)?
4. Which stream characteristics are changed most by an increase in discharge downstream (refer to Fig. 14–17)?
5. What factors affect the pattern of a stream channel?
6. Describe the effects both upstream and downstream of building a dam on a stream.
7. In what ways other than building dams might you try to control flooding by modifying a drainage system?
8. Compare and contrast the topography of arid and humid regions.
9. What type of evidence would be taken as proof of the theory of peneplanation as envisioned by Johnson and Davis?
10. Why might one's ideas about the validity of the concept of landscape evolution depend on the time span the problem considers?

SUGGESTED READINGS

Davis, W. M., 1954, Geographical Essays, ed. D. W. Johnson: Dover, 777 p.

Flemal, R. C., 1971, The attack on the Davisian system of geomorphology: A synopsis: Jour. Geol. Educ., v. xix, no. 1, 11 p.

Hack, J. T., 1960, Interpretation of erosional topography in humid temperate regions: Am. Jour. Sci., v. 258-A, p. 80–97.

Lachenbruch, A. H., 1968, Permafrost: The Encyclopedia of Geomorphology: New York, Reinhold, p. 833–839.

Leopold, L. B., and **Maddock, Thomas, Jr.,** 1953, The hydraulic geometry of stream channels and some physiographic implications: U.S. Geol. Survey Prof. Paper 252, 57 p.

Leopold, L. B., Wolman, M. G., and Miller, J. P., 1964, Fluvial Processes in Geomorphology: San Francisco, W. H. Freeman, 522 p.

Mackatschek, Fritz, 1969, Geomorphology: New York, American Elsevier, 212 p.

Morgan, J. P., 1970, Deltas—a resume: Jour. Geol. Educ., v. xviii, no. 3, p. 107–117.

Morisawa, Marie, 1968, Streams, Their Dynamics and Morphology: New York, McGraw-Hill, 175 p.

Schumm, S. A., 1966, The development and evolution of hillslopes: Jour. Geol. Educ., v. xiv, no. 3, p. 98–104.

Schumm, S. A., and **R. W. Litchty,** 1965, Time, space, and causality in geomorphology: Am. Jour. Sci., v. 263, p. 110–119.

Small, R. J., 1970, The Study of Landforms, a Textbook of Geomorphology: London, Cambridge University Press, 486 p.

15
WIND

15 WIND

INTRODUCTION

Wind is most effective where it is strong and persistent; where an abundant supply of fine particles capable of being moved by the wind is present; and where the movement of particles is not inhibited by cementing materials, moisture, or vegetative cover. Most of the effects of wind are due to near-surface movement of air where there is considerable turbulence. Only the finest dust (finely divided dry particles of organic and inorganic material that can be picked up and held in suspension in the atmosphere) particles get high enough into the air to become involved in the general planetary circulation. Turbulent flow of air occurs near the surface of the earth where obstructions set up eddies, whirlwinds, updrafts, and downdrafts. Most turbulence due to irregularities is found within a few hundred meters of the ground surface, but clear air turbulence has been encountered at altitudes of eight kilometers or more by airplanes, and turbulence is found associated with convection in thunderheads, other storms, and the jet stream. Convection, which is the cause of much high-altitude turbulence, comes from heating of the earth's surface. The layer of air just above the ground becomes heated as a result of reflected solar radiation and contact with ground warmed by absorbed solar radiation; the air expands, becomes lighter, and rises. Turbulence has the effect of reducing the forward velocity of the wind, but it plays an important role in the processes of deflation by which material from the ground surface is entrained in the air and lifted to higher altitude.

At altitudes above the level of ground turbulence the flow of air takes on characteristics of laminar flow. The velocity of the air is more steady and the wind moves in a smooth, directed stream.

The prevailing pattern of circulation of air on earth is due to the earth's rotation and the resulting Coriolis effect, plus the effects of variation of solar radiation with latitude and the configuration of continents and oceans. The location of land, water, and mountains relative to the general circulation is particularly important in determining where arid and semiarid climates are located and hence where wind action is most likely to be important. The regions of desert climate with consequent low soil moisture and sparse vegetation are shown in Fig. 15–2. These deserts, like the Gobi, Kalahari, and Death Valley, generally lie in continental interiors or where high mountains

FIGURE 15–1 Sand dunes in Death Valley. Note the ripples on the surface of the dunes. (Sante Fe Railway)

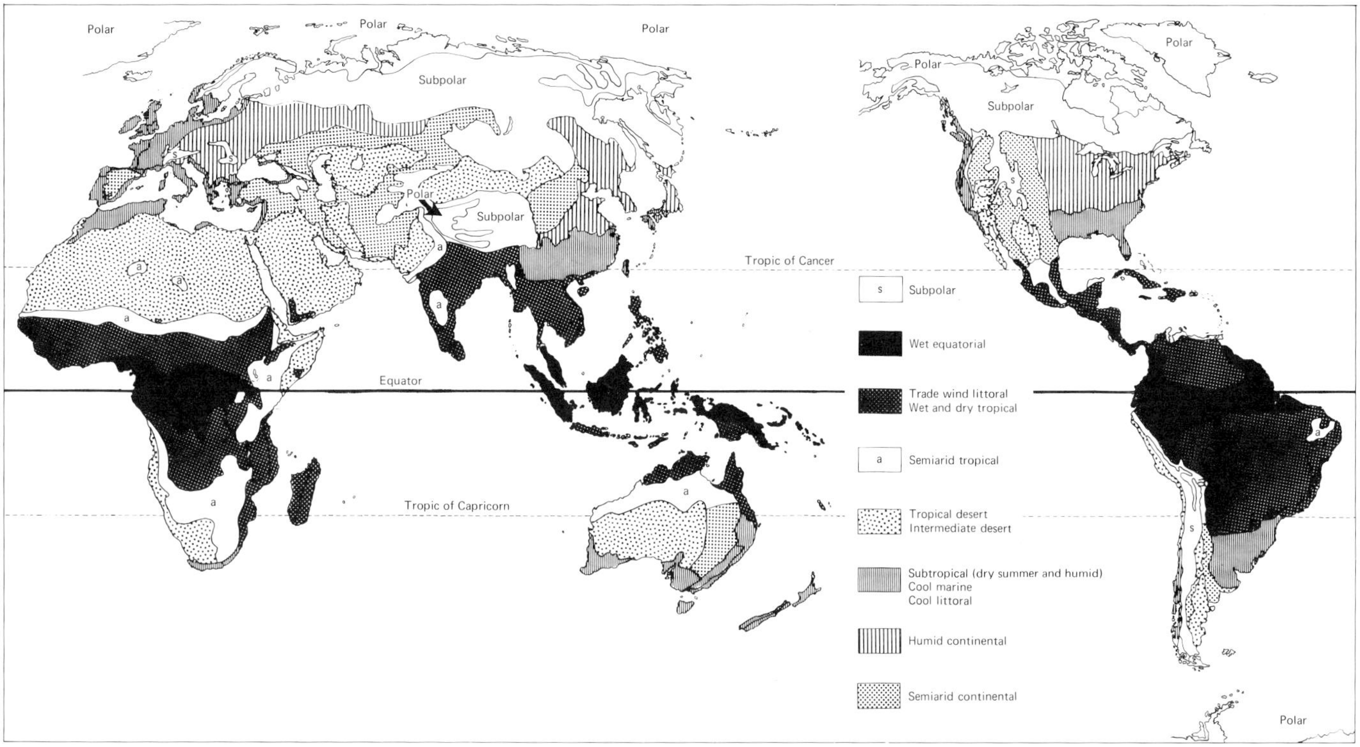

FIGURE 15–2
Climatic regions of the world.
(modified after a map by Koeppe, courtesy of A. J. Nystrom Company)

tend to intercept moisture-laden air masses and cause precipitation before the air reaches the desert. Deserts are also located in those latitudes where cold air devoid of moisture descends.

Wind action is not confined to deserts. Most coasts and many lake shores (both modern and ancient) show evidence of wind action, for example, sand dunes. Wind is able to build up to high velocities over water where there is little to break up the movement, and beaches usually provide an abundant supply of fine sediment. Even in the absence of strong prevailing winds a breeze normally blows across a coast because of differences in the rate of heating and cooling of land and water.

Strong winds analogous to the sea and land breezes of coastal regions must have existed near the edge of the continental ice sheets. In addition, air masses could develop high velocity over the ice sheets' smooth surface. Some of the widespread deposits of silt-size sediment called *loess* that originated near the ice sheets in the United States were probably caused by such winds.

SOURCES OF WIND-BLOWN MATERIAL

Potential sources of wind-blown material are found anywhere small particles come to the surface and are not held down. Plant and vegetative cover tend to prevent removal of material by wind, as does soil moisture, which acts as an adhesive. Once either of these is lost the fine particles may be blown away. Fine particle sizes are found in most soils. These are held together by root systems and soil moisture, and they are protected by plant cover in moist climates. But anything that destroys the protective devices—for example, fire, long drought, removal of cover by flood, excessive plowing or scraping—opens the way for wind action.

Beaches along coasts and around lakes usually

FIGURE 15–3 *(above)* Ancient cross-bedding. This cross-bedded sandstone is about 130 million years old. It was deposited during the Jurassic period in southern Utah.

FIGURE 15–4 Pebbles and cobbles pitted and faceted by wind-blown sand.
(M. R. Campbell, U.S. Geological Survey)

provide supplies of fine sediment. This sediment reaches the beach by longshore drifting of sediment carried into the water by streams or picked up from the sea bottom by wave action. Ancient beaches are found in and around areas formerly occupied by ice sheets. Some of these lakes formed following the melting of the glaciers. Others grew during the glacial advances when lakes formed as a result of increased precipitation and reduced evaporation. The Great Salt Lake is a remnant of such a lake system that was extensively developed in Utah and Nevada. Large lakes also covered parts of Colorado, Montana, and Wyoming. Flood plains, alluvial fans, the beds of intermittent streams, playa lakes, and pediment surfaces all provide source areas of sand, silt, and dust in arid and semiarid climates, and the exposed bedrock itself may weather to form fine particles that are blown away. Many bedrock layers are unconsolidated or semiconsolidated deposits of much older beach or sea-floor beds, and some are even ancient sand-dune deposits (Fig. 15–3). Glaciers also provide sediment sources, particularly where glacial streams lay down outwash-plain deposits of silt size.

Unlike material from other sources, large quantities of dust are placed directly in the air by volcanic explosions. Old deposits of volcanic ash are also potential sources of wind-blown material.

Man has often played an important role in making available to the wind material that otherwise would not have been exposed—for example, through poor conservation practices in farming and forestry. Plowing during years of drought accelerates evaporation of soil moisture. Moisture is lost, soil is loosened, plants die, and material small enough to be transported by the wind is blown away. The results have been disastrous in some places; fertile farms have been converted into barren areas of wind-blown sand and dust. Similar effects are produced by removal of forest cover from semiarid lands, thus exposing the soil to wind erosion.

EROSION BY WIND

The wind accomplishes erosion by *deflation* (blowing away of materials), *corrosion, abrasion,* and *impact.* Sand, silt, and dust entrained by the wind are effective tools of wind erosion; they bring about scraping or abrasion as they move over surfaces, and even more important is the long-continued effect of impact—the sand-blasting effect of wind-driven sand. Impact occurs when a sand grain is blown into or against a rock surface or into the soil. The momentum of the impact dislodges other granular material; or it may break the cement of a rock, freeing other sand grains. Man uses sand-blasting to frost glass and to cut letters in polished stones; even granites may be cut. Evidence of impact can be seen in most samples of wind-blown sand. The surfaces of grains are pitted or frosted as a result of repeated impact. Where sand and silt have been blown over rocks for a long time, the rocks become worn from repeated abrasion, achieving smooth, polished surfaces. Such stones are called *ventifacts* (Fig. 15–4).

Many desert rocks, both outcrops and pebbles, have a surface stain or crust characterized by a glistening luster and a composition of iron and manganese oxides. This surface coating, called *desert varnish,* is primarily a weathering effect, found on many ventifacts.

FIGURE 15–5
Window Rock. Note the undercutting at the base of the cliff and the holes to the left.
(New Mexico State Tourist Bureau)

FIGURE 15–6
Pedestal rocks or balanced rocks like this often result from the undercutting effects of wind-blown sand. Notice that this rock has horizontal layers. The lower layers in this case appear more susceptible to wind erosion. This rock, known as Camel Rock, is located a few miles north of Sante Fe, N. Mex.
(New Mexico State Tourist Bureau)

LAND FORMS OF WIND EROSION

Wind erosion is most effective in shifting debris that has been weathered from solid rocks. It keeps the land swept clear of debris in places and piles it up elsewhere as sand dunes, or distributes it widely as sheets of sand. Some land forms are due primarily to wind erosion. These include the following:

1. Broad, shallow depressions where soil has been blown out by repeated movement of currents in one direction. *Blowouts* of this sort are common in deserts and along sea coasts.
2. Broad, shallow cavities in the sides of hills where the impact and abrasion of sand have sculptured the lower slopes of the hill (Fig. 15–5).
3. Mushroom, table, and pedestal rocks are isolated rocks from which the base has been partially removed as a result of weathering and undercutting by wind-blown sand. Some of these develop quite unusual

FIGURE 15–7 *(at right)* Desert pavement composed of angular fragments of varnished quartzite. (C. S. Denny, U.S. Geological Survey)

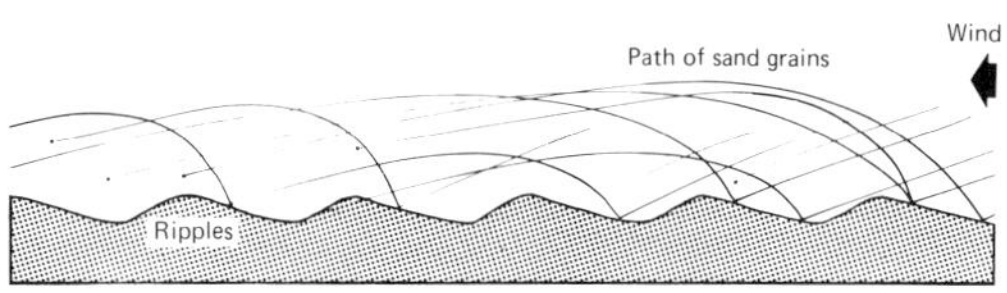

FIGURE 15–8 *(middle right)* Grains of sand in the process of saltation. Note that ripple marks are formed in the sand and that only one side of these is hit directly. The ripples are shown with vertical exaggeration. Wind ripples have height to wave-length ratios in the range of 1:6–1:10.

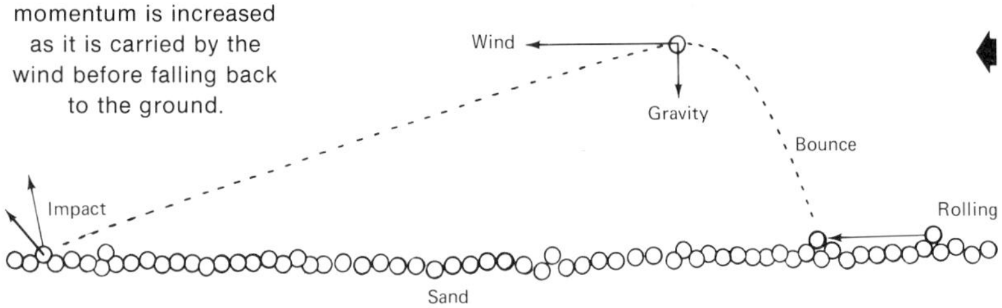

FIGURE 15–9 *(at bottom)* Sand movement. When the grain hits the sand surface the impact causes other grains to bounce and the process of saltation is begun. Sand moves initially by rolling but as a grain rolls faster it hits other grains and bounces into the air where its momentum is increased as it is carried by the wind before falling back to the ground.

forms, such as the balanced rocks. Often the top is a different rock unit from the bottom, which has been etched out because it is more susceptible to weathering and wind erosion.

Lag Gravels

When the finer size particles are removed from an originally mixed sediment, the larger and heavier fragments are left behind. These may form a thin cover of gravel, called *lag gravel,* across pediments and bajadas. An inversion of topography can occur when the coarse gravels in former stream channels remain as winding ridges where the surroundings have been lowered as a result of deflation. Lag gravels may form a sort of tight mosaic cover called *desert pavement* (Fig. 15–7), which protects underlying material from wind erosion.

TRANSPORTATION

Large-sized materials may be lifted from the ground during unusually intense wind storms. Tornadoes have been known to pick up roofs from houses, toss cars into trees, overturn railroad cars, and blow down trees along their paths, but tornadoes are not common enough to be an important agent of gradation. Hurricanes and typhoons also have localized effects, although over a period of years large areas are affected at one time or another. Such strong winds cause rapid and often dramatic erosion, but sand and silt can be moved by more ordinary winds and the cumulative effects of these less dramatic movements are great. The size of the sand moved is a function of the wind velocity. If the velocity is fairly constant, the sand is confined below a certain level above the ground. This zone is usually a few centimeters high, and only during storms does it reach more than 45 centimeters. Below this level the sand is actively moved by rolling and bouncing (Fig. 15–8) but above it the air is almost completely free of sand. Movement of sand begins when wind reaches the critical velocity necessary to overcome the inertia of the sand and grains roll along the surface. These hit other sand grains and some bounce into the air. Once in the air, they are carried forward, and their velocity is increased. The pull of gravity on a grain overcomes the upward thrust from the wind, and the grain follows an inclined path back to the ground. The increased velocity gives it more momentum (Fig. 15–9). When it hits the surface again the impact dislodges other sand grains, which start rolling or bounce into the air. Soon a chain reaction is established, and transportation is underway through a series of bounces. The process is called *saltation*. Most of the material transported in this way is sand and silt, ranging in diameter from about 2 millimeters down to $\frac{1}{256}$ millimeter. As sand is moved it tends to become rounded and sorted, and small pits, like those found on a sand-blasted piece of glass, form on wind-blown grains.

Pebbles may be rolled along the ground during high winds, but they are rarely lifted from the surface. Larger boulders may be moved by a combination of wind and mass movement. The wind may blow the finer supporting material from beneath the boulder, which then rolls in the direction in which support was lost.

Wind transports sand and silt in saltation, but the finest particles—dust, cloud particles, and smoke—are moved in a different way. Dust is often composed of flat, tabular particles with surface areas that are usually large relative to the particles' weight. This makes it possible for the force of upward motions of wind currents to more than counterbalance the downward pull of gravity. Dust settles slowly and is easily wafted up higher in the air by slight updrafts (Fig. 15–10). Large quantities of volcanic dust carried aloft during explosions of Krakatoa and Katmai stayed up for several years and were carried around the earth in upper layers of the atmosphere. Small particles like dust and ash are blown directly into the air or are lifted up by the pressure created by velocity differentials. To illustrate this effect, take a piece of paper and fold about 2.5 centimeters of the two ends toward each other. With the two ends at right angles to the rest of the paper, place the folded paper upon a flat surface so it forms an open-ended box. Then blow under the paper. If this is done properly, the paper will not be blown away, but the top will be pulled down toward the surface. The velocity of the air under the paper is greater than that above it. This creates a difference in pressure on the opposite sides of the paper. The pressure drops where the wind velocity is greatest, and the force of the pressure above pushes the top of the paper down. The same thing can take place in any fluid medium. Pressure drops of this nature are found where the wind blows over or around obstacles. Strong pressure drops occur in the centers of whirlwinds and tornadoes.

Sorting

Wind is an excellent agent for *sorting*, separation of particles according to size, shape, or weight. Wind velocity determines just what maximum size of material the wind can transport by rolling, by saltation, and in suspension; thus a distinct separation is accomplished. Smaller sizes are suspended and carried away, heavier particles lag behind on the ground, while sand and silt bounce along. Suspended and saltation loads are further separated. Only the lightest particles can be carried and maintained in suspension at extreme heights. Some parts of the load in saltation are moved more rapidly than others. Thus the wind's load becomes sorted while it is in transport.

As wind velocity drops, a *critical velocity*—the minimum wind velocity required to move a particle—will be reached for each size, and that size will settle. The result is a well-sorted sediment. The degree of sorting is usually better than that found in marine or stream deposits and, in conjunction with the pitted sand grains characteristic of wind deposits, provides a means of distinguishing between the two.

DEPOSITION

Wind deposits take two general forms—*dunes* and *sheets*. Sheets are most frequently dust sediments deposited over large areas. Within sheets, thickness varies, generally thinning away from the source. Two excellent examples of the sheetlike

FIGURE 15–10
Dust storm in New Mexico. (U.S. Soil Conservation Service)

FIGURE 15–11 *(above)* Exposure of loess, wind-deposited dust, and silt in an old road cut near Vicksburg, Miss.

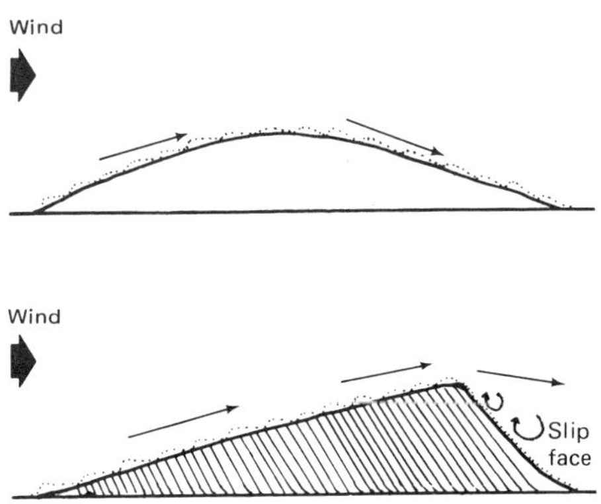

FIGURE 15–12 *(middle right)* The formation of sand dunes. These two profiles show two stages in the evolution of a dune. The top dune is smooth and rounded as the wind flows over its entire surface, but below the sand has built up until the main stream passes over the slip face where eddy currents are shown. The sand is carried up the low-sloping surface of the dune to the top where it is dropped over the side. The slip face has an angle equal to the angle of repose of the sand, usually about 35°.

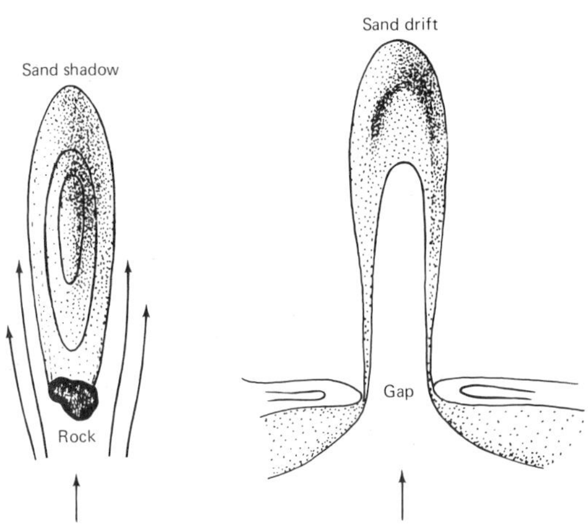

FIGURE 15–13 *(bottom right)* Sand shadows and sand drift. (based on drawings by Bagnold, 1941)

nature of some wind-blown deposits are the deposits of volcanic dust from the eruption of Katmai and the layer of loess that covers much of the central United States (Fig. 15–11). Much larger areas of the United States are covered by sheet-like deposits of loess than by sand dunes.

LOESS

Loess is an unconsolidated, unstratified, and homogeneous sediment composed of small, angular particles derived from a variety of rock types. It is porous, friable, and buff to yellowish in color. It may contain small snail shells of a type that lives on land and small tubes of calcite that apparently fill holes left after the decay of tall grasses. These could be formed if wind-blown silt and dust sifted down over fields of grass so slowly that the blades of grass were not forced down. Loess deposits are easily eroded by streams because they are unconsolidated, but the angularity of the particles makes the loess less prone to slump than might be expected. Thus small, nearly vertical cliffs are common where loess is dissected by streams, and vertical road cuts in it have remained stable for many years.

Most of the loess in the United States has apparently been derived from glacial-outwash deposits, but some loess deposits in other parts of the world are composed of materials blown from desert regions. Composition of much of the loess found in the United States is the same as that which would be produced by grinding up rocks now exposed in the northern part of the country and in Canada. Glaciers over that part of North America in the Pleistocene probably did grind these rocks into the material found in the loess. Finely ground particles were likely carried out to areas in front of the ice sheet by melt-water and deposited in the outwash plain, where winds sweeping from the ice sheet dried the loess and picked it up, carried it southward, and deposited it along the Mississippi Valley.

Loess is also extensively exposed in a long belt extending from France through Russia to China, where it forms a thick and extensive cover over vast areas in and around Shansi province. The loess in northeastern China was derived from the Gobi Desert rather than from a glacial source.

SAND DEPOSITS

Sand dunes have been defined as "mobile heaps of sand whose existence is independent of either ground form or fixed wind obstruction" (Bagnold, 1941). Thus dunes (Fig. 15–12) are distinguished from what Bagnold calls *sand shadows* (Fig. 15–13)—sand accumulations located downwind from and sheltered by an obstruction such as a boulder, grove of bushes, cliff, or the like which interrupts the streamlined flow of wind and reduces its velocity. Deposits of sand also form where a gap between obstructions (e.g., a break in a ridge) acts as a funnel from which sand deposits, termed *sand drifts,* trail out downwind.

Sand dunes and sheets cover large areas in some deserts. Over 7,800 square kilometers of the Libyan Desert is covered by a sheet of sand only a few meters thick. Dunes are best developed in deserts where they may form beautiful, symmetrical, crescent-shaped mounds isolated from one another, long ridges parallel or transverse to the prevailing winds or complex masses.

Dunes are not confined to deserts. Some of the most extensive accumulations of sand in North America are found along the eastern coast and around the Great Lakes. Sand dunes may be formed wherever the following conditions occur together: (1) strong, constant, or intermittent winds, (2) a source of sand, and (3) obstacles that can initiate dune formation.

The composition of the sand that forms dunes is highly varied. Along coast lines the most common sand is quartz sand, but in Italy it is olivine derived from volcanic rocks and in the Azores and Hawaii it is often volcanic ash; in the Arctic and Antarctic large quantities of ice sand form dunes, and in the White Sands National Monument in New Mexico the sand is composed of gypsum.

Sand-Dune Formation

Sand accumulations may start wherever an obstruction occurs in the path of the sand movement. Sand moving by saltation hits the obstacle or falls behind it where the wind velocity is below the movement threshold. The initial form of a sand dune (Fig. 15–12) may be a streamlined pile with nearly equal slope up- and downwind. Sand moves almost uniformly over the surface. But as the dune grows the streamlined pattern of flow over the dune is broken up over the crest, eddies are set up on the downwind face, and the dune begins to assume an asymmetrical form (Fig. 15–14). Sand moves up the low slope on the windward side by saltation. Just beyond the crest, wind velocity is reduced. A back slope forms from slumping and sliding of sand and assumes the characteristic angle of repose of loose sand, about 35°. This surface is barely stable and any addition of sand at the top causes small slides on the back slope. Sand on the windward side is blown over to the back slope, which is oversteepened and slumps and slides to correct its angle of repose. Once the dune is formed it may advance as sand is blown up one side to the crest, slides down the other, and eventually is blown up again. Dunes may move at rates of as much as 30 meters each year.

DUNE FORMS

The shape of any sand dune is determined by (1) how much sand is supplied to the dune, (2) the wind velocity, (3) constancy of the wind direction, (4) rate at which the sand is supplied, and (5) amount and distribution of vegetative cover. Regular-shaped dunes are not always formed. When sand is abundant, isolated and

FIGURE 15–14
View of the slip face of a large dune. The wind blows from the right to the left. This dune is located on the Oregon coast.
(Oregon State Highway Department)

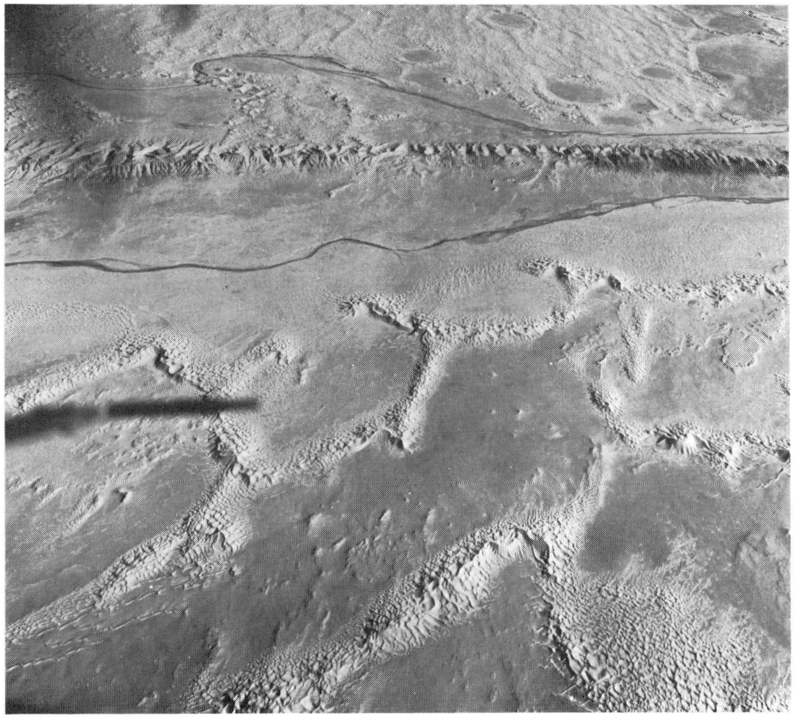

FIGURE 15–15
Complex sand dunes in North Africa. Note the two streams that flow across the desert and the various forms of the sand dunes. Across the photo between the two streams, a long outcrop of rock is visible. This may be the source of some of the wind-blown sand.
(U.S. Air Force)

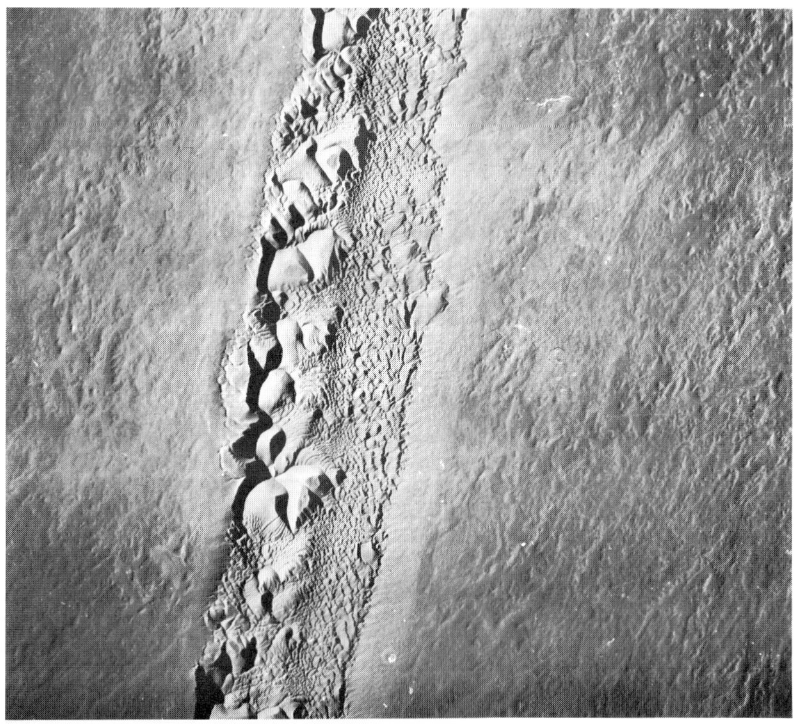

FIGURE 15–16
Long dune ridges of highly complex dune structure on sand-covered plain, Algeria.
(U.S. Air Force)

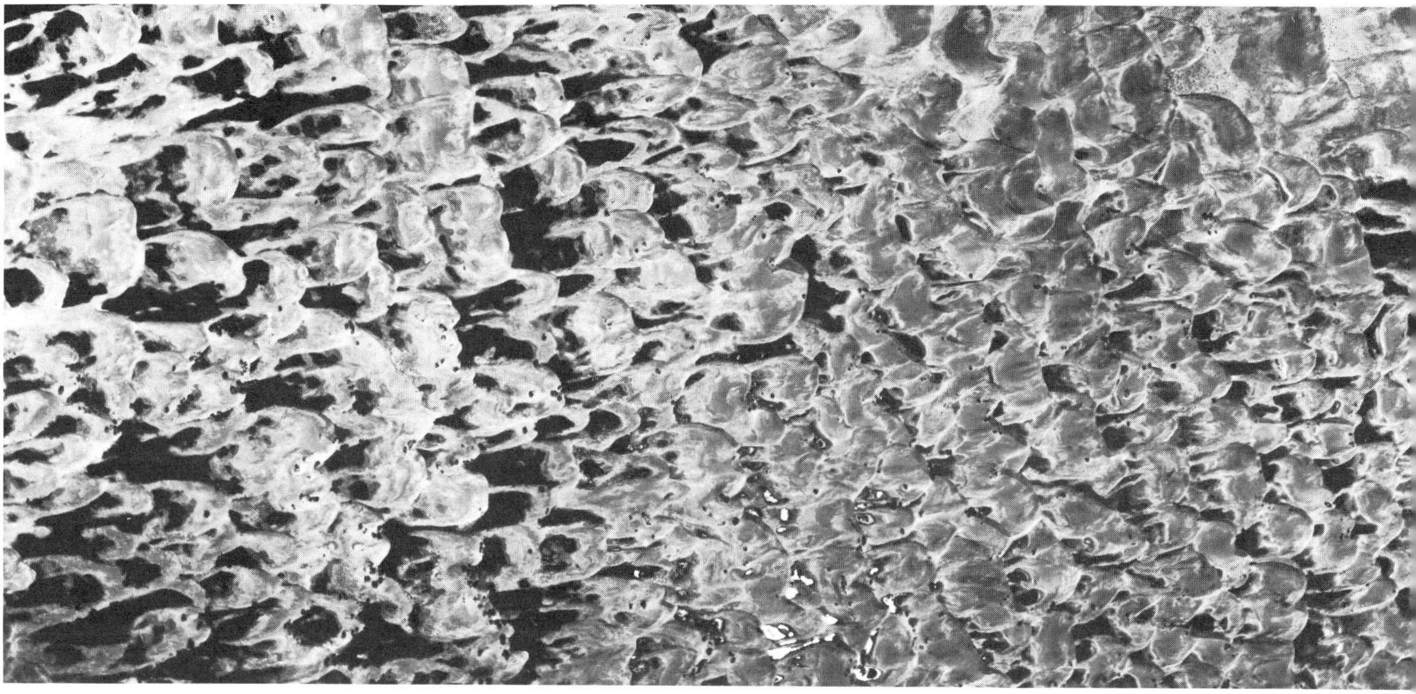

FIGURE 15–17
Vertical aerial photograph of part of the sand-dune field that dams Moses Lake, Wash. The sand moved in from the lower right. Part of the area (to the left) is flooded between dunes. These dunes were originally barchan dunes but now many of them have been altered to parabolic dunes concave to the wind direction. Scale 1:20,000.
(U.S. Geological Survey)

perfectly shaped dunes are rare; the dunes coalesce, losing their individual characteristics (Figs. 15–15 and 15–16).

Barchan Dunes

Barchan dunes (Fig. 15–18) are crescent-shaped with points or wings directed downwind. They are common where the wind direction is constant and moderate, where the surface over which the sand is moving is smooth, and when the supply of sand is limited. Wind distribution over an ideal barchan is depicted in Fig. 15–18. Steady wind sweeps the ends of the dune ahead, forming the crescent points and giving the slip face a crescent shape. Barchans are most perfectly developed where they occur as isolated mounds of sand on relatively flat surfaces. They range widely in size, achieving a maximum width of about 400 meters. The largest barchans are about 30 meters high. Bagnold indicates that a minimum height of 30 centimeters is required for the formation of a slip face.

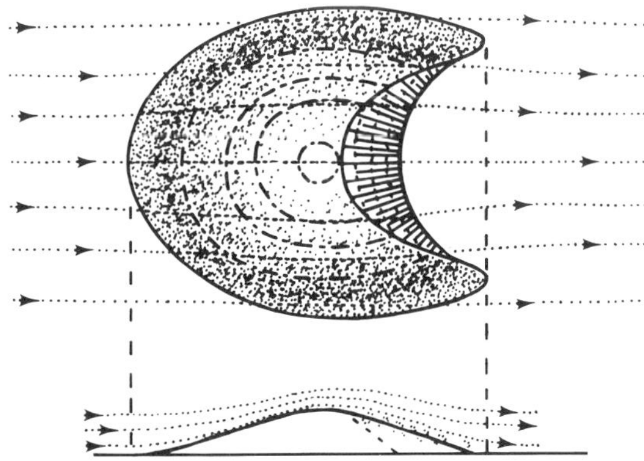

FIGURE 15–18
A barchan dune shown from above and in profile (below). Arrows indicate the approximate wind distribution. The slip face is shown by straight lines on the right side of figure.
(after Bagnold, 1941)

The rate of movement of a dune is a very important consideration where efforts to control dunes are being made. Rate is related to wind velocity, its persistence, and dune height; thus each dune must be considered individually. A rate of 1.8 centimeters per hour has been calculated for a barchan 15 meters high subjected to a wind of 13.5 meters per second (30 miles per

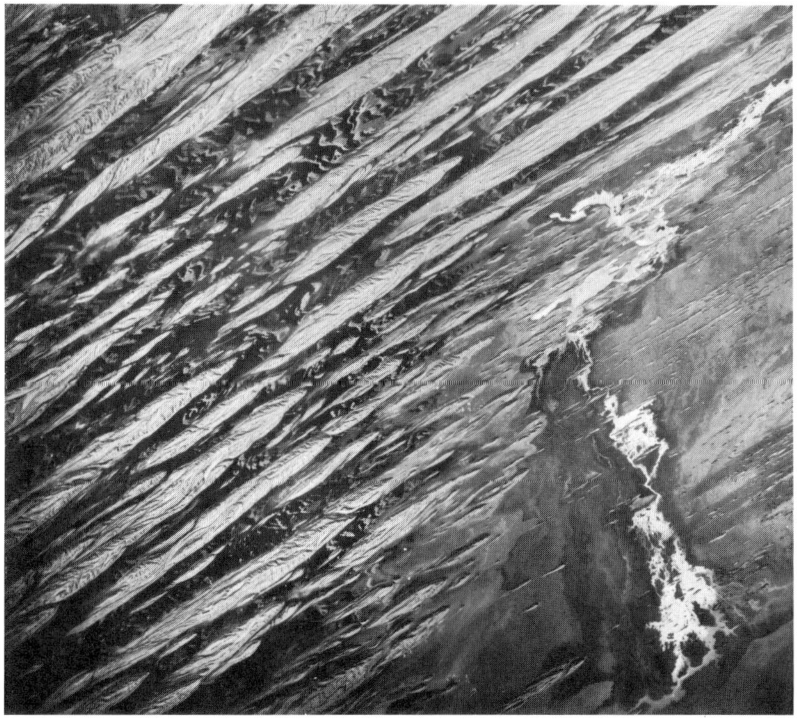

FIGURE 15–19
Northeast edge of the Shahr-e-Lūt. Beautifully developed longitudinal dunes are shown in this vertical photograph, scale 1:100,000.
(U.S. Geological Survey)

FIGURE 15–20
Complex dunes (seif ridges) with a general longitudinal arrangement in the western Sahara Desert.
(U.S. Air Force)

hour). The same wind velocity would move a dune only 40 centimeters high at a rate of 67 centimeters per hour (Bagnold, 1941, p. 217).

Longitudinal Seifs or Dunes

Many dunes are elongate in the direction of the prevailing wind, hence called *longitudinal dunes*. Some of these dunes are miles long and nearly symmetrical, like those shown in the aerial photograph of dunes in Iran (Fig. 15–19). Although the overall form is simple, the larger dunes are complex, appearing to be made up of many smaller longitudinal dunes superimposed on one another.

Longitudinal dunes are the largest dunes known. Some of those in Iran have crests up to 213 meters above their bases. They are generally about six times as wide at the base as they are high, and some are as much as 48–80 kilometers long. Once elongate dunes are formed, wind tends to be funneled between them, sweeping sand from the intervening desert floor.

Many longitudinal dunes, called *seifs*, have a distinctive shape characterized by a long, sharp ridge crest, one side of which is rounded while the other falls abruptly as a slip face. These dunes occur in isolated form but they are found more frequently as long, complex chains (Fig. 15–20). The dune chain is aligned with the prevailing wind direction, but the asymmetry of the individual dunes or ridges seems to arise as a result of temporary shifts of wind direction during which a slip face is developed along the ridges on their lee side.

Bagnold has suggested that barchans may be transformed into seif forms, as shown in Fig. 15–21, when one wing of a barchan grows more rapidly than the other. This happens when the wind flows from a direction different from that in which the original barchan developed. Eventually the extended wing begins to collect sand that is moving around the shorter wing.

Transverse Dunes

Sand dunes oriented with their long axis perpendicular to the prevailing wind are called *transverse dunes*. Like other dune types, these have the slip face on the lee side. Transverse-dune complexes like that in Pakistan, shown in Fig. 15–22, are often found in deserts that are covered by sheets of sand—the so-called "seas of sand." Slip

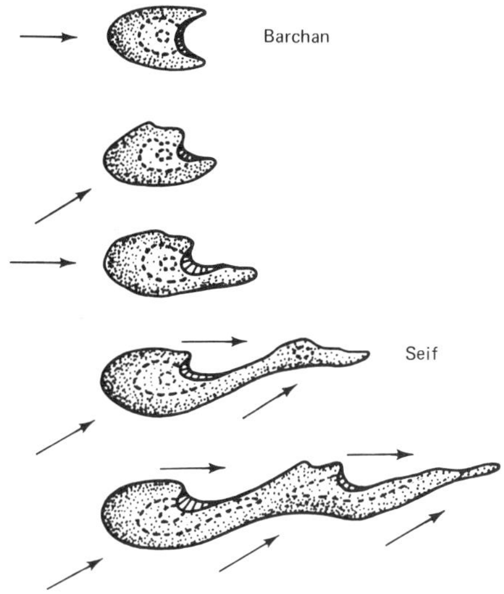

FIGURE 15–21
Transition from barchan to seif dune form due to a bidirectional wind.
(after Bagnold, 1941)

FIGURE 15–22
Transverse-dune complex in the Khārān Kalāt Desert of Pakistan. Crests of major dunes are about 1.6 kilometers apart.
(U.S. Geological Survey photograph)

faces occur on the right side of the ridges in the figure. The windward slopes are covered by dune complexes in which the individual dunes on the upper part of the slope resemble barchans in shape, and those near the troughs have longitudinal forms.

Transverse dunes may also originate where an elongate sand source is perpendicular to the prevailing wind, such as at a beach.

Parabolic Dunes

Parabolic dunes develop where vegetation is thick and at least partially covers the sand. In spots lacking in vegetation, a blowout occurs, more sand is exposed, and, consequently, sand is removed from the center of the spot. Sand movement at either side is inhibited by the vegetation. A parabolic dune with wings pointed upwind results.

CHAPTER OUTLINE

Introduction
Sources of Wind-Blown Material
Erosion by Wind
Land Forms of Wind Erosion
  Lag Gravels
Transportation
  Sorting
Deposition
Loess
Sand Deposits
  Sand-Dune Formation
Dune Forms
  Barchan Dunes
  Longitudinal Seifs or Dunes
  Transverse Dunes
  Parabolic Dunes

TERMS

abrasion
barchan dunes
blowout
corrosion
critical velocity
deflation
desert pavement
desert varnish
dunes
impact
lag gravel
loess
longitudinal dunes
parabolic dunes
saltation
sand drifts
sand shadows
seifs
sheets
slip face
sorting
transverse dunes
ventifacts

REVIEW QUESTIONS

1. What are the main sources of sand?
2. Why is wind such an excellent sorter of sand and silt?
3. What determines the slope of the slip face of a sand dune?
4. What evidence would you look for in trying to determine whether an ancient sandstone was deposited by wind action?
5. What differences might cause the formation of a sand sheet instead of dunes?
6. What does each dune type tell you about prevailing wind direction and abundance of sand?

SUGGESTED READINGS

Bagnold, R. A., 1941, The Physics of Blown Sand and Desert Dunes: London, Methuen, 265 p.

Thornbury, W. D., 1964, Principles of Geomorphology: New York, John Wiley, 618 p.

16 GROUND WATER

16 GROUND WATER

Those of us who have spent most of our lives in areas of abundant rainfall rarely think of fresh water as a valuable natural resource. At a time when industrial and agricultural demands for water are so great, when populations are increasing, and when pollution of fresh water has reached a critical stage in many streams and lakes, we should be more aware of the value of this resource, take steps to determine the extent of our holdings, and make careful plans for its future use and conservation. Many cities in the United States have already experienced periods of short water supply; each year over a thousand communities are forced to go on water rations, and we may expect more widespread shortages in the future.

That water is stored in the soil and rocks of the earth's crust is apparent from springs that issue from valley sides, from geysers, and from wells that produce much of our drinking water. So common is the experience of seeing water come out of the ground that we may be lulled into a belief that water supplies are inexhaustible.

ORIGIN OF GROUND WATER

The water contained in the ground, which man uses for drinking, for irrigation, and for industrial purposes, entered the earth from one of three sources.

Meteoric water. Meteoric water, or precipitation, includes rain, sleet, snow, hail, and other forms of precipitation. Average precipitation over continents is estimated to be 74 centimeters per year. This is taken up by plants, evaporates, and infiltrates into pore spaces in the soil and rocks; in addition, it provides most surface runoff. It is the most important source of water used by man.

Connate water. Connate water is marine or fresh water trapped in sediments when they are deposited on lake or sea bottoms. Since most sediments originate in marine water, connate water is usually salty. This is the source of much water found deeper in sedimentary units in the crust. Connate water often occurs with oil. Oil floats on it and rises upward until it is trapped.

Juvenile water. Juvenile water, also called *magmatic water,* is water produced by magmatic activity. It is hard to determine how much of this water is coming to the surface of the earth at present. Many volcanoes are located underwater, and many more are found around the margins of the oceans. More than 90 percent of all material coming out of volcanoes is steam, but much of this water may be supplied to the volcano from the surface.

WATER STORAGE UNDERGROUND

Water may be stored in soil and rocks in a variety of ways. Some underground streams do exist, but they are small in size, few in number, and restricted to limestone and volcanic terrains. Most water is stored in intergranular pore spaces (Fig. 16–1), solution pores and cavities (Fig. 16–2), and open fractures and fault zones.

The volume of pore space in a rock is called the *porosity* of the rock and is expressed as a percentage of the total volume. Some pore spaces are formed at the time of deposition, particularly in sedimentary rocks. Others are created at a later date by weathering or deformation. The amount of pore space depends on the shape, packing, size, and degree of sorting of the rock's components.

Sorting and size. It might seem likely that a boulder conglomerate would be more porous than a sandstone, but the reverse is usually true. Large particles are likely to have smaller sediments mixed in to fill up spaces between those of larger size. Thus a great assortment of sediment sizes is unfavorable for the development of maximum porosity.

Shape and packing. Great differences in porosity may result from different packing arrangement even when all of the particles are of the same size and shape. If every grain is directly above the center of another, maximum porosity results; minimum porosity occurs when the grains are offset. If the particles in a rock are elongate, the porosity of a rock in which the particles are all aligned in one direction will be very different from that of one in which they are oriented at random (Fig. 16–1). Porosity ranges of some common clastic sediments follow:

Soils	50–60%
Clay	45–55%
Silt	40–50%
Sand	30–40%
Gravel	20–40%
Sandstone	10–20%
Shale	1–10%
Limestone	1–10%

Storage in *solution cavities* is much more irregular than that in intergranular pore spaces. Caverns provide some lakes and channels that may be filled with water, but most solution pores are small. Solution cavities form along fractures or where soluble minerals have been etched out.

Fractures are present in all types of rock. Sometimes they are open, and usually water can seep down along the fracture surface. Fractures provide storage for water and are the main reservoir in some areas underlain by igneous and metamorphic rocks.

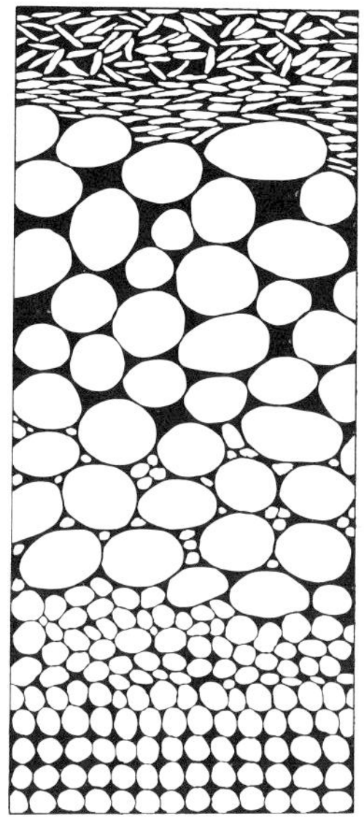

FIGURE 16–1
Variations in porosity. Water is stored in intergranular pore spaces, in fractures, and in cavities of all sorts in the earth. How much water and how readily it moves depend on the nature of the rock in which the water is stored. This illustration points out some of the variations in porosity that result from differences in size, shape, and packing of a clastic sediment.

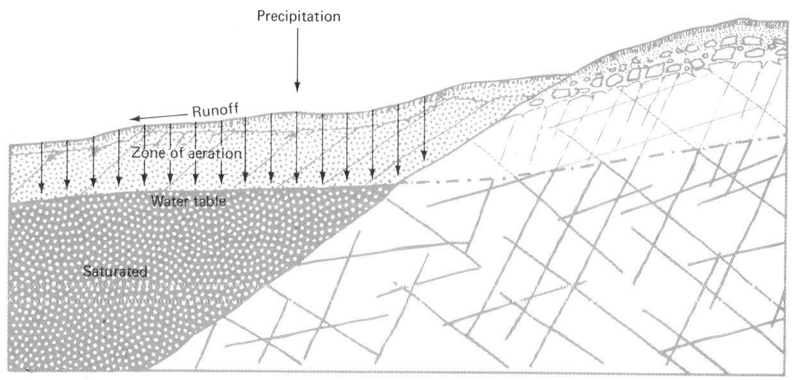

FIGURE 16–2
Water percolates into the soil, through the decayed bedrock into the pore spaces and fractures in the bedrock. In massive igneous rock the water moves mainly in fractures, but in sediments such as sandstone it can move through the pore spaces as well.

Pore spaces may be closed or filled in by natural processes. Ground water usually carries elements in solution, and these may be deposited, cementing cavities, pores, or fractures, and thereby preventing future storage of water. The most common cementing materials are calcium carbonate and silica. Man too may destroy porosity of some water-bearing sedimentary units by drawing water out of them too fast or too completely. If water is removed, some sediments, particularly those composed of clay, will become more consolidated and compacted.

Water in clay behaves quite differently from that in silt or sand. Clay colloids absorb water; therefore clay has a very high porosity, up to 50 percent, but water is held to the colloids and does not readily drain out. Thus porosity is not a reliable guide to the amount of water that can be produced. Because clay is so impervious, it does not yield much of its water content, and even in granular sediments and fractures water is held by surface tension. *Specific yield* is the percentage of the porosity that can be drained freely. The specific yield of clay is low compared with other mixtures and larger sizes of material. The specific yield of clay may be only a few percent while gravel may yield as much as 25 to 30 percent and a typical value for sand would be 10 to 20 percent.

Infiltration of Ground Water

A rock can be very porous and still not allow fluids to move through it. *Permeability* is a measure of the ability of a stratum or rock unit to transmit fluids. For movement of water, pores must be interconnected. Most well-sorted granular rocks have high porosity and high permeabil-

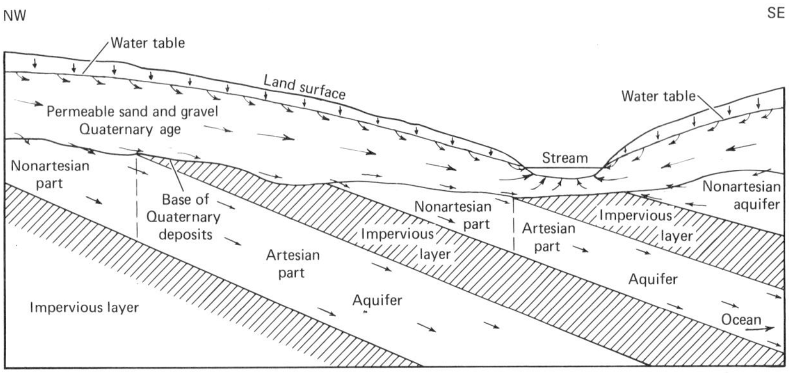

Note: Vertical scale greatly exaggerated, hence dip of aquifers and impervious layers is more nearly horizontal. Paths of ground-water movement, indicated by arrows, likewise are not to scale; vertical components of movements are greatly exaggerated (components of movement perpendicular to diagram are not indicated).

FIGURE 16–3 Idealized cross section illustrating funnel effect in recharge to Coastal Plain aquifers. (from Parker, 1964)

ity. Partially cemented rocks, those composed of clay particles, and massive igneous and metamorphic rocks tend to be less permeable unless they are fractured.

Most ground water in the upper parts of the crust comes from precipitation. This water must find its way from the surface of the ground down into the permeable rocks below. The exact way in which this movement occurs is as varied as the soils and rocks are. The water is pulled downward by the force of gravity; it is pushed down by the weight of water above it; and it is retarded by surface tension of the water as it clings to the particles of soil through which it must pass. Gradually it seeps into and through the soil. In this zone it is mixed with air, and together they bring about decay of the soil and its rock fragments. After passing through the soil the water encounters the contact between soil and unaltered rock. If the rock is porous and permeable, water will continue into the rock; if the rock is impermeable, water will either accumulate or will tend to move laterally downslope along this contact. If the rock is of homogeneous, porous, and granular material, water moving through it eventually reaches a depth below which the rock is saturated with water. That part of the soil and rock in which both air and percolating water occur is called the *zone of aeration.* The term *water table* is used to define the surface that separates a zone where the pore spaces in the rocks are not completely filled with water from a zone below in which they are saturated. The term may be somewhat misleading because the water table is not flat, nor does it maintain the same level or configuration all the time. Immediately above the water table very thin tubelets of water extend up into the pores of the overlying rock. These behave as capillary tubes, and the zone is called the *capillary fringe.*

The factors that influence the amount of water that infiltrates the ground after a rain are very important because meteroric water is the source of most of our ground-water supply. The amount of precipitation provides a guide to the maximum possible amount of infiltration, but usually only a small fraction of that total gets into the ground. Among the important factors affecting the amount of infiltration are the following:

1. Rate of precipitation and the kind of precipitation.
2. Type and condition (including porosity & permeability) of the soil.
3. Slope of the ground surface.
4. Porosity and permeability distribution of the rock underlying the soil.
5. Amount of evaporation and transpiration.
6. Amount of vegetation.

Infiltration and potential storage of ground water will be greatest when the precipitation falls at a rate slow enough to enable the water to infiltrate without building up surplus water on the ground surface that will run off, when the soil and the underlying bedrock are porous, permeable, and unsaturated, and when the rates of evaporation and transpiration are low.

Configuration of the Saturated Zones

The distribution of saturated zones in the ground is determined by the distribution of porosity and permeability and the shape of the rock body. When the structure and lithology are such that a more or less continuously connected hydraulic system is created in which water can move freely, the water is *unconfined.* When the distribution of permeability is such that water can move into layers or zones from which it cannot freely escape, the water can become *confined,* and pressure can be built up in the trapped water (Fig. 16–3).

A region underlain by a thick porous and permeable sedimentary sequence composed of gran-

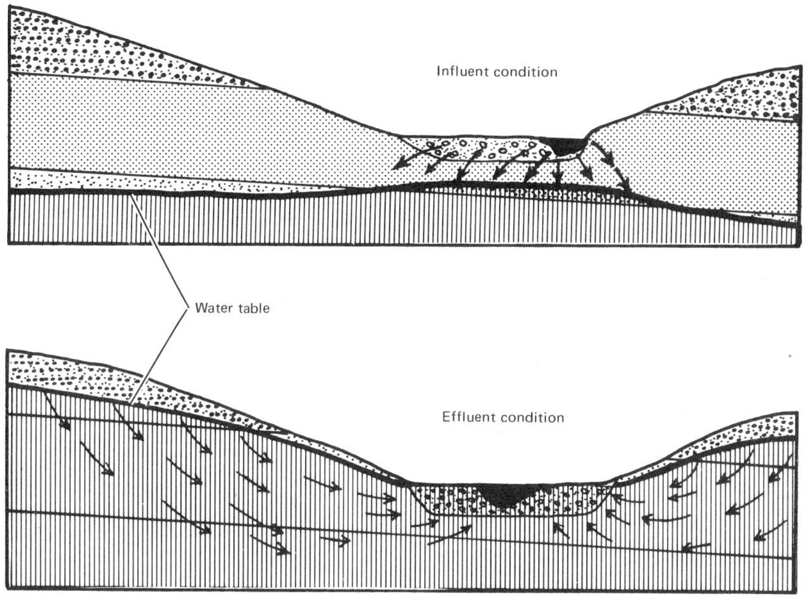

FIGURE 16–4
Cross section of an effluent stream, bottom, and an influent stream, top. The vertical-line parts of the sections indicate portions of the ground that are saturated with water. Arrows indicate the general directions of water movement.

FIGURE 16–5
Cross-sectional view showing a perched water table. Impervious layers are shown dark gray; saturated zones are shaded; dashed line indicates position of water table.

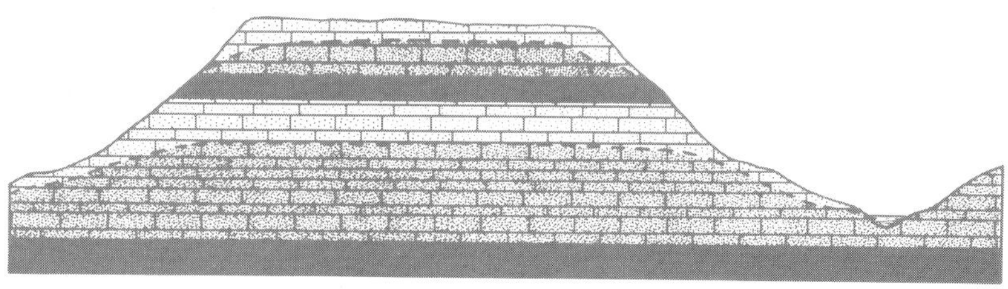

ular material provides a good example of the unconfined condition. The lower boundary of the unconfined water is formed by underlying impervious rock. The lower boundary may be bedrock composed of crystalline rock below which no water is expected, or it may be an impervious sedimentary bed below which other aquifers containing confined water may occur. The upper contact of the saturated zone in unconfined conditions is the *water table.* The shape of the water table is usually a somewhat suppressed version of the topography. It is depressed most under hills and it comes closest to the surface of the ground at the edges of lakes and along streams. Normally in regions of abundant rainfall water moves from beneath the water table into lakes and streams. When this happens an *effluent* condition is said to exist. However, during periods of long drought, in regions in which the rainfall is slight or when excessive pumping lowers the water level, the water table may become depressed below the level of the stream and lake bottoms, and water will then move from the lake or stream toward the water table. This is called an *influent* condition (Fig. 16–4) and is frequently found in arid climates. When an influent condition lasts for a long period of time the lakes and streams dry up or become intermittent, containing water only part of the time.

The shape of the water table may be represented by contour lines[1] drawn on the top of the zone of saturation. The altitude of the water table is expressed in feet or meters above sea level. Only rarely is the water table flat; the slope or gradient in a given direction is the rate of change of altitude with distance. The altitude of the water table fluctuates as the rate of supply of water varies and as water moves within the zone of saturation. Usually the water table reaches a high stage in the spring and a low stage in the fall.

Perched Water Tables

If water is moving down to an impervious layer faster than it is being transmitted laterally and faster than it is passing through the layer, then a localized water table develops above that unit. When the level of the regional water table is below this impervious unit a *perched water table* (Fig. 16–5) is said to exist above it. Water is normally lost from a perched water table through springs located where the impervious layer reaches the ground surface.

[1] See Appendix A for a description of methods used to draw and read contour maps.

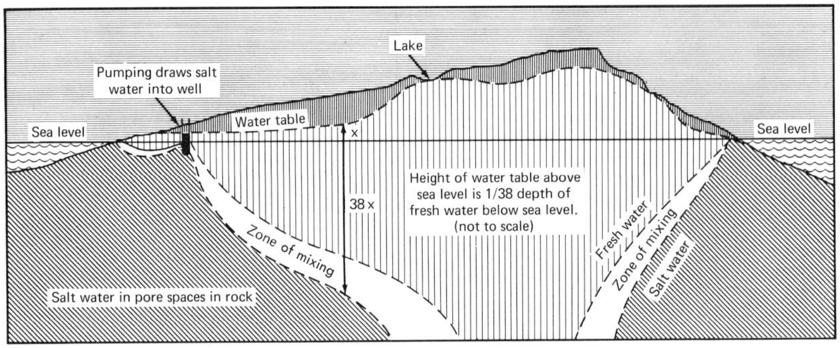

FIGURE 16–6
Storage of water in sediments in contact with the ocean. The fresh water tends to float on the more dense salt water. For every foot (0.30 meter) of elevation of the water table above sea level there are 38 feet (11.5 meters) of fresh water beneath.

FIGURE 16–7
Map of South Dakota showing isopiestic lines and areas of artesian flow of the Dakota sandstone. The lines show the shape of the piezometric surface and its position in feet above sea level.
(after Darton, from O. E. Meinzer, 1923)

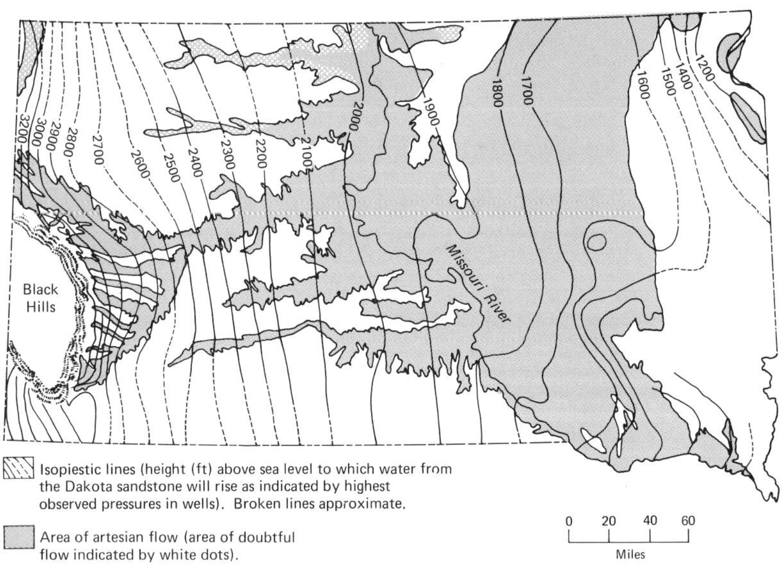

Barriers. Local barriers such as dikes or veins may prevent the free lateral movement of water underground, and these may give rise to local differences in the level of the water table in an otherwise unconfined ground-water situation.

WATER TABLES IN CONTACT WITH SALT WATER

Sea water moves into pore spaces of permeable sediment and rock below sea level along coasts. Thus it is normal to find fresh water bounded below by a contact with salt water in such rocks. Because of its lower density, fresh water floats on salt water. Consequently, a salt-water wedge extends from the ocean under the edge of the fresh water. Salt-water contamination of the fresh water occurs along this wedge, making it unsuitable for drinking purposes. The density differential between salt and fresh water is such that a balance is reached when there is about 38 feet (11.5 meters) of fresh water below sea level for every foot (0.30 meter) of elevation of the water table above sea level. This condition prevails in Florida, where fresh water is stored in porous limestones and sands that underlie the peninsula (Fig. 16–6). The fresh-water supply is a large, dome-shaped mass. Problems arise when the rate at which water is pumped from the ground exceeds the rate at which water is moving toward the area pumped. Excessive pumping brings the salt-water wedge closer to the surface and endangers the supply of fresh water.

HYDROSTATIC PRESSURE

The pressure exerted at a given point within a body of water at rest by the weight of the overlying column of water is called the *hydrostatic pressure* at that point. Most ground water moves so slowly that the pressure (called the *hydraulic pressure* for moving water) at a point within the saturated zone is nearly equal to the hydrostatic pressure that would exist if the water were at rest. The *hydrostatic level* is the level to which water will rise in a sealed well. Under unconfined conditions the hydrostatic level is the level of the water table.

The term *hydrostatic head* (*hydraulic head* when the water is in motion) is an expression of the hydrostatic pressure in terms of the height of a column of water that can be supported by the pressure.

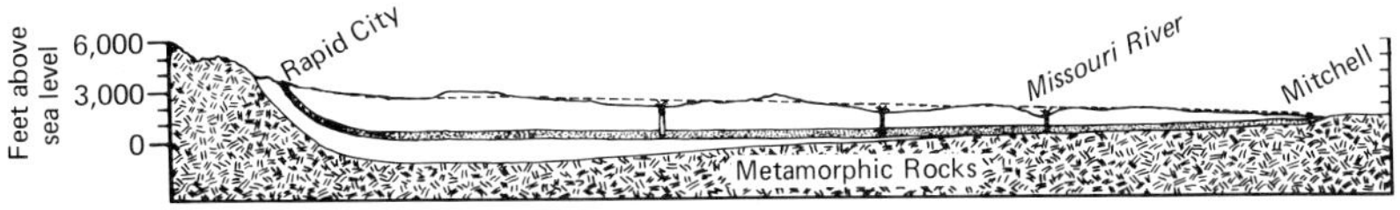

FIGURE 16–8
This section drawn from the Black Hills shows the Dakota sandstone, one of the important aquifers in the Great Plains. Note the piezometric surface, shown by a dashed line, slopes away from the Black Hills.
(after Darton, from O. E. Meinzer, 1923)

CONFINED WATER—ARTESIAN CONDITIONS

Artesian water is ground water that is under sufficient pressure to rise above the zone of saturation. An *artesian aquifer* is one that contains artesian water. An *artesian basin* is a geologic structural feature or combination of such features in which water is confined under artesian pressure.

Confined conditions arise when water is trapped within a porous and permeable unit that is overlain and underlain by impervious units of rock. Water enters such a unit where the unit is exposed at the ground surface or where it is connected with other water-bearing units and free movement of water is possible (Fig. 16–3). As water moves down into such an aquifer, the water-bearing layer is filled and hydrostatic pressure is gradually built up. If the aquifer is tapped by a well at some elevation lower than the level to which it is saturated, water will flow from it under pressure. Artesian conditions arise in extensive rock units such as the Dakota sandstone (Fig. 16–7), a porous rock unit that lies under the Great Plains and takes in water where it is exposed along the Rocky Mountain front (Fig. 16–8). Artesian water may also occur in smaller lenticular-shaped sand bodies such as buried river channels and alluvial fans.

The static level to which water in an artesian aquifer would rise if it were penetrated by a well is called the *piezometric level*. If this level is known at a number of points, the elevations of the level can be used to construct a surface that is analogous to the water table and is called the piezometric surface. The *piezometric surface* of an aquifer is an imaginary surface that everywhere coincides with the static level of the water in the aquifer.[2] It is the surface to which the water from a given aquifer will rise under its full head (Fig. 16–9). An ex-

FIGURE 16–9
Configuration of the piezometric surface of the principal artesian aquifer in Florida and southeastern Georgia as of 1960.
(from Stringfield, 1964)

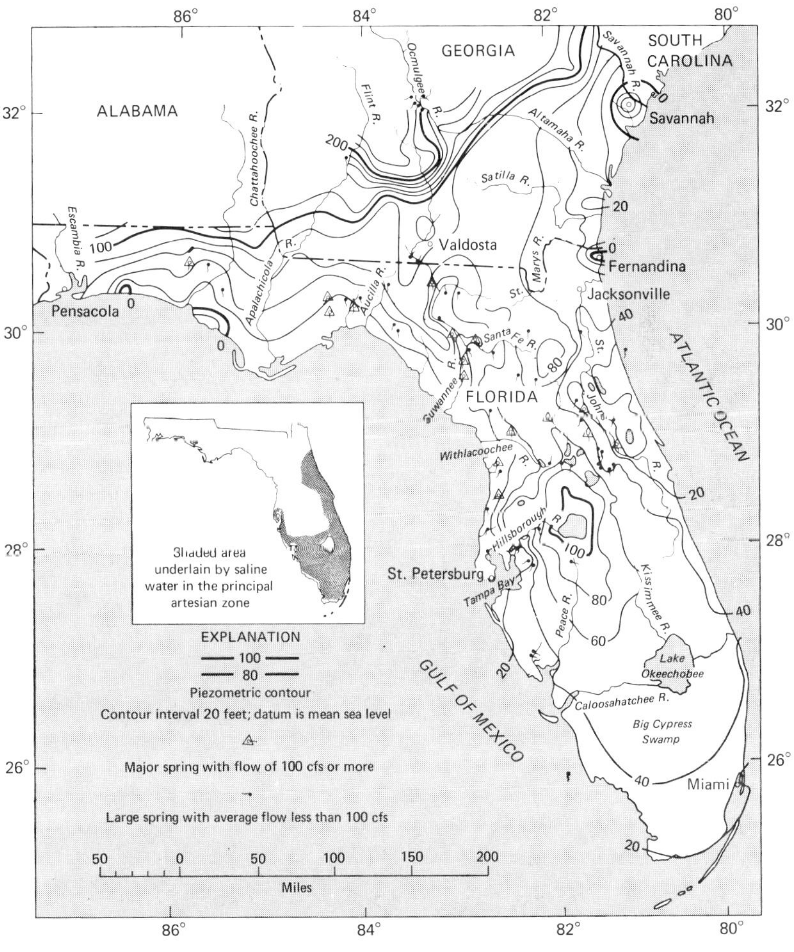

[2] Contours on a piezometric surface are called *isopiestic lines*. These are imaginary lines along which the same static level is constant. Level is usually expressed in feet or meters above mean sea level. The *piestic interval* is the vertical distance between two isopiestic lines.

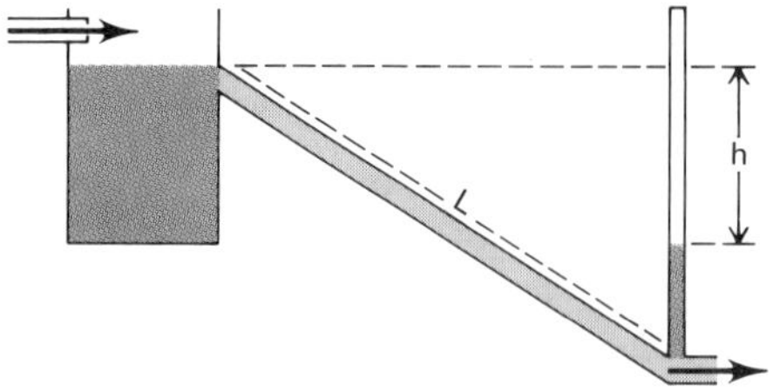

FIGURE 16–10
Experiment set up to test the validity of Darcy's Law. The law states that the velocity, *V*, of the flow of water through a unit is proportional to the permeability of the rock, *K*, and the hydraulic head, *h*, and inversely proportional to the length of the path of the flow, *L*. This law enables us to predict the amount of water which may be pumped from aquifers.

ample of such a surface is seen on the map of Florida (Fig. 16–9), which depicts the principal artesian zone. A piezometric surface may be above, coincide with, or lie below the upper surface of the zone of saturation. If the water from different depths in the aquifer will rise to different levels, the aquifer has more than one piezometric surface.

The *hydraulic gradient*, or *pressure gradient*, of an aquifer at a given place in a given direction is the rate of change of pressure head per unit of distance at that place and in that direction. The hydraulic gradient can be expressed in percentage, in meters per kilometer, and so on.

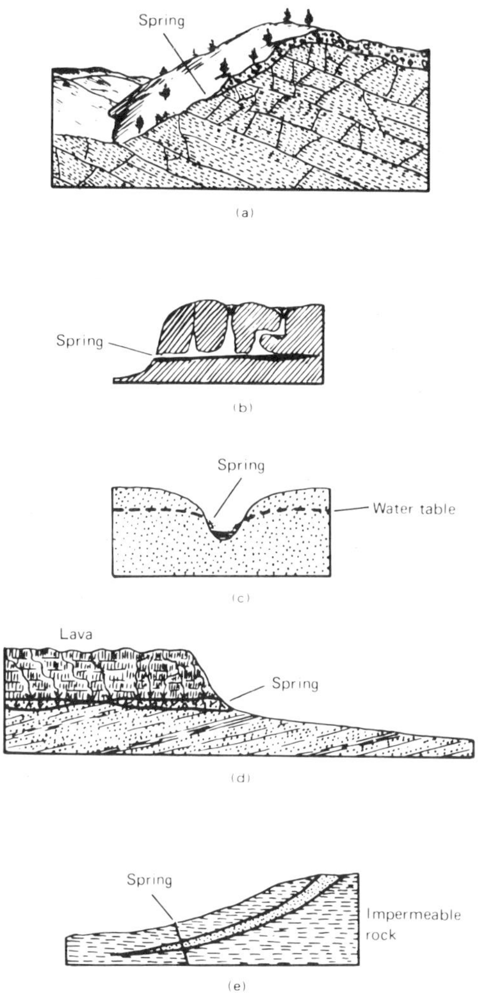

FIGURE 16–11
Cross-sectional drawings illustrating various conditions giving rise to springs: (a) fracture spring, (b) solution opening, (c) depression spring, (d) contact spring, (e) artesian spring.
(after O. E. Meinzer, 1923)

MOVEMENT OF GROUND WATER

Movement of ground water has been the subject of many experiments whose theoretical results can be applied with success to natural conditions. Of notable importance is the work of Henry Darcy (1856). He showed that the velocity of movement of water between two points on the top of the water table is equal to the product of the rock permeability and the slope of the water table. In the case of confined water the velocity is the product of the permeability and the hydraulic gradient, in this case a pressure gradient, which is equal to the difference in hydraulic head between two points divided by the length of the path of flow (Fig. 16–10). In applying this to ordinary ground-water problems it is possible to measure the length of the path (L) and hydraulic head (h), and to determine permeability (K) by making tests on the rock types through which the water passes. Thus the velocity of movements along particular paths may be calculated by Darcy's equation, $V = K\,h/L$.

Rates of water movement beneath the water table are slow. They vary considerably from one place to another according to the permeability and structure of the rocks. Rates of movement of 1–10 meters per year are frequently found but rates as high as several hundred meters per day occur in some alluvial fills, and when water moves through large solution cavities or open-fracture systems the flow approximates stream velocities. Where porosity and permeability are low the water may be almost static.

SPRINGS

Springs form where the water table intersects the land surface, where water under artesian conditions finds its way to the surface, and sometimes where the water table or artesian water rises underwater along coasts. The pressure produced by gravity acting on the water is the driving force that causes water to issue from the ground. Structure of the rock mass and topography act as controlling factors in the occurrence of springs. Several examples of structural and lithologic conditions that give rise to springs are illustrated in Fig. 16–11. Tolman (1937) proposed a classification of springs according to the type of opening from which the water issues, as follows:

1. Pervious veneer formations.
2. Thick pervious formations.
3. Interstratified pervious and impervious formations.
4. Solution openings.
5. Lava.
6. Fractures.

The first type, pervious veneer formations, occurs where a relatively thin soil or sedimentary veneer covers an impervious bedrock. The zone of saturation is near the ground surface in humid climates; and when the water table rises, springs may form on slopes where variations in permeability force the water up or in depressions in the topography. When the pervious formations are thicker, the water table may be much deeper. When permeability varies springs are likely to form where the impervious layers come to the ground surface and particularly in topographic depressions or where the contact occurs on a hillside or in a valley. Solution openings, openings in lava, and fractures give rise to some of the largest springs known. Water can move freely underground in such openings, and where they are large, vast quantities of water may accumulate and flow to the surface.

Springs are also classified according to the rate of discharge as follows:

First magnitude	100 cubic feet per second or more
Second magnitude	10–100 cubic feet per second
Third magnitude	1–10 cubic feet per second
Fourth magnitude	100 gallons per minute (0.22 cfs)—1 cubic foot per second
Fifth magnitude	10–100 gallons per minute

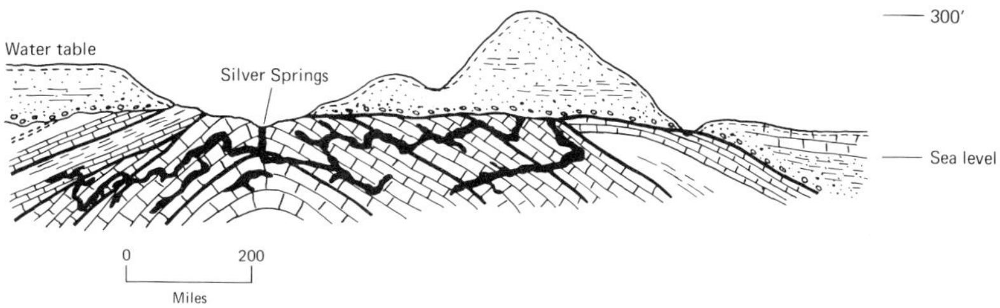

FIGURE 16–12 Generalized and diagrammatic cross section through the peninsula of Florida with great vertical exaggeration.
(after Ferguson and others, 1947)

Of the first magnitude springs in the United States, 38 are in volcanic rocks, 24 are in limestone, and 3 issue along faults.

The largest spring in the United States is Silver Springs, Fla. The spring is one of many that occur in the Ocala limestone in central Florida. The Ocala limestone is a Tertiary limestone that underlies much of the Florida peninsula. It has extremely high permeability resulting from solution cavities. Because it is overlain by impervious units in many places, the ground water in the Ocala limestone is under artesian pressure. A map of the piezometric surface for Florida (Fig. 16–9) shows a high dome east of Tampa and a ridge extending toward the north in the area where most of the springs occur (see black dots). A diagrammatic cross section through the area where Silver Springs is located (Fig. 16–12) gives an impression of the general ground-water conditions. The main pool at Silver Springs is about 250 feet (76 meters) in diameter, 36 feet (11 meters) deep, and the main opening from the Ocala into this pool is clearly visible from the surface. The opening, a solution cavity, is 65 feet (20 meters) wide and 12 feet (3.6 meters) high. The water that issues from this spring is famous for its clarity, and its discharge, 500 million gallons (1,900,000 cubic meters) per day, is sufficient to give rise to a river. Water temperature ranges from 22.2–23.3°C.

Spring deposits are usually composed of calcium carbonate and are called *tufa* if porous or cellular in structure and *travertine* if compact. Organic matter from plants and trees around a spring may be covered by the calcium carbonate deposits and become incorporated into the tufa, making it a porous and poorly consolidated material. Where it has been firmly cemented it may be cut and used as a building stone. The porous, rough texture makes it uniquely valuable for soundproofing.

Siliceous deposits, *siliceous sinter,* also form in and around hot-water springs and around geysers like those in Yellowstone National Park.

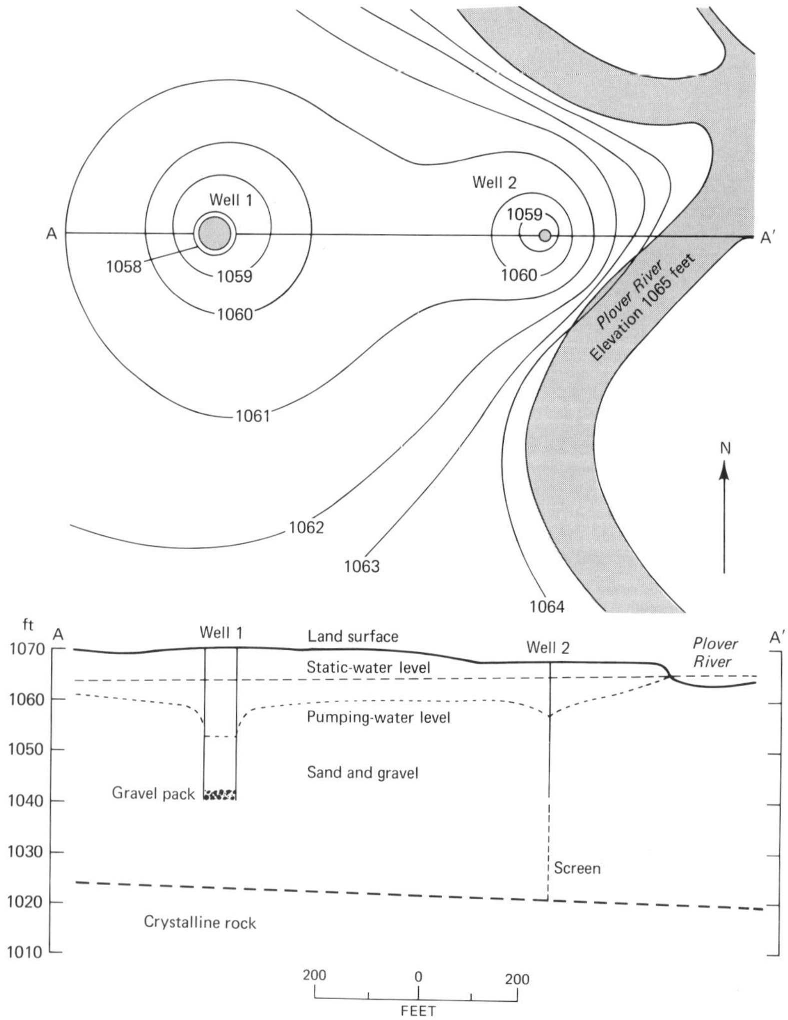

FIGURE 16–13
Sketch map of configuration of water table and cross section of part of Stevens Point, Wis., well field, showing effects of pumping from wells.
(from Holt, 1965)

EFFECTS OF PUMPING ON GROUND-WATER LEVELS

When pumping is started in a well the water table is drawn down around the well to form a cone of depression (Fig. 16–13). The cone expands and the water table is progressively lowered until a balance is achieved between the rate of flow of water to the well and the amount pumped. The results of pumping tests near the Plover River (Fig. 16–14) indicate how the cone of depression increased in size both with discharge and over a period of two months of sustained pumping. During one test 500 gallons per minute (1.9 cubic meters per minute) were produced; 1,000 gallons per minute (3.8 cubic meters per minute) were produced in a second test.

DEVELOPMENT OF GROUND-WATER SUPPLIES

Increasing attention is being given to the most effective ways of developing ground-water supplies. This is particularly true in areas where annual recharge of ground water does not equal the rate at which it is withdrawn. Such areas in the United States are located primarily in the arid and semiarid portions of the southwest and near large population centers, as in parts of Florida, Long Island, and California. Study of water in such areas has been most successfully approached on the basis of comprehensive analysis of natural basins. A *ground-water basin* may be thought of as an area bounded by natural geologic conditions that cause ground water within it to be more or less separated from adjacent basins. These conditions are met particularly in the Western States, where many structural basins exist.

Ground-water supplies of any area can be analyzed, even though the area is not an enclosed basin, if the water budget for the area can be accurately determined. The *water budget* is an expression of the total water—that introduced into the area and that used or lost through any means. An equilibrium condition usually existed before the artificial modifications caused by the activity of man. The construction of reservoirs, dams, and artificial lakes; the importation and exportation of water; irrigation; ordinary daily water consumption; construction of pipe lines, canals, and cut off of infiltration by paving and buildings all influence the water budget of an area. All these influences must be evaluated for intelligent analysis of the water supply.

The amount of water that can be withdrawn without producing undesired results is called the *safe yield* for a particular area. Undesired results include not only those that deplete or pollute the ground-water supply, but also those damaging to the area's economics. Water supply should be studied in terms of the economics of industries and agriculture dependent on water, quality of water, and legal rights to water. Undesirable effects result from long-term overdrafts that lower the water table and may cause sea-water intrusions or other contamination of ground water. It may be as undesirable to lower water quality through irrigation, which may raise salt content of the water supply in arid climates, or industrial contamination as to allow industries or agriculture to fail economically through inadequate water supply.

In more and more areas of the world people must face the fact that the supply of water underground is not unlimited. For any given area a certain amount of water is held in the ground, and a certain amount is annually recharged into the ground by natural means. To remove more water from the ground for drinking, agriculture, or industrial use than is normally recharged is to *mine* the water. When water is "mined" it may eventually be exhausted, just like any other mineral resource. Thus the amount of water in the ground, the rate at which it is naturally recharged, and the ability of man to recharge the water supply by bringing water into the area create a limit to the potential development of that area.

Understanding the geology of ground water can provide the necessary information to make intelligent use of water resources. When sufficient information is available a basin can be analyzed in terms of the *equation of hydrologic equilibrium.* An equilibrium condition is said to exist when the total amount of water coming into a basin just equals the amount taken out. Factors in hydrologic equilibrium include the amount of water in surface and subsurface storage; the quantity brought into and taken from the area by streams, underground movement, manmade devices, and precipitation, evaporation, and transpiration; and the volume consumed. The equation of hydrologic equilibrium may be stated as shown in Table 16–1.

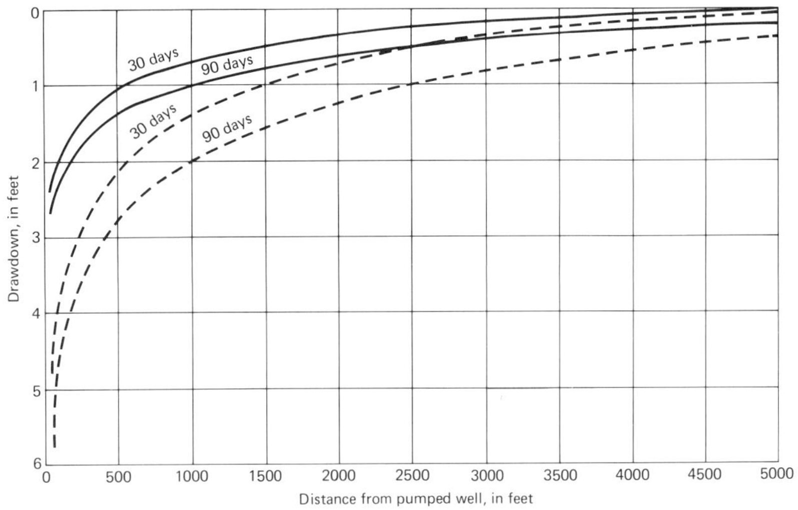

FIGURE 16–14
Theoretical distance-drawdown relation for high-capacity wells tapping the sand and gravel aquifer in the sand-plain province, Portage County. Solid line indicates well pumping 500 gallons per minute; dashed line indicates well pumping 1,000 gallons per minute. Time designates length of pumping.
(from Holt, 1965)

surface inflow through streams		increase in surface storage
+ subsurface inflow		+ increase in subsurface storage
+ precipitation		− surface outflow
+ importation by pipes & canals		− subsurface outflow
− decrease in surface storage (reservoirs)		− consumptive use
− decrease in ground-water storage		− exported outflow
balance	=	balance

TABLE 16–1
Equation of hydrologic equilibrium.

Long-range planning for any area should be based on the proposition that an equilibrium condition must be maintained. One remedy for water shortage is importation of water. It is now economically feasible to recover fresh water from sea water along coasts in arid areas, and this will undoubtedly be part of the solution to future water needs in the coastal areas of the world, but the cost of transporting water inland over mountains and long distances makes it imperative for us to develop existing ground-water supplies as completely as possible.

QUALITY OF GROUND WATER

High-quality water is needed for drinking and for many industrial and agricultural purposes. Water quality is determined by means of chemical analysis, which can reveal contamination from pesticides and other chemicals. The kinds and quantities of ions present in solution are determined and this is used as a guide to the purity of the water and the uses for which it is suitable. Concentrations of ions are expressed as parts per million, *ppm*. One ppm is one part by weight of the ion to a million parts by weight of water. When determination of the types of ions present is unnecessary, the total amount of dissolved solids can be estimated quickly by a test measuring the electrical conductivity of water, which is, in turn, a function of the salt content. Bacteriological analysis is made, particularly of drinking water, to insure that there has been no contamination from sewage. Other tests may be made to determine temperature, color, odor, and taste.

The United States Public Health Service sets standards for most drinking-water supplies. Mandatory maximum limits are set for ppm of lead (0.1), fluorides (1.5), arsenic (0.05), selenium (0.05), and chromium (0.05), and a maximum total solids concentration of 1,000 ppm. In recent years dangerously high concentrations of mercury have been discovered in some waters. Desirable limits for many less dangerous elements are also provided, along with standards for physical characteristics and bacterial quality. One of the common expressions of water quality is its *hardness*. This usually refers particularly to the amount of calcium carbonate and magnesium carbonate in solution, rather than to the quantity of all ions present.

Many industries must have extremely pure water or water containing low concentrations of certain particular ions, but these standards vary greatly from one industry to another.

Even water used for agricultural purposes must meet certain standards. Some plants are sensitive to concentrations of certain salts. For example, apples, pears, oranges, plums, peaches, and strawberries all have low salt tolerances, as do green beans, celery, and radishes. Date palms, spinach, asparagus, sugar beets, and barley can stand concentrations of three or four times the tolerable limits for the plants previously listed. This problem becomes particularly important in arid regions where irrigation is necessary, because water found in such regions is likely to have high salt content. Pumping water to the surface for irrigation may result in increasing the salt content of all the water in the vicinity. The irrigation water dissolves salts from the soil and carries the added salt load back to the zone of saturation, contaminating other water supplies, encrusting the soil, and decreasing fertility. If the same water is pumped up again its content of salts is again increased as it carries more salts downward, until eventually it is not suitable for irrigation. The possibility of producing water with a high salt content is increased as the water table is lowered. The lower water has in general been in the ground a longer time and has had longer to dissolve mineral matter. Also the cost not only of drilling the well but of pumping it increases with depth.

A great many other conditions may cause ground-water quality to be lowered. These include inadequate treatment of sewage before water is returned to the ground, dumping of sewage into streams, and dumping of wastes such as food products, oil, chemicals, and tailings from mines and mineral-processing plants. Water can be contaminated by natural causes as well. Magmatic waters contain high concentrations of salts. Water may even deteriorate as a result of the way it is brought to the surface. Excessive pumping may cause intrusion of salt water near the coasts, or water of lower quality may be drawn from one ground-water basin into another.

GEOLOGY OF GROUND-WATER OCCURRENCE

The first considerations in looking for ground water are surface and subsurface geologic conditions. Primary among these are the types of rock units present and their structure. Unconsolidated sediments such as sand and gravel are among the most important sources of water. These occur as river alluvium along modern water courses, as abandoned or buried river valleys, in plains where sediments have been eroded from adjacent mountains, in intermontane valleys where alluvial fans are deposited on flanks of mountains, in glacial-outwash deposits, and in blanketlike marine deposits from which the original salt water has been flushed. Some water is obtained from nearly all other rock types.

Porosity of most limestones is due to fracture or solution cavities and these vary considerably from place to place even in a single formation. Thus it is often difficult to predict the reservoir characteristics of limestones. Water produced from limestone is usually "hard," that is, it has a high content of calcium and carbonate ions.

Volcanic rocks also are sites of large springs. Lava flows often have high porosity, but the pores are not always interconnected. Water occurs in shrinkage joints formed when the lava cooled, in fractures caused by deformation, in zones of blocky lava, in lava tubes formed where molten lava flowed out of the early-formed crust leaving a tube or tunnel-like opening, in vesicles at the top of a flow, and between successive flows. Crystalline igneous and metamorphic rocks have the lowest porosity. Most water held in them comes from fracture systems. Clays and clay mixtures contain large amounts of water, but because they have such low permeability the water is not easily obtained.

Ground water occurs in a wide variety of near-surface sediments and bedrock types. Some of the important occurrences in the United States are the following:

1. Glacial deposits found near the surface in the northern states and Alaska.
2. Large regions of intrusive igneous and metamorphic rocks such as the Piedmont, New England, and Wisconsin highlands, and most of the central portions of mountains in the West.
3. Paleozoic sedimentary sequences, including limestones, shales, and sandstones, found in the Appalachians, in the central and eastern interior regions, and at depth in the Great Plains.
4. Mesozoic and Cenozoic sheetlike deposits of alluvial materials deposited in the Great Plains and in the basins of the Rocky Mountain region.
5. Cenozoic marine sedimentary deposits of the Atlantic and Gulf Coastal Plain.
6. Recent Cenozoic alluvial-fan deposits in the basins of the Basin and Range region of the west.
7. Volcanic eruptives that cover extensive areas of the northwestern states and Hawaii.

Ground-water provinces, large areas in which ground-water occurrence is similar, can be recognized. One such subdivision of the United States is shown in Fig. 16–15. Water occurs in a number of different ways in each of the outlined provinces. The simplification involved in this type of classification must be borne in mind.

EXAMPLES OF GROUND-WATER OCCURRENCE

The occurrence of ground water varies from place to place, reflecting changes in climatic conditions, structure and composition of the bedrock, topography, and the sources of water supplied to the reservoir rock. Sometimes marked changes in manner of occurrence are found over short distances, but other modes of occurrence apply to broad regions. Local conditions are important when questions of the quantity and quality of water supply available at a given place are considered, or when the effects of contamination of the supply by sewage, waste disposal, salt-water encroachment, and so on, are to be evaluated. The following examples have been selected to illustrate some of the more important aspects of ground-water occurrence. Although the examples give details of local situations, the principles involved are widely applicable.

Ground-Water Supply on Long Island

Long Island is located near the northern end of the Atlantic Coastal Plain. Ancient metamorphic and igneous rocks that are nearly impervious to water lie at depth under Long Island. This crystalline complex is exposed in Manhattan, Connecticut, and New York, and it slopes gently toward the Atlantic, under Long Island Sound, Long Island,

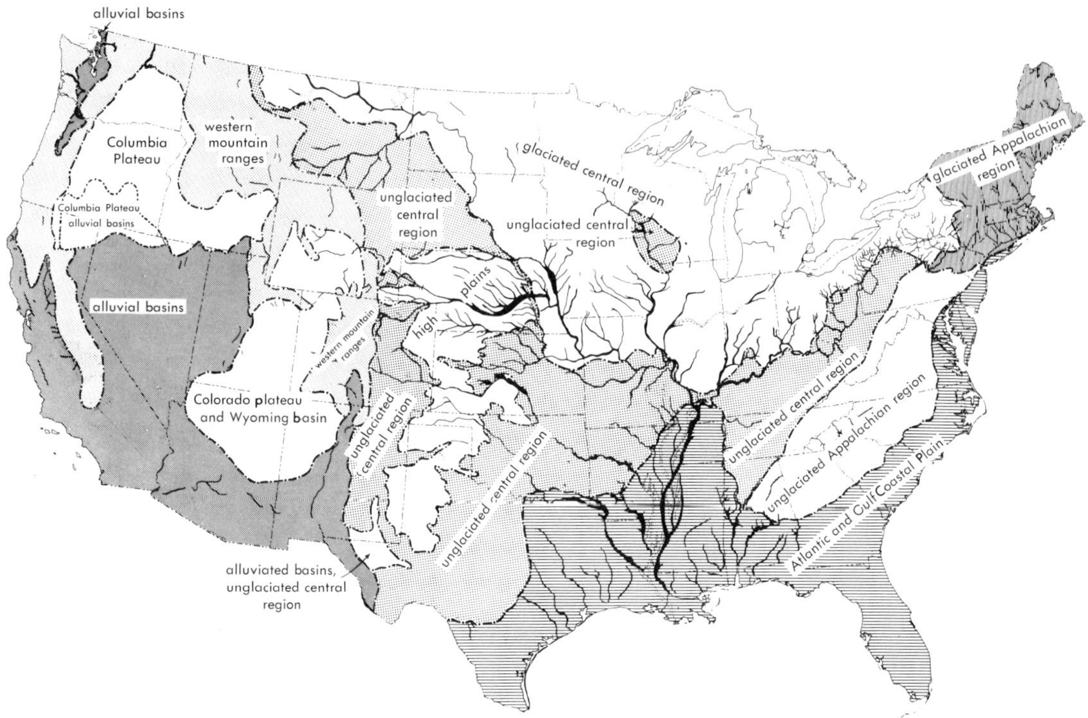

FIGURE 16–15
American ground-water provinces. The heavy black lines mark important watercourses where ground water is replenished by perennial streams. Other important provinces are alluvial basins, as marked, and glacial deposits. In the Atlantic and Gulf Coastal Plain water is produced from sedimentary units, sands and silts, of Cenozoic age. In the unglaciated Appalachian region water is held primarily in fractures. In the central region bedrock is composed of late Paleozoic and Cenozoic sedimentary units. The Columbia Plateau is largely underlain by lava flows. In the western mountains most water supplies come from surface streams. The Colorado Plateau and Wyoming basin are underlain by Cenozoic and Mesozoic sedimentary units.
(after Thomas, 1963)

and the continental shelf (Figs. 16–16 and 16–17), reaching a depth of 610 meters at the southern coast of Long Island. The crystalline basement is overlain by a sedimentary-rock section composed of semiconsolidated Tertiary units that slope toward the ocean also. The Tertiary units have been truncated by stream and glacial action at a level close to present sea level and glacial deposits (moraines and outwash) were laid down on them.

Both the Tertiary sediments and the glacial deposits on Long Island are porous and permeable. Precipitation infiltrates directly into the glacial deposits and ultimately some of it passes into the Tertiary units; salt water stands in contact with the lower density fresh meteoric water on both the Long Island Sound side and the Atlantic Ocean side. The principal water reservoirs are the glacial deposits and three of the strata in the Tertiary section. These strata contain fresh water under the island, but their seaward continuations are saturated with sea water.

The water table for the glacial deposits follows

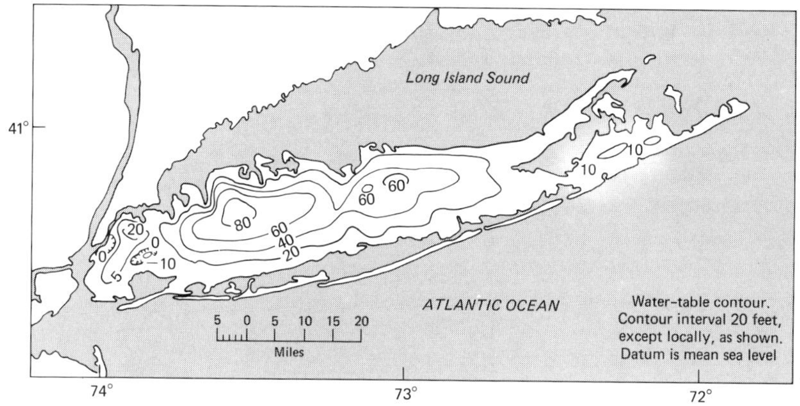

FIGURE 16–16
Generalized contours on the water table of Long Island (N.Y.) (the upper surface of the ground-water reservoir) in 1961.
(from Foxworthy and Cohen, 1966)

FIGURE 16–17
Diagrammatic section showing general relationships of the major rock units of the ground-water reservoir in Nassau County.
(from Foxworthy and Cohen, 1966)

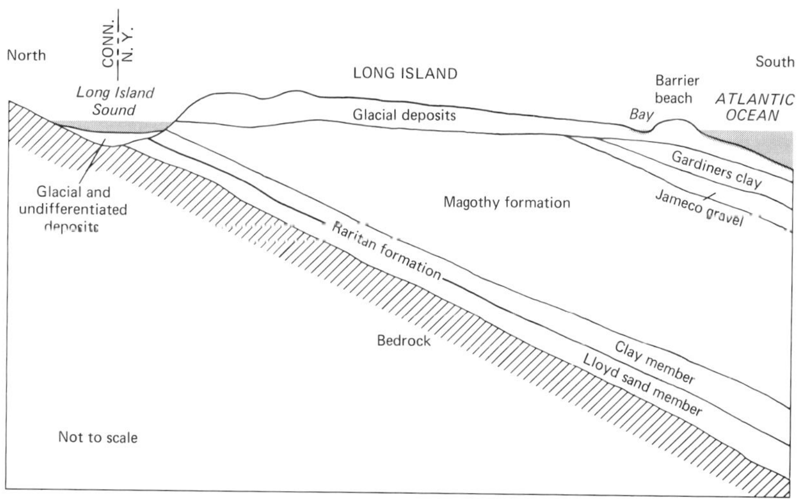

roughly the configuration of the land surface, and the piezometric surfaces for the lower aquifers, where they are under artesian pressures, are similar in shape to the water table.

Before this region became so populous, an equilibrium condition existed. The long-term average ground-water recharge and discharge were equal and the subsurface interfaces between the fresh and salty water in each of the aquifers were stable, reflecting an overall hydrologic balance. The average infiltration of rainfall, about one million gallons per day per square mile (about 1.5 million liters per square kilometer), on Long Island moved into the glacial deposits and was discharged from them into the sea or onto the ground surface. An estimated 90 percent of the total recharge was discharged from the glacial deposits into the sea, but part went down into the Tertiary rocks. The generalized ground-water conditions before development are illustrated schematically in Fig. 16–18.

As the population of Long Island grew and the demand for water increased, the number of wells put down into each of the water-bearing units increased. At first many individual wells and sewage disposal units were used. These were shallow wells drawing water from the glacial deposits, but in a short time the disposal of sewage into these same deposits created a problem of cesspool pollution of the glacial-deposit water supply, and wells had to be drilled deeper into the Tertiary section. At this stage private water supplies gave way to municipal water development. The highest population and the greatest use of water for industrial purposes are found at the western end of the island, and it was in this area that large-scale ground water withdrawal from the Tertiary-section aquifers first disturbed the position of the fresh-water–salt-water interface. As this interface was drawn farther inland the deep wells had to be discontinued, and now Long Island receives its water from the New York City supply, which is obtained from surface reservoirs in upstate New York. The intrusion of the salt-water wedge was accelerated by the gradual expansion of sewage systems designed to discharge waste into the ocean, and by the reduction of rainwater infiltration by construction of buildings, streets, and so on. In the late 1930's ground-water levels of fresh water under Brooklyn had been depressed locally to as much as 10.5 meters below sea level. This resulted in large-scale contamination of the aquifers by salt water (Fig. 16–19).

At the present time the eastern end of the island is operating as a balanced water system in which recharge is equal to or greater than withdrawal,

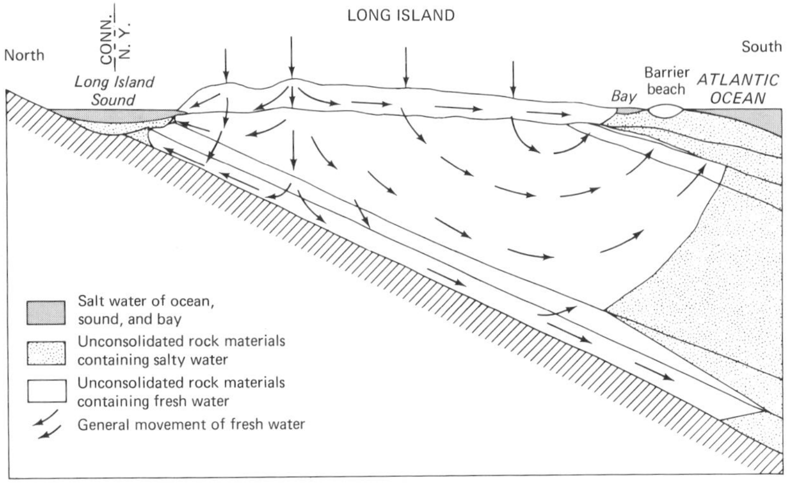

FIGURE 16–18
Diagrammatic section showing predevelopment generalized ground-water conditions.
(from Foxworthy and Cohen, 1966)

FIGURE 16–19
Diagrammatic section showing generalized ground-water conditions during phase 4 of ground-water development (deep supply wells and waste disposal through sewers to adjacent salt-water bodies).
(from Foxworthy and Cohen, 1966)

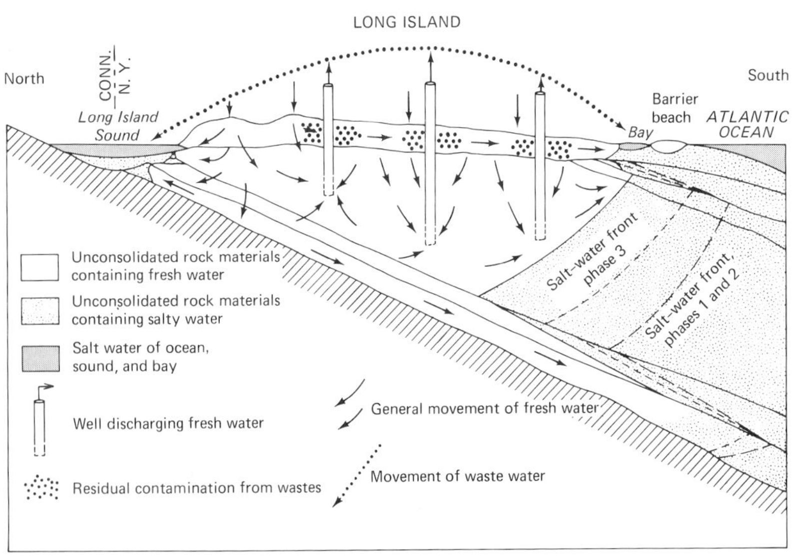

which is from shallow wells, but salt-water encroachment is a continuing problem in the central and western portions of the island.

Las Vegas Ground-Water Basin

The Las Vegas ground-water basin is located in a geomorphic province known as the Basin and Range province, most of which is characterized by an arid to semiarid climate, an average rainfall of 12.5 centimeters, and topography consisting of high mountain ranges separated by basins. The mountains rise to elevations of 3,050 meters and stand about 2,130 meters higher than the intervening basins. Many of these basins, including the Las Vegas basin, are almost completely enclosed by mountains composed of crystalline rocks and tightly cemented sedimentary rocks that are nearly impervious (Fig. 16–20). The basins have internal drainage; no surface streams drain surface water into or out of the basin from adjacent basins. Thus each basin forms a closed system in which little water moves in or out either on the surface or underground.

Water occurs in the sediments that have been deposited in the basins. The oldest sediments deposited in the Las Vegas basin are Tertiary in age. They consist of thin sand lenses and fine gravels interbedded with thick beds of clay. The sand and gravel lenses contain silt and clay and are partially cemented. These Tertiary sediments are overlain by alluvial deposits that have spread as great fans of sediment into the basin. The sediment is derived from erosion in the mountains and is transported into the adjacent valleys by intermittent streams that flow for short periods following torrential rains at infrequent intervals. The alluvial fans are composed of poorly sorted heterogeneous mixtures of boulders, gravel, sand, silt, and clay. Stringers of well-sorted gravel that is highly porous and permeable tend to accumulate in the stream valleys, and the interstream areas are covered by mudflows and sheet-wash deposits that are relatively impervious. At the toe of the fans the coarser materials give way to fine-grained sands, silts, and clay, which are interbedded with lake-bed deposits formed in intermittent lakes (playas). These alluvial-fan and playa-lake deposits have been built up over the

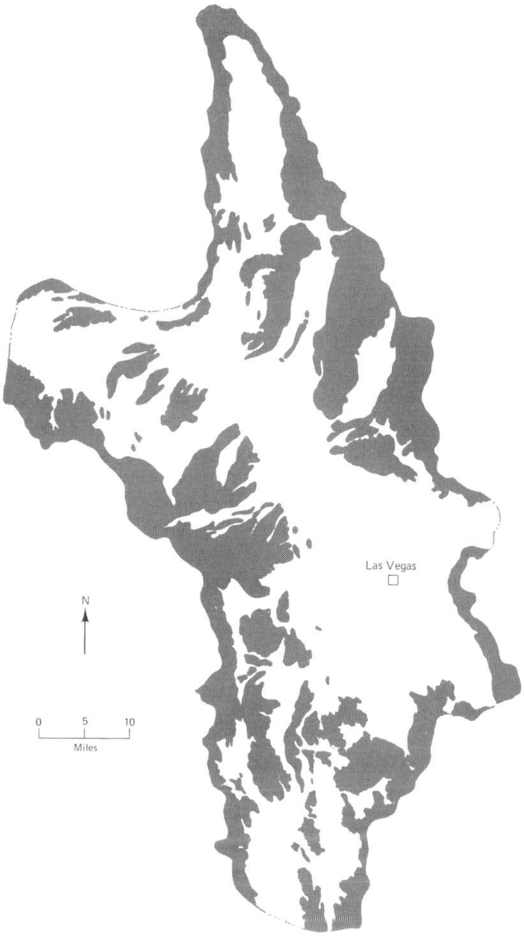

(a)

(b)

FIGURE 16–20

(*a*)

Las Vegas, Nev., ground-water basin. Solid rock exposed near the ground surface is shown in black. Outline of basin is taken at the drainage divide. White area in basin is alluvial fill.

(after Malmberg, 1965)

(*b*)

Part of the Las Vegas Valley, Nev., showing elevation of the piezometric surface in February, 1955. Bedrock is in black.

(after Malmberg, 1965)

last few millions of years (possibly 10 to 20 million) and now have thicknesses in excess of 300 meters.

During the Pleistocene only the mountains in this region were covered by ice; the basin floors were sites of lakes that formed as the "sealed" basin deposits became saturated by greatly increased rainfall and melt-water. Deposits from these lakes are impervious materials composed of clay, silt, and fine sand, and are up to 15 meters thick. Recent gravels, sands, silts, and playa deposits lie over the lake bed.

Unconsolidated alluvial deposits (Fig. 16–20) form the principal ground-water reservoir. Recharge of these beds comes largely from infiltration around the edges of the basin. Porous and permeable beds are interbedded with impervious layers (Fig. 16–21). The vertical and horizontal differences in permeability are irregular so aquifers are not continuous. Several zones of artesian conditions are present but most of the aquifers leak water into other beds, producing a more or less continuous hydraulic system. Water occurs near the ground surface over much of the lower part of the basins where evaporation and plants are responsible for transpiration of large quantities of water. Almost all of the 12.5 centimeters of precipitation in the basin is lost through evaporation and transpiration. Only in the mountains does precipitation occur in quantities sufficient to allow some infiltration. As water moves down into the aquifers it becomes confined under layers of clay or other impervious beds.

High angle faults cut the alluvial deposits, and some of the artesian layers have been severed as a result of movements on faults that cause the porous layers to be offset so that they now terminate against impervious units.

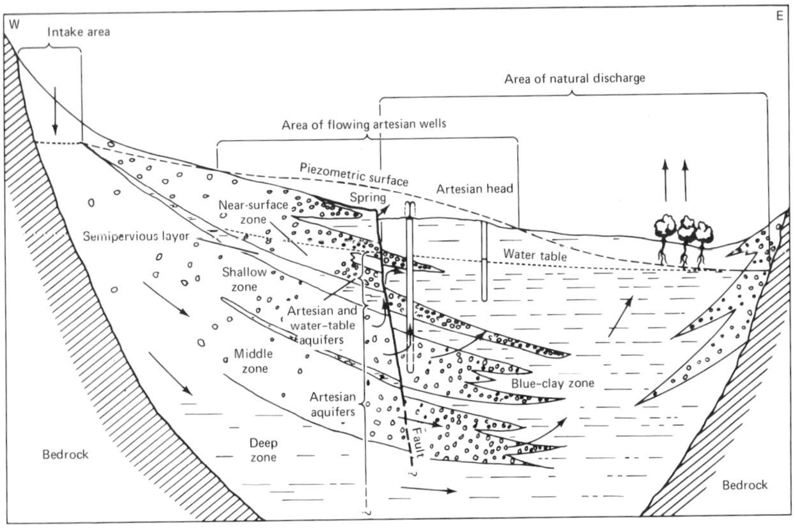

FIGURE 16–21
Las Vegas Valley, showing the principal zones of aquifers, direction of ground-water movement, and general geology of the valley fill.
(from Malmberg, 1965)

FIGURE 16–22
Regional tilting and subsidence across Las Vegas Valley resulting from the filling of Lake Mead and localized land subsidence resulting from ground-water withdrawals.
(from Malmberg, 1965)

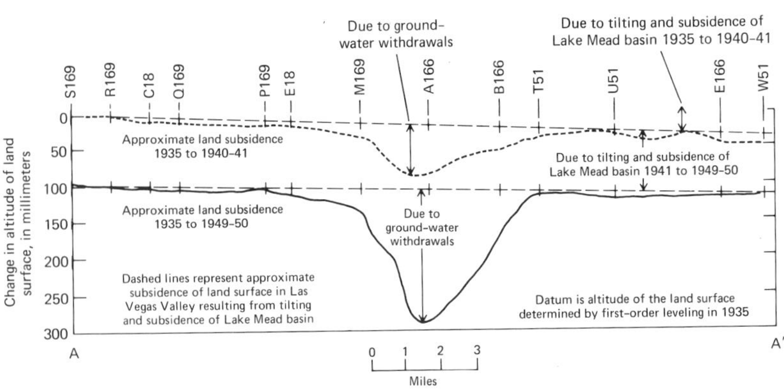

The largest withdrawals of ground water are concentrated near Las Vegas. This has caused a reduction in the artesian pressures in the aquifers and a consequent subsidence of the land surface. Additional warping of the ground surface can be attributed to the weight of the water that has been stored in Lake Mead as that reservoir has been filled. Results of careful releveling in the region (Fig. 16–22) indicate the relative contribution of each of these causes of subsidence. Water withdrawal induces compaction of the fine-grained deposits, causing subsidence of the ground. Similar compaction effects have been observed in many irrigated areas such as in the San Joaquin Valley, Calif. Careful measurements over a period of two years indicate significant compaction following periods of withdrawal of ground water (Fig. 16–23).

GEYSERS

Since its discovery in 1870, Old Faithful geyser in Yellowstone National Park (Fig. 16–24) has regularly spurted forth 10,000 to 12,000 gallons (38–45 cubic meters) of steam and water to an average height of 40 meters about once every hour. Very few geysers are this regular, but many others exist and the nature of their activity is similar. Outside of Yellowstone Park areas of geyser activity are found in New Zealand and Iceland. Geysers follow a regular cycle of activity, often similar to that of Old Faithful:

1. Immediately following an eruption water goes back into the ground at the vent or through nearby holes. A small part runs off in small streams flowing away from the cone.
2. This is followed by a period of quiet during which only a few whiffs of steam rise from the vent.
3. After about 30 or 40 minutes more, steam starts to rise from the vent and a little water appears at the surface.

4. Water and steam rise and flow out of the vent in a series of small outpourings. Some of this water runs off, but most goes back into the vent.

5. Several small eruptions occur, throwing bubbling water out of the vent to a height of several meters above the ground. This continues for several minutes.

6. The main eruption starts and a column of water and steam spurts forth, occasionally as high as 76 meters. The eruption lasts several minutes, then dies down, and the cycle is repeated.

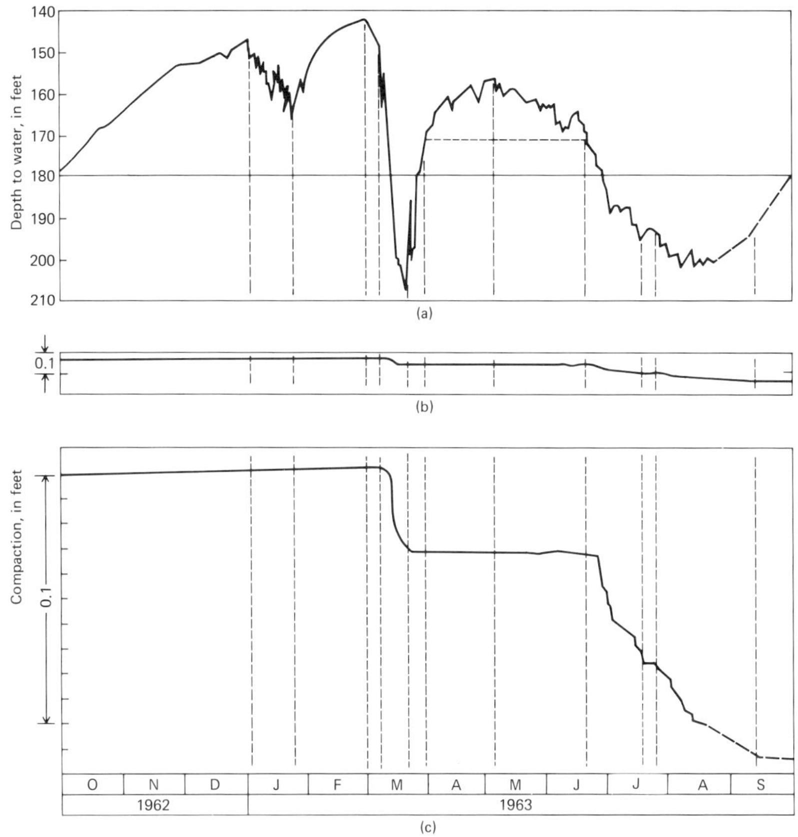

FIGURE 16–23
Correlation of compaction and water-level fluctuations near Pixley, Calif. (a) Hydrograph of well; (b) compaction in well; (c) compaction in well. (from Lofgren and Klausing, 1969)

Heat from a buried pluton rises to the surface, heating the ground water. Most of the water coming from Old Faithful and other geysers is meteoric water. It contains small quantities of common volcanic gases and solutions. Surface waters flow into the fractured and porous igneous and volcanic rocks that underlie Yellowstone Park. Beneath the surface the vent of the geysers is connected to an irregularly shaped network of fractures and cavities dissolved by the hot waters. It is from these cavity networks that the ejected water comes. After eruption, time is required for water to percolate back into the channels and refill them. The water, which has just been cooled by its ascent into the air, is warmed quickly. Since the heat in the rocks increases with depth, the water in the lower parts of the network becomes hot more rapidly than that higher. Water at the surface boils when it reaches 100°C, but water at the bottom of the system is under the pressure exerted by the weight of the column of water above it, so it must be heated to a much higher temperature before it will begin to boil. Eventually the boiling point is reached at some point in the network and steam forms. The expansion of water into steam exerts sufficient pressure to force overlying water out at the top of the vent. Just a little is forced out at first, then more, and finally enough is removed to lower the pressure on the network and permit large quantities of water to convert suddenly into steam. The main part of the eruption starts when this explosive generation of steam occurs at depth.

A large amount of silica dissolved from the underground cavities is carried in solution by the hot waters. It is precipitated at the surface when the waters cool. Usually a colloidal gel is formed. This consolidates to an amorphous form of silica and leaves a deposit known as *geyserite* (Fig. 16–25).

CAVES

Caves are naturally formed underground cavities or voids. Most of them are found in areas underlain by soluble bedrock, usually limestone, where solution by ground water is responsible for their formation. However, caves also occur in lava flows where fluid lava drained out, leaving a system of underground passageways.

Caves are abundant in the thick limestone sequences deposited early in the Paleozoic in the

FIGURE 16–24
Old Faithful in eruption, Yellowstone Park.
(Union Pacific Railroad)

region of the Appalachian Mountains and Plateau. These caves include the famous Mammoth Cave of Kentucky and the caves of Virginia. Other important limestone-cavern regions of the United States are found in central Florida, the Carlsbad Plateau of New Mexico, and the Ozarks. Most of these regions are characterized by what is called *karst topography*—topography shaped at least in part by underground solution and diversion of surface water underground in areas of limestone and dolomite bedrock. Karst topography includes such features as caves, enclosed depressions, called *sink holes* (Fig. 16–26), formed by solution, subsidence, or collapse of the surface into a cave; *sinking creeks* or rivers—streams that disappear underground through swallow holes in the channel; *natural bridges* and tunnels; and pitted and fluted bedrock caused by solution. The soil in most such areas is red and clayey and is called *terra rosa*.

Authorities generally agree that caves are formed as a result of solution by ground water, but debate about the conditions of formation continues. Probably the most important point of argument is the relation of the cave to the water table at the time that most of the solution took place. Some argue that most solution occurs above the water table by water percolating down through fractures or by streams diverted underground into a network of channels formed earlier along fractures and bedding. An alternative advanced by W. M. Davis (1930) suggests a two-cycle theory: (1) a period of solution below the water table, and (2) draining, formation of cave deposits, and diversion of surface streams. Still a third idea is that caves form very close to the water table as a result of lateral movement of water there. Convincing arguments can be made for each of these ideas, and caves no doubt differ in history as well as in geologic setting, type of bedrock, and ground-water conditions.

Proponents of the idea that caves form above the water table or near it make the following points:

1. Solution along fractures is usually evident. The pattern of passageways in some caves has a geometrical outline almost identical in plan to the alignment of fractures in the rock. These fractures and bedding surfaces are important as the prime avenues for movement of water through rock that often has almost no porosity or permeability except fractures. Blocks of rock bounded by fractures and bedding planes are found collapsed into the caves.
2. Subterranean streams are found in many caves. Some of these can be traced to the ground surface. Some are known to be streams diverted from the surface into caves only to emerge again at a lower elevation.

FIGURE 16–25
Geyserite, a deposit formed around the openings to geysers. Most of these deposits are siliceous.

FIGURE 16–26
Sink holes in limestone near Timaru, New Zealand.
(Copyright by V. C. Browne)

Streams in caves often carry sediment introduced from the ground surface in sinks or *swallow holes,* openings in a stream channel leading underground, and sediment-laden streams are effective in accomplishing erosion. Potholes, grooves, and even meanders are found in some caves.

Among the observations that favor the two-cycle theory are the following:

1. Many caves now beneath the water table and filled with water are found in the Ocala limestone of central Florida, and water-filled cavities have been encountered beneath the water table in the Appalachians.
2. The internal network of branching tubes, pockets, and irregularly shaped cavities of many caves does not possess a pattern readily explained in terms of stream action. The cavities resemble a three-dimensional, spongelike pattern with blind pockets and dead-end tubes more than a streamlike pattern. Longitudinal profiles like those of surface streams are rare in caves.
3. Some caves have crystal-lined walls. These crystals apparently grew underwater.
4. Solution cavities are found in massive unfractured rock and some cavities bear no relation to fractures.

FIGURE 16–27
Formation of cave deposits. The photograph shows stalactites, stalagmites, and columns in various stages of formation. The sequence of events in the formation of a column like that at the left is described in the text.
(James F. Quinlan, Jr., *Mineral Industries Journal*)

Cave Deposits

Moisture seeps into the cavern from the rock and drips from the ceiling and sides of the cave to form pools and streams on the floor. Points at which water drips from the ceiling become sites of calcite deposits that usually have the shape of thin straws initially. These deposits, which hang down, are called *stalactites* and may be seen in various stages of development. As a drop of water hangs from the ceiling, some of the water within the drop evaporates, leaving a thin ring of calcite or aragonite. This ring is gradually lengthened by more deposits until it becomes a long straw. Finally the inside of the straw becomes so filled with calcite that drops can no longer flow through it. Water then moves as a thin film over the surface of the straw and slowly builds up layer on layer of calcite until a large tapering cone is formed. Water that drops from the end of a stalactite falls to the floor, where it evaporates further and precipitates more calcite. Slowly a stump-shaped deposit called a *stalagmite* (Fig. 16–27) rises from the floor. Usually it has a small saucer-shaped top into which drops of water fall from the ceiling. Overflow from this saucer goes down the sides of the stalagmite, building it out. Eventually stalactites and stalagmites may grow together to form *columns*. Other formations may grow from the sides of a cavern. There droplets of water evaporate, leaving small, rounded, knoblike projections of calcite and aragonite on the walls. Calcium carbonate is also deposited from flowing water in caves, a deposit called *flowstone*. Such deposits may make up beautiful cascadelike masses on the side or floor of a cavern. Where basins are formed, rims of calcite (*rimstone*) are often built and some formations composed of flow- and rimstone give rise to large terraced deposits.

Calcite deposited in cave formations comes from dissolved limestone through which the ground water moved. The ground water was originally rainwater containing enough carbon dioxide to form weak carbonic acid, which dissolved the limestone. Precipitation of calcite as a cave deposit comes from loss of carbon dioxide and reduction in the water content when it reaches the cave. The remaining water is saturated with calcium and carbonate ions. The rate at which cave formations are built is highly variable. In active caves supplied with large quantities of water the rate is high—as much as a centimeter a year—but where the supply of water is limited the rate is slow, or formation may have ceased altogether.

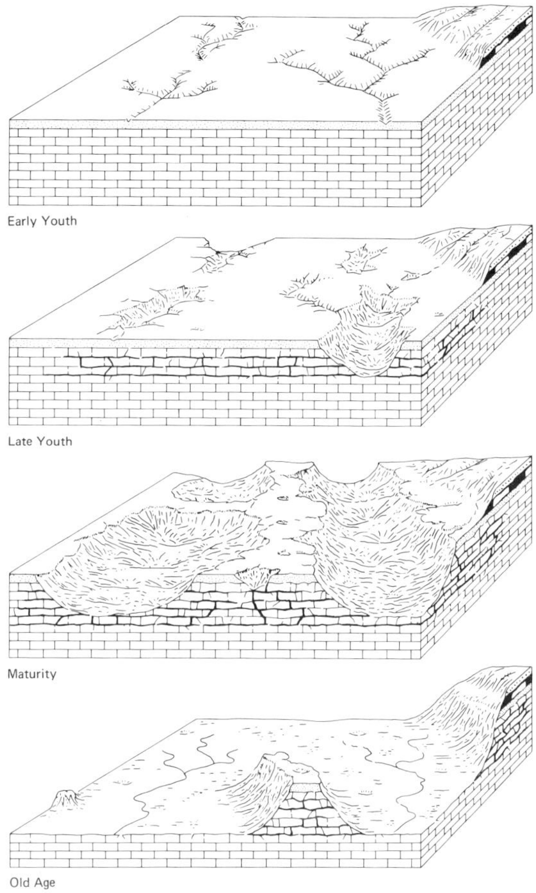

FIGURE 16–28
Stages in the erosion of a terrace in which solution is an important process in shaping the land forms.
(A. K. Lobeck, 1929)

KARST GEOMORPHIC CYCLE

Subsurface movement of water may become a dominant process in causing alteration of the landscape, but it is never the sole factor. Wherever enough water is present for formation of caverns surface drainage is present, and mass wasting invariably plays a part in degradation. *Karst topography* is likely to develop in humid climates where limestones or other soluble rocks are exposed or are near the surface of the ground. How these become exposed is important in determining the initial forms. The Ocala limestone of the Florida peninsula was broadly exposed by erosion over a large region after its uplift from the sea floor, but in the karst topography around the Mammoth Cave in Kentucky a resistant sandstone unit covers the soluble limestone. This resistant cover must be at least partially removed before karst topography can develop. In such areas, where the units are flat or tilted, the limestone is first exposed along valleys; then exposure spreads as the overlying units are eroded away (Fig. 16–28).

The erosion cycle commences with surface drainage on a limestone that is above the base level of the region. In many karst regions the cycle appears to have started with uplift that initiated downcutting of the surface streams and finally caused them to become entrenched in their present courses. At first, a large part of the precipitation runs off on the surface of the ground and a small part infiltrates the ground, causing solution along fractures, faults, bedding planes, and any other zones of weakness. Underground drainage passages begin to develop. At the same time slow percolation and solution result in solution sinks in the ground surface. Where solution beneath the surface leaves sizable cavities, collapse or subsidence forms sink holes. A few isolated sink holes exist at first, but through the years more and more of these enclosed depressions form, and a larger amount of surface drainage is trapped before it reaches main trunk streams. Small tributary streams gradually become less prominent as water originally carried by them is diverted to underground passages. This water eventually finds its way to the main streams. The water table ordinarily intersects the ground surface along the channels of these streams.

After subsurface drainage becomes well developed, caverns are abundant. Sink holes become so numerous that the sides begin to touch one another. Surface drainage becomes limited to

FIGURE 16–29
Natural Bridge, Va. The natural bridge once may have been a long tunnel. The stream, which once flowed as an underground river, is again flowing on the ground surface as a result of the collapse of the roof.
(Natural Bridge Company)

short sinking creeks. Vortexes where water swirls into cavern openings, called swallow holes, can be seen along such streams. Valleys that lead into a hillside or abruptly end where water from their streams is lost to subsurface channels are called *blind valleys.*

Subsurface drainage channels and solution cavities etch out more of the limestone and a full-scale karst topography develops. Collapse features begin to dominate the surface of the ground. *Karst windows* form where collapse of the surface over an underground stream exposes the channel. These karst windows gradually expand with the collapse of their margins and mass wasting of their sides. Eventually the depression becomes much larger, exposing the stream or caverns over a greater area. At this stage streams flow for long distances through karst tunnels and natural bridges form where tunnels partially collapse (Fig. 16–29). Mass wasting gradually causes overhanging or steep cliffs, formed where collapse has occurred, to retreat away from streams.

As the cycle nears its completion, drainage returns to the surface of the ground. The tunnels, natural bridges, and sink holes disappear. All that remains are a few more or less isolated rounded hills of limestone in which caves initially formed. These hills stand out above a relatively flat surface into which stream channels are cut. This surface may be flat if it is near regional base level, or it may be that a very resistant insoluble unit is exposed beneath the limestone. There may be more limestone, even part of the same stratigraphic rock unit, below the level of this surface on which the streams flow. In this case, further reduction has been stopped by the position of the base level, and uplift would be necessary to start the cycle over again.

CHAPTER OUTLINE

Origin of Ground Water
Meteoric water
Connate water
Juvenile water
Water Storage Underground
Sorting and size
Shape and packing
Infiltration of Ground Water
Configuration of the Saturated Zones
Perched Water Tables
Barriers
Water Tables in Contact with Salt Water
Hydrostatic Pressure
Confined Water—Artesian Conditions
Movement of Ground Water
Springs
Effects of Pumping on Ground-Water Levels
Development of Ground-Water Supplies
Quality of Ground Water
Geology of Ground-Water Occurrence
Examples of Ground-Water Occurrence
Ground-Water Supply on Long Island
Las Vegas Ground-Water Basin
Geysers
Caves
Cave Deposits
Karst Geomorphic Cycle

TERMS

artesian aquifer
artesian basin
artesian water
blind valleys
capillary fringe
caves
columns
connate water
effluent
equation of hydrologic equilibrium
flowstone
geyserite
ground-water basin
hardness
hydraulic gradient
hydraulic head
hydraulic pressure
hydrostatic head
hydrostatic level
hydrostatic pressure
influent
isopiestic lines
juvenile water
karst topography
karst windows
meteoric water
natural bridges
perched water table
permeability
piestic interval
piezometric level
piezometric surface
porosity
ppm
rimstone
safe yield
siliceous sinter
sink holes
sinking creeks
solution cavity
specific yield
stalactite
stalagmite
swallow holes
terra rosa
travertine
tufa
water budget
water table
zone of aeration

REVIEW QUESTIONS

1. Of shape, packing, and size, which is the most important in determining porosity and thus the amount of water that can be held? Explain.
2. Although clay has a porosity range 3 to 4 times as great as sandstone, a sandstone might yield more water. Why?
3. Distinguish between permeability and porosity.
4. A water table is usually a suppressed version of the local topography and may be represented by a contour map. What are some things that might affect the uniformity of the contour map locally, that is, cause large changes in a small area?
5. What are artesian water and artesian pressure? How does the latter differ from normal pressure?

6. How would it be possible for extensive irrigation in one area to contaminate the water supply for an adjacent area?

7. Why is salt-water encroachment an important problem in central and western Long Island?

8. After a cave begins to form, a sink hole may occur over it. Why is it that this speeds up the expansion of the cave into what might eventually be a large underground system?

9. Describe the concept of safe yield and explain non-geologic factors that may have to be considered in determining safe yield.

10. What steps would you recommend to help conserve water and protect ground-water supplies in the southwestern states?

SUGGESTED READINGS

Ferguson, G. E., Lingham, C. W., Love, S. K., and **Vernon, R. O.,** 1947, Springs of Florida: Florida Geol. Survey Bull. no. 31.

Heath, R. C., Foxworthy, B. L., and **Cohen, Philipp,** 1966, The changing pattern of ground-water development on Long Island: New York, U.S. Geol. Survey Circ. 524, 10 p.

Livesay, A., 1953, Geology of the Mammoth Cave National Park area: Kentucky Geol. Survey Spec. Pub. no. 7.

Meinzer, O. E., 1923, Outline of ground-water hydrology: U.S. Geol. Survey Water-Supply Paper 494, 71 p.

Overman, M., 1969, Water, Solutions to a Problem of Supply and Demand: Garden City, Doubleday, 192 p.

Todd, D. K., 1967, Ground-Water Hydrology: New York, John Wiley, 336 p.

Tolman, C. F., 1937, Ground Water: New York, McGraw-Hill, 593 p.

17 GLACIERS AND GLACIATION

17 GLACIERS AND GLACIATION

HISTORICAL BACKGROUND

Alpine glaciers attracted the attention of naturalists as early as the first part of the eighteenth century. J. J. Scheuchzer in 1723 explained the movement of ice in alpine valleys as resulting from the expansion of water in cracks in the glaciers as it freezes. Most of his contemporaries thought the only ice movement was sliding downslope. Many who lived near Alpine glaciers were familiar with the ridges of debris at the front edge of glaciers. Some recognized the *erratics,* or blocks of rock with polished and scratched surfaces that the glaciers had moved far from their site of origin. But evidence was not formally presented until 1821 when Ignaz Venetz, an engineer, read a paper at the Swiss Society of Natural History in which he detailed evidence that the glaciers in the valleys of the Swiss Alps had formerly occupied positions much farther down

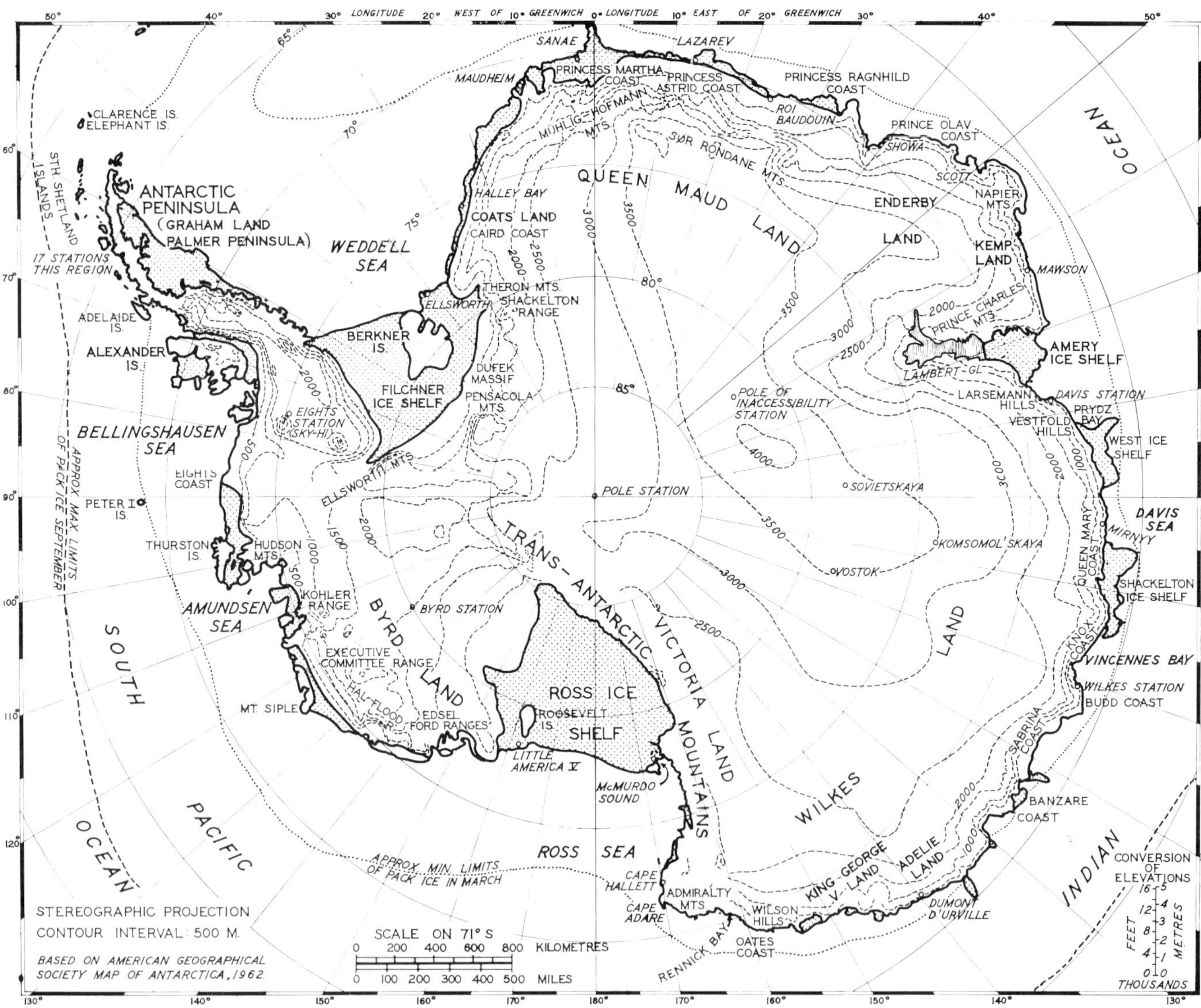

valley than their present margins, and a few years later he suggested that all of northern Europe had been covered by ice. During the next 30 years other scientists became convinced of the validity of Venetz's idea. Notable among these was a young man named Louis Agassiz, who realized the implications of the widespread glacial deposits and advanced the idea of the "ice age." K. Schimper, a friend of Agassiz, was led to the idea that thick ice had once covered large areas in Europe by the marked changes he found in fossil flora and fauna, but Schimper believed in the "iceberg theory"—that glacial deposits melted out of frozen sea water. Agassiz identified the vast extent of glacial deposits such as moraines and erratics in Europe and later in North America. The development of glacial theory was rapid in the middle nineteenth century, but it was much later before the extent of modern-day ice caps (Figs. 17–1 and 17–2) was known, and ideas concerning the cause of ice ages are still open to debate.

The present is an age of glaciation. Nearly 10 percent of the land area of the earth is now covered by ice. Almost 1.5 percent of the earth's water is now in the form of ice. Glaciers occupy valleys of many mountains in high latitudes, and some glaciers, like Kilimanjaro (elevation 5,968 meters), are located in equatorial latitudes. A thick sheet of ice covers most of one entire continent, Antarctica, and other sheets cover a large part of Greenland, Iceland, and islands in the Arctic Ocean.

For many years the glacial deposits that are so prominent in northern Europe as well as bedrock containing marine fossils were thought to be debris left by the Flood described in the Bible in the time of Noah. If we could not actually walk across these glaciers, watch them in action, and see deposits forming at their margins, it is doubtful that we would ever believe that the semiconsolidated sand, silt, and morainal deposits that cover the northern United States and northern Europe had been left there by an advancing sheet of ice. We probably would not consider accumulation of such large bodies of ice a reasonable possibility.

FIGURE 17–1
(Opposite) map of Antarctica.
(from Harrington, 1965)

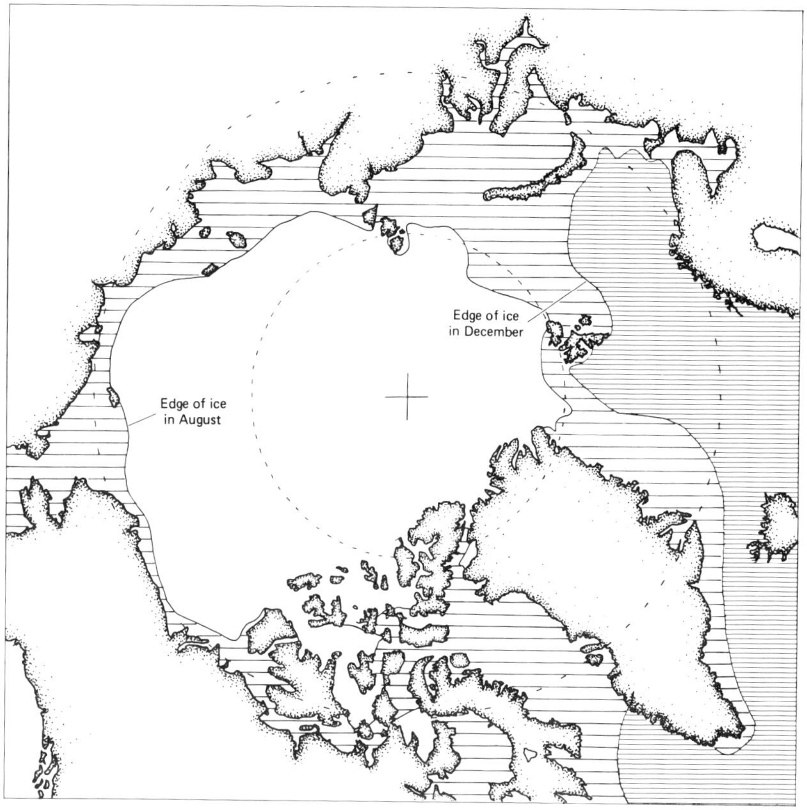

FIGURE 17–2
Ice conditions in the Arctic Ocean. The borders of the areas covered by 80 percent or more ice in early August (clear) and the areas covered by 80 percent or more ice in December (widely spaced lines) are shown.
(after U.S. Navy Hydrographic Office map)

Only four or five periods of glaciation are known during most of the history of the earth and each of these is thought to have lasted only a few million years. Thus the present is an unusual time in this respect. Although glaciers have been present during only a small part of the history of earth, they have been extremely important in producing the present configuration of our landscape. Most of the present topography of the world has formed within the past few million years, and for about the past 3 million years ice has existed in large quantities at the poles and in mountains all over the world. The last major advance of ice sheets reached a culmination about 18,000 years ago. At that time 30 percent of the land surface of earth was covered by ice, and about 8 percent of the water was in the form of ice. Deposits and erosional effects of this glacial advance have left a clear and well-documented record of the immense importance of glaciation as a process of gradation in this last period of geological time, which is called the Quaternary.

TYPES OF GLACIERS

Glaciers are conveniently classified according to their form, geography, and temperature. Glaciers may be divided into two main groups on the basis of their form and associated features: (1) *valley glaciers* and (2) *ice sheets*. The main difference between them is that valley glaciers are confined to valleys, generally in mountains, while ice sheets may reach continental dimensions and bury whole mountain ranges. Glaciers from several valleys may coalesce along a mountain front, forming an intermediate type of glacier called a *piedmont glacier*. The Malaspina Glacier in Alaska (Figs. 17–3 and 17–4) is of this type. Piedmont glaciers differ from ice sheets in that they form by joining of ice streams and are not the sources from which valley glaciers flow.

A Swedish glaciologist, Hans Ahlmann, first suggested classifying glaciers according to temperature. Temperature is an important determinant of wastage and accumulation of glacier ice, the character of the glacier terminus, and its ability to erode bedrock. Ahlmann proposed a twofold division: (1) temperate (warm) glaciers and (2) polar (cold) glaciers. The wisdom of this distinction lies in the effects temperature has on bedrock erosion by a glacier, the character of the lower end of the glacier, and the movement of the ice. Warm ice flows, yields to stress, and recrystallizes more readily than cold ice. Even the internal structure and texture of the two types are different.

Temperature of glaciers is primarily determined by climate. They are affected by insolation and radiation from the ice, by air temperature, wind, evaporation, and percolation of water. The upper temperature limit is the freezing point, 0°C, but glaciers in Greenland and Antarctica may have temperatures 10–20°C lower than that. Most glaciers are subjected to widely varying temperatures at their surface as a result of seasonal variations, but conduction is negligible once the upper few inches are at 0°C. Warming of the deeper portions of glaciers depends primarily on the percolation and refreezing of melt-water. Heat is liberated when water freezes (80 calories per gram), and this heat raises the temperature of the surrounding ice. Thus a warm glacier may be at 0°C at the surface and only slightly colder at the bottom. A cold glacier is very cold, –10°C or less at the surface, and becomes warmer at depth, but not as warm as a warm glacier.

The prime mover in all glaciation is the force of gravity. Like valley glaciers, ice sheets move downslope, but if their volume is sufficient they also flow over extensive flat regions and even over irregularities. In part, this is accomplished through the push exerted on the front of the lobe of ice by ice moving downslope behind it.

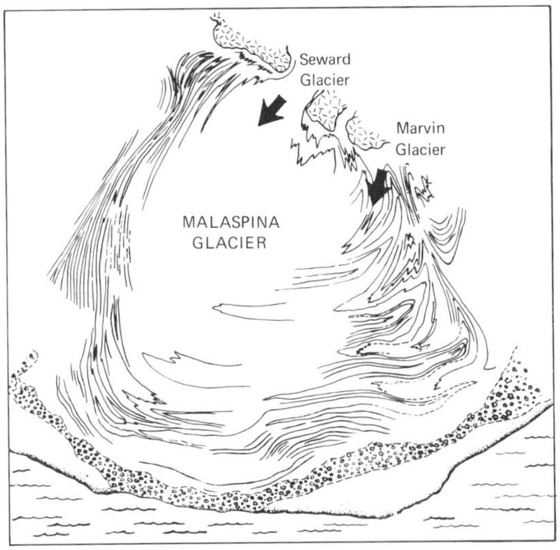

FIGURE 17–3
Malaspina Glacier, Alaska. The map view shows folds that have developed in the ice as it flowed. Note the position of Seward and Marvin glaciers that feed into Malaspina Glacier.
(after Sharp, 1960)

THE POLAR ICE MASSES

At the present time most of the land and water area covered by ice is located in polar latitudes above latitude 70° North and South (Figs. 17–1 and 17–2). The polar regions have great influence on the climate and weather of the rest of the world and the circulation in the ocean basins; they are of importance in the study of the earth's magnetic field and phenomena related to it; and they may hold valuable mineral resources. Already the edge of the Arctic basin is being explored for oil and gas. Discoveries on the northern slope of Alaska have prompted some to predict oil reserves that will rival those of Kuwait. Some geologists expect the Arctic Islands to be among the most promising oil- and gas-producing areas in the next few decades.

About 88 percent of the world's ice is contained

FIGURE 17–4
Malaspina Glacier, Alaska.
(U.S. Air Force)

in the Antarctic ice sheet. This mass of ice, about 13.5 million square kilometers, covers all but a small percentage of the area of Antarctica. The thickness of the ice is highly varied (Fig. 17–5). Most thickness determinations are made by seismic methods (see Chapter 21), but recently a method of radio sounding from low-flying aircraft promises to make future determinations much simpler. The ice in Antarctica is more than 4,000 meters thick in the thickest part. Mountain ranges, one 1,500 meters high and several hundred kilometers long, have been found buried beneath ice as much as 1,500 to 3,000 meters thick. Bedrock under a large area in the western part of Antarctica is beneath sea level. The shape of Antarctica would resemble that shown in Fig. 17–6 if the ice were instantly removed. The lowest point would be 1,800 meters below sea level. The eastern part of the continent consists of a buried plateau that stands about 2,000 meters above sea level.

Glacier ice from the Antarctic ice sheet has flowed out into the sea, producing ice shelves, some reaching 1,000 meters in thickness. The Ross

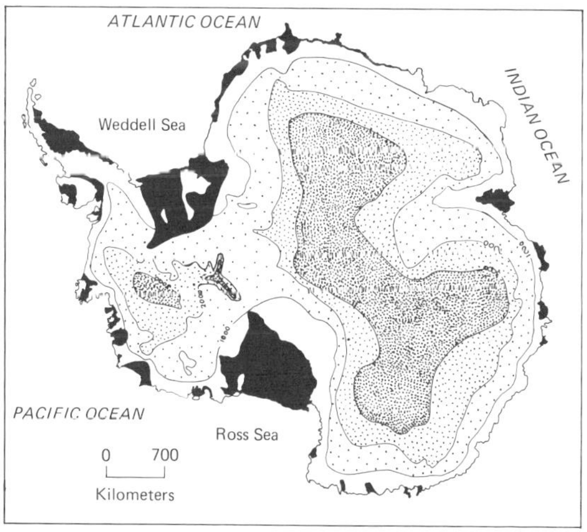

FIGURE 17–5
Generalized map of ice thickness in Antarctica. Black areas show shelf ice. Thicknesses are indicated as follows: white, 0–1,000 meters; lightest stippling, 1,000–2,000 meters; medium stippling, 2,000–3,000 meters; heavy stippling, 3,000+ meters.
(after a map by Thiel, 1962)

and Filchner seas are covered in this manner. In addition *pack ice,* floating ice consisting of pieces of ice driven together (Fig. 17–1), in the Antarctic Ocean covers most of the sea surface above 70°S latitude. The maximum limit of pack ice at the

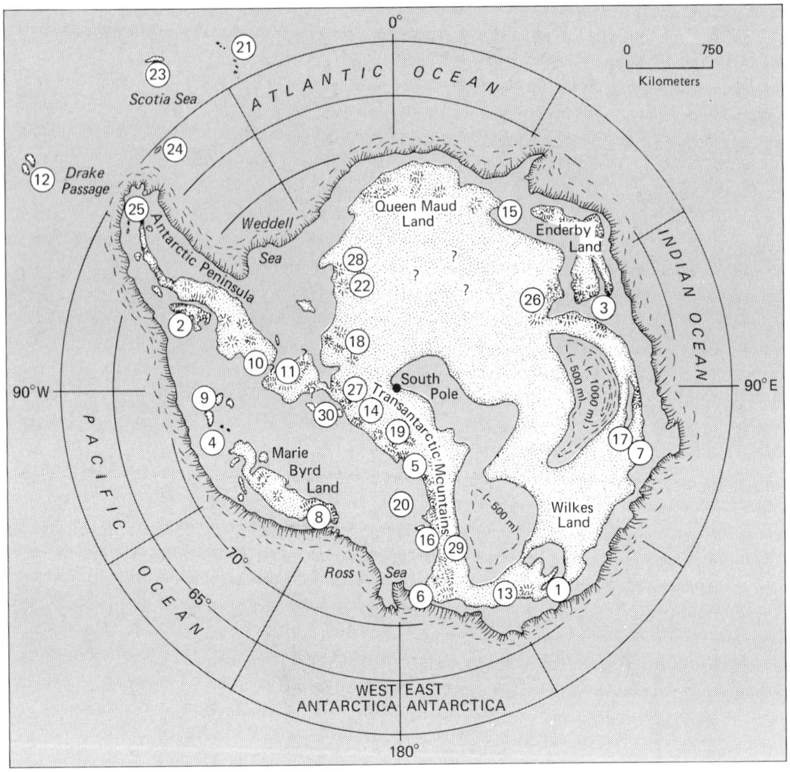

FIGURE 17–6
Physiographic diagram of Antarctica with ice cover removed. This map is slightly modified from a U.S. Geological Survey Map (1962) of the bedrock surface, unadjusted for isostatic uplift. Hachures show continental slope and major mountain areas; stippled regions are above sea level. In alphabetical order, numbered features are as follows: (1) Adelie Coast, (2) Alexander Island, (3) Amery Ice Shelf, (4) Amundsen Sea, (5) Beardmore Glacier, (6) Cape Adare, (7) Denman Glacier, (8) Edsel Ford Ranges, (9) Eights Coast, (10) Ellsworth Land, (11) Ellsworth Mountains, (12) Falkland Islands, (13) George V Coast, (14) Horlick Mountains, (15) Lutzow-Holm Bay, (16) McMurdo Sound, (17) Mount Sandow, (18) Pensacola Mountains, (19) Queen Maud Range, (20) Ross Ice Shelf, (21) Scotia Arc, (22) Shackleton Range, (23) South Georgia, (24) South Orkney Islands, (25) South Shetland Islands, (26) Southern Prince Charles Mountains, (27) Thiel Mountains, (28) Theron Mountains, (29) Victoria Land, and (30) Whitmore Mountains.
(from Ford, 1964)

end of winter in September covers most of the ocean south of 65°S latitude and projects above 60°S latitude north of the Palmer Peninsula. Pack-ice conditions in the Arctic Ocean are illustrated also in Figs. 17–2 and 17–7.

ICE-BUDGET CONCEPT

The ice budget is of great interest in Antarctica because we may predict the trend of the world's climate from it. The budget concept is an attempt to account for the balance between moisture supplied to a glacier or ice cap and that lost from it. The budget may be balanced or show a net loss or gain. A gain of snow and ice signifies accumulation and growth of the glacier; a loss signals wastage and decay. Estimates of the supply and loss of ice and snow over such a vast area are extremely difficult. However, it has been estimated that the total annual accumulation of ice and snow over the Antarctic continent is 1,100 cubic kilometers (in water equivalent). That is a low value compared with the water supplied to other continents. The highest accumulation rates (the equivalent of about one meter of water) occur along the coastal areas. The losses are due only in minor part to melting, as we might expect in an area with a mean annual temperature of −30°C. Ice is lost by sublimation from the ice surface, but the greatest loss is from the breaking away of ice from the ice sheet where it protrudes into the sea. Icebergs (Figs. 17–8, 17–9, and 17–10) are formed in this way. Ice is also lost as a result of melting at the bottom of the ice shelves.

The recent estimates indicate that the Greenland ice sheet has a nearly balanced budget (−100 cubic kilometers per year water equivalent) and the Antarctic ice sheet is growing (+1,100 cubic kilometers per year). This is something of a surprise since sea level has been rising in recent geologic time (see Chapter 9).

FORMATION OF GLACIERS

A sheet of ice covered most of Canada and over a third of the United States during the last glacial advance. It extended southward into the United

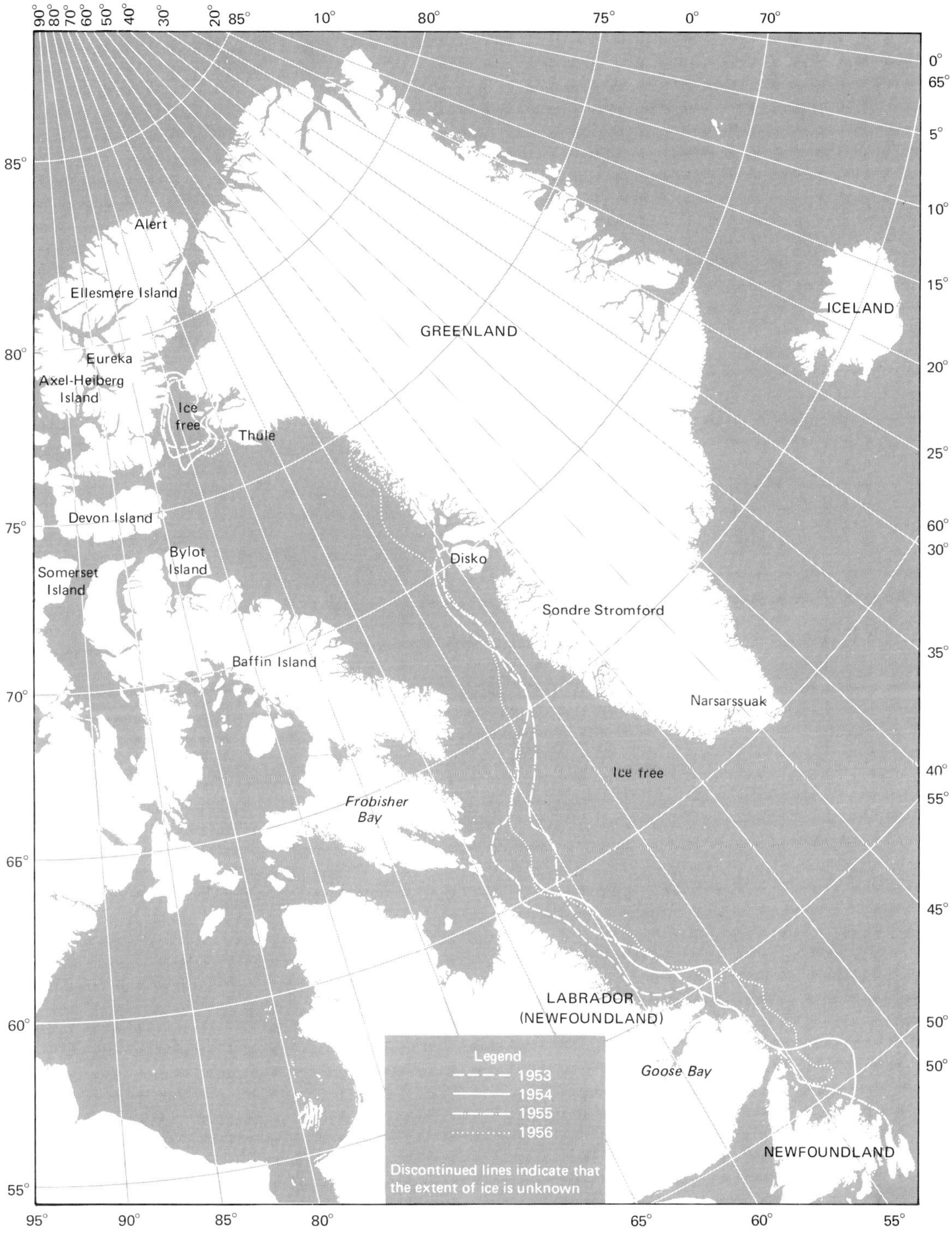

FIGURE 17–7
Comparison of ice-pack boundaries in Baffin Bay and Davis Strait in mid-May for the years 1953 through 1956.
(U.S. Navy)

FIGURE 17–8
Icebergs frozen in sea ice.
(U.S. Navy)

FIGURE 17–9
Icebergs calving from the margin of a continental glacier in the Antarctic. Notice how these are breaking out along a set of perpendicular fractures.
(U.S. Navy)

States, where its terminus is usually marked by ridges of debris deposited near the front of the advancing ice; these are called *moraines.* Moraines form a ridge on Long Island, cross Staten Island into New Jersey, and features formed at the margin of the ice can be traced across Pennsylvania to the Ohio River. The southern margin of the ice cap is marked approximately by the position of the Ohio and Missouri rivers. Because no ice sheet occupies central Canada today, we are prompted to wonder in what way conditions differed when ice did cover such extensive areas from conditions today, and if there is any chance that glaciers might return. We do not have a definite answer to the second question, but it appears that the only condition necessary for the growth and expansion of glaciers is that the amount of snow and ice accumulation must exceed the amount of ice wastage through melting and sublimation for an extended period of time. Growth of glaciers is accelerated by increased precipitation in the form of snow and sleet and temperature low enough for the ice to be preserved throughout the year. The extremely low temperatures at Antarctica, which average about −30°C, are certainly not necessary for formation and growth of continental glaciers. In fact, glaciers are most likely to grow when temperatures are only slightly below freezing, because air at that temperature can hold much more moisture than can air at extremely low temperatures. If the average annual temperatures over the face of the earth dropped several degrees, it would almost certainly bring about formation of large masses of ice on the continents. These would form first in high mountains, where temperatures are now nearly low enough to promote large-scale mountain glaciation. Accumulation occurs where snow is permanent throughout the year. The lower limit of this area of perennial snow is called the *snow line.* The altitude of the snow line fluctuates over a period of years, reflecting changes in temperature and amount of snowfall.

The area above the snow line where snow re-

FIGURE 17–10
Left, icebergs. Most of the ice is beneath the water. In order to float the ice must displace a volume of water equal in weight to that of the ice.
(U. S. Air Force)

mains from year to year is called a *snow field* (Fig. 17–11). Snow fields of ice sheets are vast, with accumulations of snow and ice completely burying all topographic features. This must have been the condition during the recent ice ages in central Canada where relief is low. In mountainous regions, however, the snow fields may lie below the summit levels of peaks. This is true of the present snow fields in the northern Rockies and in some of the mountains of Antarctica. Peaks or isolated hills that protrude through the snow fields or glaciers are called *nunataks*.

Formation and Movement of Glacier Ice

Snow is a crystalline form of water and is, consequently, a mineral. Snowflakes (Fig. 17–12) display an almost infinite variety of delicate and beautiful hexagonal patterns. If the temperatures remain below freezing after a snowfall, individual flakes recrystallize. The extremities of the delicate patterns melt and the water moves toward the centers of the flakes until all the ice is recrystallized as small grains. This partially consolidated snow and granular ice is called *firn*. As more snow is added, pressure of the overlying snow and firn compacts the ice grains together at depth. The rounded grains of firn tend to rotate and move downslope under the weight of the overlying snow. Large crystals of ice may grow as grains of ice become rotated until the atomic arrangements within adjacent grains have the same orientation. Then one grain fuses into another, becoming a single crystalline mass. Once the thickness of overlying snow and firn reaches about 60 meters, the weight is so great that the firn compacts into a solid mass of ice. Movements of the ice layers are accomplished in part by slippage between layers of atoms, primarily by gliding on a crystallographic plane within the crystals. This gliding mechanism is accompanied by recrystallization. Movement is also accomplished by shearing along planes through the ice. Thus deformation deep in a glacier is thought to involve both external and internal movements of the individual ice crystals and recrystallization. The net effect is flowage with characteristics similar to the flowage of plastics and of viscous liquids, a type of behavior referred to as *plastico-viscous* or *pseudo-plastic*.

FIGURE 17–11
Tongues of ice descending from an ice cap into a large U shaped valley at the head of a fiord
(U.S. Air Force)

FIGURE 17–12
Enlargement of a snowflake.
(U.S. Weather Bureau)

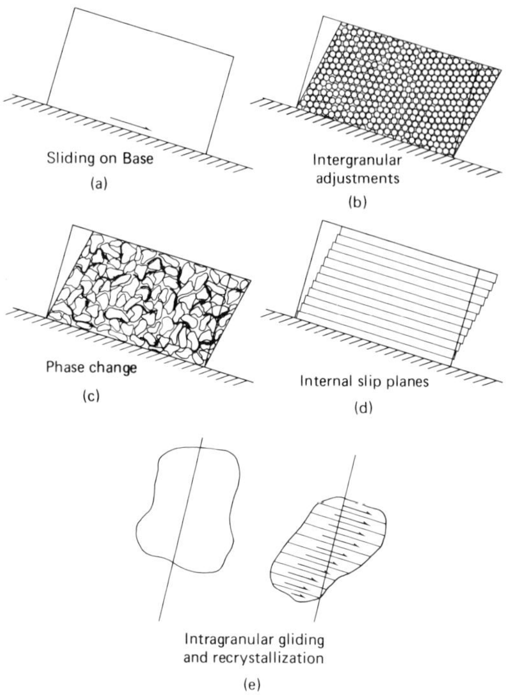

FIGURE 17–13 Sketches illustrating various possible mechanisms of glacier movement. (from Sharp, 1960)

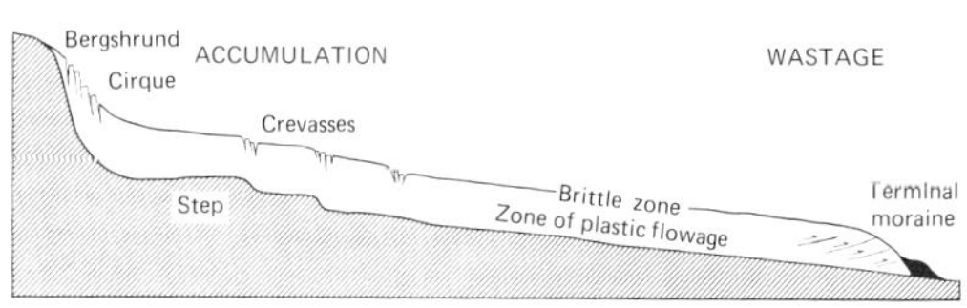

FIGURE 17–14 Longitudinal cross section of a glacier. This section has a schematic representation of the zones of movement, accumulation, wastage, and other features associated with glaciers.

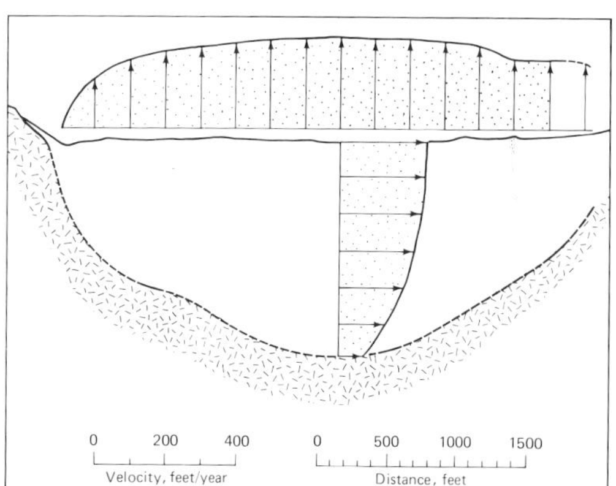

FIGURE 17–15 Measured transverse surface-velocity profile (horizontal component) on Saskatchewan Glacier and calculated vertical velocity profile. (after M. F. Meier, 1960)

Thus movement of snow and ice in a glacier is the combined effect of sliding of the ice over bedrock, intergranular rotations of the firn, movement of water as it recrystallizes into ice, phase change, and slip, both within ice crystals and by shearing through the ice (Fig. 17–13).

The main characteristic of a plastic is that it will deform continuously once a certain threshold value of load is applied to it. Ice behaves as a brittle, elastic solid until the pressure, or load, on it is equal to the weight equivalent of 60 meters of ice. At that load it will flow continuously. The ice sheets that covered Canada accumulated to a great thickness, probably comparable to that in the Antarctic where the ice is several thousand meters thick over large areas. Once the thickness became great enough, pressure on lower layers from the weight of the overlying mass caused pseudo-plastic flowage near the bottom of the ice sheet. Thus the sheets slowly spread out over irregularities.

Because the behavior of ice changes as load is increased, a typical glacier has an upper brittle crust 45 to 60 meters thick and a lower pseudo-plastic zone (Fig. 17–14). Fractures and crevasses form in the upper brittle portion of a glacier as it moves. These crevasses open and close as the ice is sheared, compressed, and extended.

If snow fields form in mountains pre-existing stream drainages become the course of least resistance for the ice as accumulations reach the thickness necessary to initiate pseudo-plastic flow. Then tongues of ice begin their slow movement down the valleys as valley glaciers. How fast and how far these glaciers move depend on the rate of accumulation of snow in the snow fields, the temperature, the shape of the surface over which the ice moves, the thickness of the ice, and, of course, the rate of wastage at the terminus of the glacier.

Movement Patterns in Valley Glaciers

Many existing valley glaciers are moving at slow but measurable rates. One early technique used to study this motion was to drive a series of stakes across the surface of the valley glacier in a straight line and come back periodically to observe the position of these stakes. The distribution of movement is often similar to that in a stream. Frictional drag along the contact between ice and the rock at the sides and the bottom retards movement there. Ice at the center of the glacier moves along much more rapidly (Fig. 17–15). The maximum velocity of most glaciers is near the surface. Frictional drag of air on ice is negligible. The variation of velocity with depth has been determined for some glaciers by measuring the rate of deflection of an originally vertical pipe placed in them. Some glaciers move so slowly that trees and other plants grow in the debris on top of them; others move at rates as high as 10 meters per day or, in exceptional cases, over 30 meters per day. Extremely rapid movements of glaciers are called *surges.* Hassanabad Glacier in the Karakoram advanced 10 kilometers in 2.5 months—nearly 130 meters per day. That the margins of ice sheets are in motion is shown by the continual splitting off of masses of ice where the sheet extends into the sea. This process is called *calving.* The big blocks of ice broken off in this way form *icebergs.*

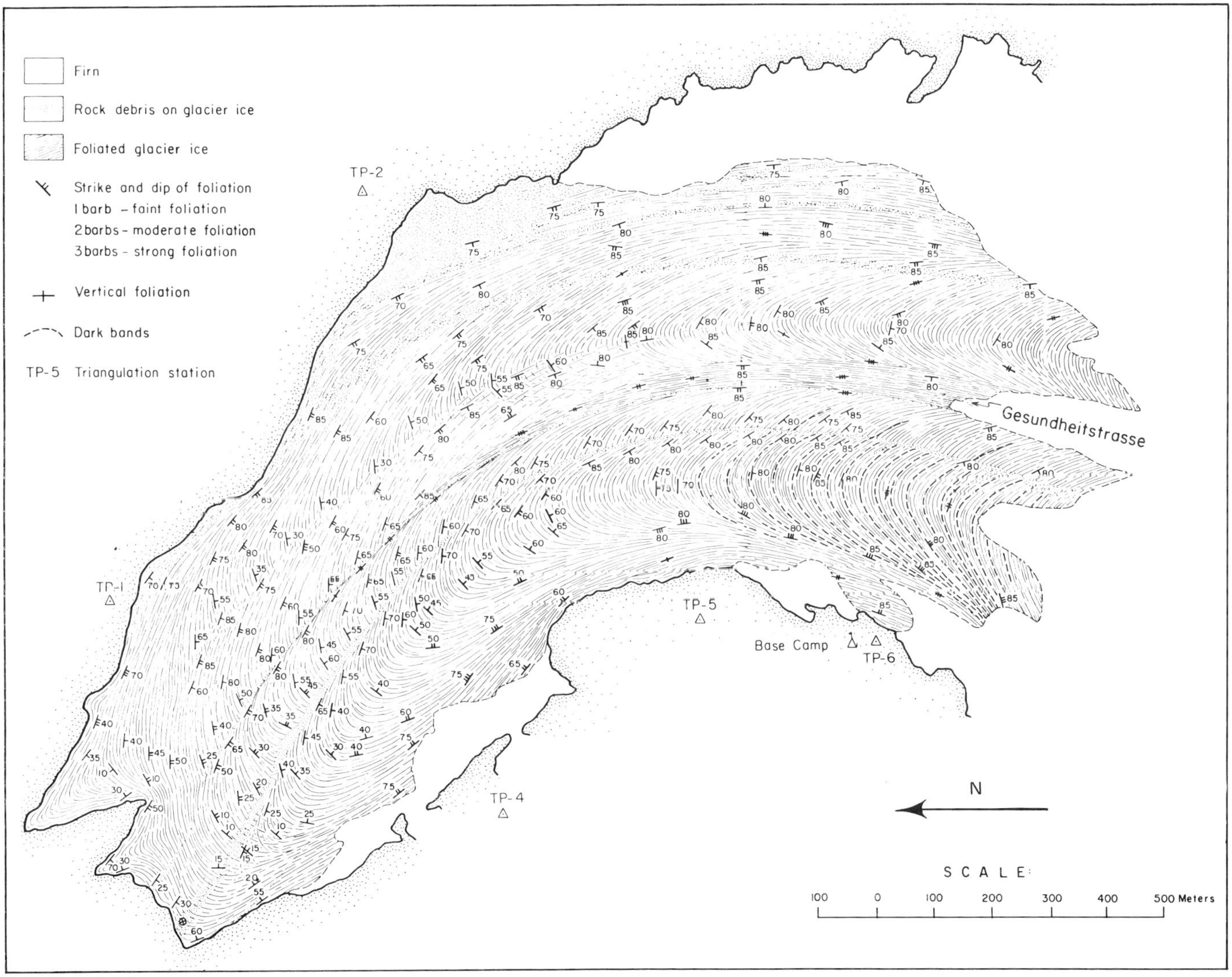

FIGURE 17–16
Structures exposed on surface of Blue Glacier below the firn edge. Numbers in glacier area indicate angles of dip.
(from Sharp, 1960)

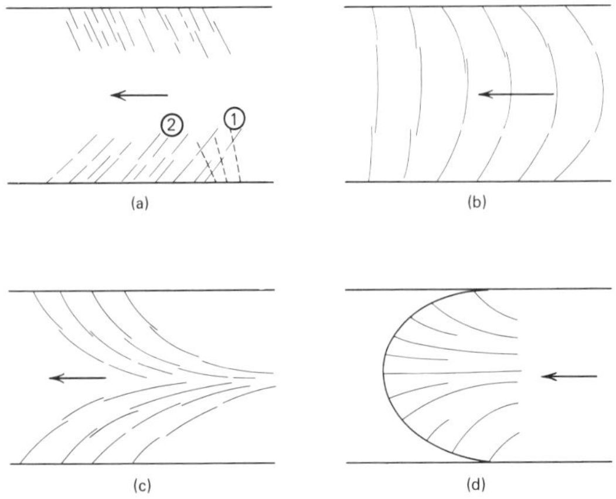

FIGURE 17–17
Principal types of crevasses in valley glaciers. (a) Marginal, (1) old rotated crevasses, (2) newly formed crevasses; (b) transverse; (c) splaying; (d) radial splaying. Arrow indicates flow direction.
(from Sharp, 1960)

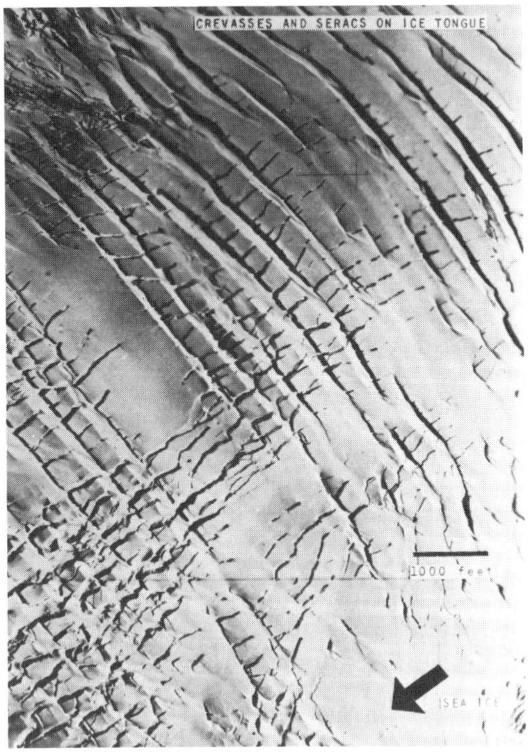

FIGURE 17–18
Crevasses developed on an ice tongue in the Antarctic.
(U.S. Navy)

Other indications of movement pattern in glaciers are provided by the arrangement of crevasses, the pattern of foliation in glacier ice, and the form taken by debris that gets onto the glacier surface, such as lateral and medial ridges. Flowage of the ice of the Malaspina Glacier, Alaska (Fig. 17–3), is beautifully depicted by the pattern assumed by lateral and medial ridges of debris that flow into the glacier, a piedmont glacier, from its tributary valley glaciers such as the Seward and Marvin. These debris ridges assume zigzag patterns as the ice in the Malaspina Glacier spreads laterally, unconfined by valley sides.

The foliation pattern in glacier ice of the Blue Glacier of Washington has been mapped (Fig. 17–16) by Robert Sharp. The pattern suggests the streamlined flow of ice with higher rates of movement in the center of the two valleys clearly shown.

The principal types of crevasses in valley glaciers are depicted in Fig. 17–17. These are rarely found together. Most of them result from extension of the ice; the break forms at right angles to the direction of extension of the ice mass (Fig. 17–18).

EROSION BY GLACIERS

Evidence of the great erosive capacity of glaciers is seen in the large quantities of debris deposited from them, in the changes they cause in glaciated terrains, and in the abrasion, plucking, and polishing of bedrock surfaces they have moved over.

Incorporation of debris in glaciers. Most glaciers carry heavy bed loads and erode the bedrock below them. Glaciated valleys are deepened and irregularities are smoothed out and removed under the ice. Debris becomes incorporated in the base of the ice as a result of freezing of debris into the ice, frictional drag of blocks under the ice, wedging of rock by water freezing in cracks in bedrock, and because ice near the bottom of the glacier is pseudo-plastic, it can flow into irregularities around rocks. The processes by which pieces of bedrock are lifted out of place are called glacial *plucking* or *quarrying*. Fractures and bedding in bedrock play an important role in this process. Fractures are opened by frost wedging and they make removal of bedrock blocks possible.

Abrasion. Bedrock surfaces in glaciated regions are usually abraded and polished. The surfaces may bear striations and grooves aligned in the direction of flow of the ice, *chatter* marks caused by the removal of chips of bedrock, *crescentic fractures* concave downstream, and *crescent-shaped gouges* where pieces of rock have been removed (these gouges may be lunate—concave downstream—or crescentic—concave upstream). Movements of ice sheets are indicated by alignments of striations over a large region as indicated for the northeastern United States in Fig. 17–19.

The combined effects of abrasion and removal of debris and bedrock produce streamlined topographic features like those shown in Fig. 17–20 in New Zealand. The alignment of lakes on the bedrock surface in Canada (Fig. 17–21) reveals the direction of glacier movement in that area.

Mass movement. Valley sides are scraped and blocks broken off are frozen into the ice and carried away by valley glaciers. This leads to undercutting of the sides of the valley and produces unstable slopes. Slumping, sliding, and debris avalanches bring great quantities of debris onto the top surface of the glacier.

Scraping and pushing. The front edges of glaciers may push and scrape material in front of them. This is particularly effective in removing soil and semiconsolidated sediments.

FIGURE 17–19
Striation orientations in the northeastern United States. End moraines are shown in black. (after the Glacial Map of North America, 1945)

METHODS OF TRANSPORTATION

Mass wasting is accelerated by increased frost action accompanying glaciation. Large quantities of talus fall on the margins of the valley glaciers. Much of this load is carried on the surface of the ice. Some of the debris may remain in the ice from the head to the terminus of the glacier, but much of it slowly works its way down, particularly through crevasses, toward the bottom of the ice. This load plus the debris frozen into the bottom of the glacier ice leads to accumulation of a heavy load near the bottom of the ice mass. This part of the load is responsible for the abrasive action of the glacier. The remainder of the load is retained in the body of the glacier, and some load is pushed as the advancing ice overrides new ground.

FIGURE 17–20
Topography grooved and smoothed by movement of glacier, South Island, New Zealand.

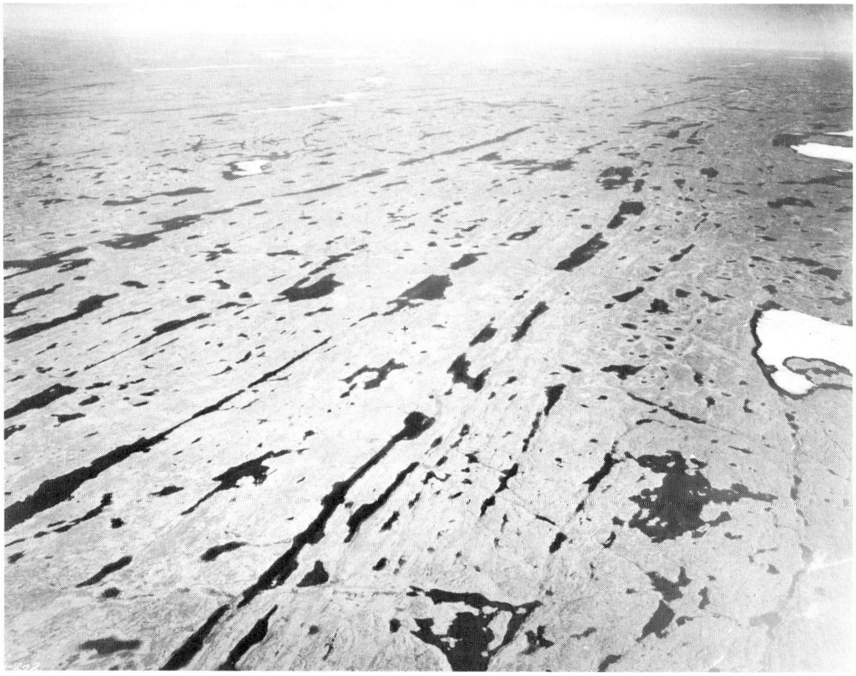

FIGURE 17–21
Air photograph showing a surface of Precambrian crystalline rocks eroded by continental glaciation in Canada.
(Royal Canadian Air Force)

FIGURE 17–22
The island in the foreground is a drumlin.
(Royal Canadian Air Force)

Continental glaciers carry little surface load because once the sheet has become thick enough to flow over flat areas its top surface is usually above all but the highest mountain peaks. Thus valley glaciers carry a much greater load per unit volume of ice than do ice sheets.

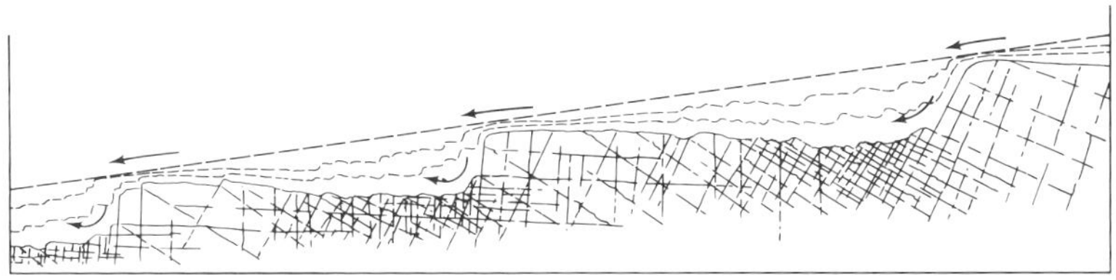

FIGURE 17–23
Longitudinal profile of a glacial valley showing selective quarrying due to fracture intensity as postulated by F. E. Matthes, 1930.
(from Hill, 1964)

FEATURES PRODUCED BY EROSION

Many small-scale erosional features are common to continental and valley glaciers. Striations, grooves, and polished surfaces are found both on the bare rock over which the ice moved and on the rocks carried in the bottom of the ice.

Large-scale erosional features produced by ice sheets are not as distinctive as those produced by valley glaciers. Ice sheets tend to streamline preexisting topography, removing irregularities like peaks and sharp ridges (Fig. 17–20). Rounded hills that have been streamlined by glacial abrasion and scouring, called *roches moutonnée,* are one of the features found where valley or, more commonly, continental glaciers have passed. These hills, often largely solid rock, are asymmetrical with a smoothed and streamlined *stoss* side and another steeper and more angular side produced by glacial quarrying. The effect is to repress any land forms that might stand out. *Drumlins* are one of the streamlined features shaped by continental sheets (Fig. 17–22). They are hills shaped like the back of a whale; the steeper slope faces the direction from which the ice moved. Drumlins are composed of *glacial till*—that is, unconsolidated materials such as soil and rock mixtures deposited by glaciers—and are shaped by movement of the glacier over the till. They are products of combined erosion and deposition, but the relative importance of erosion and deposition is still debated.

FIGURE 17–24
High oblique view of the Beartooth Mountains, Mont. Note the plateau surface, 11,000 feet (3,300 meters) high. This ancient, nearly flat erosion surface has been dissected by glaciers, but now mass movement is one of the main processes tending to change its appearance. High piles of talus may be seen on the sides of the valleys. These converge at the bottom of the valleys to form rock glaciers that slowly move downhill. The tops of the plateaus are strewn with frost-heaved boulders.
(U.S. Air Force)

Erosional Features of Valley Glaciers

Cirques. The heads of most valley glaciers occupy hollowed-out amphitheater-shaped features called *cirques* (Figs. 17–23 and 17–24). The cirque is the site of glacial quarrying, or plucking, where the glacier starts its movement down the valley. Walls of cirques produced by this quarrying action become precipitous and may drop almost vertically

FIGURE 17–25
Sketches of some of the important geomorphic features that result from erosion and deposition.

thousands of meters before they curve out into the floor of the valley. Cirques usually form at the head of the glacier near former drainage divides of streams that predated the glaciers. Some such valley heads were undoubtedly sites of steep slopes and steep stream gradients before glacial erosion started. The glacier steepens these slopes further by removing soil, talus, and bedrock. But cirques also can develop on the valley sides where no stream valleys preceded them. The bottom of the cirque is often gouged out to produce a depression in the valley floor. After the glacier melts, a small lake, called a *tarn,* usually occupies this depression.

The natural course for a glacier to follow is the old drainage pattern through the mountains. Thus glaciers begin by following the line of least resistance, but the flow and erosion methods of streams and glaciers are so different that stream valleys are not suited for optimum movement of glacier ice. Where ice is forced around sharp curves, mountain spurs that extend into the major valley between stream tributaries are cut back. These features are called *truncated spurs* (Fig. 17–25). Ice streamlines the valley, cutting off or smoothing out curves and removing obstacles. It cuts out a U-shaped valley (Figs. 17–27 and 17–28) instead of the V-shaped valley characteristic of mountain streams.

The longitudinal profile taken along the valley floor of many valley glaciers shows a steplike pattern (Fig. 17–23). The cirque itself may show up as a step if the valley glacier is one flowing off of a high plateau (Fig. 17–24), but other less prominent steps are common. The origin of these steps has been interpreted in several ways. F. E. Matthes (1930) suggested that steps are formed where the intensity of fracturing in the bedrock of the valley floor is greatest, facilitating removal of rock by plucking and quarrying processes.

Valleys of tributary streams also become sites of glaciers, but these tributary valley glaciers are smaller than the major glacier, accomplish erosion at a slower rate, and do not cut their valleys as deeply as that of the glacier to which they are tributaries. When the glaciers recede, valleys of the tributary glaciers are exposed. If they open into the larger valley above the valley floor, they are called *hanging valleys.* Waterfalls or cataracts

FIGURE 17–26
Snowfields, cirques, valley glaciers, arêtes, cols, lateral and medial debris on the glacier, and horns are depicted in this photo.
(Royal Canadian Air Force)

FIGURE 17–27
The Sierra Nevada Mountains, Calif. The large valley in the center is a glaciated valley.
(U.S. Geological Survey)

FIGURE 17–28
Modern lake created by glacial moraines. Note smooth valley sides. South Island, New Zealand.

FIGURE 17–29
Glaciated valley showing fiord in foreground and U-shaped valley, Fiordland, New Zealand.

FIGURE 17–30 Sharp ridges and cols cut into the northern Rocky Mountains in Canada by valley glaciers.
(Royal Canadian Air Force)

(Fig. 17–27) develop where renewed streams flow out of the hanging valley into the larger valley.

Fiords. *Fiords* (Fig. 17–29) are glaciated valleys, the floors of which are submerged below sea level, some as much as several thousand meters. They are formed where a valley glacier flows from the mountains into the sea. Glacier ice is able to carve its valley below sea level before the ice is melted in the sea water or breaks away to form icebergs. Usually the valley becomes shallower and a threshold is found near the coasts. Today these large valleys on the continental shelf are filled with sea water. This is due partly to rises in sea level since the last major glacial advance and partly to the glacier's cutting its valley below sea level. The coasts of Norway, Greenland, British Columbia, Patagonia, and southwest New Zealand are famous for their spectacular fiords.

Arêtes (Fig. 17–30). *Arêtes* are sharp ridges produced by glacial erosion. Rugged ridges form along the crest of a mountain range, between two mountains, and between mountain spurs that separate cirques. Plucking and quarrying at the head of the glacier may lead to enlargement of the cirque and movement of the cirque walls toward the divide between adjacent valleys. Eventually the cirque walls may come together closely enough to produce an arête.

Cols. A *col* (see Fig. 17–25) is a pass formed where two cirques converge, cutting into the same wall and lowering it below the level of the remaining summit area.

Horns. The sharp peaks that project above the surrounding area and are shaped by glacial erosion are called *horns* (Fig. 17–31). The Matterhorn in Switzerland is a classic example. Horns are usually produced where several valley glaciers head in the same area. As the cirques are enlarged and converge, an isolated peak more or less surrounded by cirque walls is produced. Horns are formed during a late stage in glaciation after the summit of the preglacial mountains has been reduced to a few isolated peaks. Horns are often aligned, since they may be remnants of an arête.

FIGURE 17–31
The Matterhorn, the classic example of a horn. Isolated peaks such as this give us an idea of how high the surrounding country was before erosion cut it down.
(Trans World Airlines, Inc.)

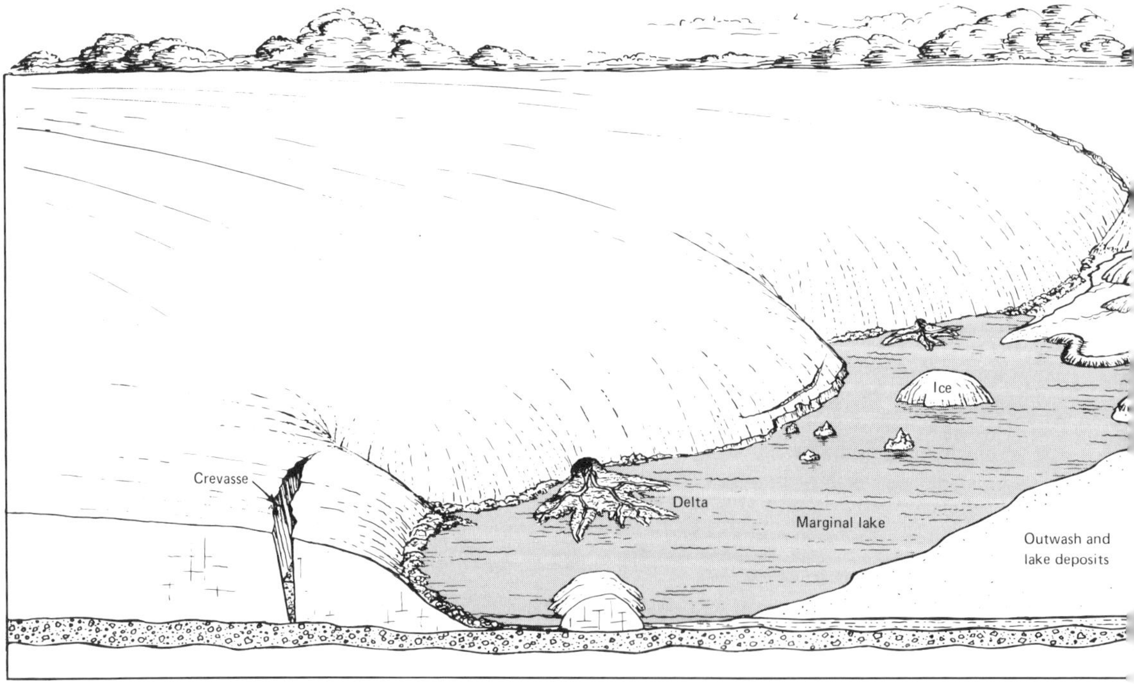

FIGURE 17–32
Sketch of some of the important features, such as moraines, kettles, outwash, and eskers, formed by erosion and deposition by continental glaciers.

DEPOSITION

Glaciers carry as great an assortment of rock types as crop out in the terrain over which the ice moves. Soil, alluvium, colluvium, and any other unconsolidated materials encountered by the moving ice will be incorporated into the ice along with fresh solid rock plucked from bedrock and products of abrasion and gouging. Deposition takes place when the ice melts. Thus deposition from ice occurs most rapidly at its terminus, where melt-water carries the smaller particles of the load out beyond the ice. The terminus may move slowly forward as melting occurs; it may remain in one position with the amount of *ablation* (melting and sublimation) just equal to the amount of ice moving to the end of the glacier; or melting may exceed the rate of movement, causing the terminus of the glacier to recede. The position of the edge of an ice sheet or the terminus of a valley glacier is determined by the balance between wastage and supply of ice. Glacial deposits, called *drift,* may be classed as two types, (1) stratified, primarily glaciofluvial deposits, and (2) unstratified (till). Each of these two types may occur in a number of topographic forms, depending on the location and conditions under which they are deposited.

Unstratified Deposits

Unsorted, usually unstratified or poorly stratified glacial drift is called *till.* All till is of glacial origin, but two distinctly different processes of deposition are involved and these two types can generally be recognized.

Debris that has been in transport near the base of the ice or under the ice is subjected to considerable pressure. As this debris becomes freed where the ice is wasting it may become pressed, plastered, or "lodged" in the deposits under the glacier. These till deposits are called *lodgment tills.* They show no sorting, but elongate rocks are usually aligned with their long axis in the direc-

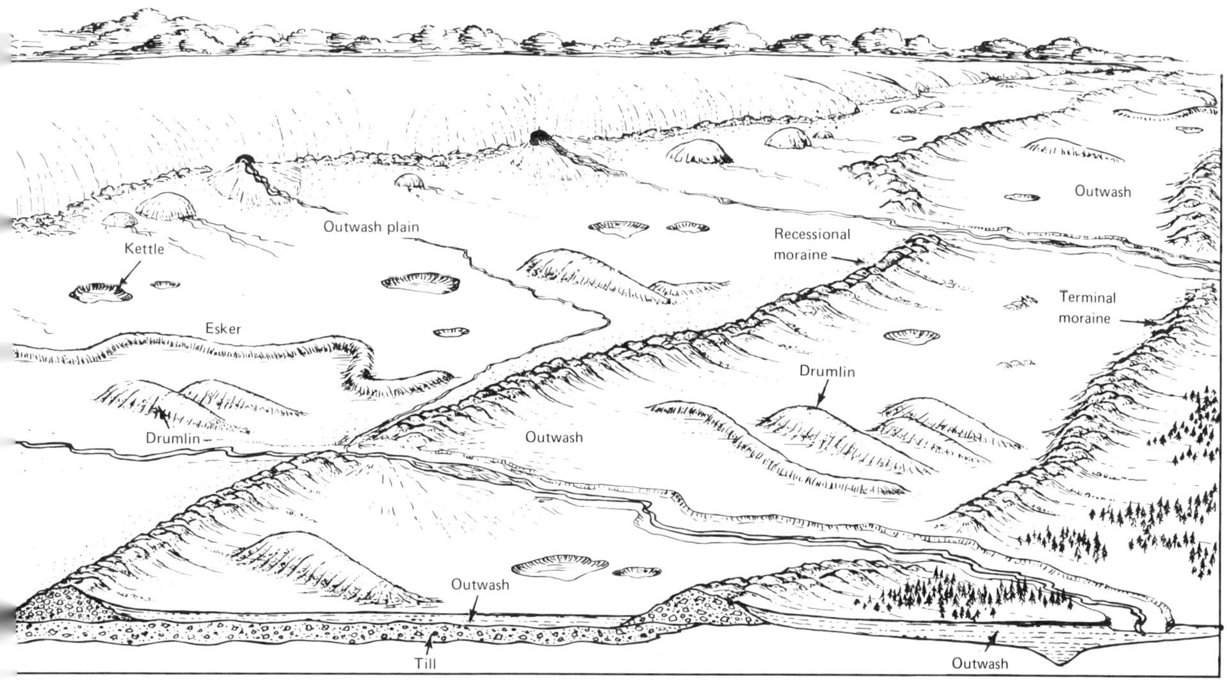

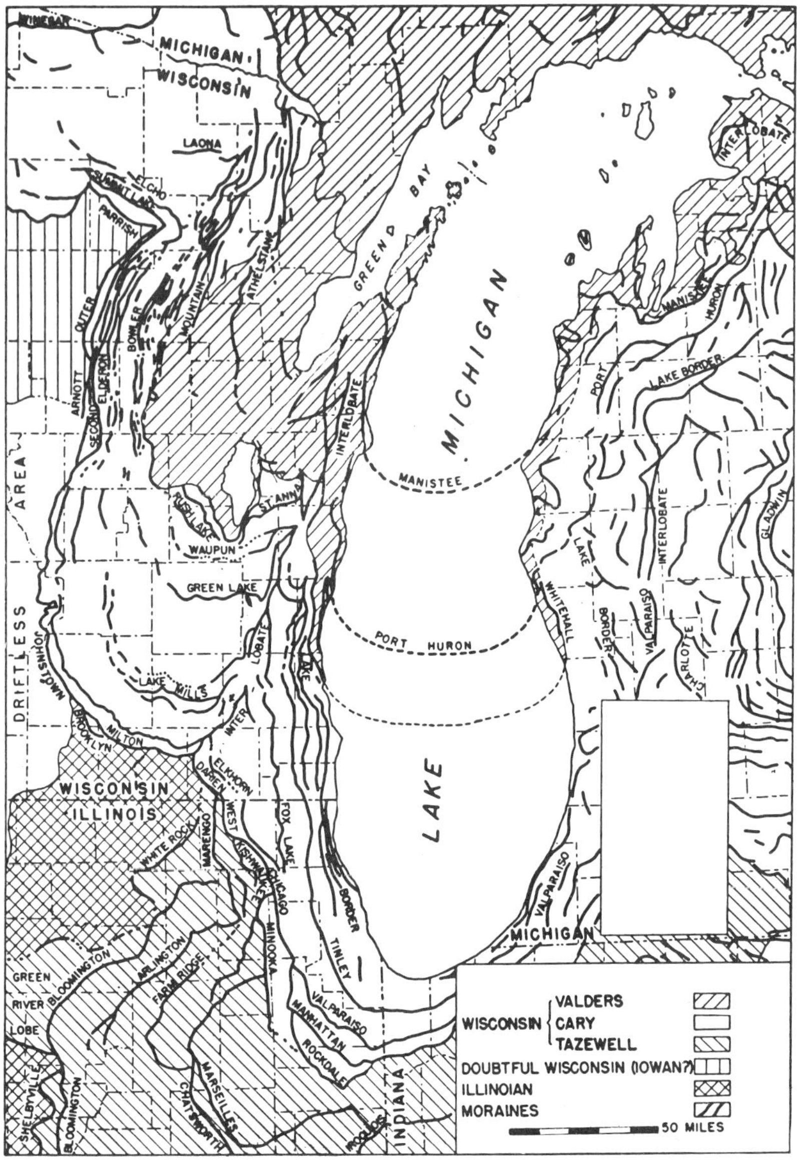

FIGURE 17–33
Details of the location of morainal debris left by advancing ice sheets from the north.
(from Thwaites and Bertrand, 1957)

tion of ice movement; particles may be crushed and the deposits may have a shalelike fissility.

The name *ablation tills* is given to drift deposits formed from the debris that is transported within or on the ice and gradually lowered as ablation causes the glacier to shrink. The resulting deposit is loose and shows no signs of pressure effects such as those in *lodgment tills.*

Stratified Deposits

Deposits laid down under and immediately beyond the glacier by melt-waters are called *glaciofluvial deposits.* These and till differ from most other sediments in that their components are largely fresh; they consist of fragments and particles of rock that have been broken down by disintegration, not decomposition. Hornblende, feldspar, and other easily decomposed minerals are often present. Except for their composition, glaciofluvial deposits have the same characteristics as other water-deposited sediments. They are stratified, sorted, and may have cross-bedding or other primary sedimentary structures. A milky-colored water usually issues from streams flowing out from under both valley and continental glaciers. The material in this water, a powdery silt-size substance, is called *rock flour.* The streams carry rock flour beyond the end of the ice where it gradually settles out to become part of the outwash sediment. This is the probable source of surficial wind-blown deposits, called *loess,* that cover the central part of the Mississippi Valley.

Land Forms of Glacial Deposits

Terminal moraines. *Terminal moraines* (Fig. 17–32) are the ridges of drift (largely till) deposited along the edge of a glacier at the position of the glacier's furthest advance. They form as debris continually melts out of the glacier while the balance between ice supply and wastage causes the edge of the ice to remain in one location. Melt-waters flow down the edge of the ice and over the terminal moraine while the glacier is close to it, but when the glacier begins to retreat, the resulting great quantity of melt-water usually forms a lake.

Eventually the lake tops the dam made by the moraine, cutting narrow valleys through the moraine, and the water drains out, carrying finer debris beyond the terminus. Broad, low alluvial fans may grow from the notches cut into the moraine; beyond, the streams typically flow as braided streams subject to seasonal freezing as long as the ice is nearby.

FIGURE 17–34
Sharp ridge in foreground is composed of lateral moraine and kame deposits. A wasting valley glacier covered by debris is located in the bottom of the valley. Near Mount Cook, New Zealand.

Recessional moraines. If the margin of the ice remains in one position for any period of time, a ridgelike mass of drift accumulates from the debris dropped by the glacier. The formation of this ridge, a *recessional moraine* (Figs. 17–32 and 17–33), is similar to that of a terminal moraine—they differ only in position. Marginal moraines continue to grow as long as the ice is thick enough to move forward, bringing a new supply of load to the melting margin. Older moraines are usually destroyed if the glacier readvances, and the till from them is redeposited in a new position.

Lateral moraines. A large proportion of the load of a valley glacier may be carried along the margins on the top surface of the glacier, forming ridges on the surface of the ice. These ridges are made up of debris that accumulates on the glacier surface as a result of the normal downslope movement of talus and colluvium from the sides of the valley. The amount of such debris is greatly increased as a result of freezing, thawing, and undercutting by the glacier ice. When the ice melts it is laid down along the edge of the valley in ridgelike *lateral moraines* (Fig. 17–34).

FIGURE 17–35
Convergence of debris-covered valley glaciers. Note the irregular surface of the debris in the foreground. Ice is melting under it. Near Mount Cook, New Zealand.

Medial moraines. One or more long ridges of debris may be formed on the surface of a valley glacier parallel to the lateral ridges but located toward the middle of the glacier surface. These medial ridges of debris are derived from the confluence of two or more valley glaciers. As the two join, the lateral debris on the adjacent sides of the two glaciers merges to form a single mass that will continue down valley as a single medial ridge. When the glaciers melt, these medial ridges are lowered and may be recognized as ridges of till called *medial moraines* (Fig. 17–35). This merging of glaciers occurs wherever tributaries of a glacier join, so a number of medial ridges may be formed as shown in Fig. 17–26.

Ground moraines. When melting of a glacier has progressed far enough or when the supply of new ice decreases, the glacier ice ceases to move forward. As melting continues the debris on the surface, that within the ice, and that caught near the bottom of the ice are slowly lowered to the base of the ice. The resulting deposits are called *ground moraines.* These moraines have low relief and consist of till or mixtures of till and water-laid deposits.

Eskers. Both ice sheets and valley glaciers partly melt during summer even while glacial advance is in progress, and melt-waters are abundant as the conditions giving rise to the glacier begin to wane. Water flows on the surface of the ice, seeps down along the margins, and even moves through holes in the ice. The water accumulates on the surface and beneath the glacier and forms streams. These streams are different from ordinary streams, however, in that their banks are composed of ice. All types of material in the transported load of the glacier may get into the stream. The small fractions of the load are carried easily downstream, but the coarser boulders and blocks are not. After the glacier melts, a sinuous ridge of this debris may remain, marking the former stream channel; such a ridge is called an *esker* (Fig. 17–36).

Eskers may also form where a stream beneath the ice cuts into underlying sediment or weak rock beneath the glacier, leaving deposits of the coarser part of its load. If the sediment on either side of the stream deposits is more easily eroded than the material in the channel, the stream deposits may be etched out to stand as a ridge.

Kames and kettles. The term *kame* is widely used to describe glacial deposits laid down in contact with ice, but so many different features have been called kames that the term lacks precise meaning. The American Geological Institute's *Glossary of Geology* suggests the following definitions, among others: (1) "a hill of stratified drift deposited, usually as a steep alluvial fan, against the edge of an ice sheet by debouching streams of sediment-laden meltwater," or (2) "a low, steep-sided hill of stratified drift, formed in contact with glacier ice." It is also the name given glaciofluvial deposits formed by sediments washed into depressions on the top surface of a glacier by surficial melt-waters. As the ice melts, the material that formerly filled depressions on top of the glacier is dropped and makes small hills. The ice near the margin of a glacier melts faster than that near the center, causing lakes and streams to flow along margins of valley glaciers or along trough-like depressions on continental ice sheets, and melt-waters to concentrate debris in these lakes and channels. If the water cannot carry the entire load, part is dropped. Terraces, called *kame terraces* (Fig. 17–25), are built in this way along the sides of the valley. Kames may also form beneath the ice where sediment washed down through crevasses and holes in the ice accumulates in small hills.

Frequently debris in the lower part of a continental ice sheet and on top of a valley glacier completely covers the ice in the late stages of the melting. As the ice disappears, blocks of ice may be trapped and protected from melting by the covering debris. As these last remnants of ice melt away an enclosed depression called a *kettle* is left. Most kettles are formed in a marginal position near the edge of the ice.

Outwash plains. When a glacier first begins to melt the melt-water is incapable of transporting all the debris present. Part of it is thus deposited as broad, flat alluvial fans or, in the case of ice sheets, as a series of coalescing alluvial fans called the *outwash plain* (Fig. 17–37). Here rock flour is dropped along with most of the sand and clay carried from the glacier. Because the streams are heavily loaded, and because they are shifted by often-changing conditions, braided-stream patterns characterize the outwash-plain streams. Later, if the rate of melting begins to increase, the streams may establish more permanent channels through which large amounts of the debris are moved beyond the immediate region of the glacier.

Sand and gravel deposits laid within a valley and down valley from the end moraine of a glacier are called *valley trains.* Multiple superimposed valley trains frequently are formed as the supply of sediment fluctuates with each phase of glacier advance and retreat.

Erratics. The name *erratic* (Fig. 17–38) is applied to any rock that is carried by a glacier and laid down on a different type of rock. They may be blocks up to about 30 meters high.

MAJOR FEATURES OF PERIODS OF CONTINENTAL GLACIATION

Judging from the scarcity of glacial deposits in the rock record of past times, periods of glaciation are very unusual. The remains of corals and other animals that ordinarily favor warm waters are found in rocks of Paleozoic age in the Arctic and Antarctic. What has caused glaciers to form on earth is an intriguing question that has given rise to a great deal of speculation. Scientific literature abounds in hypotheses that try to answer the question. Some of them are based on assumptions that have proved fallacious, and the validity of others remains uncertain. No definitive answer to the question can yet be given.

For any hypothesis to be acceptable it must explain, or be compatible with, the observed nature of glaciers and the major features of glaciation, or at least it must not contradict these observations. For example, a proposed hypothesis must take into account the history of glaciation and the record of climatic changes. A number of quite different types of climatic indicators for times past have been recognized. The most frequently used of these are the following (Broecker, 1965):

1. The temperature of the near-surface waters of the oceans as indicated by oxygen-isotope and paleontological studies made on cores of marine sediment.
2. Fluctuations in sea level used as a guide to the amount of water tied up in the form of ice on land.
3. The size of the continental ice sheets estimated from the extent of their margins as indicated by the position of glacial deposits of any given age.
4. Large lakes formerly occupied many en-

FIGURE 17–36
An esker in northern Canada.
(Royal Canadian Air Force)

FIGURE 17–37
Pitted outwash plain. The depressions are kettles.
(Royal Canadian Air Force)

FIGURE 17–38
Recessional moraine with kettles and erratics in foreground. Fine outwash sediment is in the distance. Note also the glacially smoothed and terraced valley walls on the mountainside. South Island, New Zealand.

closed basins such as the Great Basin of the western United States. Thus the arid conditions there are modern.

5. The distribution of different types of plants that are temperature and moisture sensitive as indicated by the presence of plant pollen in sediments.

The occurrence of till containing striated and grooved pebbles assures us that several periods of glaciation predate the one in which we now live. The oldest of these took place in the Precambrian, and it is possible that more than one period of glaciation occurred during the Precambrian. The oldest ice sheets occupied an area in the southern part of the Canadian Shield. Glacial sediments are also found in widely separated areas in rocks that lie just beneath the Cambrian system. These tills are more than 600 million years old. They are scattered over the Canadian Shield, in Utah, Norway, Australia, eastern Greenland, and India. The third major period of glaciation occurred during the late part of the Paleozoic era in the Carboniferous and Permian periods. At this time, about 230 million years ago, glaciers covered parts of South Africa, South America, Australia, Antarctica, and India. In Australia five separate sheets of till that are interbedded with coal are of this age. This relation indicates cold periods of glacial advance separated by warm, moist interglacial periods during which plant life flourished. The distribution of glacial deposits in areas of modern temperate and tropical climates has been best explained in terms of continental drift and movement of the crust relative to the geographical pole. The late Paleozoic glaciation may have taken place when the areas affected were close enough together to allow coverage by a single continental glacier, as suggested in Fig. 25–3. The orientation of glacial grooves and striations on these now widely separated areas is compatible with the hypothesis of a common ice center when the position of the continents is restored as shown.

We have a relatively complete understanding of the events that have taken place during the most recent glacial period, the Pleistocene, and our knowledge imposes important limitations on the theories of glaciation. Pleistocene glaciation started about 3 million years ago. Since its beginning, at least four major intervals of glacial advance and three interglacial ages are indicated.

Soil profiles developed in glacial drift have been used as an effective means of identifying tills. Depth of oxidation, amount and kind of chemical weathering, or extent of decay may be used as means of estimating relative ages of different tills. The most recent sheet of glacial till is fresh in comparison with older sheets. Studies of these till sheets and the interglacial deposits have led to the following subdivision of the Pleistocene, in order from youngest to oldest:

Wisconsinan glaciation. Defined on the basis of deposits in Wisconsin.

Sangamon interglacial. For deposits in Sangamon County, Ill.

Illinoian glaciation. For deposits in Illinois.

Yarmouth interglacial. For deposits in Yarmouth, Iowa.

Kansan glaciation. For sections in northeastern Kansas.

Aftonian interglacial. For deposits in Afton Junction, Iowa.

Nebraskan glaciation. For deposits in Nebraska.

The last great resurgence of ice reached a maximum about 18,000 years ago. During periods of advance, the ice sheets have covered most of Canada, the northern United States (Fig. 17–39), Greenland, northern Europe, northern Siberia, Antarctica, and high mountains throughout the world. Each of these advances took place essentially simultaneously throughout the Northern Hemisphere. This suggests a relationship to latitude such as we now see reflected in mean-temperature distribution. Thus, whatever caused the glaciation must have affected this huge region at about the same time. Even within each of the glacial advances fluctuations in the position of the ice sheet and minor advances and retreats of relatively short duration took place. These advances and retreats were not periodic, and we lack sufficient data to demonstrate conclusively whether or not the major glacial periods occurred at regular intervals. Consequently, we cannot be sure if the cause of glaciation is one that occurs at fixed intervals.

Ice sheets grow and decay over periods of a few thousand years—a short interval in geologic time. Most glaciologists now believe that ice sheets decay nearly five times as fast as they grow.

Variations in temperatures on earth of between 5° and 10°C should be sufficient to cause glacial advances and retreats. Even today a lowering of only a few degrees in the mean annual temperatures would certainly give rise to a great increase in the land area covered by ice and snow throughout the year, and it might be sufficient to trigger a new resurgence of ice sheets across North America.

OXYGEN-ISOTOPE STUDIES AS TEMPERATURE INDICATORS

Oxygen-16 (O^{16}) is the type of oxygen in the air and in water, but a small amount of the heavy isotope of oxygen, O^{18}, occurs with it. About 0.2 percent of the oxygen in natural water is O^{18}, but the ratio of O^{18} to O^{16} varies widely. This ratio can be measured by use of a mass spectrometer. The great significance of this isotope is that the ratio varies with temperature; thus it varies with altitude and from season to season, providing a valuable tool in the study of glacier ice. It is possible to identify the annual layers of snow by use of these ratios and in this way analyze the rate of accumulation of glacier ice.

Marine organisms take oxygen from sea water to construct their shells. Thus the shells of such abundant microorganisms as the planktonic foraminifers have shells made of $CaCO_3$ in which the ratio of O^{18} to O^{16} (which increases 0.02 percent per 1°C temperature drop) is a function of the temperature of the near-surface water masses where these organisms live. A strong correlation is found between species assemblages and water temperature. Both of these clues to water temperature have been used to analyze cores of Pleistocene sediments for the purpose of determining variation in water temperature. Emiliani (1955, 1964) used oxygen-isotope methods of analysis on several cores to derive a temperature curve for the past few hundreds of thousands of years, and Ericson and his co-workers (1964) have derived a temperature curve primarily on the basis of the species assemblages (Fig. 17–40). Both used estimated sedimentation rates as a guide to age of the intervals in the lower part of the core. Radio-

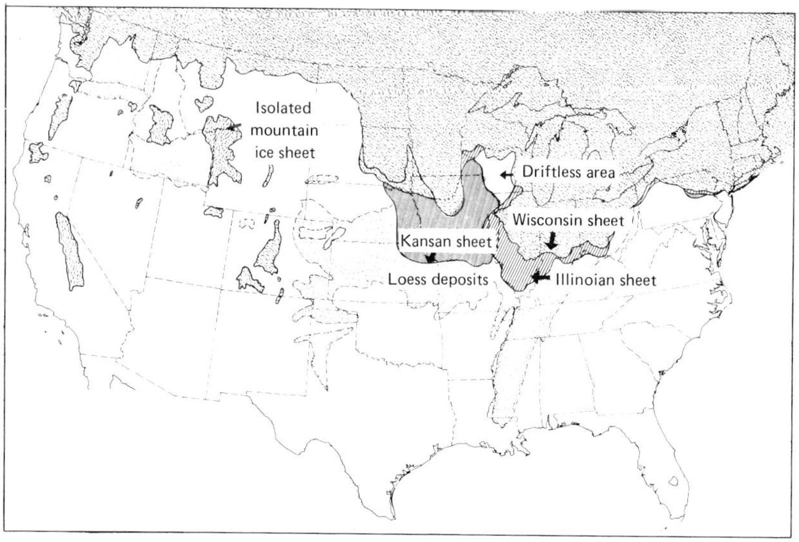

FIGURE 17–39 The margins of the ice sheets that have advanced into the United States during the Pleistocene.

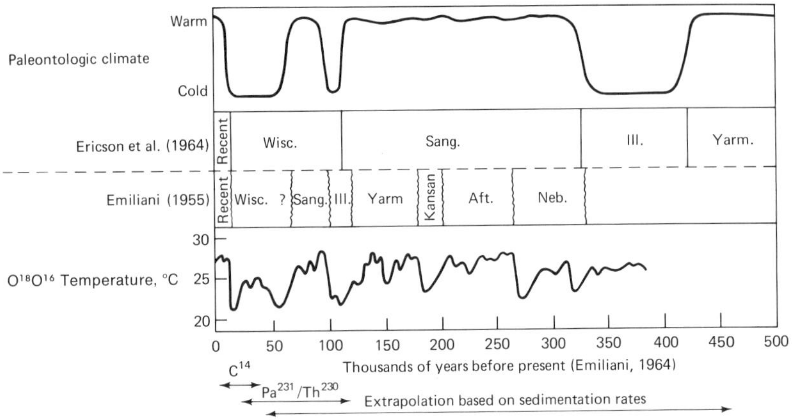

FIGURE 17–40 Comparison of climatic curves of Emiliani (1955, 1964) and Ericson and others (1964). Although the two generally agree in their assignment of ages to various levels in the core, they seriously disagree as to (1) the climate characterizing the period from 120,000 to 300,000 and (2) the correlation with the continental glacial sequences. Ericson's curve is based on the types of fauna at various levels in the core. Emiliani's curve is based on temperature as determined by oxygen-isotope ratios. The abbreviated subdivisions of the Pleistocene are Wisconsinan glacial, Sangamon interglacial, Illinoian glacial, Yarmouth interglacial, Kansan glacial, Aftonian interglacial, Nebraskan glacial.

metric dating was used for the upper hundred thousand years. Major discrepancies in the two curves occur at 120,000 years before the present. A number of factors may be involved in these discrepancies. One possible error is use of rates of sedimentation as a guide to age.

SIDE EFFECTS OF PLEISTOCENE CONTINENTAL GLACIATION

Emphasis here has been placed on the formation, movement, and wastage of ice on land physically in contact with ice and melt-waters and the consequent erosional and depositional features. Continental glaciation has had numerous indirect effects of major importance that occur far removed from the ice. Of these, we have already discussed the effect of glaciation on sea level (see Chapter 9). Taken on a world-wide basis sea level has been rising for the last 15,000 years (see Fig. 9–31), with that rise starting from a level about 130 meters below modern sea level and rising at slower rates for the last 5,000 years than for the previous 10,000 years. If all the present ice melted, sea level would rise by an amount conservatively estimated as being between 25 and 60 meters. Changes in sea level have had pronounced effects on the evolution of coastal land forms and they have dominated the recent history of the continental shelves.

Changes in climate, patterns of atmosphere circulation, precipitation, and water supply have had dramatic effects over vast areas beyond the limits of the ice sheet. Lakes such as glacial Lake Agassiz in Canada, which covered more area than the present Great Lakes, were formed as the retreating ice margins left natural dams in the form of recessional moraines. The Great Lakes owe their presence to moraines that prevent their drainage toward the south and to deepening due to scouring and removal of rock and debris by the glaciers. Lakes in the western United States were deepened by as much as 300 meters by glacial processes. Other lakes formed beyond the termini of the glaciers, partly due to melt-water but primarily as a result of increased precipitation and lower evaporation rates. The Great Salt Lake is a remnant of one such lake, Lake Bonneville, located far from the edge of the ice and covering approximately 80,000 square kilometers. Such lakes were common in the present arid regions of the southwestern United States, the Sahara, Arabian, Kalahari, Atacama, and Patagonian deserts. These areas were moist, perhaps fertile, and well watered during these times, called *pluvial* ages, when patterns of atmospheric circulation were different and precipitation much increased (Fig. 17–41). These deserts are now located in areas crossed by trade winds or in *horse latitudes.* The air in the trade-wind belts is warm and often low in water content, so high rates of evaporation are possible; the horse latitudes are areas of calm winds where dry air descends from high altitudes. During the pluvial ages these areas were probably located differently relative to the pattern of circulation. Possibly the equatorial belt of sluggish circulation, called the *doldrums,* was wider and shifted southward over some of these desert regions. The doldrums are characterized by warm, moisture-laden air and high precipitation.

GLACIAL HYPOTHESES

Many possible causes of glaciation have been suggested. These hypotheses may be classified broadly into three groups: (1) cosmic, (2) planetary, and (3) geological and geophysical.

Cosmic. Hypotheses that involve such ideas for lowering temperature on earth as variation in the heat produced by the sun and the intersection of radiant energy from the sun by cosmic dust clouds.

Planetary. Hypotheses based on periodic changes in motions of the earth, tilt of the earth on its axis, or position of the earth relative to the sun to reduce the amount of solar radiation and cause cooling of the earth's atmosphere.

Geological and geophysical. Hypotheses that include such ideas as movements of the earth's crust relative to the poles; changes in oceanic circulation; and variations in the amount of volcanic dust in the atmosphere.

Cosmic Theories

The temperatures on the surface of the earth might be lowered enough by fluctuations in the amount of solar radiation, leaving the sun to bring about a general glaciation. At the present time the average amount of energy reaching the earth's surface is about 2 calories per square centimeter each minute. This value has fluctuated by about 3 percent since 1918. Thus fluctuations of the amount of energy coming to the earth from the sun actually do occur, but the question of whether the variations are ever sufficient to cause glaciation is still open. And even if these fluctuations are the cause of glaciation, the question of which—increased or decreased solar radiation—is responsible for glaciation remains. Decreases in solar radiation would cool the earth and cause increased snowfall and accumulation of snow and ice in mountains and in high land areas until ice sheets developed and began to advance. Then, when the amount of radiation began to increase, the ice would melt and the glaciers would retreat.

However, one can also reason that increased solar radiation would cause glaciation. If solar radiation did increase, the increase would vary with latitude, being greatest at the equator and least at the poles. More atmospheric convection and stronger winds would occur in equatorial latitudes. This would cause increased evaporation of sea water, more clouds, and consequently more precipitation. This precipitation would fall as snow in the polar regions in the early stages of increased radiation. Thus more snowfall would account for more snow accumulation and an ice age. If temperatures continued to rise, the ice would melt and the glaciers would retreat and disappear. This would initiate a warm, wet, interglacial period such as those that are known between ice advances of the Pleistocene. If the amount of radiation began to diminish again, the temperature would fall and the amount of precipitation would decrease, so a dry, cold, glacial period might follow.

Planetary Hypothesis

Movements of the earth and the earth-moon system cause periodic changes in the position of the earth relative to the sun.

1. Once every 92,000 years the earth's position is at a maximum distance from the sun because of eccentricity of the earth's orbit.
2. The angle that the earth's axis makes with the plane in which it circles the sun (the ecliptic) varies through 3° every 40,000 years.
3. The axis of the earth wobbles because of the gravitational effect of the sun, moon, and planets. The period of this movement is 21,000 years.

Conditions would be most favorable for glaciation of the poles when they are farthest from the

FIGURE 17–41
Comparison between fluctuations of the mid-continent glacier and the Great Basin pluvial lakes.
(from Broecker and Kaufman, 1965)

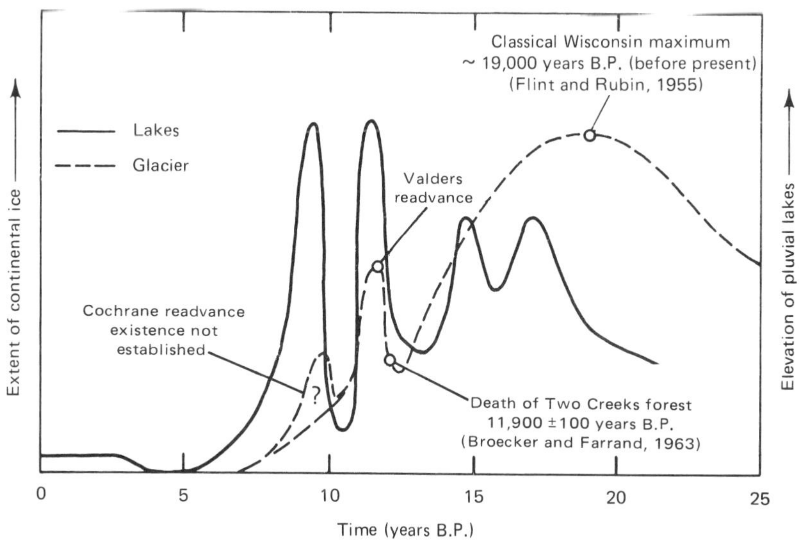

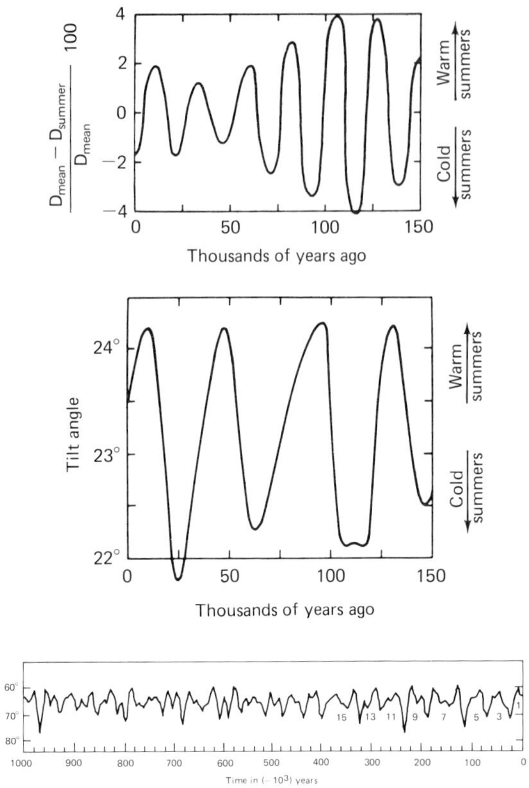

FIGURE 17–42

(*a*) Variation of the earth's tilt and of the June 21st sun-earth distance over the last 150,000 years. High tilt angles and short June 21st sun-earth distances lead to increases in summer insolation in the Northern Hemisphere. The cyclic variations in the June 21st sun-earth distances are due to the earth's precession. The variation in the magnitude of this cycle results from changes in the eccentricity of the earth's orbit. (from Broecker, 1968)

(*b*) The Milankovitch curve of variations of solar radiation for the Northern Hemisphere expressed in terms of equivalent latitude with 65°N being the normal. The Emiliani core stages are indicated. (from van Woerkom, 1953)

sun; that is, when the earth's location at a maximum distance in its orbit coincides with maximum tilt of the axis.

The planetary hypothesis was developed most completely by Milankovitch (1938). He constructed curves showing the fluctuations of solar radiation at various latitudes that should have taken place as a result of the effects of tilt of the earth's axis and precessional wobble of the axis for the last 150,000 years (Figs. 17–42a, b). These two curves may be combined to give a record of the net effect of warming and cooling. The variation in summer insolation represented by these changes amounts to about 1 percent of the annual heat budget and would lower the mean summer temperature by about 3°C (according to Broecker, 1968). Broecker compared the combined Milankovitch curves with ages established by geochronological method (radiometric dating techniques) for sea-level fluctuations and temperature of the Atlantic Ocean (Fig. 17–43) and finds the following strong evidence for correlation of the two:

1. The last two abrupt warmings that mark the end of oceanic cold periods are coincident with the two most prominent warm insolation peaks.
2. The present high sea stand and the last three high sea stands all fall just after the four prominent warm insolation peaks.
3. The major glaciation maximum that occurred 18,000 years ago closely follows the prominent insolation cold peak 22,000 years ago.

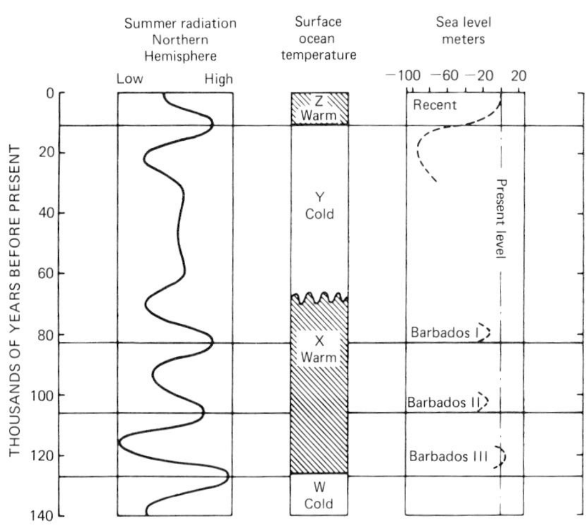

FIGURE 17–43 Correlation between insolation curve, changes in surface ocean temperature, and changes in sea level over the past 140,000 years. Sea-level stands determined in Barbados are indicated. (from Broecker, 1968)

Geological and Geophysical Theories

Volcanism. A connection between volcanism and glaciation seems unfounded on the basis of the association of volcanic and glacial sediments. The present is a time of great volcanism and glaciation, but no glacial deposits correspond in age to other periods of widespread volcanism.

Solar-topographic theory. Professor Richard Flint (1957) emphasized the importance of uplift of land areas in conjunction with variations in output of solar energy as a cause of glaciation. When large areas of land are uplifted by mountain-building processes (see Chapter 25), patterns of atmospheric circulation are changed, temperature drops, and more land area is raised above the snow line, encouraging the accumulation of ice and formation of glaciers. The late Tertiary was a time of extensive mountain building in many parts of the world. The Alps, Sierra Nevada, and Himalayas have been uplifted 2,000 meters or more during late Tertiary and Pleistocene time. Mountain building also took place about the time of late Paleozoic glaciation.

Uplift may be important in initiating glaciation but crustal movements cannot account for the relatively rapid and repeated growth and decay of glaciers during the ice ages.

Effect of an Ice-Free Arctic Ocean

Ewing and Donn (1956; Donn and Ewing, 1968) emphasize the importance of the Arctic Ocean in controlling glacial and interglacial stages in the Northern Hemisphere during the Pleistocene. At the present time northern Canada and Siberia, areas that were covered by ice during the last glacial advance, are desertlike in their low precipitation (10–25 centimeters per year). The presence of ice sheets in Greenland indicates that temperatures are low enough at high latitudes to support glaciers. They are rare in those latitudes because of insufficient precipitation.

Ewing and Donn suggest that an ice-free Arctic Ocean would provide the necessary source of precipitation for high-latitude glaciation. As these ice sheets grew and temperatures dropped, the Arctic Ocean would become frozen. Any additional expansion of the ice sheets would depend on sources of precipitation from lower latitudes. A glacial stage would be terminated by the failure of the moisture supply in the Arctic. Long time lags would occur between the start of glaciation on land in the high latitudes and temperature drop of the ocean water. Similarly, a long lag would occur between the time the glaciers began to recede and the time sea temperatures rose again. Glaciation was initiated in the Arctic, according to this theory, as a result of the migration of the earth's geographic pole into the Arctic with the consequent reduction in temperature that results from lower solar radiation at high latitudes. Polar wandering is discussed in Chapters 19 and 25.

Ice-Shelf Theory

A. T. Wilson (1964) offered a new glacial hypothesis in which he postulated that the ice shelves around Antarctica become unstable and grow rapidly to cover large areas of the southern oceans. The enlarged ice shelf would increase the reflection of solar energy sufficiently to induce extensive glaciation elsewhere in the world.

The Ross Sea shelf ice is a good example of the type of shelf ice described. It is about 250 meters thick at the seaward edge and extends nearly 1,000 kilometers from land. Shelf ice would grow rapidly if the ice sheet on land became unstable; that is, if the pressure near the bottom of the ice approached the melting point, reducing resistance to movement. The Kutiah Glacier in northern India became unstable in 1953 and advanced at a rate of 4.7 meters per hour. During an interglacial period such as the present the ice sheet slowly builds up. The bottom of the ice is heated by normal geothermal heat flow and eventually the pressure melting point is reached and the ice yields by both thawing and plastic yield. Once initiated, the movement causes heating and the weight of the ice sheet drives continued movement.

An Antarctic ice shelf might be expected to grow out to the Antarctic convergence where warmer water would break up and melt the ice shelf. An ice shelf of this dimension would greatly increase reflection of solar radiation and decrease the heat on the earth as a whole by 4 percent, causing ice sheets in the Northern Hemisphere to grow.

Eventually colder ice would move into the rock-ice interface under the Antarctic ice sheet and rapid movement would stop. Calving and melting at the edge of the shelf ice would exceed the rate of expansion and gradually the shelf ice would decay and the glacial period would end.

CHAPTER OUTLINE

Historical Background
Types of Glaciers
The Polar Ice Masses
Ice-Budget Concept
Formation of Glaciers
Formation and Movement of Glacier Ice
Movement Patterns in Valley Glaciers
Erosion by Glaciers
Incorporation of debris in glaciers
Abrasion
Mass movement
Scraping and pushing
Methods of Transportation
Features Produced by Erosion
Erosional Features of Valley Glaciers
Cirques
Fiords
Arêtes
Cols
Horns
Deposition
Unstratified Deposits
Stratified Deposits
Land Forms of Glacial Deposits
Terminal moraines
Recessional moraines
Lateral moraines
Medial moraines
Ground moraines
Eskers
Kames and kettles
Outwash plains
Erratics
Major Features of Periods of Continental Glaciation
Oxygen-Isotope Studies as Temperature Indicators
Side Effects of Pleistocene Continental Glaciation
Glacial Hypotheses
Cosmic Theories
Planetary Hypothesis
Geological and Geophysical Theories
Volcanism
Solar-topographic theory
Effect of an Ice-Free Arctic Ocean
Ice-Shelf Theory

TERMS

ablation
ablation tills
arêtes
calving
chatter (marks)
cirques
cols
crescent-shaped gouges
crescentic fractures
doldrums
drift
drumlin
erratic
eskers
fiords
firn
glaciofluvial deposits
ground moraines
hanging valleys
horns
horse latitudes
ice sheets
icebergs
kame terraces
kames
kettle
lateral moraines
lodgment tills
loess
medial moraines
nunataks
outwash plain
pack ice
piedmont glacier
plastico-viscous
plucking (glacial)
pluvial
pseudo-plastic
quarrying (glacial)
recessional moraines
roches moutonnée
rock flour
snow field
snow line
surge
tarn
terminal moraines
till
truncated spurs
valley glaciers
valley trains
volcanism

REVIEW QUESTIONS

1. Summarize the data that must be known about an ice sheet to make an accurate determination of the ice budget of the sheet.
2. What erosional features are similar for valley glaciers and ice sheets?
3. How might you distinguish two moraines of different ages that occur in the same area?
4. What are the differences in the behavior of warm and cold glaciers?
5. What are the sources of information about sea-level stands during the Pleistocene?
6. What data would you seek from areas now covered by ice sheets to determine Pleistocene history?
7. What is the effect beyond the ice terminus on streams during a glacial advance?
8. What effects would large ice sheets have on climates near the edge of the ice?
9. Explain why the size of pluvial lakes might increase during short periods of growth of ice sheets but not during a glacial maximum (refer to Fig. 17–41).
10. Explain the combined effects that total melting of ice on earth and isostatic adjustments would have on sea level. If effects on continents and ocean basins would be different, explain.

SUGGESTED READINGS

Davies, J. L., 1969, Landforms of Cold Climates: Cambridge, M.I.T. Press, 200 p.

Embleton, Clifford, and **King, C. A. M.,** 1968, Glacial and Periglacial Geomorphology: New York, St. Martin's, 608 p.

Ericson, D. B., and **Wollin, Goesta,** 1964, The Deep and the Past: New York, Alfred A. Knopf, 292 p.

King, P. B., 1965, Tectonics of Quaternary Time in Middle North America: The Quaternary of the United States, ed. Wright and Frey: Princeton, N.J., Princeton University Press, p. 831–871.

Sharp, R. P., 1960, Glaciers: Eugene, Ore., University of Oregon Books, 78 p.

IV
PHYSICS OF THE EARTH

18 THE EARTH'S GRAVITY FIELD

18 THE EARTH'S GRAVITY FIELD

FIGURE
(preceding page)
A sharp downward pitch in an anticline (Big Horn Mountain, Wyo.). Such folded structures indicate the magnitude of forces deforming the earth.
(U.S. Geological Survey)

THE LAW OF GRAVITATIONAL ATTRACTION

Since the time of Newton, scientists have been much concerned with the earth's gravity field, and their ability to measure forces of gravity has become increasingly precise. Results of these measurements have yielded insights into such diverse problems as determining the shape of the earth, mapping distribution of bodies of high- and low-density rock beneath the surface, and understanding the structure of mountain ranges. Gravitational studies have led to the discovery of oil, gas, and metals. They have also provided the basis for the theory of *isostasy*, an important concept concerning the strength of the earth's interior and how this affects deformation of the crust.

Sir Isaac Newton (1642–1727) lived at a time when science was flourishing. Gilbert was making great contributions to the understanding of magnetism; Huygens was exploring the nature of light and centripetal force, experimenting with the pendulum clock, and had already invented the reflecting telescope; Roemer had measured the speed of light; Hooke was studying and formulating principles of elasticity; and Agricola had written his famous book on mineralogy.

Until the time of Newton, the chief concern of astronomers had been the determination of planetary paths. Now their concern turned to another question: What forces must act on a body to make it move in an elliptical orbit? Newton showed that by postulating that all bodies in the universe exert forces of gravitational attraction on one another he could derive the motions of all bodies in the solar system. In *Principia* he set forth the mathematical explanations of the forces of gravitational attraction between bodies and applied these to the motions of the planets, comets, the moon, and the tides.

He established first that any body moving in accordance with Kepler's Second Law, the law of equal areas [planets traveling in elliptical paths sweep out equal areas in equal periods of time (Fig. 18–1)], must be acted on by a central force. It had been shown that each planet does move in accordance with Kepler's Second Law, and that the net force on each planet must be directed toward one focus of the elliptical path. Newton established the magnitude of the force as well as its direction. And he was able to show that the force upon each planet must at any moment be inversely proportional to the square of the distance from the center of the body (if it is spherical) to the focus of the ellipse. Newton saw in the attraction of objects on the earth to the earth the solution to the origin of these forces—all objects attract one another with a gravitational force. Thus, he reasoned, the central force on a planet was the gravitational attraction to the sun. Similarly, the force on the moon must be the gravitational attraction between the earth and the moon. Newton found that the magnitude of the force is also proportional to the masses of the two objects. Thus his law of universal gravitational attraction can be stated as follows:

$$F = G\frac{M_1 M_2}{R^2}$$

The value of the constant, G, was later established by Cavendish (1731–1810) to be 6.670×10^{-8} dynes cm^2/g^2.

Newton tested his theory by using these hypotheses to calculate the period of revolution of the moon about the earth. He calculated a value of 27 days, compared with the actual value of $27\frac{1}{3}$ days, and this result showed that he was on the right track.

MEASURING THE FORCE OF GRAVITY

Gravitational force may be thought of as a pull exerted between two objects. It acts along a line connecting the centers of mass of the two objects, and it has magnitude. When several objects are mutually attracting one another the effect on any one of them is the resultant of the effects of each of the others.

Instruments designed to measure gravity are among the most sensitive yet developed. Meters for measuring earth tides are read to 1 part in a billion, and instruments used for field measurements are accurate to 1 part in 100 million. One of the earliest instruments used for this purpose was the pendulum. The period (time of one complete swing) of a pendulum is a measure of the force of gravity at the place where the pendulum is set up. Other methods of measuring gravity have involved timing the rate of fall of a body in a near-vacuum, the use of supersensitive spring balances, torsion balances, and measurement of air pressure.

An ideal pendulum completes its full swing in a time (period) T, given by the following equation:

$$T = 2\pi\sqrt{\frac{L}{g}}; \qquad g = \frac{(2\pi)^2 L}{T^2}$$

L in this equation is the length of the pendulum and g is gravity. Thus the gravitational attraction at a given location can be determined by measuring the length of the pendulum and the period of the swing. (Note that the dimensions of g are those of accelerations, or distance/time^2.) In order to compare the acceleration due to gravity at two stations with a pendulum the following simple relation is used:

$$\frac{g_1}{g_0} = \frac{T_0^2}{T_1^2}$$

Because this method is time-consuming, because the apparatus is difficult to make, and because the instruments are so heavy that they are not readily portable, pendulums have been largely replaced by *gravimeters*[1] designed especially for measuring the difference in gravity fields. Pendulums are still used for absolute gravity measurements, however.

Gravimeters are sensitive weighing devices. The weight of a constant mass varies with any variation in the gravitational field. Thus, if the change in weight can be measured, differences in gravity can be obtained. In a gravimeter of the spring-balance type, a mass is delicately suspended from a spring and the elongation of the spring is measured. This elongation is a measure of weight. Changes in the length of the spring are so small that an enormous magnification of the elongation is needed. Optical, mechanical, and electrical methods are used to obtain these magnifications. One of the standard procedures for checking the calibration is to take readings with the instrument on various floors of a building. The readings are checked against the theoretical changes arising from differences in elevation. Modern gravimeters are lightweight, portable, and can be read directly in a short time. Thus they are widely used for gravity studies.

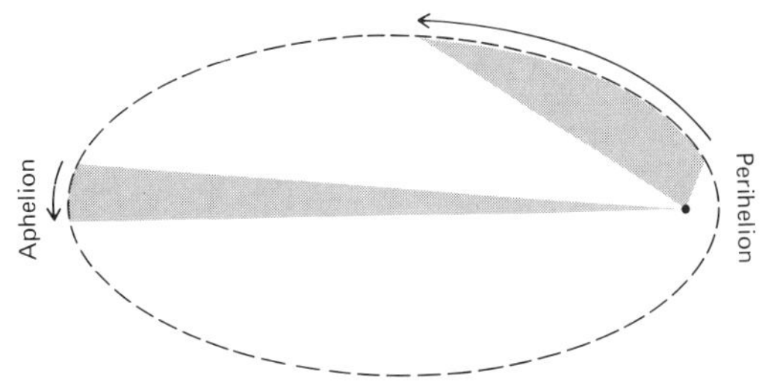

FIGURE 18–1
Kepler discovered that planets traveling in elliptical paths sweep out equal areas (shaded) in equal periods of time. Thus a planet at perihelion must travel more rapidly than one at aphelion.

Absolute measurements of gravity are expressed as *gals* (a unit named in honor of Galileo) —1 gal is an acceleration of 1 cm/sec^2; 1 milligal is $\frac{1}{1000}$ gal. Accuracy of 0.1 milligal is considered necessary in modern studies. Precision of 0.01 milligal is possible with a gravimeter as compared with 1–2 milligals with a pendulum.

MASS AND DENSITY OF THE EARTH

Acceleration of falling bodies is also used to determine the value of gravity at a point. The acceleration of gravity, g, is also the acceleration imparted to a body by its own weight (w = mass $\times g$). This is known as Newton's Second Law (force = mass × acceleration, $F = ma$). The acceleration of free-falling bodies in a vacuum is approximately 980 cm/sec^2 or 32.15 ft/sec^2 on earth. Using this value and knowing the radius of the earth it is possible by using Newton's Law of Gravitation to determine the mass and density of the earth.

$$mg = G\frac{mM}{R^2}$$

Where M is the mass of the earth and R is the distance to the center of the earth,

$$g = \frac{GM}{R^2}$$

$$M = \frac{gR^2}{G}$$

If R = 6,380 kilometers,

$M = 5.98 \times 10^{27}$ grams

[1] The Gulf Oil Company used pendulums extensively for oil exploration from 1932–36 until they were superseded by gravimeters. Pendulums are now used primarily to establish precise values along "calibration lines" to aid in calibration of gravimeters.

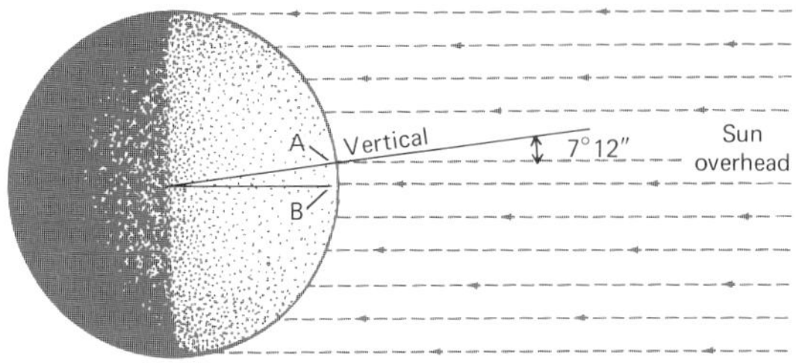

FIGURE 18–2 Eratosthenes measured the diameter of the earth by observing the angle a vertical pole made with the incident sunlight at *A*, which was a known distance from *B* where no shadow was cast. The angle between *A* and *B* at the center of the earth is 7°12″, the same as the angle between the vertical and the sun. Through knowledge of this angle and of the distance from *A* to *B*, the radius of the earth could be calculated.

The volume of the earth is $V = \frac{4}{3}\pi R^3$ approximately. The weight-density of the earth (mass/volume) can be calculated, and this can be converted to specific gravity, resulting in the conclusion that the specific gravity of the earth is about 5.5.

SHAPE AND SIZE OF THE EARTH

The Pythagorean brothers, a monastic brotherhood founded in one of the Greek colonies in southern Italy about 550 B.C., taught that the earth is a sphere. The fact that other celestial bodies could be seen to be spherical suggested to them that for reasons of symmetry the earth also might be spherical. They argued that an earth of any other shape would fall in on itself. The concept did not gain wide acceptance, however, until the time of Plato (428–348 B.C.), who argued that the sphere was the most perfect shape for a body and hence the earth, at the center of the universe, must be a sphere. Aristotle (ca. 340 B.C.) pointed out a more convincing argument: During an eclipse of the moon, when the earth comes into a position directly between the sun and moon, the shadow cast by the earth on the moon is circular. He also pointed out that a man who travels a short distance northward or southward will find a changed star pattern. These observations plus the familiar experience of seeing a ship disappear over the horizon laid a firm qualitative basis for the idea of a spherical earth early in the history of science.

A remarkably accurate estimate of the size of the earth was made by Eratosthenes about 230 B.C. He observed that at noon in midsummer a vertical stick at Syene (Aswan), Egypt, did not cast a shadow, whereas at the same time a vertical stick at Alexandria, 800 kilometers north, did cast a shadow. If the stick at Syene is considered to be in a truly vertical position, the shadow cast by the stick at Alexandria indicates that the sun there is displaced by 7°12″ from the vertical (Fig. 18–2). It was either known or assumed that Alexandria and Syene are on the same longitude (Alexandria is due north of Syene). If the distance from Alexandria to Syene could be accurately measured, the circumference of the earth, and in turn its diameter, could be calculated. Eratosthenes determined that the earth's diameter was 12,560 kilometers, which is only 112 kilometers less than it is estimated to be today. Eratosthenes gave the distance in "stadia," a measure whose actual length is now debated by historians, so he may have been much further off. Many modern astronomers have doubted the accuracy of Eratosthenes' calculations, and some have thought that by good fortune he may have incorporated some compensating errors. The method used by Eratosthenes was much like modern methods involving triangulation (a surveying method that relies on the location of points by lines of intersection).

To a first approximation the earth is a sphere, and, as such, we might expect a uniform radius and that the lengths of all one-degree arcs measured along any great circle path would be identical. There was great interest in measuring the size of the earth in Newton's time, and in 1669 J. Picard measured an arc at Paris. This work was extended by Cossini to the south into Spain. Much to the surprise of those involved, the length of a one-degree meridian arc (along a line of longitude) turned out to be 111,017 meters to the north and 111,284 meters to the south. Two explanations appeared possible: (1) that the earth is not spherical but is flattened toward the north; or (2) that there is some anomaly in France. To settle this question geodetic expeditions were sent to Peru to measure a meridian arc near the equator and to Lapland to measure an arc at high latitudes. The results showed that the earth is flattened toward the poles and is not spherical. Flattening is expressed by

$$F = \frac{a - c}{a}$$

where a is the radius near the equator and c is

the radius at the poles (Fig. 18–3). The flattening determined by the French expeditions gave a value of $\frac{1}{216.8}$. Thus the earth's shape comes close to being an *oblate spheroid*, as explained theoretically by Newton in *Principia*.

Refinement of the earth's shape is possible by (1) measuring meridian arcs on the earth's surface, (2) measuring gravity over the earth's surface, (3) determining the motion of artificial satellites around the earth, and (4) measuring precession of the earth's axis.

Departures from the Oblate Spheroidal Shape

The earth's surface is a maze of irregularities to the man walking over its surface, but these irregularities of the earth's surface must be smoothed out in making an approximation of the shape of the earth as a whole. To accomplish this, use is made of a smoothed figure of the earth called the *geoid*. The *geoid* is the shape of a *sea-level* surface. Level is determined on earth by use of *plumb bobs* that hang *normal to level*. Sea level is the mean level of the sea after tidal and wind effects are removed. The geoid is defined as the level the sea would have if extended across land areas by means of imaginary tunnels or canals.

If the earth were homogeneous and perfectly spherical, the geoid would also be a sphere. Flattening, caused by the earth's rotation, was detected early in the history of *geodesy*; thus a geoid with an oblate spheroidal shape is closer to the actual condition than the spherical shape. Irregularities in shape due to irregular mass distribution near the surface also affect the geoid. For example, the surface of the ocean is irregular as a result of the gravitational attraction between rock masses and water. Water is pulled toward the continents and toward major relief features such as the midocean-ridge systems and the Hawaiian Island swell (Fig. 18–4). Thus the geoid consists of many irregularities superimposed on a larger scale shape. Distortions of the geoid result not only from the effects of continental rock masses on water but also from variations in density distribution within the crust and mantle. These variations are of great significance in geology, but they tend to be relatively limited in extent and of small amplitude when compared with the radius of the earth.

As early as 1743 Clairaut deduced a mathematical expression to show how gravity should vary from the equator to the poles on a flattened spheroid (for derivation see Garland, 1965). The *reference spheroid* is a mathematical model that represents an idealized geoid. The geoid would be a spheroid if no lateral variations in density occurred within the earth. Clairaut's work is valid only as a first-order approximation, but the relation between shape and gravity is now one of the most powerful tools of geodesy. The formula adopted by the International Union of Geodesy and Geophysics is derived from Clairaut's work. The formula gives the value of g (γ_0) on the reference spheroid for any geographic latitude (ϕ).

$$\gamma_0 = 978.0490(1 + 0.00529 \sin^2\phi - 0.0000059 \sin^2 2\phi)$$

Note that no longitude term occurs in this equation because it was set equal to zero in Clairaut's equation.

Satellite Observations

Satellites would travel in an elliptical path within a fixed plane passing through the earth's center of gravity if the earth were a true uniform sphere, but because the earth's gravity field is uneven, the satellites are drawn into perturbed paths. The

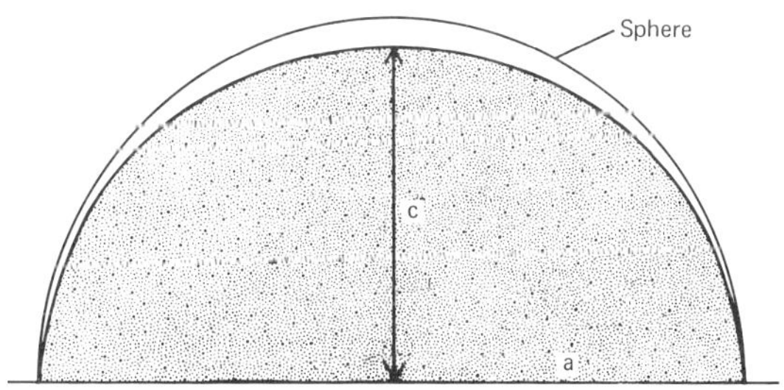

FIGURE 18–3 Relation of a flattened spheroid to a sphere of radius *a*.

FIGURE 18–4 (*Below*) Comparison of a spherical and a sea-level geoid and the effects on plum bobs.

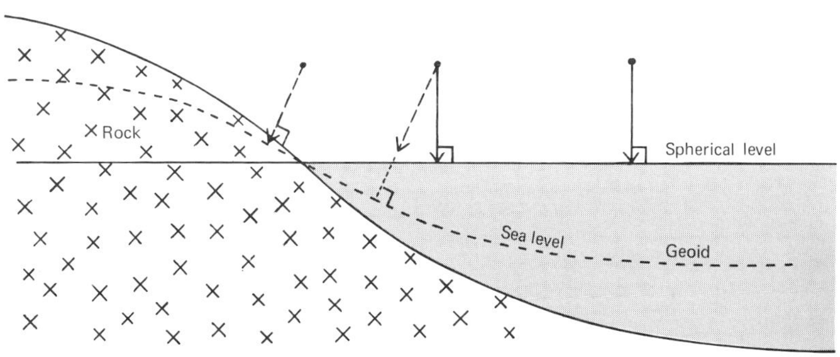

same types of observations can be made on the moon but the effects of the earth's gravity field on the satellites is nearly a million times greater. The perturbing forces are the drag of the atmosphere on the satellite and the effects of the nonsphericity of the earth, the largest component of which is the equatorial bulge. The bulge has the effects of causing the orbital plane to slowly rotate around the earth's axis of rotation and of causing the orbital ellipse to rotate within its own plane. The actual paths of satellites can now be observed and plotted with high precision and these paths are related to the flattening of the reference spheroid. Observations on Sputnik 2 showed that the amount of flattening was less than that predicted earlier. The field determined by the satellites indicates large regional bulges of the geoid in polar latitudes. These bulges have an amplitude of some 80 meters and are the basis for calling the earth pear-shaped; however, they are minor when compared with the overall flattening of some 21,000 meters between the equator and the poles.

It is more difficult to determine the amount of deviation of the equatorial plane of earth from a circle. It appears that the equatorial plane is approximately elliptical, as are planes through the poles. As the earth rotates through the orbital plane of the satellite, effects of variations from all longitudes are combined, but this combined effect can be avoided by putting the satellite in a stationary orbit. Results of some of the first studies of this type indicate that a triaxial ellipsoid is required and that the a-axis at longitude 33°W exceeds the b-axis by nearly 400 meters (Cook, 1961).

The same type of analyses of the paths of moon orbiters have resulted in knowledge of the moon's gravitational field. Some areas showed relatively high anomalies postulated to come from concentrations of mass; hence they were named *mascons*.

CAUSES OF VARIATION IN THE EARTH'S FIELD

Gravity should have a constant value everywhere on the surface of an isolated, homogeneous, perfectly spherical earth, but the earth is none of these. It is possible, however, to calculate what the force of gravitational attraction would be on an ideal earth model if the earth were isolated from the pull of other planets, the sun, and the moon, and if the mass of the earth were incorporated in a homogeneous, spherical body without irregularities on the surface. The following factors cause deviation from such a theoretical gravity field.

1. Deviation of the earth's shape from that of a sphere. Earth's radius is 21 kilometers longer at the equator than it is at the poles. Also, the earth is slightly pear-shaped. The earth behaves as a fluid balanced between gravitational forces, tending to make it spherical, and centrifugal forces, tending to flatten it. The force of attraction at the equator is less than at the poles for two reasons:
   - **a.** The surface of the earth at the equator is farther away from the center of mass.
   - **b.** Centrifugal acceleration of the earth at the equator is outward and opposite in direction to gravitational acceleration. Centrifugal acceleration is zero at the poles because they do not rotate. This results in a variation in gravity of about 5 gals equator to pole, or about 1.0 milligal per mile north-south. Effects of the deviation of earth's shape from a sphere can be calculated as described earlier.
2. Attraction of extraterrestrial bodies. Every body in the universe exerts some attraction on the earth, but since the force of attraction varies as the square of the distance between the bodies, it is unnecessary to consider the gravitational effect of bodies outside the solar system. Effects of the sun and the moon are the most notable. They produce tides in the oceans and also distort the shape of the "solid" earth. Likewise, they exert a pull on the instrument being used to measure gravity. This variation (*earth tides*) changes from hour to hour and with the seasons as positions of the moon and sun change. The range of earth tides is seldom more than 0.3 milligal and usually much less.
3. Variation with elevation. A point on a mountain is farther away from the center of mass of the earth than one at sea level. Thus gravitational attraction on objects on earth is a function of elevation. This gradient is about 0.31 milligal per meter in elevation.

4. Variation due to attraction of surface material. The mass of material between sea level and a gravity station influences measurements. This is known as the *Bouguer effect* and, depending upon density used, is about 0.1 milligal per meter in elevation. Distribution of mass in the surrounding topography also exerts an attraction on the instrument.

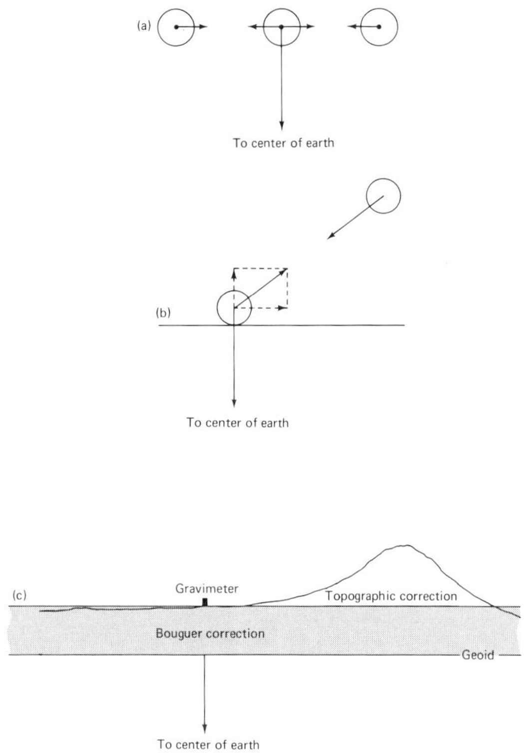

FIGURE 18–5
Gravity corrections. In (a) three masses are shown. If the gravimeter were located at the center position the attraction on the meter by the two other masses would be equal and directed in opposite directions. The two attractions would balance one another. In (b) a second mass is shown in a position above one on the ground. The solid arrows represent the direction along which the forces act. The effect on the mass at the ground surface is the same as it would be if two forces shown by the dashed arrows were acting on it. Thus part of the pull by the earth on a meter in this position would be offset by the upward force due to the second mass. This situation is essentially that caused by the sun, moon, or segments of the topography located above the position of a gravimeter. In (c) a gravimeter is shown set up at some elevation above the geoid. In this case the reading must be corrected for the effect of the topography, the distance of the meter above the geoid, and the attraction of the plate of rock between the geoid and the gravimeter.

REDUCTION OF GRAVITY OBSERVATIONS TO REVEAL ANOMALIES

Insight into the structure of the earth's crust and interior comes from interpreting deviations of actual gravitational acceleration from acceleration on an ideal earth model. Ideal in this sense means the value of g calculated by the earlier mentioned international gravity formula (see p. 427), which is based on an idealized reference spheroid that is flattened and within which there is no lateral density variation. For purposes of geological interpretation we are primarily interested in explaining deviations from ideal accelerations, called *anomalies,* in terms of lateral variations in density distribution. These lateral variations are caused by the spatial distribution of rock bodies. Anomalies may be obtained by making corrections in the observed value for the effect of latitude, terrain, Bouguer effect, *free-air* (elevation above sea level), and attraction of the sun and moon (Fig. 18–5). The observed value is thus reduced to what it would be at the geoid if all the terrain above sea level were removed and the pull of the sun and moon were nullified. If the resulting corrected value still deviates from the theoretically calculated value, the difference in the two values is a gravity anomaly and is positive if the observed value after correction (reduction) is greater than that calculated by the international formula for that latitude.

Free-Air Reduction

Gravitational attraction varies with the square of the distance between the centers of mass of the earth and the gravimeter. Thus gravimeter readings reflect changes in elevation as well as lateral changes in density of rock. Therefore if lateral variations in density are to be analyzed, meter readings must first be corrected or "reduced" to a uniform level of datum. The level used for this purpose is the geoid. The value of gravitational acceleration at sea level, g_0, is a function of the gravitational constant, G; the radius of the earth, R; and the mass of the earth, M.

$$g_0 = G\frac{M}{R^2}$$

The gravitational acceleration at some height, h, above the geoid on a 1-gram mass is:

$$g_h = G\frac{M}{(R+h)^2}$$

These two values can be related as follows:

$$\frac{g_0}{g_h} = \frac{(R+h)^2}{R^2} = 1 + \frac{2h}{R} + \frac{h^2}{R^2}$$

because R is so large the last term can be neglected and the expression becomes:

$$g_0 = g_h\left(1 + \frac{2h}{R}\right)$$

$$g_0 = g_h(1 + 0.00031h) \quad (h \text{ in meters})$$

The difference between the value g_0 and the theoretical value of gravity, γ_0, defined earlier, is called the *free-air anomaly*.

$$\text{free-air anomaly} = g_0 - \gamma_0$$

This anomaly is widely used for purposes of interpretation in gravity studies at sea. It involves the smallest number of assumptions and is consequently easy to use with other geophysical data.

Free-air correction is always positive over continents. The magnitude of the correction may be estimated as follows:

Elevation of 10 meters requires a correction of +3.1 milligals.

Elevation of 1,000 meters requires a correction of +308.6 milligals.

Elevation of 5,000 meters requires a correction of +1,543.0 milligals.

The elevation correction can be reduced to a constant. It is 0.3086 milligal per meter of elevation above the geoid.

Bouguer Correction

In making the free-air reduction it was assumed that the gravimeter reading was taken in the free air with no material between the instrument and the geoid. Actually, of course, rock material fills this space, and this rock exerts a pull on the instrument. Thus the reading of g_0 is too high by the amount of this attraction. Bouguer recognized this error and approximated the effect by computing the attraction on the instrument under the assumption that the mass of rock between the station and sea level is a plate of uniform thickness extending on a horizontal plane to infinity. Depending on the density of this plate, its attraction is nearly one-third of the free-air correction. The Bouguer correction amounts to 0.00004ρ gal per meter. When the free-air and Bouguer corrections are combined the gravity at sea level with the two corrections becomes

$$g_0 = g_h + h(0.00031 - 0.00004\rho) \text{ gals}$$

where ρ is density in grams per cubic centimeter and h is in meters.

Topographic Correction

The Bouguer correction is approximately correct if the assumption of the plate's density value is nearly correct and if topographic effects can be neglected. It is often necessary to correct for high topography. This can be done, if adequate topographic control is available, by drawing circles with successively increased radii around the station and dividing these into zones by lines radiating out from the station. An average elevation is determined for each zone, a density assigned, and the effect calculated. Some idea of the effects of the Bouguer reduction with topographic correction can be judged from the correction over Pikes Peak, 48 milligals, and over Mont Blanc, France, 123 milligals.

When the theoretical value of gravity for a given latitude calculated by the international formula is subtracted from the observed gravity at a station corrected for free-air effect, Bouguer effect, and topography, any residue is called a *Bouguer anomaly*.[2]

Anomalies are the keys to geological interpretation since they are due to local mass distributions not accounted for in the reduction. Of course, anomalies can arise from incorrect assumptions in the reduction formulae (e.g., the density of the rocks in the topography may be different from the assumed value). Woollard (1959) points out four causes of gravity anomalies:

1. Variation in near-surface densities.
2. Structure of the surface of the crystalline basement.
3. Structure of the dense sub-basement.
4. Existing degree of isostatic equilibrium.

INTERPRETATION OF GRAVITY ANOMALIES

Gravity anomalies do not provide unique solutions to subsurface structure; that is, an anomaly of a particular shape and magnitude may be explained by more than one density distribution at

[2] If no topographic corrections have been made it is usually referred to as the *simple Bouguer anomaly*, as distinguished from the *complete or terrain-corrected Bouguer anomaly*.

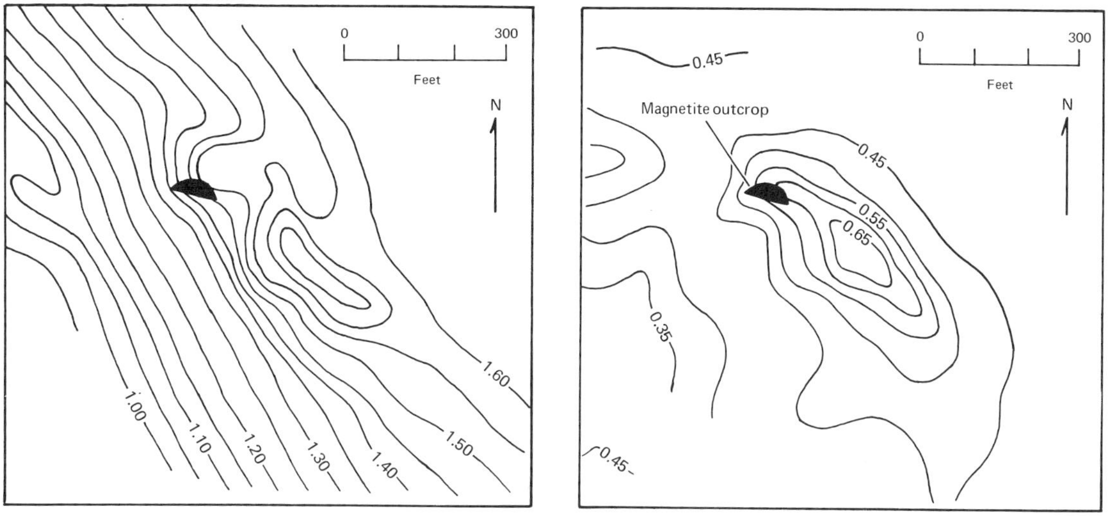

(a) (b)

FIGURE 18–6 Gravitational force over a magnetite body, Iron Mountain, Tex. (a) Before subtraction of regional; (b) after subtraction of regional.
(after Barnes and Romberg, 1949)

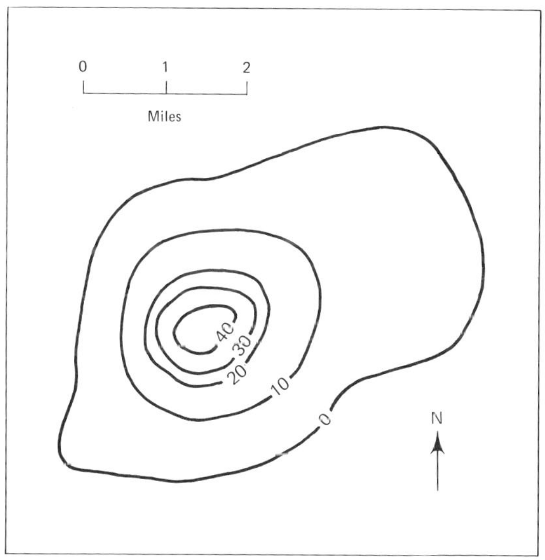

FIGURE 18–7 Contours of gravity observations over the Grand Salina salt dome. Corrections for free-air, latitude, Bouguer, and regional trends have been applied.
(after Peters and Dugan, 1949)

depth. For example, a small body of high density buried at a shallow depth might create an anomaly very similar to the anomaly of a much larger but more deeply buried mass. Despite this limitation gravity analysis is one of the most important methods of exploration for buried mineral deposits. Many of the more than two hundred salt domes in the Gulf Coastal Plain and continental shelf were discovered by gravity surveys. Salt has a lower density than the enclosing rock; thus negative anomalies are found over bodies of salt. Salt domes and associated structural features have been prolific oil-producing traps and are well known from well data. The typical plug is about 1.6 kilometers in diameter, located a few hundred to a few thousand meters deep, and shaped somewhat like a carrot, tapering downward. Gravity surveys are the most economical and rapid methods of exploring for salt domes. With new improved gravimeters it has become possible to use gravity surveys in ocean areas; and many domes have been located in the Gulf of Mexico.

Positive anomalies are found where metallic ore bodies are located in less dense country rock. Other anomalies help locate faults or folds in which high-density rocks have been offset.

Interpretation of gravity anomalies relies heavily on other types of data, both quantitative and qualitative, that limit the number of possible solutions. The interpretation must be reasonable in terms of knowledge of regional geology, the types of structures that might be expected, and the densities of the rocks that are most likely involved. Knowledge of the geology at the ground surface in land areas imposes some rigid limitations as does well-control data. Surface geology

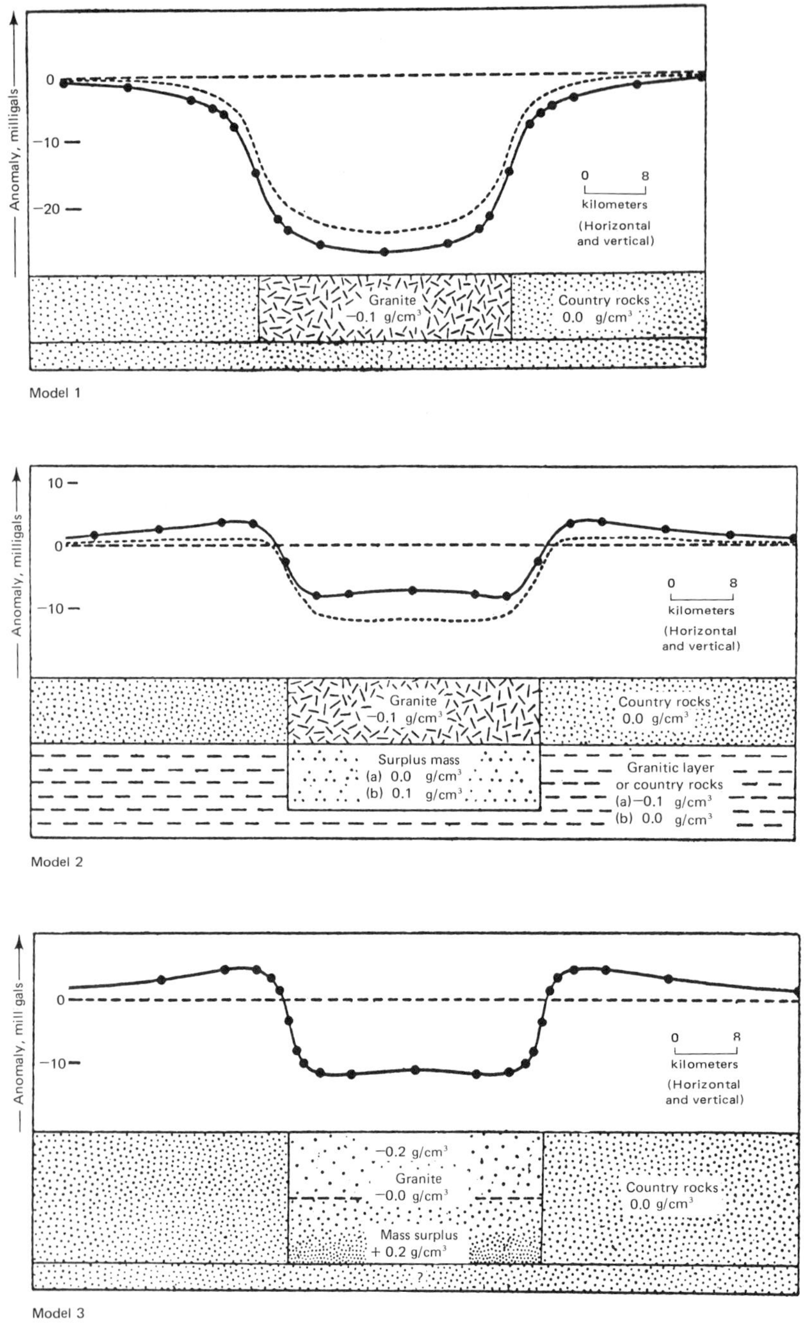

FIGURE 18–8
Models to explain anomaly pattern over granite plutons. (Model 1) A rectangular mass deficiency without a compensating surplus mass. The field across the diameter of a vertical cylinder is also shown (dashed curve). The curves are similar in shape. (Model 2) A rectangular mass deficiency with an equivalent mass surplus directly beneath as shown. The field for a vertical cylindrical model is again shown by a dashed line. (Model 3) A rectangular mass contrast in which the density is constant over each horizontal plane but changes linearly with depth from −0.20 gram per cubic centimeter at the surface to 0.20 gram per cubic centimeter at the bottom (16 kilometers down). The total relative mass deficiency is zero.
(from Bott, 1956)

can often be projected to depths of several thousand meters with a reasonably high level of confidence. Seismic methods may provide control on density distributions at depth and give position of the surface of rock bodies.

Use of models. Models are often used in gravity interpretation. A model of density distribution is constructed and the anomalies that would be produced by that model are calculated and plotted. Then a comparison is made between the calculated anomaly pattern and the observed pattern. The model must incorporate all control data and still produce an anomaly pattern close to that observed if it is to be acceptable. This method is illustrated in Fig. 18–8 as it has been applied to granite plutons. The shape and magnitude of the anomaly (Fig. 18–8) are sensitive to changes in density distribution at depth. Compare this with the observed and calculated curves over the Mascoma granite dome of New Hampshire (Fig. 18–9). Both positive and negative Bouguer anomalies are found over plutons depending on the relative density of the pluton and the enclosing country rock. A positive Bouguer anomaly is found over the Crazy Mountains, Mont. (Fig. 18–10), where diorite, plutons, and mafic dikes are intruded into tuffaceous sediments. A strong regional gradient in this area must be removed to isolate the anomaly due to the pluton.

Gravity surveys are used to interpolate between locations where seismic profiles have been made. A gravity survey is made along the line of the seismic profile, which provides the control needed to extend the interpretation.

One important conclusion about the earth's crust comes from analysis of Bouguer anomalies from all over the world. Positive Bouguer anomalies are generally found over ocean basins and negative anomalies over continents. To calculate Bouguer anomalies in water-covered areas the water must be replaced with material of average density (usually 2.67–2.80 grams per cubic centimeter) or topographic features dominate the anomaly picture (Fig. 18–11). A Bouguer-anomaly map resembles a smoothed out topographic map and the amount of the anomaly is statistically proportional to the elevation. Thus high mountains tend to be associated with larger negative anomalies than are low areas (Fig. 18–12). Free-air anomalies are often smaller if the attraction of material above sea level is neglected. This is interpreted to mean that the surplus mass present in high regions is compensated by the presence of low-density rock at depth.

The relationships between topography, crustal rock type, structure, and free-air, Bouguer, and isostatic gravity anomalies described later can be seen in profile (Fig. 18–13) across the United States from Virginia to California. The geologic cross section is based on well data and projections at depth from surface geology. The Bouguer anomaly was calculated using an average density value of 2.67 grams per cubic centimeter for crustal rocks.

ISOSTASY

The theory of isostasy postulates a system for large-scale distribution of material in the earth's crust and interior. The theory is based on, and explains, observed gravity values. We can, of course, measure the density of rocks exposed at the surface and in wells, but these measurements give an idea of the density of only the outermost part of the crust. Rocks making up continents are generally found to be of lower density than those in the ocean basins. The difference is essentially that between basalt (2.8–3.1) and granite (2.5–2.8). This does not answer the important question: How are continents supported so high above the level of the ocean floors? Is the earth's mantle rigid and do continents simply rest on a solid platform, or is this weight of the continents buoyed up in some way? This problem concerned Leonardo da Vinci in the sixteenth century. He concluded that the mountains were lighter than the plains and that this was in part responsible for their relative elevations. Pierre Bouguer, who was a member of a French expedition sent to Peru in 1749 to establish by survey the length of a meridian arc, reported that the Andes have less mass than had been expected. An explanation offered soon after this was that mountains have less density because they represent material that has risen as a result of thermal expansion of rocks at depth.

The next major advance came a hundred years later, in 1850. Another survey was being made to establish the length of a meridian arc. This time the survey was being conducted in India and one of the geodesists was a man named George Everest, for whom Mount Everest was named. Two methods, surveying and astronomical observation, were used to determine the difference in latitude between two cities. When the work had advanced far enough, it was discovered that a difference in latitude measurements made by these two methods amounted to $5\frac{1}{4}$ seconds of latitude in a distance of only 600 kilometers. No errors could be found in either set of data to explain such a large difference. One of the men studying the data was J. H. Pratt, who concluded that the error had been introduced because the leveling system, a plumb bob, used to set up the telescope to make the astronomical observations had been thrown off by the mass of the Himalayas. He calculated what the deflection should be at each of the two stations, taking into consideration the mass of the nearby mountains. He concluded

FIGURE 18–9
The Mascoma dome, New Hampshire.
(after Bott (1956); adapted from Bean, 1953)

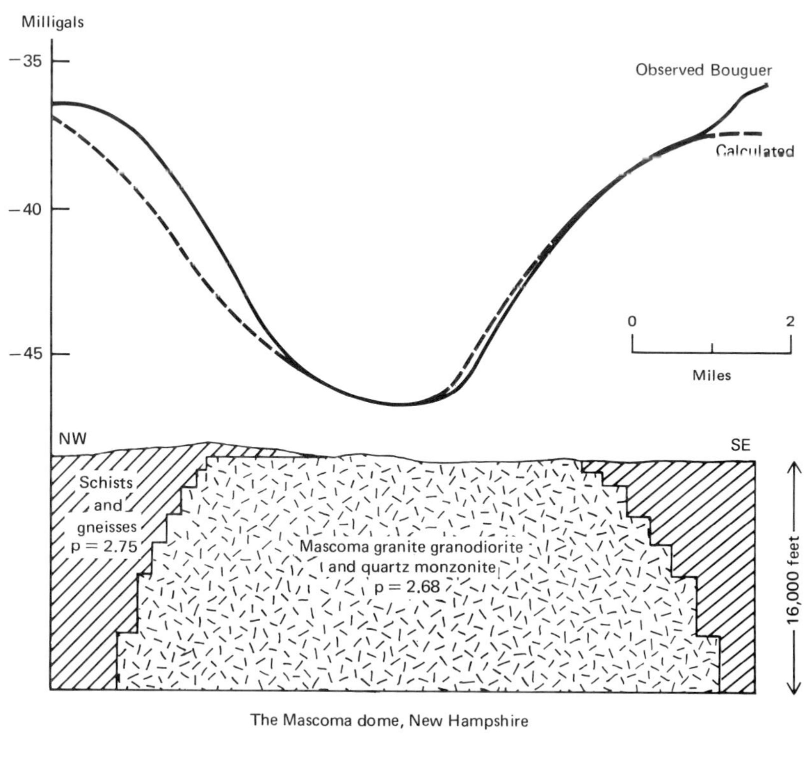

The Mascoma dome, New Hampshire

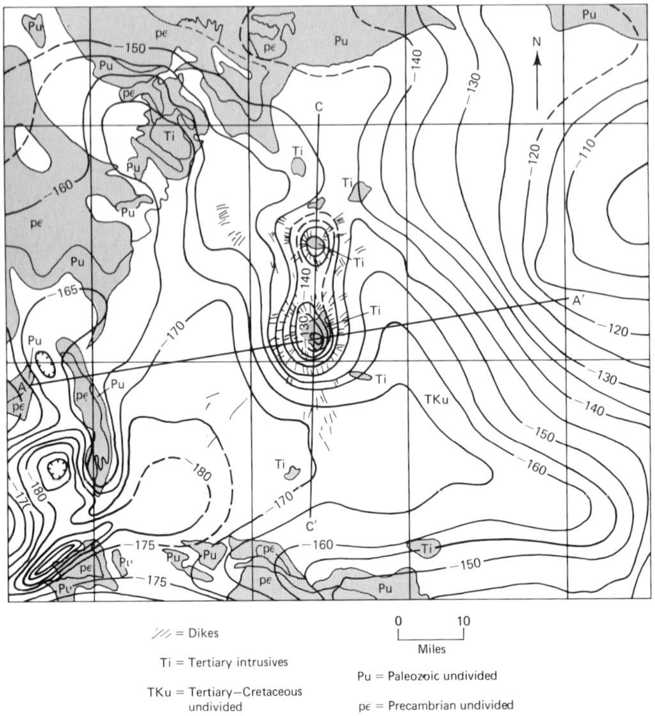

FIGURE 18–10
Bouguer-anomaly map, profiles, geologic cross sections, and gravity model for one profile across the Crazy Mountains, Mont.
(from Bonini, 1967)

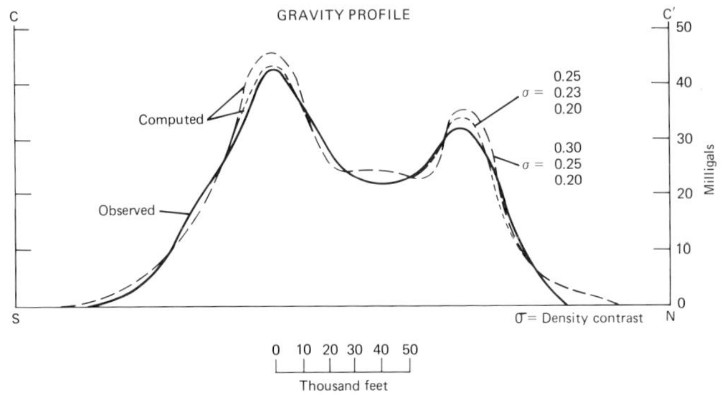

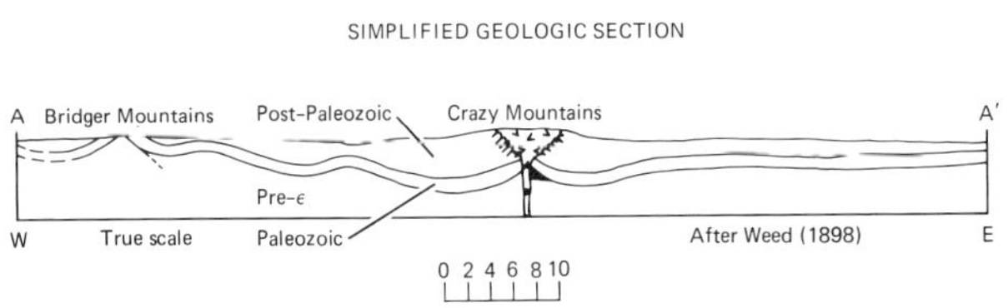

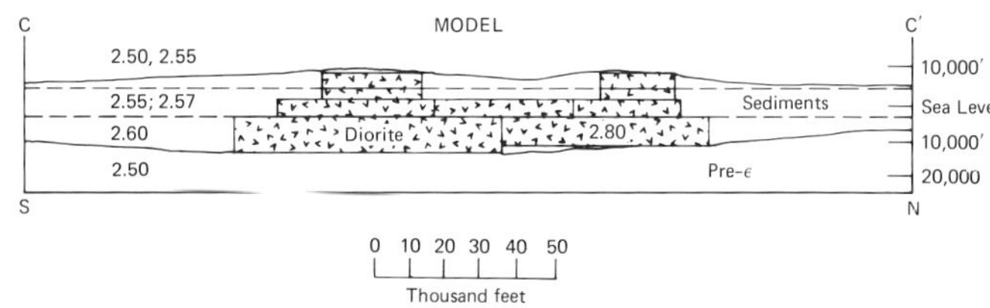

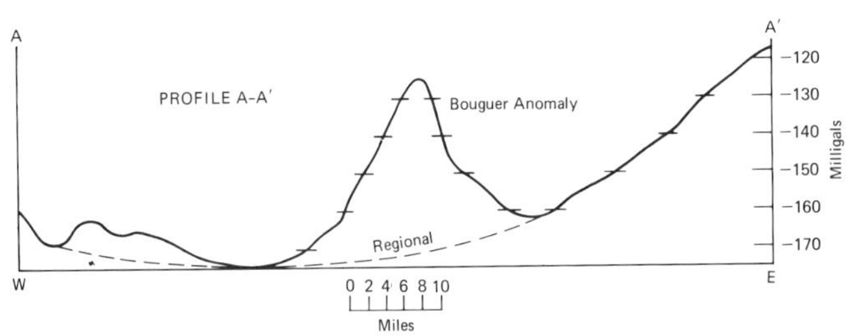

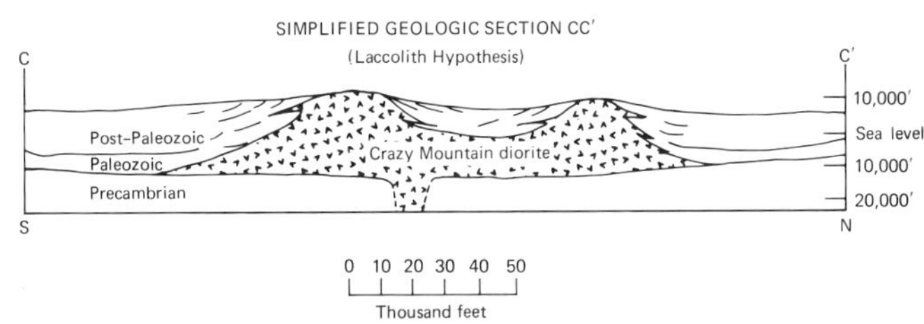

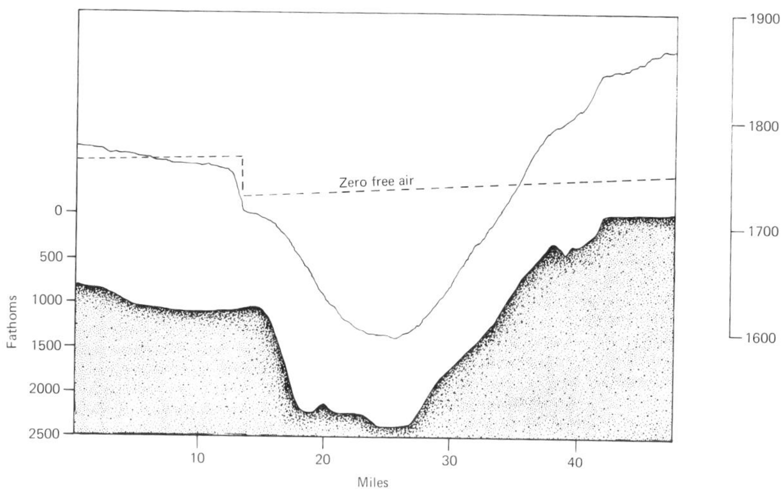

FIGURE 18–11
Record made by the Graf-Askania sea gravimeter. The difference between the dashed line and the recorded trace gives the free-air anomaly at any point. Note the close correspondence of gravity to topography (shown at the bottom of the figure). The slope of the zero free-air line reflects the change of gravity with latitude and the step at the left is caused by gravity correction resulting from change in the heading of the ship.
(after Talwani, 1964)

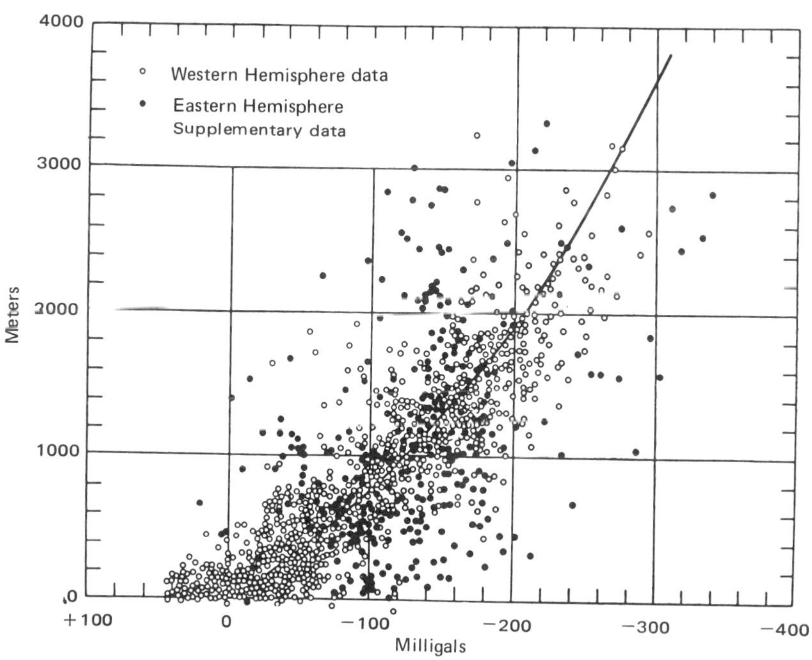

FIGURE 18–12
Relation of Bouguer gravity anomaly to surface elevation. Supplementary data are given for the Eastern Hemisphere.
(from Woollard, 1959)

that the two readings should have been off by nearly 16 seconds instead of $5\frac{1}{4}$ seconds. This surprising discovery—that the mountains did not cause as much deflection of a plumb bob as they should—led him to believe that part of the attraction must be compensated for within or under the mountains by a mass deficiency. Pratt developed the idea that mountains are caused by expansion of high-density rocks at depth. This model allowed the mountain range to have lower density, accounting for the observed deflection of the plumb bob (which was deflected toward the mountains, but not as much as it should have been). He held that this thermal expansion gradually decreased at depth under the mountains in such a way that eventually a depth was reached at which density would be uniform. He calculated that at a depth of about 150 kilometers the surficial differences in gravity would be evened out. This depth would thus be a depth at which compensation occurs.

Within a few months of the time Pratt submitted his paper to the Royal Society of London, another scientist, G. B. Airy, submitted a paper (1855) on the same subject. Airy had a different idea of the mechanism of density distribution.

> I conceive that there can be no other support than that arising from the downward projection of a portion of the earth's light crust into the dense lava; the horizontal extent of that projection corresponding rudely with the horizontal extent of the tableland, and the depth of its projection downwards being such that the increased power of flotation thus gained is roughly equal to the increase of weight above from the prominence of the tableland. It appears to me that the state of the earth's crust lying upon the lava may be compared with perfect correctness to the state of a raft of timber floating upon the water; in which if we remark one log whose upper surface floats much higher than the upper surfaces of the others we are certain that its lower surface lies deeper in the water than the lower surface of the others.
>
> It will be remarked that the disturbance depends on two actions; the positive attraction

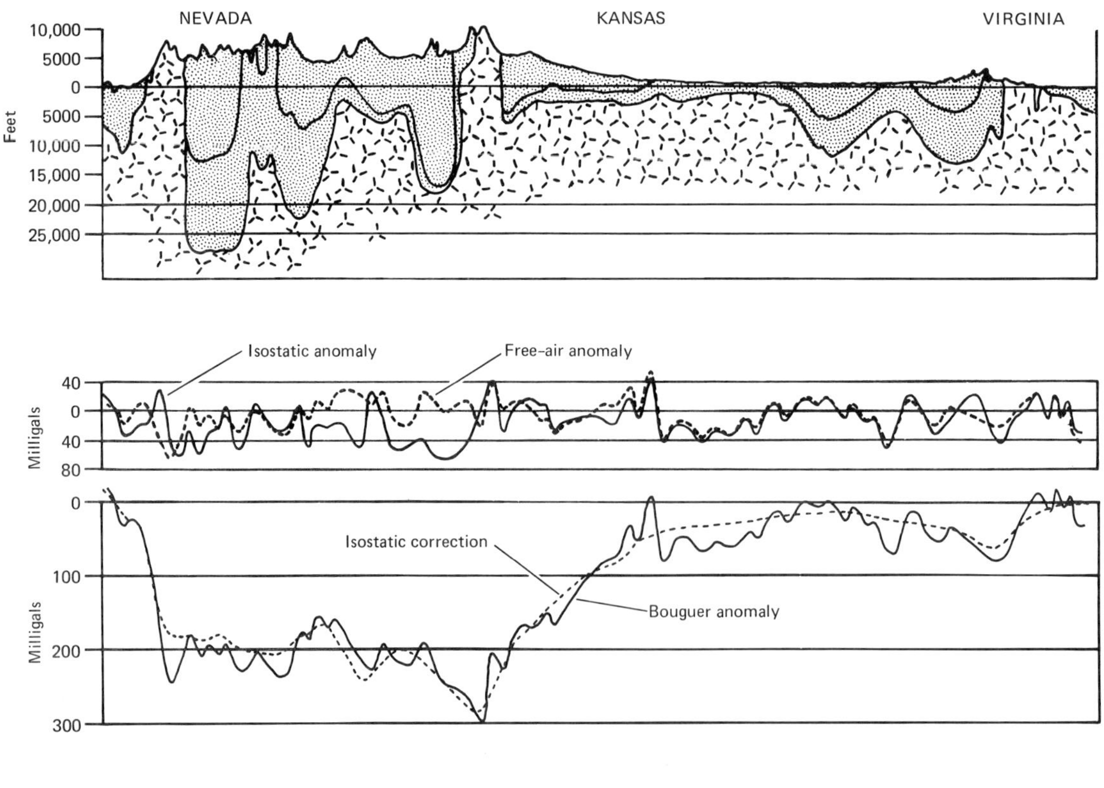

FIGURE 18–13
Geologic cross section (top); free-air, Bouguer, and isostatic anomalies along a profile across the United States from Virginia to California.
(after Woodward, 1966)

> produced by the elevated tableland; and the diminution of attraction, or negative attraction, produced by the substitution of a certain volume of light crust (in the lower projection) for heavy lava.
>
> The diminution of attractive matter below, produced by the substitution of light crust for heavy lava, will be sensibly equal to the increase of attractive matter above.

Thus was originated the idea that mountains have roots.

About thirty years later C. E. Dutton defined the term *isostasy* as follows:

> If the earth were composed of homogeneous matter its normal figure of equilibrium . . . would be a true spheroid of revolution; but if heterogeneous, if some parts were denser or lighter than others, its normal figure would no longer be spheroidal. Where the lighter matter was accumulated there would be a tendency to bulge, and where the denser matter existed there would be a tendency to flatten or depress the surface. For this condition of equilibrium of figure, to which gravitation tends to reduce a planetary body, irrespective of whether it be homogeneous or not, I propose the name isostasy. . . . We may also use the corresponding adjective, isostatic. An isostatic earth, composed of homogeneous matter and without rotation, would be truly spherical. . . .
>
> But if the earth be not homogeneous—if some portions near the surface are lighter than others—then the isostatic figure is no longer a sphere or spheroid of revolution, but a deformed figure, bulged where the matter is light and depressed where it is heavy. The question which I propose is: How nearly does the earth's figure approach to isostasy?

Modern measurements show the answer to Dutton's question to be that the earth as a whole closely approaches the condition of *isostatic equilibrium.* The original models of Pratt and Airy have been modified, but regardless of which of these models is used, the conclusion is reached that the earth is essentially in a condition of isostatic equilibrium.

Pratt assumed that the boundary between upper light material and lower dense rocks is at a uniform depth, called the *depth of compensation.*

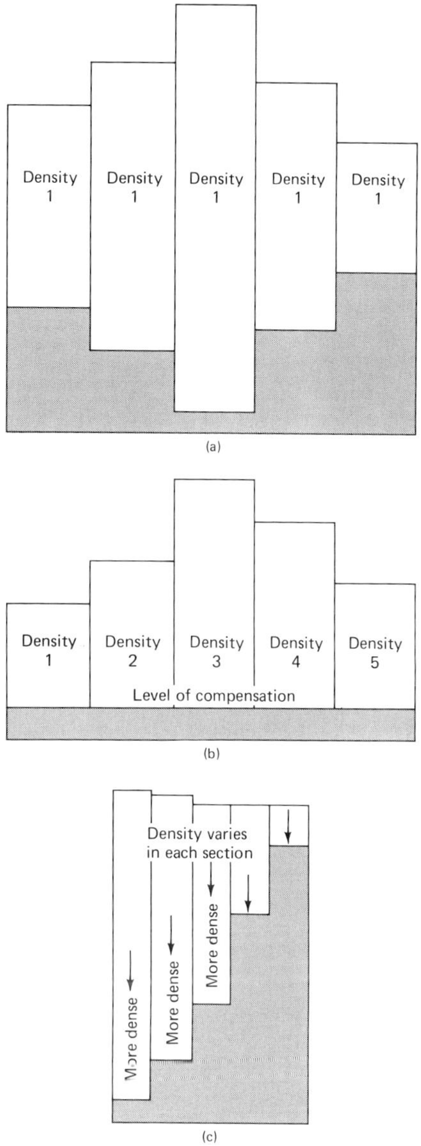

FIGURE 18–14
(a) Airy's theory of isostasy. This illustrates one way in which isostatic equilibrium might be obtained if the crust of the earth consisted of an essentially uniform density unit of different heights. Airy would account for the mountains by the presence of deep roots of crustal material extending down into the high-density mantle. This actually represents a case of flotational equilibrium. The mantle rock is plastic and can flow.

(b) Pratt's theory of isostasy. Pratt recognized that the crust of the earth is not made up of materials of uniform density. He visualized isostatic equilibrium as being the result of density. The low-density materials would represent the continents and the high-density materials, the ocean. He postulated a depth or level of compensation at which the effects of different densities above were balanced.

(c) Heiskanen's theory of isostasy. He combines the assumptions of both Pratt and Airy. According to his theory density varies both between columns and within each. The higher densities are represented by the shorter columns and toward the bottom of each column. This theory accounts for the roots of mountains, and for the variations in density in different parts of the crust.

He further postulated that variations in the density of the lighter layer are related to the elevation of the surface. Lighter material lies under mountains, and heavier material under oceans. The weight of columns of rock extending from the surface to the depth of compensation in different parts of the earth is thus the same. The United States Coast and Geodetic Survey uses as the depth of compensation 113.7 kilometers. This figure is obtained by calculating isostatic anomalies for stations at a large number of different elevations, using different depths of compensation. The depth that gives the smallest average anomaly is considered the most probable actual depth of compensation (Fig. 18–14).

Heiskanen's Modifications

Airy postulated that columns of rock are nearly the same in density, floating like blocks in a fluid. Pratt suggested that different segments of the earth are of different densities, but that the differences are compensated at a certain depth. A third hypothesis has been formulated by Heiskanen (1933). He combined the assumptions of both Airy and Pratt (Fig. 18–14). It has been observed that rocks at sea level are more dense on the average than those at higher elevations (2.76 grams per cubic centimeter at sea level down to 2.70 grams per cubic centimeter at elevations in high mountains). Other evidence of an increase in density with depth comes from the field of seismology. Heiskanen assumes that this change continues downward, tending to make deeper rocks more dense than shallower ones in all sections of the earth's crust. Heiskanen's theory is based on knowledge of density variations obtained by direct measurement, and it yields low isostatic anomalies.

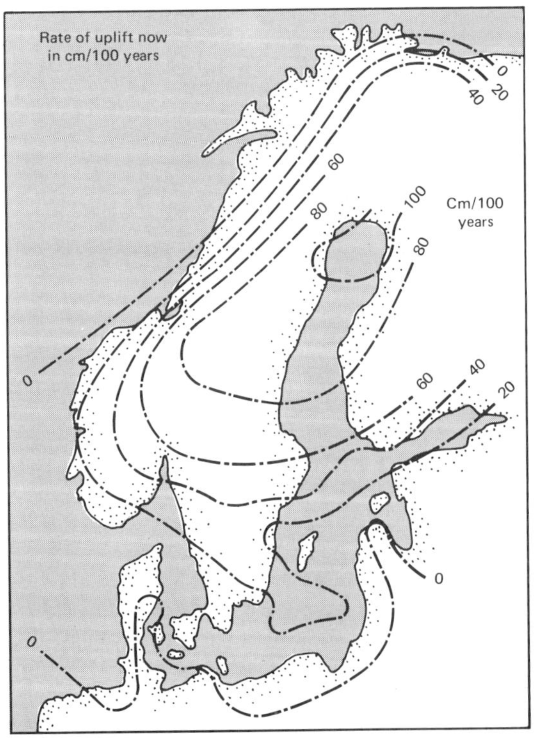

FIGURE 18–15
Rate of uplift in Scandinavia. This map shows the present rates of uplift in centimeters per 100 years.
(after Beno Gutenberg, 1941)

Application of Isostasy

According to all three of these theories the earth's crust is in a state of approximate equilibrium, meaning that few large imbalances in the distribution of mass occur. Most geologists and geophysicists support either Pratt's or Heiskanen's theory or some modification thereof. If either is correct, then we have an extremely valuable tool that may be used to explain large-scale crustal features and to predict behavior of the earth's crust in response to changes in the distribution of mass brought about by such processes as volcanism, mountain building, and erosion. Undoubtedly the crust and interior of the earth down to the depth of compensation have some strength and rigidity that enable them to support loads of rock for at least short periods of time, but over long periods of time a state near compensation appears normal.

The theory of isostasy holds meaning both for the interior and for the crust of the earth. If a zone of compensation does exist, then the rock below that zone must be in a plastic or pseudoviscous state in order to maintain compensation as loads on the crust are shifted from one place to another. Tops of mountains are eroded rapidly, and this debris is transported away by mass movement, running water, and glaciers. As more and more mass is moved, the column of rock through the crust under the mountain is lightened in relation to other columns in the crust. Since pressure at the zone of compensation is due to the weight of the column of rock over it, the pressure under the mountain would be lessened. This would cause material below the zone of compensation to move toward that position and raise the column of rocks under the mountains again. Thus mountains tend to rise as they are eroded away.

Though most of the crust is in a state of isostatic equilibrium, the 3 percent that is not provides some interesting applications of the theory. Among the most striking of these are the isostatic anomalies. An *isostatic anomaly* is that gravity value remaining after all earlier corrections have been made and an isostatic correction is applied.[3]

[3] An isostatic correction is similar to the Bouguer correction. The effect of a slab of material with some estimated density and a thickness from the geoid down to the depth of compensation is calculated.

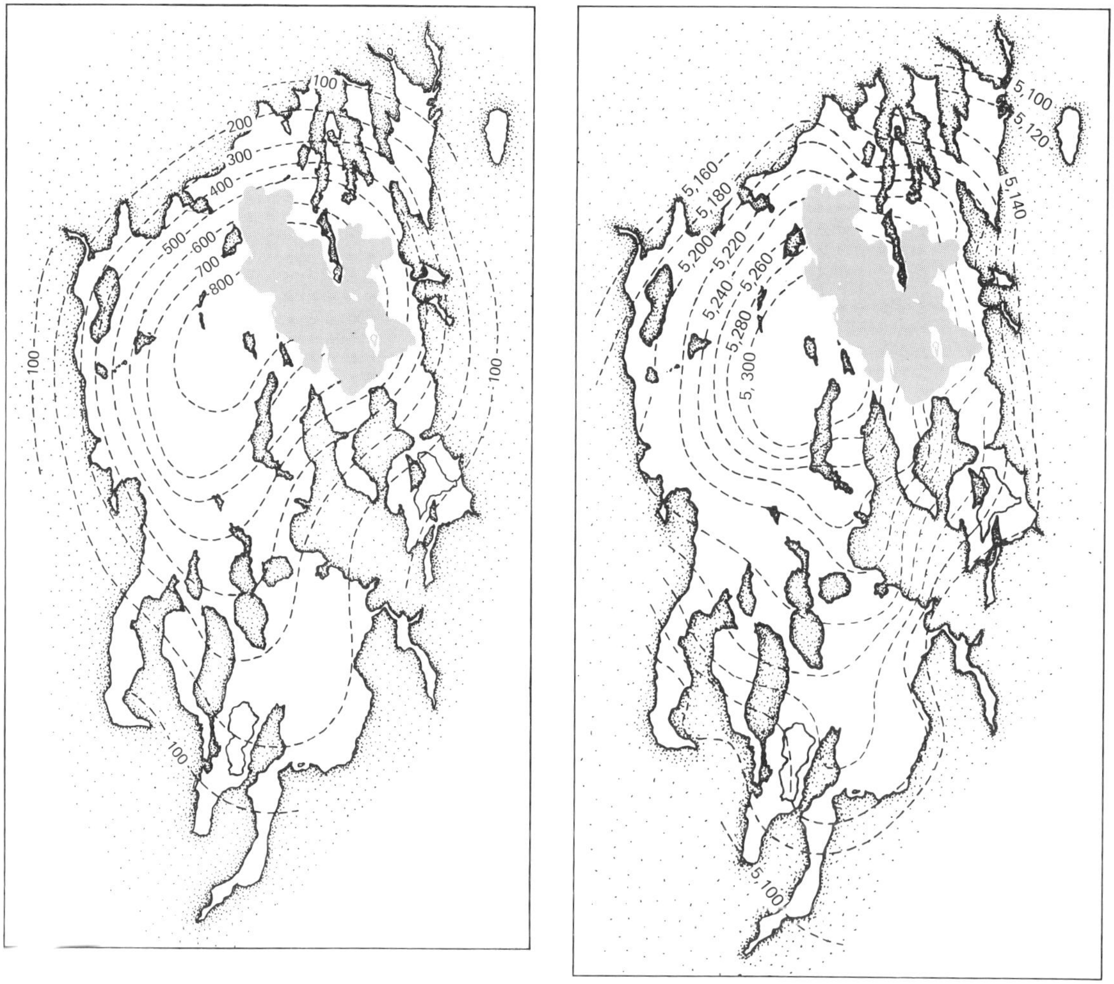

FIGURE 18–16
Isostatic deformation of Lake Bonneville. Lake Bonneville was one of the great periglacial lakes of the Basin and Range province. Great Salt Lake (dark gray) is a remnant of it. These two diagrams depict the present position of one of the former shore lines of Lake Bonneville. The left diagram shows contours for the depth of water in the old lake, averaged over 40-mile-radius circles. As the water has drained out and slowly evaporated, the lake has been uplifted, deforming the former shore lines; the right diagram shows the present elevation of the deformed shore line. The shore lines have been elevated most in those areas formerly covered by the greatest depth of water.
(after Crittenden, 1963)

The isostatic correction is derived by determining the gravitational effect of the less dense "root." Thus a mountain range in isostatic equilibrium, such as the Rocky Mountains, has a Bouguer anomaly of −300 milligals and the isostatic correction for the less dense root is equal to +300 milligals, giving an isostatic anomaly of zero. Negative isostatic anomalies are associated with the regions that during the ice ages were sites of thick accumulations of ice, in some cases more than 1.6 kilometers thick. Such areas include parts of the Canadian Shield and the Fennoscandinavian Shield. Careful surveys have been made across both areas in recent years. Results show that portions of Scandinavia are being uplifted at rates as much as a centimeter each year (Fig. 18–15). Marks made at sea level in the region hundreds of years ago confirm that this uplift has been going on a long time.

These same regions are sites of negative gravity anomalies signifying a deficiency of mass. This can be interpreted in terms of movement of subcrustal rocks. When the ice was present, plastic flowage in the subcrust took place out and away from the centers of ice accumulation. The ice melted more rapidly than compensation by return flow in the subcrust took place. Flowage back into the areas is still going on. The size of the anomalies is an indication of how much more movement must take place before the areas are compensated again. One estimate is that parts of Scandinavia will be uplifted by close to 200 meters before compensation is complete. In Canada we can predict on this basis that Hudson Bay will slowly disappear as the continent rises. Similar adjustments have taken place around the Great Salt Lake (Fig. 18–16).

Positive anomalies are often found over volcanoes. Such an anomaly indicates that the volcano has a surplus of mass and that it is not underlain by a void. It also indicates that the crust is locally strong enough to hold up the mass of the volcano.

Another example is found in the Mississippi Delta, where large quantities of sediment are being deposited annually on the continental shelf. These deposits might be expected to increase the mass of that segment of the crust, but measurements in the region reveal no positive anomaly. This is a region where compensation is taking place.

CHAPTER OUTLINE

The Law of Gravitational Attraction
Measuring the Force of Gravity
Mass and Density of the Earth
Shape and Size of the Earth
- Departures from the Oblate Spheroidal Shape
- Satellite Observations

Causes of Variation in the Earth's Field
Reduction of Gravity Observations To Reveal Anomalies
- Free-Air Reduction
- Bouguer Correction
- Topographic Correction

Interpretation of Gravity Anomalies
- *Use of models*

Isostasy
- Heiskanen's Modifications
- Application of Isostasy

TERMS

anomalies
Bouguer anomaly
Bouguer effect
depth of compensation (isostatic)
earth tides
free air
free-air anomaly
gal
geodesy
geoid
gravimeter
isostasy
isostatic anomaly
isostatic equilibrium
mascons
oblate spheroid
plumb bobs
reference spheroid
sea level

REVIEW QUESTIONS

1. Normal sea level is used as the surface of the geoid, although even this is subject to irregularities. Why?
2. The earth's gravitational field would be uniform if it were a perfect, isolated, homogeneous sphere. However, it is not. What are some of the most important considerations affecting gravitational differences?
3. What is a Bouguer anomaly? How is it determined?
4. Although gravity anomalies are not subject to unique interpretation, they are useful in interpreting subsurface density changes. What additional information is needed in order to make a unique interpretation of gravity data?
5. In general, a map of Bouguer anomalies shows positive anomalies over ocean basins and negative anomalies over continents; it also approximates a

rough topographic map. What does this imply about the earth's crust in general?

6. In the earliest studies it was discovered that mountains generally did not possess as much mass as was assumed. How did Pratt's study in India indicate this?

7. Compare Airy's concept of isostasy with that of Pratt and then draw a schematic and simple diagram depicting your idea of their conceptual differences.

8. In the Fennoscandinavian area and Canadian Shield, rebound is taking place. Explain the mechanism involved here and the use of gravity anomalies and isostatic theory as applied in making predictions of future amounts of rebound.

9. What does it mean to say that a region has a positive isostatic gravity anomaly?

10. Under what conditions will gravity studies be most effective in exploration for economically important materials?

SUGGESTED READINGS

Dobrin, M. B., 1960, Introduction to Geophysical Prospecting: New York, McGraw-Hill, 446 p.

Garland, G. D., 1965, The Earth's Shape and Gravity: New York, Pergamon Press, 183, p.

Heiskanen, W. A., and **Vening-Meinesz, F. A.,** 1958, The Earth and Its Gravity Field: New York, McGraw-Hill, 470 p.

Jacobs, J. A., Russell, R. D., and **Wilson, J. T.,** 1959, Physics and Geology: New York, McGraw-Hill, 424 p.

Sumner, J. S., 1969, Geophysics, Geologic Structures and Tectonics: Dubuque, Iowa, Wm. C. Brown, 115 p.

19 THE EARTH'S MAGNETIC FIELD

19 THE EARTH'S MAGNETIC FIELD

MAGNETS AND MAGNETIC FIELDS

The Chinese knew of some phenomena of magnetism as early as 2000–3000 B.C. The Greeks knew of the magnetic effect of lodestone by the sixth century B.C. and primitive compasses were in use by the Arabs and the Chinese by A.D. 1000. William Gilbert (1540–1603), founder of the field of geomagnetism, demonstrated that the earth as a whole behaves like a great bar magnet. Among the observations that had been made by Gilbert's time were the following:

1. Certain rocks, known as lodestone and composed of the mineral magnetite (Fe_3O_4), are natural magnets.
2. Several other materials, including steel, may be made into magnets by being held close to lodestone. When a bar-shaped magnet of steel is placed in a pile of iron filings, the filings will be attracted to the steel, with more being attracted to the ends of the bar than to the middle. The points where the most filings are held are called the *poles* of the magnet.
3. If a steel needle is magnetized and placed carefully on the surface of water so that it floats, it will rotate until one end points north and the other south. The two ends are known as the *north* and *south poles* of the magnet.
4. If two such magnetic needles are brought close together, opposite poles attract one another and similar poles repel one another.
5. The force of attraction or repulsion between two magnetic poles is directly proportional to the product of their pole strengths and inversely proportional to the square of the distance between them.

The force of attraction between two poles of strengths p_1 and p_2 separated by a distance d is given by

$$F = \frac{1}{\mu}\frac{p_1 p_2}{d^2}$$

where μ is a constant known as the *magnetic permeability* of the medium in which the poles are located; it has a value of 1 for air. The unit of pole strength, a *unit pole,* is of such strength that it will repel a similar pole of equal strength with a force of 1 dyne when the two are separated by 1 centimeter in a vacuum. The area around a magnet within which the influence of the magnet can be detected is called the *magnetic field.* The field for the earth is very great indeed, but for a small magnet the field loses strength rapidly with distance. A magnetic field at a point may be described in terms of (1) the orientation a very small bar magnet at that point would assume if it were free to move and (2) the intensity or magnetic strength of the field at that point. *Magnetic-field strength* (H) is the force (in dynes) that would be exerted on a unit pole at the point.

$$H = \frac{p}{\mu d^2}$$

The force is designated in terms of dynes per unit pole strength, called *oersteds.*

Field strength may be visualized in terms of "lines of force" in the field. The orientation of the lines indicates the direction of the magnetic field and the number of lines in a cross-sectional area indicates the field intensity. The number of lines of force per square centimeter is taken as equal to the number of dynes with which the field would act on a unit pole. Thus one line of force per square centimeter indicates a strength of one oersted. (The name *gauss* is sometimes used instead of oersted. The two are numerically equal for $\mu = 1$.) The total magnetic field of the earth is normally $\frac{1}{2}$ oersted. Instruments designed to measure magnetic-field strength are capable of detecting variations of a very small fraction of an oersted; thus, a unit called the *gamma* ($= 10^{-5}$ oersted) is used.

MEASUREMENT OF THE EARTH'S MAGNETIC FIELD

The total intensity and the direction of the lines of force of the geomagnetic field are the values most frequently measured in studies of earth magnetism. The direction of the field is determined by measuring the direction of horizontal and vertical components of the field. Both of these can be found by use of a compass needle; however, in practice sensitive measuring instruments are used. A compass is designed to align itself in the direction of the horizontal component of the field, and a *dip needle,* in which the magnetized needle is mounted to rotate on a horizontal axis, will measure the *inclination* (a measure of the slope of a plane or plunge of a line). If a magnetized needle free to rotate in any direction could be suspended from a point it would align itself in the field.

Early dip needles, used to determine local anomalies due to buried bodies of magnetic mineral such as iron ore deposits, were mounted so that they rotated freely on a horizontal axis and the south end of the needle had a counterweight mounted on it for use in the Northern Hemisphere. In this procedure, the counterweight is moved along the needle until the needle is horizontal—a point at which the gravitational torque on the south end of the needle is equal to the magnetic torque acting on the north end. Then any anomalies within a local region can be detected by deviation of the needle from the horizontal. A survey is made by establishing a line or network of stations. As the dip needle is moved over an ore body, the needle is deflected from the normal position by the anomalous field in the vicinity of the ore body.

The Schmidt-type *magnetic balance* has been widely used for prospecting purposes, although it is now being replaced by the *flux-gate magnetometer.* The magnetic balance is similar in principle to the dip needle, but is more sensitive. It employs a magnet that is pivoted slightly off the center of mass. A balance is reached between magnetic and gravitational torques around a pivot point; the angle of inclination, which depends on the strength of the field, is measured. Measurements to within 1 gamma are possible. Two magnetometers are used, one for the vertical and one for horizontal components of the field.

The flux-gate magnetometer consists essentially of two cylinders made of materials with high permeability and wrapped with identical but oppositely wound coils. These coils are connected to an alternating current that magnetizes the cores first one direction of polarity, then the other. The earth's field will add to the flux or field strength in one coil and decrease the other. This causes a voltage drop between the two coils, which is a measure of the earth's field strength. These instruments are accurate to about 1 gamma.

Proton-precession magnetometers are widely used to determine the total field intensity. They are suitable for use on ships and airplanes as well as on land. This type of magnetometer consists of a bottle of fluid (water or an oil) suspended within a wire coil. The protons in the water act like small magnets that have a tendency to align themselves in the earth's field. When a current is passed through the coil a much stronger magnetic field is created within which the protons become aligned.

When the current is switched off after a few seconds, the protons begin to precess in a path around the earth's field. *Precession* is a motion like that described by a spinning top when the axis is not vertical. The axis of the top spins and also slowly changes orientation, describing a cone around a vertical line. The spinning of the protons generates an electromotive force in the coil that has the same frequency, f, as the precession and can be measured.

$$f = \frac{\gamma_p H}{2\pi}$$

where γ_p is an experimentally determined constant and H is the total field.

The similarity among magnetic, static electric, and gravitational phenomena is striking. In 1820 Hans Oersted made public his discovery of an important relation between electric and magnetic fields. He had placed an electric wire in which a current was moving over a compass needle and discovered that the needle rotated until it was oriented at right angles to the wire. He showed that electric charges not only have electric fields associated with them, but also that when the charge is in motion it sets up a magnetic field oriented perpendicular to the direction of motion of the charge. This connection between electric and magnetic fields set the stage for modern ideas about the origin of the earth's magnetic field. But before turning to modern developments, we will examine the nature of the earth's field (Fig. 19–1).

Gilbert thought that the earth behaved as a magnet because a large body of permanently magnetized rock like lodestone was buried in the earth. Such a bar as Gilbert visualized would not pass through the center of the earth, however; it would miss the center by almost 960 kilometers. The *north magnetic pole* is not situated at the

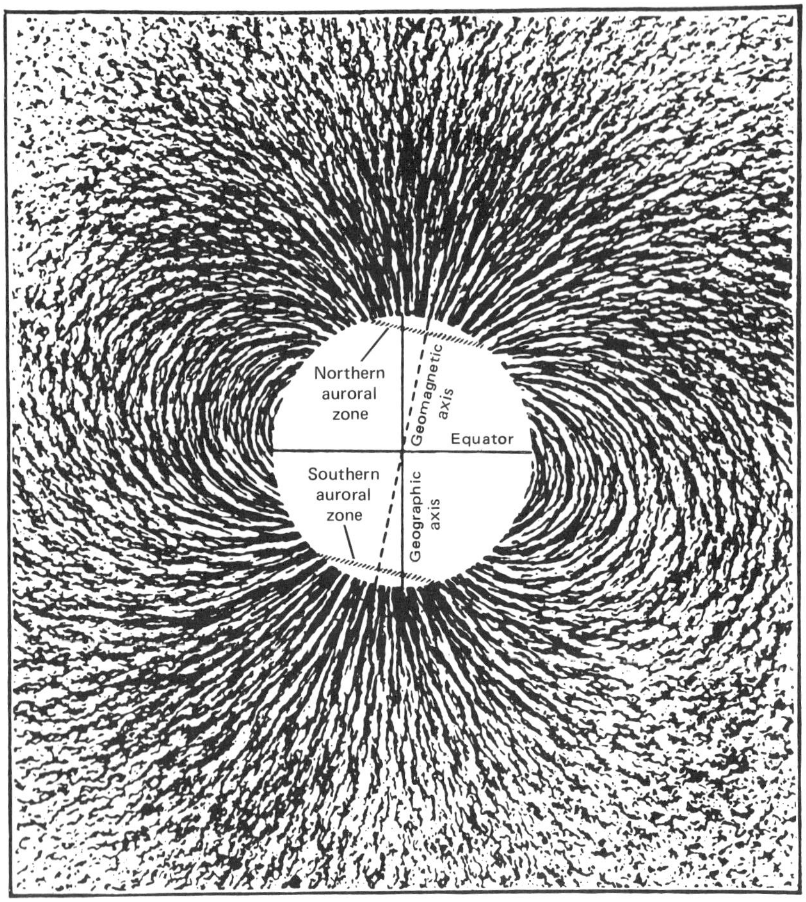

FIGURE 19–1
The magnetic field of the earth.
(from Vestine, 1960)

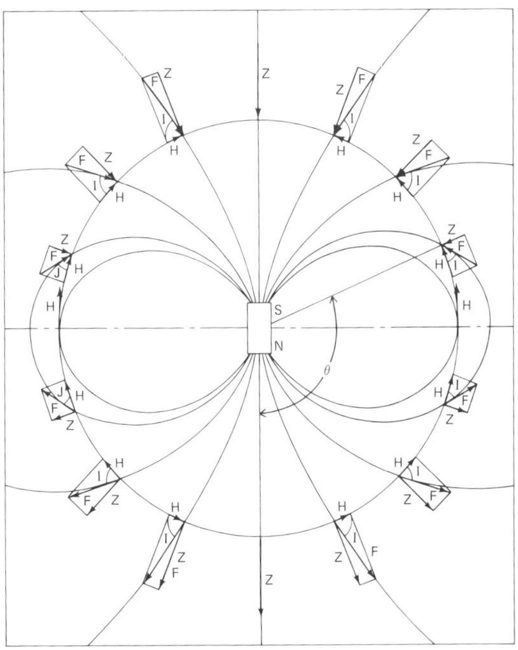

FIGURE 19–2
A simplified model of the earth's magnetic field at various latitudes.
(after McComb; courtesy of U.S. Coast and Geodetic Survey)

north *geographic* pole, which is a pole of *rotation*, but rather in northern Canada at about latitude 75°N. The *south magnetic pole* is located at latitude 68°S, south of Australia. Also the positions of the magnetic poles are known to shift. Some shifts are periodic, but others represent permanent drifting.

DESCRIBING THE EARTH'S MAGNETIC FIELD

The nature of the field is depicted by maps showing lines of equal inclination of the field in relation to the earth's surface, lines of equal intensity or strength, and lines of equal *declination* (the horizontal angle at a locality between the direction to magnetic north and true [geographic] north) of the magnetic pole from the geographic pole (Figs. 19–2 to 19–5). Figure 19–6 shows a hypothetical scheme of these lines. The total intensity or strength of the earth's field is determined by finding the force of attraction at a given place. As expected, regions near the poles have the highest magnetic intensity, but the south magnetic pole is slightly stronger than the north, suggesting that the center of the earth's field is somewhat closer to the surface at the south magnetic pole.

Although Gilbert was wrong in thinking that the earth's field was produced by buried magnets, it is now known that rocks differ in their magnetic properties. Some rocks contain large proportions of magnetite, and a great many other minerals are slightly magnetic. When sufficiently accurate maps of the earth's field are prepared, irregularities—magnetic anomalies—resulting from distribution of these rocks of differing magnetism show up. This fact has proved to be a most useful tool in exploration for economically important iron ore and other deposits that have magnetic minerals associated with them. When a measurement of the magnetic field is made, the instrument used is affected by a force field that includes various components: magnetic fields external to the earth; those arising from deep internal causes; and others caused by near-surface variations in the distribution of magnetic minerals. This last component of the field, which is of great importance to the geologist who seeks data on the location of mineral deposits or the pattern of sub-

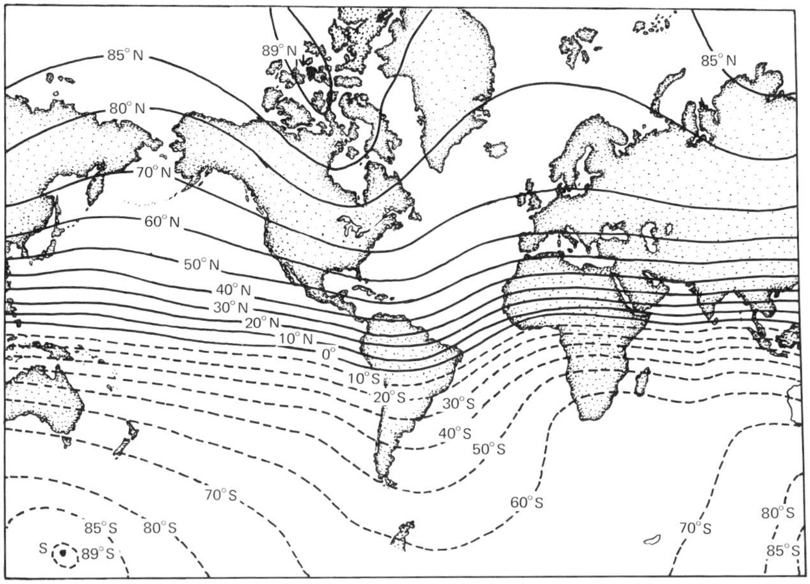

FIGURE 19–3
Lines of equal inclination of the earth's magnetic field.
(after E. H. Vestine and others, 1947)

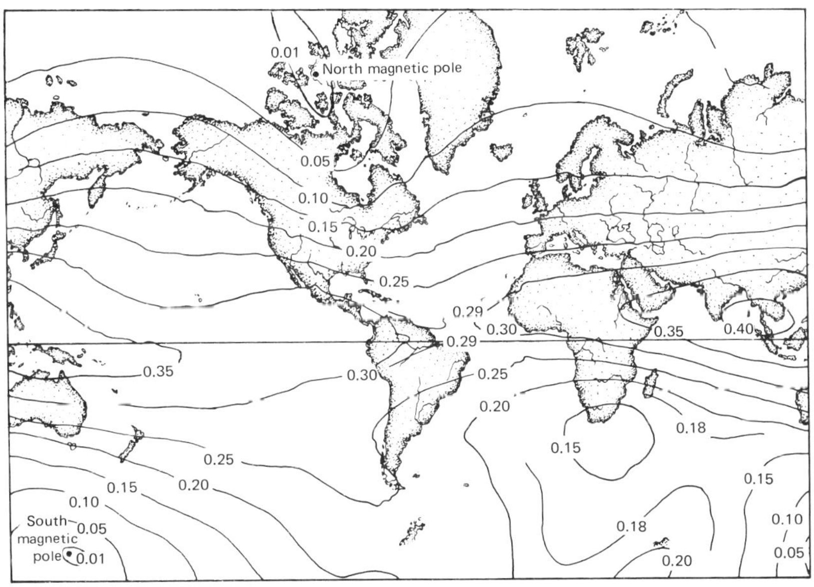

FIGURE 19–4
Lines of equal horizontal intensity of earth's magnetic field expressed in cm/g/sec.
(after E. H. Vestine and others, 1947)

surface structure, is relatively constant, but variations of the field due to other causes are well known.

VARIATIONS IN THE FIELD THROUGH TIME

Detailed study of the magnetic field has been in progress for over ninety years. Data collected show clearly some of the variations that take place through time. Some of these variations are periodic, others are irregular. Variations that occur with a period of less than one day are related to solar radiation; they vary from night to day and with the season of the year. Monthly and annual cycles related to motions of the sun and moon are also recognized. These variations occur with

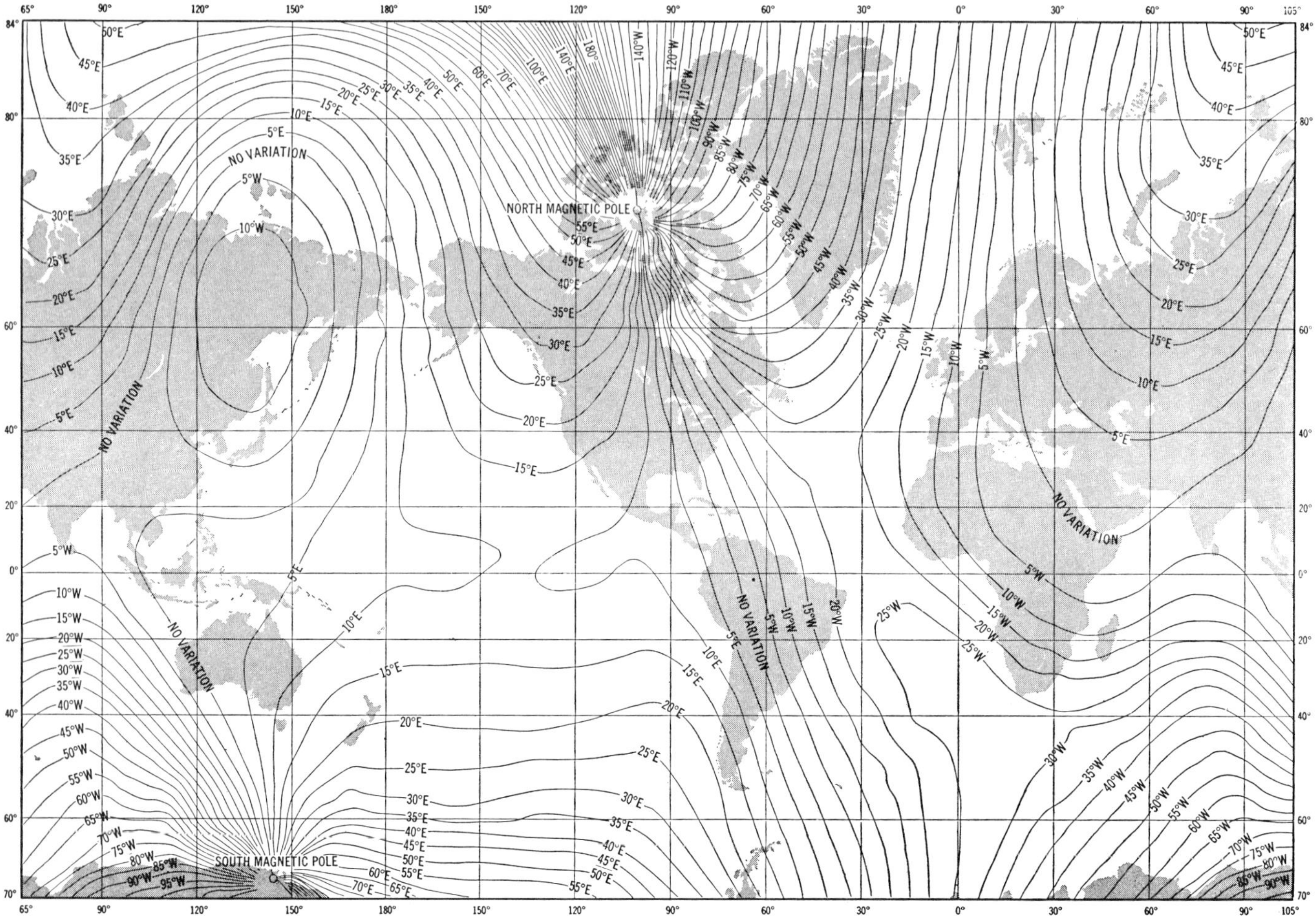

FIGURE 19–5
World map of magnetic declination; lines connect points of equal variation of compass from geographic north.
(U.S. Hydrographic Office)

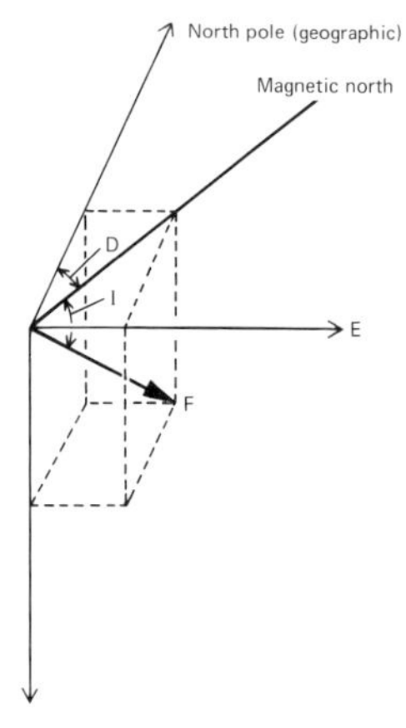

FIGURE 19–6
Force vector F for the earth's magnetic field. D is declination of magnetic north from true north, and I is inclination of the field.

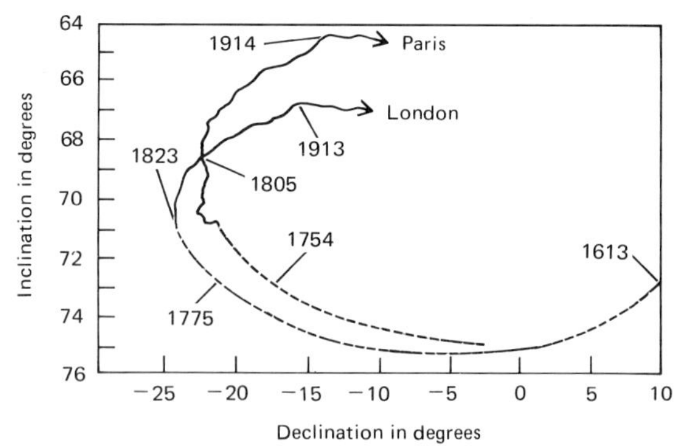

FIGURE 19–7
Secular variation of the direction of the geomagnetic field in Paris and London.
(after Gaibar-Puertas, 1953)

fixed periodicity, can be predicted with accuracy, and must be accounted for in magnetic surveys where local variations in the field are to be detected. Diurnal variations may be smooth, regular, and of low amplitude (called *quiet days*), and reflect the predictable variations due to the movements of sun and moon. At other times, *disturbed days*, the records show great variations that are not regular; these are called *magnetic storms*. Variations of as much as 1,000 gammas may occur during storms. Magnetic surveys for purposes of ore exploration and structural analysis cannot be conducted during magnetic storms because the variations are so great and unpredictable. Irregular variations are due to such phenomena as *sunspot activity*, during which high-energy particles from the sun bombard the earth; at these times auroral activity at the poles is at a maximum and may extend even to the middle and low latitudes.

Long-term changes have also been noted. Gellibrand (1635) observed that the position of the north magnetic pole was shifting. It was later suggested that the magnetic pole may be moving in a circle about the geographic north pole, but such a simple explanation does not satisfy the observation that the period of this motion is different at different locations on the earth (Fig. 19–7). Indeed, it appears that the magnetic field is influenced by local or regional conditions. Such regional characteristics probably reflect differences in the physical properties of the material deep in the earth's interior.

These long-term variations are known as *secular variations*. They are observed throughout the earth as progressive changes in declination, intensity, and inclination, and they show rates of change that vary with time at any given point. Such changes can be plotted as *isoporic* maps (Fig. 19–9) for a large region. An *isopor line* is one that connects points of equal annual change. The maps typically show lines of no change and points, called *foci*, where high rates of change occur. The positions of the foci change with time and they may die out as new foci form. This continual change of the field is probably related to processes going on within the earth's core.

Halley noticed in 1692 that some of the nonaxial features of the field drift westward at a rate of about 0.5° per year. This is seen in such observa-

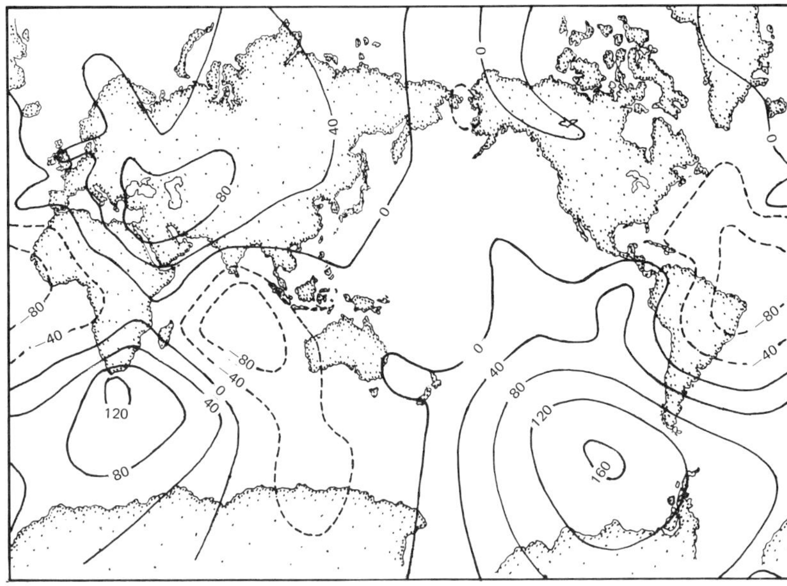

FIGURE 19–8
World map showing the secular variation of the vertical component of the magnetic field for July, in gammas per year.
(after E. H. Vestine and others, 1947)

FIGURE 19–9
Isoporic chart for the vertical component of the earth's magnetic field in gammas per year for the year 1960.
(modified after Nagata and Sawada, 1963)

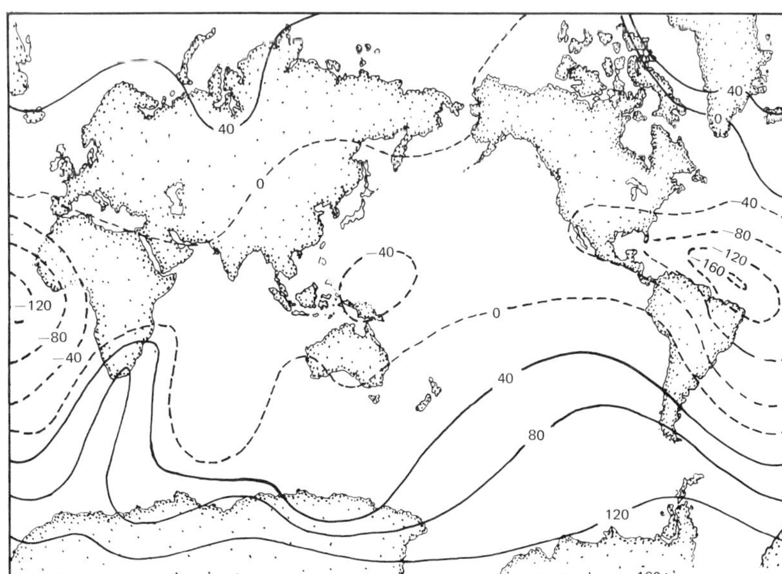

tions as the westward shift of the point of zero declination, shifts of isoporic foci, and movement of the region of maximum declination (it moved westward a distance of 25° in 250 years). The westward-drift rate is not uniform for the earth nor is it the same as the secular variation.

ROCK MAGNETISM

"The magnetism of rocks is the vector sum of the *induced magnetization* [italics added] dependent on an external field—and *remanent magnetization* independent of any external field" (Lindsley and others, 1966).

Natural materials may be classified into groups that exhibit several different types of magnetic properties. *Ferromagnetic* substances are those that reach a magnetization intensity of the same order of magnitude as iron; *paramagnetic* substances become weakly magnetized in the direction of the magnetic field; and *diamagnetic* substances are those that tend to move from stronger to weaker fields and, if free to rotate, will become oriented normal to the field.

Magnetic poles do not occur independent of one another; thus the simple representation of a magnet is a *dipole,* which consists of two opposite poles of equal strength separated from one another by a distance, L. The product of the pole strength, P, and L is called the magnetic moment, M:

$$M = PL$$

One theory of induced magnetism is that all materials that can become magnetized consist of large numbers of very small molecular magnets that are arranged at random when the material is not in a magnetic field. When the material is magnetized, the small magnets become oriented so that north poles point one direction, south poles point the other. The degree of alignment depends on the field strength. The effect is that poles are induced on the surface (within the magnet the north and south poles are so close together they cancel one another out). This type of induced magnetization, called *polarization,* acts in the direction of the external field and has a strength that is proportional to that of the external field.

If the external field is homogeneous and of strength H, and if it makes an angle ϕ with a normal to the surface of the magnetic material, the induced pole strength per unit area, I, is equal to the following:

$$I = kH \cos \phi$$

where k is a constant, called *susceptibility,* and depends on the nature of the material. Cos ϕ becomes 1 when the field is normal to the surface of the material, and the susceptibility becomes

$$k = \frac{I}{H}$$

Examples of magnetic susceptibilities of some common natural materials are shown in Table 19–1.

Rocks vary so greatly in iron content it is probably more useful to consider the variation in susceptibility with regard to magnetite content; however, many nonferrous minerals are also magnetic. Most minerals are paramagnetic with positive or negative susceptibilities in the range of 10^{-6} cgs (centimeter-gram-second) units and their susceptibility is independent of the applied field, but ferromagnetic minerals (iron and nickel) show a variation in the intensity of magnetization that is a function of the applied field and their susceptibility approaches unity as field strength is increased.

Magnetic susceptibility of rock is influenced by composition, texture, temperature, and pressure. Temperature is particularly significant. If the temperature is increased, a point will be reached at which the susceptibility vanishes and the rock is no longer magnetic. This temperature is characteristic for the rock type and is called its *Curie point* (e.g., Curie points for some common minerals: magnetite = 578°C; hematite = 675°C; pyrrhotite = 300°C; nickel = 358°C).

USING MAGNETIC SURVEYS TO DETAIL SUBSURFACE STRUCTURE

Magnetic surveys are frequently used to supplement other types of geologic field studies. Magnetometers can be operated from airplanes, boats, and over areas where soil, ice, water, swamp, or other cover obscure outcrops. These surveys are relatively inexpensive. Magnetic anomalies have been used to locate magnetic ore bodies and to outline or trace other rock bodies (e.g., dikes) that contain iron minerals. Magnetic anomalies have also been useful in petroleum exploration to predict depth to basement rocks as well as to reveal buried topography, locate salt domes, and to help trace out covered rock units.

Variations in the magnetic susceptibility of rocks make it possible to analyze subsurface distributions of rock masses in much the same way that density variations are used in gravity surveys. Magnetic studies also face many of the same difficulties—for example, the necessity to make corrections for several magnetic field components and general failure to obtain a unique solution. Corrections applied to readings of magnetic intensity include the following:

1. *Diurnal correction.* Curves showing diurnal variations are prepared by the Coast and Geodetic Survey. These may be used where fine precision in a survey is not required, but where detail is necessary the usual procedure is to return with the magnetometer to a base station periodically during the day so that a variation curve can be constructed for the immediate area.
2. *Regional correction.* Large-scale regional variations in magnetic fields are usually encountered and the recognition and removal of these presents a significant problem in magnetic surveys. If the area of the survey is small—for instance, a mine—the regional variations may be neglected, but over larger areas a correction is necessary. One method used to apply this correction is to contour the field intensity over the region using data points located outside the area of a particular magnetic anomaly. Then the difference between these smoothed-out contours and observations within the area of survey is the residual anomaly. Another method consists of plotting the total field along a section and fitting a smooth curve to the data (Fig. 19–10). For the United States, published maps of regional values constructed from the network of stations located about 16 kilometers apart are available.
3. Some magnetometers are sensitive to temperature changes. These instruments must be calibrated for this purpose and a correction applied to the reading.

TABLE 19–1

	$k \times 10^6$	Field strength H in oersteds
magnetite	500,000±	0.5 oersted
ilmenite	135,000	1.0
peridotite	12,000	1.0
granite	2,700	1.0
basalt	700	1.0
sandstone	17	1.0

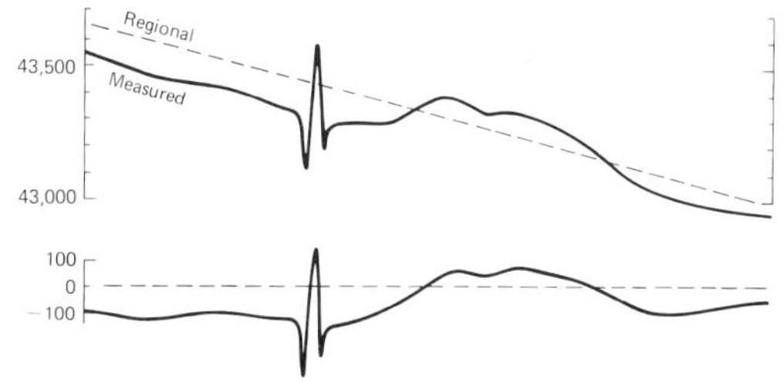

FIGURE 19–10
Examples for the application of regional fields on the corrected values of a magnetic measurement.
(after Bentz, 1961)

The values left after the above corrections are called the *residual magnetic anomalies.* Several methods are used to interpret these anomalies. Some important discoveries have been made in surveys of the oceans simply by examination of the anomaly pattern obtained from a regional survey of closely spaced observations (see Chapter 25). Use may be made of models, as in gravity work. The anomaly that would be produced by a model of a particular geometry and prescribed magnetization is calculated. A recent method involves a spectral analysis of the total magnetic intensity along a specific line of profile. The short wave lengths in a spectrum are found to correspond to sources that are shallow (in the crust or upper mantle) and long wave lengths to sources much deeper in the mantle and core.

REMANENT MAGNETIZATION

The remanent (permanent) magnetization of rock depends on the quantity and composition of the ferromagnetic minerals they contain, the environment (heat, pressure, chemical) in which they have been located, and the history of environmental changes. Rocks may have received their

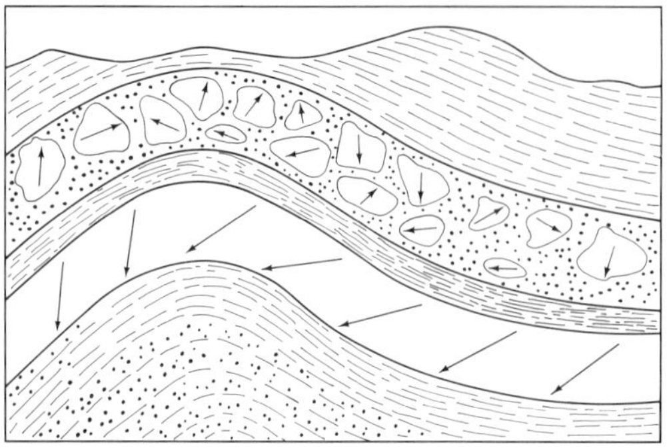

FIGURE 19–11
Bedding tilt and conglomerate tests. Directions shown in fold and in conglomerate cobbles indicate a stable predeformation magnetization.
(after Cox and Doell, 1960)

permanent magnetization during (1) crystallization, (2) temperature change, or (3) deposition.

Most igneous and many high-temperature metamorphic rocks acquired permanent magnetization on cooling. The temperature at which materials lose their spontaneous magnetization, as mentioned on p. 450, is called the Curie point. It is about 570°C for ferromagnetic minerals. Magnetization of lavas has been studied experimentally. Lava is reheated within a zero magnetic field until the Curie point is reached and then cooled within a magnetic field similar in strength to that of the earth, and the magnetic moment of the lava is observed as temperature decreases. The intensity of magnetization in the specimen is found to increase as temperatures are lowered below the Curie point. When the cooling lava reaches normal temperatures the magnetization is permanent and is unaffected by changes in the external magnetic field.

Magnetization of sediments was first noted in studies of varved clays that were initially deposited in glacial lakes. The clays exhibited a remanent magnetization now commonly ascribed to preferred orientation of iron oxide grains within the geomagnetic field that existed at the time of deposition. The grains behave much like small compass needles. The theory is easily tested by dispersing the clay in water and allowing redeposition in the laboratory. The clays become magnetized in the general direction of the field acting at the time they are deposited. The orientation of a rock sample to be analyzed for paleomagnetic pole direction must be recorded when the sample is collected. The alignment of the remanent magnetic field is determined, and on the basis of its inclination and direction the paleomagnetic latitude of the sample locality is determined. The *paleo-pole location* can then be found.

Red sandstones and siltstones have also been thoroughly studied. They are more strongly magnetized than other sediments and sometimes show a marked magnetic stability. A test of this was made by determining the direction of magnetization of pebbles derived from a Precambrian rock now cemented in a much younger (Triassic) conglomerate. The directions were found to be random. Thus the ancient remanent magnetization had been stable for millions of years. Similarly, it has been found that the direction of magnetization of some sandstones that have been folded does not agree with those of unfolded rocks of the same age until the directions are rotated by the amount of the folding (Fig. 19–11).

Rocks may become magnetized in a number of other ways. *Chemical magnetization* is probably active in a broad range of geological processes. When magnetite is formed by the reduction of hematite it may acquire a remanent magnetization. Chemical magnetization may be acquired in several ways: (1) oxidation, (2) reduction, (3) recrystallization, (4) chemical precipitation, or (5) exsolution at temperatures below the Curie point. An example is found in hot humid climates where some of the hematite in surface layers becomes hydrolyzed. The hydroxide may later decompose to hematite again, taking up a magnetization parallel to the geomagnetic field at the time. This would be particularly important in the case of porous sediments. Lightning may produce a strong magnetization when it strikes a rock. When we consider the large number of possible causes of magnetization, it is not surprising that the rocks deposited or formed during any one period of geologic time do not indicate the same position for the pole location for that time, but if the spurious results can be eliminated we will have a powerful tool that can be used to help unravel the history of the earth.

Pole Reversals

When remanent magnetization is determined experimentally for a rock, the direction and polarity of the field are found. The orientation of the specimen must be carefully measured before it is removed from the outcrop so it is possible to relate the measurements made on the specimen in the laboratory to the earth's present magnetic field and the surrounding geology. One of the most surprising results of these paleomagnetic studies has been that the polarity of some of the specimens from particular rock masses is the reverse of the present geomagnetic field. Reversals of this sort have been found in Tertiary lavas of the Massif Central of France, the Columbia Plateau of the northwestern United States, Iceland, and Japan. Reversals by now have been found in sediments of most geologic ages. Two explanations appear possible: (1) Either the earth's field has periodically reversed, or (2) some process produces a reversal in rocks that is independent of the geomagnetic field.

Nagata (1955) suggested a process by which reversals may occur. He studied a piece of pumice that became magnetized in the opposite direction from that of the field in which it crystallized. As it cooled, very small intergrowths of two magnetic minerals occurred. The mineral with the higher Curie point became magnetized first in the direction and with the polarity of the geomagnetic field. As the second component reached the Curie point, the geomagnetic field within the mineral became overwhelmed by the field of the first component. From this experiment Nagata concluded that if the reversed magnetization of the second component is greater than the magnetization of the first component, the net result is a reversed direction for the specimen. Although this process of reversal is known, it does not appear to outweigh the observations that Tertiary lavas do not possess this self-reversal property. Also, analysis of magnetization direction of dikes and the immediately surrounding country rock that was heated above the Curie point during intrusion of the dikes indicates that both the dikes and the altered country rock have the same direction and polarity. If real reversals of the geomagnetic field do not occur, we might expect that the dikes and country rock would, at least occasionally, have opposite directions of magnetization. Actually, both have identical direction regardless of whether it is the same or the reverse of the present geomagnetic field, and about half the time it turns out that both are the reverse of the present field.

The earth's field has undergone frequent reversals. The times of these reversals can be dated when radiometric ages as well as polarity can be established for rock samples. The geomagnetic time scale has changed as more data are analyzed. The scale for the last 4 million years as it looked in 1967 is illustrated in Fig. 19–12, and a longer scale, extending back 80 million years, is shown in Fig. 19–13. The long-term weakening of the geomagnetic field has been well documented. This may be an indication of what happens. The field may weaken gradually until it fades out and reappears reversed, gaining in strength to another maximum. If the field did disappear, the earth's surface would become subjected to a vastly increased bombardment of particles from the sun, which are normally funneled into the polar regions. Marked changes in the fauna of the earth have taken place at times that coincide with times of reversal. Thus we may have an indication of one important process by which mutation rates are sharply increased.

Wandering of the Poles

Long-term secular changes in the position of the magnetic poles have been documented through observations considered earlier (see p. 449). With the development of methods to determine polarity and direction of magnetization for rocks we have a means of extending observations of pole position location back in time through millions of years. We will assume that the geomagnetic field of the past acted in much the same way and had the same effect on rocks that it now has, and thus that sediments formed at a particular time have actually tended to become magnetized in the direction dictated by the geomagnetic field of that time and at that locality. Similarly, cooling and crystallizing rocks have taken on the direction of

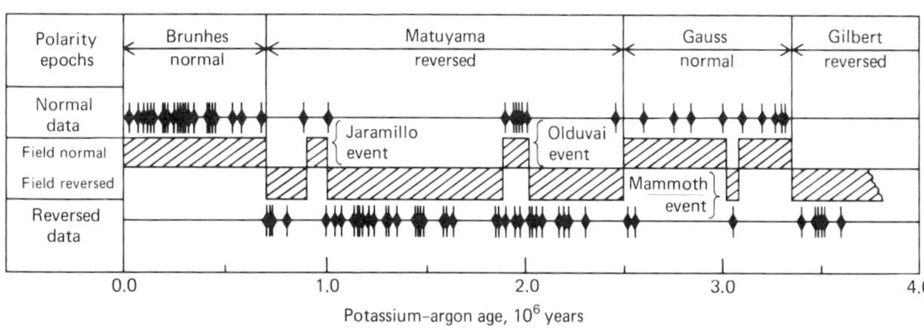

FIGURE 19–12
Time scale for reversals of the earth's field. The observed magnetic polarities and potassium-argon ages of volcanic rocks are shown as diamond-shaped symbols. Only the more reliable primary group of data is included. The inferred time scale is shown by the bar at the center.
(from Cox and Dalrymple, 1967)

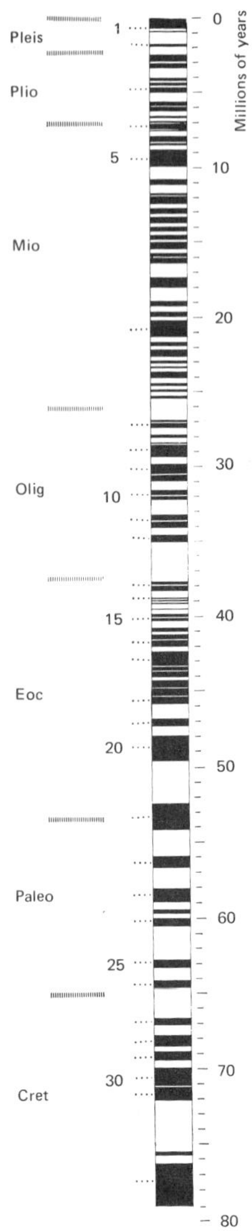

FIGURE 19–13
The geomagnetic time scale. Cretaceous to present time scale for geologic eras, numbers assigned to bodies and magnetic anomalies; geomagnetic field polarity with normal polarity periods colored black.
(Heirtzler and others, 1968)

the field when they reached their Curie point. If these assumptions are valid, we should be able to reconstruct the geomagnetic field for any time in the past for which an adequate rock record is left, recognizing that care must be taken to remove effects of later folding and that anomalous results may be found due to effects of lightning, chemical changes, or reheating.

When a large number of Recent or Pleistocene rocks are studied and the direction of their magnetization determined, it is found that the pole positions cluster around the present geographic poles. Magnetic and geographic pole positions have not coincided, but the poles during the last few thousand years have remained within the same general area. When pole positions are determined for more ancient times, some rather remarkable features are found. First, the clustering of pole positions does not coincide with present geographic poles. Secondly, the pole positions for any given period of geologic time, other than the most recent, tend to be different for rocks collected from different continents. A great deal of effort has gone into attempts to document the path along which the pole position has wandered through geologic time for North America, Great Britain, and Australia, and many geologists feel that reasonably accurate paths have been determined. The fact that these paths are different for different continents may be taken as evidence that the continents themselves have moved.

ORIGIN OF THE EARTH'S MAGNETIC FIELD

Those who investigate the origin of the geomagnetic field are beset with problems. The earth's interior, where the major part of the field is thought to originate, is known only indirectly, and most conclusions regarding the composition, state of motion, and distribution of compositional variations are based on assumptions and theory. The magnetization of the near-surface layers is not nearly as great as it would be if the earth were uniformly magnetized. In addition, we have seen that rocks lose their permanent magnetization when heated to the Curie point, and temperatures should be high enough to cause such a loss for all rocks including iron at depths of about 25 kilometers; that is, within the continental crust and only slightly (15 kilometers) below the oceanic crust. (It is possible that the Curie point increases with extreme pressure, but there is no experimental validation of this idea). It has been suggested that the field is caused by the rotation of the earth, but the magnetization that would arise from such causes is negligible. Whatever the cause of the geomagnetic field, any theory must explain the following: (1) the secular drift of the field, (2) the failure of the field to line up along the axis of rotation of the earth, (3) the reversals of the polarity of the field, and (4) the large anomalies in the field. Many small anomalies are caused by distribution of magnetic materials within the crust. Intermediate scale anomalies that might be attributed to the mantle appear to be missing. The larger anomalies and the main part of the field are attributed to the core and inner core.

Earth as a Dynamo

The most widely favored hypotheses for the origin of the geomagnetic field are those involving electromagnetic phenomena. Sir Joseph Larmor (1919) was the first to suggest that the magnetic field of the sun might be somewhat like a self-exciting dynamo, and his theory was later applied to the earth. It has already been shown that the outer core of the earth behaves as a liquid and that the liquid interior may be a nickel-iron compound of some type. Both of these qualities would promote the possible existence of electric currents in the core. The inner core appears to be a solid, probably composed of ferromagnetic material but much hotter than the Curie point. The remaining condition necessary for the development of a magnetic field would be to start such an electric current and then to sustain it in motion. Electric currents might have begun in the interior of the earth as a result of (1) chemical processes similar to those in a battery, (2) slow movements due to slight differences in chemical composition within the melt, or (3) the chance that the core has not yet reached a condition of stability. Once electric currents started they would be amplified if movements in the interior were taking place. Motion would logically arise from heat differences between the inner and outer parts of the core. As a result of these differences in temperature, *convection* (a slow, boilinglike motion) should occur within the molten material. The secular variations of the geomagnetic field could reflect eddies within the core. W. M. Elasser (1939) recognized this possible relationship and laid the basis for the present theory. Few earth scientists doubt the existence of some form of motion in the core and lower mantle, but the depth, amount, and intensity of movement and its role in causing the geomagnetic field are continuing fields of investigation. As yet the dynamo theory is not proved. There are, however, other possible causes of electric currents in the core. Chemical and thermoelectric processes generate electromotive forces that are most likely to be found along the contacts of materials of different electric properties. The boundaries between the mantle and core and between the core and inner core would be likely places, and Elasser has suggested thermoelectric currents might be due to inhomogeneities in the core that are created and regenerated by turbulent motion. Unfortunately, it is difficult to prove or disprove such hypotheses.

How well do such explanations fit the observations? First, they certainly agree with our knowledge of the interior derived from seismology. Furthermore, they explain the cause of the variability of the earth's magnetic field. If many small, eddylike convection currents are present, and if electric and magnetic fields are associated with these, then we should find corresponding local variations in the observed field at the earth's surface, as we do. That these many small magnetic fields do not show up more distinctly may be due to the rotation of the earth as a whole. This would impart a general pattern of symmetry to the entire field and cause the alignment of the fields of convection cells; thus what we observe at the surface is the net overall effect of the fields, with some minor variations superimposed.

FIGURE 19–14
Auroral displays taken at the University of Alaska.
(Vic Hessler, ESSA photograph)

The Aurora

Most dramatic of the effects of the earth's magnetic field are the *auroral displays* (Fig. 19–14). Sometimes visible as far south as the southern United States, they are consistently seen in the Arctic and Antarctic; displays are most common and most pronounced in belts located 23 degrees from the earth's axis in what are called the *auroral zones.* Auroras take on many patterns and colors, but the most common are pale glows that are greenish, or yellow, with occasional brighter colored greens, yellows, reds, and even blues. The aurora may take the form of arcs of diffuse color slowly rising from the horizon in the direction of the magnetic pole and moving toward the zenith along lines approximately parallel to lines of the magnetic field. Several of these arcuate clouds of color may succeed one another through the night. At other times, the arcs break up into rays or even into bands of rays that resemble folded draperies or curtains. Usually these rays move about and pulsate, and generally they indicate the breaking up of the auroral display.

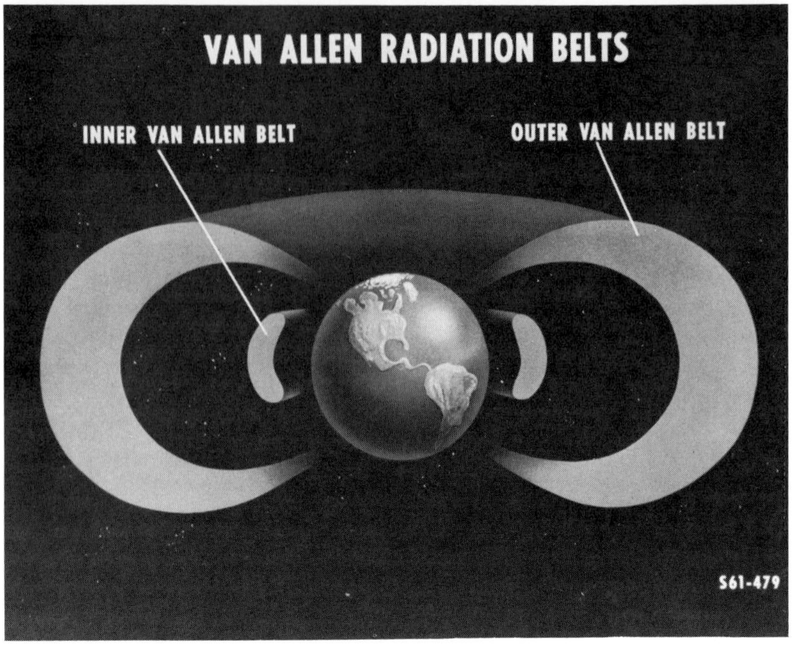

FIGURE 19–15
Schematic representation of the Van Allen radiation belts and a representation of the relations between the sun and the earth's magnetic field.
(U.S.N.C.–I.G.Y., N.A.S.)

FIGURE 19–16
Schematic representation of the interactions between the sun and the earth's magnetic field.
(U.S.N.C.–I.G.Y., N.A.S.)

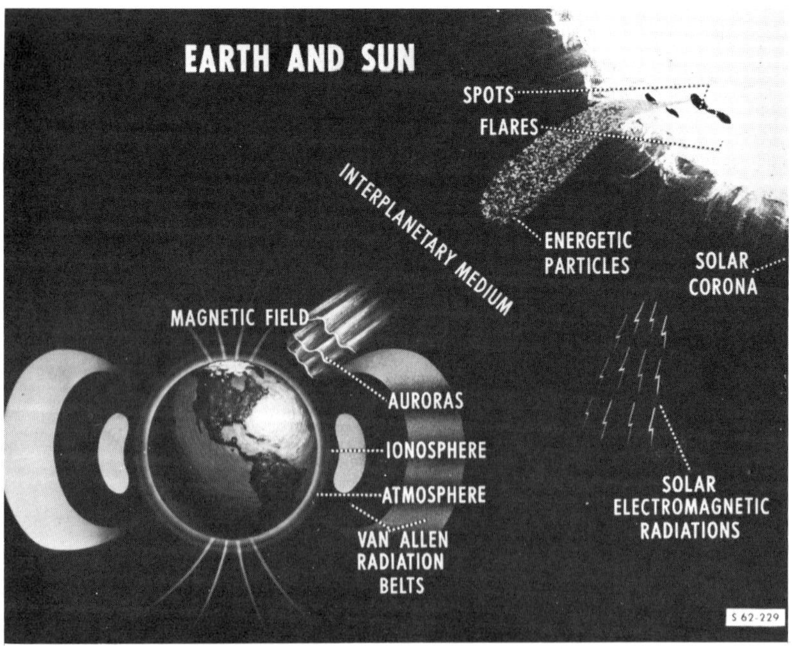

The aurora is related to the magnetic field of the earth because the auroral zones are located where the lines of the earth's magnetic field converge and dip into the earth toward the core (see Fig. 19–1). It is also well-known that fluctuations in intensity of the magnetic field accompany the waxing and waning of the aurora. The most widely accepted explanation of the cause of the aurora also connects it with solar activity and the earth's magnetic field. Protons and electrons from the sun bombard the atmosphere, exciting oxygen and nitrogen atoms in the upper atmosphere to a state of luminescence. That this activity is connected with the sun is shown by the 11-year period of increasing and decreasing auroral display that parallels the 11-year cycle of sunspot activity. It has also been observed that the most prominent auroras follow periods of solar flares. There is a slight lag, suggesting that the greater part of the ionized particles move from the sun to the earth at a speed of about 1,600 kilometers per second, taking almost a day to reach the earth. Some invisible clouds of particles, known as *plasma clouds,* continue to arrive for several days. Information about these clouds is now being derived largely from artificial earth satellites. When one of these clouds of ionized particles reaches the outer parts of the earth's magnetic field it causes the field to be deformed, bending it in toward the earth. This accounts for the presence of auroral displays at low latitudes on such occasions, and accounts for some of the magnetic storms on earth that interrupt radio communication.

VAN ALLEN BELTS

Two large belts of intense radiation surrounding the earth have been discovered through the use of artificial earth satellites. These were unknown until rockets were put to use to investigate the space immediately surrounding the earth. The first indication of radiation belts came from high-altitude balloon and rocket studies in 1952–53, when James van Allen and a number of co-workers were studying the cosmic-ray intensity in the atmosphere of the Arctic region. A few of their records showed abnormally high readings for the radiation in a zone starting at an altitude of about 48 kilometers. Several years later it was established with the use of rockets that a similar belt of high-intensity radiation existed in the Antarctic. At about the same time, Carl McIlwain (1958) discovered that the radiation includes protons (hydrogen nuclei) and electrons. The greatest advances after this came when a number of satellites were put into orbit about the earth. The most detailed pictures of the radiation belts were obtained by Explorer IV and Pioneer III. Data from these showed that the earth is surrounded by two belts of radiation shaped somewhat like hollow doughnuts, one inside the other (Fig. 19–15). These concentric doughnuts are oriented so that they are in the plane of the earth's magnetic equator. The inner belt lies above altitudes of 800 kilometers and extends to about 34° north and south of the equator. The peak in this belt occurs at about 3,200 kilometers above the earth. The peak of the outer belt lies at about 16,000 kilometers. Beyond that the intensity drops off and falls to a low level at 64,000 kilometers. The intensity of radiation in the belts is about 25,000 counts per second on a Geiger counter, which is estimated to represent bombardment by some 40,000 particles per square centimeter per second.

The explanation advanced to account for this radiation is that the particles originate from the sun; they are injected into the earth's magnetic field and are deflected into corkscrew paths that follow to some extent the lines of the magnetic field. Once in the field they are trapped, but not permanently. As the number of particles increases, some are lost, particularly in the auroral zones, giving rise to the displays observed there. Some normally escape, causing the phenomenon known as *airglow*—a faint illumination of the night sky. An alternative theory is that neutrons released in the upper atmosphere through bombardment of atoms by cosmic rays decay into protons and electrons, which are trapped in the magnetic field. It is agreed that this process takes place, but some doubt that a sufficiently high number of particles could be produced in this way.

CHAPTER OUTLINE

Magnets and Magnetic Fields
Measurement of the Earth's Magnetic Field
Describing the Earth's Magnetic Field
Variations in the Field through Time
Rock Magnetism
Using Magnetic Surveys To Detail Subsurface Structure
Remanent Magnetization
- Pole Reversals
- Wandering of the Poles

Origin of the Earth's Magnetic Field
- Earth as a Dynamo
- The Aurora

Van Allen Belts

TERMS

airglow
auroral displays
auroral zones
chemical magnetization
convection
Curie point
declination
diamagnetic
dip needle
dipole
disturbed days
ferromagnetic
gamma
gauss
hysteresis
inclination
induced magnetism
isopor line
isoporic maps
magnetic field
magnetic-field strength
magnetic permeability
magnetic storms
north and south poles
oersted
paleo-pole location
paramagnetic
plasma clouds
poles
- geographic (earth)
- magnetic (earth)
- of a magnet

precession
quiet days
remanent magnetism
residual magnetic anomalies
secular variations
sunspot activity
susceptibility
unit pole

REVIEW QUESTIONS

1. In what way or ways is it possible to describe the intensity of a magnetic field?
2. At least four things affect the intensity of the earth's magnetic field to produce both long-term and short-term changes. Describe their effects briefly.
3. What is susceptibility? What are the variables involved in its determination?
4. What is a residual magnetic anomaly? What are some problems encountered in determining this anomaly?
5. In the interpretation of magnetic anomalies, models may be used—especially with the help of computer techniques. However, the derived model may vary considerably from what actually exists. What factors do you think could produce such divergent results?
6. What are some evidences that remanent magnetization remains constant, that is, permanent through time in sedimentary rocks?
7. How could evolutionary trends be related to changes in the earth's magnetic field?
8. In studying causes for the earth's magnetic field, the Curie point at depth is an important consideration. Why?
9. Do the hypotheses concerning the origin of the geomagnetic field fit the observations we make at the surface as well as those derived from seismology? How might the rotation of the earth affect the geomagnetic field?
10. Why are auroral displays concentrated around the poles?

SUGGESTED READINGS

Cox, A., and **Doell, R. R.,** 1960, Review of palaeomagnetism: Geol. Soc. America Bull., v. 71, p. 645–768.

Cox, A., Dalrymple, G. B., and **Doell, R. R.,** 1967, Reversals of the earth's magnetic field: Sci. American, v. 216, no. 2, p. 44–54.

Runcorn, S. K., 1959, Rock magnetism: Science, v. 129, p. 1002–1012.

20 EARTHQUAKES

20 EARTHQUAKES

EARTHQUAKES

Earthquakes are dramatic evidence that the earth is a dynamic body within which processes are actively taking place. More people have been killed by earthquakes and earthquake-induced phenomena than by all other geological processes combined. The numbers of casualties in earthquakes over the past 1,000 years has been compiled (Hill, 1965) and includes seven earthquakes in which the death toll exceeded 100,000. An estimated 800,000 people were killed in the 1556 Shenshi, China, earthquake. By comparison, the death tolls of 5,600 in Chile in 1960 and 114 in the 1964 Alaskan earthquake seem small.

The position on the earth's surface directly above the region where motion is initiated, the *focus,* is called the *epicenter* of the earthquake. Waves may be seen moving across the ground near the epicenter; buildings are turned on their foundations; clay, brick, and stone structures disintegrate; the ground cracks open; mass movements occur in marginally stable bodies of ice, soil, and mountainslopes; tsunami may be generated along coasts; and volcanoes may start to erupt. Of course, which of these things happen depends on the local geologic setting, the magnitude of the earthquake, the depth of the focus, and the distance of the features from the epicenter.

Ground motion near an epicenter is complex, as is shown by the trace made by a dresser leg on asphalt tile during the 1964 Alaskan earthquake (Fig. 20–1). Studies of such movements are critically important in the design of safe structures for use in heavily populated earthquake belts.

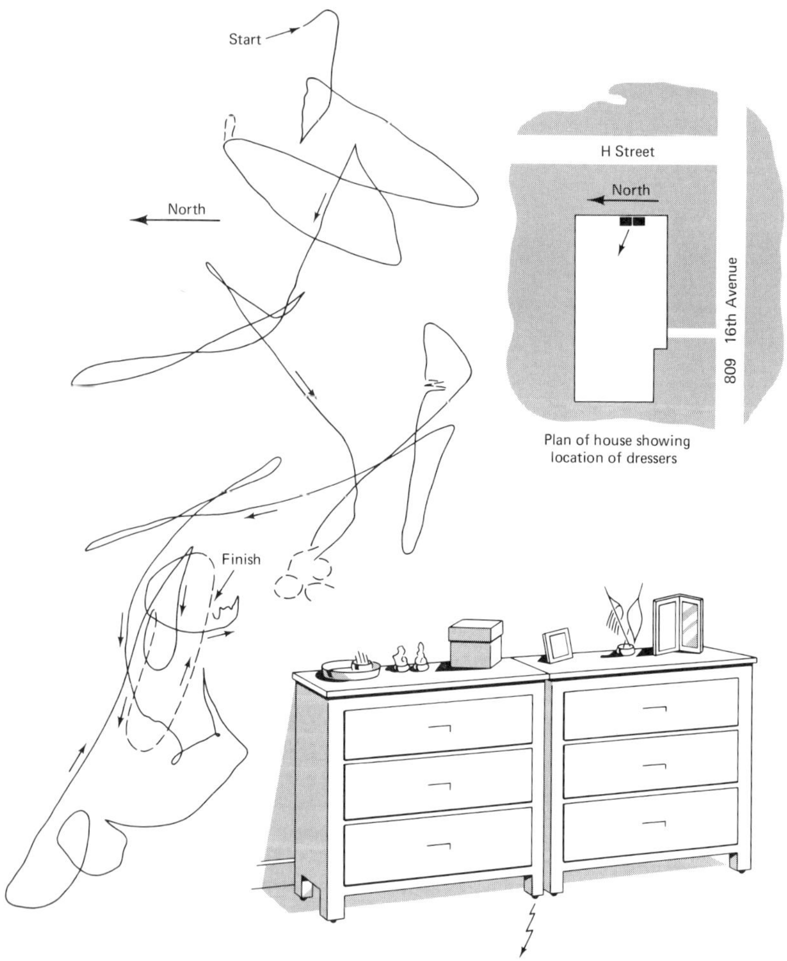

FIGURE 20–1
Horizontal trace on asphalt-tile flooring left by dresser leg during the 1964 Anchorage, Alaska, earthquake.
(from W. R. Hansen, 1965)

MODIFIED MERCALLI INTENSITY SCALE

Mercalli made studies of regional effects of earthquakes between 1890 and 1901. At the time of his study there were few seismograph stations; thus Mercalli devised a scale of intensities based on observations he could expect to be reported with a fair degree of objectivity. He found that some regions suffered more intense shocks than others.

His original scale has been modified, but it is still in use today. However, an earthquake magnitude scale—a measure of the amount of energy released at the focus—developed by C. Richter is used more frequently (see page 489). The following is the modified Mercalli scale.

I. Not felt except by a few under especially favorable conditions.

II. Felt by only a few persons at rest. Delicately suspended objects may swing.

III. Felt quite noticeably under favorable circumstances. Standing automobiles may rock slightly.

IV. Felt by many or most persons. Some awakened. Dishes, windows, doors disturbed, walls crack. Sensation like heavy truck striking building.

V. Some dishes, windows, and so on, broken; a few instances of cracked plaster; unstable objects moved.

VI. Felt by all; many frightened and run outdoors. Some heavy furniture moved; a few instances of fallen plaster or damaged chimneys.

VII. Everyone runs outdoors. Damage negligible in buildings of good design and construction; slight to moderate in well-built ordinary structures; some chimneys broken. Noticed by persons driving cars.

VIII. Panel walls thrown out of frame structures. Fall of chimneys, factory stacks, columns, monuments, walls. Heavy furniture overturned. Sand and mud ejected in small amounts.

IX. Buildings shifted off foundations, ground cracked conspicuously. Underground pipes broken.

X. Ground badly cracked, rails bent, landslides considerable from river banks and steep slopes. Water splashed over banks.

XI. Few masonry structures remain standing. Bridges destroyed. Broad fissures in ground. Earth slumps and land slips in soft ground.

XII. Damage total. Waves seen on ground surface. Lines of sight and level distorted. Objects thrown into the air.

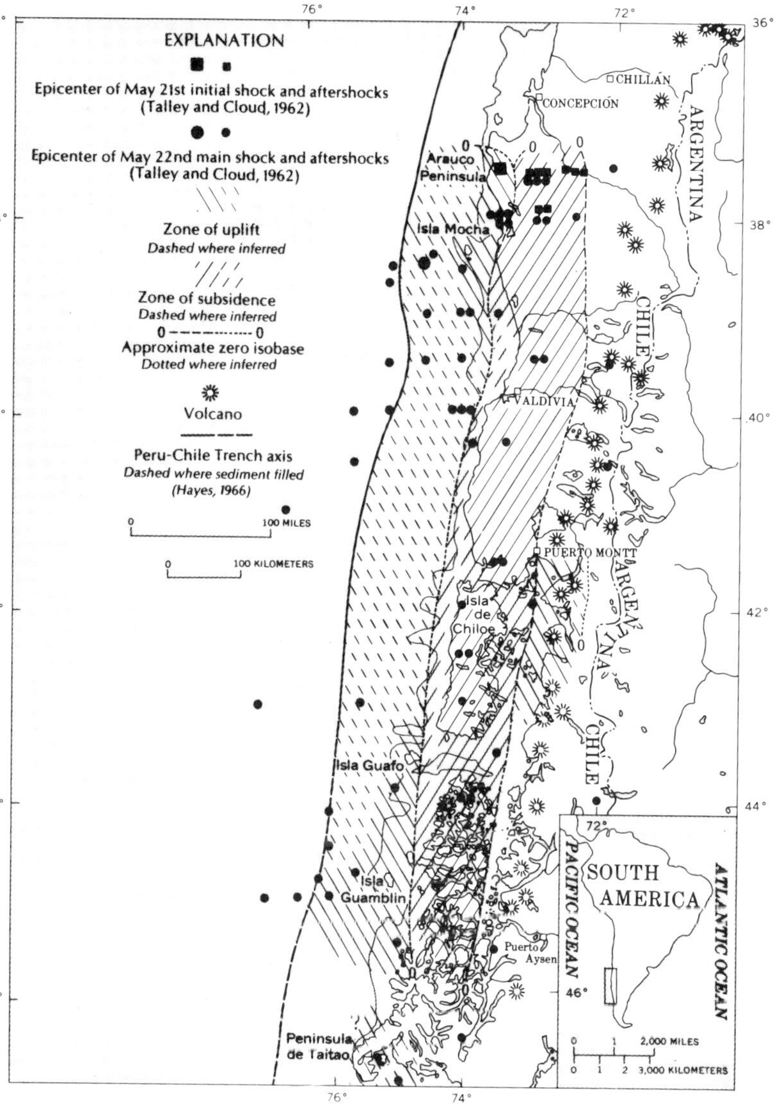

FIGURE 20–2
Spatial distribution of epicenters and zones of land-level change associated with the 1960 Chilean earthquake.
(from Plafker and Savage, 1970)

CASE HISTORIES

Chilean Earthquake of 1960

At 6:00 A.M. on May 21, 1960, a sharp earthquake was felt in the region around Concepción, Chile. Considerable property damage was caused by this movement and most of the people in the region affected were so frightened that they ran out of their houses. This earthquake, which was later determined to have originated at a depth of 50 kilometers, was followed by small aftershocks all the next day. At 3:00 P.M. the second day a second sharp shock came, followed 15 minutes later by the most severe shock of what was to be a long period of violent earth movements, several of which were rated Richter magnitude 7.5 to 8. The center of this shock was almost 300 kilometers south of Concepción (Fig. 20–2), but it affected an area 150 kilometers wide and 1,600 kilometers long along the coast of Chile. People ran out of their houses during the first shock, and when the second shock came waves were visible in the

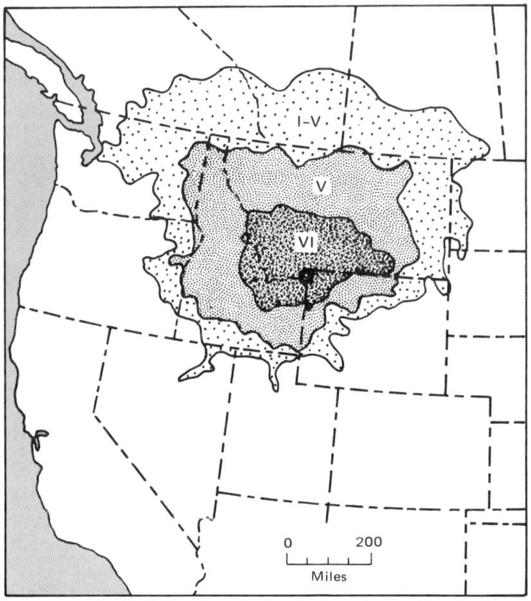

FIGURE 20–3
Isoseismal map of Hebgen Lake, Mont., earthquake of August 17, 1959, 11:37 P.M., M.S.T.
(from Steinbrugge and Cloud, 1962)

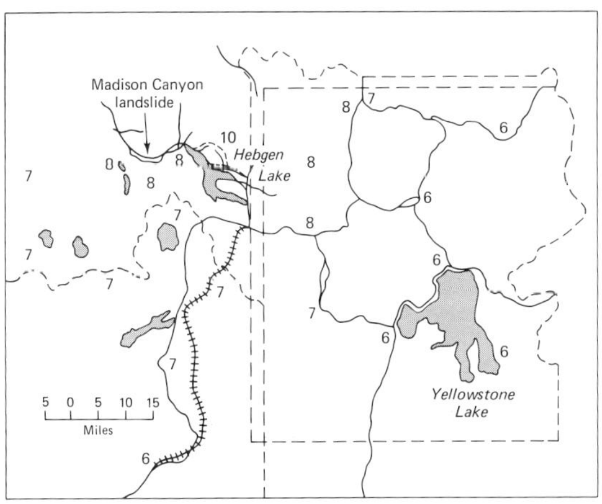

FIGURE 20–4
Intensities at various locations in and near the epicentral area of the Hebgen Lake, Mont., earthquake.
(from Steinbrugge and Cloud, 1962)

streets and ground. Trees and telephone poles swayed back and forth, loud noises came from the ground, and cracks formed in the earth. Cars were tossed back and forth across the street and some fell into fissures. Almost all manmade structures in this region were damaged and many were totally destroyed. Landslides caused by the shocks buried parts of a number of cities.

The strongest shocks were felt at Isla de Chiloe where trees were snapped off, the ground cracked open and highways were split, road fills gave out, and soils became liquid and flowed.

After the main shocks the sea began to retreat well below low tide and came back 20 minutes later as a tsunami between 4 and 11 meters high, destroying parts of many towns. Five hundred people were drowned in one town alone. Tsunami were set up by subsidence of parts of the coast, in places amounting to 2 meters. The wave motion caused in Chile by this subsidence crossed the Pacific and destroyed towns in Hawaii and in Japan.

On May 24th an explosion occurred on the side of Mount Puyehue, one of the large active volcanoes in Chile. Ash, steam, and finally lava poured from a fissure about 100 meters long. Shocks continued in Chile for several months and were responsible for billions of dollars in property damage and the loss of more than 5,000 lives in what may have been the most severe earthquake of this century.

Shocks and damage associated with this sequence of earthquakes were so widespread and extensive that it is difficult to relate the activity to any particular geological structure. Observers generally agree that the most intense activity was located along the coast and along a line of active volcanoes inland (Fig. 20–2), with a region of less activity in between. The coastal strip seems to have been lowered, perhaps along concealed faults.

The lower reaches of many rivers were flooded immediately after the May 22nd shock, and releveling surveys indicate that a coastal strip 30 kilometers wide and 600 kilometers long was bent or dropped down as much as 1 to $1\frac{1}{2}$ meters.

The Hebgen Lake, Mont., Earthquake

One of the most severe earthquakes in the United States took place near midnight on August 17, 1959. It was felt by persons in nine states and three Canadian provinces, as shown on the *isoseismal map,* a map with lines connecting points of equal intensity (Fig. 20–3). Intensities of 10 (Richter magnitude 7, see page 489) were assigned in the epicentral area on the basis of the extensive topographic changes that accompanied the earthquake, including the formation of steep fault scarps 6 meters high, warping of the ground surface, and a number of large landslides, among them the Madison Canyon landslide described in greater detail in Chapter 13.

In the immediate vicinity of the earthquake epicenter people were thrown out of bed, the ground was set in violent vibration, mortar work near Hebgen Lake broke, buildings were lifted and dropped back, tilted and knocked askew, brick chimneys in West Yellowstone were twisted on their foundations by ground motion, and bricks fell out of walls and chimneys as far as 100 kilometers away from the epicenter. Rumbling and grinding sounds were heard as a roar near the epicenter and some reported hearing the earth movements at distances of 80 kilometers. A long series of aftershocks followed the main shock in an irregular fashion, but with a general decrease in frequency and severity.

The geysers in Yellowstone National Park, which lies within the intensity VI zone (Mercalli scale) were unaffected. Prominent changes did occur in the area around Hebgen Lake (Fig. 20–4). The most spectacular of these is the formation of several fault scarps, particularly the Hebgen and Red Canyon fault scarps (Fig. 20–6), which can be traced for a distance of nearly 32 kilometers. These scarps occur in soil and *colluvium* (a mixture of soil, stream deposits, and talus) on the mountain flanks north of Hebgen Lake. Although the displacement on the scarps amounts to as much as 6 meters over long distances, solid rock is not exposed; thus it is difficult to demonstrate that the faults continue at depth into solid rock. Many of the faults may have originated as shown in Fig. 20–7, by a slipping downslope of the colluvium over the much harder, more compacted, and tightly cemented bedrock; but Witkind (1964) points out that these new scarps lie near old fault traces, faults that date back some 70 million years to the Cretaceous period when the modern Rocky Mountains were uplifted. These old faults are nearly vertical or inclined slightly to the south and the south side is relatively down. Thus it is probable that these old faults were reactivated. Some of the most spectacular features occurred in the colluvium along the new fault scarps where the colluvium on the side of the fault that is relatively down has subsided into the fault zone (Fig. 20–5). At either end of the fault, and particularly well shown at the eastern end, the faults pass into a flexure (a *monocline,* meaning dipping in one direction) as the displacement on the fault decreases and dies out.

Changes in the topography near the epicenter caused by regional tilting were immediately obvious because much of the area is occupied by an artificial lake, Hebgen Lake. Parts of the lake margin were submerged, and other areas were raised so that piers were elevated completely out of water (Figs. 20–8 to 20–10). Detailed resurveying using points of previously established level such as the profiles along the highway located on the north side of the lake made it possible to outline the regional changes in elevation (Fig. 20–6). Major elongate depressions formed that are

FIGURE 20–5
Grabenlike subsidence on downthrown side of fault formed during the Hebgen Lake earthquake, 1959.

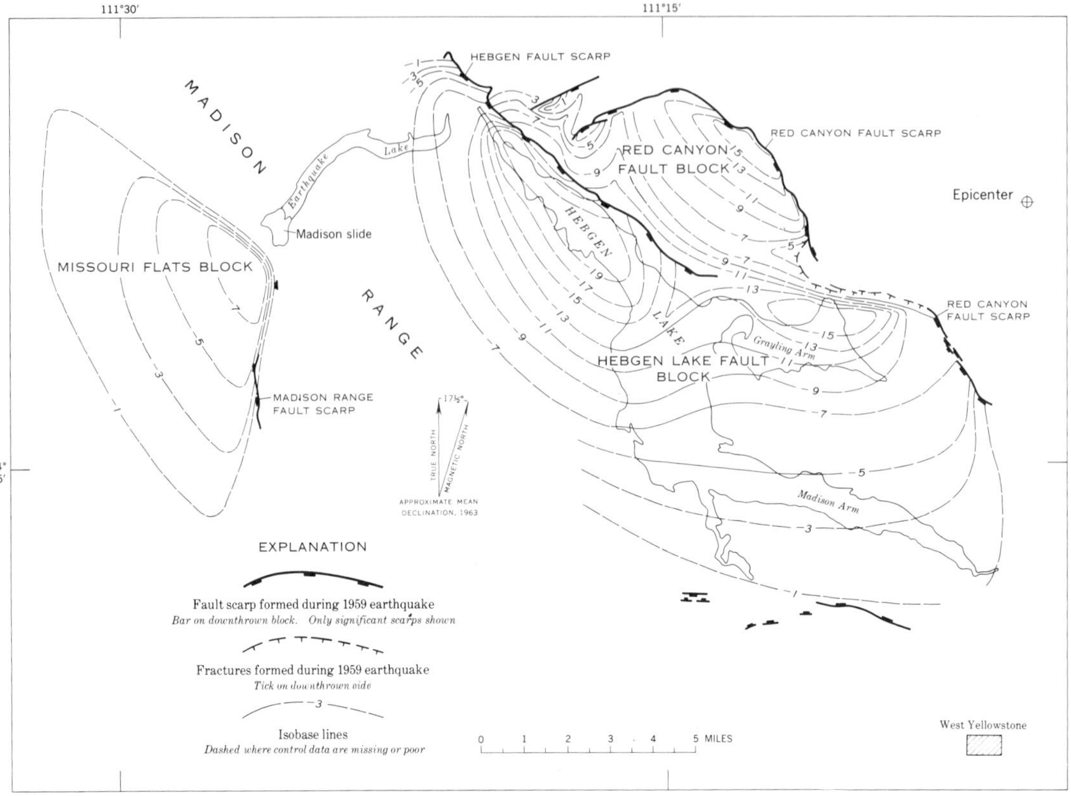

FIGURE 20–6
(*Above*) Map illustrating the dual-basin, fault-controlled pattern of deformation at Hebgen Lake, Mont. (isobase lines are labeled in feet).
(from G. D. Fraser and others, 1965)

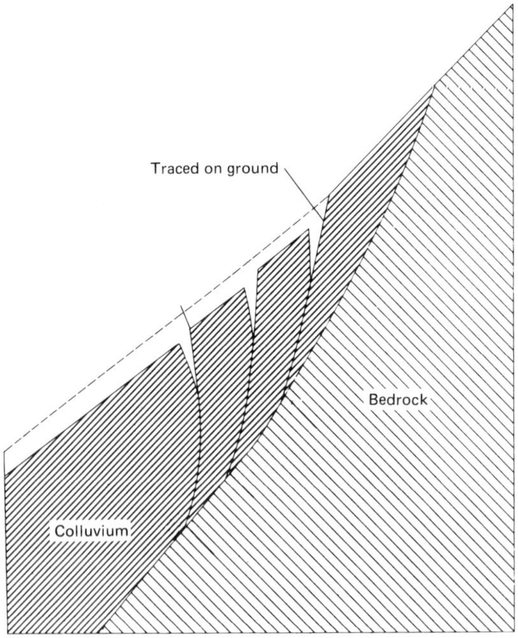

FIGURE 20–7
Suggested mode of formation of step faults in the Hebgen Lake area.
(from I. J. Witkind, 1964)

FIGURE 20-8
A fault scarp formed during the Montana earthquake of 1959.

FIGURE 20-9
Landslides, a side effect of severe earthquakes. This slide of 35 million cubic yards of debris was set off by an earthquake in Montana in 1959.
(U.S. Geological Survey)

FIGURE 20-10
Tilted lake basin. As a result of faulting, the Hebgen Lake basin tilted, leaving these docks 1.8 meters out of water.
(U.S. Geological Survey)

bounded on the north and northeastern sides by fault scarps.

Assam Earthquake

The Indian earthquake that occurred in the range of mountains south of the Himalayas and north of the Bay of Bengal in 1897 is probably the most violent in recorded history. The shocks occurred when the mountains, which are formed of large blocks of the crust tilted along faults, were shifted northward and uplifted as much as 10 meters in places. Faults that opened in the surface were traced for more than 32 kilometers into the jungle. Drainage of some streams was stopped, lakes formed, and large areas were flooded. In other areas lakes and swamps were drained. Many fissures opened in the ground. Almost all masonry buildings in the mountains were leveled. The region of complete destruction covered nearly 23,000 square kilometers; extensive damage was done in an area of 390,000 square kilometers; and the earthquake was felt over an area of 2,000,000 square kilometers.

FIGURE 20–11
Setting of the observed and inferred vertical displacements of the Kodiak group of islands with respect to regional tectonic deformation in south-central Alaska.
(from Plafker and Kachadoorian, 1966)

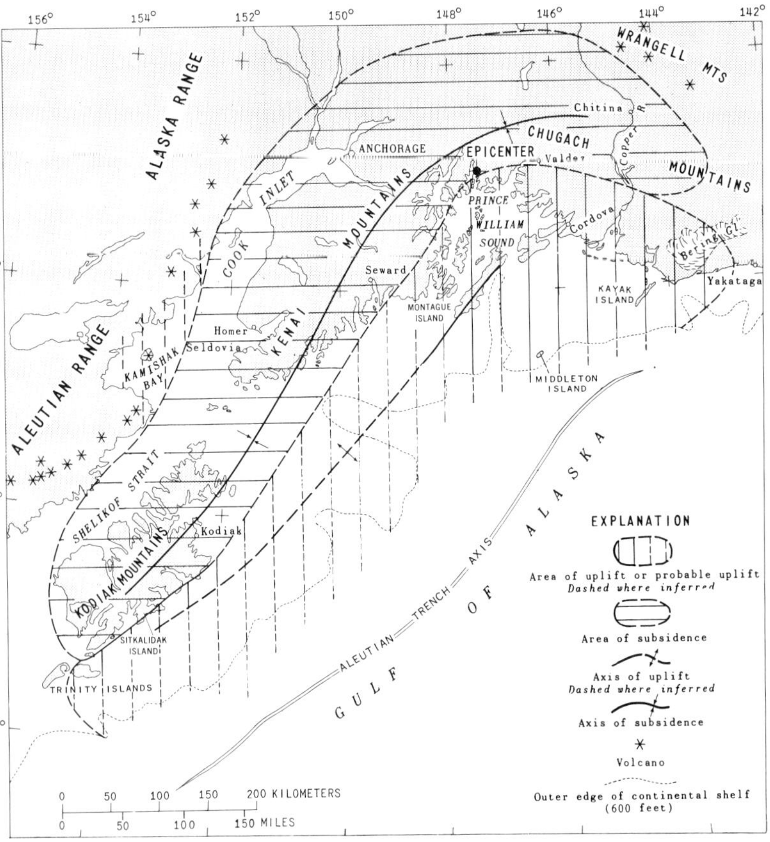

Kwanto Earthquake

Another of the most intense earthquakes on record occurred about 80 kilometers from Tokyo, Japan, on September 1, 1923. Estimates of loss of as many as a quarter of a million lives have been made. The earthquake occurred shortly before noon, and much of the damage was caused by fires that swept through the ruins afterward. Fires were started because so many families were cooking over open flames at the time of the earthquake. Buildings collapsed and ignited; water mains were broken, making it impossible to extinguish the fires. Half a million buildings were shaken down in the region around Sagami Bay, and much of this damage was concentrated in the heavily populated cities of Tokyo and Yokohama. Hydrographic resurveys of the bay shortly after the earthquake revealed changes of such magnitude in the shape and depth of the bay's bottom that most geologists attribute part of the changes to errors in earlier surveys. If the figures of surveys both before and after are taken at face value, some areas of the bay were lowered more than 200 meters and other areas were raised more than 245 meters. Even allowing for errors in surveying, the magnitude of the changes was exceptionally great. Most of the sediments in the bay came from eruptions of the famous volcano Fujiyama. It is likely that these unconsolidated sediments shifted or slumped during the earthquake, which was caused by movement along a major fault. Waves on the surface of the bay attained a height of 11 meters during and shortly after the earthquake, and more than 1,000 aftershocks of minor intensity followed the main shock, which lasted only a few minutes.

Alaskan Earthquake of March 27, 1964

One of the greatest geotectonic events of our time occurred in southern Alaska late in the afternoon of March 27, 1964. Beneath a leaden sky, the chill of evening was just settling over the Alaskan countryside. Light snow was fall-

> ing on some communities. It was Good Friday, schools were closed, and the business day was ending. Suddenly without warning half of Alaska was rocked and jarred by the most violent earthquake to occur in North America this century (Hansen and Eckel, 1966).

The effects of the Alaskan earthquake have been studied more intensively than those of any other earthquake. Much of the damage from this shock was caused by mass movement (see Chapter 13) and tsunami (see Chapter 9), but the regional geologic setting and the *tectonic* (large-scale structural features) effects of this earthquake are of unusual interest.

The shocks that accompanied the Alaskan earthquake were of high intensity (magnitude estimates range from 8.3–8.7), of unusually long duration (estimates made with wrist watches ranged from 1 to 7 minutes), and they were particularly destructive. The epicenter has been located in the high mountainous region near the head of Prince William Sound, about 128 kilometers east-southeast of Anchorage (Fig. 20–11), and the focus is estimated to have been between 20 and 50 kilometers deep, but it is probably wrong to think of this series of shocks as having originated from a point source. A long series of aftershocks followed the main shock, and these aftershocks were distributed over a region of about 260,000 square kilometers, mainly along the continental margin of the Aleutian Trench (Fig. 20–11) from Prince William Sound to Kodiak Island. Twenty-eight aftershocks of magnitude 6 or more followed within the first day and about 12,000 shocks of magnitude 3.5 occurred within the first 70 days after the main shock.

This earthquake lies within one of the most seismically active belts in the world, the nearly continuous belt that circumscribes the Pacific. The seismicity of the Alaskan part of this belt is indicated by the density of epicenters of major earthquakes recorded between 1898 and 1961 (Fig. 20–12). Many of these earthquakes are concentrated along the Aleutian Islands, an arcuate chain of active volcanic islands that extends from the Alaskan Peninsula to Kamchatka. Most of the epicenters lie north of the axis of the deep trench in the sea floor that is located south of the Aleutian Islands. Most of the remaining epicenters are associated with the mountain system that includes the Alaska Range and extends down the western coast of Canada. These mountains are high and they are the sites of modern crustal deformation.

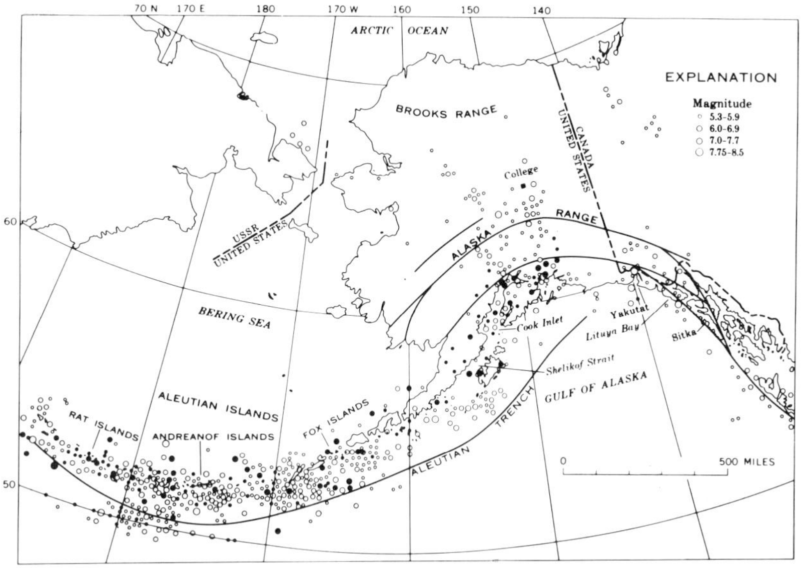

FIGURE 20–12 Earthquake epicenters in the Aleutians and Alaskan region. Earthquakes of various magnitudes are indicated by the size of the circles, and focal depths greater than 50 kilometers are indicated by filled-in circles. (from W. R. Hansen and others, 1966)

One of the most notable aspects of the Alaskan earthquake of March 27th was the vast area over which the level of the land changed. The zone extends from Prince William Sound to Kodiak Island (Fig. 20–11). An extensive program of releveling was undertaken in this area. Bench marks and triangulation stations provide the most accurate measurements of changes in level, but these are widely scattered; thus it was necessary to use the displacement of intertidal *sessile* (attached to the bottom) marine organisms such as barnacles along the coasts, and resurveys of highways. The results show that a zone southeast of Kodiak Island was elevated whereas a zone to the northwest was depressed. Maximum uplifts of 8 meters were found on Montague Island and maximum depression of about 2 meters was recorded in the Kenai Mountains. These movements are best described as a regional warping or tilting; only two faults were found and one was on Montague Island. Interpretation of the resulting uplift, subsidence, and distribution of the epicenters is illustrated in Fig. 20–13. Plafker (1965) favors the view that deep-seated movement occurred on a fault that comes to the surface along the Aleutian Trench and dips back under the island arc. These movements are shown near the zone of contact between the oceanic (basaltic) and continental (granitic-andesitic) crust. He concludes that the continental crust is moving out over the oceanic crust in this region.

San Francisco Earthquake of 1906

Frequent predictions of impending disastrous earthquakes in California and the grim reality of the 1971 Los Angeles earthquake have focused attention on the seismicity of the West Coast and the history of the 1906 earthquake, which affected large areas in California and devastated a large part of San Francisco.

The 1906 earthquake resulted from movement on the San Andreas fault, which runs subparallel to the West Coast and passes 12.8 kilometers from San Francisco. Features on opposite sides of that fault were displaced laterally as much as 4.5 to 6 meters with little vertical displacement. Railroads, paths, orchards, and streams were displaced. Movements were recorded over a distance in excess of 300 kilometers; movement was parallel to the fault trace and the west side moved north relative to the east side.

Damage was extensive over a large part of California but the greatest losses were in San Francisco and nearby towns where an estimated 500 to 700 people perished and where approximately 20 percent of the city was destroyed. Brick and masonry structures were totally destroyed and many frame buildings were lost in the fire that swept the city and raged for days until it was finally extinguished by means of dynamiting buildings in its path. Water mains were disrupted by the earthquake.

The San Francisco earthquake had a magnitude of 8.3 on the Richter scale but intensity was found to vary greatly depending on the type of soil or bedrock; and extent of damage was more closely related to foundation materials than to proximity to the fault. Intensities of 9 and 10 were recorded in areas of construction over manmade fills in the bay area whereas intensities were lower in areas of solid bedrock.

This experience should have served as adequate warning about building on unconsolidated materials, but the 1971 Los Angeles earthquake again caused extensive damage to buildings constructed on weak materials. Buildings tipped over and collapsed, reservoir dams were damaged, and overpasses fell, despite the low magnitude of the 1971 earthquake.

FIGURE 20–13
Schematic sections across the Aleutian arc, showing approximate (solid) and inferred (dashed) land-level changes, aftershock distribution, and geologic interpretations along the section line. Heavy dots are earthquake foci (U.S. Coast and Geodetic Survey) within 30 kilometers of the section line.
(after Plafker, 1965)

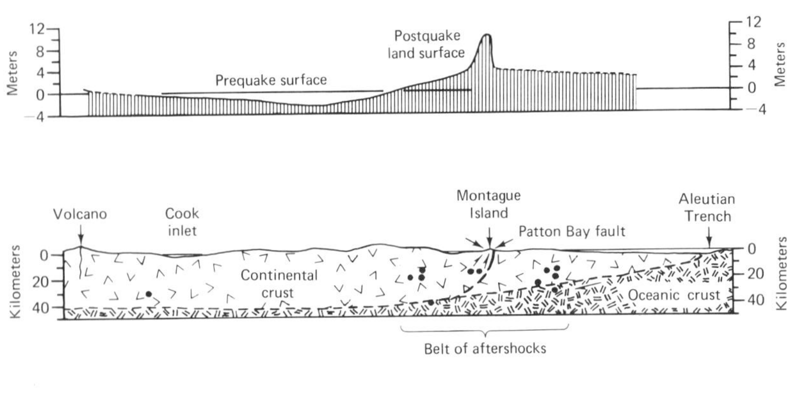

DISTRIBUTION OF EARTHQUAKES

It might appear that the distribution of earthquakes on the face of the earth is a simple matter. However, the picture one gets of the seismicity of the world depends on what factors are taken into consideration when the data are compiled. Different patterns are obtained if each of the following is plotted:

1. Distribution of the number of earthquakes that were felt or recorded.
2. Distribution of deep or shallow earthquakes.
3. Distribution of quakes of different intensities.

Seismograph stations are most abundant in heavily populated areas, and records have been kept in some areas for a much longer period than in others. Thus, if a summary of all available data is used, a false idea of the activity of different parts of the earth results. Detailed studies of the distribution of earthquakes have been made by the late Professor Gutenberg and Professor Richter at the California Institute of Technology. The results of their work are shown in Figs. 20–14 and 20–15. Richter (1958) summarizes the major fea-

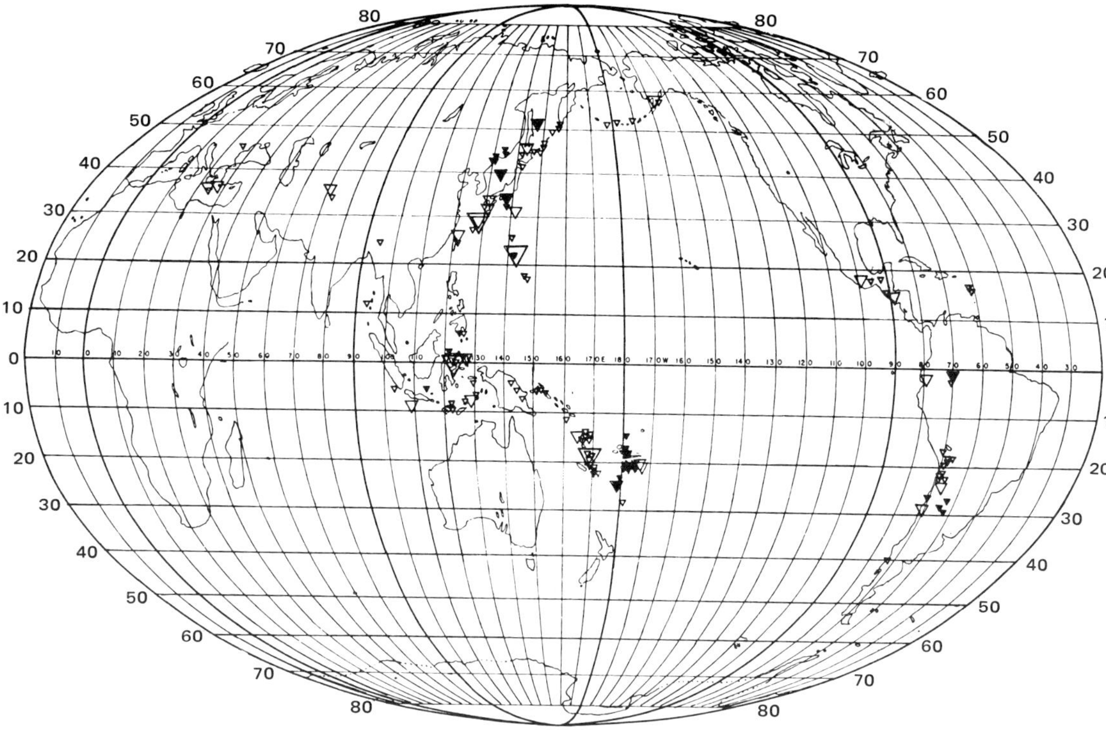

Modified Mollweide projection

FIGURE 20–14
Occurrence of large intermediate and large deep shocks.
(from Richter, 1958)

Large intermediate earthquakes, 1903–56

h = 70 — 300 km
▿ Magnitude 7.5–7.8
▽ Magnitude 7.9–8.5
▽ Magnitude 8.6–8.7

Large deep earthquakes, 1903–56

h = 300 km and over
▾ Magnitude 7.5–7.8
▼ Magnitude 7.9–8.4

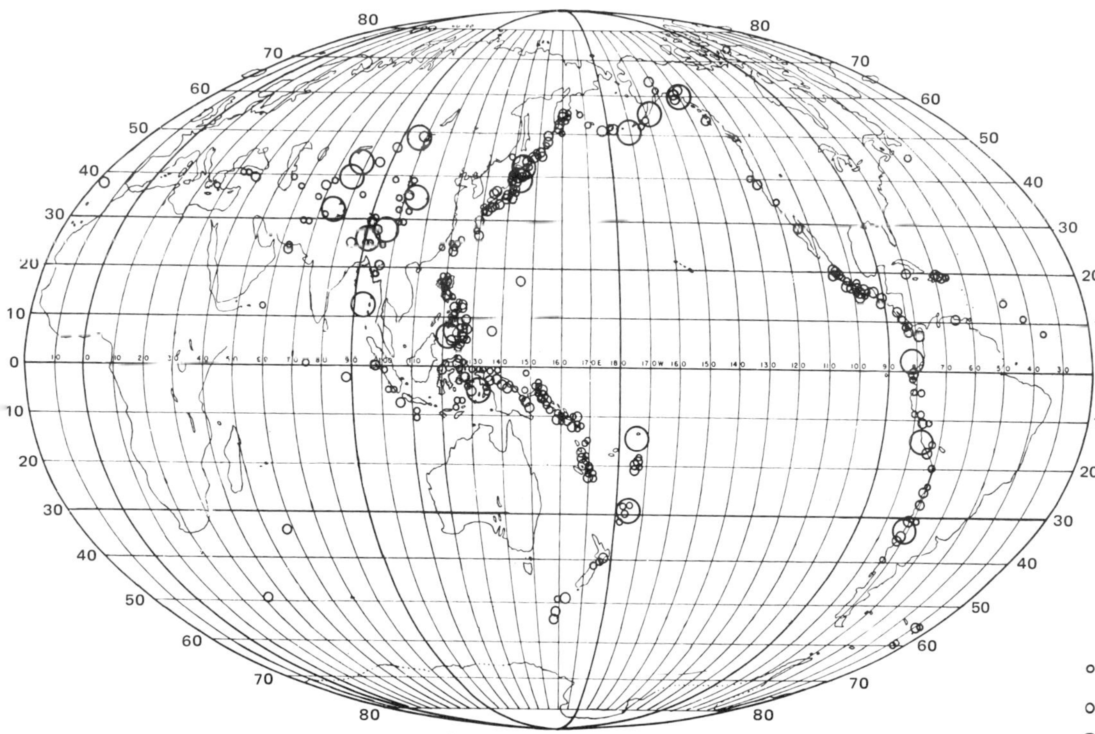

Modified Mollweide projection

FIGURE 20–15
Occurrence of large shallow shocks.
(from Richter, 1958)

Large shallow earthquakes

○ Magnitude 7.5–7.8, 1918–56
○ Magnitude 7.9–8.4, 1897–1956
◯ Magnitude 8.5–8.9, 1897–1956

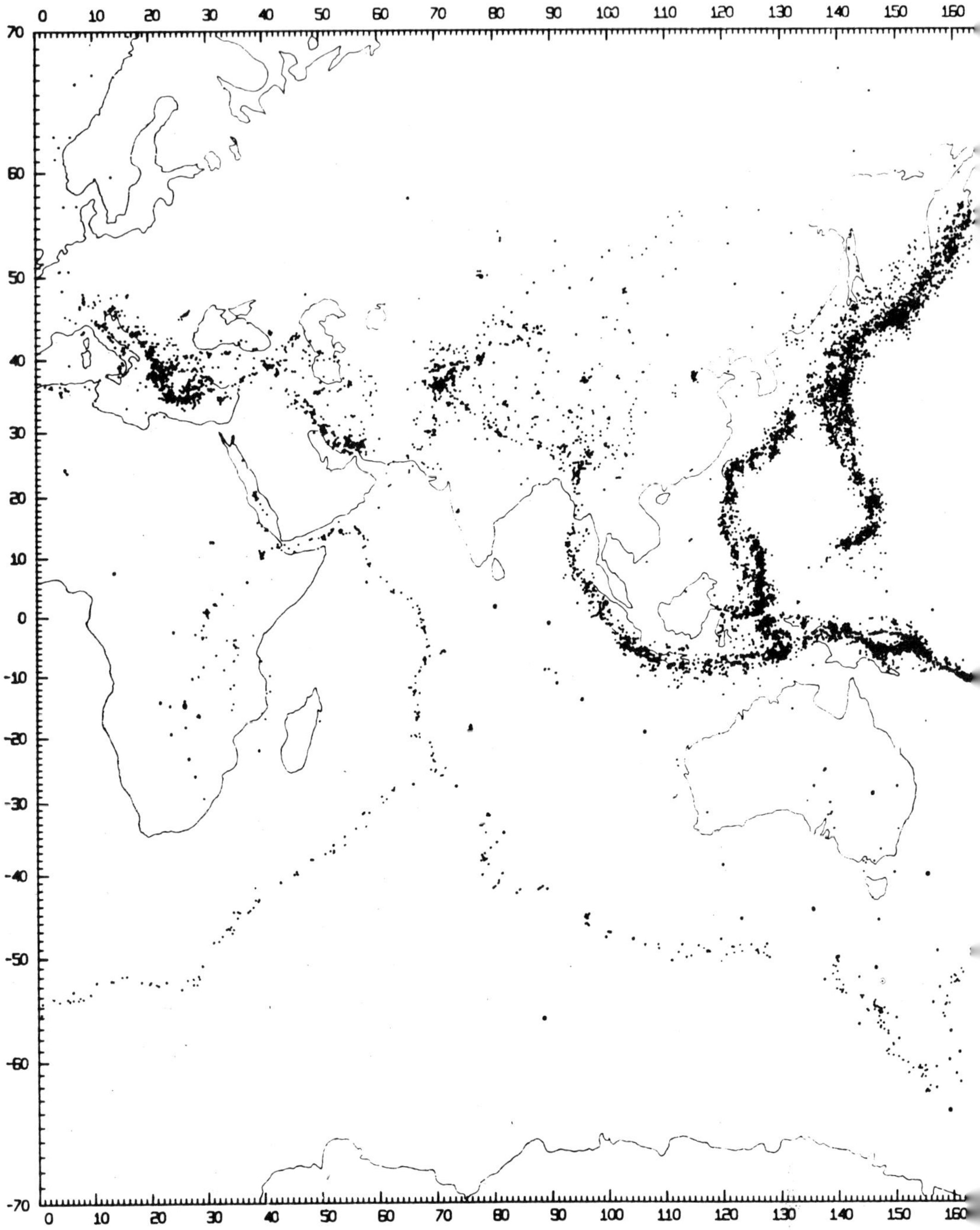

FIGURE 20–16
World-wide distribution of all earthquake epicenters for the period 1961 through 1967 as reported by U. S. Coast and Geodetic Survey.
(after Barazangi and Dorman, 1968)

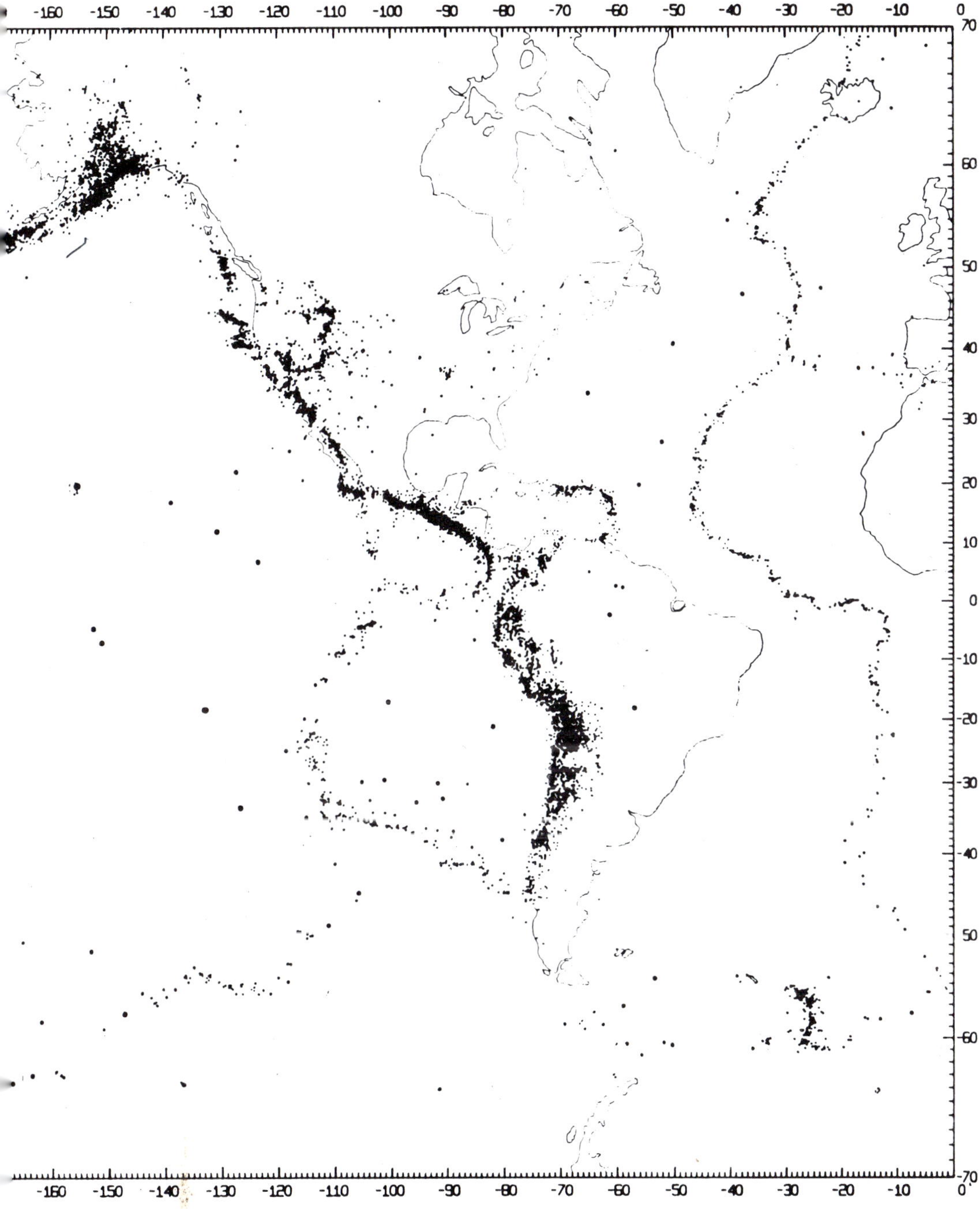
-160
-150
-140
-130
-120
-110
-100
-90
-80
-70
-60
-50
-40
-30
-20
-10
0
70
60
50
40
30
20
10
0
-10
-20
-30
-40
50
-60
-70

tures of the seismicity of the earth as follows. Epicenters occur chiefly in a few narrow belts or zones. Certain wider areas show fairly general moderate activity. Seismologically speaking, the most important subdivisions of the earth's surface are the following:

1. The circum-Pacific belt, with many branches and complexities.
2. The Alpine belt of Europe and Asia; this may be considered an extension of one of the main branches of the circum-Pacific belt.
3. The Pamir-Baikal zone of central Asia.
4. The Atlantic-Arctic belt.
5. The belt of the central Indian Ocean, with branches.
6. Rift zones, notably those of East Africa.
7. A wide triangular active area in eastern Asia, between the Alpide belt and the Pamir-Baikal zone.
8. Minor seismic areas, usually in regions of older mountain building.
9. The central basin of the northern Pacific Ocean; almost nonseismic except for the Hawaiian Islands.
10. The stable central shields of the continents, also nearly nonseismic.

The stable areas are generally known as *shields.* They include the following: (1) the Canadian Shield (Greenland being considered a detached part); (2) the Brazilian Shield; (3) the Baltic Shield; (4) the Angara Shield, located in northern Asia; (5) the African Shield (Arabia and Madagascar being detached parts); (6) the Australian Shield; (7) the Antarctic Shield; and (8) Peninsular India.

The circum-Pacific belt contains most of the shallow- and intermediate-depth earthquakes and almost all deep-focus earthquakes. The belt running through the mountain system that contains the Alps and Himalayas contains most of the remaining shallow and intermediate quakes, and two deep quakes have been recorded in it in Spain and Italy.

A world-wide network of about one hundred stations has recently been set up to provide much more uniform coverage of the earth's seismicity. The instruments in all these stations are standard; thus we may expect more refined determination of seismicity in the future. Already the increase in data is dramatic. Approximately 5,100 earthquakes were reported for the period 1904–52 compared with 29,553 (about 4,300 per year) for the period 1961–67.

EARTHQUAKE MECHANISM

The intensity of near-surface earthquakes leaves little question about the great amounts of energy that are released during earthquakes. The use of magnitude scales gives us a precise estimate of the amount of energy involved. It is also evident that a broad spectrum of elastic waves is generated by earthquakes, but we do not know with certainty the exact mechanism(s) by which these waves are generated or the manner in which the potential energy held at the focus is released to become kinetic energy of motion.

Certainly some earthquakes are generated by volcanic activity, but most shallow earthquakes appear to be related to faulting. In general earthquakes associated with volcanoes are more localized, both in extent of damage and in intensity of the waves produced, than are those associated with faults. In some regions, particularly in the circum-Pacific belt, volcanic activity and active faulting are closely associated. Nevertheless, it is possible to distinguish some earthquakes as being caused by activity due to volcanic phenomena not associated with major breaks in the crust of the earth. A shock may be produced by any of the following mechanisms:

1. Explosion of the volcano upon the release and expansion of gases and lavas.
2. Faulting within the volcano resulting from pressures in the chamber of molten rock.
3. Collapse of the center of the volcano into the space formed by the extrusion of gases and molten matter.
4. Vibrations set up by the movement of magma up the pipe or fissure.

Most American geologists favor the hypothesis that shallow (having focal depths within the crust) earthquakes originate as a result of some mechanism associated with faulting. Faults are breaks within the solid rock of the crust along which displacements occur.

Macelwane (1947) summarizes the character of the majority of slight earthquake shocks as follows:

1. The area of greatest intensity often lies along a known fault zone.
2. They occur far from any volcano.
3. Even in the neighborhood of an active volcano they are often not correlated with any particular sign of volcanic activity.
4. They often associate themselves in groups in which the center of intensity of successive earthquakes migrates parallel to a fault zone.

The California earthquake of 1906 is often taken as a model by American geologists. H. F. Reid (1910) made a study of U.S. Coast and Geodetic Survey records of triangulation surveys across the fault over a period of years before 1906 and concluded that movement of the west side toward the north had been going on gradually over a long period of time with an accompanying build up of strain across the fault. He theorized that the strain accumulated until it became so great that the fault broke and moved to relieve it.

The surveys showed that the strain was localized near the fault. Possible mechanisms for generation of shock waves by faulting are the following:

1. Vibration of the strained region when release occurs.
2. Grating along the fault surface due to frictional drag.
3. Frictional effects due to breaking at one point followed by a progressive increase in stress along the fault and steplike breaks along the fault.

These theories, all of which go by the name *elastic rebound,* derive support from the manner in which most earthquakes are followed by a long series of aftershocks. This may be depicted by the release of energy over a period of time (Fig. 20–17), accompanying faulting.

The mechanism of deeper earthquakes, those in the mantle, is even more difficult to assess. These deeper earthquakes occur below the depths at which brittle failure of rock is possible. Experimental rock-deformation studies have shown that rocks do not exhibit brittle fracture of the type with which we are familiar at the surface when they are subjected to confining pressures of the order of magnitude found below the continental crust. The following are among the suggested mechanisms for these depths:

1. Impulsive movement of magma.

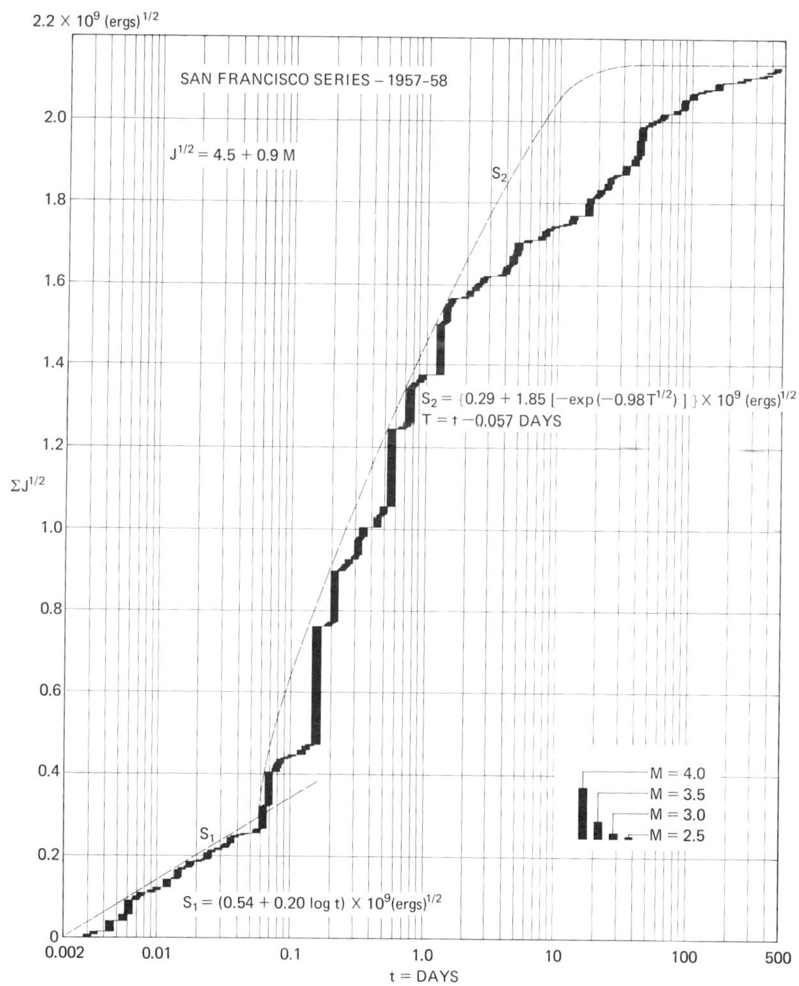

FIGURE 20–17
A graph depicting elastic rebound. The graph shows a plot of the cumulative energy release in a series of earthquakes against time. M is the earthquake magnitude and is drawn to scale for each earthquake in the series. $\Sigma J^{1/2}$ is the square root of the cumulative energy; S_1 and S_2 are two types of strain-release curves derived from the plot. (California Division of Mines)

2. Implosions or explosions due to sudden changes of phase of minerals (some phase changes involve 12 percent or more volume increase).

3. A fault movement consisting of a *stick-slip* type similar to a phenomenon encountered in plastic materials when they are compressed and sheared.

CHAPTER OUTLINE

Earthquakes
Modified Mercalli Intensity Scale
Case Histories
- Chilean Earthquake of 1960
- The Hebgen Lake, Mont., Earthquake
- Assam Earthquake
- Kwanto Earthquake
- Alaskan Earthquake of March 27, 1964
- San Francisco Earthquake of 1906

Distribution of Earthquakes
Earthquake Mechanism

TERMS

colluvium
elastic rebound
epicenter
focus
isoseismal map
monocline
sessile
shields
stick-slip
tectonic

REVIEW QUESTIONS

1. What is the Mercalli scale and what is it based on?
2. It is difficult to prove that the Hebgen and Red Canyon fault scarps at Hebgen Lake are more than shallow structures. Why? What relation do they have to any pre-existing structures?
3. Depending on what is plotted, different results may be obtained in trying to discover the most seismically active areas of the earth's crust. Explain why.
4. Seismic activity occurs chiefly in several narrow belts or zones. Where are the most active areas? Are they geologically similar in other ways?
5. What caused the most damage and loss of life in the earthquakes described in this chapter?
6. What zoning ordinances should any city subject to earthquakes have to protect property and population?

SUGGESTED READINGS

Hansen, W. R., 1965, Effects of the earthquake of March 27, 1964, at Anchorage, Alaska: U. S. Geol. Survey Prof. Paper 490, 196 p.

Richter, C. F., 1958, Elementary Seismology: San Francisco, W. H. Freeman, 768 p.

U.S. Geological Survey and Cooperating Agencies, 1964, The Hebgen Lake, Montana, earthquake of August 17, 1959: U.S. Geol. Survey Prof. Paper 435, 242 p.

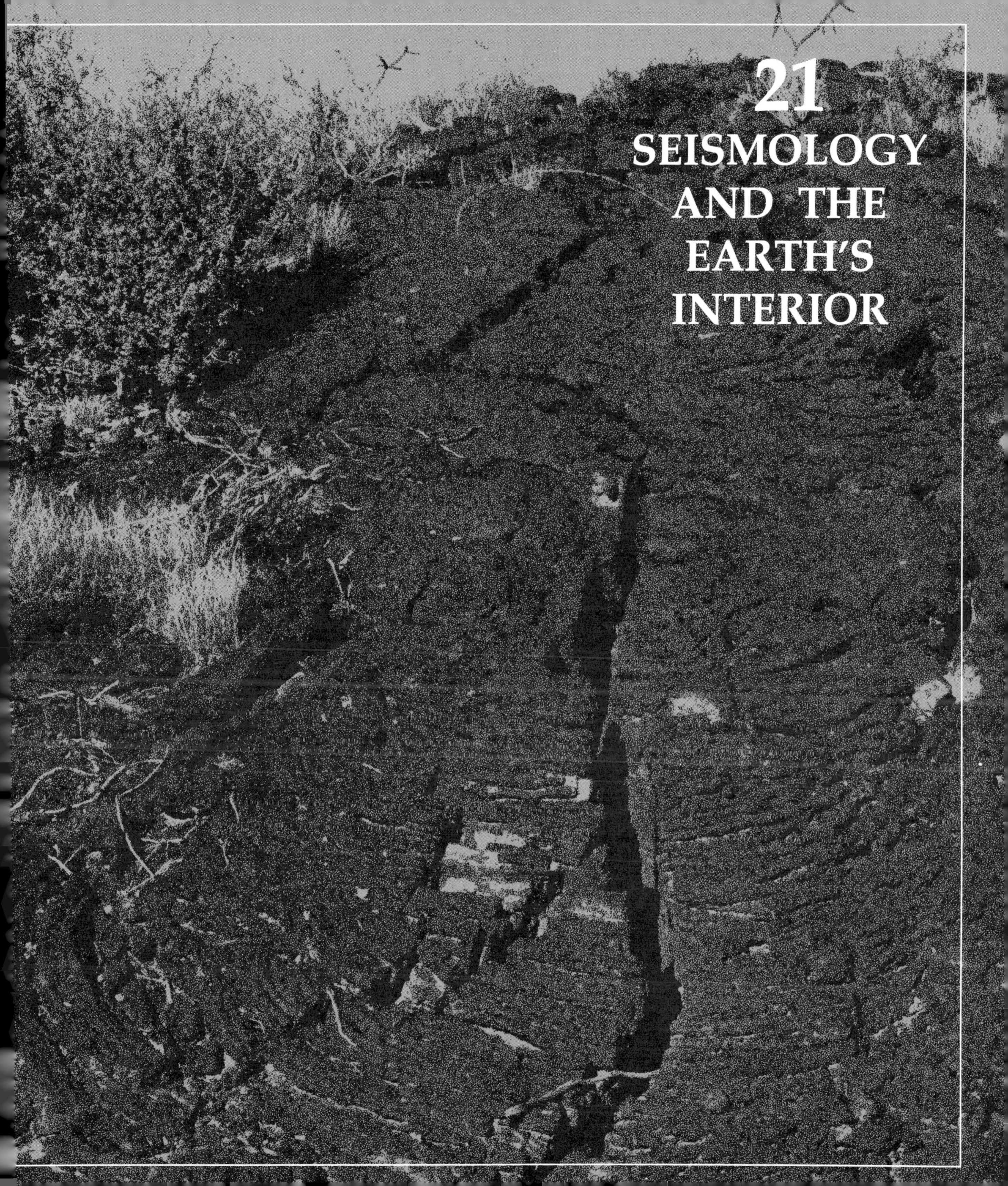

21 SEISMOLOGY AND THE EARTH'S INTERIOR

21 SEISMOLOGY AND THE EARTH'S INTERIOR

FIGURE
(preceding page)
Basaltic lava flows in central New Mexico.
(New Mexico State Tourist Bureau)

PROPAGATION OF SHOCK WAVES[1]

When energy is suddenly released within an elastic medium, it is propagated through the medium. Explosions caused by atomic bombs, TNT, and volcanic eruptions are accompanied by this type of energy release; artificial explosions are valuable in many types of seismic work. Energy is also released when *brittle failure* occurs in solid rock; this is thought to be the most common cause of shallow earthquakes, those within the upper 10 kilometers of the earth's crust where most rock is brittle. Other mechanisms may provide the energy released in earthquakes at greater depths, from 10 to 700 kilometers. Sudden changes in volume, as in a volcanic explosion or a sudden slip along a fault that sets rock in vibration, initiate elastic waves that are then propagated through the rock. Some waves travel throughout the rock mass just as a sound made underwater will travel out in all directions in ever-expanding, nearly spherical shells. Because the wave front is continually expanding as it moves farther away from the source the energy is distributed over an expanding surface, and the amount of energy in any part of the wave front is continually diminishing. The energy is said to be *attenuated.* The spherical shape of this expanding shell of energy is maintained as long as the medium through which it is being propagated is *isotropic*—that is, it has the same properties in all directions. Numerous physical discontinuities exist within the crust of the earth, and the outside surface is itself another discontinuity. Waves that are propagated throughout a three-dimensional continuum are called *body waves. Surface waves* are those elastic waves propagated near the surface of a body. Like the waves set up when a stone is dropped into a still body of water, they are propagated in a two-dimensional medium and the energy is distributed over an ever-expanding circle or near-circle. Surface waves are generated by earthquakes that occur near the surface of the earth and by manmade explosions. Explosions generate fewer surface waves than shallow-focus earthquakes, and the differences are used to distinguish the two.

[1] See Appendix C for discussion of the elasticity of rocks.

TABLE 21–1
Seismic velocities (mostly laboratory determinations) for some common rock types, in km/sec.
(from Clark, 1966)

	Normal Pressure		Higher Pressures (all V_p)		
Rock Type	V_p	V_s	1000 bars	4000 bars	10,000 bars
Soil	0.1–0.2	—			
Wet clay	1.5–2.5	—			
Sandstone	2.0–4.5	—	5.3	5.6	5.8
Salt	4.4–6.5	—			
Limestone (Solenhofen)	6.0	2.9			
Granite	5.6	2.9	6.3	6.4	6.5
Basalt	6.4	3.2	6.7	6.8	6.9
Gabbro	6.5	3.4	7.1	7.14	7.2
Dunite	7.4	3.8	7.9	8.1	8.3
Serpentine	—	—	6.7	6.8	7.0
Gneiss	3.5–7	3.4–3.6			

Two types of body waves—dilational and shear—are generated if the energy release occurs in rock. *Dilational waves,* also known as *primary, P-, compressional,* or *longitudinal waves,* involve changes in the volume of the rock through which they are transmitted. Primary was used originally because this wave travels faster and is the first, or primary, arrival observed on earthquake seismographs. Sound waves are examples of this type. Sound travels through any compressible medium such as water, air, or rocks. A propagating vibration or sound is set up in the medium by motion of an object, as in the motion of the surface of a drum that has been hit. As the face of the drum moves out on each vibration it forces the air away from it; the air is compressed. The layer of compressed air in turn compresses the layer of air next to it. Each time the drum face moves in, the air next to it is pulled as if it were under tension. This pull is transmitted to the next layer of air and follows the compressed layer outward as an expanding spherical shell. The waves are thus transmitted as a succession of compressions and rarefactions. The direction of vibration is along the path of travel of the wave front—hence the name longitudinal.

Shear waves, also called *secondary, S-,* or *transverse waves,* involve a shearing of the material; they vibrate perpendicular to the direction in which the wave front is moving. No volume changes are associated with them, but other distortions of shape are. They may be thought of as the waves transmitted along a bar when the bar is twisted or the motion of a string connected to a wall and wiggled. Waves can be seen to propagate to the wall, but any point on the string moves normal to the length of the string. Unlike compressional waves, those involving a shearing motion cannot be transmitted through liquids. These waves travel out from the source at different velocities.

Lord Rayleigh was the first to recognize the special types of wave motion that are concentrated at the surface of an elastic solid, and one of these wave types is named for him. Wave motion is vertical and in the direction of the movement of the wave front in *Rayleigh waves* (Fig. 21–1). Each particle involved in this motion follows an elliptical orbit in retrograde motion (along the bottom of the ellipse the motion is in the direction of propagation and at the top it is in the opposite direction). Rayleigh waves are superficially like ocean waves, but they differ in that the restoring force is elasticity instead of gravity, motion is retrograde elliptical instead of prograde circular.

A. E. H. Love later demonstrated the existence of another type of surface wave, now called *Love waves,* which travel with transverse displacements of the surface in the direction of motion of propagation of the wave front (Fig. 21 1).

Velocity of Elastic Waves

The velocity of propagation of seismic waves through rock is determined by the elastic properties[2] and the density of the rock (Table 21–1). Equations giving the velocity of elastic waves transmitted through a homogeneous, isotropic, and perfectly elastic material are given below:

Compressional waves: $$V_p = \sqrt{\frac{k + \frac{4}{3}\mu}{\rho}}$$

Shear waves: $$V_s = \sqrt{\frac{\mu}{\rho}}$$

where ρ is the density of the rock, μ is the modulus of rigidity (shear modulus), and k is compressibility.

[2] See Appendix C.

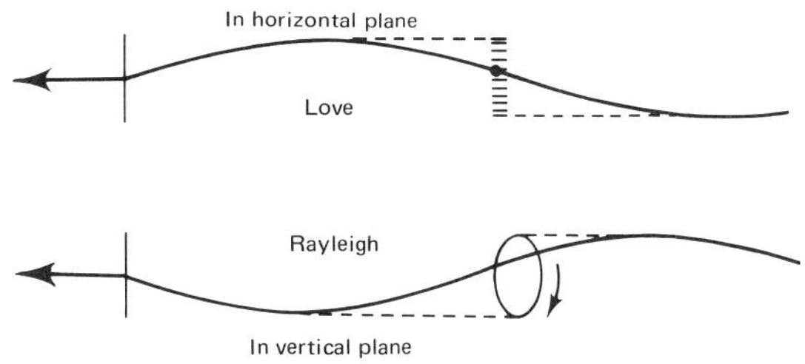

FIGURE 21–1 Earth-particle motions in surface waves. The upper diagram shows the horizontal earth particle motion transverse to the direction of propagation in the case of Love waves. The lower diagram shows the retrograde character of the earth-particle motion (in the vertical plane) in the case of Rayleigh waves.
(from Neumann, 1966)

These formulae indicate that the velocity of P-waves is increased as the rocks become more rigid and less compressible, and that the velocity is decreased as the rocks become more dense. Actually more dense rocks have higher values of compressibility, shear modulus, and Poisson's ratio so the velocities of seismic waves do increase in more dense rocks.

The velocity of the shear wave increases with increased rigidity and decreases with increased density. Shear waves cannot travel through media that cannot be sheared, such as liquids, whereas the P-wave does continue through liquids because it depends on the compressibility of the medium. These effects of density should be especially noted since density and seismic velocity increase with depth in the earth. This must mean that rigidity of the rocks increases faster than density.

Poisson's ratio, σ, for any rock can be determined from measurement of V_p and V_s.

$$\frac{V_p}{V_s} = \sqrt{\frac{k}{\mu} + \frac{4}{3}} = \sqrt{\frac{1 - \sigma}{\frac{1}{2} - \sigma}}$$

$$\sigma = \frac{(V_p/V_s)^2 - 2}{2(V_p/V_s)^2 - 2}$$

This relationship allows us to determine this important physical property of rocks even when they are too deeply buried to sample.

Wave Paths: Bending, Reflection, and Refraction

Because the earth is neither isotropic nor homogeneous, wave fronts of body waves do not travel out in spherical shells. Among the main causes of inhomogeneity on the earth are the following:

1. The strong discontinuity at its surface.
2. Variations in density and rigidity with

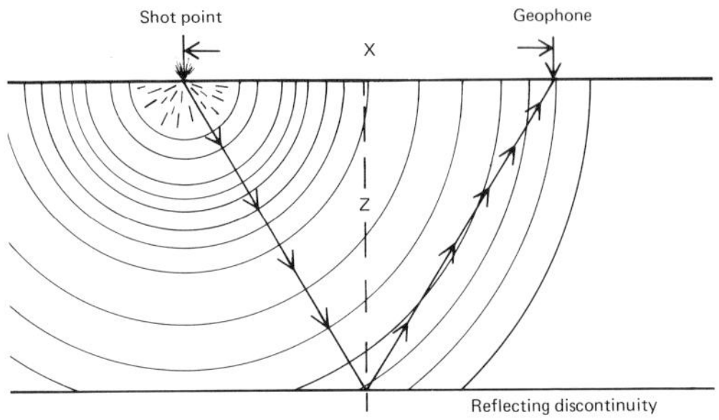

FIGURE 21–2
Wave reflected from a single interface with constant velocity, V.
(after Dobrin, 1960)

depth due, in part, to weight of the overlying rock.

3. Variations in physical properties caused by changes in rock type, due to variation in composition.
4. Variations in the state of the material may occur at great depth due to high temperature and high pressure.

It is useful to consider the path of the wave in terms of rays, lines drawn so that they are everywhere perpendicular to the wave front. Waves are reflected and refracted when they reach the boundaries between layers with different elastic properties or densities. Part of the energy may be reflected and part refracted (bent) into the adjacent layer when a shock wave reaches a discontinuity in physical properties. The boundary between two layers of different types of rock illustrates this point. Reflection takes place so that the angle at which a ray is reflected from a discontinuity surface is equal to the angle of incidence of that ray (Fig. 21–2). Each of the layers has a characteristic seismic velocity for the P- and S-waves if each layer is isotropic and homogeneous. So long as the wave moves within a given layer it moves with the same velocity, but when energy travels across the contact of the two layers it takes up the velocity of the other layer. This change in velocity causes the direction of propagation of the wave to be modified according to Snell's Law, which states that the sines of the angles between any incident ray and the normal to the interface are in direct proportion to the velocity of the respective wave in the medium (Fig. 21–3). This can be written

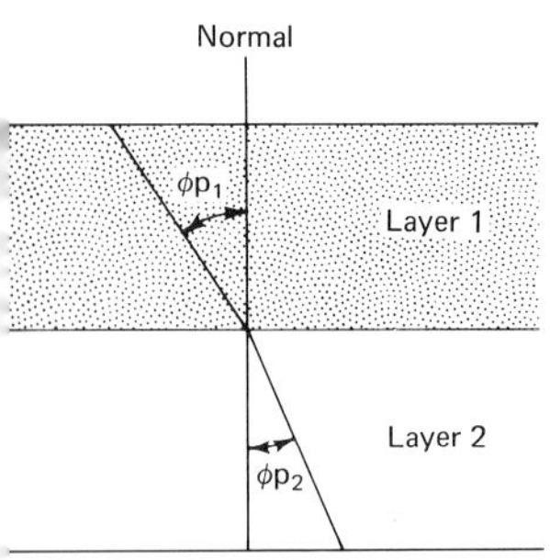

FIGURE 21–3
Refraction, bending, of a ray on passing through a physical discontinuity.

$$\frac{V_{s_1}}{\sin \phi_{s_1}} = \frac{V_{s_2}}{\sin \phi_{s_2}} \qquad \frac{V_{p_1}}{\sin \phi_{p_1}} = \frac{V_{p_2}}{\sin \phi_{p_2}}$$

where the subscripts s and p indicate P- and S-waves and subscripts 1 and 2 denote the two layers.

If a wave approaches a discontinuity at right angles, along the normal to the surface, the angle between the incident ray and the normal is zero, and the energy passes through the discontinuity without being bent. If the incident wave approaches a discontinuity at some other angle, refraction occurs. When the wave passes from one layer to another of lower rigidity (lower velocity), the wave front (as shown by a ray) is bent down (toward a line perpendicular to the contact between the two layers, the normal). The reverse occurs when the waves pass from a less rigid to a more rigid layer. Thus the gradual increase in the rigidity of rocks with depth causes the paths of the shock waves to be curved upward.

The curved path of rays in a material in which rigidity increases with depth may be thought of as a multilayered case with very thin layers, each of which is more rigid than the one above. The wave is bent on passing through each interface and the bending is away from the normal (Fig. 21–4).

REFLECTION SEISMIC METHODS

It is instructive to consider the methods of exploratory seismic studies in which the location, depth, size, and time of occurrence of the generation of the shock wave are all known before examining the more complicated problems of earthquake seismology. Seismic methods are used by oil companies, engineering firms, and others who need to determine the depth and position of rock layers.

The method of *reflection seismology* is quite simple in principle. A signal from an explosion or sound generator is propagated through the ground. When energy from this source encounters discontinuities of various types in the earth part is refracted into the units below and part is reflected back toward the surface. The part that comes back to the surface can be detected by *geophones*, and the signal recorded. The time at which the signal was generated is known and the time at which each reflection reaches the geophones can be measured. If the velocity of propagation of the signal through the various units in the earth is known or can be determined, then the depth to any reflecting horizon is a function of the travel time of the signal. The depth can be calculated by a simple geometrical relation (Fig. 21–2). In practice the application of reflection techniques has become sophisticated, and the principles are known to be somewhat more complicated than outlined above.

The oldest method of generating a signal is through explosion of dynamite. If the soil and mantle of decayed rock are shallow, a hole is drilled deep enough to set the charge off beneath the overburden. At sea the charge is usually set off at the surface. Newer methods involve use of large vibrators on land to generate the signal. With these the frequency range of signals can be limited and various frequencies tried, if the first one does not yield a satisfactory record. At sea new methods involving use of a spark-discharge gas exploder (a mixture of propane and oxygen detonated by a spark) and air guns in place of explosives have been developed. In this method the ship moves at 1.5–2.5 meters per second and spark discharges are timed between one and four seconds. Geophones towed behind the ship pick up reflections as the ship moves, giving a continuous profile of reflections (Fig. 21–5). On land the geophones are set out in some stationary pattern, which is then moved between shots. Various types of geophone placement patterns are used. A common one involves placing twelve geophones in a line on either side of the shot point. The reflected signals are picked up by each of these and recorded on a drum recorder that shows the twenty-four traces. Usually the signals are also recorded on magnetic tape. Numerous systems of modifying the recorded signals from tapes have been developed. They are designed to strengthen the returning signal and make interpretation of the record easier (Figs. 21–6 and 21–7).

One of the most important advantages of the reflection technique is that reflections are obtained from horizons at several depths. In addition, the depth of the reflection can be determined (if velocity can be obtained by another method). Thus a unique point in the subsurface can be located by reflection. The important question for the geologist is: What does that point or surface

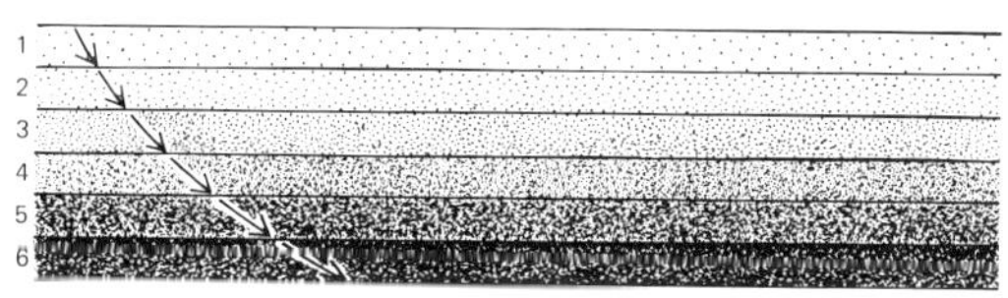

FIGURE 21–4
Gradually increasing rigidity with depth causes bending of the rays. Taken far enough, the ray will be directed back toward the surface.

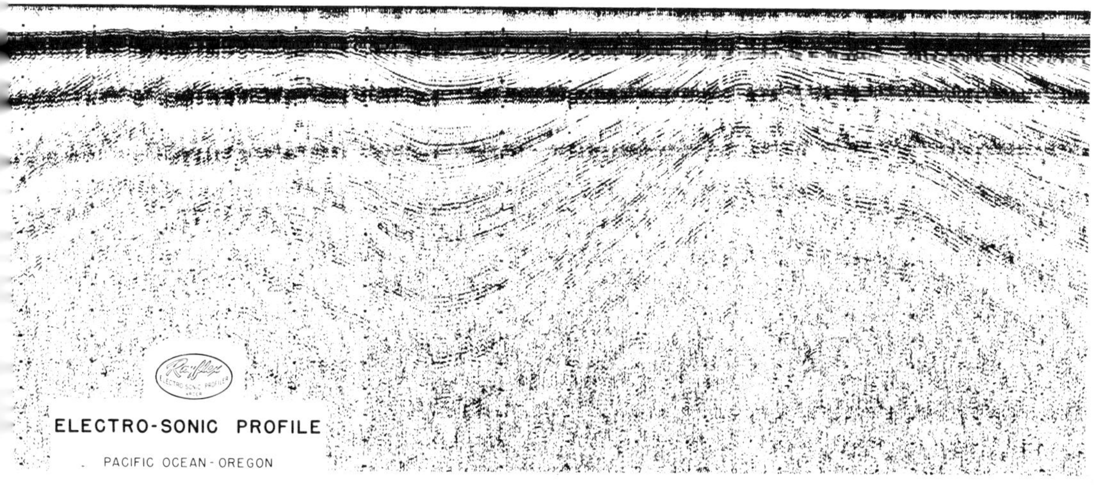

FIGURE 21–5
Cross section off Oregon coast made by continuous electro-sonic methods.
(Rayflex Co.)

represent? The physical properties of rock, specifically the density and rigidity, determine the rate at which shock waves are propagated through it. A change in these physical properties would necessarily cause a change in the velocity of propagation. It is from such changes in physical properties that reflection may be expected to occur. Thus, if a marked change in the properties of rocks occurs between one formation and another, the contacts of that formation may be useful marker horizons, and it may show up prominently on the records. The obvious way to check the existence of such a horizon is to have a well in which the formation is penetrated and to see if the contact is the reflector. Many reflections that do not represent formation contacts appear on records. Changes in physical properties may be changes in porosity or the degree and type of cementing. Phantom horizons that do not have any direct correspondence to the stratigraphic sequence in the area where the reflections occur may appear. This happens because the reflecting energy is reflected back down through the section as well as up toward the surface. An interference pattern is created; some signals are reinforced, others negate one another. For this reason reflection surveys are most useful when they can be correlated directly with well logs or surface sections through the same formations.

FIGURE 21–6
Vibratory seismic galvo-trace uncorrected cross section composed of records from Delaware Basin, western Texas.
(from R. S. Finn and W. O. Heap, 1962)

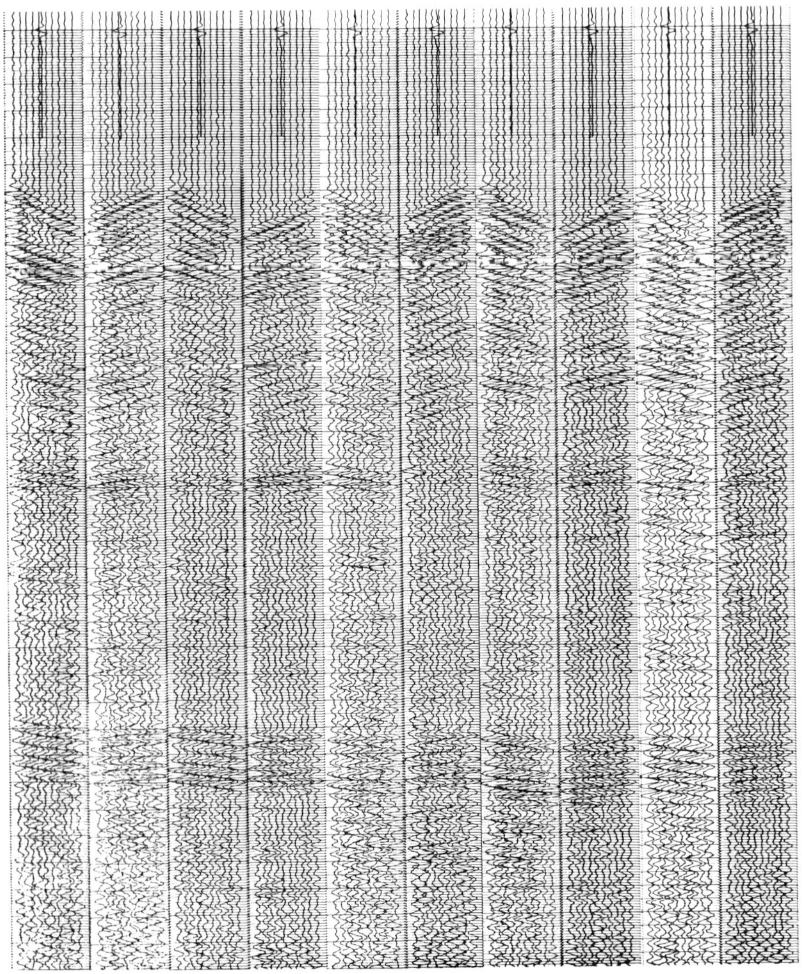

Determination of the attitude of a reflecting horizon can be made from one or more profiles on that horizon. Consider first a horizontal discontinuity. The first shock received at each geophone is the wave that travels in a straight line close to the ground surface. The first reflected wave should arrive at the closest geophones on either side of the shot at about the same time. This reflection will have traveled progressively farther to reach each of the other pairs of geophones. The time lag appears on the record and it should be uniform for geophones at identical distances from the *shot point* on either side.

Two profiles at high angles to one another are used to determine strike and dip. Assume that the profile line is perpendicular to the strike of the formation. The distance traveled and the travel time from the shot point to the horizon and back to geophones located at identical distances on either side of the shot point will be different when the horizon is dipping. The time difference is a measure of the dip of the horizon. The dip may be calculated by the following formula:

$$\text{sine } \alpha = \frac{V(t_1 - t_2)}{x}$$

The sine of the angle of dip is equal to the velocity of propagation within the units above the horizon multiplied by the time difference in arrival of the reflection at two geophones spaced equal distances from the shot divided by the distance between the geophones. [See Dobrin (1960), page 108, for derivation of the equation.]

Reflection records may be used directly for structural interpretation provided the structure is

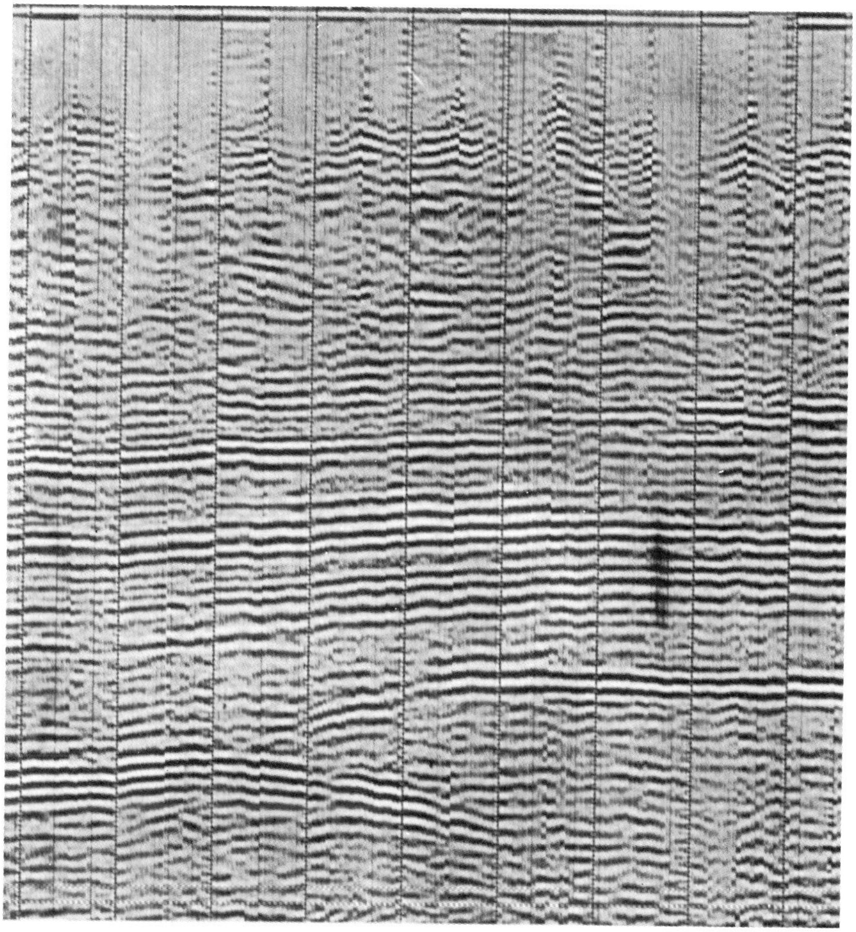

FIGURE 21–7
Example of variable-density cross section from central Oklahoma from vibratory seismic records. Presence of fault is shown.
(courtesy of Continental Oil Co., from R. S. Finn and W. O. Heap, 1962)

not complicated, but generally it is best to make interpretations from data that have been corrected for elevation and depth of weathering, and in which times have been converted to depths. The depths for dipping units must be moved so that the depth lies under the correct point along the profile. This point is not vertically below the shot point when the horizon is dipping (Fig. 21–8). These data may be plotted on a profile (Fig. 21–9) with depths to a selected horizon indicated. From this a structure contour map may be drawn. More commonly, the data are studied in the form of a profile or cross section. Results from a series of overlapping profiles may be used for continuous correlation along the profiles. If wells are available along the line of the profile, data from the well may be drawn to the same depth scale as the seismic section and used for even more exact correlation.

Records on magnetic tapes have proved particularly valuable for interpretation of seismic profiles. The tape method allows corrections to be programmed and also makes it possible to modify the traces. Common modifications are to cut off the peaks and troughs that have amplitudes over some determined value, "folding" back peaks and troughs into spaces in between, and to vary the density of the reproduction so that peaks appear black and troughs white.

Even when records are excellent their inter-

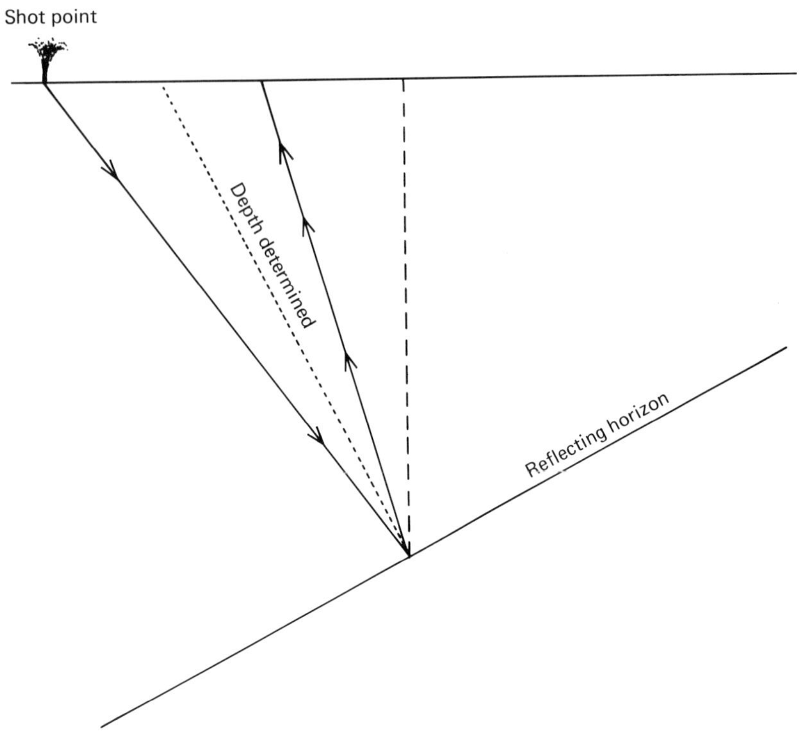

FIGURE 21–8
Ray path of shock reflected from a dipping horizon.

FIGURE 21–9
Cross section constructed from a plot of dip shooting results (short lines indicate dips in this profile).

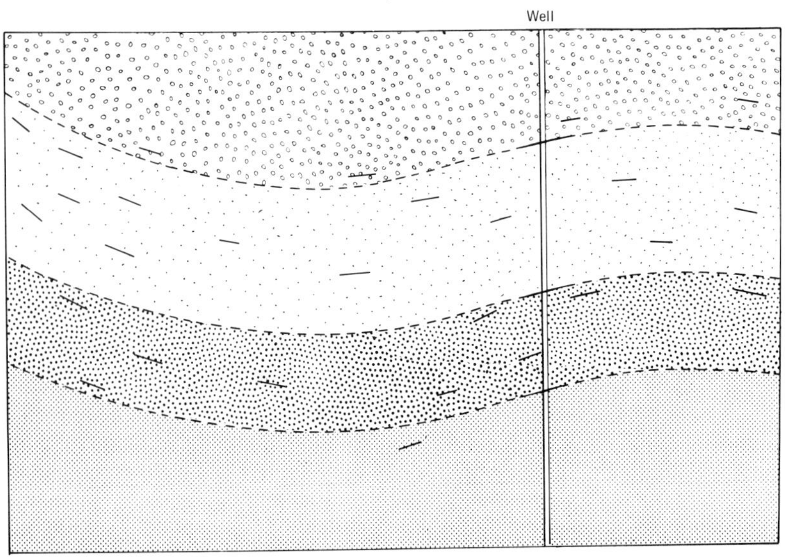

pretation requires a thorough knowledge of the geophysical techniques and a broad background in possible stratigraphic and structural configurations, particularly those to be expected in the region under investigation.

REFRACTION SEISMIC METHODS

When a shock wave encounters a discontinuity in a sequence of stratified rocks, part of the energy is refracted through the discontinuity into lower rock units. The velocity of transmission of the shock waves differs in the two rock units if they have different physical properties. At some distance away from the shot point the wave front will impinge on the discontinuity at such an angle, known as the *critical angle,* that the refracted wave will travel along the discontinuity instead of through it. As the shock wave moves along the discontinuity it is continually refracted back up through the overlying rock. Energy that was refracted through the first discontinuity may be returned toward the surface after being refracted along a second discontinuity, but it is likely to be recognizable only if the velocity in the lower unit is higher than that in the overlying unit. Modern techniques have emphasized use of these later arrivals.

To detect these shock waves a set of geophones are placed on the ground surface. Various patterns are used, but the geophones are often simply placed in a line extending away from the shot point. To obtain the same depth of penetration the geophones must be placed much farther away from the shot point than in reflection work because the first wave that can be refracted back to the surface is at a distance from the shot point determined by the geometry and velocity of the rock unit. Spacing of geophones depends on the depth to which solutions are desired. Usually spreads equal to about six times the desired depth are used. The time of the explosion and the time at which the first arrivals reach each of the geophones are recorded. The data are then plotted; the travel time of the first arrival is plotted against the distance from the shot point to the geophone (Fig. 21–10).

The travel time vs. distance plot, called a *travel-time curve,* is a straight line if all of the first arrivals have traveled along the same discontinuity and if the velocity in that unit is uniform. Sometimes the velocity increases continuously with depth in a unit. This is likely in semiconsolidated sediments where density increases with depth as a result of compaction. When this happens the line connecting points marking first arrivals is curved as is the ray path followed by the shock waves. When the velocity of propagation along the first discontinuity is faster than that through the surface material, the first arrival is a wave that has actually traveled a longer path. That path is from the shot point down to the discontinuity, along the discontinuity, and then back up to the receiving geophone. In a layered sequence with more than one discontinuity and in which velocity is higher in each successively lower layer, at some distance from the shot the wave traveling a path that goes down to the second discontinuity will outrun the one above it. For geophones at that critical distance, and beyond, the travel time will be proportionally less than it was for the closer geophones. This will show up on the plot by a deviation from the slope of the initial straight line (Fig. 21–10). First arrivals from other lower discontinuities will be represented on the graph as additional lines with ever-decreasing slope. Refraction studies have also been used in studies of crustal thicknesses. One such crustal section (Fig. 21–11) was made in eastern Montana.

One of the critical disadvantages of *refraction seismology* is the difficulty of detecting layers in which velocity decreases from what it was in overlying layers. One advantage of it over reflection methods is that it does provide information about the velocity of layers through which the shock wave passes.

Refraction profiles have been used to find faults. When waves refracted along a discontinuity strike a fault, there is a time lag before arrival of the waves from the same discontinuity on the other side of the fault. Thus the straight line from first arrivals from this discontinuity is broken and offset, but unless the dip of the discontinuity changes across the fault the slope of the line on the travel-time graph remains constant.

It is also possible to determine the strike and dip of an inclined surface through refraction methods. As with reflection seismology, two lines of data (often shot at right angles) are used. The dip can be calculated [see Nettleton (1940) or Dobrin (1960) for explanation of the methods].

DETECTION OF SEISMIC WAVES

Motion of the ground surface resulting from an earthquake depends on how deep and how far away the *focus* is located. Near the *epicenter* in zones of high intensity, ground motion may turn buildings on their foundations or throw objects into the air, but at great distances from the epicenter the motion consists of small displacements less noticeable than the slight jar resulting from blasting in a quarry or building site located several miles away. In fact, most seismic waves are not felt at distances of more than a hundred miles from the epicenter.

The *seismograph* is an instrument designed to detect the amplitude and direction of these minute displacements. Any quantity that has amplitude and direction can be resolved into three mutually perpendicular components. For this reason most seismograph stations employ

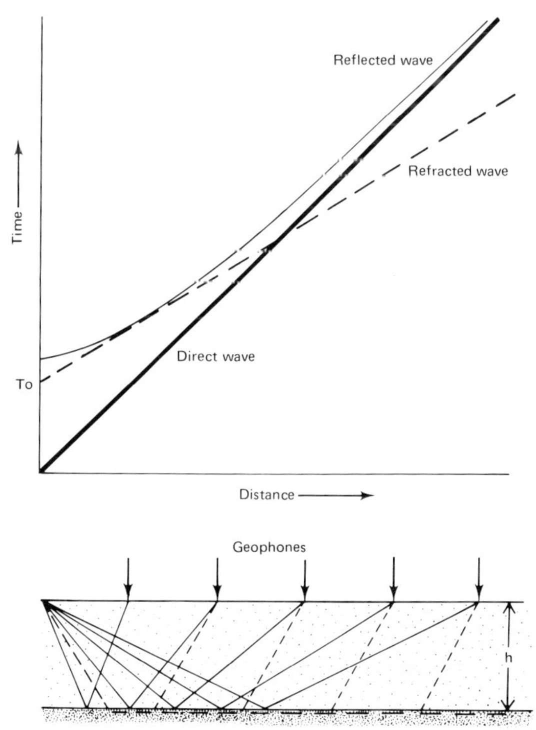

FIGURE 21–10
Cross section showing ray paths between an explosion and geophones for reflected and refracted waves, below. Travel-time curve for waves, above.
(after Dobrin, 1960)

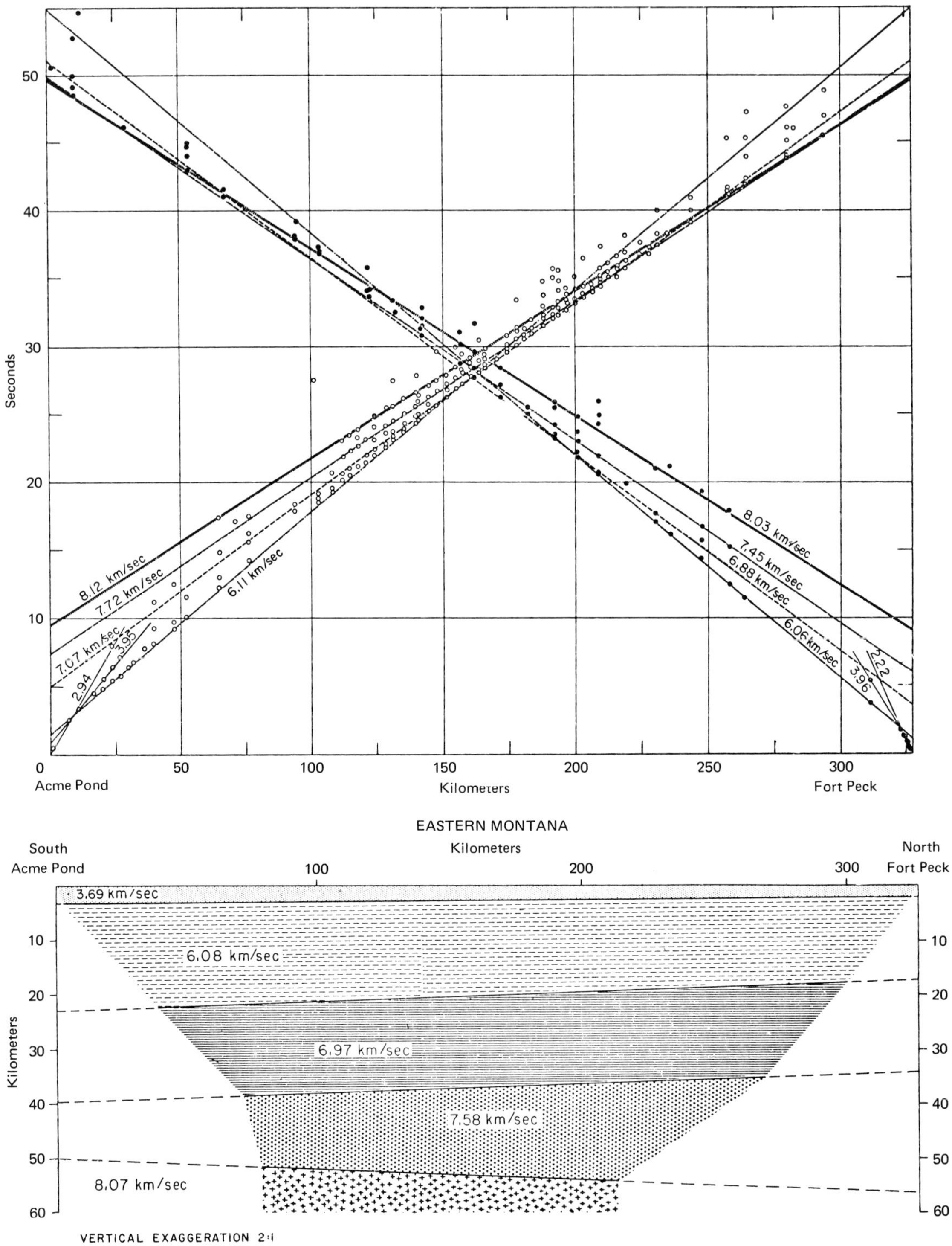

FIGURE 21–11
Reversed seismic profile, top, and interpreted cross section, bottom, from an area in eastern Montana.
(from Myer, Steinhart, and Bonini, 1961)

three instruments to record the components of ground motion along three mutually perpendicular lines. These are horizontal motion along north-south and east-west lines and motion perpendicular to the earth's surface. The only difference in the three components is the manner of suspension of the mass and the orientation of the instrument. The principle on which most seismographs operate is simple. The perfect seismograph is one in which it is possible to isolate a mass (as a point of reference) from the ground motion and measure the motion of the ground in relation to that mass.

Two types of seismographs are widely used. Both measure motion of the ground by recording the relative motion between the ground and the center of oscillation of some form of pendulum. The two seismographs differ depending upon whether the relative motion of the pendulum is recorded directly on paper through a simple mechanical or optical lever, or whether it is measured indirectly by generating an electrical current that depends on the motion of the pendulum. The second type has proved more effective for modern work. Some of these are capable of magnification of ground motion 100,000 times or more, but they usually operate in a magnification range of 500 to 20,000. A mass is delicately suspended by springs so it is relatively unmoved during the quake. Relative movement between the ground and a suspended mass is measured. A magnet is used for the mass; a coil of wire is placed around the magnet and anchored to the ground so it moves past the magnet when the ground is in motion. The two are isolated so that they do not touch one another. During an earthquake the coils move through the magnetic field set up by the magnet. This induces an electrical current in the coils. The magnitude of the current is a function of how fast the coils move and thus of the amount of ground motion. The current is amplified and fed into a mirror galvanometer, causing a deflection of the small mirror mounted in the galvanometer. Again the amount of deflection of the mirror is a measure of the ground motion. A beam of light is directed through a slit onto the mirror in such a way that it is reflected to hit the surface of a rotating drum (Fig. 21–12). Thus the signal is greatly amplified. A piece of photographic paper covers the surface of the drum. The drum is set to rotate at a constant rate and mounted in such a way that it moves along its axis as it rotates. Thus if the beam of light hitting the drum were perfectly still it would describe a spiral path on the drum with no two lines crossing or overlapping. A radio at the station receives time signals from the Naval Observatory or other stations broadcasting Greenwich Standard Time (this is used at seismograph stations throughout the world to avoid confusion in timing the occurrence of an earthquake). This time signal is converted into a small impulse of current sent to the galvanometer each minute, causing a deflection of the beam and thus marking the exact time on the seismogram. It is also possible to record the electrical signal from the seismograph on magnetic tape. This has advantages in analysis of the data since various computer frequency filters can be used as well as other computer operations.

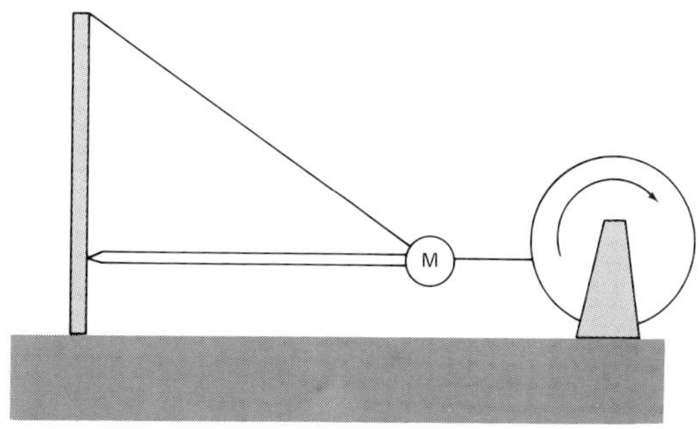

FIGURE 21–12 A simple seismograph. The mass suspended on the wire tends to remain stationary while the support and drum move during an earthquake. This instrument measures motion in the direction perpendicular to the page.

INTERPRETATION OF SEISMIC WAVES

While the seismograph is in operation, the trace indicating the earth's motion on the seismogram is rarely at rest. Movements of trees in the wind, waves breaking along the seashore, and even trucks passing on nearby highways are picked up by the more sensitive instruments. These minor movements are called *microseisms*. With the arrival of the first shock wave from an earthquake the trace is suddenly displaced from the rather sinuous path it was following. This break is indicated by a sudden change in the direction of movement and increased amplitude of the trace's swing (Fig. 21–13). The key to identifying the arrival of subsequent waves is also change in

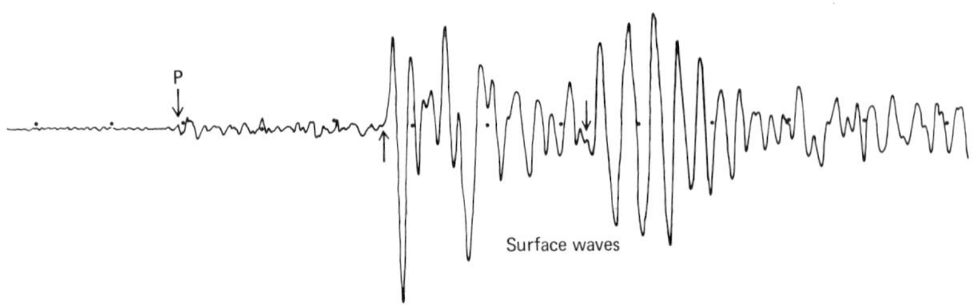

FIGURE 21–13
(*Above*) A seismogram. The minor fluctuations in the trace are called microseisms. They are received almost continuously. The first break in the trace marks the arrival of the wave that traveled the most direct path with the greatest velocity. This is the P-, or compressional, wave. The shear wave, which arrives some time later, started at the same instant as the P-wave and travels the same path but at a lower velocity. The spread between the arrival of these two waves is a measure of the distance to the focus.

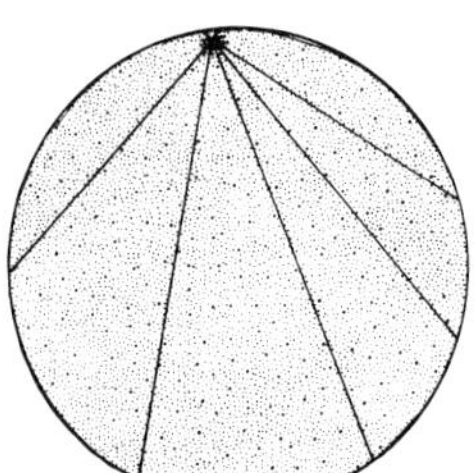

FIGURE 21–14
Cross section of isotropic sphere showing straight-line seismic-wave paths.

direction and amplitude of the trace's movement. The time of arrival of each wave at the station varies according to the path traveled by the wave and the velocity of the propagating wave. The first wave to arrive is the one that travels the most direct path with the highest velocity. That wave is the P- (compressional) wave.

If the earth were isotropic and homogeneous the waves would follow courses indicated in Fig. 21–14; all rays would be straight lines; and all wave velocities would be constant. The outer crust of the earth is neither isotropic nor homogeneous and this has proved true of the interior as well. The methods of exploratory seismic work are sufficient for detailing discontinuities in the upper few kilometers of the earth, but the energy necessary for deeper penetration of shock waves comes from earthquakes and nuclear explosions. The structure of the interior has been inferred largely from analysis of travel-time curves. The times of arrival of waves at various seismograph stations around the world are interpreted in terms of models of velocity distribution (depths of velocity discontinuities) until a model is constructed that fits all or most of the data.

Seismic Discontinuities

The first major seismic discontinuity is the *Moho discontinuity* (*M-discontinuity*), which is named for A. Mohorovičić who first recognized it. This discontinuity has been well documented by methods of both exploration seismology and earthquake seismology, and it is now accepted as defining the base of the *earth's crust.* It occurs at varying depths but generally at about 10 kilometers in deep ocean basins, from 30 to 40 kilometers on continents, and may reach a depth of 60 kilometers or so under mountains. There is a sharp break in velocities from about 6 km/sec to 8.2 km/sec for P-waves, and S-waves increase from about 3.5 km/sec to about 4.5 km/sec at *M*. The zone below the crust is known as the *mantle* of the earth. Shock waves originating in the crust are reflected from the *M*-discontinuity and refracted through it into the mantle.

The way inference of discontinuities is made is easy to understand if we consider the travel times of the first seismic waves to arrive at a hypothetical series of stations located on a *great circle* (the circle defined by the intersection of the earth's surface and a plane passing through the center of the earth) that includes the epicenter (Fig. 21–15). The distance of stations from the epicenter is defined by the angle between the focus, center of the earth, and the station. The location of the focus is not initially known, but it can be determined by methods we will consider later.

Assume that an earthquake occurs at shallow depth. P- and S-waves travel out in all directions at high velocities; much slower waves spread over the surface. Because rigidity and velocity increase downward in the mantle, P- and S-waves travel bent paths that cause them to return to the surface in less time than the P- and S-waves that travel the entire distance at the same depth. Thus the first arrival should be a P-wave that has traveled a curved path. P-waves are reflected from the earth's surface and *M*-discontinuity back down into the mantle as either P- or S-waves. Therefore, a P-wave that has traveled a path consisting of a series of reflections from the surface can reach the station. Such waves are referred to as PP, PPP, SS, SSS, and so on (Fig. 21–15), and if a wave travels part of the path as a P-wave and part as an S-wave it may be designated PS, SP, PPS, and so on.

Stations located between 15° and 22° receive initial arrivals of P- and S-waves earlier than might be expected. The velocity of P appears to have increased to 9.1 km/sec and S to 5.3 km/sec. Sir Harold Jeffreys interprets this as a change brought about by passage of the waves through a zone in which the rate of change of velocity with depth increases. He postulates that the depth of this zone is about 400 kilometers.

Beyond 22° the velocity of P- and S-waves increases continually, and apparently the waves do not pass through strong discontinuities in this zone. This observation holds for all stations located up to a distance of 103° from the epicenter. Beyond 103° the amplitude of P- and S-waves decreases, or they are not recorded at all. This decreased amplitude is observed by stations between 103° and 143° distant. Then beyond 143° the amplitude increases again for the compressional waves, but direct shear waves have disappeared altogether. These observations lead to the identification of the second major discontinuity, the outer boundary of the earth's core. Only diffracted P-waves are obtained between 103° and 143° because of the presence of the core; waves coming to the boundary of the core along a line tangent to the core are refracted into the core, come out, and are recorded at stations between 143° and 180° distant. Thus the zone between 103° and 143° is known as the *shadow zone* (Fig. 21–16).

Some P-waves have been identified in the shadow zone. Lehmann (1936) suggested that they might be due to a small inner core with a higher velocity than the outer core. This has now been confirmed by other observations of the travel times for waves passing through the core and is generally accepted. The inner core has a radius of 1,300 kilometers and is more rigid than the outer core. The core as a whole has a radius of 2,900 kilometers.

Waves are both reflected and refracted at the prominent discontinuities found at the surfaces of the outer and inner core. The predicted travel times of waves traveling various reflected and refracted paths are observed as wave arrivals at stations located at the predicted distances, and this is taken as confirmation that such paths have been followed. Examples of these paths are illustrated in Fig. 21–15, and wave paths are designated as follows:

P = path on which a wave travels as a compressional wave.

S = path on which a wave travels as a shear wave.

c = reflection at outer surface of core (e.g., *PcP, PcS*).

I = path within the inner core (e.g., *PKIKP*).

K = path within the core (e.g., *PKP, PKKP, PKKS*).

KK = a reflection within the core from the core–mantle boundary.

Travel-time curves (Fig. 21–17) are constructed on the basis of the theory described above, and they are now used widely to identify wave arrivals. Travel-time curves will depend on the depth of focus of the earthquake. Waves from a focal

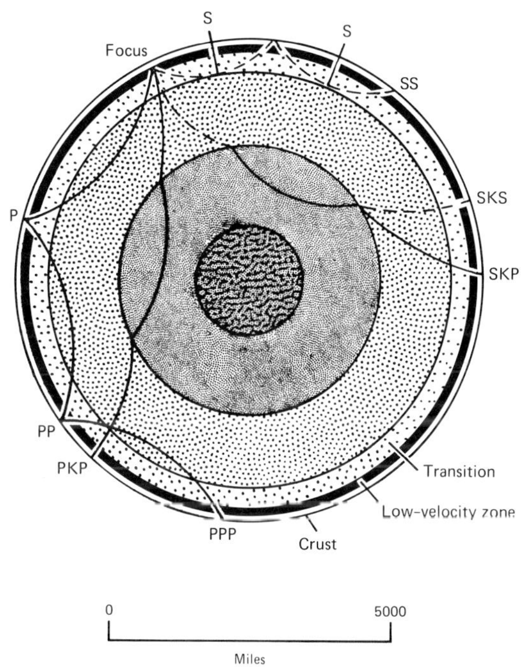

FIGURE 21–15 Cross section of the earth showing major discontinuities and selected wave (ray) paths.
(after E. C. Robertson, 1966)

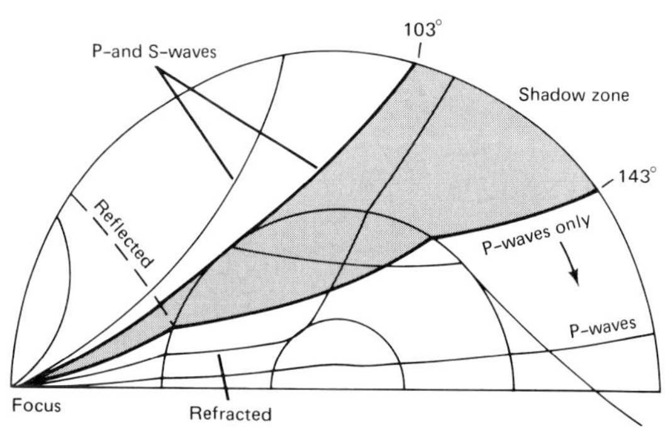

FIGURE 21–16 Ray diagram showing some paths followed by waves through the earth's interior. Stations located more than 103° away from the epicenter do not receive shear waves, and stations located between 103° and 143° away do not receive either type of wave. For this reason this zone is called the shadow zone. Waves arriving in the shadow zone have been refracted and reflected from the inner core or have been reflected in the mantle.
(after Gutenberg and Richter, 1939)

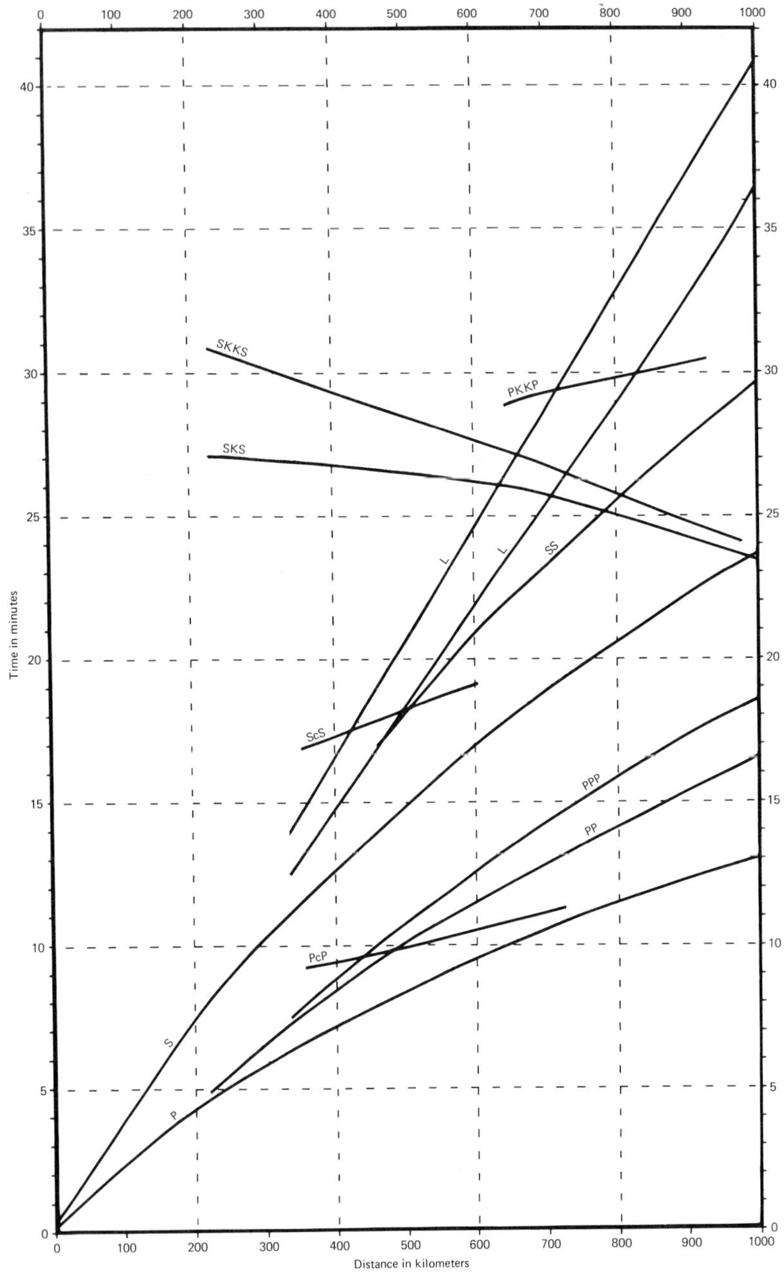

FIGURE 21–17
Travel-time curves for earthquake waves. ("L" designates Rayleigh and Love waves.) (after U. S. Coast and Geodetic Survey charts based on curves prepared by Gutenberg and Richter)

depth of 400 kilometers or more (some are as deep as 720 kilometers) will arrive at the station earlier than those from a shallow event.

When the surface waves reach a station the seismometer is set in long-period vibrations that generally obscure identification of additional arrivals.

Location of the Epicenter

The position of the epicenter and time of an earthquake are known initially only if the earthquake occurs in a populated area, and often people within the area of a violent earthquake lose communication facilities. The first indication of the earthquake at a station some distance away is the beginning of the arrival of the wave train. If it is possible to identify wave arrivals then it is a simple problem to determine the distance to the epicenter. The P- and S-waves travel the same path but at different velocities. Thus the distance to the epicenter is a function of the time lag between these two arrivals. Because each arrival cannot always be positively identified and because the velocities vary along the path, the method actually used to determine the distance to the epicenter is based on the travel-time curves. Distance determinations from three stations are required to locate the epicenter (Fig. 21–18).

Magnitude and Intensity

The modified Mercalli intensity scale was described in Chapter 20 (p. 460). The intensity scale refers to the violence of the earthquake motion and is based on effects directly observable. *Earthquake magnitude* is a measure based on the amount of energy released at the focus. "The magnitude of an earthquake is the logarithm of the maximum trace amplitude expressed in microns with which the standard short-period torsion seismometer would register that shock at an epicentral distance of 100 kilometers" (Neumann, 1966).

The magnitude and intensity of earthquakes at focal depth of 18 kilometers have been compared by Gutenberg and Richter as shown in Table 21–2.

The amount of energy released in earthquakes is enormous. A magnitude 5 earthquake is estimated to represent the release of 10^{21} ergs of energy, about the same as an atomic bomb or 20,000 tons of TNT. The strongest earthquakes release about 10^{25} ergs.

Magnitude	2.2	4.0	6.0	8.0	8.5
M–M Intensity*	1.5	4.5	7.8	11.2	12.0

* M–M = Modified Mercalli

TABLE 21–2
Magnitude and intensity of earthquakes at 18 kilometers focal depth.

MODELS OF THE INTERIOR

Seismology provides the most valuable method of studying the earth's interior, but important contributions have come from other indirect methods of making observations, particularly gravity and magnetic studies. Among the main objectives of these studies are determination of the nature of the materials inside the earth and their distribution. We would like to know the composition, structure, pressure, temperature, and state (solid, liquid, etc.) as well as variations and distribution. The things we come closest to observing directly are seismic velocities that depend on elastic properties (compressibility and rigidity) and density. The character of the material must be inferred from these with consideration given to the independently determined characteristics such as mass, moment of inertia, magnetic properties, nature of crustal geology, and results of experimental studies on the behavior of rock at high temperature and confining pressure.

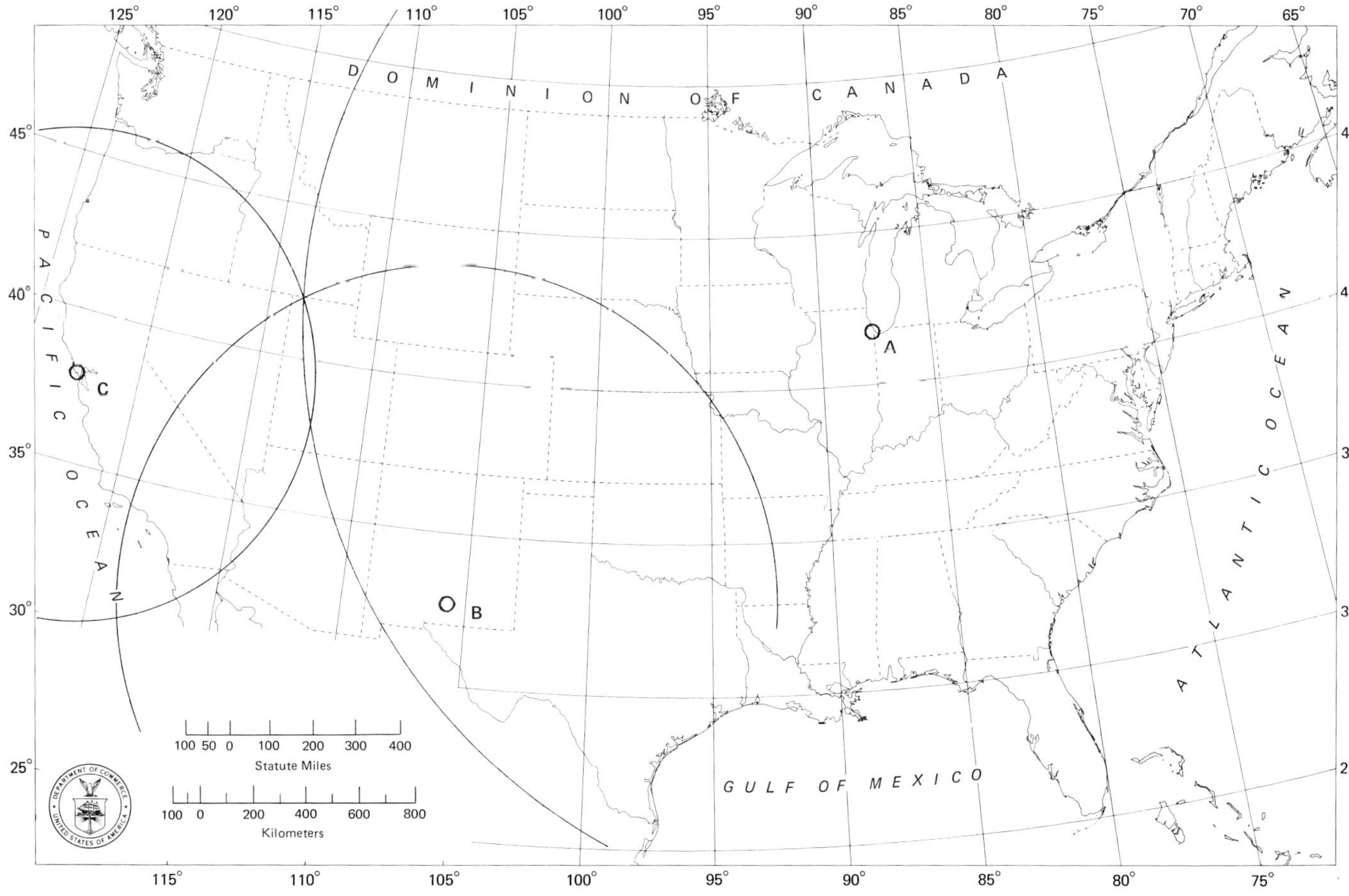

FIGURE 21–18
Location of epicenter of an earthquake is obtained by use of distance determinations from three seismograph stations.

Region	Depth (km)	Features of region
———	—	———
A		Crustal layers
	33	———
B		Steady positive P- and S-velocity gradients
———	413	———
C		Transition region
———	984	———
D		Steady positive P- and S-velocity gradients
———	2898	———
E		Steady positive P-velocity gradient
———	4982	———
F		Negative P-velocity gradient
———	5121	———
G		Small positive P-velocity gradient
———	6371	———

TABLE 21–3 Bullen's (1963) model of the earth's interior showing regions, depth, and features of the wave velocity in each region.

One model of the interior is that suggested by K. E. Bullen (1963) and H. Jeffreys. The interior is divided into regions as shown in Fig. 21–19a, based on velocity distribution and the seismic characteristics shown in Table 21–3.

Additional seismic discontinuities, particularly in the upper mantle, have been recognized in recent years. These discontinuities show up as breaks in the travel-time curves. The data do not yield unique solutions. The interpretation favored by D. L. Ánderson (1967) is shown in Fig. 21–19b. The *M*-discontinuity and a low-velocity channel are apparent, but other discontinuities or rapid increases in velocity occur at depths of 320 kilometers and 640 kilometers. Additional discontinuities have been proposed from studies of waves generated by nuclear explosions.

State of Rocks in the Regions of the Interior

The materials in the crust and mantle behave as elastic solids when subjected to stresses whose time of duration is that found in seismic-wave propagation. This does not mean that these materials are elastic solids like rocks at the ground surface. All evidence suggests that mantle materials will flow (somewhat like asphalt) in response to stresses of long duration.

S-waves have not been observed to be transmitted through the core, but P-waves are transmitted, although at a lower velocity. These are characteristics we would expect if the wave motion were transmitted through a liquid, or a material that exhibits low rigidity for stresses of the durations involved. P-waves traversing the inner core encounter a higher velocity path. This has been widely interpreted as meaning that the inner core is solid, but until we have a better understanding of the behavior of materials at extremely high pressures the interpretation remains hypothetical.

Models of Density Variation

Most rocks outcropping at the earth's surface have a density in the range of 2.5 to 3.5 grams per cubic centimeter. The overall density of the earth is in the range of 5.5 grams per cubic centimeter, which means that the density of subcrustal material must be much higher than that of the crust. P- and S-velocity distributions are functions of ratios between incompressibility and density, and between rigidity and density. These relationships are the essential factors in density estimates at depth. Rigidity is taken as zero at the outer edge of the core; thus the above ratios are observationally determined to that depth with a high degree of accuracy. Bullen (1961) determined a model of density variation (Fig. 21–20) for the interior based on these ratios and modified to agree with calculations based on the moment of inertia of the interior.

Composition of the Interior

Indications of the composition of the layer immediately below the *M*-discontinuity point to olivine basalt or peridotite, two similar rocks. Diamond pipes and similar deep-seated intrusions are peridotites, and this rock type is brought up to the surface from depth as blocks and extruded with basaltic lavas where the crust is thin. Studies of P- and S-velocities in the mantle provide an indication of the density and elastic properties in the subcrust. These confirm the prediction that the subcrust is rich in olivine ($(Mg,Fe)_2SiO_4$) or peridotite. The material in the subcrust is either olivine, or like olivine, in that it has a relatively low density and a high velocity for the compressional waves. A well drilled through the crust would have solved this problem, but it seems relatively safe to predict that peridotitic or olivine basalt will be found.

Bullen's density model (Fig. 21–19a) indicates a chemically homogeneous layer, B, between 33 and about 400 kilometers. More recent models have been constructed to correspond to seismic-velocity data (see Fig. 21–19b). Some modern models have numerous density breaks and gradients in the upper 1,000 kilometers of the mantle. Many students of the problem think that these breaks correspond to changes in the crystal structure or form of the minerals at depth.

Sir Harold Jeffreys (1936) suggested that the changes in velocity near the top of layer C might be due to the transformation of ordinary olivine to a high-pressure modification. Francis Birch and A. E. Ringwood have done much of the significant experimental work to determine the feasibility of such a transformation. Ringwood concluded the following:

1. Fayalite (Fe_2SiO_4) will transform to spinel with a 12 percent density increase at pressures predicted at 120 kilometers.
2. Forsterite (Mg_2SiO_4) will transform to spinel with an 11 percent density increase at pressures predicted at a depth of about 400 kilometers.

Thus layer C could be spinel, and the increasing density of mantle material with depth could reflect an increase in iron content with depth. The density model is also compatible with a mantle composition consisting of dunite and peridotite with up to 10 percent garnet.

Region D is thought to be relatively homogeneous chemically with probable composition of spinel. Birch suggested closely packed oxides of magnesium, iron, and silicon. Ringwood describes this region as a homogeneous spinel phase consisting of a disordered solid solution.

It is often assumed that the outer part of the earth's core is a liquid because no shear waves

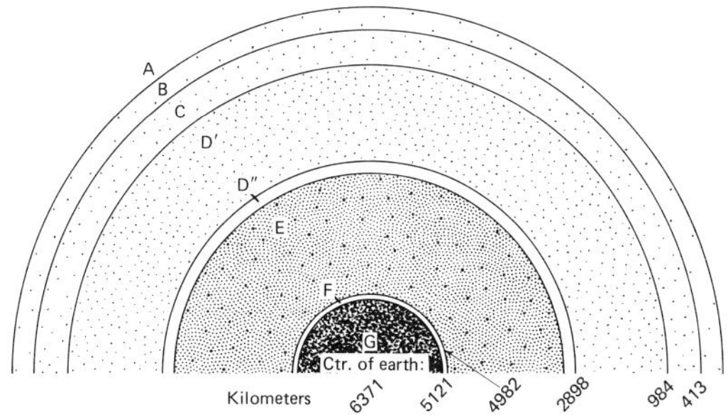

FIGURE 21–19a Earth's cross section is divided into distinct layers through which seismic waves travel at different velocities. See text for definition of the layers designated A–G. D″ is a thin transition zone between the core and mantle. (after Bullen, 1963)

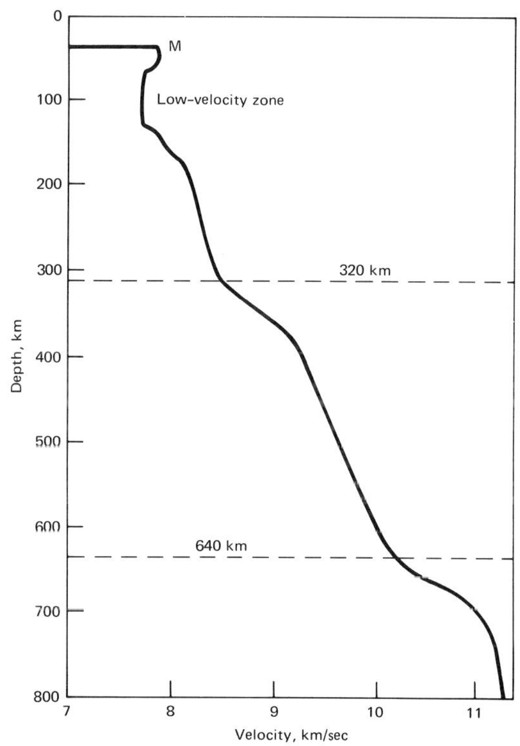

FIGURE 21–19b Anderson's model of upper-mantle velocity-depth relationships. (after Anderson, 1967)

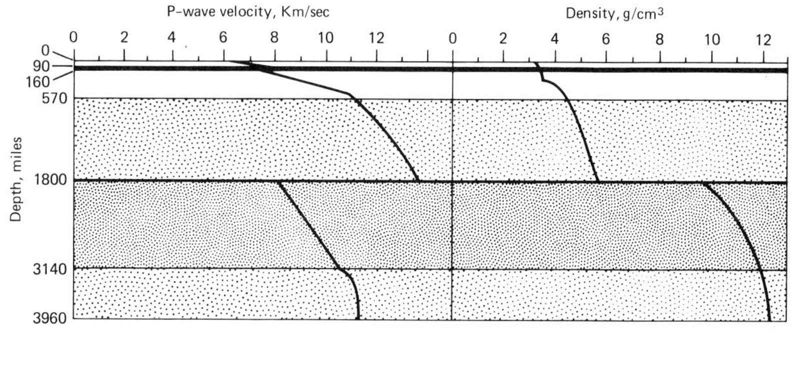

FIGURE 21–20 (*Below*) Variations of P-wave velocity and density with depth in the earth and major seismic discontinuities are shown. (after E. C. Robertson, 1966)

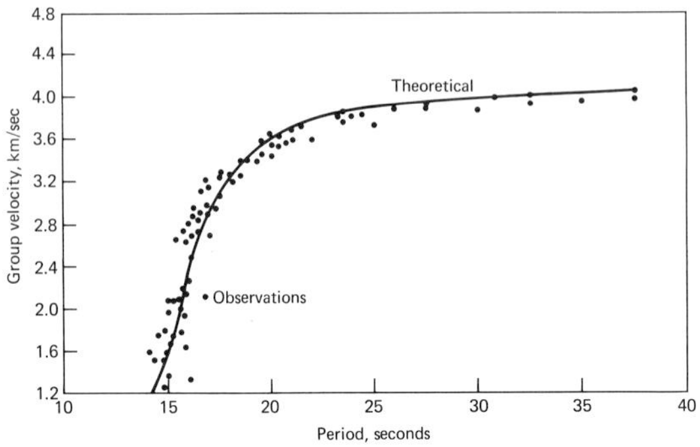

FIGURE 21–21 Theoretical and observed Rayleigh wave dispersion for oceanic paths. (from M. Ewing and F. Press, 1956)

are propagated through it. Iron in a molten state was preferred as the conjectured composition for this layer for many years. Support for this point of view is found in the nickel-iron composition of many meteorites that were presumably formed at the same time that the earth formed. Iron and several oxides are the only common minerals that are heavier than silicates, and the density of the core is so great that the types of silicate minerals found in the crust are excluded. Oxides are excluded because they would dissolve the silicates (olivine) in the mantle, oxides and silicates being mutually soluble.

Another possibility suggested by Ramsey (1948) is that the core is not chemically distinct from the mantle but rather consists of a high-density modification of mantle material. According to this idea the density would be largely a function of pressure alone except right at the core boundary where a large change is necessary. One argument against accepting iron is that the calculated density of the core is somewhat higher than that of iron. It should also be noted that many meteorites are not nickel-iron in composition but instead contain a high proportion of olivine. Meteorites of this type are called *stony meteorites* or *chondritic meteorites.* Ringwood has suggested that as much as 20 percent of the core may be silicon, an estimate based on the composition of chondrites.

Still another hypothesis is that the core consists of iron alloyed with lighter elements, possibly even metallic hydrogen. It has been shown that at the pressures found at a depth of 1,600 kilometers hydrogen probably changes to a metallic state with a sudden increase in density (0.4–0.8 g/cm^3).

The composition of the inner core is in even greater doubt. It may be a solid, as indicated by the increased velocity of P-waves passing through it, but until we have a better understanding of the nature of the changes that can take place in materials under extreme pressures some doubt will persist.

SURFACE-WAVE INTERPRETATION

So far we have discussed the interpretation of body waves that penetrated deeply into the earth; surface waves are also yielding significant information about the outermost layers—the crust and upper mantle. When surface waves start to arrive at a station the body waves are masked by the surface movements, which have much greater amplitudes and much longer wave lengths. Surface waves recorded on the vertical-component seismometer are predominantly Rayleigh waves; those recorded on the horizontal components are Love and Rayleigh waves. Surface-wave motion often continues to arrive at a station for hours; surface waves set off by the most violent earthquakes have been detected after having traveled around the earth up to seven times, with significant surface motion lasting as much as a whole day.

Unlike body waves, which arrive suddenly and pass through a station quickly, surface waves travel as a wave train. The exact time of the first arrival is often difficult to pick and once wave motion starts it continues for a long period of time. The seismograph traces are set in long-period motion. The *period* of a wave is the time between the arrival of two adjacent wave peaks (or troughs). Periods of surface waves fall in a range of 15 to over 3,000 seconds; waves used in most studies, however, are in the range of 15 to 75 seconds. The periods of waves in the train are not constant. Generally the earlier arrivals are of longer period than the later arrivals. This alone would suggest that what we are observing is the arrival of many different waves that originated at about the same time but traveled with different velocities—velocities related to the period of the wave motion. This continual spreading out of an initial disturbance into a train of waves is called *dispersion.* The original motion is sorted out into *groups* of simple harmonic waves, and each of these groups has a particular wave length, period, and travels forward with a particular *group velocity*. Thus it should be possible to identify each group by measuring the period of each surface wave and if the time and place at which the earthquake originated are known, then the velocity of each group can be simply calculated. An easy graphical method for plotting these data is a plot of group velocity vs. group period, as shown in Fig. 21–21. (It turns out to be more convenient to plot this on logarithmic graph paper because the range of periods is so great.) When plots of this type are made at one station for several different epicentral locations it becomes clear that the *dispersion curves* are not identical—surface-wave dispersion is thus related to the particular path followed by the waves, an observation suggesting this is a tool that may be used to interpret crustal structure.

The key to the use of surface waves for interpretation of crustal structure lies in the fact that the depth to which a wave penetrates the medium through which it moves is a function of its wave length (also period or frequency). Long-period waves penetrate more deeply; we say they "feel" or "see" to greater depth. Those waves that penetrate deeply into the upper mantle are influenced by higher velocities. Now it should be clear why the longer period waves arrive first—they have traveled with greater velocity. It follows that the dispersion curves should reflect differences in the distribution of physical properties (hence structure) of the crust along the path. The prominent differences between the structure of the crust (depth to *M*-discontinuity) under oceans and continents is clearly defined by dispersion curves; the curves join to form a single curve for groups with periods of about 100 seconds. The longer period groups are influenced by the mantle in which great difference in properties along different paths does not appear to exist; this may be taken as evidence of relative homogeneity. Similar results are found when Love waves are studied.

Additional information can be obtained from the analysis of *phase-velocity* curves for Rayleigh waves (Press, 1956). Phase velocity is the velocity of a particular spectral component of a wave form. This technique may be used where stations are so closely spaced that it is possible to correlate individual waves from station to station. When this is possible the velocity of a given wave of specific period can be calculated between stations. A phase-velocity curve may be prepared by plotting the phase velocity of a number of waves of different period so that a plot of velocity vs. period is obtained. Interpretation of this in terms of crustal structure is then possible by calculating what the phase-velocity curve would be for specific models of crustal structure. The variables in such models are layer thickness and elastic moduli. A correct model should yield a phase-velocity curve that coincides with the observed curve.

GUTENBERG LOW-VELOCITY LAYER

The *M*-discontinuity is widely recognized and it has been accorded great significance, but another, deeper zone may well prove to be of greater importance in actively influencing processes taking place near the surface of the earth. Beno Gutenberg was the first to recognize a *low-velocity layer* in the upper mantle from the *M*-discontinuity to a depth of about 200 kilometers. He noticed that the amplitudes of P-waves coming to the surface at distances of 600 to 1,500 kilometers from an earthquake epicenter were much lower than the amplitudes of those reaching the surface beyond about 1,500 kilometers (Fig. 21–22). Low-velocity layers are difficult to detect by seismic methods because the lower velocity causes waves that are refracted into the layer from above to be bent downward toward the normal.

Gutenberg's original evidence for the low-velocity layer was the existence of a shadow zone around earthquake epicenters (Fig. 21–23). This evidence was first confirmed on the basis of predictions regarding the way the shadow zone's shape would be affected by variations in the depth of focus of the earthquake. The zone should be widest when the focus is right at the upper boundary of the zone; it should appear when the focus is above the zone; and it should not be detected when the focus is below the low-velocity layer. Subsequently abundant confirmation has come from other sources.

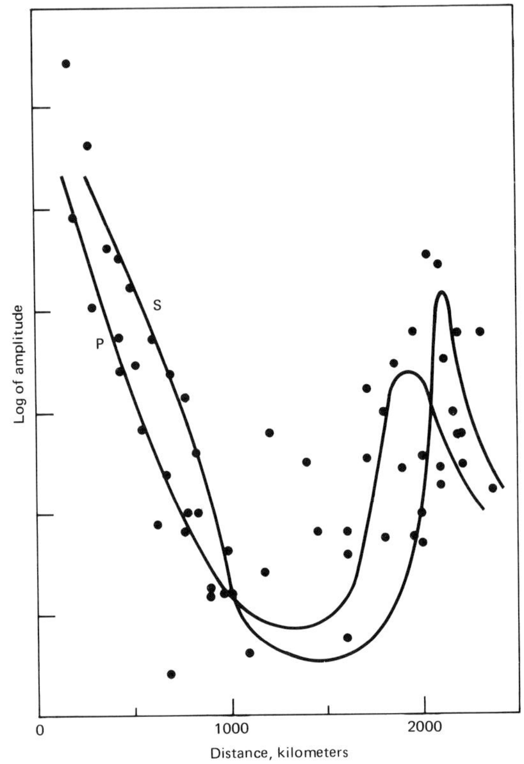

FIGURE 21–22
Curves plotted by Gutenberg showing change in amplitude of earthquake and explosion waves with distance from focus.

1. Underground nuclear explosions have provided the carefully controlled conditions needed to test this idea. These confirm Gutenberg's findings but indicate the zone lies between 60 and 250 kilometers deep.

2. Analysis of the dispersion of surface waves (Fig. 21–21) indicates that the material in the low-velocity zone is less rigid than that above and below. Surface-wave studies have shown that the layer is world-wide in its extent, and that under continents it is slightly different from under ocean basins.

3. Plots of focal depths of large numbers of earthquakes show a gap within the depth range of the low-velocity zone.

4. A layer of lower rigidity is necessary to explain free oscillations (vibrations) of the earth as a whole, such as those that are caused by the most intense earthquakes.

He calculates the depth of the zone to be centered at about 150 kilometers and to be on the order of 100 kilometers thick. Within this zone velocities are decreased by about 6 percent from velocities just below the *M*-discontinuity.

The low velocities observed in this zone strongly suggest that material in this layer is less elastic (less rigid) than that above. It follows that this zone is probably more plastic and that it has less strength than zones above and below. The most probable explanation for the existence of this zone lies in temperature and pressure effects on the behavior of the material at this depth. Both temperature and pressure are thought to increase steadily with depth, but the composition of material in the low-velocity zone must be such that its behavior is different from that above and below. This could mean that the composition is different or that the material undergoes some change of state in this zone.

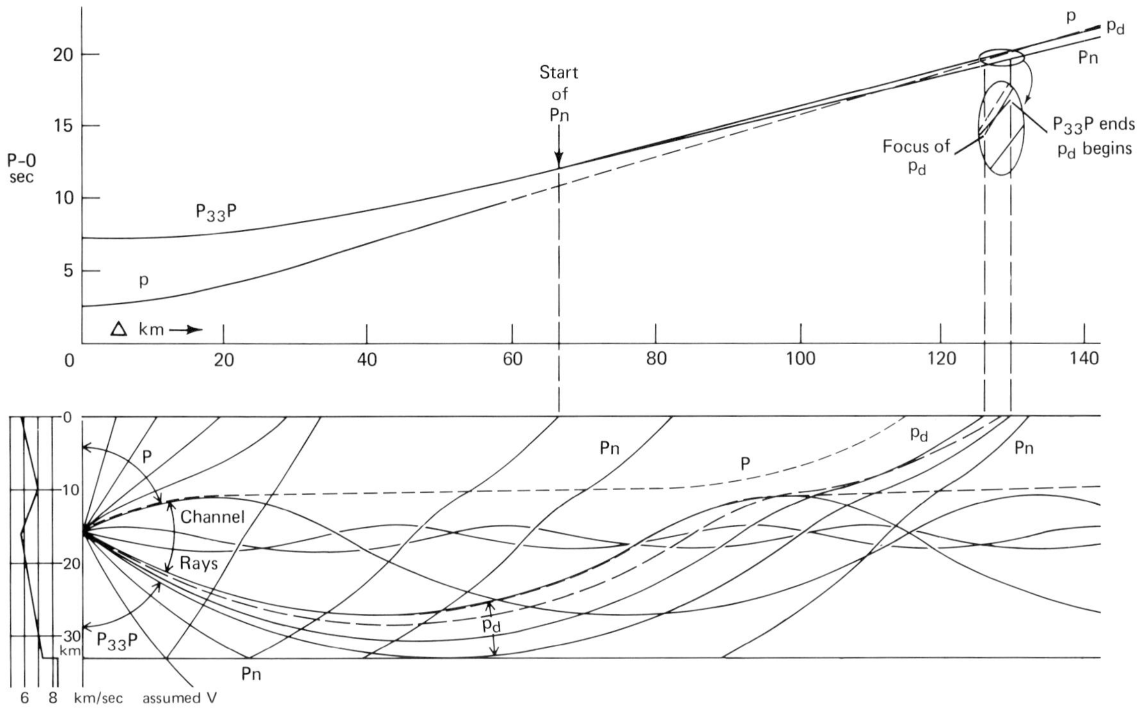

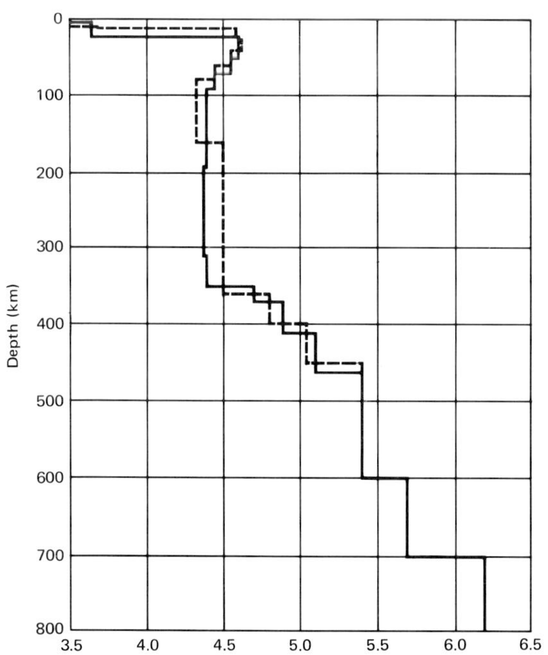

FIGURE 21–23

(*Above*) Wave paths of longitudinal waves in the lithosphere channel (also called the Gutenberg low-velocity channel) and corresponding travel times. $P_{33}P$ is a wave reflected from the bottom of the low-velocity zone, Pn is refracted along the bottom of the zone, and Pd is a direct wave bent along the low-velocity zone

(from Gutenberg, 1954)

FIGURE 21–24

Shear-velocity distributions for two crustal models.

(from Toksöz and Anderson, 1966)

CHAPTER OUTLINE

TERMS

attenuated
body waves
brittle failure
chondritic (stony) meteorites
critical angle
dilational waves (also primary, P-, compressional, and longitudinal)
dispersion curve
earth's crust
epicenter
focus
geophones
Great Circle
isotropic
low-velocity layer
M-, or Moho, discontinuity
magnitude (earthquake)
mantle
microseism
period
phase velocity
reflection seismology
refraction seismology
seismograph
shadow zone
shear waves (also secondary, S-, or transverse)
surface waves
travel-time curve

REVIEW QUESTIONS

1. Both compressional and shear waves increase in speed with depth. What does this indicate about rocks at depth? Why?
2. A wave front on water travels in nearly perfect circles. However, seismic waves do not travel out in circles. Why not?
3. When refraction methods are used, sometimes a refracted wave will arrive at a geophone before a wave traveling above it near the surface. Why does this happen and what are some of the implications.
4. What is the *M*-discontinuity? What is its significance and what is its effect on seismic waves?
5. What is the shadow zone and why does it occur?
6. Earthquakes release huge amounts of energy. The quantity of energy can be expressed as a function of the magnitude. Differentiate between magnitude and intensity of an earthquake.
7. K. E. Bullen developed a model for the earth's interior. Briefly explain it and what it is based on.
8. Because surface waves have different velocities they may be used to form dispersion curves that give certain indications about the earth's crust. What are some of these indications and why are the velocities different? (What are they dependent upon?)
9. According to findings from underground nuclear tests, where does the Gutenberg low-velocity zone lie? What is its significance and what are its implications in regard to the structure of the earth?
10. What is initial movement and why is it an important consideration?

SUGGESTED READINGS

Benioff, Hugo, 1954, Orogenesis and deep crustal structure—Additional evidence from seismology: Geol. Soc. America Bull., v. 65, no. 4.

Gutenberg, Beno, 1954, Low-velocity layers in the earth's mantle: Geol. Soc. America Bull., v. 65.

Hodgson, J. H., 1964, Earthquakes and Earth Structure: Englewood Cliffs, N.J., Prentice-Hall, 166 p.

Press, Frank, 1961, The earth's crust and upper mantle: Science, v. 133, no. 3463, p. 1455–1463.

Richter, C. F., 1958, Elementary Seismology: San Francisco, W. H. Freeman, 768 p.

Stacey, F. D., 1969, Physics of the Earth: New York, John Wiley, 324 p.

Sumner, J. S., 1969, Geophysics, Geologic Structures and Tectonics: Dubuque, Iowa, Wm. C. Brown, 115 p.

Talwani, Manik, 1964, A review of marine geophysics: Marine Geol., v. 2, p. 29–80.

22 STABILITY AND INSTABILITY OF THE EARTH'S CRUST

22 STABILITY AND INSTABILITY OF THE EARTH'S CRUST

DEFINITION OF THE EARTH'S CRUST

The crust is defined as the outer part of the lithosphere between the ground surface and the *M*-discontinuity (Moho), recognized by the sudden increase from about 6.5 to 8 kilometers per second in the velocity of the compressional shock waves. It marks a sudden change in the density of rock and probably a change in either composition or phase. The configuration of the underneath side of the earth's crust is almost a mirror image of the surface (Fig. 22–1).

GROSS FEATURES OF THE EARTH'S CRUST

The most important divisions of the crust are the continents and the ocean basins. These differ in many respects, and they are separated by zones of transition having characteristics that are different from those of either the continents or the ocean basins.

The average elevation of continents is almost 0.8 kilometers above the level of the modern-day seas, but the ocean floors average almost 4.8 kilometers beneath sea level, producing about a 4-kilometer difference in surficial elevation between the two.

The thickness of the crust beneath oceans is strikingly different from its thickness beneath continents. The crust is between 8 and 10 kilometers thick under the oceans, counting the ocean water; but it is 20 to 40 kilometers thick on the continents, and it reaches a maximum under some mountains where it may be 40 to 60 kilometers thick.

The crust under the oceans has a bulk composition that differs from crust under the continents. The continental crust is made of rocks of granitic composition (NOTE: a mixture of limestone, sandstone, and shale has a bulk composition similar to granite). These rocks are rich in the elements silicon and aluminum, and the abbreviation *sial* (*si* for silicon and *al* for aluminum) is used to denote this type of composition when more specific information is unknown or unnecessary. The upper part of the continental crust is thought to be sialic, but the oceanic crust contains only a thin upper layer of sialic material, if any at all. Holes drilled into the oceanic crust encountered basalt, and basaltic rocks constitute the main rock type erupted from volcanoes in the ocean basins. Thus the conclusion that the minerals of the oceanic crust are those that compose basalt seems warranted. They are rich in silicon and magnesium (abbreviated *sima*).

Extrusions and intrusions in oceans tend to be basaltic in composition, and those in and on the continents tend to be granitic or intermediate (andesitic) in composition. The extent of this difference is partially dependent on the ultimate determination of just how many of the granite batholiths known on continents are really of igneous origin and how many of them are metamorphic rocks. No granitic batholiths have been found in the deep ocean basins.

The oceans and continents are similar in a few respects. Both have broad plain areas that are nearly flat; both have high mountain ranges that extend for thousands of kilometers. However, these mountains and plains are fundamentally different in structure and origin. Both oceans and continents have extremes of relief that comprise a relatively small amount of the total area. High mountain peaks such as Everest, 9,000 meters, on continents have extreme deep-sea counterpart trenches up to 11,400 meters deep.

The crust may be subdivided into six major subdivisions on the basis of rock types present, structure, crustal thickness, topography, and stability:

1. Continental shields	Continental crust
2. Young folded mountain belts	
3. Shelflike continental margins	Transitional
4. Island arc systems	
5. Deep oceanic crust	Oceanic crust
6. Midocean ridges	

The character and origin of these major divisions is the subject of Chapter 24.

STABILITY AND INSTABILITY OF THE EARTH'S CRUST

The brevity of recorded history, the lifespan of man, and the slow rate of most processes of crustal deformation tend to obscure the extent to which the crust exhibits long-term instability. Yet the evidence proves conclusively that most of the crust has been elevated and depressed at various times in the earth's history and some parts of the crust have been intensely deformed. Perhaps the most convincing evidence of modern instability is the ground movement that accompanies large earthquakes. The main lines of evidence concerning major crustal instability come from the following sources:

1. Study of relative positions of sea level and features produced at or below sea level (e.g., wave-cut terraces, marine deposits, etc.).
2. Detailed surveying in which leveling is carried out with great precision.
3. Observation of structural features that cannot, or are not known to, originate in the absence of deformation.

Deformation of the earth's crust is given the general name *diastrophism,* the processes by which the crust is deformed and the results of this deformation. Two essentially different types of deformation are in evidence. One type is characterized by broad uplifting or downwarping of a large area, called *epeirogenic* movement or deformation. These broad movements of uplift or subsidence affect all or a large part of a continent or sea bottom and are indicated by uplifted or drowned beaches. Strata are not crumpled or strongly folded by this type of deformation, although they may be tilted or warped.

The second type of deformation of the crust is called *orogenic,* a general term applied to the processes of mountain building. Strata are crumpled and folded; they may become broken by low-angle faults along which the rocks are displaced laterally many miles. Large-scale regional uplift usually accompanies orogenic deformation. Orogenic movements affect long, narrow belts of the crust, which are called *mobile belts.* The central portions of these belts are generally metamorphosed, suggesting a connection between orogeny and thermal activity as well as uplift and compression. Orogeny differs from epeirogeny in intensity of the deformation and in the general nature of structural features produced; the two are caused by different although related processes.

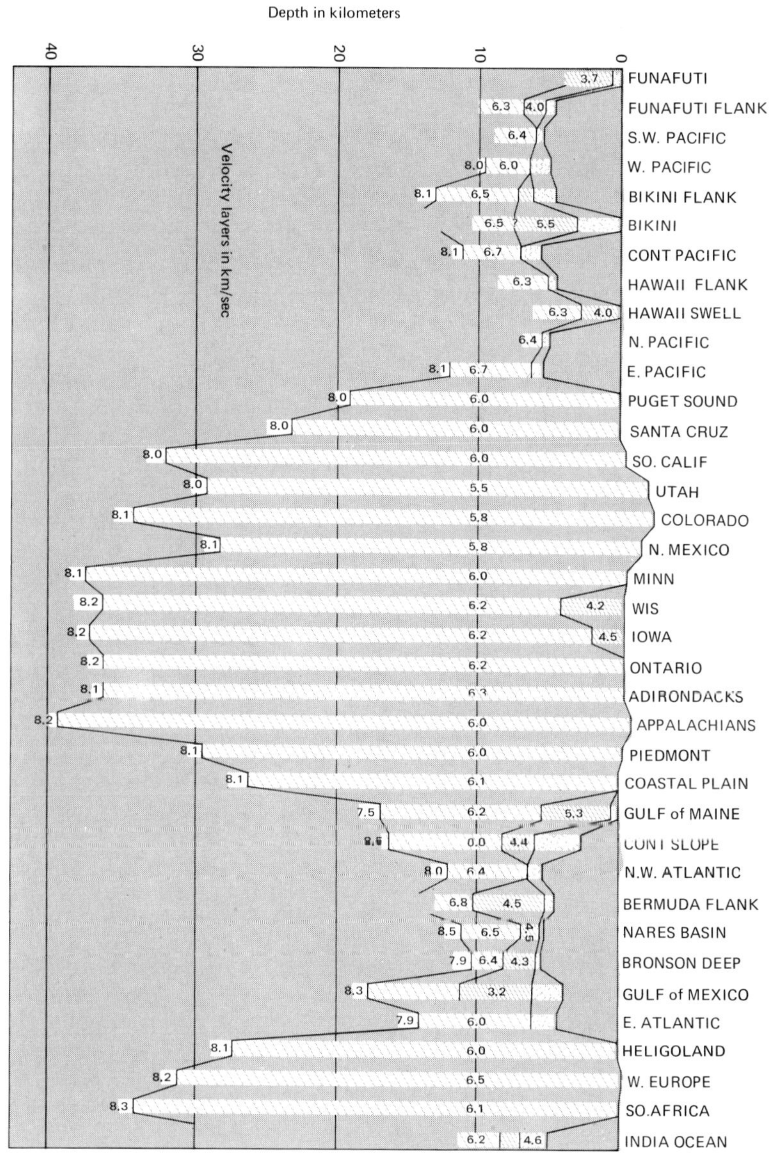

FIGURE 22–1
Crustal structure. The thickness of the crust of the earth is shown for a number of locations. Note the great thickness under continents as compared with the thickness under ocean basins. The solid black represents the top of the mantle.
(from G. P. Woollard, 1959)

EVIDENCE OF EPEIROGENY

Changes in Level Relative to Sea Level

In the sixth century B.C. a large and rich city was built on the shore of the Bay of Taman on the northern shore of the Black Sea. The remains of this city, named Phanagoria, are being excavated. These excavations reveal that some of the streets of the city now lie under the sea. Thus it appears that the region has been lowered in relation to sea level since the city was built. Similar evidence is found near Naples, Italy, in the ancient Roman public market known as the Temple of Jupiter Serapis. Here marine gastropods have bored holes in columns 5.4 meters above the floor of the temple. The remains of some of these gastropods are still found in the holes. It is evident that the shore was submerged beneath the sea after the Romans built the temple, and then raised above sea level where it stands today. Crustal movements in the Bay of Naples have been studied carefully. The floor of Serapaeum had been sinking at a rate of about 13 millimeters per year since 1538 when an eruption took place at "Campi Flegrei" (fiery fields) near Naples. Uplift has been taking place recently—about 80 centimeters has occurred since October of 1968. New volcanic fissures opened in 1969 and 1970; masonry walls were cracked and parts of this area were evacuated.

Sea level is the level to which most crustal movements must be related. It is the basis for comparison of modern altitudes and is useful for documenting past levels because the conditions above, at, and below sea level leave a clear impression on the rock record. Marine sediments are deposited below sea level; wave action and reef growth are concentrated within a relatively narrow range of water depth, and subaerial processes act above sea level. Thus a reasonably close approximation of sea level is obtained where reefs, wave-cut and other marine terraces, or littoral marine sediments are found.

Although ample evidence indicates relative positions of sea level and land for many times in the past, it does not follow that each disparity in level between some older stand of sea level and present sea level can be explained in terms of epeirogenic or orogenic crustal movements.

Fluctuations in sea level have resulted from changes in the relative amounts of water stored in the sea and as ice on the continents. The last glacial maximum was reached about 11,500 years ago, and at that time sea level was several hundred meters below its present position (see Chapters 9 and 17). Enough ice remains to allow a rise of about 60 meters in sea level simply by a warming of the earth's climates. Many fluctuations in sea level occurred as a result of glacial and interglacial periods during the past 2 to 3 million years. An almost continual addition of water to the surface from the interior of the earth through igneous and volcanic activity has been postulated by most students of oceanic history. The estimated rates vary, but all agree that the increase has taken place. Uplift and subsidence in ocean basins or on continents that affect the volume (size or shape) of the ocean basins would cause relative changes in level, and sea-level fluctuations have also been attributed to global expansion in which the size of the oceans has increased although the size of the continents has remained about constant. Changes in level must be carefully considered against this background of likely causes; it is not always possible to isolate exactly what has been the cause, but it is possible to demonstrate that some changes are caused by epeirogenic uplift.

Some wave-cut terraces stand at elevations higher than sea level would be if all ice melted. Raised marine terraces are prominent along the western coast of North America. A number of different levels can be identified, as on profiles of the Palos Verdes hills area southwest of Los Angeles (Fig. 22–2), where some terraces stand 400 meters above present sea level. In addition to the terraces, banks and scattered deposits of modern marine shells are found as high as 900 meters. The western coast of North America lies within a region of modern orogenic deformation, so it is difficult to distinguish the relative effects of local warpings caused by folding from the regional uplifts and depressions. Possibly the lower terraces and the submerged terraces are due pri-

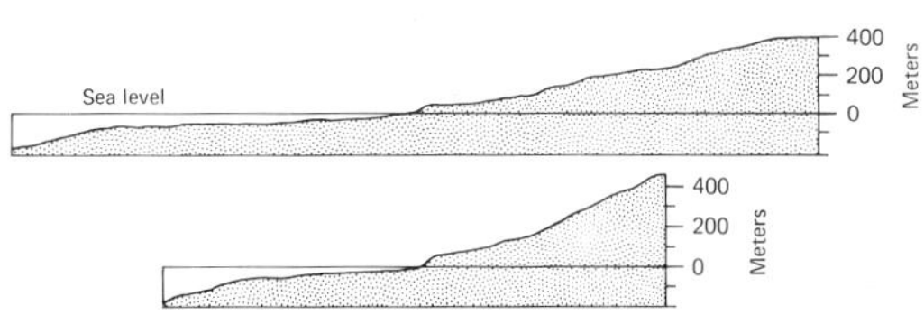

FIGURE 22–2
Profiles of Palos Verdes hills and offshore area southwest of Los Angeles to show emerged and submerged marine terraces.
(after King, 1965)

marily to eustatic changes in sea level because of glaciation.

Guyots

Guyot (Gē'yō) is the name H. H. Hess (1946) applied to flat-topped seamounts that he discovered while making profiles of the sea floor in the Pacific basin. The flat tops of guyots range up to about 10 kilometers across. Dredging on these flat-topped mountains yielded rounded pebbles and cobbles of basalt, which supports the idea that the mountains were originally volcanoes that were eroded until the tops were nearly flat and close to sea level. Such erosion might be envisioned as partly subaerial and partly due to wave action at sea level. Most of these mountaintops are now between 1 and 2 kilometers below sea level. Reef faunas of Cretaceous and Tertiary age have been found on some guyots now submerged nearly 2,000 meters. Such reefs are now alive only within the upper 100 meters of the sea; the mountains must have been drowned by either a phenomenal rate of sea-level rise or as a result of subsidence accompanied by relatively minor additional effects of eustatic changes of sea level.

Drilling and seismic work on modern atolls have yielded significant data on this problem; 1,400 meters of coral was penetrated on Eniwetok Atoll and 1,000 meters of coral is indicated at Funafuti Atoll. The ages of the coral can be identified from the particular faunas present. The top of each Tertiary unit and the top of the basalt occur at different elevations on the different atolls, indicating that changes of sea level alone are not sufficient to account for the elevations. Guyots topped with sediment containing faunas of different ages occur at nearly the same level (e.g., some topped with Lower Cretaceous deposits are at the same level as others topped with Eocene deposits). Thus actual differential crustal subsidence must also be involved.

It will be possible to document actual times and areas of subsidence and uplift of the ocean-basin floor as more data are obtained that reveal the depth to layers of each age under atolls and the age of formation of the truncated tops of guyots. It is notable that most guyots and atolls are located in the Pacific Ocean. The position of a now-collapsed oceanic rise in the southwestern Pa-

FIGURE 22–3

(*Above*) An aerial view of the sea coast south of Kaikoura, South Island, New Zealand, showing elevated marine terraces.

(National Publicity Studios, New Zealand)

(*Below*) Uplifted coral terraces. Unlike the wave-cut terraces, these were formed by the growth of reefs to a position just below sea level. Like uplifted wave-cut terraces, these indicate a relative elevation of the land with respect to the sea.

(U.S. Air Force)

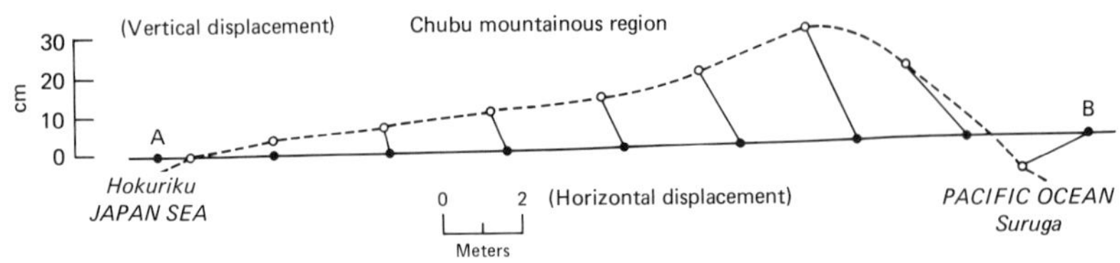

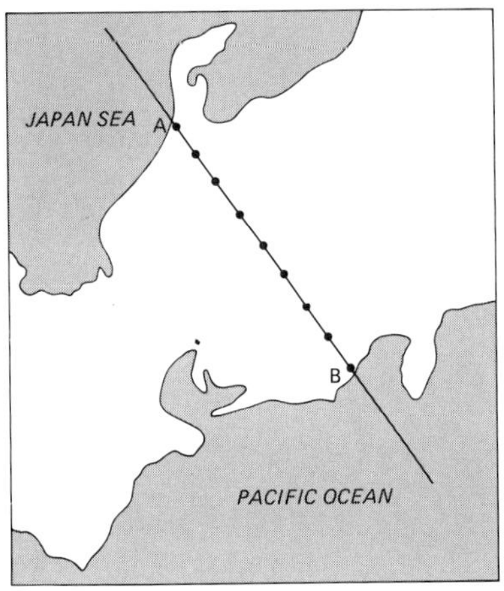

FIGURE 22–4
Profile of land deformation for about 60 years across Chubu District, Japan, in northwest-southeast direction.
(from Eiji Inoue, 1960)

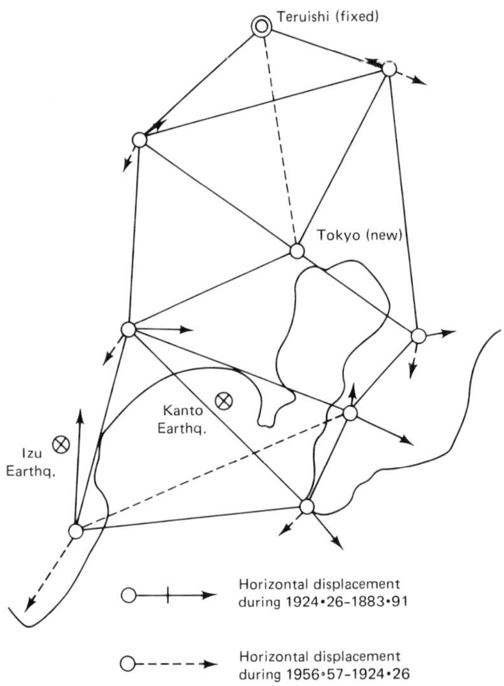

FIGURE 22–5
Horizontal displacements at first-order triangulation points in Kanto District, Japan.
(from Eiji Inoue, 1960)

cific, the Darwin Rise, has already been postulated on these types of data.

Leveling and Modern Epeirogeny

Aristotle and other Greek philosophers noted evidence of large-scale movements of the crust, but one of the first efforts to measure the rates of such movements came in 1731, when special marks were made on rock cliffs along the coast of Sweden at sea level. These marks are now more than 2 meters above sea level. Such a rise is extremely rapid. A region would be uplifted 1.6 kilometers in less than 150,000 years at this rate. Because this is isostatic uplift, however, it is unlikely to continue for that amount of time.

Modern techniques of leveling and surveying allow such precise measurements that it is possible to detect minute changes of elevation and position by carefully resurveying the same area. The simplest method involves setting up a series of survey markers, usually *bench marks*. Bench marks may be concrete posts inlaid with brass markers, solid rock into which holes have been drilled and a brass plate cemented in place, a concrete floor containing a marker, and so on. Each bench mark must be set where it is not subject to soil movement, foundation subsidence, or any other type of movement. Then a survey is made to determine the elevation of each point. This technique of surveying is called *leveling* (Fig. 22–4). Since the survey is run over a short period of time, it enables a careful comparison of the elevation of each marker relative to every other marker. It is thus possible to detect any changes in the relative elevation between stations over the period of time between surveys. The relative movements between stations ultimately must be related to stations that are stable or to mean sea level in order to determine absolute movements (which stations actually moved and which way they moved). If the network of stations is large enough so that some of the stations are located in areas that undergo no change, then the regional patterns of uplift and subsidence can be determined.

If the aim of the surveying is to determine movements that include horizontal components, a method known as *triangulation* is used. A base line is carefully laid out between two points. The length of this line is measured precisely. A new triangulation station visible from both ends of the base line is located by taking a bearing (compass direction) from each end of the base line to the new point, which is located where these two bearing lines cross. A network of stations (Fig. 22–5) is set up in this fashion. If horizontal movements occur within the network between resurveys, the bearings between stations will be changed. A combination of leveling and triangulation makes it possible to determine relative movements with precision.

The margins of lakes are important natural level-surface markers. A lake level must be stable long enough for a beach to form or for wave-cut or wave-built terraces to be developed. Such markers have been used to establish former levels of the Great Salt Lake (Fig. 22–6) and to document the uplift that has taken place in Canada since the last (Wisconsinan) ice sheets occupied central Canada. The shore lines of a successive series of lakes in the Great Lakes region (Fig. 22–7) have been tilted. The tilting has been cumulative so that older beaches have been tilted more than younger ones. The lake shores are nearly horizontal toward the south and show increasing elevation toward the north. Following the last ice advance the sea invaded central Canada in the Hudson Bay region about 8,000 years ago. Shore-line features have been identified over a vast region. By selecting the highest shore-line features of this age (commonly determined by radiometric dating), the level of the seashore is

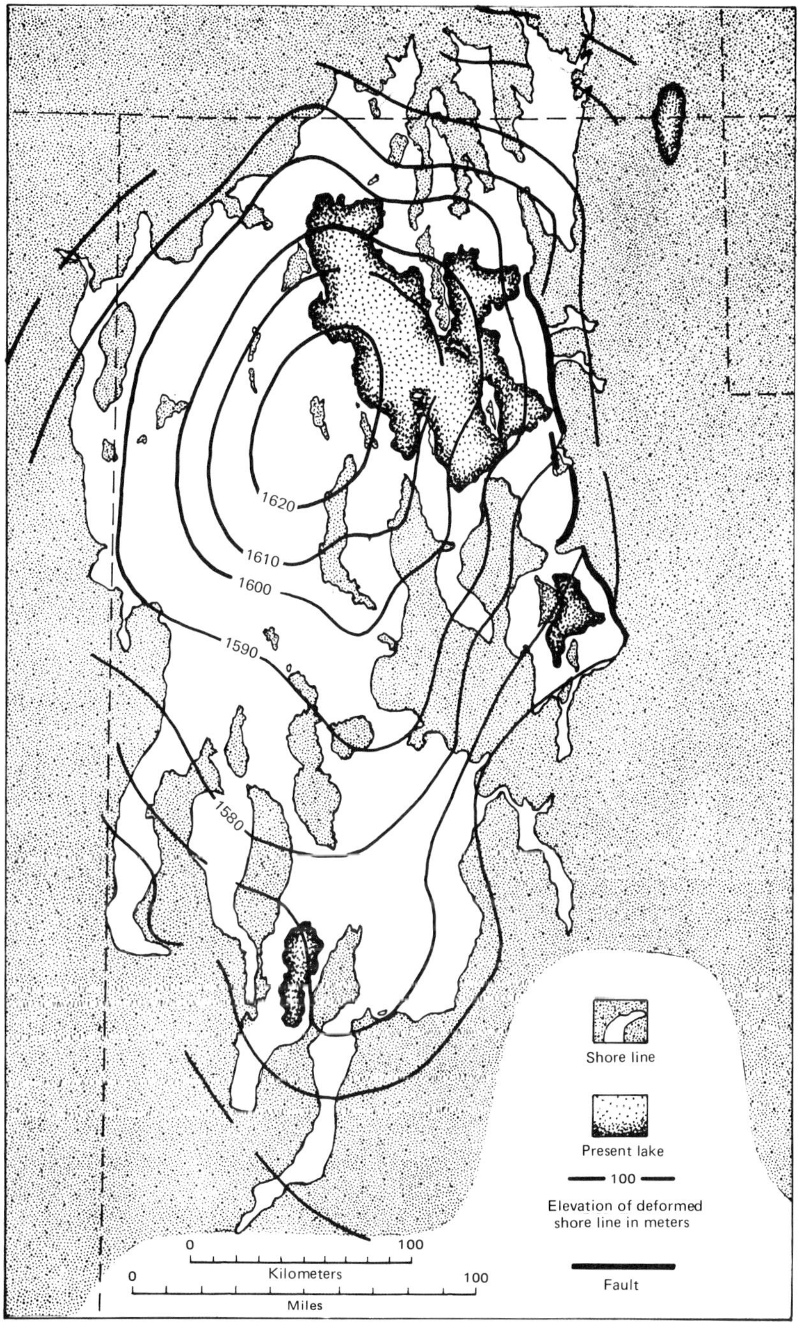

FIGURE 22–6
Map of northwestern Utah showing maximum area of Lake Bonneville and deformation of its highest shore line.
(from Crittenden, 1963)

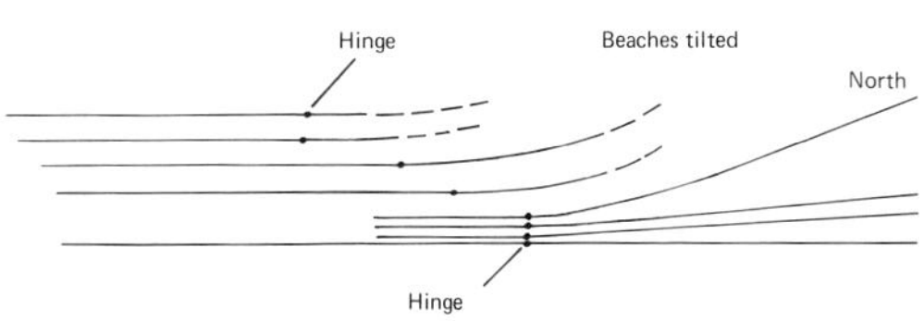

FIGURE 22–7
Generalized diagram across the central part of the Great Lakes region, showing tilting of shore lines of successive lakes and northward displacement of hinges, or zero isobases. Older lakes are above and to left, younger lakes below and to right.
(after Taylor, 1927)

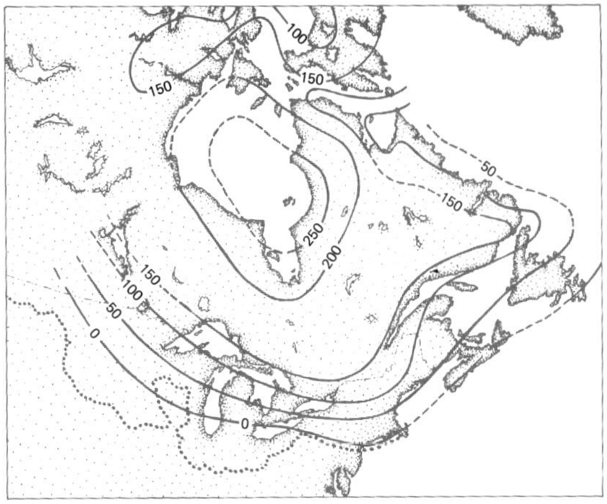

FIGURE 22–8
Map of northeastern North America illustrating postglacial uplift. Contours on the highest observed marine shore lines indicate the minimum amount of postglacial uplift.
(compiled from Farrand and Gadja (1962) and other sources; from King, 1965)

FIGURE 22–9
Rate of land emergence in Juneau region, Alaska. Contour interval, 0.4 cm/year.
(from Hicks and Shofnos, 1965)

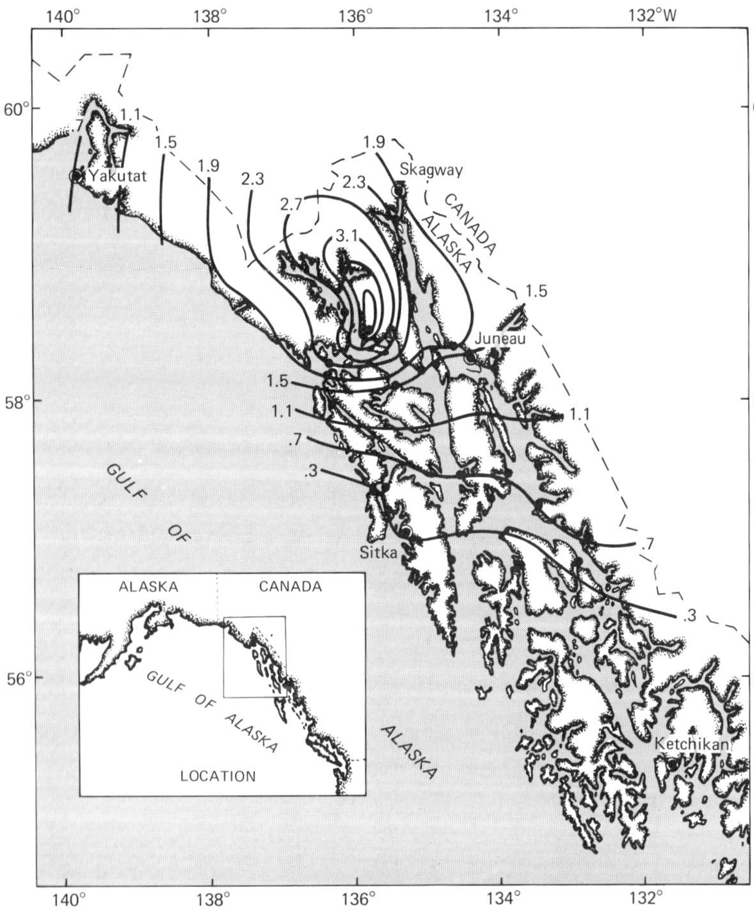

approximately established. This level (Fig. 22–8) is now elevated to an estimated 250 meters near Hudson Bay, and precise leveling studies and gravity anomalies indicate that uplift is still in progress at rates on the order of 1 meter per century in the Hudson Bay area.

Other studies of rates of epeirogenic movement indicate that the Gulf Coast of the United States is subsiding at the rate of 6.6 mm/year, the coast of Newfoundland is rising 0.8 mm/year, the shield of Scandinavia is rising about 7 mm/year, the eastern coast of Australia is rising at the rate of 0.4 mm/year, and the eastern coast of India is subsiding at the rate of 1.2 mm/year.

Particularly rapid rates of emergence have been found in the region near Juneau, Alaska (Fig. 22–9). Sea-level observations have been made in this region since 1887 and good control has existed for the last 40 years. The maximum observed rate of emergence is nearly 4 cm/year in the vicinity of Glacier Bay and a 2 cm/year rate covers an extensive area from the coast to Juneau and Skagway. This uplift, like that in central Canada, is attributed to rebound resulting from deglaciation, but the rates near Juneau are much greater and the western coast of North America lies within an orogenic belt.

Marine Invasions of the Continental Interior

Epeirogenic movements are by no means confined to the present, nor even the recent periods of geologic time. One of the most remarkable records of epeirogenic movement is found in the section of strata exposed in the Grand Canyon of the Colorado River in Arizona (Fig. 22–10). Near the bottom of the canyon Precambrian rocks, more than half a billion years old, are exposed. These rocks have been intensely deformed, folded, intruded by magma, and metamorphosed. A long period of erosion followed the mountain building and the mountains formed during the orogeny were reduced to a nearly flat plain. Such reduction would take place only above sea level where the mountains could be eroded by running water and other geomorphic agents. Then marine sedimentary layers were deposited across the edges of the older eroded rocks. A second period of deformation subsequently took place

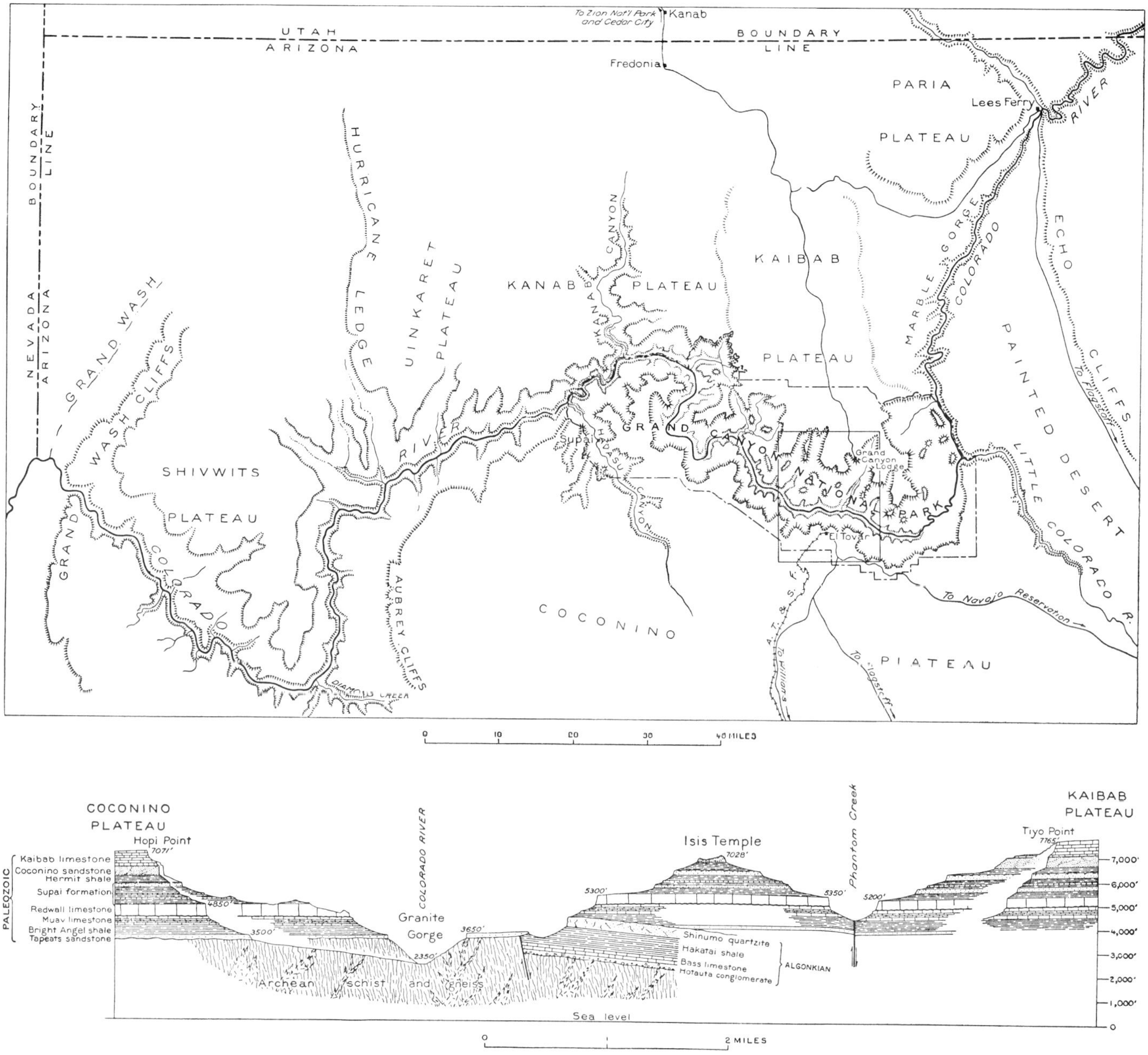

FIGURE 22–10
Map and geologic cross section of the Grand Canyon.
(from F. E. Matthes, U.S. Geological Survey Topographic Map)

during which these units were faulted, tilted, and undoubtedly uplifted. This was followed by another long interval of erosion before the first Paleozoic seas occupied the region. The oldest of these layers is about half a billion years old. During the millions of years represented by the strata exposed in the canyon, the region of the southwestern United States was beneath or just

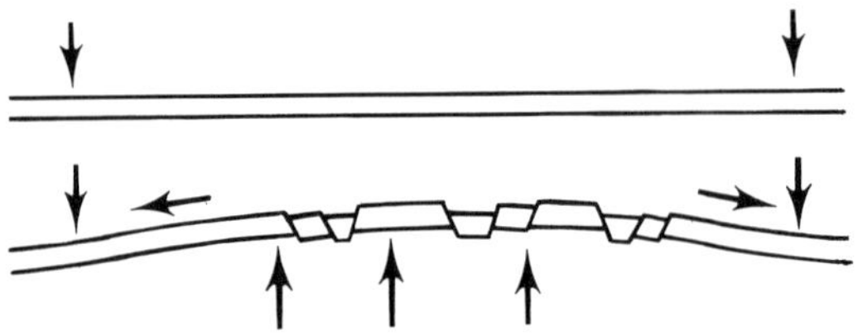

FIGURE 22–11
Origin of graben and horst structure by uplift.

at sea level most of the time. Occasionally the region was uplifted, and some of the sediments were eroded off, but during most of that time new sediments accumulated on top of the older ones, which continued to subside slowly. Today the marine sediments found in the walls of the canyon nearly 1.6 kilometers above sea level are as flat-lying as when they were initially formed on the floor of the ocean. This segment of the earth's crust has been uplifted more than 1.6 kilometers without intense deformation of the rocks within the block during the millions of years since these strata started to form. Other epeirogenic movements preceded the one that brought the region to its present elevation, but at no time in that half a billion years was the region of the Colorado Plateau intensely folded. The Colorado Plateau lies within the orogenic belt of western North America, and it is flanked by high-angle faults and strong flexures, but this crustal block more or less escaped the orogenic deformation that surrounded it.

Evidence of epeirogenic movement similar to that in the Grand Canyon is found on every continent. Seas have transgressed across continental margins and occupied the interiors of every continent. Such seas were particularly widespread during the Early Paleozoic. Deposits from these seas are now widely exposed in the Appalachian Mountains and North American interior lowlands, but they often lie buried under younger sediments in the Western States.

Faulting Associated with Epeirogeny

Broad epeirogenic uplift or downwarping is likely to cause fracture of brittle crustal rocks, and if the movements are primarily vertical, stretching of the crust occurs, accommodated by development of faults (Fig. 22–11). These faults, usually high-angle, undergo movements that allow extension of the brittle crustal layers. Faults, presumably of this nature, abound in the Basin and Range geomorphic province (Nevada, Utah, southern California, Arizona) where many of them have been active within the period of recorded history.

Epeirogeny and Isostasy

The theory of *isostasy* provides the best explanation for the cause of epeirogenic uplifts. The earth's crust is known to be of variable thickness, composition, and density. The crust is constantly undergoing change through the shifting of loads of rock on the surface as a result of weathering, erosion, transportation, and deposition. Over a long period of time mass is shifted mainly from the high mountains to the lower parts of the surface. Volcanoes and lava plateaus are constructed through igneous activity, and mountains are formed through orogenic processes. These processes build up large masses of rock.

According to the theory of isostasy, loads on the crust are approximately balanced. If mass is removed from a certain area on the earth's surface, equilibrium is upset, and mantle rock, behaving like a plastic, flows and exerts a pressure on the crust, tending to elevate the area from which mass has been removed and to restore what is essentially a flotational equilibrium.

The rates of epeirogenic uplift of the Canadian and Fennoscandinavian shields may be explained in this manner. During the ice ages both of these areas were sites of accumulation for large glaciers. It appears probable that parts of Canada were covered by ice ranging from 2 to 4 kilometers in thickness. The weight of this ice upset isostatic equilibrium, causing that segment of the earth's crust to subside slightly into the somewhat plastic zones of rock beneath the crust. Since the earth is a solid sphere, the material displaced by the downward extension of the area under the Canadian Shield had to move. This movement was toward crustal segments that were becoming lighter, such as the ocean basins, which had lost water (to form the ice), since otherwise it would have caused a general upward exertion of pressure throughout the subcrust. When the ice melted away as the glaciers retreated, the weight was reduced on the shields, and equilibrium is now being re-established by compensating uplift of those areas.

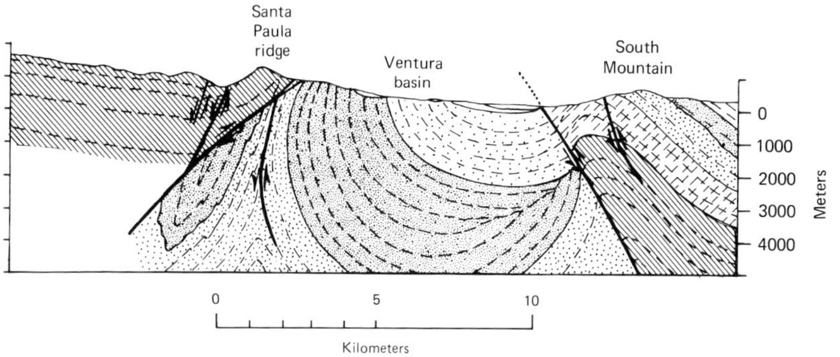

FIGURE 22–12
Section illustrating mid-Pleistocene deformation in Ventura basin, California. This section is across the basin, about 20 kilometers east of Ventura, showing structure of Tertiary and Quaternary rocks.
(after Bailey and Jahns, 1954)

EVIDENCE OF OROGENY

The existence of long belts of mountainous terrain is proof of orogeny. It is evident from examination of the exposed rock masses that processes other than epeirogenic uplift followed by dissection are involved in the formation of these mountain belts. Sedimentary layers are folded and often repeated by faults that have caused a piling-up of strata, and by the alteration of sedimentary rocks to crystalline metamorphic rocks, indicating high temperature and pressure. These features of modern high mountain belts such as the Alps, Himalayas, Andes, and Rockies are also found in the much older and lower belts like the Ural, Appalachian, and Ouachita mountains. The same types of structures also have been found in the subsurface where the mountain system has been lowered and covered by more recent deposits, and they are found in Precambrian rocks where the mountain systems have been eroded to a level so that no relief remains to attest to their original grandeur.

The metamorphism usually found in the central parts of orogenic belts and the abundance of igneous rocks in such belts indicate the importance of heat in orogeny. The folds and low-angle faults (see Fig. 24–8) suggest the action of directed stresses of a magnitude sufficient to exceed the strength of the rocks. Because folded layers and repetition of beds by faulting (see Fig. 23–31) imply a shortening of the beds, it has been widely assumed that orogeny takes place as a result of strong crustal compression.

Orogeny is going on today. Most of the earthquake belts of the world (see Fig. 20–16) coincide with young, folded mountain belts or with areas in which orogeny is thought to be going on. These include the circum-Pacific belt of volcanism, seismicity, and diastrophism. Drilling for oil in southern California's Ventura basin has provided control at depth, which makes it possible to construct an accurate cross section of the basin (Fig. 22–12). Sedimentary sequences in the basin, in-

FIGURE 22–13
Vertical displacement for the last several decades.
(from Fiji Inoue, 1960)

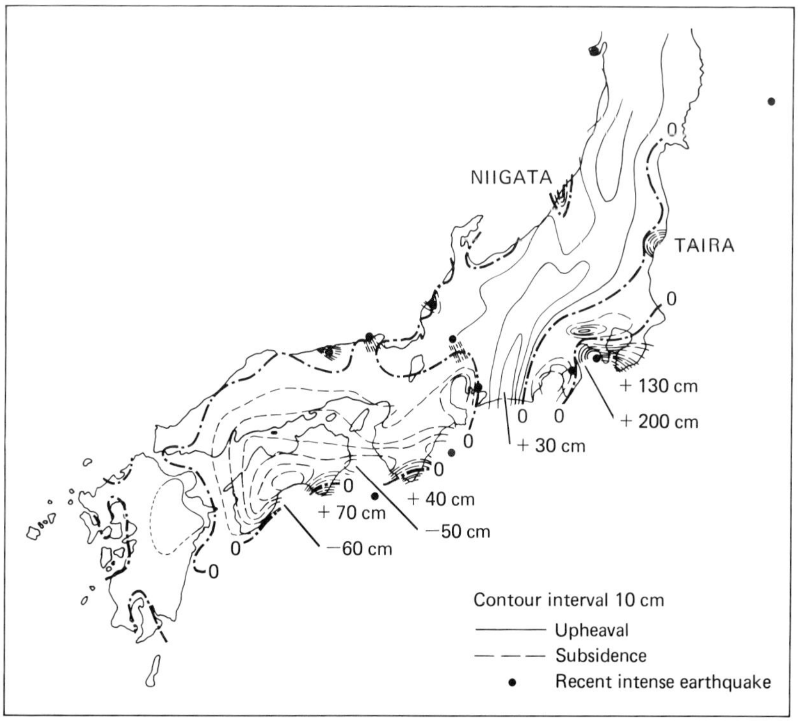

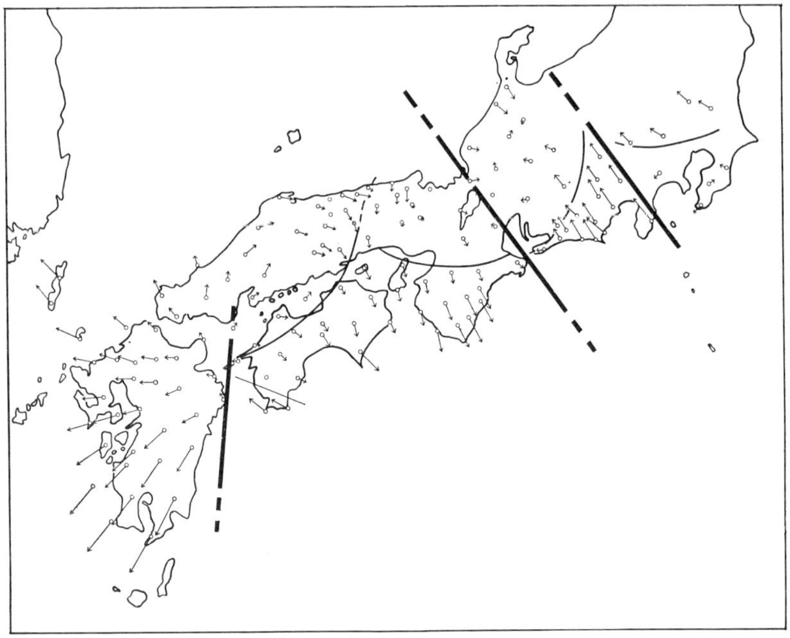

FIGURE 22–14
Discontinuous boundaries of horizontal displacements in the southwestern part of Japan.
(from Eiji Inoue, 1960)

cluding Pleistocene units, are folded and faulted as shown. Thus this region of active seismicity also bears evidence of recent orogenic deformation.

Japan is a site of modern orogeny where the network of surveying stations and leveling is sufficient to document both horizontal and vertical movements. A summary of vertical displacements over the past several decades (Fig. 22–13) indicates a far more irregular pattern than that which characterizes the epeirogenic deformation of Canada or Scandinavia. Horizontal displacements also exist (Fig. 22–14) on a regional scale.

CHAPTER OUTLINE

TERMS

bench marks
diastrophism
epeirogenic
guyot
isostasy
leveling
mobile belts
orogenic
sial
sima
triangulation

REVIEW QUESTIONS

1. Man's life span is brief compared with the geologic record. Thus it would appear, with the notable exception of catastrophic phenomena, that the earth's crust is fairly stable. However, this is not the case over long periods. What are some indications of this instability?
2. Differentiate between epeirogeny and orogeny.
3. What are some evidences indicating changes in sea level during recorded history? What are some more general indications that geologists might use?
4. Other than epeirogenic or orogenic movement, what processes have most assuredly caused changes in ocean levels?
5. What is a guyot and what are its implications geologically?
6. There are evidences of great orogenic movement in the Grand Canyon and the rocks exposed at its base are over half a billion years old. Yet most of the sediments exposed in the canyon walls are fairly flat-lying and undeformed. Explain why the higher rocks are undeformed, particularly in view of their high elevation.
7. What sort of faults are caused by upwarping during epeirogeny?
8. What is flotational equilibrium? How does it work?
9. Why is heat an important form of energy associated with orogeny?
10. Describe the modern crustal movements that have taken place in Japan. How were they determined?

SUGGESTED READING

King, P. B., 1965, Tectonics of Quaternary Time in Middle North America: The Quaternary of the United States: Princeton, N.J., Princeton University Press, p. 831–870.

V THE STRUCTURE OF THE EARTH

23 ROCK DEFORMATION AND THE STRUCTURAL FEATURES OF DEFORMED ROCKS

23 ROCK DEFORMATION AND THE STRUCTURAL FEATURES OF DEFORMED ROCKS

STRUCTURAL FEATURES

The term *structure* is applied to the form or the arrangement of parts of a body. In geology, structure refers to form and orientation of the parts of rock masses. Here we are concerned with the internal and external arrangements of rock particles resulting from movements that occur after deposition and often after the rock is consolidated. These are therefore *secondary structural features;* the initial particle arrangements are products of the processes of deposition or crystallization and are *primary structural features.* Secondary structural features are usually recognized and described by the distortion of primary features. The most widespread primary structural feature of sedimentary rocks is stratification. Most sediments are deposited on flat areas or on slopes of a few degrees (Fig. 23–1). This provides a valuable guide to initial or primary shape and orientation.

If the original shape and position of a rock are known, it becomes possible to recognize changes in shape or position that occur later. It may be difficult to distinguish structural features formed during compaction and consolidation from those formed later. Unconsolidated and semiconsolidated sediments, particularly those composed of clay, silt, or sequences in which thin beds of different lithology are interbedded, are easily deformed. They may become folded and faulted as a result of downslope movements such as submarine slides or slumps while deposition is taking place. These features may resemble structural features that result from deformation of consolidated rocks subjected to directed pressures.

In the following chapters we will consider some of the large-scale processes that may cause rocks in the earth's crust to be subjected to strong, directed pressure (*stress*); that are important in the formation of mountain ranges; that cause metamorphism; and that produce most secondary structural features. Here we consider the smaller scale structural features that can be seen in hand specimens, in rock outcrops, or that can be traced over larger areas. Structural features most frequently encountered are the following:

Fractures. Planes (breaks) through a rock across which cohesion has been lost (Figs. 23–2 and 23–3).

Faults. Shear zones along which rocks on either side have been displaced.

Folds. Bends or distortions of rock masses most easily recognized when bedding is deformed (Fig. 23–4).

Structural features of metamorphic rocks (cleavage, foliation) were discussed in Chapter 5.

Scale of Structural Features

Secondary structural features range from submicroscopic size to whole mountain systems. It is convenient to think of features in the following size categories:

Microscopic. A size most readily studied with the aid of a microscope.

Mesoscopic. The size of an outcrop, ranging from a few meters to a few hundred meters across.

Regional. Features may be many kilometers across (Fig. 23–5).

FIGURE 23–1
Remnants of a formerly extensive layer of flat-lying sandstone in southern Utah.
(Esther Henderson)

Tectonic. The large-scale structural features such as mountain ranges, mountain systems, island arcs, and so on, which are complexes of many smaller scale features.

The study of microscopic features has proved valuable as a means for understanding the internal movements that occur between and among the grains or crystals during deformation. This approach has shown how micaceous minerals of slate and schist are aligned along cleavages; how calcite in marbles and limestones has deformed by slippage along certain glide planes through the individual crystals; and indicates that movements in many sandstones and quartzites have taken place by rotation and granulation of individual grains. The microscope has made it possible to measure distorted oölites and fossils as well as the way crystals of pyrite or garnet have been rotated (Fig. 23–6) during deformation of the enclosing rock. Microscopic approaches have been particularly valuable as a method for comparing experimental laboratory tests on small specimens with tests on naturally deformed rocks.

Most of what we know about regional and tectonic scale features is ultimately derived from observations made at the outcrop from mesoscopic scale features or from subsurface data. Fractures, faults, folds, deformed fossils, rotated structures, extended features—all abound at this scale. The orientation in space of each linear and planar feature can be determined; photographs, sketches, and diagrams are used to describe the features at an outcrop.

To describe regional structural features, it is necessary to compile data obtained at many outcrops or drill holes. Observations from outcrops are plotted on maps so that regional relationships may be seen. Geologic maps are prepared that show the areal distribution of rocks of different types and ages. Structural maps may be compiled on which the elevation of a marker surface such as the top of a given stratum is shown by use of contour lines like those used to represent the elevation of the land surface. Contour maps of this sort drawn on a structural feature are called *structure contour maps*. Other maps, called *isopach maps*, are prepared to show the thickness variation of a given stratum or sequence of strata. Cross sections may be prepared from all of the

FIGURE 23–2
Rectangular jointing in the upper surface of a coal seam, Yorkshire, England.
(Crown Copyright)

FIGURE 23–3
Fractures in limestone. These fractures have been filled with calcite (white) precipitated from ground water.

FIGURE 23–4
Sharp, asymmetric syncline at Sandy Mouth, Cornwall.
(Crown Copyright)

FIGURE 23–5
Belt of folded mountains in Africa (Atlas Mountains). These mountains have been deeply dissected. Erosion products of the mountains lie on the flanks of the ridges.
(U.S. Air Force)

above maps to show the structure as it appears at depths. Subsurface data obtained from drilling, seismic reflection or refraction profiles, and other methods may be used to aid in the interpretation of the subsurface configuration of strata.

Analysis of regional studies in terms of the largest scale features of the earth's crust is covered by tectonics. Regional structures are synthesized and interpreted in terms of our knowledge of rock deformation obtained from experiments, theory, analysis of the history of the earth's crust, and from our understanding of the earth's interior.

DESCRIPTION AND CLASSIFICATION OF STRUCTURAL FEATURES

Positions of planes and lines in three-dimensional space can be described by simple measurements of bearings and inclinations. The compass direction of a horizontal line or the horizontal projection (i.e., N 20° E) and its inclination or *plunge* (i.e., the angle, measured in a vertical plane, between a line and the horizon) define the position of a line (Fig. 23–7). Planes (e.g., bedding, fractures, faults, cleavages are all often planar) are described by *strike* and *dip* measurements.

Strike. The compass direction (bearing) of any horizontal line on the plane.

Dip. The angle between the plane and the horizon, measured in a vertical plane at right angles to the strike (Fig. 23–8); the *direction of dip* is the bearing of the direction toward which the plane is inclined.

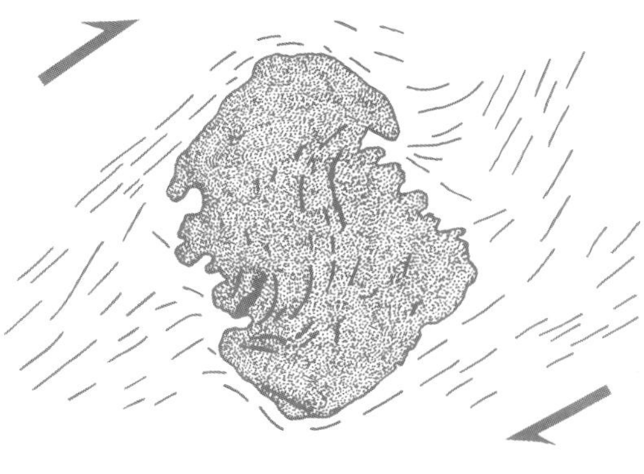

FIGURE 23–6
Sketch of a rotated garnet crystal. The black lines indicate alignment of hornblende and mica in the rock around the garnet.
(drawn from a photograph by Alan Spry)

FOLD NOMENCLATURE

Parts of a Fold

Limbs. The flanks or sides of a fold.

Crest. The highest part of a fold form.

Trough. The lowest part of a fold form.

Axial surface. An imaginary surface that approximately bisects the fold into symmetrical halves.

Axis (hinge). An imaginary line formed by the intersection of the axial surface with a bed-

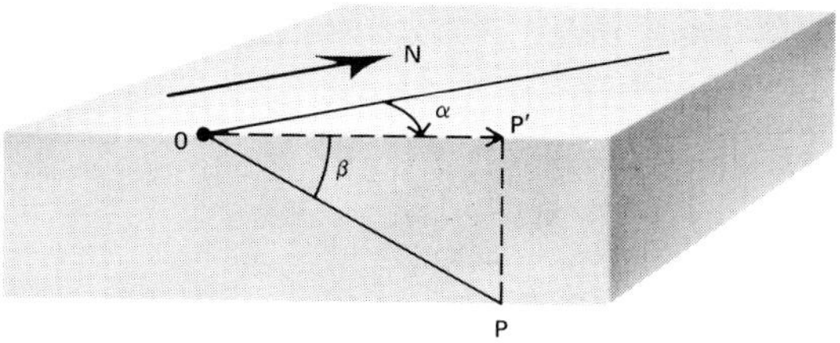

FIGURE 23–7
Block diagram showing bearing (angle between north and $OP' = \alpha$) of the horizontal projection (OP') of line OP and the plunge of OP (β). The block has been drawn so that OP is located in one side of the block.

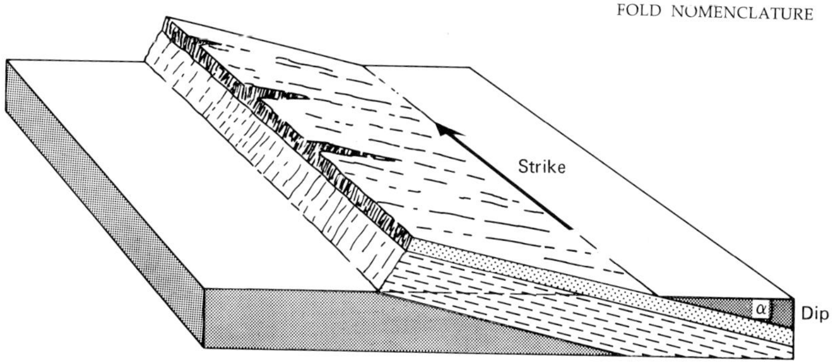

FIGURE 23–8
Strike of a rock unit is the compass direction of horizontal lines of the unit. Dip is the angle between the horizon and the surface of the unit measured at right angles to the strike.

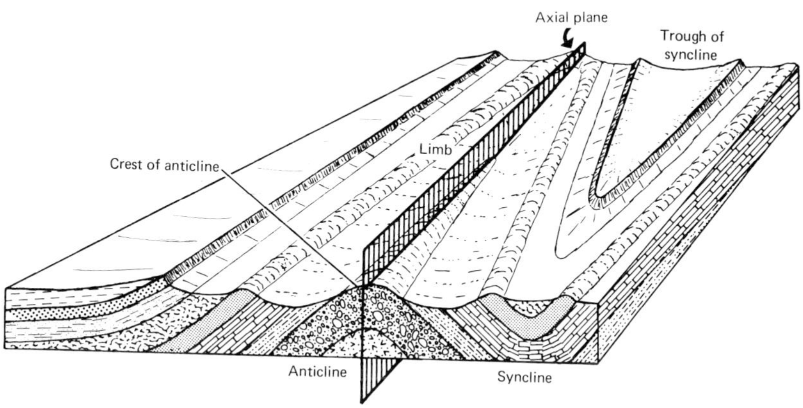

FIGURE 23–9
Block diagram illustrating open, symmetrical plunging anticline and syncline and showing the axial plane of the anticline.

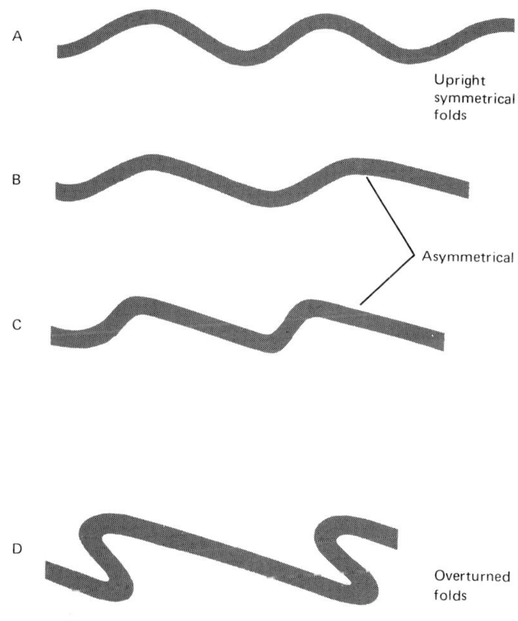

FIGURE 23–10 (*right*)
Cross sections across fold systems showing progressive shortening.

ding surface. The hinge is located where the curvature of the bed is greatest.

Plunging axis. When the axis is not horizontal, the angle of inclination measured from the horizon is the *plunge*.

Fold Symmetry

Symmetrical fold. When two limbs of a fold are nearly mirror images of one another.

Asymmetrical fold. When the limbs are not symmetrical.

Overturned fold. When the axial surface is inclined so one limb is over part of the other limb.

Recumbent fold. When the axial surface is nearly horizontal.

Types of Folds

The principal fold types are defined on the basis of shape, direction of inclination of limbs (Fig. 23–10), and relative positions of layers of different

FIGURE 23–11
A structural dome. The top of this dome has been planed off through long-continued erosion. The oldest rocks are exposed at the center and successively younger units are exposed away from the center.
(U.S. Air Force)

FIGURE 23–12
A structural basin. A series of hogback ridges are exposed around the margins of the structure. The youngest units are exposed near the center of the structure.
(U.S. Air Force)

ages. Mechanisms of folding—for example, flowage, flexure, slip, and the geometry of the fold—are also used in classifications. Thus several names may be appropriately applied to any fold.

Anticline. An upfold in which the limbs dip away from the fold axis. Older beds are located in the core of an anticline; thus, if the crest of an anticline is breeched by erosion, older rocks are exposed (Fig. 23–9).

Syncline. A downfold in which the limbs dip toward the axis. Younger beds are exposed along the trough of a syncline.

Dome. A circular or elliptical uplift; older beds lie in the center (Fig. 23–11).

Basin. A circular or elliptical downwarp or structural depression, with younger beds in the center.

Tilted beds and flexures (Fig. 23–13) are also classed with folds.

Homocline. Where the beds have a single, relatively uniform direction of dip (one limb of a large fold may be described as homoclinal).

Monocline. A single inclination; a flexure in which the direction of dip does not change.

Structural terrace. A flexure that results in a flat, nearly horizontal surface.

It is usual to classify folds according to their geometry (Fig. 23–14). Among the terms used to describe fold geometry are the following:

Concentric (parallel) folds. Those in which the thickness of the individual layers remains constant (measured perpendicular to the bedding) with depth.

Similar folds. Those in which the shape of the fold remains the same from one layer to another in the fold, but the thickness varies from place to place.

Disharmonic folds. A term applied where the geometry of the fold form varies in an irregular way with depth.

It is desirable to classify folds on the basis of the manner in which they originated, a genetic classification, but this is often very difficult to do.

When it is possible to observe or infer the manner in which the fold formed, the following genetic classification applies:

Flow fold. A fold formed by rock flowage.

Shear fold (slip fold). A fold formed as a result of the minute displacement of beds along closely spaced fractures or cleavage planes.

Flexural folds. Where flow or slip is restricted by the layer boundaries, the layering exercises an active control on the deformation; the resulting folds represent a true bending of layers. Flexural slip is analogous to the movement involved between cards in a flexed deck of cards.

Passive folds. Where flow or slip crosses the layer boundaries, the layering exercises little or no control on the deformation; the layering is passive, and layer boundaries serve merely as markers, parts of which are displaced relative to other parts to produce an apparent bending.

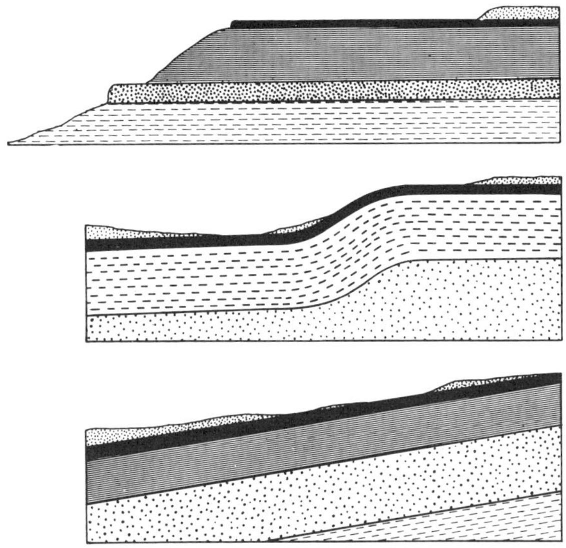

FIGURE 23–13
Schematic cross sections of flat-lying strata, top; a monocline, middle; a homocline, bottom.

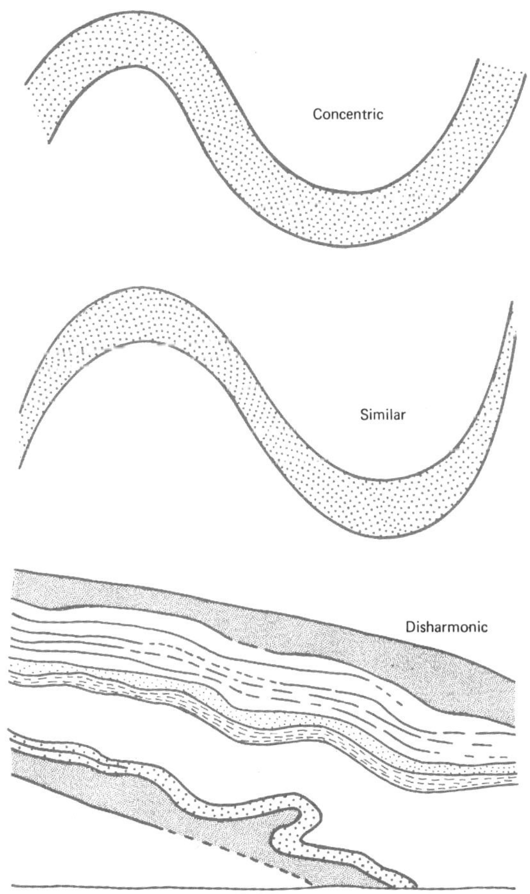

FIGURE 23–14
Fold geometry. Beds in a concentric fold retain constant thickness, but beds in a similar fold show thickening in crest and troughs.

FAULT NOMENCLATURE

Faults are shear zones along which displacement has occurred. Faults may be exposed in valley walls or along road cuts (Fig. 23–15); but because the rock in the fault zone is broken and sheared, the zones are eroded rapidly and often appear as topographic lows. Fault zones may be recognized by the following criteria:

1. *Displacement of geomorphic features.* Streams, mountains, hills, and other topographic features may be offset across faults.
2. *Fault scarps.* Small escarpments often mark the position of displacements where one side moved up in relation to the other. Such scarps are most frequently found along recent faults or when rocks of different resistance to erosion are brought in contact along the fault.
3. *Displacement of geological structures.* The outcrop of uniformly inclined strata may be displaced, or a structural feature such as a fold may be offset across a fault.

Most faults are recognized through geologic

FIGURE 23–15
Overthrust that passes into bedding at both ends.
(Crown Copyright)

FIGURE 23–16
Block diagram showing a normal fault. (Hanging wall is down relative to footwall.) Slickensides indicate dip-slip movement. Distance *AB* is called the displacement. In this instance the displacement is equal to the dip separation.

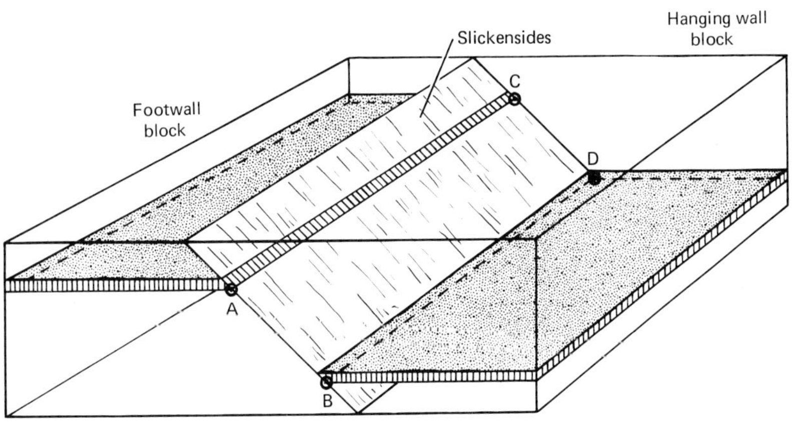

mapping when it is found that rocks of a different age are juxtaposed across narrow zones of breccia or when sharp changes appear in structural attitude of the strata across the zone.

General Terms

Fault plane. The zone along which the two sides have moved is called the fault plane or zone. It may be a smooth plane produced by a sharp break, or it may be a wide zone of distortion hundreds of meters thick.

Breccia. Crushed and brecciated rock is usually present within the fault plane. Sometimes crushing is so complete that breccia is reduced to a powdery-textured material called *gouge* when soft or *mylonite* when it is lithified.

Slickensides. Striations and grooves formed where rocks on either side of the fault have moved across one another.

Fault blocks. Faults locally divide the crust into blocks. If the fault is not vertical, one of these blocks must lie above the fault. The upper block is called the *hanging wall,* a mining term applied by a miner standing in a mined-out fault zone to the block hanging over his head. He stands on the other block, called the *footwall.*

Classification of Faults by Relative Movement of Blocks

Normal faults. Steeply inclined faults along which the hanging wall has moved downward in relation to the footwall (Fig. 23–17).

Reverse faults. Steeply inclined faults along which the hanging wall has moved up in relation to the footwall.

Thrust fault. A fault inclined at a low angle along which reverse-type movement has occurred. Many thrusts involve movements parallel to bedding.

Classification of Faults by Displacement

The straight-line distance, after movement, between two points that were originally adjacent on opposite sides of the fault is called the *displacement* or *net slip.*

Unfortunately, it is often impossible to identify features that were originally adjacent. Linear features that were cut by the fault and offset afford the best opportunity to identify and measure displacement. Linear features such as fold axes, lines of intersection of two dikes or a dike and a bed, or an old stream channel provide satisfactory markers (Fig. 23–18).

If it is possible to identify and measure displacement (net slip), the amount and direction of movement necessary to account for the displacement can be determined and the fault can be classified on the basis of its type of movement. This movement is described in terms of its relationship to the strike and dip of the fault if the movement is translational (blocks do not rotate during movement).

Strike-slip fault. If the direction of displacement is horizontal along the strike of the fault.

Dip-slip fault. If the direction of displacement is up or down the dip of the fault.

Oblique-slip fault. If the displacement is oblique across the fault plane.

Rotational fault. A scissorslike motion is also possible, involving rotation of the fault blocks relative to one another. Displacement directions and amounts vary from place to place in the fault.

Classification Based on Separation

A classification of faults based on *separation* (the distance measured in some specified direction between the displaced parts of a marker horizon, e.g., bed) is preferable to the "slip" classification (Fig. 23–19). Separation can almost always be determined; slip is difficult to determine.

Dip separation. Measured straight down the dip of the fault.

Strike separation. Measured straight along the strike of the fault.

Vertical separation. Measured in a vertical line.

The direction of strike separation is specified as *left-lateral* (if one stands on one side of the fault, markers are displaced to the left across the fault) or *right-lateral.*

Special Names Applied to Fault Structures

Graben. Derived from the German word meaning "grave", applied to an elongate block, bounded by normal faults on either side, which has moved down in relation to the sides (Fig. 23–25).

Horst. Applied to an elongate block bounded by normal faults on either side when the block has moved up in relation to the sides.

Klippe. An isolated part of a thrust sheet separated from underlying rocks by a fault. Usually the fault is part of a thrust fault, and usually, though not always, rocks in a klippe are older than those underlying them (Fig. 23–20).

Nappe. A large mass of deformed rock that

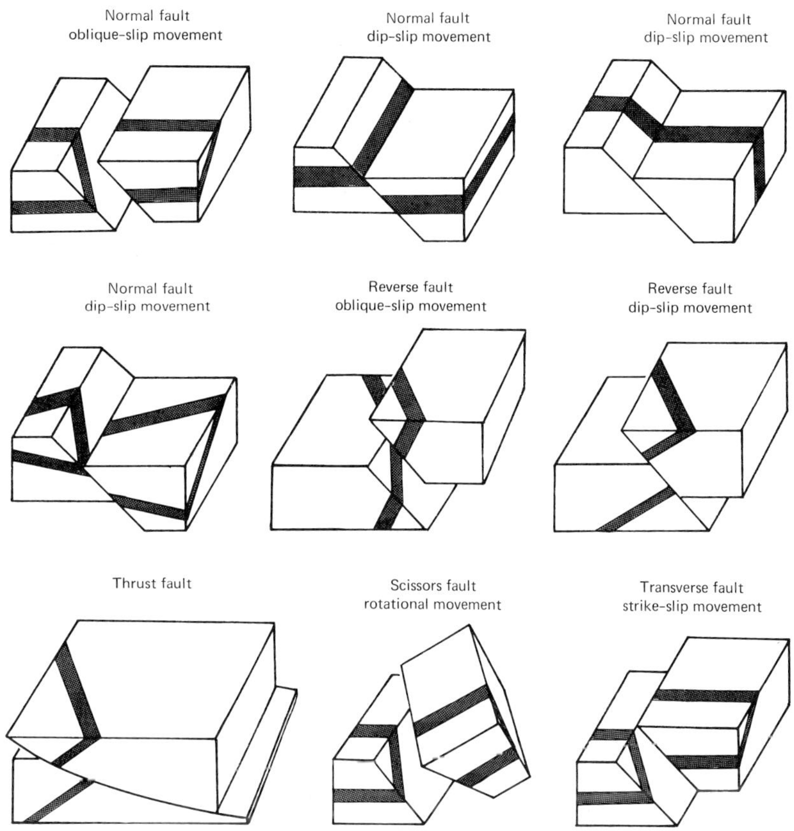

FIGURE 23–17
Fault blocks illustrating various combinations of types of movement and strike and dip of strata.

FIGURE 23–18
Schematic diagram showing the line of intersection of two dikes. This type of feature when offset by a fault can be used to determine displacement.

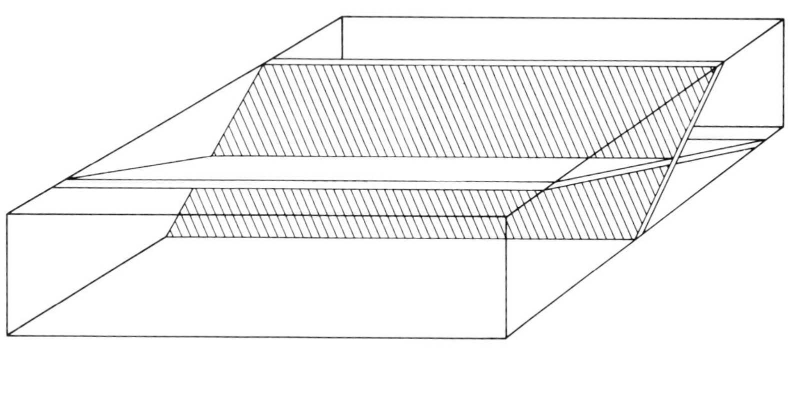

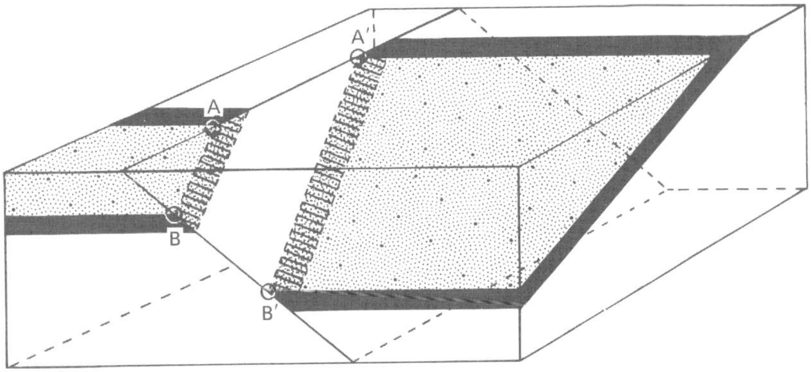

FIGURE 23–19
Block diagram showing a stratum that has been displaced across a fault. The distance *AA′* is called the strike separation; *BB′* is the dip separation.

FIGURE 23–20
Schematic block diagram of structural features of the Central Switzerland nappes. Root zones are located at the right. Nappes are numbered I and II, and K designates several klippes.
(after Arbenz, 1912)

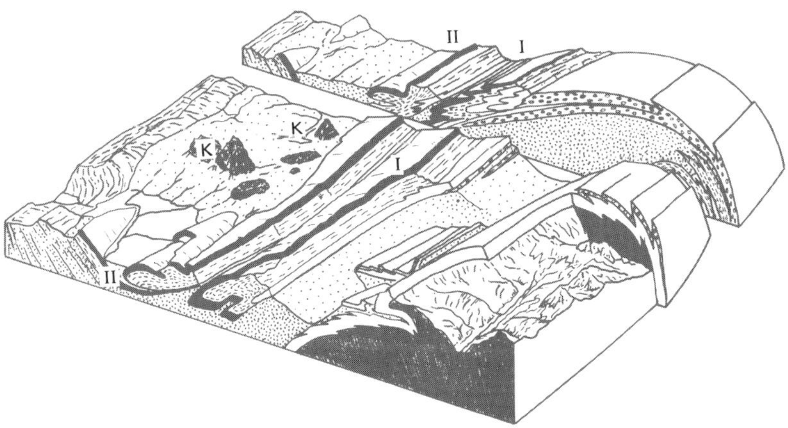

has been moved a long distance from its point of origin. Most nappes are parts of strongly folded strata in which folding has proceeded until one limb is overturned and the nappe is displaced by thrusting and folding (Fig. 23–20).

OCCURRENCE AND ORIGIN OF ROCK FRACTURES

Fractures are planar breaks in rock along which no displacements have occurred. The rock has lost cohesion across these breaks, and thus they impart a weakness to the rock. Usually fractures occur as parallel sets (Fig. 23–21) with one or more orientations. For this reason the rock mass may be very weak in some directions but strong in others. The existence of fractures as well as their orientations and spacings may be crucial factors, therefore, in the design of dam or building foundations or in the construction of tunnels; we have already seen (Chapter 13) how fractures may lead to slope instability. Fractures are important also because they account for the porosity of many igneous and metamorphic rocks. Water is frequently stored in fractures, and oil and gas are produced from fractures in a few fields.

Some fractures are caused by the contraction that results from the drying out of sediment or the cooling of rock melts. Others are clearly related to larger structural features like folds and faults. Such fractures result from brittle failure of the rock when directed stresses (see Appendix C) are applied and cause the folding or faulting. Still other fractures originate after formation of the larger structures. Thus a fold may form under temperature and pressure conditions that cause the rock to be ductile; later the entire mass may be fractured when the rock is brittle.

Fractures Caused by Volume Reduction

When molten rock begins to crystallize its volume is reduced through reduction in thickness and usually through reduction in surface area as well. The latter is accomplished in part by development of breaks, cooling cracks, along smooth planar or curved surfaces. Cooling cracks form a number of different patterns (Fig. 3–9); some are radiating; others are irregular, intersecting, curved surfaces; and still others are hexagonal or polygonal. Fractures formed in slightly deformed sedimentary rocks may likewise originate through contraction caused by compaction and shrinkage as sediments are consolidated into sedimentary rocks.

Fractures Caused by Diastrophism

The relationship of fractures to other structural features is determined by measuring fracture orientations that occur on folds and near faults and comparing these results with results derived from experimentally produced fractures. Experiments also make it possible to relate the fractures to particular types of stress applications.

Two general classes of fractures can be distinguished: (1) those that result from *extension* and (2) those that result from *shearing*. Extension may be expected where originally flat-lying beds are

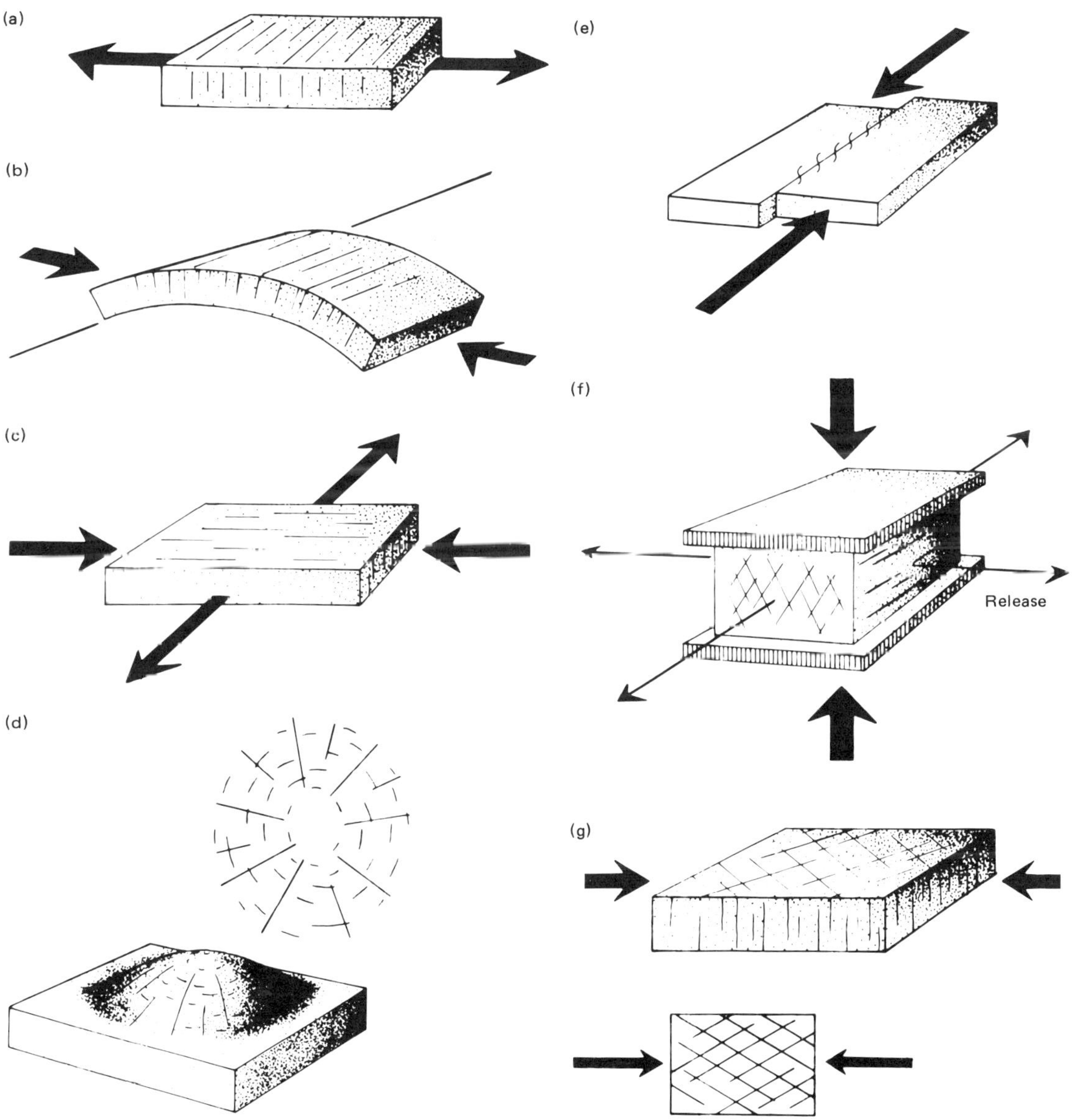

FIGURE 23 21
Fracture orientations. Extension fractures formed by extension of block (a), by flexure (b), by compression of brittle material (c), by doming (d). Gash fractures due to the action of a couple (e) are usually open and often filled with secondary deposits. Shear fractures (f and g) are shown as formed in a press (f) and as they may appear in a rock layer subjected to horizontal compression (g).

domed or folded into an anticline or syncline (Fig. 23–21b). Folds frequently have a set of extension fractures oriented parallel to the fold axis, as shown. A set of radial fractures and a second set of concentric fractures are often found on circular domes. Extension fractures may also develop oriented perpendicular to the direction of extension (Fig. 23–21d).

Extension fractures may develop as a result of rotation due to the shifting of blocks, as shown in Fig. 23–21e. Here the initial fractures form in an *en echelon* arrangement and may be found in the sedimentary layers over deep faults that have undergone lateral displacement.

The above-described conditions can be approximated in simple experiments with clay cakes, wax, or brittle materials. If a block of brittle material is compressed in a vise, the first fractures to form are generally a set oriented parallel to the direction of compression (Fig. 23–22). These are extension fractures generated as the material expands in one of the unconfined directions.

Shear Fractures

Fractures that cannot be explained in terms of extension are also found in rocks. Frequently these fractures have slight displacements, and the fracture surface is slickensided, indicating that the rock was under pressure that forced the surfaces together. Relatively simple experiments can be performed to generate shear fractures. As an illustration of a brittle material's behavior, a block of soap, wax, or rock compressed in a vise will usually break initially along extension fractures that are oriented parallel to the axis of the compression. But if the material is not brittle, fractures oriented between 30° and 40° to the axis of the vise (Fig. 23–22) form and undergo a shear displacement as deformation continues. Often these compose two sets of shear fractures that intersect at angles of 60°±, and the angle between them is the direction of maximum compression.

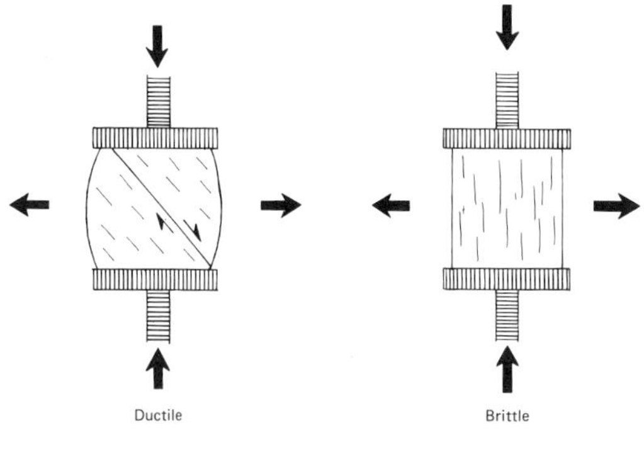

FIGURE 23–22 Extension fractures (right) formed in brittle material and shear fractures (left) formed in ductile material in response to compression in a vise.

ROCK CLEAVAGE

Rock cleavage (see Fig. 5–18) is another form of fracturing. At least two different types of rock cleavage can be recognized. One type, like that in slates and some metamorphic rocks, is due to the parallel alignment of micaceous minerals. A second type, consisting of closely spaced fractures, is called *fracture cleavage*. Both *slaty* or *flow cleavage* and fracture cleavage may occur together. They are common in fine-grained rocks but less so in sandstones. Rock cleavage is a plane of weakness, but it should not be confused with mineral cleavage (Fig. 2–30), which is determined by atomic arrangements and not by deformation.

FAULTS

Most earthquakes are associated with active faults. Movements on faults have been responsible for many disasters; large dams and other foundations have failed as a result of movements on faults that passed under them or nearby. Even when faults are inactive their presence is important where a knowledge of subsurface conditions is needed: for example, in mining, in exploration for oil, gas, water, or mineral resources; in tunneling, dam, highway, and building construction; and in underground waste disposal. Regional and urban planning, the selection of construction sites, and zoning should always be done with consideration given to the location and nature of faults in the area.

Normal and Reverse Faults

High-angle normal and reverse faults are found both inside and outside the orogenic belts. Some mountain ranges, called *fault-block mountains,* owe their high elevations to uplift on such faults. The Sierra Nevadas and the Tetons are fault-block mountains, the blocks having been tilted up along faults on their east side. Many of the blocklike mountain ranges in Nevada, Utah, and Arizona are horsts separated from one another by grabens. Prominent grabenlike features form the Connecticut Valley (Fig. 23–23) and other lowlands in the Appalachians. These are much older features than those in the western United States. The typical graben is a straight-edged, gravelike depression bounded on each side by one high-angle fault or a series of steplike faults. Examination of the geometry of a graben (Figs. 23–24 and 23–25) leads to the conclusion that normal faults, grabens, and horsts are caused by extension of the crust. Such extension might be caused locally by arching of layers or by actual extension over a large area (Fig. 23–26).

Grabens also are found in sedimentary layers over salt domes along the Gulf Coast. Here the origin is clear. The salt domes have risen, causing doming of the layers over the salt. The grabens and step faults have developed in response to extension over the dome (Fig. 23–27).

Thrust Faults

Thrust faults are confined to orogenic belts, providing one of the key identifying features of these belts. What we see and can map of a thrust fault depends on how it is exposed in the topography through erosion. Thus it is not surprising that we know much of such faults in the Alps where glaciers have enlarged and modified deep valleys and in the Appalachians where many millions of years of denudation have taken place.

Some thrusts have only slight displacements and these may be nearly planar faults, but other thrust faults have displacements measured in kilometers. Major thrust faults are often zones of intense shearing and brecciation; some have breccias hundreds of meters thick. The overriding hanging-wall blocks in these cases may exhibit extremely complex folding. Many thrusts can be traced into faults that extend along a bedding plane (see Fig. 24–7). Some of the great thrust sheets of the Alps, called *nappes,* have become detached from their original sites.

Figure 23–28 shows a map of a thrust plate in the central Appalachians. The fault trace is irregular because the fault is nearly flat, so its trace is affected by the topography. Parts of the thrust

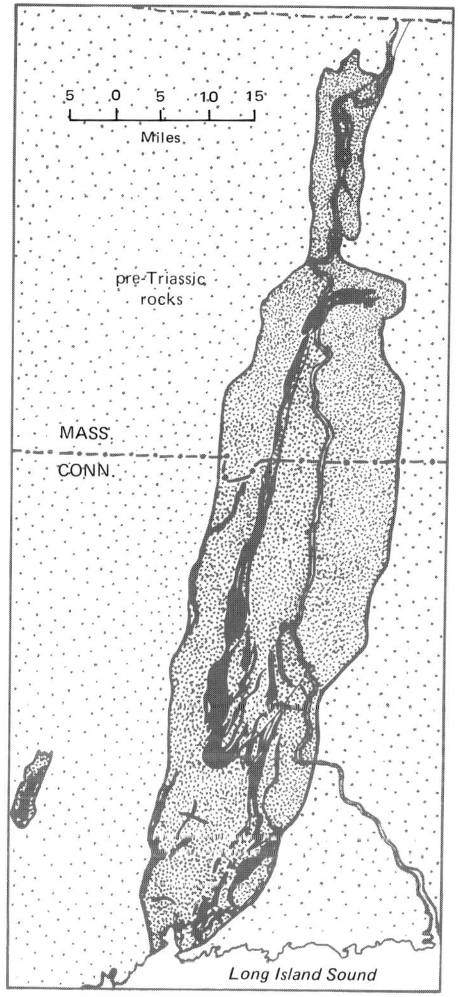

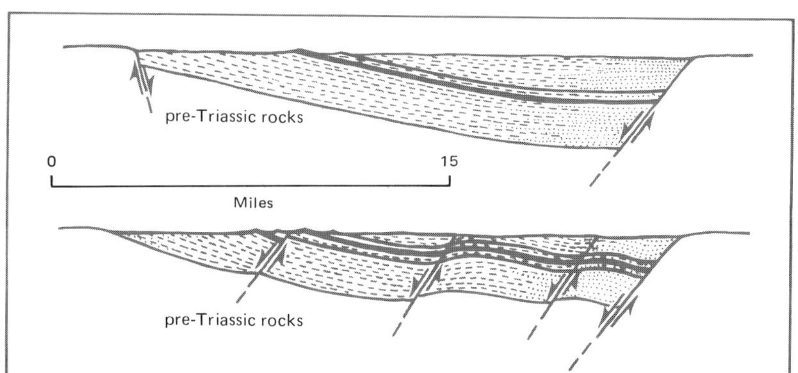

FIGURE 23–23
Map and cross sections of the Triassic deposits along the Connecticut Valley.
(after Longwell, 1933)

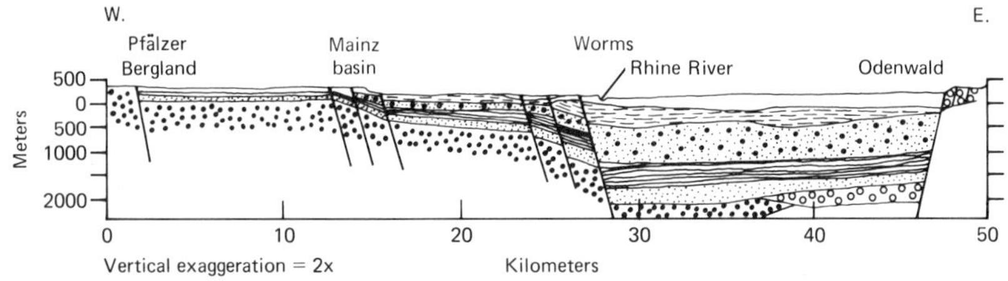

FIGURE 23–24
Cross sections across the Rhine graben. The Rhine River flows down a down-dropped troughlike depression called a graben. Normal faults form both sides of the graben.
(from Billings, 1960; after Dorn)

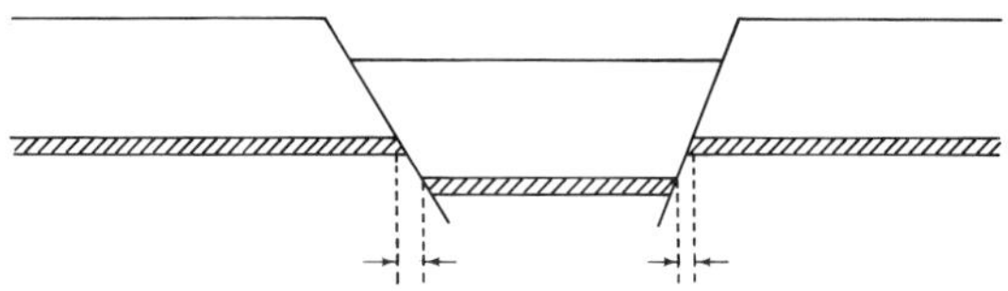

FIGURE 23–25
Schematic cross section of a graben showing amount of horizontal extension that accompanies the faulting.

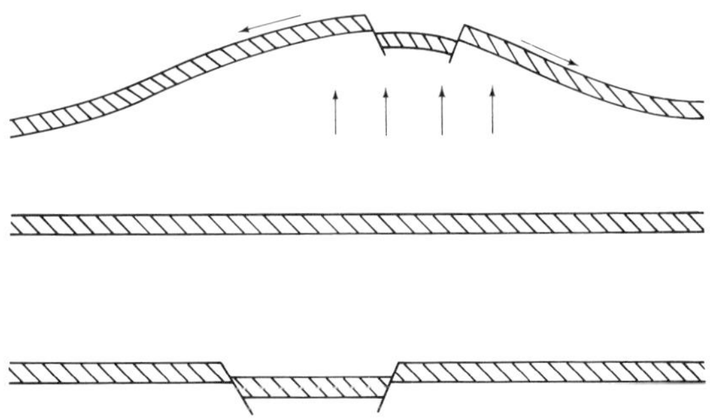

FIGURE 23–26
Schematic cross section showing an undeformed layer, center; graben formed over a dome, top; graben formed as a result of horizontal extension, bottom.

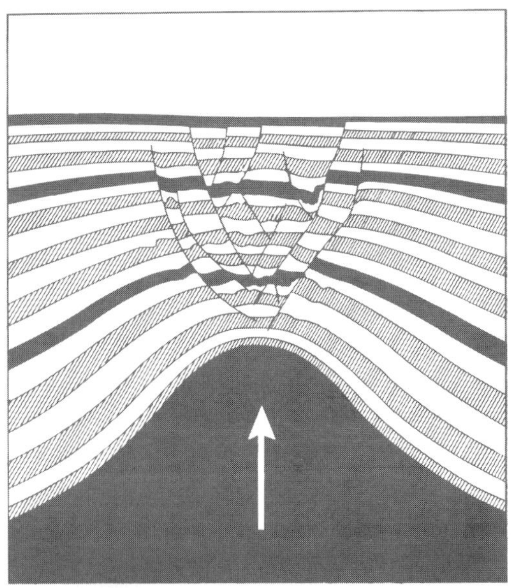

FIGURE 23–27
Graben fault pattern formed in model experiment.
(modified after Currie, 1956; redrawn from Hughes, 1960)

plate may be isolated as a result of erosion, forming klippes, and at other places the fault is breached by erosion to form a window. Both compression and gravity sliding may occur in orogenic belts.

The origin of low-angle thrusts of the sort described here is a controversial subject. The traditional view has been that they originate as a result of deep crustal compression that shortens the crust and causes the overthrusting. The alternate view is that the thrusting is caused largely by gravity sliding of sedimentary layers along bedding planes. The sliding is possible only because frictional drag is reduced. The favored mechanism for reducing friction is to build up pressure in water in the pore spaces of a porous and permeable layer. This requires an overlying impervious layer. As pressure in the water increases, the internal sliding friction is reduced.

Strike-Slip Faults

Many major faults that have essentially strike-slip displacement are now known both in the crust of continents and in ocean basins. These faults have straight traces that cut across topographic features and show little or no deviation, even where they cut across valleys. This indicates that they are nearly vertical zones. Within the zone intense crushing and brecciation occurs. These crushed zones may be hundreds or even thousands of meters wide.

Evidence of strike-slip faulting is strong in the case of such faults as the San Andreas. The San Andreas is still active, and right-lateral offsets are evident where the fault has cut through roads and orchards. Even streams are offset along the San Andreas. To document the total amount of displacement it has been necessary to use other criteria. Offset dikes provide a good measure along some related faults. Figures estimating more than 320 kilometers of movement (the total slip accumulated over nearly 70 million years) have been suggested for the San Andreas fault. These figures are based on the offset of major plutons on either side of the fault and on offsets of shore-line sedimentary beds that were formed millions of years ago. Most geologists have been convinced by the evidence that strike-slip of this magnitude has occurred. Such a fault must be very deep, probably extending through the continental crust.

Relation of Fault Orientations to Stress Applications

Faults are shear zones across which there has been displacement. In this sense all faults are closely related to shear fractures in mode of origin. If they are related to shear failure, then we must explain why the orientations differ so greatly. The answer appears to lie in the application of nonuniform stress. In the absence of directed stress a material buried in the earth is subject to all-sided compression, called *lithostatic pressure*. If a directed stress is applied (e.g., an east-west compression), the pressure in different directions is nonuniform. The vertical pressure is the weight of the overburden; the maximum stress becomes the east-west compression. The north-south pressure may be equal to the vertical pressure or it may be some other value. We have seen that shear fractures tend to form at an angle of about 30° to the direction of maximum compression. Similar experiments show that the line of intersection of the two shear fractures or faults lies in the direction of the intermediate stress. Thus fault orientations bear the relationship shown in Fig. 23–29 to the directions of maximum, intermediate, and minimum stress.

FIGURE 23–28
Block diagram showing thrust faults. Note the trace of the thrusts on the topography. Diagram depicts area on the Little River near the sinks in Tennessee.
(after King, 1964)

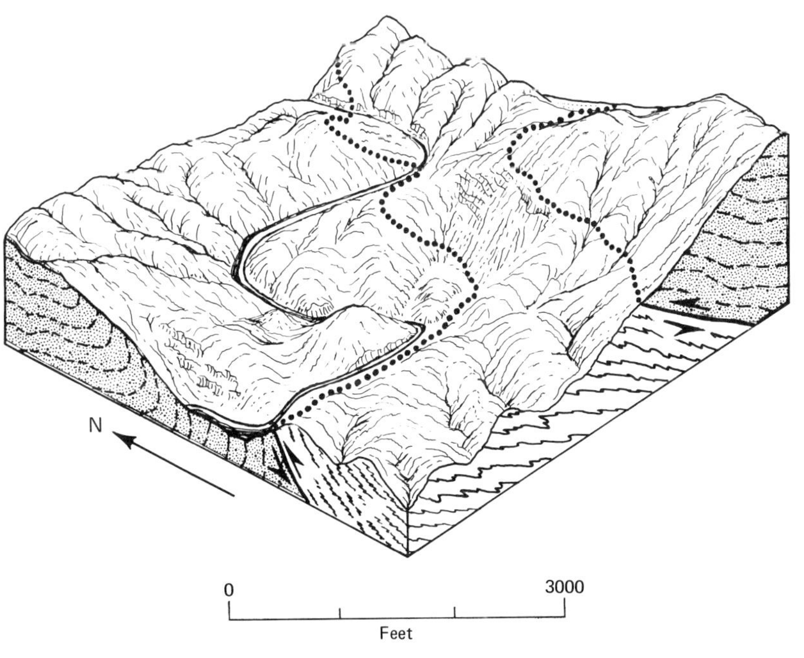

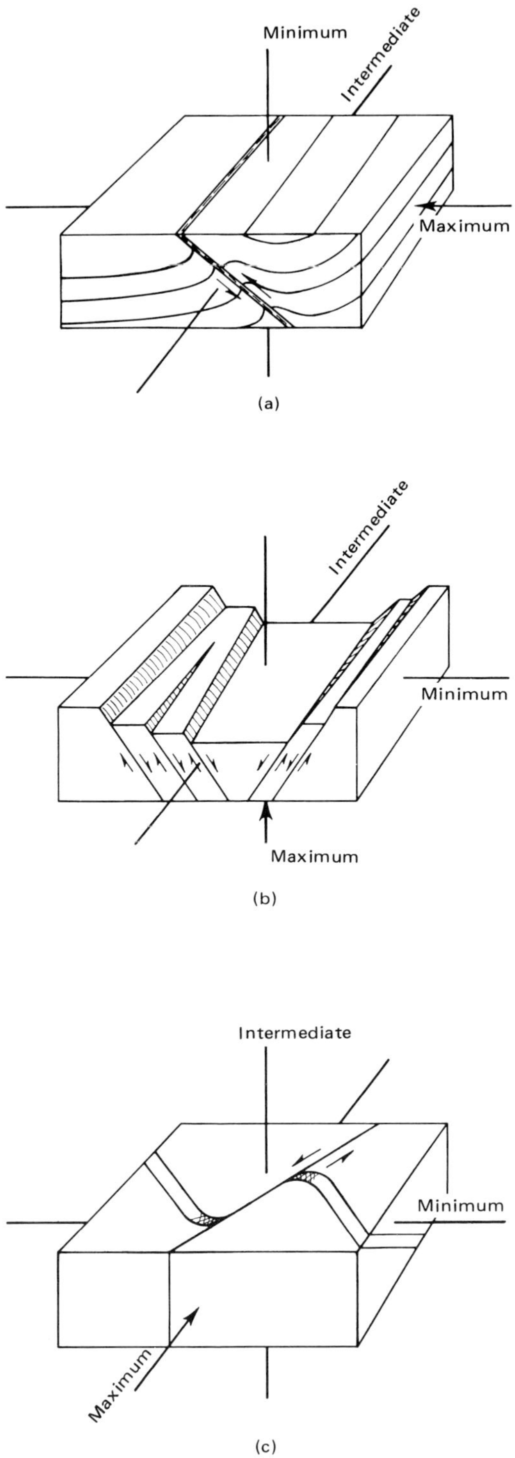

FIGURE 23–29
Relationships between principal stresses and common fault orientations for (a) thrust, (b) normal, and (c) strike-slip faults.
(after Anderson, 1951)

FOLDS

Flexures and folds, usually most easily seen in stratified rocks, occur widely and under very different circumstances. For example, they may form (1) during the deposition of sedimentary layers; (2) while the sediment is still soft; (3) as a result of vertical uplift; (4) with faults; (5) as a result of lateral compression; and (6) as a gravity-glide phenomenon (Fig. 23–30).

If the surface on which sediment is being deposited is irregular, that irregularity will impart an initial irregularity to the sedimentary layers. The submergence of a hilly topography would create such an irregular surface, as would a reef or lava flow on the sea floor. The bending of the layers is accentuated as the sediment becomes compacted. Folds also form as a result of the differences in the amount of compaction possible in various types of sediment. Sand can be compacted very little, while the reduction in volume of clay is great. Thus a sand lens in a clay layer can cause the layering in the clay to become folded as compaction occurs.

Soft sediment is highly plastic when the clay content is great, and pore spaces in sand layers are usually filled with water. Sediments can slide and slump where deposited on slopes of more than a few degrees. Beds may become detached, roll up, and become complexly folded.

Vertical uplift results in several kinds of folds. Strata are found draped over the edge of some faults where the fault did not continue up into the layer and the sediment yielded by stretching rather than failure. Nearly circular domes are formed in sediments over salt domes, which are abundant in the Gulf Coast region. These domes are typically 1.6 kilometers in diameter and underlain by salt plugs that have risen from depth, causing bending of the overlying layers. Other domes are formed by intrusion of magma and where basement rock is raised or lowered by other processes.

Long systems of plunging folds (Fig. 23–5) are characteristic of parts of most major mountain belts. These folds are often associated with thrust faults (Fig. 24–9) in the central portion of the mountain belt, but they occur beyond the thrust-faulted areas as well. Their origin has been strongly debated. Shortening of a layer is usually necessary to produce a fold (Fig. 23–31). Thus fold belts have been interpreted to mean shortening in those areas of the earth's crust; but in recent years an alternative explanation has been advanced, citing vertical uplift accompanied by sliding of the layers due to gravitational stresses as the source (Fig. 23–32).

The Little Dome anticline, Wyoming, (Fig. 23–33) is a beautiful example of a doubly plunging anticline. It is located in the Bighorn Basin. Notice how erosion has lowered the crust, exposing lower and older strata in the core of the fold. This is one of many folds located around the margin of a large intermontane basin. The floor of the basin has been lowered, and this fold probably developed largely as a result of basinward movement of the sedimentary veneer.

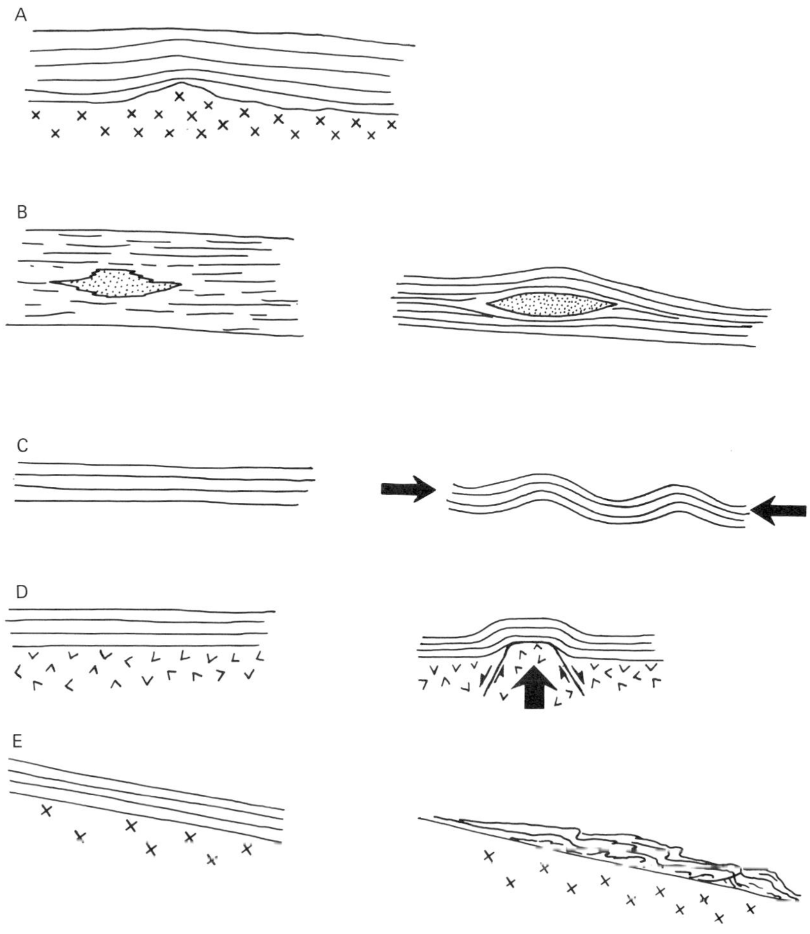

FIGURE 23–30
Folds formed as a result of different geologic processes: Compaction over buried topography, A; compaction fold in shale as a result of differential compaction, B; folding due to lateral compression, C; folding due to bending of layers over uplifted block, D; gravity gliding, E.

Behavior of Rock

Rocks like granite, gneiss, sandstone, quartzite, and limestone appear to be hard, durable materials capable of sustaining great loads. They can be broken when high stresses are applied rapidly, and they usually break in the manner of brittle materials. The development of fractures in slightly deformed rocks is compatible with these observations. Fractures result from failure of rock when stresses exceed rock strength. Yet, beds of all types of sedimentary and metamorphic rocks can be found folded in strongly deformed orogenic belts, and often the folded rocks show no evidence of fracturing—they did not behave as a brittle material. Limestones and many metamorphic rocks are found folded into patterns (Fig. 23–34) that resemble flowage in a viscous liquid. Thus rocks that are brittle under some conditions are ductile under others, and it is not difficult to postulate some of the conditions that influence their behavior.

Rock type. The behavior of a rock must reflect in large degree the behavior of the component minerals, and the physical properties of minerals vary greatly. For example, minerals vary in strength so that some deform under much smaller applied stresses than others.

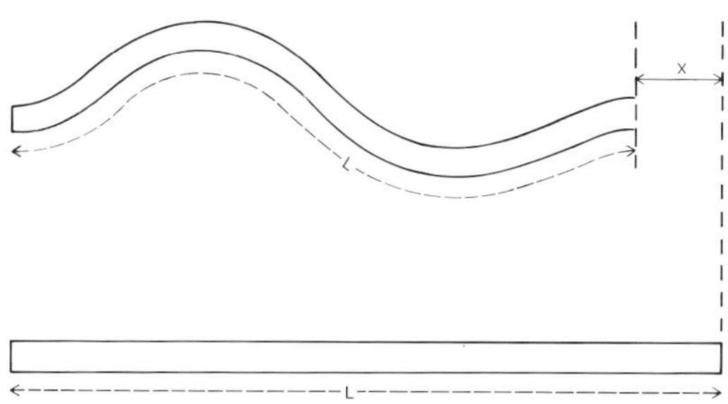

FIGURE 23–31
Schematic cross sections showing unfolded layer, below, and a fold and local shortening, above; L is the original length of the layer and X is the apparent shortening.

Confining pressure. Lithostatic pressure increases with depth, and it reaches high magnitudes

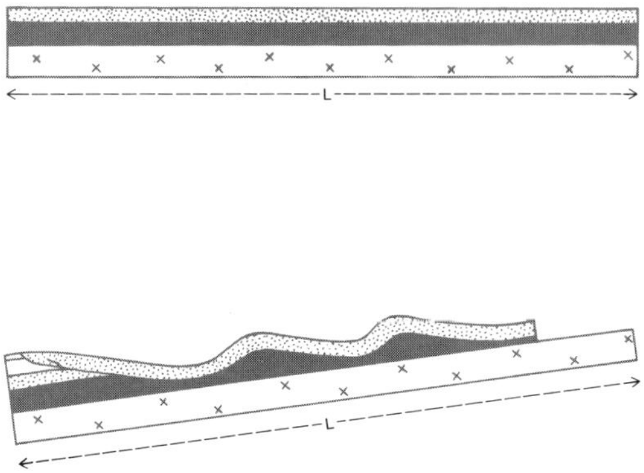

FIGURE 23–32
Schematic cross section showing how apparent shortening, thrust faulting, and folding can occur as a result of sliding of layers on an inclined rigid basement.

FIGURE 23–33
Vertical air photograph of Little Dome, Wyo. This structure is best described as a doubly plunging anticline. The dips dipping away from the axis are Cretaceous shales and sandstones.
(photograph by the U.S. Geological Survey)

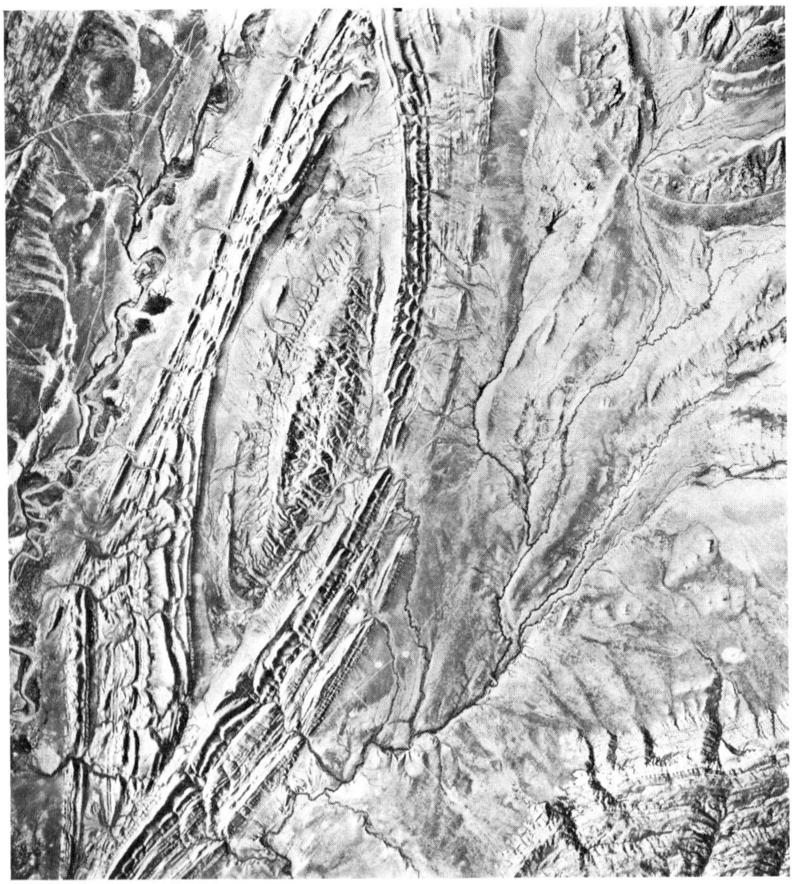

within the crust. Experimental deformation studies show that rocks are more ductile under great pressure.

Temperature. Temperature increases with depth and this, too, affects rock behavior. When temperatures reach high levels, metamorphic effects result and ultimately magma may be produced; but below these temperature levels mobility of material may be predicted. Evidence of ductile behavior of rocks is more frequently found among metamorphic rocks than other types, but such behavior is not confined to metamorphism.

Solutions. Solutions in pore spaces promote movement of materials. Solutions accelerate and facilitate movement of ions and lubricate mechanical movements such as grain rotation and slip on fractures.

Time. The effect of time in rock deformation is the hardest factor to evaluate in orogeny. Marble slabs supported only at the ends are found at some of the older universities and in old graveyards. It is not uncommon to find that the center of a slab that has been standing for hundreds of years has sagged. The amount of sag is sometimes pronounced and it is easily measured. When these marble slabs were originally cut, they were smooth and flat, so the sags must have formed in the time since the slabs were mounted. Here, then, is evidence that solid rock can be deformed under a confining pressure of 1 atmosphere with nothing more than the force of gravity acting as a stress, if the time involved is great enough. This suggests that the contortions in the earth's crust may result from small directed stresses applied over very long periods of time.

Deformation is time-dependent even among some materials with which we are familiar, such as asphalt, stitching wax, and "silly putty." These materials are brittle under sudden, impulsive deformation, but they flow like viscous liquids under low stresses if given sufficient time. The history of orogenies suggests that most rock deformation takes place over long periods of time. Millions of years may be involved in the formation of some of the structural features now exposed in mountain belts.

Behavior of Rocks: Experimental Evidence

Most experiments in the investigation of rock behavior are conducted on small specimens, and the duration of the experiments is generally short. Equipment has been designed that allows specimens to be subjected to high confining pressures and temperatures at the same time that a directed stress is applied. Typically, the specimen is about 2.5 centimeters in diameter and several centimeters long. It is inserted in a heavy steel cylinder in which a hole slightly larger than the specimen has been bored. Confining pressure is built up within an oil that surrounds the specimen, and the specimen is then deformed by means of directed stresses applied through a piston arrangement. Both compression and extension experiments can be performed.

The usual experimental procedure is to control all variables, holding all but one constant. Then that one is varied. The following are examples of common procedures:

1. Hold confining pressure and temperature constant and observe the way shortening (for compression test) varies as the amount of the applied directed stress is increased (Fig. 23–35).
2. Hold confining pressure and temperature constant and observe the amount of directed stress needed to produce a given rate of shortening (for compression test) or lengthening (for extension test).
3. Hold confining pressure and temperature constant and observe the way the amount of strain varies with time for a given constant applied surface stress.

After a series of tests are run, it becomes possible to draw curves that show the way strain varies with applied stress or the way the amount of strain varies with time. Then one of the variables that was held constant can be varied and another series of experiments run. Thus the effects of each variable can be evaluated.

As an example of the application of these methods, consider the experimental results obtained by the deformation of the Solenhofen limestone (Fig. 23–35); two series of tests designed to evaluate the effects of confining pressure were used. During the first series of tests the confining pressure was held at 5,000 atmospheres and temperature was held constant while the amount of directed stress was varied and the amount (measured in percentage) of strain was measured. Confining pressure was then raised to 7,500 atmospheres in the second series of experiments (run on another specimen of the same rock).

FIGURE 23–34
Folds in Precambrian gneisses. These complex patterns of folding resulted from deformation of the rocks under high temperatures and high confining pressures. The patterns suggest viscous deformation.

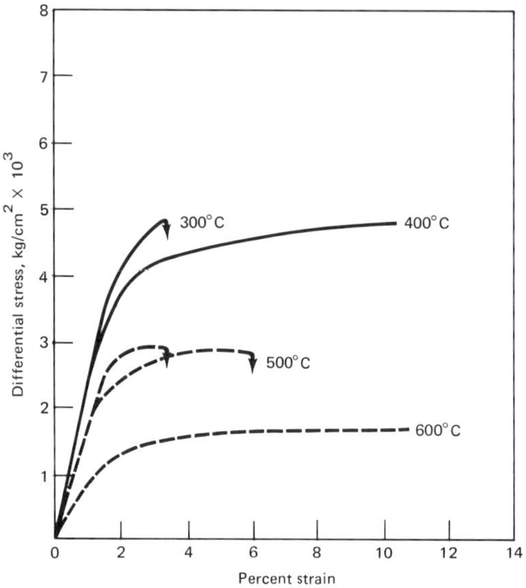

FIGURE 23–35 Extension stress-strain curves for Solenhofen limestone at various temperatures and confining pressures. Solid lines are for test run at 5,000 atmospheres confining pressure; dashed lines are for test run at 3,000 atmospheres confining pressure. (from Heard, 1960)

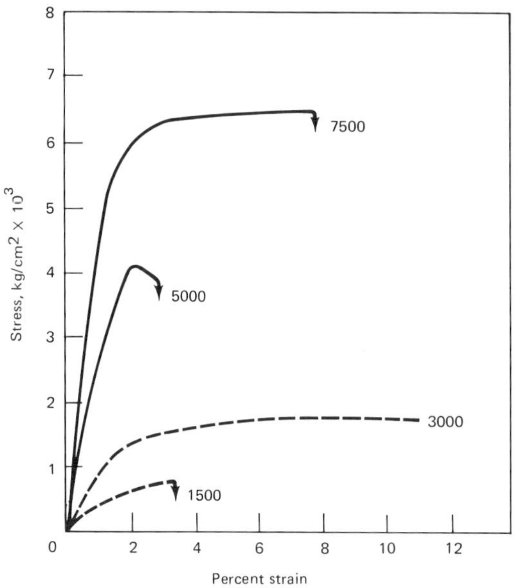

FIGURE 23–36 Extension stress-strain curves for Solenhofen limestone at various confining pressures. Solid lines are for test run at 25°C; dashed lines are for test run at 600°C. (from Heard, 1960)

Notice that a linear relationship exists between stress and strain for both series of tests up to a certain level (about 3 kg/cm² at 7,500 atmospheres). After that linear relationship ceases to exist, the specimen begins to undergo large amounts of strain for small increments of stress. This condition continues until the specimen fails. The higher confining pressure caused the limestone to sustain much greater strain before failure, and a higher stress was needed to cause a given amount of strain at higher confining pressures. The stress level that corresponds to the highest point on the curve is called the *ultimate strength* of the rock under the given conditions. Strength of the limestone was increased by increasing confining pressure.

The effects of temperature are evaluated for the same limestone, as shown in Fig. 23–36. Much greater amounts of strain are possible at higher temperatures, and less stress is needed to produce a large strain after some initial level of stress has been applied (about 4 kg/cm²). Curves of this general character have been determined for a wide range of rock types.

The general strain–time relationships and effects of temperature and confining pressure (Fig. 23–37) have been defined on the basis of long-term experiments. Conclusions warranted from this experimental evidence are the following:

1. When temperatures and confining pressures are low, the amount of strain for a given applied stress reaches a certain level and does not increase with time.
2. The amount of strain increases with time as the temperature, confining pressure, or amount of directed stress is increased; and when the temperature or stress is very high, the strain rate becomes high and remains high until the specimen fails.

Continuous strain under constant, applied stress is important in causing effects of rock flowage so frequently found in metamorphic rocks. Experimental studies point to a transition in the behavior of rocks from that of brittle material at low temperatures and low confining pressures to that of ductile material at high temperatures and pressures. This is evident in the way the material deforms, that is, whether it fractures or deforms continuously without fracture.

Rock Behavior: Theoretical Models

A number of simplified models have been used for purposes of describing the differences in behavior of different types of materials. These include three basic models: (1) elastic, (2) viscous, and (3) plastic behavior, as well as combinations of these. Each of the three basic models may be defined in terms of the way the material strains when it is subjected to directed stress, and the way strain varies with time.

Elastic model. The ideal elastic material is one in which the amount of strain is directly proportional to the amount of stress applied. The graph of the relationship is a straight line up to some particular stress, called the *elastic limit.* Once the elastic limit has been passed, the material no longer behaves as a purely elastic substance. A spring is an excellent mechanical model of an elastic material. Strain remains constant for a given stress level over any amount of time, provided the stress level is within the elastic limit of the material.

Viscous model. The ideal viscous material is one in which strain takes place continuously for any amount of applied stress. No increase in stress is needed to cause strain to increase, and strain will continually increase with time for any applied stress.

Plastic model. The ideal plastic material is one in which behavior is initially elastic. As the amount of stress applied is increased, the amount of strain increases proportionally. This continues up to a certain stress level, called the *yield point,* beyond which the material strains continuously with no additional stress. The behavior may be likened to that of a mass of clay. When weight is added to the clay, strain occurs, after which continued deformation stops until a certain amount of weight is added. Once the yield point is reached, the clay starts to flow and deforms continuously.

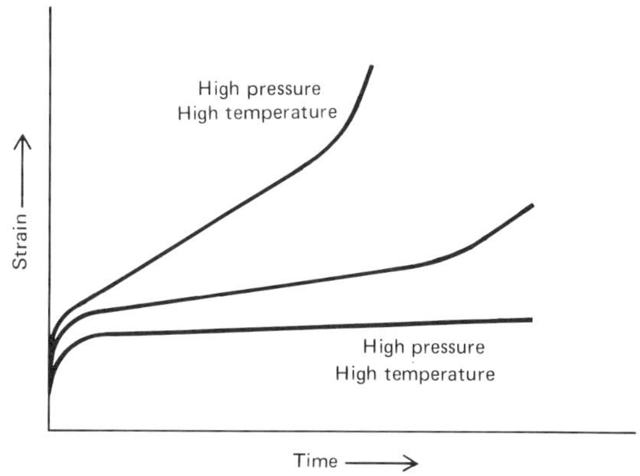

FIGURE 23–37 Generalized strain-time curve. See text for description.
(after Nadai, 1950)

It is instructive to compare the theoretical models with the experimentally determined stress–strain and strain–time curves obtained for the limestone specimens (Fig. 23–35). When this is done it is apparent that the limestone behaved initially as an elastic material, but as the stresses were raised its behavior changed and became more like that of the plastic model. The behavior of rocks is best described as a combination of the ideal models. Rocks are often termed *elastico-viscous* materials. Such substances exhibit the characteristics of elastic solids under some conditions and those of a fluid under others. One of the most commonly known elastico-viscous materials is "silly putty." Its behavior is very much like that of the crust of the earth except for the time involved. Rocks and "silly putty" have these things in common: (1) If subjected to a sudden, sharp impact, both will rupture; (2) if subjected to rapid, small deformations, both are elastic; (3) if subjected to slow, continuous stress, both will flow.

CHAPTER OUTLINE

Structural Features
- Scale of Structural Features

Description and Classification of Structural Features

Fold Nomenclature
- Parts of a Fold
- Fold Symmetry
- Types of Folds

Fault Nomenclature
- General Terms
- Classification of Faults by Relative Movement of Blocks
- Classification of Faults by Displacement

Classification Based on Separation
Special Names Applied to Fault Structures
Occurrence and Origin of Rock Fractures
Fractures Caused by Volume Reduction
Fractures Caused by Diastrophism
Shear Fractures
Rock Cleavage
Faults
Normal and Reverse Faults
Thrust Faults
Strike-Slip Faults
Relation of Fault Orientation to Stress Applications
Folds
Behavior of Rock
Rock type
Confining pressure
Temperature
Solutions
Time
Behavior of Rocks: Experimental Evidence
Rock Behavior: Theoretical Models
Elastic model
Viscous model
Plastic model

TERMS

anticline
asymmetrical fold
axial surface
axis
basin
breccia
concentric folds
crest
dip
dip separation
dip-slip fault
disharmonic folds
displacement
dome
elastic limit
elastico-viscous
en echelon
extension
fault blocks
fault plane
fault scarps
faults
flexural folds
flow fold
folds
footwall
fracture cleavage
fractures
gouge
graben
hanging wall
homocline
horst
isopach maps
klippe
limbs
lithostatic pressure
monocline
mylonite
nappe
net slip
normal faults
oblique-slip fault
overturned fold
passive folds
plunge
plunging axis
primary structural features
recumbent fold
reverse faults
rock cleavage
rotational fault
secondary structural features
separation
shear fold (also slip fold)
shear fractures
slaty cleavage
similar folds
slickensides
stress
strike
strike separation
strike-slip fault
structure
structural terrace
structure contour maps
symmetrical fold
syncline
tectonic features
thrust fault
trough
ultimate strength
vertical separation
yield point

REVIEW QUESTIONS

1. On the basis of your earlier study of mineral structure would you expect a rock composed largely of quartz to deform in the same manner as one composed of calcite or mica? Explain.
2. If fault orientations are related to the principal stress directions, as suggested in the text, what types of faults would you expect to develop at shallow depths as a result of (a) extension of a region along N–S lines and (b) vertical uplift?
3. What type of information can we gain from experi-

mental deformation of small cores that can be used to help understand the origin of tectonic features?

4. What factors influence the transition from brittle to ductile behavior in most natural materials?

5. Would you expect concentric folding to occur more readily under conditions of flexural folding or passive folding? Why?

6. Why is separation a more practical means of measuring fault movements than slip?

7. A klippe and a horst may appear to be similar topographic features. How would you distinguish between the two structurally?

8. In what structural features other than domes would you expect to find tension cracks? In what part of the feature in particular?

9. Rock cleavage and mineral cleavage may appear to be similar physical features but in fact are two different things. Explain the difference.

10. Compare the studies shown in Figs. 23–35 and 23–36. Summarize the effects of temperature and confining pressure on rock deformation in extension.

SUGGESTED READINGS

Hills, E. S., 1953, Outlines of Structural Geology: New York, John Wiley, 182 p.

Ragan, D. M., 1969, Introduction to concepts of two-dimensional strain and their application with the use of card-deck models: Jour. Geol. Educ., v. xvii, no. 4, p. 135–141.

Spencer, E. W., 1969, Introduction to the Structure of the Earth: New York, McGraw-Hill, 597 p.

Wilson, J. T., 1965, A new class of faults and their bearing upon continental drift: Nature, v. 207, p. 343–347.

24 CONTINENTAL SHIELDS, MOUNTAIN SYSTEMS, AND CONTINENTAL MARGINS

24 CONTINENTAL SHIELDS, MOUNTAIN SYSTEMS, AND CONTINENTAL MARGINS

CONTINENTAL SHIELDS

Large exposures of Precambrian rocks occur within the boundaries of each continent (Fig. 24–1). The term *shield* is applied to Precambrian areas exposed at the surface or shallowly buried that have not been folded, thrust faulted, or complexly deformed since the end of the Precambrian. Their stability is exhibited by scarcity of igneous, volcanic, diastrophic, and seismic activity; however, they are not totally void of activity. Subsidence has created regional basins in which sediments have accumulated. Shallow seismic activity is occasionally recorded, and high-angle faults that have been active since Precambrian time are sometimes found. Although the shields have been relatively stable for over half a billion years, the areas they encompass were not stable in Precambrian time. Structural evidence of old mountain systems (Fig. 24–2) is found in every shield. Rocks were strongly deformed and metamorphosed, and large-scale plutonic activity occurred in each shield at one or more times during the 4 billion years of the Precambrian. Precambrian rocks must lie under the rest of the continents, but these have yielded to deformation, or at least to subsidence, to a greater degree than the shields.

Most Precambrian rocks and all early Precambrian rocks have been metamorphosed. Large portions are granitic in composition, and, although some of these are now known to be granitized sedimentary rocks, others are of igneous origin. All of the Precambrian orogenic belts have been reduced by erosion to surfaces of low relief. The terrain is that of gently rolling hills and isolated geomorphic features etched out by many years of erosion. Often shield margins are stable portions of the continent that have subsided since the Precambrian, allowing sediment deposition. These sediments thicken away from the shield toward young folded mountain belts as in the Appalachians and the Rocky Mountains.

A shieldlike area appears to continue southward into the United States as far as the Ouachita Mountains in Arkansas. This large central area of the country contains some regions where shallow sediment covers Precambrian rocks; in other areas, like the Michigan and Illinois basins, many thousands of meters of Paleozoic sediment lie over the Precambrian rocks.

Some shields extend to the edge of the continental margin, as in South America, Australia, Africa, and India. The absence of orogenic belts around such shields supports the theory of continental drift and argues against the once-popular theory that continents grow by the welding on of marginal orogenic belts.

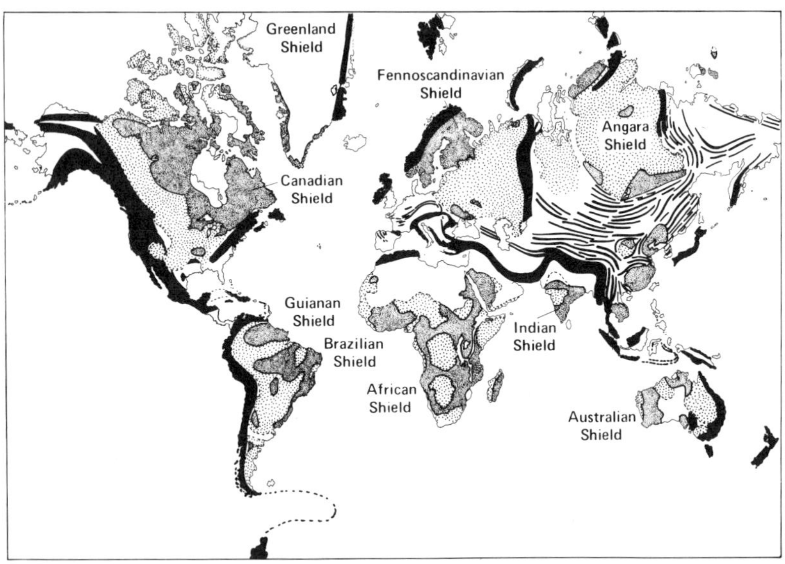

FIGURE 24–1 Generalized world map showing shields (fine stipling); areas where Precambrian basement is covered by little deformed strata or extrusives (dot pattern); and post-Precambrian orogenic belts (solid black and heavy black lines). (simplified after Weeks, 1952)

YOUNG FOLDED MOUNTAIN BELTS

One or more mountainous belts in which the sedimentary units are folded and thrust faulted are found on every continent. These belts have exhibited mobility since the Precambrian. The term "young" is used in a strictly relative sense. Only geologists or astronomers would ordinarily think of anything half a billion years old as being young. The term "elongate" is hardly sufficient to describe some of these belts, which are several thousand kilometers long and 160 to 320 kilometers wide.

Stresses in the crust sufficient to lift masses of rock of this size must be included in any acceptable theory of orogeny. Consider the weight of a single section across the Himalayas 1.6 kilometers wide, 320 kilometers long, and averaging 4,900 meters in elevation. The amount of rock in this section, which has been uplifted 4,900 meters

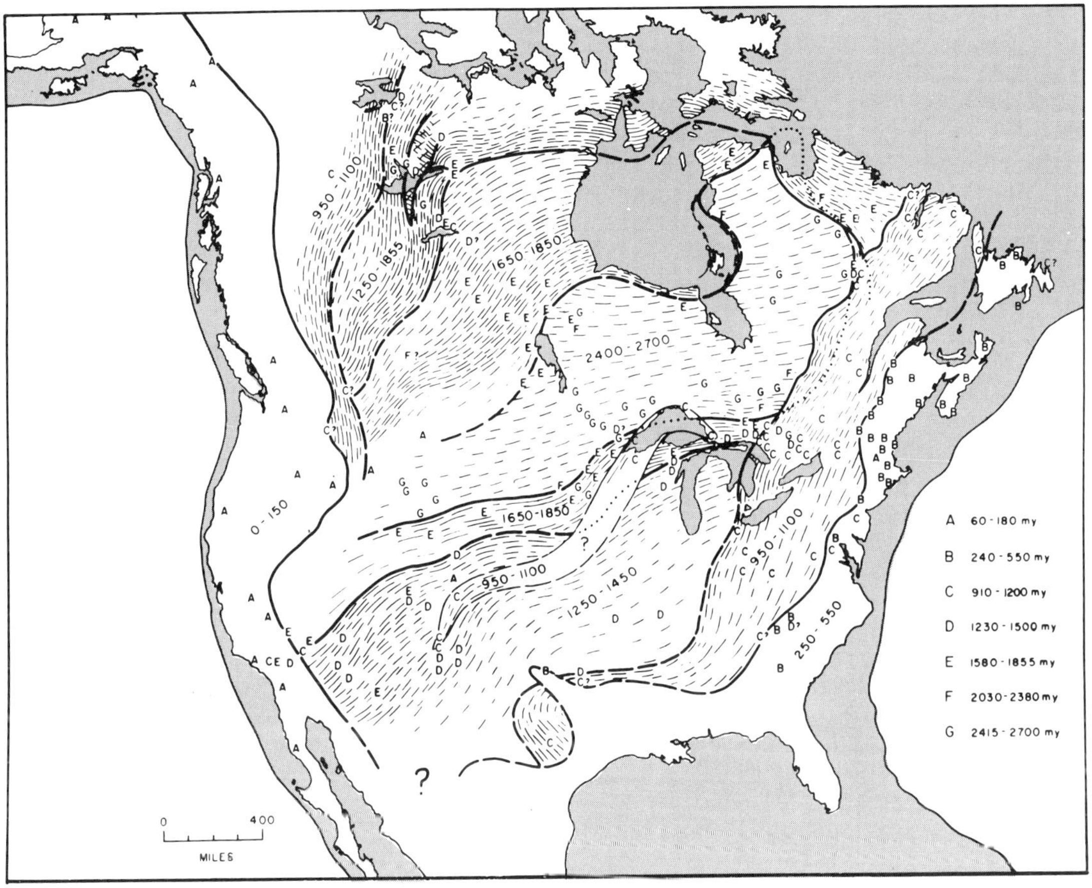

FIGURE 24–2
Precambrian orogenic belts of North America. Age ranges for each belt, obtained by radioactive dating methods, are indicated in millions of years. Many of the dates outside central Canada are from areas where most Precambrian rocks are largely covered by younger sediments.
(courtesy of Dr. Gordon Gastil)

above sea level, is approximately 2,500 cubic kilometers. The term "mountain belt" is not restricted to mountains that are topographically high today. Mountains may be eroded down to a relatively flat surface within a period of 20 to 40 million years. Thus mountains high at some point within the past half billion years may not be high today. Once the crust of the earth is disturbed and deformed into a folded mountain system, the structures brought into existence at that time leave a relatively permanent imprint in the crust, and although the top part of the mountains are eroded off, the lower and marginal structures may be preserved indefinitely.

Some mountain belts have been deformed several times since the Precambrian; others have experienced only one major deformation. The most recently deformed mountains are those that stand as high mountains today. Older mountains are usually lower topographically and the oldest ones may not be visible at all. However, old mountain systems have often been partly rejuvenated during later orogenies. For example, the Rocky Mountains are located in part where early Precambrian mountains once stood.

Location of Some Young Folded Mountain Belts

The Alps, Himalayas, Rocky Mountains, Appalachians, Urals, and Caledonides are among the best known young orogenic belts.

The highest and most complex system of mountains today is that of which the Alps and Himalayas are a part (see Fig. 24–1). It extends from Indochina northward through China across Tibet

FIGURE
(*page 539*)
A section of the Sierra Nevada.
(U.S. Geological Survey)

and the northern part of India; the Kunlun Mountains of Russia are a part; westward the system is made up of the Caucasus Mountains between Turkey and Russia, the mountains of Greece, the Alpine system, and extensions from it into North Africa, the Atlas Mountains, and into Spain, the Pyrenees. This system was strongly deformed during the Cenozoic era.

The Rocky Mountain system extends northward through western Canada and into Alaska, and southward through Mexico, continuing southward as the Andes along the western margin of South America into the Palmer Peninsula of Antarctica. Deformation has taken place in this system during the Mesozoic and Cenozoic eras and continues today.

The Appalachian folded belt of eastern North America is covered by younger rocks at its southern end in Alabama, but it extends northward into and along the eastern border of Canada in Nova Scotia and Newfoundland. The Ouachita and Marathon mountains are possibly a southwestern continuation of the Appalachians. The Appalachians were deformed several times during the Paleozoic. The last orogeny took place near the end of the Paleozoic era. The Urals of Russia and the Caledonian system that extends through Scotland and northward along the coast of Norway are also of Paleozoic age.

Pattern of Mountain Belts

Mountain chains are relatively continuous structures, although their paths are by no means straight and structures may differ within the same chain. They are usually curved and some almost double back on themselves. The overall pattern of some mountain belts may be drawn as a connected series of arcs. The North American Cordillera is a system that starts in Alaska. The ranges first trend northeast, then curve southward and extend through western Canada, the United States, and Mexico. Connections through Mexico are unclear, but an orogenic belt extends through Cuba and follows the island arc system of the Caribbean, coming back to land in Venezuela, thence across northern Colombia along the west coast of South America in Peru and Chile. At Cape Horn the belt bends eastward again and runs through the South Sandwich Islands and then into the South Shetland Islands and the Palmer Peninsula of the Antarctic before disappearing under the ice cap. This sinuous belt extends over roughly half the circumference of the earth.

Geosynclinal Theory

The idea that folded mountain belts are formed in long, relatively narrow sections of the crust where subsidence and long-term sedimentation produce great thicknesses of sediment was first proposed by James Hall, an early American geologist. Hall studied the central part of the Appalachian Mountains in New York, Pennsylvania, and Virginia where the thickness of the Paleozoic sequences is on the order of 9,000–12,000 meters. Such long narrow belts of subsidence, sedimentation, and folded strata are called *geosynclines,* and these characteristics have been found in many other folded mountain belts. Hall noticed that the present Appalachian Mountains are located in a belt that had been the site of thick sedimentary accumulations in the Paleozoic and that the sediment had accumulated in shallow water. The water could remain shallow only if the geosyncline subsided as sediment accumulated. Thus the geosyncline was not a deep trough that became filled gradually; it was a belt in which subsidence and sedimentation took place together. This led Hall to think that the weight of the sediment caused the subsidence. But the crust is strong enough to support thick sedimentary piles; thus the cause of the subsidence lies primarily in some process that forces the crust down or causes it to subside.

Hall also suggested that the folds in the folded mountains are caused by the sliding and slumping of the sediment toward the center of the trough during sedimentation. He did not explain why mountains are high, and for this reason his theory has been called a theory of mountain building with the mountains left out. Later work has shown that soft-sediment deformation is frequently present in geosynclines but the major folds and faults are not thought to be of this origin because folding in most areas came many millions of years after the sediments were deposited.

The thickness of accumulation in a geosyncline before an orogeny often reaches 15,250 or more meters, as compared with the 600 to 900 meters of sediment that might have accumulated outside the geosyncline during the same period of time. Geosynclines subsided during the time before the orogeny because shallow-water sediments usually occur all through the sequence. Geosynclinal sediment accumulations lie off the eastern and Gulf coasts of the United States under the continental shelves and continental slopes.

A famous German geologist, Hans Stille, suggested that many geosynclines consist of two subparallel troughs that can be distinguished by the types of sediments they contain. The inner trough, closer to the continent, he called the *miogeosyncline.* The miogeosyncline typically contains thick sequences of sandstone, shale, and limestone. The outer belt, called an *eugeosyncline,* contains the same sedimentary types as the miogeosyncline, but in addition it usually has thick deposits of graywacke (an arenaceous sediment with fragments of other rocks, often of igneous or volcanic origin). Eugeosynclines also contain considerable quantities of volcanic debris and lava flows. Many of these flows are pillow-shaped masses of lava, thought to form when lava is extruded underwater. In addition to the shallow-water sediments, some deep-sea sediments are also found. These include such sediments as the radiolarian oozes, which are siliceous in composition and form now only in very deep water.

Marshall Kay (1951) applied this concept of geosynclines to North America and concluded that the origin of the sedimentary rocks in both

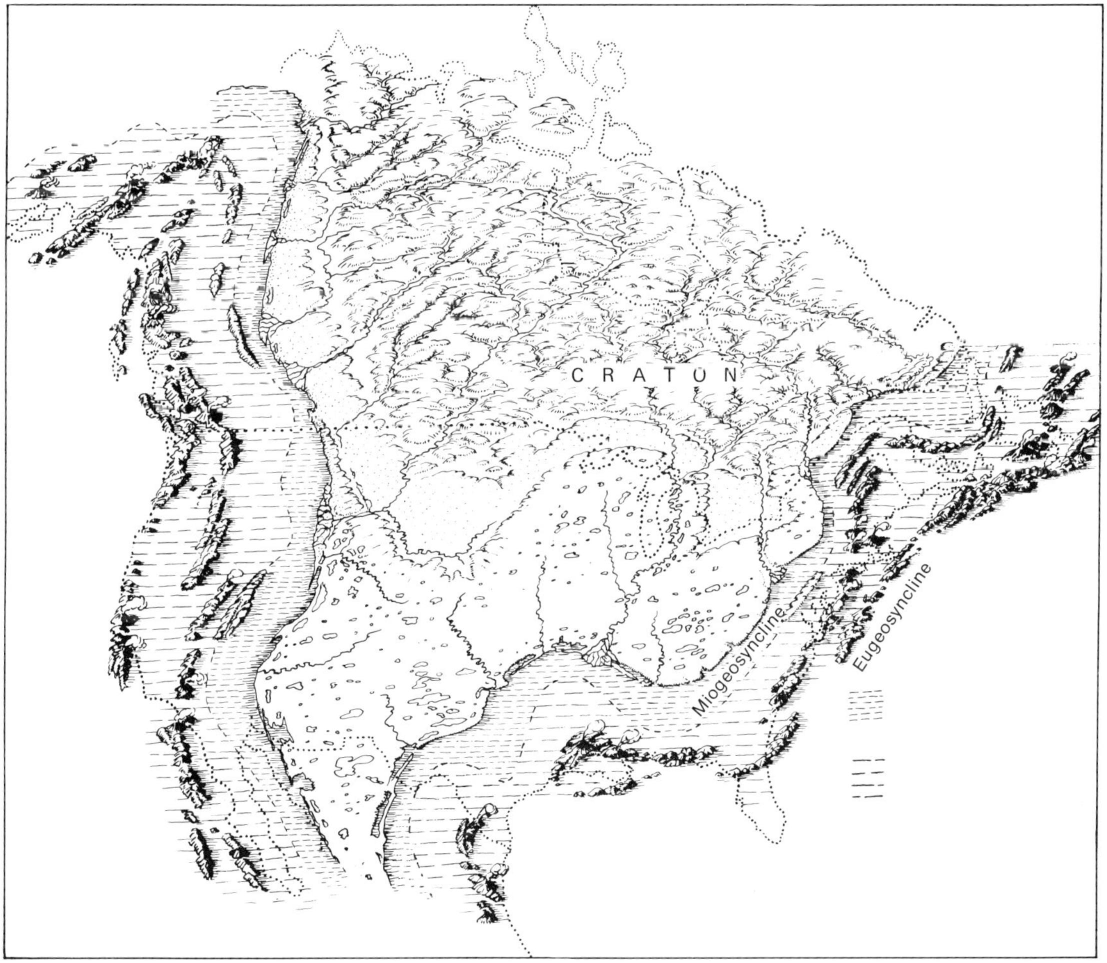

FIGURE 24–3
North America reconstructed as it might have appeared in the Ordovician period, according to the theory that the *craton,* the stable continental interior, was separated from the deep-ocean basins by miogeosynclinal belts in which sandstones and limestones accumulated, and the eugeosynclinal belt in which great thicknesses of graywacke, sandstones, lavas, limestones, and shales accumulated.
(after Marshall Kay and E. Raisz, 1951)

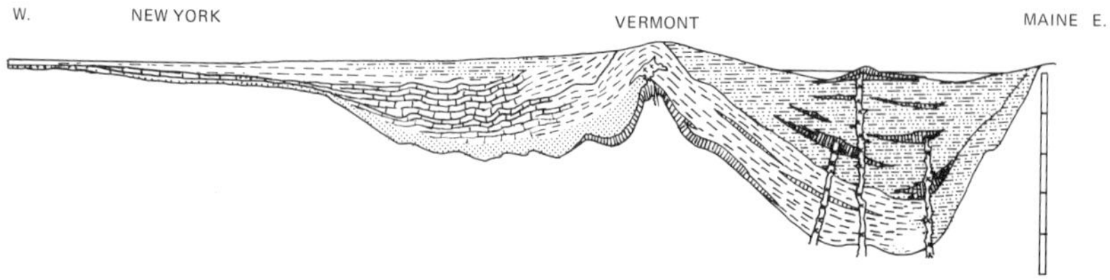

FIGURE 24–4
New York to Maine. Section across the eugeosynclinal belts from New York to Maine in the middle of the Ordovician period.
(after Marshall Kay, 1951)

FIGURE 24–5
Cross section across the Catskill Mountains. Notice how the units grade into one another. It was once thought that each unit was of a different age from the others. More detailed study has shown that they were laid down almost simultaneously.
(after P. B. King, 1950)

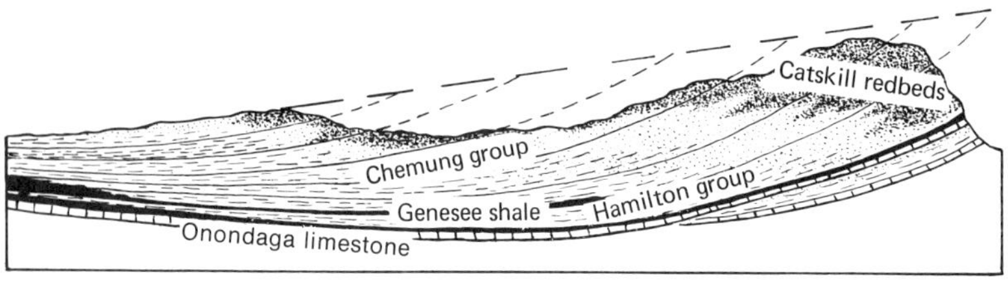

the Appalachian-Ouachita geosyncline and the Cordilleran geosyncline of the western United States could be best explained in terms of an inner miogeosyncline (Appalachian Valley and Ridge area; Cordilleran Rocky Mountain region) and an outer eugeosyncline (Appalachian piedmont and New England highlands; Cordilleran western belt). Kay likened the eugeosyncline to modern volcanic island arcs, as shown in a suggested restoration of North America for part of the Ordovician period (Figs. 24–3 and 24–4).

Concept of Sedimentary Orogenic Cycle

The history of a region can be deduced from the type of sedimentary materials deposited in the surrounding areas. The sediments provide a key to the environment in which they form. On this basis, several concepts of geosynclinal sedimentation have been advanced that relate the sediment to the orogenic activity in the geosyncline. One such cyclic concept, proposed by F. J. Pettijohn (1957), recognizes four stages in an ideal cycle:

Preorogenic stage. Sediments deposited in the geosyncline are clean sandstones and limestone. The types of sediment now being deposited on the shelf off the eastern coast of the United States are typical of this stage.

Euxinic stage. The initial indications of orogenic activity are the development of deeper basins in which circulation of water is not good. Black shales and limestones may be deposited in such basins.

Flysch stage. At this stage uplift in the core of the orogenic belt has started and deposits of sandstone and shale are deposited by turbidity currents in deep water.

Molasse stage. In this final stage the mountains are high and coarse sediment is shed into the sedimentary basins. These sediments may contain abundant feldspar, as in arkose; they may be extremely coarse; the iron may be oxidized; and they can often be traced into terrestrial deposits.

Of course the type of sediment deposited at any given time is a function of location as well as stage in the orogenic cycle. Thus it may be possible to trace coarse terrestrial arkosic sediments into coarse marine deposits to shale, and eventually the shale may grade into marine limestones at great distance from the uplift (Fig. 24–5).

Igneous and Metamorphic Complexes

The core of most highly deformed mountains is a metamorphic complex usually intruded by magmas that have been partially assimilated and largely altered (Fig. 24–6). These cores are characterized by recrystallization of original rock, flowage, and distortion. The extensive alteration of the rocks is indicative of the depth to which the rocks found there had been buried before the deformation and reflects the high temperatures and pressures to which the rocks have been subjected. Every orogenic belt contains ultrabasic intrusions such as peridotites, dunites, or serpentines. Extensive outcrops of these are rare, but the small isolated bodies found are located near the central axis of the mountain system; sometimes they are aligned on either side of the axis. They occur so persistently in orogenic belts and not outside them that they must be related to orogeny. They may best be explained as having been parts of the upper mantle that became emplaced in the eugeosynclinal sedimentary pile during deformation.

Most batholiths are found within young folded mountains and in the island arcs. Granite, other intrusives, and extrusives are common but not confined to these belts.

It is easy to explain the presence of plutons in the eugeosynclinal portions of geosynclines if the analogy to volcanic island arcs is valid. The plutons would be magma sources for the lavas and volcanic ash in surrounding sediments. Precise dating of igneous activity and folding has shown that the two are not always synchronous nor do they always occur in the same locations. A causal connection between the two may not be warranted, but both still may be expressions of orogenic activity. Uplift and folding can occur with plutonic activity and plutons can be emplaced with accompanying folding and uplift. Both tend to occur most frequently in those parts of the continental crust that are orogenic belts

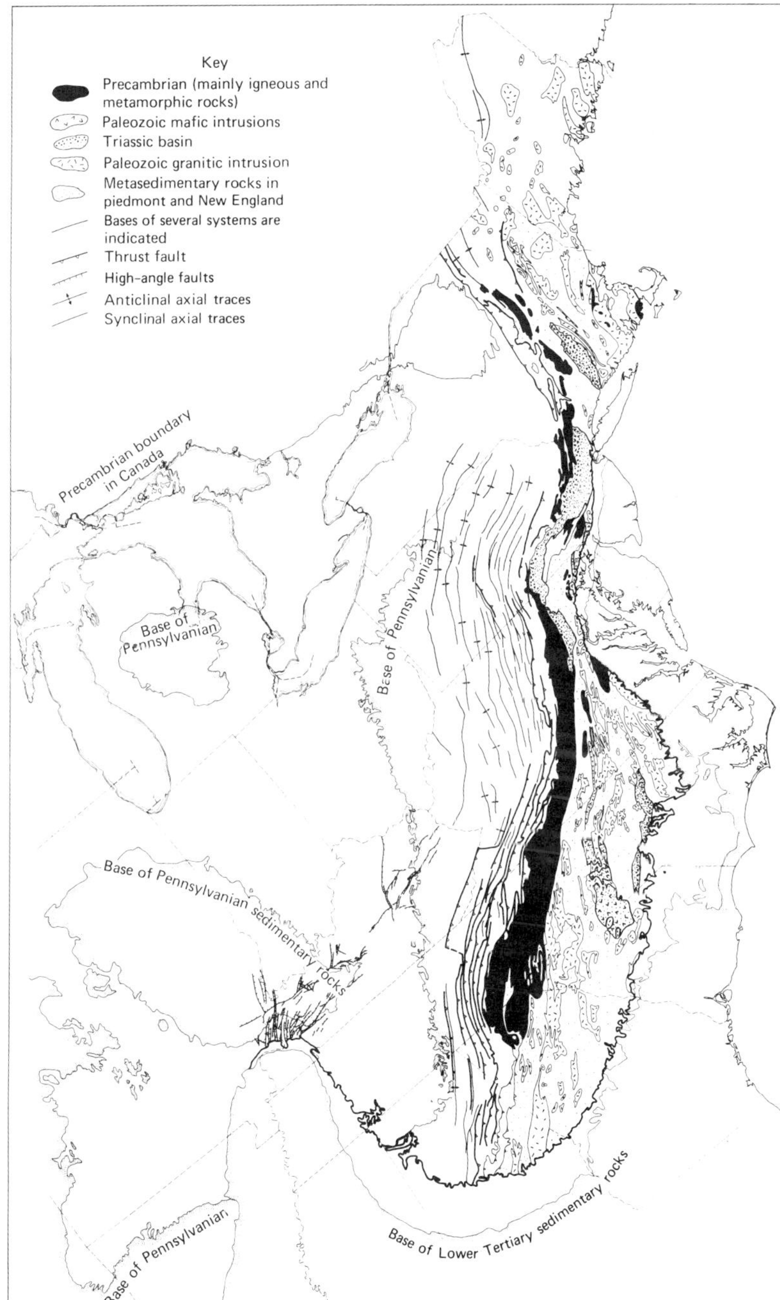

FIGURE 24–6
Sketch map of the Appalachian region of the United States.

Structural Features

Structurally, orogenic belts are characterized by long folds or systems of folds. It is frequently possible to trace the folds from the margins of a belt, where broad arches are present, into zones of increasingly intense deformation. These may culminate in the formation of recumbent folds (Fig. 24–7). Thrust faults, also prominent in these belts, represent part of the accommodation of the crust locally to the stresses that were acting in the crust at the time of the deformation. The direction of movement indicated by folds and faults is usually away from the central part of the mountain system toward the margins.

The fact that many of the youngest mountain ranges lie along the continental borders may be very significant, particularly since the island arcs occupy this position. Island arcs (see Fig. 25–1) are the sites of present-day orogeny. For that reason they are of special interest and importance in the theory of orogeny.

Origin of Folds and Thrusts

Folding and thrusting are characteristic features of every orogenic belt. Located on the flanks of the core, they are called *foreland folds*. The greatest uplift coincided with the areas where the oldest rocks or the most highly metamorphosed rocks are now exposed.

Folds and faults may be created in several ways. They could result from shortening of the crust of the earth. Measurements in the Appalachians show that 100 kilometers or more of shortening is necessary to create the folds found there. This figure is arrived at by measuring the length of folded units in a cross section drawn perpendicular to the general trend of the structures (northwest–southeast in the Appalachians) and comparing it with the horizontal distance across the section. Shortening was favored originally because the earth was thought to be contracting. Some of this shortening, although striking in its surface expression, may actually be only "skin deep" (Fig. 24–8). Skin-deep deformation could be produced by slippage of a sedimentary veneer over crystalline basement rocks or along faults that parallel original bedding surfaces. Such slippage, referred to as *décollement*, may be primarily caused by the force of gravity. Slip is facilitated by the presence of clay, gypsum, shale, or other soft and weak materials on which sliding could occur. Such sliding may take place after uplift in major orogenic belts. One of the best examples of the shallow depth of deformation is found in the Pine Mountain overthrust of the Cumberland Mountains in the Appalachians. Here a large section of sedimentary rocks has been thrust to the northwest along a fault that lies flat and nearly parallel to sedimentary rocks below, as shown in the cross section.

Skin-deep deformations that do not involve the deeper parts of the crust may be caused by sliding of a comparatively thin sedimentary veneer under the pull of gravity. This movement is downward from the central uplifted core of the mountain belt.

FIGURE 24–7
Sketch of the valley walls in the Glarner Alps showing thrust sheets. The labels indicate two mountains.
(after Oberholtzer, 1933)

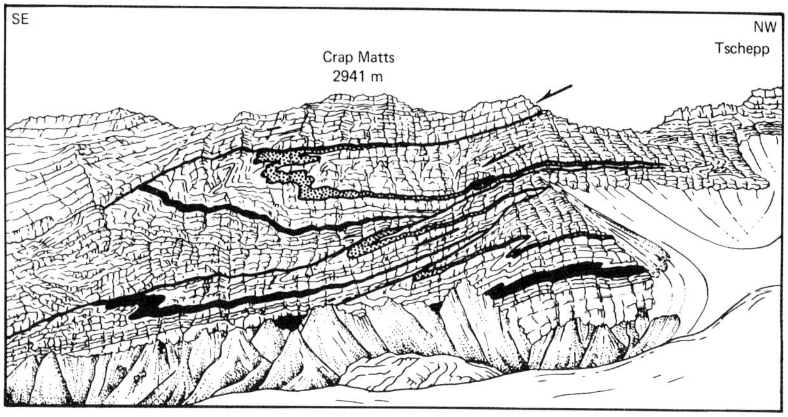

Orogenic Deformation: Summary

Orogenic deformation appears to have followed a similar course in many mountain belts. The initial phase is formation of a geosyncline—subsidence and the deposition of a thick accumulation of sedimentary rocks (Fig. 24–9). Some geosynclines have both a miogeosyncline and a volcanic island arc system in a eugeosyncline. The orogenic de-

FIGURE 24–8 (*below*)
Structure of the Kubore-Bismarck Cordillera and the Papuan foreland folded belt, composite section based on the *décollement* principle of foreland folding.
(after Smith, 1965)

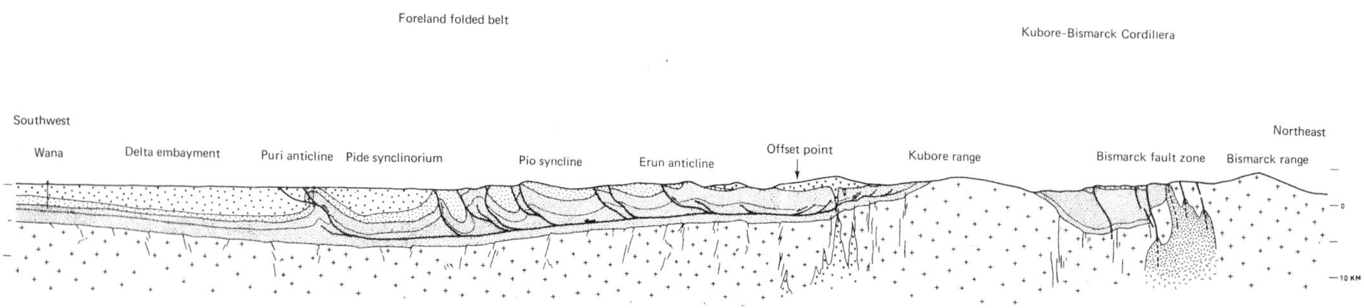

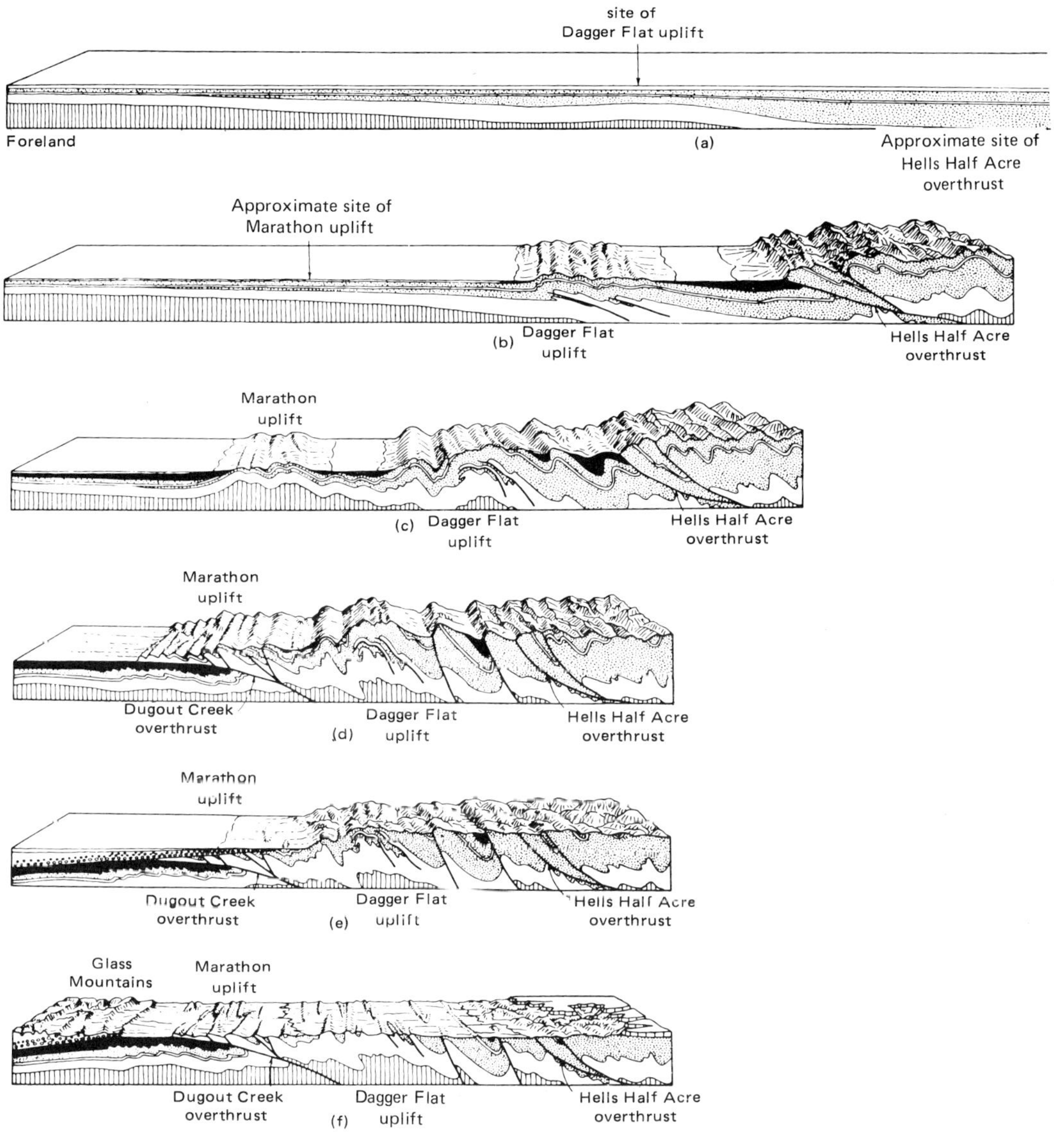

FIGURE 24–9 Hypothetical block diagrams showing progressive development of structural features in the Marathon region of Texas. (after P. B. King, 1937)

formation involves uplift and the formation of folds and thrust faults in the foreland. This deformation may or may not be accompanied by igneous activity and may involve compression as well as uplift. Movement indicated by folds and faults is generally from the continental margins toward the continental interior. As the uplifted mountains are eroded a central core of metamorphosed rocks and plutons is exposed. Sediment from the mountains is spread into nearby basins and seas.

Some orogenic belts have been block-faulted following the folding and uplift. Graben and horsts were found in the Appalachians in the Triassic, and terrestrial sediments eroded from the high mountains were deposited in basins including the present Newark basin and the Connecticut Valley. Grabens and horsts are also widely developed in the Basin and Range province of the western United States. These structural features formed in the Cenozoic and many of the faults are still active.

CONTINENTAL MARGINS

The continental margins mark the zone of transition from crustal structure of the continents to that of the ocean basins. Several distinctively different types of transition zones are found. Particularly strong contrast can be seen by comparing of the margins of the Atlantic Ocean with those of the Pacific. The Pacific is encircled by a broad zone of young folded mountain belts, modern diastrophism, volcanism, and seismicity. The northern and western margin of the Pacific is characterized by an almost continuous chain of island arcs starting with the Aleutians in the north and extending to the Tonga-Kermadec Islands on the south side of the basin. The long and almost continuous belt of young folded mountains comprised of the Cordilleran system of western North America and the Andes of South America lies near the continental margin all along the eastern side of the Pacific. These great structural belts are parallel to the margins along most of that distance with the notable exception of the Central American portion. These mountain systems are high and young, and deformation is continuing along much of the belt.

The structure of the ocean floor off this eastern Pacific margin is highly varied. Deep trenches lie parallel to the continental margins of Chile, Peru (the Chile Trench), and along Mexico and Central America (the Mid-America Trench). The East Pacific Rise, which is located west of these trenches, trends northward into the Gulf of California where it disappears under the Cordilleran mountain system. A northern continuation of this rise reappears at Cape Mendocino and runs parallel to the coast of British Columbia before it disappears again in the Gulf of Alaska. A number of major faults have been traced in the Pacific basin and across the East Pacific Rise. They are prominent topographic features in the sea floor, and they displace linear magnetic anomalies that are parallel to the rise system.

The North Atlantic is quite different from the South Atlantic. The Mid-Atlantic Ridge lies in a central north–south position through the Atlantic, marking almost the exact center of the ocean basin. The geology of the Atlantic continental margins is strikingly different from that of the Pacific. The young folded mountain systems around the Atlantic do not closely parallel the continental margins except in the case of the Appalachian system. The Atlas Mountains of North Africa, the Pyrenees of Spain and France, and the Caledonide belt of Scotland and Ireland all trend toward the Atlantic Ocean basin and appear to end at the coast. They do not continue into the ocean basin. It seems probable that they have been rifted apart and separated as the Atlantic Ocean basin grew larger. Ancient Precambrian shields of South America and Africa lie across from one another in the South Atlantic, and these margins show no evidence of later folding that might be related to the continental margins. Only two island arc systems are located in the Atlantic, and these are positioned quite differently from those of the Pacific. The Caribbean arc and the South Sandwich Islands arc are both situated between continental plates and not in front of them, as is true of most Pacific arcs. The continental margins of the Atlantic are quiet. Little evidence of Cenozoic volcanism, seismicity, or diastrophism is present. Long stretches of the continental margins around the Atlantic Ocean are characterized topographically by a broad continental shelf, a continental slope, and a continental rise from the continent into the deep abyssal plains.

Continental Margins of Eastern United States

The American continental margins are better known through geophysical studies than are most other continental margins in the world. Intensive studies of the eastern margin have been undertaken in recent years, but it remains to be seen if this margin may be considered typical of the margins around the Atlantic. An obvious difference is that a folded mountain belt parallels this margin whereas no other long margin of the Atlantic has such a feature.

The major geomorphic provinces of the eastern United States margin are illustrated in Fig. 8–9. The continental shelf is essentially a seaward continuation of the Atlantic Coastal Plain, which is an area of very low relief underlain by Cenozoic marine sedimentary rocks that dip gently and gradually thicken toward the coast. The newly developed technique of continuous echo sounding by means of high-energy spark discharges and low-level explosions has made it possible to produce continuous cross sections across the margins (Fig. 8–11). These indicate that the Cenozoic sediments continue out to the edge of the shelf as fairly flat-lying units; they continue beyond the shelf down the slope, but many of the layers have broken off, sliding downslope and piling up where the slope levels off. The orderly progression from shelf to slope to rise, which is so perfectly seen north of Cape Hatteras, is broken up in the south where two sharp topographic breaks are found—one at the edge of the continental shelf where the sea floor drops down to the Blake Plateau, and a second at the Blake escarpment. Cross sections based on seismic studies reveal the structure of the margin across the Blake Plateau (Fig. 11–12). This southern part of the continental margin is broken by major faults that have dropped down large blocks of the crust toward the oceanward side. If similar block faulting occurs to the north, it is probably deeper.

Other seismic studies have led to the remarkable discovery that the total sediment thickness of the continental margin under the continental slope and rise is very great (Figs. 24–10 and 24–11). The sediment is over 6,000 meters thick in places—a modern geosynclinal accumulation. The age of this sediment is yet to be determined. It may be an eastern part of the Appalachian geosyncline, or it may be a Mesozoic geosyncline that formed as the Atlantic Ocean began to open. This great sediment accumulation (Fig. 24–11) thins out under the Blake Plateau, and it does not continue across Florida into the Gulf of Mexico where another modern geosynclinal accumulation is known to exist beneath the Gulf Coastal Plain and Shelf.

The continental rise of the eastern Atlantic is notable for the absence of linear magnetic anomalies such as those found on either side of the Mid-Atlantic Ridge and those found off the coast of California. A number of small circular positive anomalies are found associated with the Kelvin Seamount group; similar anomalies presumably lie over buried seamounts of similar origin. These are thought to be extinct volcanoes. They are aligned roughly along an east-west lineament at latitude 40°N, a lineament that shows on the map of sediment thickness under the rise. It seems probable that this is a major fault in the oceanic crust that, like faults of the Pacific, trends perpendicular to the continental margin.

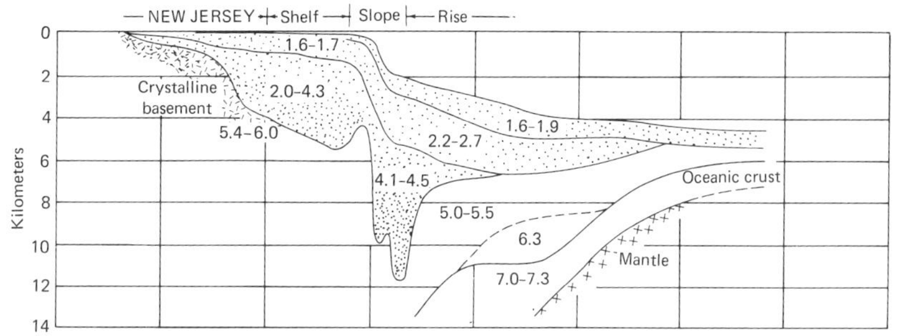

FIGURE 24–10
Seismic refraction section across the continental margin east of Cape May, N.J.
(from Drake, Ewing, and Stockard, 1968)

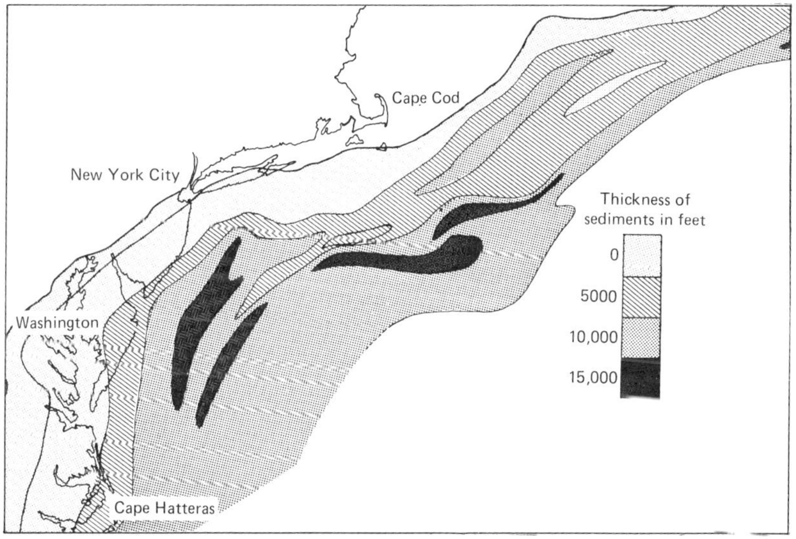

FIGURE 24–11
Sediment thickness off the eastern coast of the United States. The inner margin is the coastal plain. The black areas are those in which more than 15,000 feet of sediment and sedimentary rocks overlie the basaltic layer of the ocean floor.
(modified after C. L. Drake, Maurice Ewing, and G. H. Sutton, 1959. Reprinted with permission from *Physics and Chemistry of the Earth*, vol. 3, 1959, Pergamon Press, Inc.)

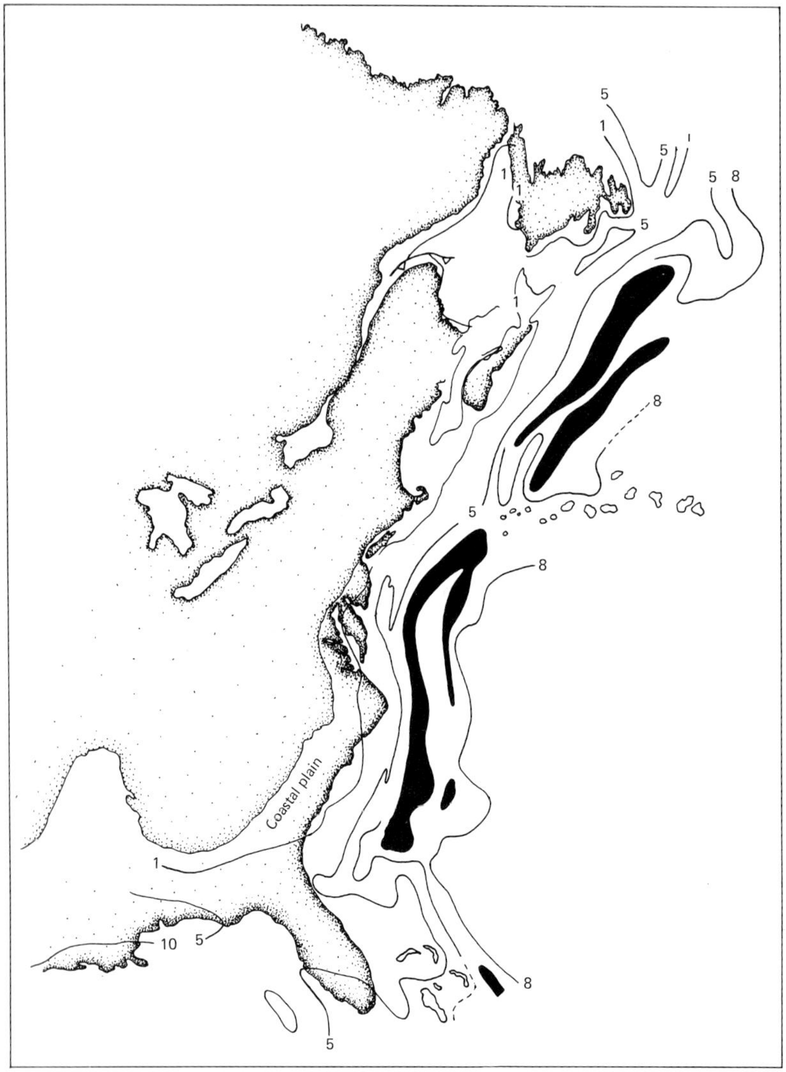

FIGURE 24–12
Contours on the basement in eastern North America. Depths of 10 kilometers or more are shown black.
(after Drake and Kosminskaya, 1969)

Pacific Margin of North America

The most prominent difference between the Atlantic and Pacific margins of North America is the apparent relationship of the continental edge to the oceanic rise. The Mid-Atlantic Ridge is located in the center of the Atlantic Ocean; but the East Pacific Rise appears overridden by the western continental margin. The Mid-America Trench and the Guatemala basin on the western edge of Mexico separate Mexico from the crest of the East Pacific Rise, but farther north the crest of the rise passes under the Gulf of California (Fig. 25–1).

Block faulting is prominent in Baja California and along the edges of the Gulf of California. Transverse faults trending northwest cut across the peninsula. It appears that the gulf opened by *rifting* and movements of the blocks between these faults (Fig. 24–13). Gravity profiles across the gulf in several places show pronounced positive anomalies over the gulf, indicating that high-density material lies beneath the gulf floor. This may be in the form of dikes of mantle material, or the top of the mantle itself at shallow depth under the gulf. High magnetic anomalies and high heat-flow through the floor of the gulf are also found. Seismic sections across the gulf and Baja California support this conception of the structure. The gulf appears to have originated as a splitting apart of a part of the continental margin over the crest of the East Pacific Rise (Fig. 24–13).

Block faulting seems a feasible explanation for the irregular topography of the shelf west of southern California where deep abyssal basins lie close to high blocks that rise near the surface of the ocean. The strong transverse faulting indicated for Baja California continues northward along the coast. The San Andreas fault is part of this system. But the structure of the sea floor along the continental margin appears much simpler between Los Angeles and Cape Mendocino than it does to either the north or south. This stretch of coast located between the Mendocino and Murray escarpments in the Pacific basin floor has a narrow continental shelf, a slope, and a rise that lead progressively deeper into the deep-sea floor.

North of Cape Mendocino (Fig. 24–14), the sea floor is not as deep, and a complex system of submarine ridges and deep enclosed basins indi-

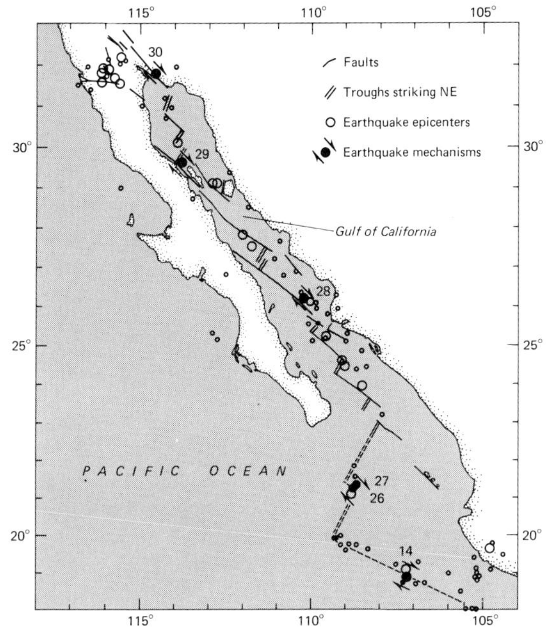

FIGURE 24–13
Structural features of the Gulf of California. (after Sykes, 1968) Relocated epicenters of earthquakes for the period 1954 to 1962.
(from Isacks and Oliver, 1968)

FIGURE 24–14
Physiographic provinces in northeast Pacific Ocean.
(from McManus, 1967)

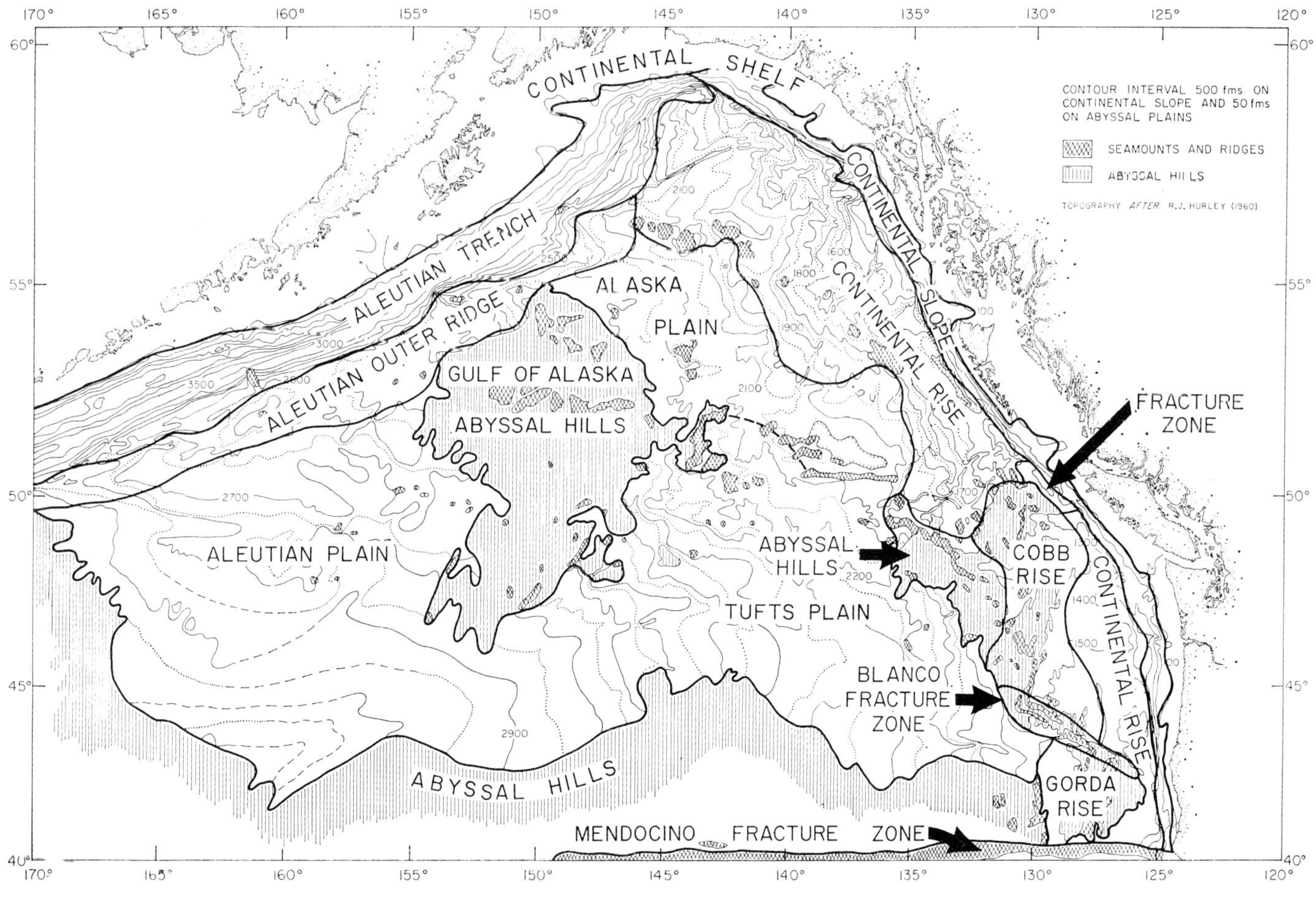

cates structural features in the sea floor. Numerous seamounts are located in this region, all the way into the Gulf of Alaska. The ridges and adjacent basins are thought to be located along faults that cut across and displace a northern continuation of the East Pacific Rise. Shallow-focus earthquakes in this region are located along these faults and along the crest of the rise.

CHAPTER OUTLINE

Continental Shields
Young Folded Mountain Belts
- Location of Some Young Folded Mountain Belts
- Pattern of Mountain Belts
- Geosynclinal Theory
- Concept of Sedimentary Orogenic Cycle
- Igneous and Metamorphic Complexes
- Structural Features
- Origin of Folds and Thrusts
- Orogenic Deformation: Summary

Continental Margins
- Continental Margins of Eastern United States
- Pacific Margin of North America

TERMS

continental margins
cordillera
craton
décollement
eugeosyncline
euxinic stage
flysch stage
foreland folds
geosynclines
miogeosyncline
molasse stage
preorogenic stage
rifting
shield

REVIEW QUESTIONS

1. Describe the distribution of mountain belts and shields (Fig. 24–1).
2. What does the distribution described above suggest about the processes of mountain building?
3. Compare the age and deformation of rocks on the East Coast of the United States with the age and deformation of those on the West Coast.
4. What is the significance of thick sections of feldspathic sediment such as arkose or graywacke?
5. How would you recognize an ancient geosyncline in an area eroded to low relief?
6. What evidence supports the idea that mountains are formed primarily as a result of processes in the earth's interior?
7. How would isostasy affect the formation of mountains?
8. Why would heat in the earth's crust facilitate mountain building?
9. Summarize evidence for the origin of folds as a result of shallow deformation.
10. Compare the continental margins of eastern North America with those of the Aleutians or Japan (refer to Figs. 24–1; 24–10, 11, 12; 24–14; 8–14, and 8–17).

SUGGESTED READINGS

Billings, M. P., 1960, Diastrophism and mountain building: Geol. Soc. America Bull., v. 71, p. 363–397.

Geological Survey of Canada Dept. of Mines and Technical Surveys, 1965 (1967), The world rift system: Canada Geol. Survey Bull. Paper 66–14, 471 p.

Geological Survey of Canada Dept. of Mines and Technical Surveys, 1965 (1966), Continental margins and island arcs: Canada Geol. Survey Bull. Paper 66–15, 486 p.

Johnson, Helgi, and **Smith, B. L.,** 1970, The Megatectonics of Continents and Oceans: New Brunswick, N.J., Rutgers University Press, 282 p.

Phillips, O. M., 1968, The Heart of the Earth: San Francisco, Freeman, Cooper, and Co., 236 p.

25 CONCEPTS OF GLOBAL TECTONICS

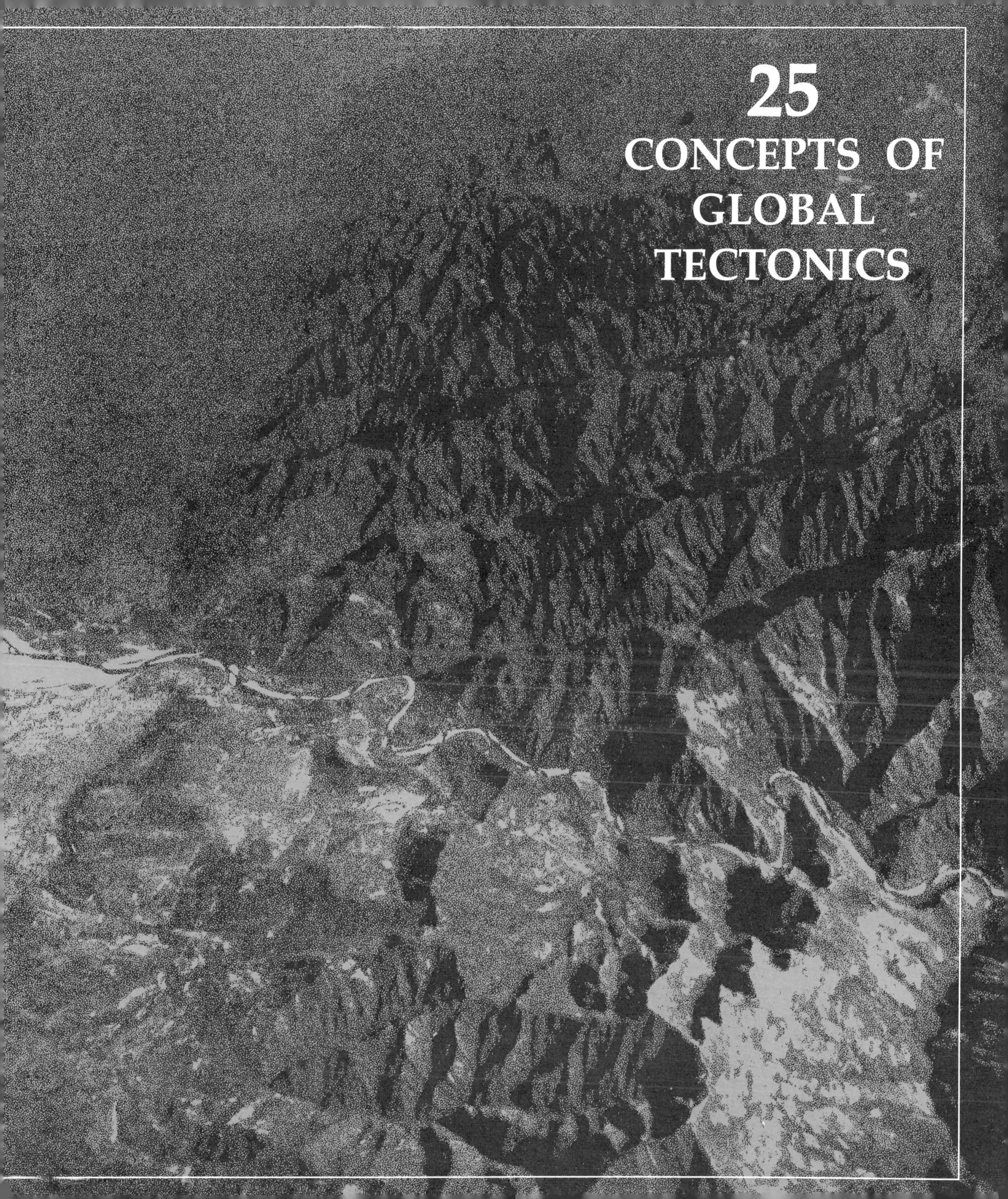

25

CONCEPTS OF GLOBAL TECTONICS

FIGURE
(previous page)
High mountains in China; these are eroded by running water.
(U.S. Air Force)

Tectonics refers to formation and deformation of the earth's crust resulting in large-scale structural features, such as orogenic belts, continents, island arcs, and midocean ridges. The origin of features on this scale and the relationships among them provide some of the most intriguing problems in earth science today.

How permanent are the continents and ocean basins? Has the present distribution of continents and ocean basins persisted since the continents first formed, or are the ocean basins actually young? Have the continents grown by the welding on of successive orogenic belts, as the theory of *continental accretion* postulates? Or did the continents originate as one or perhaps two continental plates that subsequently were rifted apart, moved, and modified by orogeny to their present form, according to the theory of *continental drift?* What causes orogeny to take place? Has the size of the earth changed during its history? Is it contracting, expanding, or has there been little change? These are among the questions being investigated today. Dramatic discoveries in the last decade have made tectonics one of the most rapidly evolving fields of geology.

Speculation and hypotheses about the origin of the earth's large-scale features preceded the emergence of geology as a science and have played a crucial role in its development. Hypotheses stimulate and give direction to research. Many field studies, observations, and experiments are designed to test their validity. The hypotheses are then modified to better explain observational data, or they are discarded. Several major hypotheses stand out as unifying concepts in the study of tectonics: (1) the contracting earth concept, (2) continental drift, (3) convection currents in the earth's interior, (4) sea-floor spreading, (5) and global expansion. They are hypotheses to which many other theories and data can and must be related. The way geologists use multiple working hypotheses to obtain a better understanding of the earth is illustrated by the way tectonic hypotheses have evolved.

THEORY OF THE CONTRACTING EARTH

Lord Kelvin was the first scientist to advance the idea that the earth is cooling. He viewed the earth as a heat engine that is slowly running down. Great quantities of energy are expended in volcanic activity, diastrophism resulting in uplift of huge crustal segments, and in seismic activity and faulting. The source of this energy is not immediately apparent although it must come from the earth's interior. Volcanic activity suggests that the source might be heat energy. Support for this point of view was found in the *dynamic encounter hypothesis* advanced by Buffon in 1749 in which it was postulated the earth originated as a mass of hot gases derived from the sun as a result of a collision with a comet. The earth condensed, according to this theory, into a rotating sphere of gases that contracted to a ball of magma and, on further cooling, a crust formed over the magma. The earth's core is the remaining liquid from this original magma. Heat lost from the cooling of the interior might thus provide the energy expended in volcanism and orogeny. Accordingly, the crust represents the already cooled portion of the earth, and the hot interior is part of the original molten mass. Élie de Beaumont (1829) saw in this cooling a simple way to account for the folded and thrust-faulted orogenic belts by crustal compression. Compression would explain the apparent shortening in folds and thrust faults (see Figs. 23–31 and 23–32), uplift of orogenic belts, and igneous and metamorphic rocks in the uplifted cores. According to this hypothesis, the volume of the earth's interior decreases as it cools, and the outer crust is forced to accommodate itself to a sphere of smaller radius. This is accomplished by wrinkling or folding of the crust.

The contraction theory gained widespread acceptance among geophysicists when Sir Harold Jeffreys (1950) discovered a seismic discontinuity at a depth of between 70 and 100 kilometers. This discontinuity coincided in depth with the level previously predicted to be a level of "no strain," according to the contraction hypothesis. The level of no strain is the level at which crustal compression decreases to zero.

The idea that the orogenic belts originated as a result of crustal compression caused by contraction of the earth was the most popular theory of orogeny until the advent of the convection current hypothesis in the 1930's. However, the fundamental assumptions of the theory have been subject to re-examination during the last two decades. With the discovery of great quantities of radioactive minerals in the crust, the question arose as to whether the interior is actually cooling and contracting, or heating and expanding, for radioactive decay is accompanied by the release of heat.

The discovery of many structural features such as grabens and horsts (Fig. 23–24) that are best explained as arising from extension of the crust also poses a problem for the contraction hypothesis. Continental grabens such as the East African valleys, the Rhine graben, and the Basin and Range province were recognized early, but extensional features were not really apparent until studies of the midocean ridges in the 1950's revealed a globe-encircling zone of grabens along the crest of the oceanic ridges and rises (Fig. 25–1). It is difficult to reconcile the existence of such widespread extensional features with the concept of crustal compression.

ORIGIN OF THE THEORY OF CONTINENTAL DRIFT

Credit for conception of the continental drift theory appears to belong to an American, A. Snider, who explained the origin of the Americas in an article published in 1858. Carozzi (1970) describes Snider's theory as follows:

> He assumes that before the Deluge, dry land formed a single mass which, through the action of underground fire, became increasingly hardened . . . dry land was split by incipient cracks lined with volcanoes, . . . During the Deluge, increased expansion due to the volcanoes led to a further splitting of dry land, mainly along its major fissure, and the island of Atlantis, accompanied by its volcanoes, drifted westward during a single violent commotion until planetary equilibrium was again reached, becoming America. A proof of this displacement is the similarity of the opposite coasts of the Atlantic.

Snider published a map showing South America joined to Africa and North America joined to Europe. He made this reconstruction in order to explain the similarities of fossil plants found in Carboniferous coals of both Europe and North America. The modern debate about the idea started with papers published by F. B. Taylor in 1910 and by Alfred Wegener in 1915. Since the theory was first advanced, the question of permanence of continents and ocean basins has been intensely debated. Wegener advanced the idea of continental drift to explain anomalous distribu-

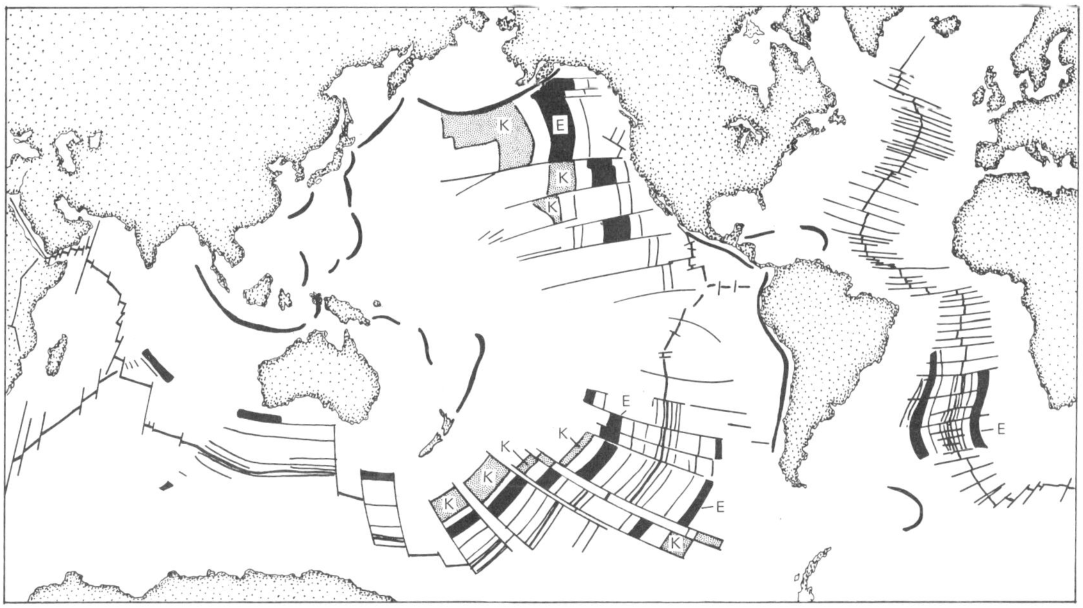

FIGURE 25–1
Major structural elements of the oceans showing the midocean-ridge crest and transverse faults across it, deep-sea trenches (black), and belts of crust of various ages (K, Cretaceous, 65 million years, dots; E, Eocene, 38 million years, black; boundaries indicated are 2, 7, 26, 38, 54, 65, and 136 million years) are shown.
(after Heirtzler and others, 1968)

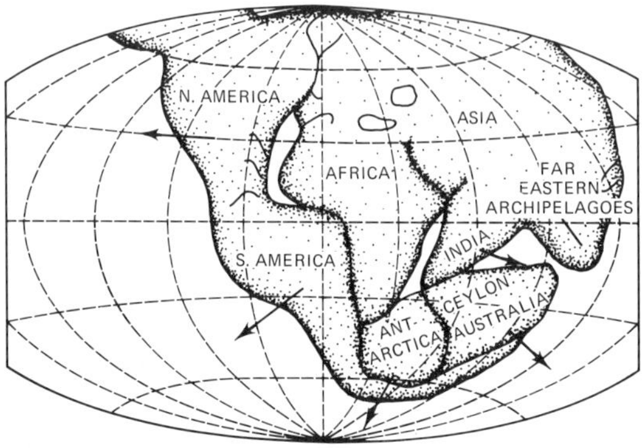

FIGURE 25–2 Hypothetical arrangement of the world's continents according to Wegener. From a single mass like this the continents supposedly drifted apart. (after Wegener, 1915)

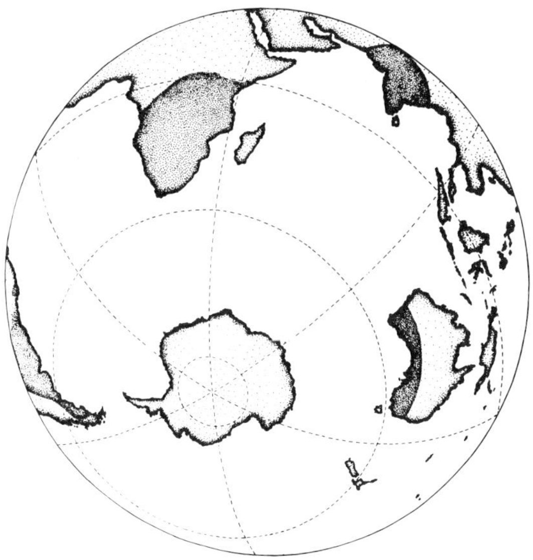

FIGURE 25–3 Areas covered by ice in the late Paleozoic glaciation are shown shaded. A single ice cap is needed if these continents were together at that time.

tion of ancient climatic belts. Wegener attempted to fit all continents into a single protocontinent called *Pangaea* (Fig. 25–2), and the close fit of the coasts on either side of the Atlantic remains a convincing line of evidence for continental drift. Other predrift reconstructions have favored the existence of two primitive continents—*Gondwanaland,* which included Africa, South America, India, Australia, and Antarctica, and *Laurasia,* which was composed of the continental masses in the Northern Hemisphere.

Gondwanaland

Evidence of Carboniferous-Permian continental glaciation is found in the southern parts of South America, Africa, India, Antarctica, and Australia (Fig. 25–3). If we accept the present disposition of the areas that were glaciated in the Paleozoic, then ice sheets existed at the equator. Why were there glaciers in India but not in areas farther north toward the poles? The geographic distribution of evidence for late Paleozoic glaciation is difficult to understand if we accept the idea that the continents are fixed in position.

There is evidence of more than a single period of glaciation on each of the continents. Two *tillites* (sedimentary rocks composed of cemented till) and associated varves separated by interglacial deposits occur in South America and India; several tillites are found in parts of Africa; and 11 glacial stages are indicated for parts of Australia. As in the Pleistocene, there were repeated resurgences of the ice sheets in the late Paleozoic.

The bedrock surfaces over which the ice sheets moved have been uncovered in some places, exposing glacial striations, erratics, and ground moraines. If erratics can be identified it is often possible to determine where they originated and hence, indirectly, the movement of the ice. Glacial erratics of a rock type common to South Africa have been discovered in South America, where the rock type is not known to exist. Striations also indicate the orientation of ice movement. This type of evidence indicates that the ice sheets centered in the south of India, to the east of South America, and south of Australia. Ice movements in Africa are generally directed to the west or south from a center located in Rhodesia. The overall pattern of movement and geography of the glaciated areas are much easier to explain if these continents were at one time located close to one another in a pattern like that shown in Fig. 25–4. The restoration of the southern continents to form what is known as Gondwanaland has long been a compelling idea to many geologists.

The theory of continental drift did not receive the support of most American geologists despite the evidence shown by the geographic fit of Africa and South America, late Paleozoic glaciation, and similarities of fauna and flora in the Southern

Hemisphere. The main objection was the lack of a mechanism that could move continents through what appeared to be a solid and strong sea floor.

EXPANDING EARTH HYPOTHESIS

The theory that the earth is expanding, first proposed in 1933, has been revitalized recently by the discovery that new oceanic crust is forming along the central parts of the oceanic rise systems. O. C. Hilgenberg (1933) was the first person to advocate global expansion, but his notion was largely ignored because he suggested the unsubstantiated hypothesis that the interior of the earth has increased in mass as well as volume. A few years later the astronomer J. K. E. Halm proposed that the earth's original density was much higher (9.13) than it is now (5.5). Halm pointed out that atomic packing can vary and that exceedingly high densities are found among some stars, within which it is thought that electrons are lost from atoms with a consequent compaction of matter into very dense forms (densities of some white dwarfs—compacted, white-hot stars—exceed a hundred thousand times that of water).

L. Egyed (1955) suggested that the core of the earth is composed of an ultra-high pressure phase of material that is unstable, and that it is being transformed into a lower density phase with accompanying increase in volume. Egyed devised a geologic test of this hypothesis based on the assumption that the quantity of water on earth has either remained the same or increased through geologic time. All evidence points to the validity of this assumption. Egyed reasoned that if global expansion is taking place, it is occurring in the ocean basins since no great amount of expansion of the continental plates is geologically feasible. Furthermore, if the oceans have enlarged while the continents have remained about the same size, then the area of the continental plates covered by marine seas should have decreased through time. Egyed used paleogeographic maps prepared for each of the geologic periods since the Cambrian, and he measured the area of the continental plates that was shown covered by water for each period. A plot of these data (Fig. 25–5) shows fluctuations that might result from diastrophism and glaciation. A general decrease through time of continental area covered by water is also indicated. This supports global expansion even if the amount of water has remained constant; if the amount of water has been increasing, the amount

FIGURE 25–4 Reconstruction of continents suggested by Carey (1958).

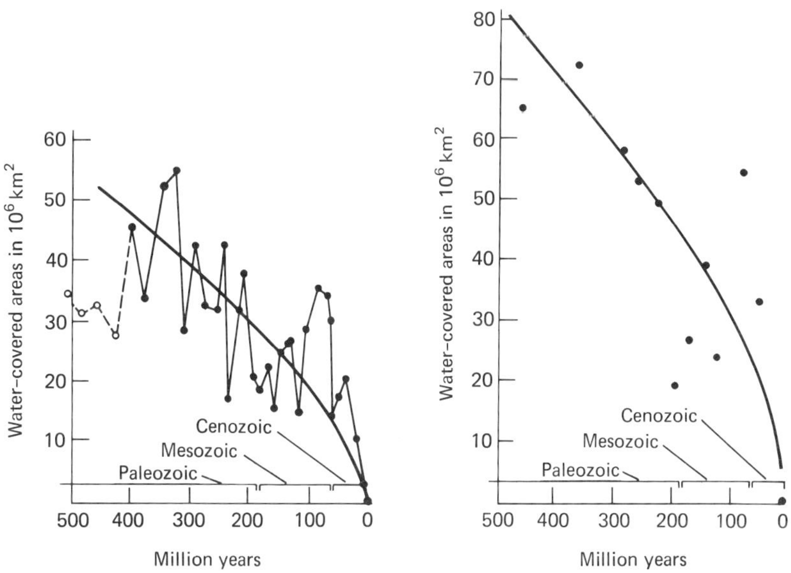

FIGURE 25–5 *(below)* Progressive emergence of continental massifs with time. (after Egyed, 1957)

of expansion is greater than indicated. The drawback of this test is that it is based on the assumption that the paleogeographic reconstructions are sufficiently accurate for this purpose, that the evidence now concealed beneath the continental margins and oceans would not tend to raise the curve. The curve would be raised if former land areas have been submerged during the Cenozoic.

When S. W. Carey completed his reconstruction of the continental plates, allowing for the effects of continental drift (Fig. 25–4), he found that they would not fit together perfectly unless he assumed that the earth's radius was smaller before drifting started. A diameter three-fourths the size of the earth's at the present time provided a near-perfect fit. This would suggest that the earth was considerably smaller at the beginning of the Mesozoic era before fragmentation of the crust and drift started. Carey also points out that the direction of continental drift is toward the Pacific basin for every major continent; yet the Pacific is the largest of all oceans. Carey's analysis of the tectonics of the Pacific's perimeter suggests that it has been increasing in size also. Effects of crustal extensions are in agreement with the theory of global expansion (although they can be explained in terms of convection and vertical tectonics as well).

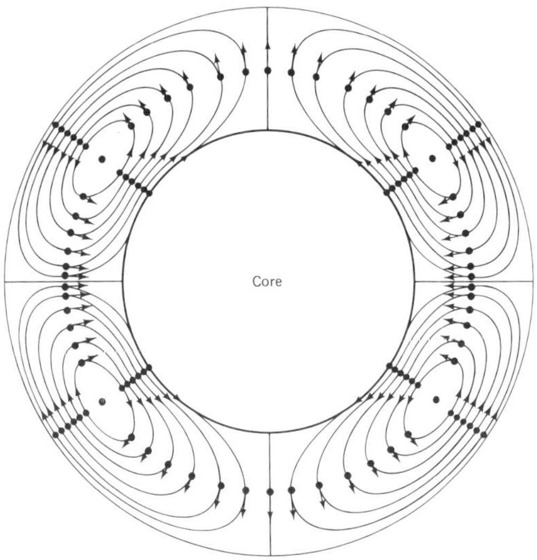

FIGURE 25–6
Convection cell pattern. Length of the arrows indicates relative rates of movement.
(after Pekeris, 1935)

Some astronomical evidence also supports the idea of global expansion. Astronomers have found that the length of a day is increasing at the rate of about 2 seconds per 100,000 years, amounting to over 3 hours since Precambrian time; this corresponds to a reduction of the rate of rotation by that amount. If this rate of rotation is due to an increase in the diameter of the earth, then it would correspond to an increase of radius of about 0.6 millimeter per year.

The idea of global expansion must be regarded as an interesting and highly provocative hypothesis. The best hope for proof of global expansion is that we will someday be able to determine the location of points on the crust of the earth with such precision that it will be possible actually to measure any changes in distance between them. If the earth is expanding at a rate of as much as half a millimeter per year, then points at different places on the earth should be dispersed from one another in much the way that points on a balloon are dispersed as the size of the balloon increases.

FIGURE 25–7 (*below*)
Regional isostatic negative gravity anomalies. Vertical-line pattern marks areas with 0 to −50 milligal anomaly. Cross-hatched pattern covers areas with −50 to −200 milligal anomalies.
(after Vening-Meinesz, 1948)

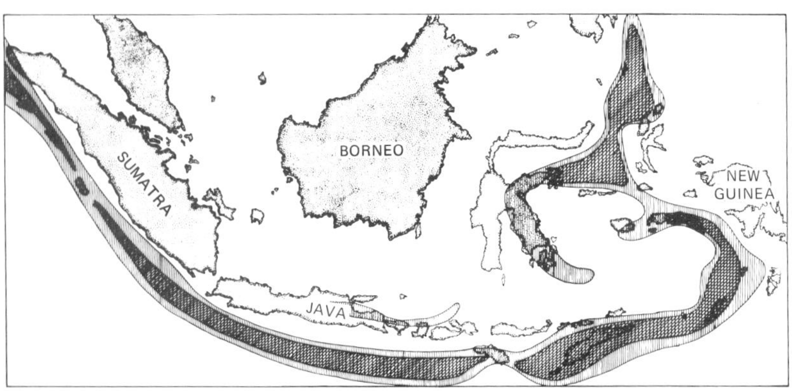

CONVECTION CURRENT HYPOTHESIS

Convective overturn of materials within the earth's core and mantle has been postulated for many years. Convection currents form in fluids when temperature gradients occur. The interior of the earth is hotter than the surface and high temperature gradients may be established locally by concentrations of radioactive minerals. Convection could not exist in the earth's crust because it

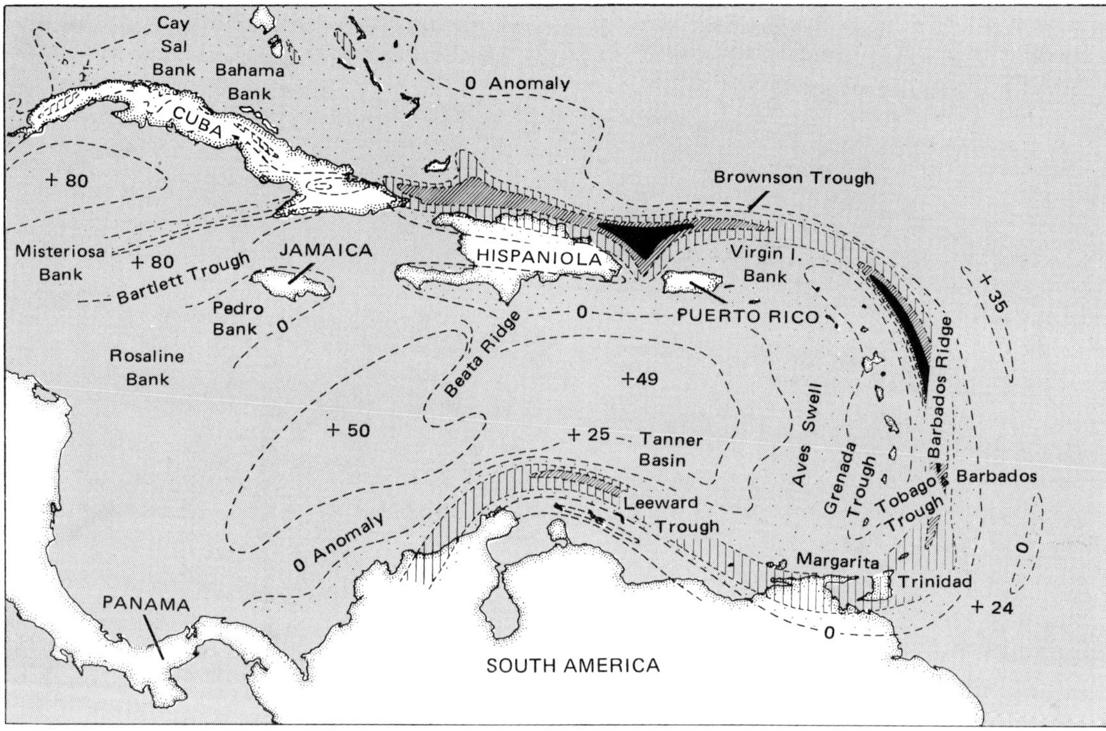

FIGURE 25–8
Gravity anomalies in the Caribbean. The black and diagonally lined areas are negative anomalies where there is a mass deficiency. Such negative anomalies are associated with deep-sea trenches in the island arc systems of the Pacific also.
(after Hess, 1939; from Eardley, 1957)

is too rigid, but within the mantle, where the material is thought to be plastic (pseudoviscous) as a result of high temperature and confining pressure, convection is possible, and it is likely in the outer core which is thought to be fluid.

A model of convection in the earth (Fig. 25–6) was postulated by Pekeris (1935). Only one of several possible models is illustrated. Unfortunately, the location, shape, and size of convection cells cannot be detected directly in the earth. However, the nature of the earth's magnetic field and temperature gradients strongly supports the idea that convection exists in the earth.

Gravity Surveys and Development of the Convection Theory of Orogenesis

A Dutch geophysicist, Vening-Meinesz, made gravity surveys in the southwestern Pacific Ocean, particularly around Java and Sumatra, in the 1930's. He found large negative gravity anomalies over the deep-sea trenches (Fig. 25–7). Later, such anomalies were found to be associated with most trench systems and to persist after corrections were made for differences in density between rock and sea water in the trenches. In fact, they exist even over some islands and submarine ridges that are aligned with the trenches (see Barbados Ridge, Fig. 25–8). Negative anomalies remain even after isostatic reductions are made. Vening-Meinesz concluded that these anomalies resulted from projection of low-density rocks down into the denser subcrust under the island arcs. He postulated that lower density rocks, the sialic part of the crust, had been pulled into a downbuckle, later called a *tectogene,* by convection currents. According to his hypothesis, convection within the mantle exerts a frictional drag on the solid crust, and in areas where two cells converge and move down, the crust is pulled down as a tectogene into the mantle. Initial downwarping of the crust produces a geosyncline in which great quantities of sediment may accumulate. As the downbuckle is pulled farther down,

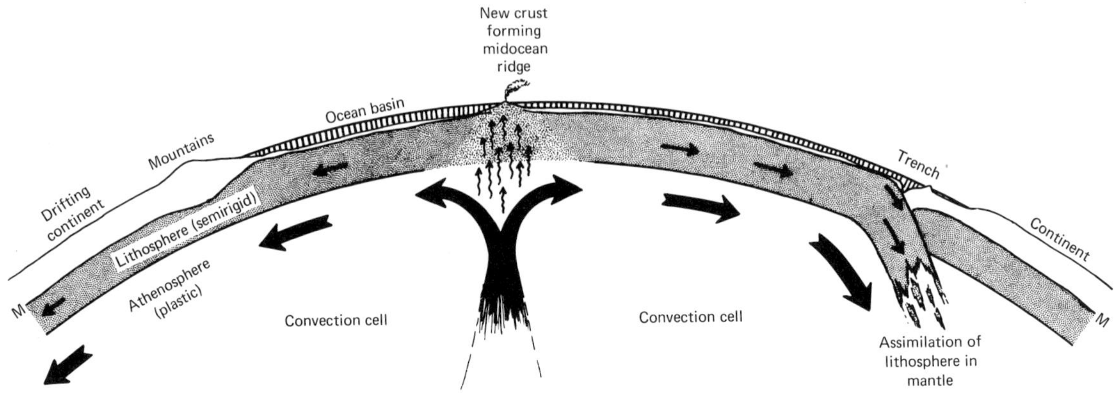

FIGURE 25–9
Generalized and schematic drawing (not to scale) showing rise of convection cells under midocean ridge where new crust is formed. The continent at the left is riding passively as the lithosphere moves as a result of convection and development of new crust. The lithosphere at right moves laterally and sinks beneath a trench where the lithosphere is assimilated back into the mouth.

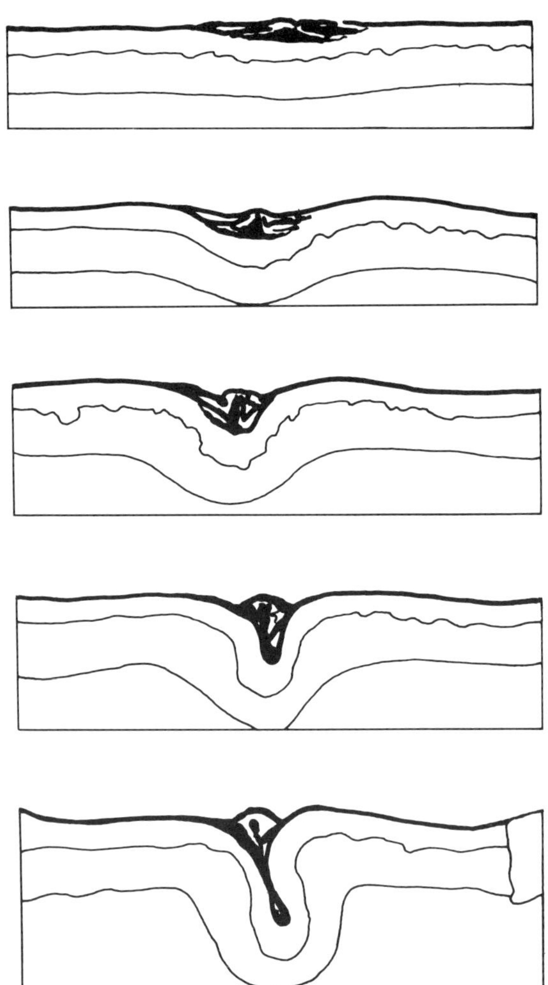

FIGURE 25–10
Experiment in the formation of a tectogene. A floating plastic layer is shown as it is compressed. According to Vening-Meinesz, the downbuckle may be formed in the earth by the drag of convection currents on the crust. These drawings, representing five successive stages, are based on photographs taken during an experiment by Ph. H. Kuenen.

the deeper portions are heated, deformed, and metamorphosed; magmas are created and intrude the mass. Eventually the downbuckle (Fig. 25–10) may be pressed together so that part of the material in it is forced upward to become a folded mountain range. Uplift is also sustained by isostatic adjustments that arise from the low-density crust's having been pulled down into a high-density mantle. As the uplift occurs, a metamorphic complex is brought close to the surface where it may be exposed later by erosion in the central part of the orogenic belt, and foreland folding takes place. Possibly the marginal folding is derived from the upward and outward push of the mountain root, or it may be the result of gravity sliding from the uplifted core. Thus Vening-Meinesz' hypothesis successfully explained many observed features of folded orogenic belts.

The idea of the tectogene is illustrated clearly in the experiments made by Ph. H. Kuenen (Fig. 25–10), in which a plastic layer is buckled and drawn down into a viscous sublayer.

Advocates of convection have been forced to modify some of the initial ideas because of more recent discoveries in the ocean basins. For example, the idea of a granitic crust under the oceans has been disproved by seismic studies, which reveal that the crust under the ocean basins is distinctly different from that under continents (Fig. 22–1). Thus the tectogene cannot be a downbuckle of sialic crust as Vening-Meinesz thought. Subsequently it has been suggested that the downbuckle takes place in the zone of transition between continental- and oceanic-type crusts.

Fischer and Hess (1963) suggested that oceanic crust consisting of basalts and water-laden sediment on serpentinized peridotite is carried into the mantle by convection until partial melting occurs.

Measurements of the rate of heat flow from the interior of the earth to the surface have supported the convection hypothesis. The rate of heat flow is abnormally low in island arcs; it is postulated that convection cells converge here and move down. But the rate of heat flow is abnormally high in the vicinity of the midocean ridges, which suggests that the midocean ridges are located over rising convection currents that bring warm mantle material close to the surface. This upwelling could also account for the high topography along the ridges and rises, and crustal extension in the form of grabens along the crest as well as the presence of volcanoes, as along the Mid-Atlantic Ridge (Fig. 25–11).

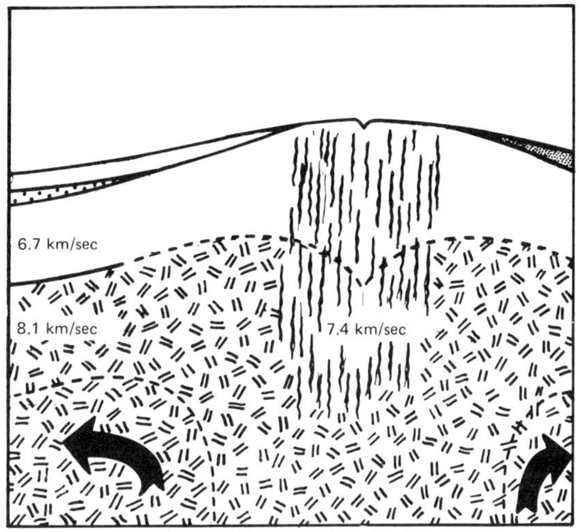

FIGURE 25–11 Schematic cross section across a midocean ridge showing apparent overlapping of ocean sediments on the ridge, dike swarms arising from an area of hot and low-velocity mantle under the ridge, and directions of convection movement deep in the mantle. (modified after Hess, 1962)

TOPOGRAPHIC AND GEOLOGICAL FIT BETWEEN AFRICA AND SOUTH AMERICA

A number of attempts have been made to fit the continents back together using their present shapes. Most attempts have met with limited success, but then it would be rather unusual if the lines along which the original continental plate broke were faithfully preserved during the time of separation and drifting and through all subsequent changes. In fact, it is remarkable that the fit between South America and Africa is so close. S. W. Carey first demonstrated the exactness of this fit (Fig. 25–12) by using maps drawn on a spherical surface. The best fit is obtained when a connection is drawn using the 2,000-meter isobath. This fit is so perfect that it stood as the most convincing line of evidence throughout the long debate over the validity of the theory of continental drift. Good fits are also found across the Gulf of Aden, the Red Sea, the Gulf of California, and between Africa and Madagascar and Africa and India. Perhaps the reason for the good fit between Africa and South America is that the break between these two occurred through a Precambrian shield in which the older orogenic belts had been completely welded together by metamorphism. The complex Paleozoic orogenic belts of North America, England, and eastern Europe lie along the line of rifting farther north (Figs. 25–23 and 24–1).

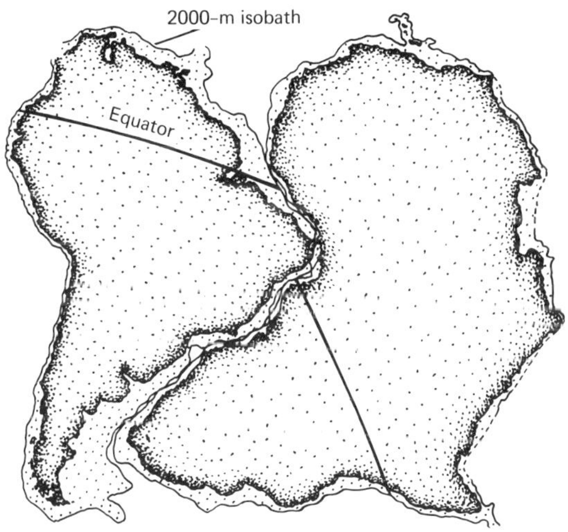

FIGURE 25–12 Fit between South America and Africa at the 2,000-meter isobath as constructed by Carey (1958).

Many attempts have been made to match the geology (as it appears on geologic maps) on one continent with that of another. These have often met with little success and it is easy to see why. If the fit should be made at the 2,000-meter isobath, the geologic provinces now exposed on land were originally separated by the width of the continental shelves including part of the continental

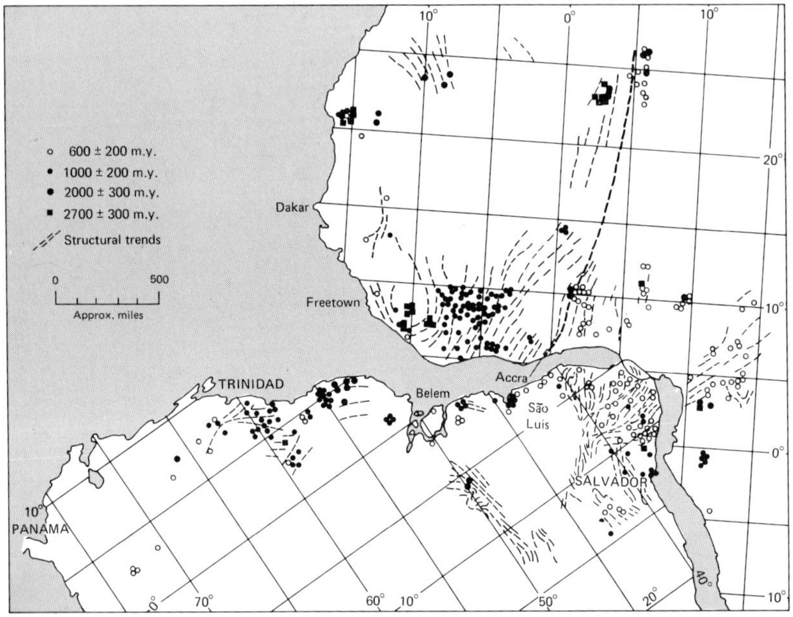

FIGURE 25–13
West Africa and South America shown fitted together according to the reconstruction of Bullard and his associates. In West Africa the 2,000-million-year Eburnean age province (solid circles) adjoins the 550-million-year Pan-African age province (open circles); the boundary between them is shown by the heavy dashed line. If Africa and South America were once joined together, this line would have entered Brazil near São Luis. The age measurements for Brazil are in accordance with those in West Africa, with the boundary at the predicted location. There may be a similar correlation between West Africa and the east coast of Brazil north of Salvador.
(from Hurley and others, 1967)

slopes on either side of the ocean. Thus only approximate fits can be obtained. For example, it was suggested (Kay, 1967) that the Appalachian Mountains, whose northern extremity is at Newfoundland, may be connected with the Caledonide Mountains of Ireland, Scotland, and Norway. Restoration by closing the North Atlantic leaves a covered shelf 320 kilometers wide between Newfoundland and Ireland on which no exposures can be seen.

P. M. Hurley and others (1967) studied the geology on either side of the Atlantic. A test of the fit was made at a point where the predrift reconstruction between West Africa and Brazil, based on topography, places the land areas close together. A large number of samples were taken on each continent from coastal localities (Fig. 25–13). Trends of folds and faults are shown along with the radiometric ages of the rocks. Results show a close correspondence between structural trends and ages when the continents are restored to their predrift positions.

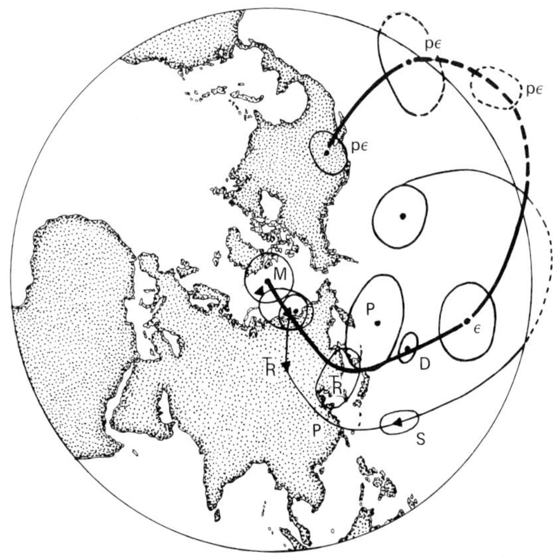

FIGURE 25–14
Polar wandering paths for Europe and North America. Solid triangles show American rocks. (pє, Precambrian; є, Cambrian; M, Miocene; Ʀ, Triassic; S, Silurian; D, Devonian; and P, Permian).
(after Runcorn, 1959)

EVIDENCE OF DRIFT FROM POLAR WANDERING

The discovery of remanent magnetism and the development of techniques by which the orientation of the magnetic field and its strength can be determined make it possible to reconstruct the magnetic field as it existed at times in the past (see Chapter 19). Many studies have been made of remanent magnetism for rocks of various ages and rocks of the same age located in different areas. Results indicate that the position of the poles has not remained constant through time. This is shown by calculation of pole positions for rocks of different ages on a single continent. For example, studies in North America indicate that the position of the north pole has shifted along the line indicated in Fig. 25–14 from a Precambrian location in the central Pacific to a Triassic position in Siberia to its present location. This indicates a shifting of the pole in relation to modern geographical distribution of land.

When paleomagnetic data from Europe are plotted on the same diagram, a similar shift of poles is indicated, but the lines are not coincident, as they should be if the geographic positions of

Europe and North America had remained fixed relative to one another. The two polar wandering paths can be made to coincide by shifting the geographic positions of Europe and North America back to the suggested predrift position shown in Fig. 25–4. Studies of the same type from other continents likewise reveal shifting of continents in relation to one another. Figure 25–15 shows the position of Australia relative to the pole at various times in the last 600 million years as determined by remanent magnetism.

The determination of polar wandering paths in the 1950's was an important factor in rekindling interest in the theory of continental drift. Here was an independent line of evidence of drift, and though it did not resolve the problem of drift mechanics, it did stimulate many earth scientists to look for new ways to evaluate the drift hypothesis.

CRUSTAL STRUCTURE

Even from the earliest stages of development of tectonic theories the importance of composition, structure, and processes in the earth's interior has been recognized. But relatively little was known of the interior until seismic and other geophysical techniques were applied. Some of the significant discoveries described earlier included the following:

1. The theory of isostasy and its implication that the outer part of the earth is essentially in a state of flotational equilibrium (Chapter 18).
2. Mohorovičić's discovery of the seismic discontinuity that separates the crust from the mantle (Chapter 21) and led to the first clear-cut distinction between continental and oceanic crustal structure (Fig. 25–16).
3. Gutenberg's discovery of the low-velocity zone at about 100 kilometers depth (Chapter 21, Fig. 21–23).

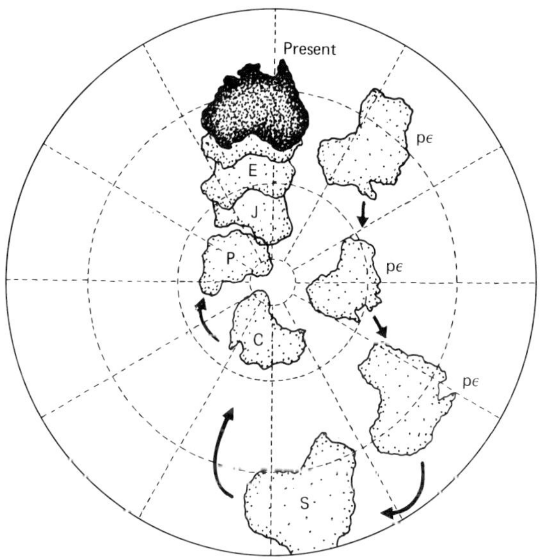

FIGURE 25–15 Stereographic projection showing position of Australia relative to the pole. (E) Lower Tertiary; (J) Mesozoic, probably Jurassic; (P) Permian; (C) Upper Silurian; (pє) Precambrian.
(after Runcorn, 1959)

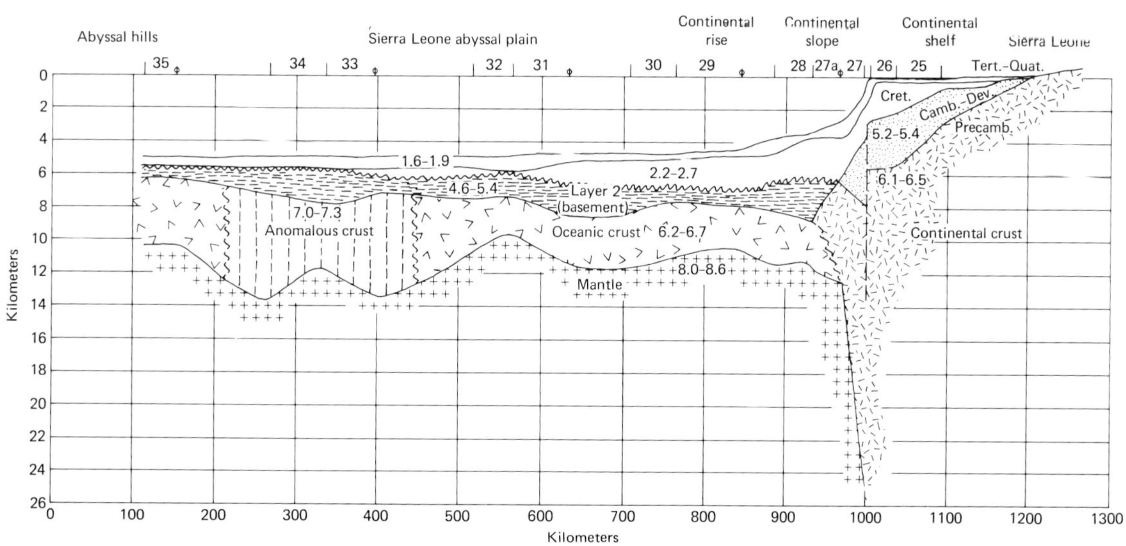

FIGURE 25–16 Interpreted structure section based on seismic refraction data off Sierra Leone. Vertical exaggeration, 20:1.
(from Sheridan and others, 1969)

Seismic Shear Zone—Benioff Zone

Other important discoveries have also played key roles in the development of our new global tectonics. One such study was made by Hugo Benioff (1949), who compiled records of the locations of earthquake foci and demonstrated that these foci are concentrated in distinct patterns. The main zone of deep earthquakes coincides with the circum-Pacific belt, where earthquakes are strikingly related to island arcs and young folded mountain belts. Benioff plotted the foci in these belts and showed that the foci define seismic zones that emerge near the deep-sea trenches and dip at about 50-degree angles back under the volcanic islands and toward the continents (Fig- 25–17). A similar zone is found dipping east under the Andes from the Peru Trench. It appeared that the island arcs and continents were riding out over the deep-sea floor. This *seismic shear zone*, as it is sometimes called, could be traced to great depth, 700 kilometers.

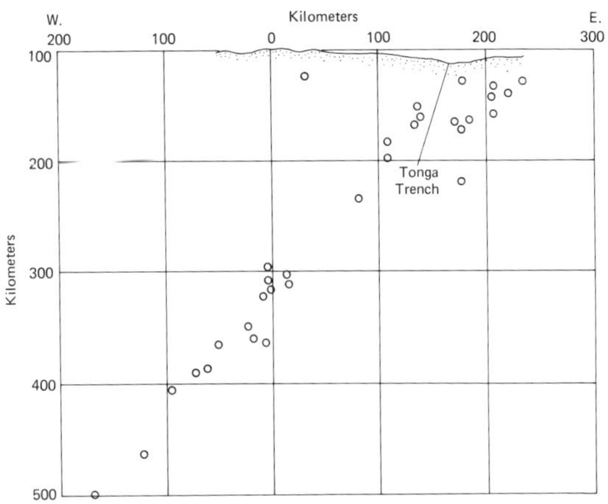

FIGURE 25–17
Plot of earthquake foci against depth in a cross section across the Tonga Trench and island arc. The topography has been exaggerated vertically.
(after Isacks and Oliver, 1968)

FIGURE 25–18 (*below*)
Generalized cross section across the Mid-Atlantic Ridge based on seismic and gravity data. Figures are seismic velocities in km/sec. Vertical exaggeration, 40:1.
(after Talwani and others, 1965)

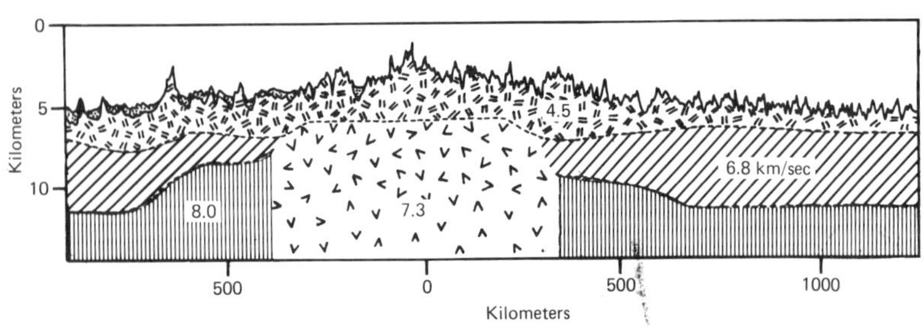

MIDOCEAN RIDGES AND ISLAND ARCS

Cross sections of crustal structure of the midocean ridges and island arcs prepared from seismic refraction profiles and gravity surveys first appeared in the early 1950's and have now been made across many arcs, continental margins, and ridges. Cross sections of the Mid-Atlantic Ridge, for example, show the following thicknesses (Fig. 25–18):

1. 0.5 kilometer of sediment.
2. 3 kilometers of rock with velocity of 5 kilometers per second (about that of basalt).
3. A deep substratum of velocity—7.3 kilometers per second.

The most significant distinction between this and sections of the deep oceanic crust (Fig. 22–1) is the substratum with velocity 7.3 kilometers per second. This velocity is intermediate between that of the crustal rocks typical of the oceans (6.7 km/sec) and those of the mantle (8.1 km/sec), below the *M*-discontinuity. The material with 7.3-kilometers-per-second velocity may be serpentine. Mantle rock could be altered by solutions to form serpentine. An alternative theory is that the 7.3-kilometers-per-second velocity occurs in a zone where the mantle and crust are mixed as a result of volcanism and igneous intrusions.

Crustal sections across island arcs have consistently shown thin oceanic crust under the trench and a projection of material (possibly serpentine) 15 to 30 kilometers down into the mantle. Material under the islands has both lower velocity and lower density than mantle rock. These anomalous velocities and densities are too high to be granitic rock and probably represent altered basalt or ultrabasic rock of the upper mantle.

BIRTH OF THE CONCEPT OF SEA-FLOOR SPREADING

A unifying concept of global tectonics emerged in papers by R. S. Dietz (1961) and H. H. Hess (1962). No single discovery can be credited with prompting this new conceptual scheme. The preceding decade had been one of great achievement in the search for knowledge of the sea floor and the structure of the earth's crust and upper mantle.

The convection current hypothesis, greatly strengthened by heat flow measurements over ridges, was widely accepted. Knowledge of crustal sections of continents, continental margins, the sea floor of the deep basins, island arcs, and oceanic ridges had been obtained from gravity and seismic studies. Continental drift had come back into prominence largely because of paleomagnetic studies of polar wandering. Gutenberg had discovered the low-velocity layer, and its significance as a zone of plasticity was recognized. These were certainly among the important considerations when Dietz wrote the following:

> The concept proposed here, which can be termed the 'Spreading of the sea floor,' is largely intuitive, having been derived through an attempt to interpret sea floor bathymetry. . . . The assumed model [of crustal structure] is as follows:
>
> 1. Large scale thermal convection cells, fuelled by the decay of radioactive minerals, operate in the mantle. They must provide the primary diastrophic forces affecting the lithosphere.
> 2. The sequence of crustal layers beneath the oceans is markedly different from that beneath the continents and is quite simple [see Fig. 22–1].
> 3. . . . The outer 70 km of the Earth . . . is moderately strong and rigid even over time spaces of 100,000 years or more; this outer rind is the *lithosphere*. Beneath lies the *asthenosphere* separated from the lithosphere by the level of no strain . . . ; it is a domain of rock plasticity and flowage where any stresses are quickly removed.
> 4. . . . the continents are low-density tabular masks of sial—a "basement complex" of granitic rocks about 35 km thick. . . .
>
>   . . . the oceanic "crust" is almost wholly coupled with the convective overturn of the mantle creeping at a rate of a few cm/yr. Since the sea floor is covered by only a thin veneer of sediment with some mixed in effusives, it is essentially the outcropping mantle. So the sea floor marks the tops of the convection cells and slowly spreads from zones of divergence to those of convergence. . . .
>
> The gross structures of the sea floor are direct expressions of this convection. The median rises mark the up-welling sites or divergences; the trenches are associated with the convergences or down-welling sites. . . .

H. H. Hess made many contributions toward the development of modern tectonic theories. He outlined some basic assumptions and conclusions in 1962 as follows:

1. The mantle is convecting at a rate of 1 cm/yr.
2. The convecting cells have rising limbs under the midocean ridges.
3. The convecting cells account for the observed high heat flow and topographic rise.
4. Mantle material comes to the surface on the crest of these ridges.
5. The oceanic crust is serpentinized peridotite, hydrated by release of water from the mantle over the rising limb of a current. In other words, it is hydrated mantle material.
6. The uniform thickness of the oceanic crust results from the maximum height that the 500°C isotherm can reach under the midocean ridge.
7. Seismic velocities under the crests of ridges are 10–20 percent lower than normal for the various layers including the mantle, but become normal again on ridge flanks. This is attributed to higher temperature and intense fracturing with cooling and healing of the fractures away from the crest.
8. Midocean ridges are ephemeral features having a life of 200 to 300 million years (the life of the convecting cell).
9. The Mid-Pacific Mesozoic Ridge (extends from the Marianas to Chile) is the only trace of a ridge of the last cycle of convecting cells.
10. The whole ocean is virtually swept clean (replaced by new mantle material) every 300 to 400 million years.
11. This accounts for the relatively thin veneer of sediments on the ocean floor, the relatively small number of volcanic sea-

mounts, and the present absence of evidence of rocks older than Cretaceous in the oceans.

12. The oceanic column is in isostatic equilibrium with the continental column. The upper surface of continents approaches equilibrium with sea level by erosion. It is thus axiomatic that the thickness of continents is dependent on the depth of the oceans.

13. Rising limbs coming up under continental areas move the fragmented parts away from one another at a uniform rate so a truly median ridge forms, as in the Atlantic Ocean.

14. The continents are carried passively on the mantle with convection and do not plow through oceanic crust.

15. Their leading edges are strongly deformed when they impinge upon the downward moving limbs of convecting mantle.

16. The oceanic crust, buckling down into the descending limb, is heated and loses its water to the ocean.

17. The cover of oceanic sediments and the volcanic seamounts also ride down into the jaw crusher of the descending limb, are metamorphosed, and eventually probably are welded onto continents.

18. The ocean basins are impermanent features, and the continents are permanent although they may be torn apart or welded together and their margins deformed.

19. The earth is a dynamic body with its surface constantly changing. The spherical harmonics of its topography show unexpected regularities, a reflection of the regularities of its mantle convection systems and their secondary effects.

FIGURE 25–19
Summary diagram of total magnetic-field anomalies southwest of Vancouver Island. Areas of positive anomaly are shown in black. Straight lines indicate faults offsetting the anomaly pattern. Arrows show the axes of the three short ridge lengths within this area; from north to south they are the Explorer, Juan de Fuca, and Gorda ridges.
(from Raff and Mason, 1961)

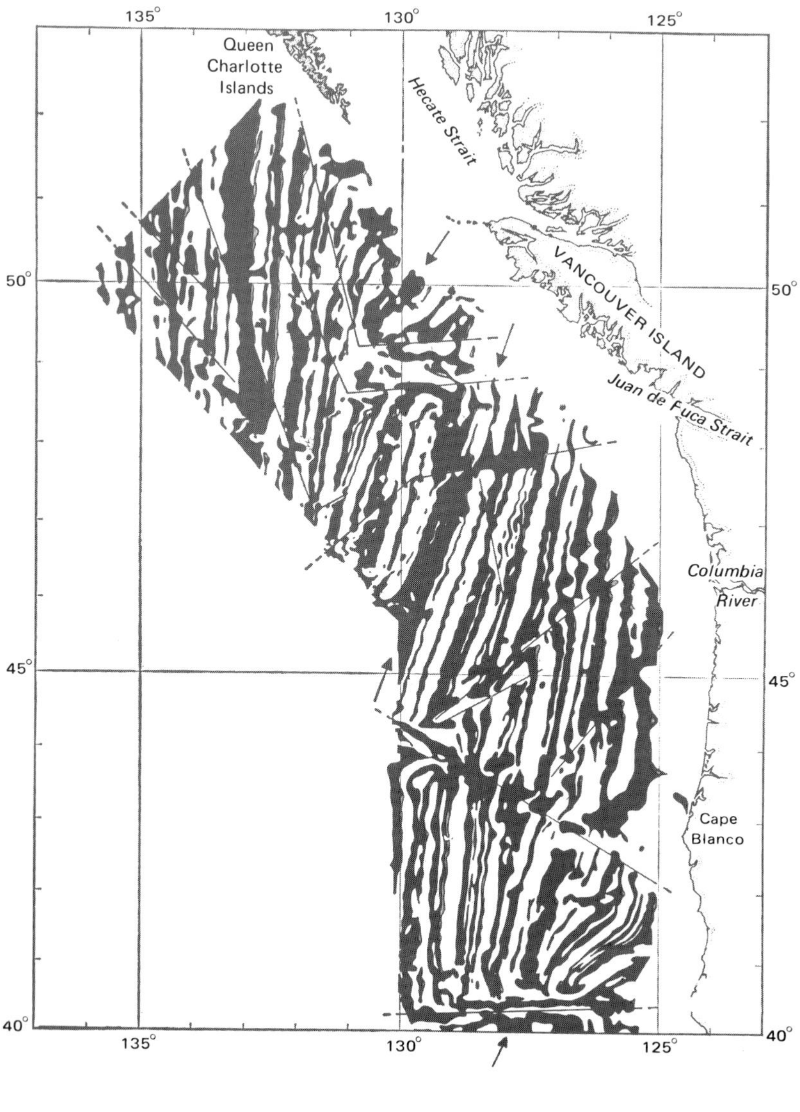

Magnetic Anomalies: A Key to Sea-Floor Spreading

A. D. Raff and R. G. Mason (1961) compiled maps that show a striking anomaly pattern of high and low magnetic-field intensities even before sea-floor spreading was proposed (Fig. 25–19). The anomalies are roughly parallel to the west coasts, but show offsets where major faults in the Pacific sea floor cut east–west across the belts of anomalies. Surveys of magnetic-field strength revealed a similar pattern of belts of anomalies symmetrically distributed on either side of the crest of the Mid-Atlantic Ridge. In 1963, F. J. Vine and D. H. Matthews suggested that these variations in intensity are caused by changes in the strength and polarity of the earth's magnetic field that had become "fossilized" in the oceanic crust. As new crust is formed along the crest of the oceanic ridges it cools, passing through the Curie point at which the magnetic-field strength and polarity are "frozen" in. Thus over a period of time the sea floor spreads apart from the crest of the ridges as new crust is formed (Fig. 25–20). Not only does this explains the anomaly belts, but it indicates a way to measure the rate of spreading. If the anomalies are in fact caused by reversals in the polarity of the field, then the anomalies can be correlated with the reversals on continents where the age of the reversals can be determined by radiometric methods (Figs. 19–12 and 19–13). The distance of an anomaly belt of any known age from the crest of the ridge can be used to calculate the rate of spreading. Plots of anomalies against distance (Fig. 25–22) indicate that the sea floor is spreading at a rate of between 2 and 4 centimeters per year. Spreading rates of 8 to 12 centimeters per year have been calculated for parts of the East Pacific rise.

Some of the anomaly belts have been traced throughout much of the length of the midocean-ridge system. Once the traverses are spaced closely enough together to allow exact correlation of anomalies, it becomes possible to restore the continents to their positions at the time a particular anomaly formed. Figure 25–23 shows the Atlantic restored to the shape it had when anomaly 31 was located at the crest of the Mid-Atlantic Ridge.

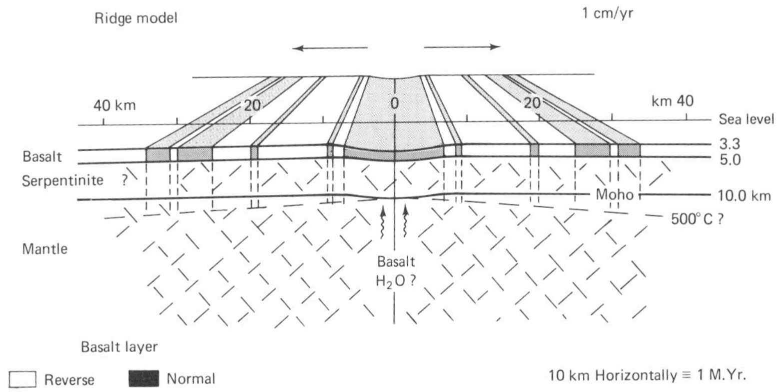

FIGURE 25–20
Diagrammatic representation of the oceanic crust at a midocean-ridge crest, assuming active spreading at a rate of 1 centimeter per year per ridge flank, and the geomagnetic reversal time scale of Cox and his associates (1967).
(from Vine, 1969)

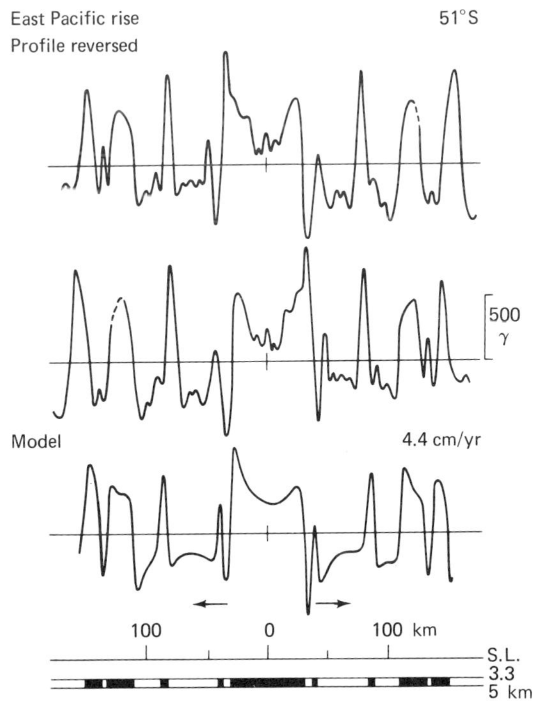

FIGURE 25–21
An observed magnetic anomaly profile across the East Pacific rise (upper curves) compared with a computed profile for this area, assuming the model of Fig. 25–20 and a spreading rate of 4.4 centimeters per year. (observed profile from Pitman and Heirtzler, 1966) The computation assumes an intensity and dip for the earth's magnetic field of 48,700 gammas and −63°, and a magnetic bearing for the profile of 102° (1 gamma = 10^{-5} oersted). Normal or reverse magnetization is with respect to an axial dipole vector. Effective susceptibility assumed = ±0.01, except for central block (+0.02). S.L. = sea level.
(from Vine, 1969)

Sediment Distribution in the Oceans

Rates of modern sedimentation in the oceans have been estimated in many places in the ocean basins from measurements of cores. The measurements indicate accumulations of between 5 millimeters and 2 centimeters per thousand years. If the smaller figure is used, the ocean basins should have accumulated approximately 8.5 kilometers of compacted sediment. Instead, seismic studies indicate an average thickness of only 1.3 kilometers in the deep-ocean basin. Long before the concept of sea-floor spreading was advanced this discrepancy led some geologists to conclude that the oceans must be relatively young.

A corollary of the spreading hypothesis should

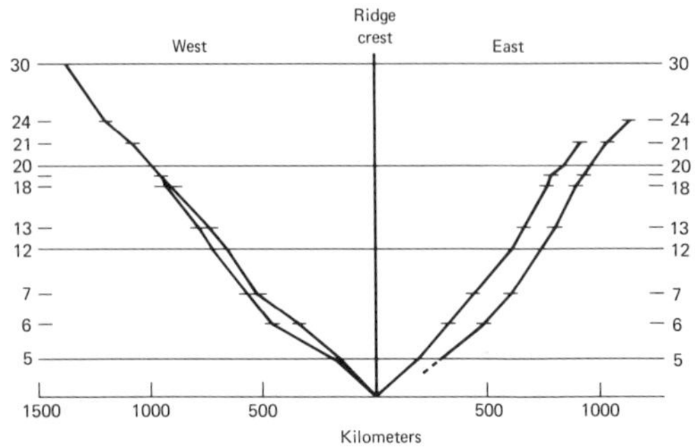

FIGURE 25–22
Plot of magnetic anomaly numbers on either side of the Mid-Atlantic Ridge in the South Atlantic against distance from the ridge crest.
(modified after Dickson and others, 1968)

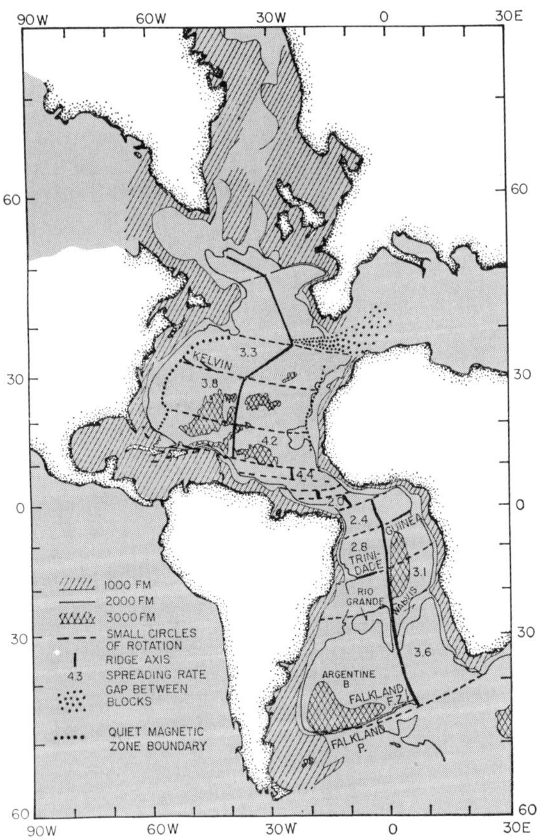

FIGURE 25–23
The Atlantic at the time of anomaly 31. The possible locations of major Mesozoic fracture zones, shown by dashed lines, were obtained by assuming that the pre-Cenozoic drifts can be described for South America and North America by a single rotation each. The average spreading rate in centimeters per year for each block is obtained by assuming a constant rate of spreading between 120 and 70 million years ago. The bathymetry is not shown for the Caribbean and Gulf of Mexico. Notice that the trends of the Falkland plateau and fracture zone, Rio Grande and Walvis ridges. Trinidade ridge, Guinea ridge, and Kelvin Seamount chain agree well with the predicted trends.
(from Le Pichon, 1968)

be that sediment thickness increases away from the crest of the midocean ridge and that older sediment should be found at progressively greater distances from it. Graphs showing sediment thickness on five traverses across the midocean ridge (Fig. 25–24) confirm the predicted distribution of thickness.

Discovery of Underthrusting in Island Arcs

The seismic shear zone under island arcs first identified by Benioff (1954) has been confirmed by more recent studies of seismicity. Many geologists who have worked in the island arcs have interpreted the fold and thrust faults there to be products of uplift and horizontal crustal compression; however, some suggested that the folding and lateral transport of large overturned folds and thrust sheets were caused entirely by gravity sliding after uplift. Those who favored the idea of crustal compression in the arcs interpreted the seismic shear zone as a major fault zone in the crust and mantle along which the *relative* motion is such that one crustal plate is moving over another. Seismic studies have confirmed that most arcs are under compression and that movement on the Benioff shear zone is thrusting.

Bryan Isacks, Lynn Sykes, and Jack Oliver (1969) initiated studies of waves generated by deep earthquakes in the seismic shear zone under the Tonga-Kermadec island arc. They found that waves that traveled up along the seismic shear zone were different from those that traveled out of the zone and through the mantle under the deep-sea floor. This analysis demonstrated the existence under the arc of an anomalous zone in the upper mantle, estimated to be 100 kilometers thick and bounded on its upper surface by the seismic shear zone. The model derived to explain the seismic character of this zone (Fig. 25–26) shows the oceanic crust and upper mantle (lithosphere) moving down and under the island arc. This anomalous zone is traced to a depth of 700 kilometers when it is inferred that the lithosphere is assimilated into the mantle.

The concept of sea-floor spreading is greatly strengthened by the discovery of the anomalous crust and underthrusting in the island arcs. The island arcs appear to be "sinks" where old oceanic crust goes back into the mantle as new crust is generated at the midoceanic ridges. If such sinks did not exist we would be forced to accept the idea that the earth is expanding in order to accommodate new sea floor.

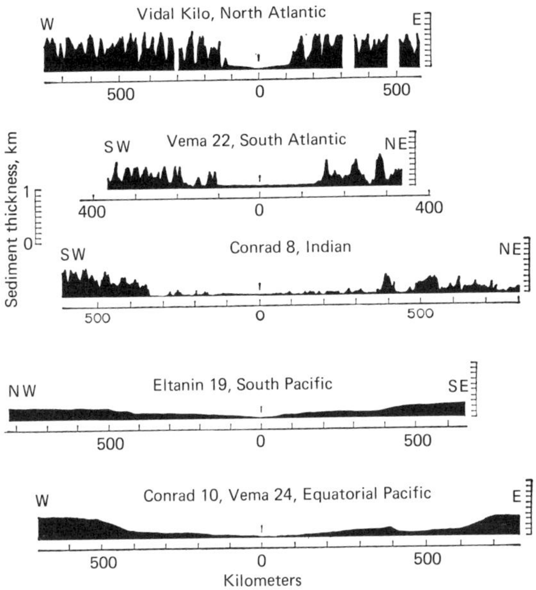

FIGURE 25–24

Graphs of sediment thickness versus distance from the ridge crest for five traverses. The bottom one is a composite made from three traverses in the Pacific where the belt of thick equatorial sediments intersects the East Pacific rise. On the eastern side the pattern is complicated by fractures (transform faults) or structure associated with the Galapagos Islands. The names of the cruises made to map the traverses are given above each section.

(from Ewing and Ewing, 1967)

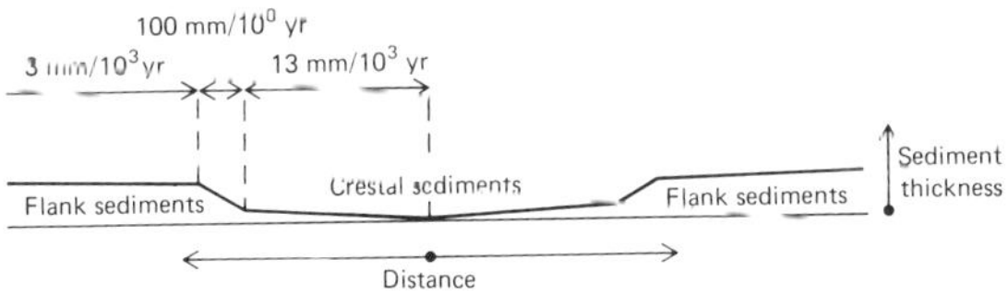

FIGURE 25–25

Generalized sketch of crestal and flank sediment thickness based on data from the equatorial Pacific. The slopes give accumulation rates as indicated if spreading is assumed constant at 4.5 centimeters per year.

(from Ewing and Ewing, 1967)

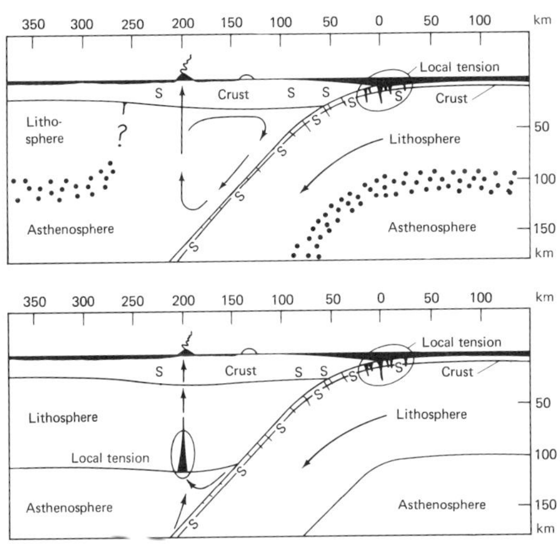

FIGURE 25–26

Vertical sections through an island arc indicating hypothetical structure. Both sections shown down-going slab of lithosphere, seismic zone near surface of slab and in adjacent crust, and tensional features beneath the ocean deep where slab bends abruptly and surface is free. (In both sections, S indicates seismic activity.)

(Top)

A gap in mantle portion of lithosphere beneath island arc and circulation in mantle associated with crustal material of the slab and with adjoining mantle.

(Holmes, 1965)

(Bottom)

The overriding lithosphere in contact with the down-going slab and bent upward as a result of overthrusting. The relation of the bending to the volcanoes follows Gunn (1947). No vertical exaggeration.

(from Isacks and Oliver, 1968)

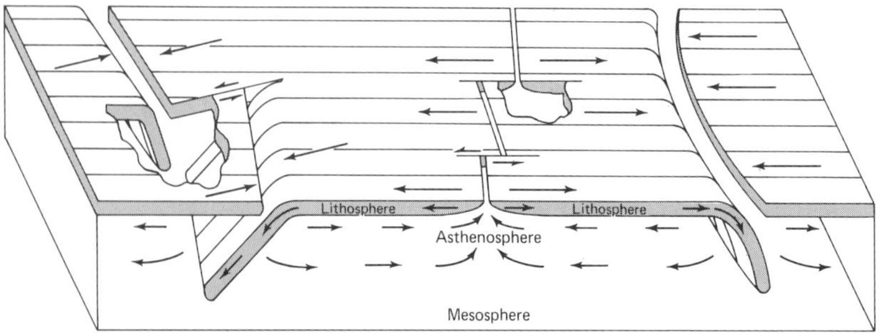

FIGURE 25–27
Block diagram illustrating schematically the configurations and roles of the lithosphere, asthenosphere, and mesosphere in a version of the new global tectonics in which the lithosphere, a layer of strength, plays a key role. Arrows in lithosphere indicate relative movements of adjoining blocks. Arrows in asthenosphere represent possible compensating flow in response to downward movement of segments of lithosphere. One arc-to-arc transform fault appears at left between opposing zones of convergence (island arcs), two ridge-to-ridge transform faults along ocean ridge at center, and simple arc structure at right.
(from Isacks and Oliver, 1968)

Focal Mechanism and Movements in the Crust

The association between earthquakes and faulting has been long established. Seismologists began to study the focal mechanisms of earthquakes hoping to find a relationship between the direction of initial slip on a fault and the movement direction of the first seismic wave to arrive at a station. They succeeded in establishing this relationship, making it possible to determine the initial slip direction on faults by study of seismograms. The World-Wide Standard Seismograph Network was established and provided the data necessary to analyze initial movements on faults in the seismic shear zone as well as on faults that cut across and displace the midocean ridges (Fig. 25–1). The initial movement directions are consistent with the idea of crustal compression in most of the island arcs (Figs. 25–27 and 25–28).

FIGURE 25–28
Motion directions in the Aleutians. Pacific plate is moving northwesterly with respect to Alaska and the Bering Sea. These motions are inferred from the mechanisms of earthquakes located along the inner margin of the trench. Events in the deep part of the trench are characterized by large components of normal faulting with the axes of tension nearly perpendicular to the local strike of the trench.
(from Stauder, 1968)

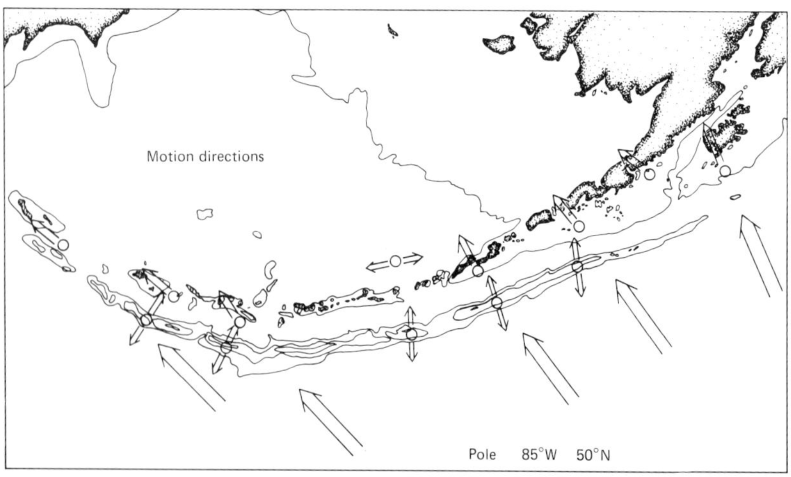

Transform Faults

Major fault zones were identified in the ocean-basin floor soon after detailed topographic studies began. The long east–west trending faults in the eastern Pacific were the first zones to be studied in detail. These faults make prominent scarps in the sea floor, some thousands of meters high. Seismic and gravity surveys reveal that these faults extend deep into the sea floor through the crust. When magnetic surveys were made (Fig. 25–19) the magnetic anomalies were found to be offset along the faults. Similar faults cut across the Mid-Atlantic Ridge (Fig. 25–1).

The displaced magnetic anomalies along these faults seemed to afford an excellent way to determine the strike-separation along the fault, but when attempts were made to correlate the magnetic anomalies across the faults serious problems arose. The separation appeared to be left-lateral at one end of some of the faults and right-lateral at the other. In 1965 J. T. Wilson offered an explanation of this dilemma. He concluded that the faults could be interpreted in terms of a mobile sea floor and that some of these faults separated portions of the sea floor that are spreading at different rates (Fig. 25–29). He called faults of this type *transform* faults.

PLATE TECTONICS

The crust of the earth has been shown to possess long, often irregular, narrow belts where high crustal mobility and other signs of instability such as volcanic and plutonic activity, high seismicity, and diastrophism are concentrated. These belts include the modern orogenic belts, the island arcs, and the midocean-ridge system and may be used as a basis for subdividing the crust into a number of large, relatively stable plates (Fig. 25–30). It now appears possible that modern orogeny, sea-floor spreading, and crustal deformation in general can by analyzed in terms of the movements of these plates. The direction of movement of the plates can be deduced from the first-motion analysis of seismic waves. The movement picture delineated in this way is shown in Fig. 25–30.

The concepts of convection in the mantle, continental drift, and sea-floor spreading can be easily combined to make a unified theory of tectonics, which has been called the *new global tectonics*. Key aspects of this theory are the following:

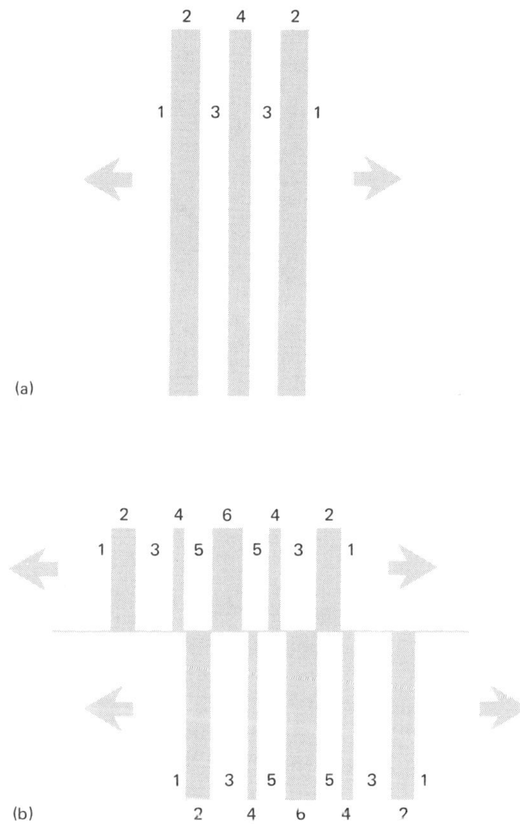

FIGURE 25–29
Transform fault. Schematic drawing showing distribution of belts of magnetic anomalies due to field reversals along a ridge with uniform spreading rates (a), and along a ridge broken by a transverse fault that separates two parts of the ridge that have different spreading rates (b).
(after J. T. Wilson, 1965)

1. Convection currents in the mantle moving below a depth of 100 kilometers probably provide the motive force for diastrophism.
2. Convection currents rise beneath the midocean-ridge system where mantle rock is altered by solutions.
3. Systems of dikes and plutons of basaltic composition are emplaced in the vicinity of the ridge crest, and these intrusions take on the magnetizations of the earth's field at the time of crystallization.
4. Rates of formation of new sea floor vary along the crest of the ridge and transform faults separate areas with different spreading rates.
5. The crust and solid upper mantle, called the lithosphere, ride passively on the top of the spreading mantle rock, which is plastic and capable of flowage.
6. Continental plates are carried passively on the moving lithosphere. In some places the continental plates are located over rising convection cells. The Gulf of Aden and the Red Sea mark one such place. The crest of the East Pacific rise passes under the western part of North America, entering at the Gulf of California and emerging at about the latitude of Oregon. The continental plates are frequently split by graben structures where the ridges pass under the continental plate.
7. The oceanic plates move slowly away from the midocean ridges where new crust is being formed, and this oceanic crust eventually goes back into the mantle in what are called *zones of subduction*. The island arcs are examples, but subduction may also occur under continental margins and under continents as well.
8. Orogeny is related to the movement of crustal plates in several possible ways. Compression in zones of subduction could account for folding and thrusting both in island arcs and along continental margins. As relatively lighter material is carried

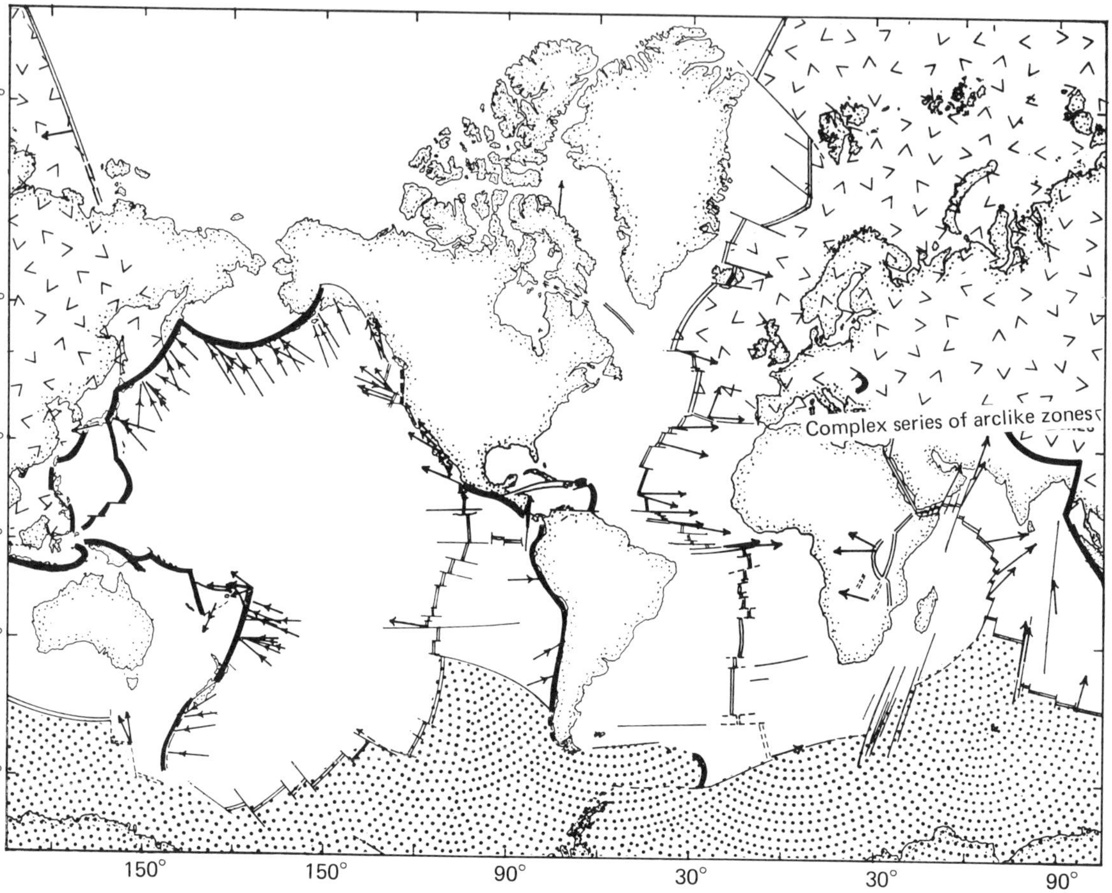

FIGURE 25-30
Division of the earth's crust into plates. Borders are seismically active. Directions of movement are indicated by arrows. Plates are bounded by trenches, oceanic ridge crests, and orogenic belts.
(modified after Le Pichon, 1968)

down into the mantle and metamorphosed, it becomes buoyant, causing uplift in the overlying crust. Orogeny may also occur where there are differential rates of movement between plates or where two plates are forced together. The west coast of South America may be viewed as an orogenic belt formed where the continent is moving westward relative to the east side of the East Pacific rise. The Urals and the Himalayas appear to have formed where movement of plates forced two continental shields together, compressing, deforming, and uplifting the sediments formerly deposited in the seas between them.

These new theories of sea-floor spreading and plate tectonics no doubt will be refined and modified in coming decades, but so far they have provided the most satisfactory hypotheses yet devised for studying relationships of large-scale crustal features.

CHAPTER OUTLINE

TERMS

asthenosphere
continental accretion
continental drift
contracting earth theory
convection current hypothesis
dynamic encounter hypothesis
expanding earth hypothesis
Gondwanaland
Laurasia
lithosphere
new global tectonics
Pangaea
tectogene
sea-floor spreading
seismic shear zone
tillite
transform faults
zones of subduction

REVIEW QUESTIONS

1. Compare Figs. 25–1 and 25–30 with Figs. 20-16, 17, and 18, and describe the relationships you observe.
2. What is the significance of the observation that free-air gravity anomalies lie over some islands and submarine ridges on a trend with deep-sea trenches (see Figs. 25–7 and 25–8).
3. Summarize the directly observed data and the inferences drawn from those data that are used to support the concept of sea-floor spreading.
4. How could continental drift be proved by direct observation?
5. What information is needed to reach a conclusion about rate of sea-floor spreading?
6. How are magmas thought to originate according to the tectogene hypothesis and sea-floor spreading? Does this adequately explain the origin of all magma? If not, give examples of volcanic activity it does not explain.
7. What objections are there to the idea that continents move over and through the mantle and oceanic crust?
8. If you were free to plan a series of deep wells to plumb the mantle in search of evidence concerning continental drift, where would you have them drilled and what data would you seek?
9. Explain the way in which a transform fault differs from most other strike-slip faults.
10. Summarize the way folded mountains may originate as a result of plate tectonics.

SUGGESTED READINGS

Dietz, R. S., 1961, Continental and ocean basin evolution by spreading: Nature, v. 189.

Geological Survey of Canada Dept. of Mines and Technical Surveys, 1965 (1967), The world rift system: Canada Geol. Survey Bull. Paper 66–14, 471 p.

Geological Survey of Canada Dept. of Mines and Technical Surveys, 1965 (1966), Continental margins and island arcs: Canada Geol. Survey Bull. Paper 66–15, 486 p.

Hess, H. H., 1962, History of ocean basins, *in* Petrologic studies: A volume in honor of A. F. Buddington: Geol. Soc. America, p. 599–620.

Howell, B. F., Jr., 1959, Introduction to Geophysics: New York, McGraw-Hill, 399 p.

Isacks, B., and **Oliver, Jack,** 1968, Seismology and the new global tectonics: Jour. Geophys. Research, v. 73, no. 18, p. 5855–5899.

Johnson, Helgi, and **Smith, B. L.,** 1970, The Megatectonics of Continents and Oceans: New Brunswick, N.J., Rutgers University Press, 282 p.

Menard, H. W., 1964, Marine Geology of the Pacific: New York, McGraw-Hill, 271 p.

Munyan, A. C., ed., 1963, Polar wandering and continental drift: Soc. Econ. Paleontologists and Mineralogists Spec. Pub. 10, 169 p.

Phinney, R. A., ed., 1968, The History of the Earth's Crust, A Symposium: Princeton, N.J., Princeton University Press, 244 p.

Takeuchi, H., Uyeda, S., and **Kanamori, H.,** 1967, Debate about the Earth: San Francisco, Freeman, Cooper, and Co., 253 p.

Vine, F. J., 1969, Sea-floor spreading—New evidence: Jour. Geol. Educ., v. xvii, no. 1, p. 6–16.

APPENDICES

A. TOPOGRAPHIC AND GEOLOGIC MAPS

B. CHEMISTRY

C. ELEMENTS OF GEOLOGICAL PHYSICS

D. SOURCES OF GEOLOGICAL INFORMATION

E. OIL AND GAS

A

TOPOGRAPHIC AND GEOLOGIC MAPS

A *topographic map* is a means of accurately representing the configuration of the land on a plane surface. Generalized impressions of land forms may be obtained by use of shading or hachure lines, but a topographic map is constructed so that a very close approximation of the elevation of any point on the map is possible. This is accomplished by using lines, called contour lines, which are map projections of lines connecting points of equal elevation on the ground (see Fig. A-1). These lines are drawn at set, but arbitrarily determined, intervals. The contour interval might be five feet in the coastal plain but one hundred feet in a mountainous region. An interval is selected to provide an optimum amount of information without having the lines merge. Normally the elevation of every fifth contour is given on the map and these lines are heavier lines.

Contours have certain properties as a result of their definition. They connect points of equal elevation. Thus points falling between two adjacent contours normally have elevations intermediate between the elevations of the lines to either side. This generalization fails along the top of a ridge or in a valley where the elevation of intermediate points may be above or below that of two contours of identical elevation. Contours may be closed, as they are around a hill that is entirely shown on the map, or they may run off the margin of the map. A single map may have more than one contour of the same elevation. If there are two hilltops, each may be ringed by a contour of the same elevation. Contours never cross one another unless they are superimposed over a vertical cliff or along an overhanging ledge.

Topographic maps are made by the United States Geological Survey, the Coast and Geodetic Survey, and by some states and private firms. They are also frequently constructed by geologists and surveyors for small areas. The government series of topographic maps are most commonly printed at a scale such that 1 inch of map distance is equal to 1 mile of horizontal ground distance (1:62,500) or at a scale of 1:24,000. These maps cover areas of 15 or 7½ minutes latitude by 15 or 7½ minutes longitude respectively, and are called 15 minute and 7½ minute quadrangles. In addition to contour lines quadrangle maps provide the following information:

1. Scale in miles, feet, and kilometers. The scale is written as a fraction also, that is, 1:24,000, meaning that, regardless of the units of measure used, one of those units measured on the map is equivalent to 24,000 of the same units measured horizontally across the ground.
2. Cultural features such as roads, highways, towns, houses, churches, cemeteries, ranches, camps, irrigation ditches, trails, township lines, county lines, park boundaries, and state boundaries.
3. Physical features including rivers, lakes, marshes, levees, beaches, and mountain peaks.
4. The name of the map area and the names of those that adjoin at corners and along sides, printed at map margins.
5. Exact elevations of points that have been determined by ground surveying. These include bench marks (BM with elevation given) and triangulation stations (triangle with elevation). Exact elevations of some peaks and other points may be given also.

Topographic maps are widely used by geologists for interpreting land forms, for studying drainage systems, and especially as base maps on which to record other information. They are also of great value in surveying, in highway planning, and in regional and city planning.

GEOLOGIC MAPS

Geologic maps show interpretations of the areal distribution of one or more of the following:

1. Rock types such as granite, basalt, schists, and sandstone.
2. The rocks of a certain age, those that were formed during a certain interval of geologic time.
3. Units of rock, called formations, that can be recognized and mapped (e.g., a sequence of sandstone beds, etc.).

Other specialized maps are *structure-contour maps,* which show the elevation of a particular stratum, isopach maps, which indicate the thickness of strata, or paleogeographic maps, which show ancient geography. Geologic maps may be of any scale. Some are done in detail using a scale where a few feet ground measure equals an inch on the map. More frequently the scale is between 500 feet to an inch and 1 mile to an inch, but others cover entire states or countries. The scale selected should allow enough room to show the amount of detail required by the purpose for which the map is being made.

Several types of data are usually presented on geologic maps:

1. Many geologic maps are printed on topographic map bases to make possible location and correlation of land forms and geology.

2. The same cultural features that appear on topographic maps may appear on geologic maps.

3. Some geomorphic features such as streams, lakes, coasts, and mountain peaks are indicated.

4. Lines of latitude and longitude are usually shown.

5. The contacts between the units being mapped are drawn whether these are contacts between rocks of different ages or of different rock types. The surface between two bodies of rock (e.g., between two layers or between an igneous intrusion and the country rock) is called a *contact*. This surface of contact intercepts the ground surface to form a line of contact. To determine where these contacts are located on the ground surface and to show their location on a map are primary objectives of most geologic mapping. Contacts are drawn as solid lines where the bodies of rock on both sides of the contact are exposed at the surface. Contacts are shown dotted or dashed where soil, ice, or sediment covers the rock or where the contact is inferred.

6. Faults are shown on the map. The lines are solid where the fault is exposed, dotted or dashed where it is not exposed. Symbols are used to indicate the direction of relative motion of the two sides of the fault. (Arrows indicate strike-slip, *U* and *D* signify upthrown and downthrown sides, barbs or *T* indicates a thrust fault.)

7. Strike and dip symbols are used to indicate the local attitude of the plane of contact between two rock units; they may also appear on faults and fractures or to indicate the orientation of platy minerals. It is highly desirable to have this information when making interpretations of the map or when drawing cross sections.

8. A legend is provided. Along one margin of the geologic map a column or vertical series of boxes contains the color pattern or symbols used to represent each of the rock units on the map. Standard procedure is to separate the three types of rocks (igneous, metamorphic, and sedimentary). Within each of the types the youngest rock unit is placed at the top of the series with successively older units farther down in the column. The basic rock units used for the purposes of mapping are called *formations*.

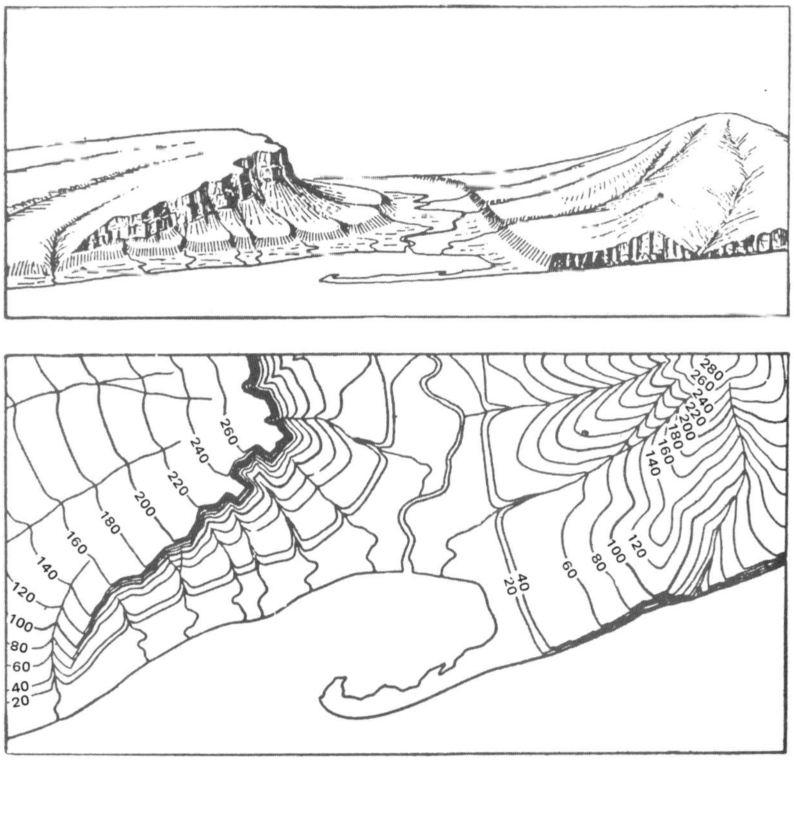

FIGURE A–1
An example of a topographic map.
(U.S. Geological Survey)

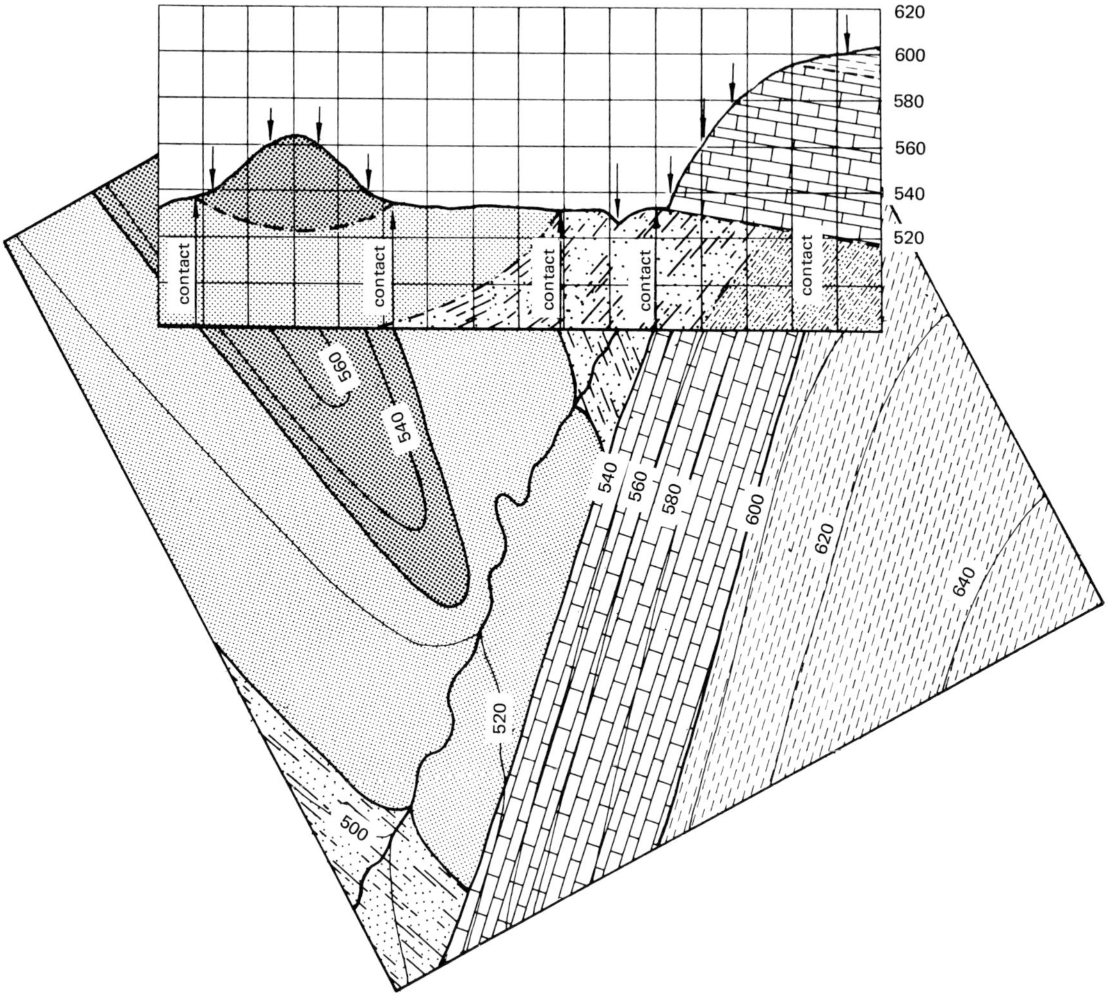

FIGURE A–2
Preparation of a cross section. First, a topographic profile is prepared along the line of the section. Then the positions of contacts are located along the profile; the dip of the contacts is drawn in and projected at depth to complete the cross section.

A formation is a body of rock, usually sedimentary rock, that has certain characteristics that make it recognizable and a mappable unit in the field. Most formations are approximately of the same age throughout their extent. The ages of the rock units are also indicated in the legend.

PREPARATION OF CROSS SECTIONS

Once the line along which a cross section is to be made has been determined, place a piece of paper (preferably graph paper) along that line on the geologic map (Fig. A-2). A scale for the vertical dimension must be selected. The vertical scale will be the same as the horizontal scale of the map unless vertical exaggeration is desired. A vertical exaggeration of two to three times often gives more satisfactory results. If the map has topographic contours on it, use these to draw a profile of the land surface along the line of the cross section. Do this simply by noting where each contour line intersects the line of the cross section, observing the elevation of the contour line, and then making a point at the corresponding elevation on the graph paper. A smooth line connecting these points gives a fair approximation of the land profile. Now, with a profile on the graph paper, go back and mark each contact between rock units and the places where each fault intersects the line of the cross section. If the strike and dip of the rock units and faults are given, lay these off on the cross section with a protractor, measuring dips from the horizon and drawing in the formational boundaries and faults as an initially straight line.

After the projections of the lines have been drawn it will be obvious that some of the lines should be connected to form folds, but a great deal of experience may be needed to make the correct decisions each time. Once these connections have been made, the section is complete. Certain patterns of outcrops provide hints needed to make the correct subsurface connections.

OUTCROP PATTERNS ON GEOLOGIC MAPS

Some geologic maps are very simple, making the ordered nature of the distribution of rock units quite apparent, but others are complex. A few observations will serve to demonstrate the significance of the outcrop patterns. Sedimentary rock units represented on maps are usually assumed to be platelike in shape. They are approximately uniform in thickness, and they extend over large areas. Like most generalizations, this is not always a valid assumption, but it simplifies the problem of map interpretation to make this assumption at first. Outcrop patterns are determined by the following five factors:

1. Thickness of the strata (Fig. A-3).
2. Slope of the ground surface (Fig. A-4).
3. Attitude of the sedimentary strata (Fig. A-5) and the igneous and metamorphic rock units.
4. Structure of the rock units.
5. Sequence of events in the area.

If strata are flat-lying in a plain, the rock unit exposed at the surface may cover the entire map, but if the area has an irregular land surface, the contacts between strata will be exposed. Their

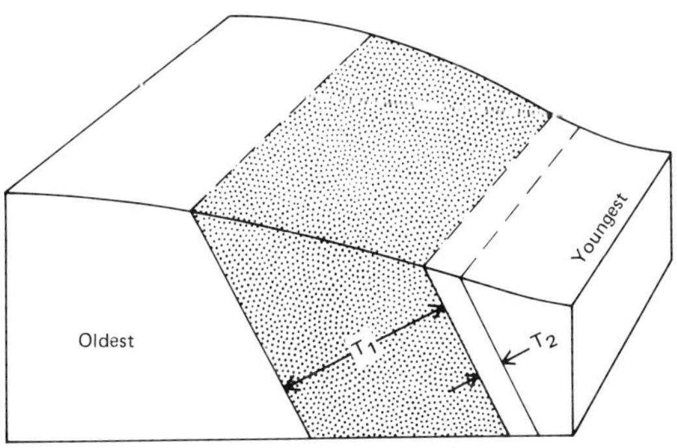

FIGURE A–3
Thickness (*T*) of a rock layer is an important determinant of its outcrop width on the ground and on a map.

FIGURE A–4 (*below*)
Cross sections showing effect of slope (θ) of ground surface on width (*GW*) of outcrop belt.

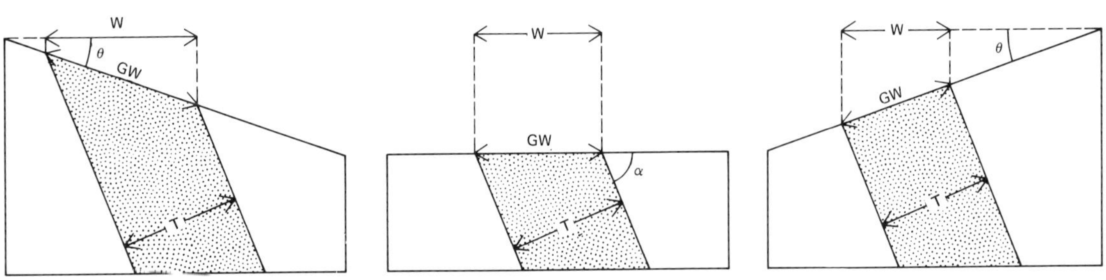

FIGURE A–5 Block diagrams showing effect of dip of bed on outcrop width. One marker is shown vertical (top). As dip (α) decreases, the width of outcrop belt increases (bottom).

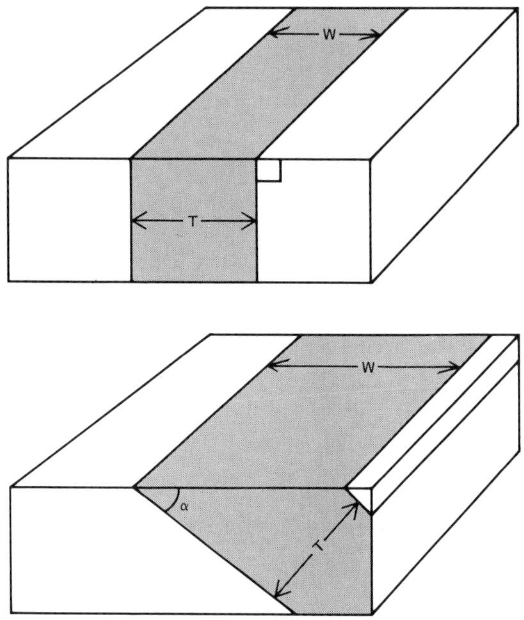

FIGURE A–6 (*below*) Geologic map showing dendritic outcrop pattern formed by flat-lying strata exposed in a region where there is a dendritic stream valley. Three layers (white, top; stippled, middle; and black, lower) are shown. (simplified from the State Geologic Map of Pennsylvania)

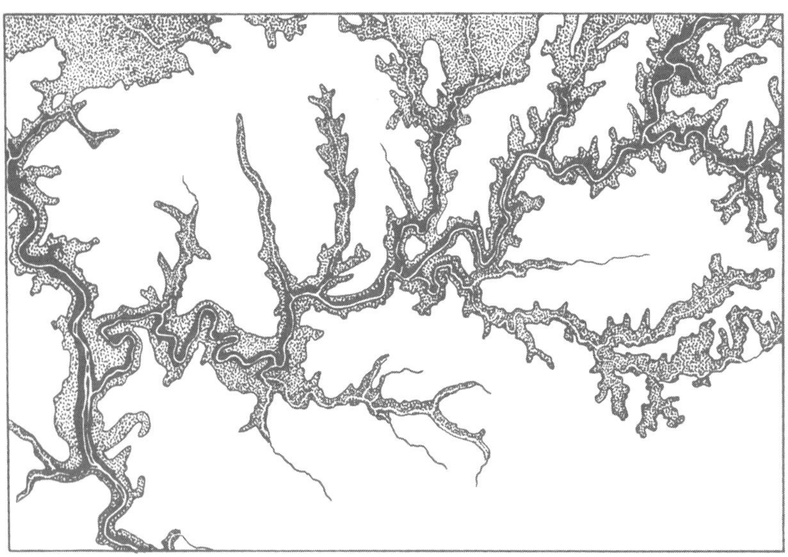

pattern will resemble that of contour lines (Fig. A-6). Many contacts are exposed if differences in elevation are great and the thickness of individual strata in the sequence is slight.

If a sequence of strata is uniformly tilted, the width of outcrop depends on the amount of the dip and the amount and direction of the ground slope. Note that the youngest strata are always in the direction of the dip and that the outcrop pattern is a sequence of bands (Fig. A-3).

If a bed is dipping vertically, the outcrop of the contacts will form straight lines across the map regardless of the topography, and the thickness of the outcrop pattern will be equal to the thickness of the bed (Fig. A-5).

The Law of V's

If the direction of dip of a stratum is not given by a symbol on the map, it is possible to determine this by noting the outcrop pattern where a stream valley cuts across the contact. A V-shaped pattern is usually produced. The V points *upstream* under the following conditions:

1. In horizontal strata.
2. In strata that dip gently upstream.
3. In strata dipping downstream if the slope of the stream is greater than the amount of the dip.

The V points *downstream* if the strata are dipping downstream at an angle greater than the slope of the stream valley. No V is formed if the strata are vertical. Figure 6 shows V's, most of which point upstream.

Metamorphic rocks of sedimentary origin can be mapped in much the same fashion as sedimentary strata. Although metamorphic rocks are frequently banded, contacts around igneous rock bodies may have very irregular shapes. Their outcrop patterns may be as regular as that of a stratum, as in the case of sills or dikes, but they also may be very irregular, appearing as patches on the map, as is shown in the illustrations in Chapter 5.

STRUCTURE ON GEOLOGIC MAPS

Folds

The identification of folds on geologic maps is usually easy because processes of erosion remove rock from the land surfaces. As a dome is raised, the top of the dome tends to be eroded more than are the margins. This brings about removal of the topmost layer of sediment from the top of the uplift, exposing older rock units (Fig. A-7). The result is a series of concentric contacts and exposure of successively older rock units toward the center. The rims of structural basins are subjected to erosion first, exposing a series of concentric contacts, but the youngest rock units are now found in the center. Likewise, anticlines and synclines may be identified by applying the same sort of reasoning. Erosion is fastest on the upfolded parts. Thus the axial zone of an anticline is truncated first, exposing older units, but the limbs of a syncline are truncated before the axial zone, leaving younger units along the axis. It is possible to find folds in which the mountains are anticlines and the valleys are synclines. The term "cigar-shaped mountain" aptly describes such a doubly plunging anticline; similarly, "canoe-shaped mountain" suggests the shape of a syncline. But it is important to remember that structural features may or may not be directly reflected in the topography.

Faults

Faults often can be recognized by the abrupt changes in strike of beds or the pattern of contacts across the fault. It is important to remember that the upthrown side of a fault is usually subjected to erosion and is consequently lowered faster than the downthrown side. If contacts between beds are not vertical, the contact will migrate in the direction of dip of the contact as the ground surface is lowered (Fig. A-8).

Fault traces, like contacts between beds, will appear as straight lines even across rough topography if the fault plane is vertical. The dip of a fault plane can be inferred using the Law of V's. When a fault is horizontal its trace parallels contour lines.

Unconformities

Breaks in the history of sedimentary sequences bring unconformities into existence. These unconformities simply indicate prolonged interruptions in the process of deposition of sedimentary rocks; they may have been brought about by uplift of the sediments from the sea bottom to the surface where they were attacked by the agents of weathering and erosion and partially removed;

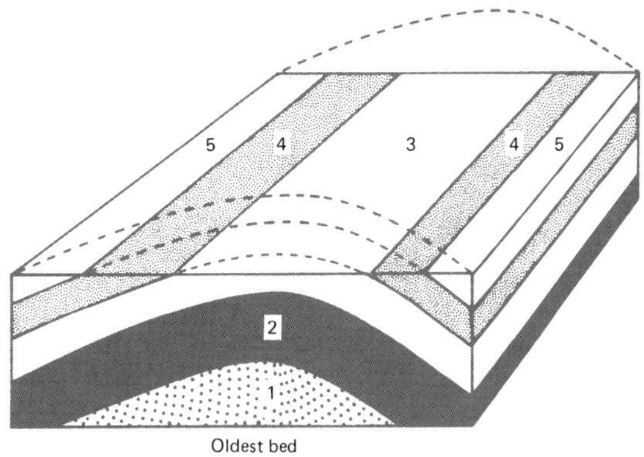

FIGURE A-7 Block diagram showing geologic map and cross sections of an anticline with horizontal axis. Note that the oldest stratum crops out along the axis.

FIGURE A-8 (*below*) Block diagrams and maps showing the effect of faulting and erosion on the map pattern of four uniformly dipping strata.

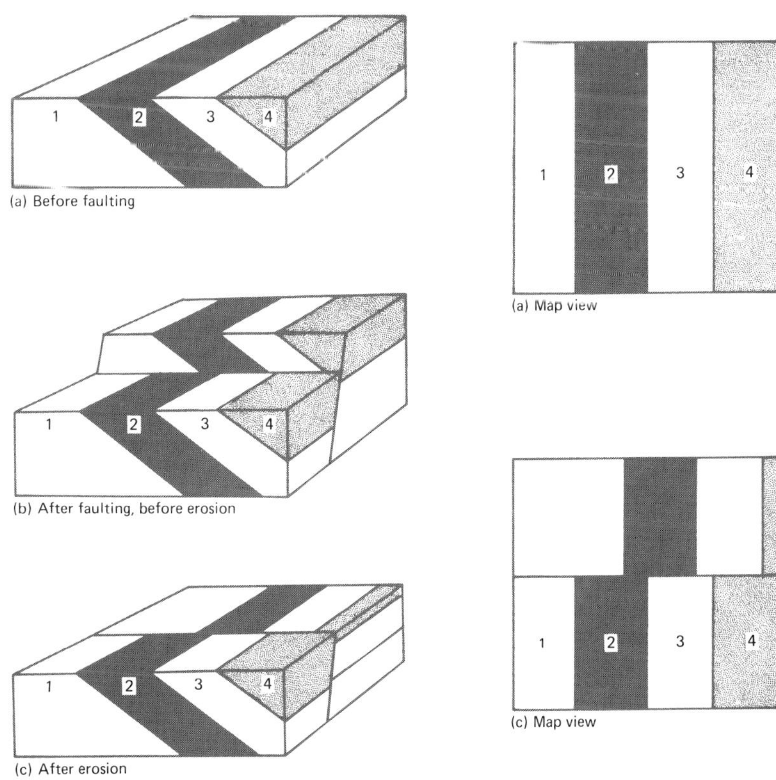

or the unconformities may represent a major break in the sequence during which uplift, folding, faulting, and erosion all took place before sedimentation was resumed and more units of rock were deposited. The first two types of unconformities affect map-outcrop patterns only in that the expected sequence of units is not found. Some beds either were not deposited or were eroded away. The third type, angular unconformities, may greatly influence the outcrop pattern. For example, the break in the record may have been preceded by deposition, folding of the strata into synclines and anticlines, and erosion of these before new strata were formed; the new strata are now probably horizontal or tilted (see Fig. 1–9).

B

CHEMISTRY

SYMBOLIC REPRESENTATION

The following examples illustrate the basic symbolic representation of elements in chemistry:

$$^{234}_{92}U \quad \text{or} \quad {}_{92}U^{234}$$

The symbol for the element uranium is U. The *bottom number* (92) is the *atomic number* of the element, that is, the number of positive units in the atom's nucleus. The *upper number* (234) is the *atomic weight* (see pp. 26–27).

The symbol for the element sodium is Na. The sodium ion is expressed as follows:

$$Na^{+} \quad \text{or} \quad Na^{+1}$$

The + or +1 indicates that this is an ion with one positive charge, that is, one electron has been lost from the outer orbit of the atom.

THE PERIODIC CHART

The periodic chart is an outgrowth of work by Dmitri Mendeléev, who formulated what he called the Periodic Law about 1868:

> If all the elements be arranged in the order of their atomic weights a periodic repetition of properties is obtained. This is expressed by the law of periodicity; the properties of the elements, as well as the forms and properties of their compounds, are in periodic dependence.

The structure of the chart has changed over the years as new elements and isotopes have been discovered and as more accurate weights and other data are accumulated. The chart contains an array of rectangles, one for each element, arranged in rows and columns. Each column is designated as a *group*. The elements in each group exhibit similar behavior (e.g., the inert gases occur in a group at the right of the chart). In general, the column number corresponds to the number of electrons in the outer orbit of the element. This is in turn reflected in certain properties of the atom. The rows in the chart are called *periods* or *series*. Each series contains one new set of electronic shells. All elements in group 1 have one electron in their outer orbit; all elements in series 2 have two orbits, and so on.

Iron and nickel are *transition elements* that essentially belong to group 8. A full description of these elements, as well as a fuller description of the periodic chart, can be found in basic chemistry texts.

In the simplified periodic chart on the back endpapers, basic data about selected elements are given within rectangles. For more complete data refer to the "Geochemical Table of the Elements for 1959," by J. Green, *Bulletin of the Geological Society of America,* volume 70, 1959. This table includes the following data: abundances of elements in various rock types, minerals that contain the element, ionic radii, coordination, volume of ions, natural isotopes, and some physical properties.

Table B-1 lists the half lives of radioactive isotopes, some of which are useful in geological dating.

Isotope	Half life
Potassium	
K^{40}	$1.3 \cdot 10^{9}$ years
Rubidium	
Rb^{87}	$4.3 \cdot 10^{10}$ years
Beryllium	
Be^{7}	53.9 days
Be^{10}	$2.7 \cdot 10^{6}$ years
Lead	
Pb^{210}	20 years
Pb^{211}	36.1 minutes
Pb^{212}	10.6 hours
Pb^{214}	26.8 minutes
Radium	
Ra^{223}	11.6 days
Ra^{224}	3.6 days
Ra^{226}	1,620 years
Ra^{228}	6.7 years
Thorium	
Th^{227}	18.2 days
Th^{228}	1.9 years
Th^{230}	$8 \cdot 10^{4}$ years
Th^{231}	25.6 hours
Th^{232}	$1.4 \cdot 10^{10}$ years
Uranium	
U^{234}	$2.5 \cdot 10^{5}$ years
U^{235}	$7.1 \cdot 10^{8}$ years
U^{238}	$4.5 \cdot 10^{9}$ years

TABLE B-1
Half lives of some important radioactive isotopes
(from Green, 1959)

C

ELEMENTS OF GEOLOGICAL PHYSICS

STRESS AND STRAIN

Structural features of deformed rocks are caused by the application of forces to the rocks. Forces are of two general types: (1) *body forces* that act throughout a material as does the force of gravity; (2) *surface forces* that act on the body *surface*. Consider a cube resting on a surface. Initially the only forces acting on the cube are the pull of gravity directed downward toward the center of mass of the earth (it acts throughout the cube) and a counterbalancing force pushing upward on the surface on which the cube is resting. If an external force, a surface force, is applied to one side of the cube (that is, if we push it from one side), unless the cube is constrained it will move in the direction toward which the force acts. If the cube is constrained so it cannot move, it will change shape.

Force is defined as the product of mass and acceleration ($F = ma$). The state of motion of a body that is free to move is therefore changed when a force is applied to that body. If the body is not free to move, the effect of the force applied to the body's surface is to distort or deform the body. Force is a *vector quantity*—that is, it has both direction and magnitude—and can be represented by a line drawn in the direction along which the force acts and with a length scaled so that it is proportional to the magnitude of the force. When a force is applied to the surface of a body, the amount of force per unit of surface area is called *stress,* designated as a *directed stress* unless the stress acts uniformly from all directions, in which case it is called *hydrostatic stress.*

When a force or stress is applied to a body (e.g., mass of rock, cube, etc.), the body responds either by changing its state of motion or by becoming strained. Strains may be either *dilational* (change of size) or *distortional* (change of shape). The type of strain that occurs depends on the way the stresses are applied. Dilational changes occur when uniform pressure is applied, as when rocks are buried. Dilational strains in rock are difficult to detect because the original size of a dilated object is rarely known. Most of the strains with which we deal are changes in shape; these are readily identified because primary shapes such as ripple marks, oölites, cross-bedding, fossils, mineral crystals, and sedimentary layering are well known. Thus layering often provides a marker useful to describe folds, measure displacements, and otherwise determine strain.

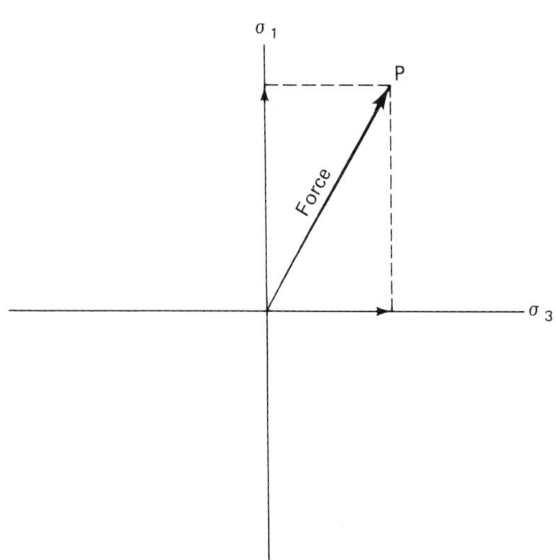

FIGURE C–1
Resolution of a force, P, onto two reference axes, σ_1 and σ_3, the latter of which is the least principal stress direction.

REPRESENTATION OF STRESS CONDITIONS

The conditions of stress at a point can be analyzed if the forces acting at that point are known. Because forces are vector quantities, the effects of several different forces of different magnitude and direction can be added at a point and a resultant stress system determined.

Analysis is simplified by resolving all forces into components on two axes (for plane-stress analysis) as shown in Fig. C–1, or three (for three-dimensional analysis) mutually perpendicular axes. The effect of the two forces acting along the axes on a body is the same as if the single force P acted alone. If several forces were acting, each could be resolved into components acting along the references axes; those acting along any given axis could then be added to summarize the stress condition.

Consider the stress conditions on the small cube in Fig. C–2. The stresses on that cube can be described in terms of the stresses perpendicular to the faces of the cube in three mutually perpendicular directions, called the maximum, intermediate, and least *principal stress directions.*

The three principal stress directions may be

equal in magnitude (the condition of hydrostatic pressure); two may be equal and one either greater or less; or, in the most general case, one may be maximum (σ_1), one intermediate (σ_2), and one least (σ_3).

A number of different combinations of forces can produce the same set of principal stresses. Usually we do not know the exact forces acting in the earth or the specific stresses that acted in the past to produce rock structures, but strains in rock can be related to the forces that cause them in terms of principal stresses. Of course, it is possible to subject materials to known stresses in experimental work and much has been learned about rock deformation in this way.

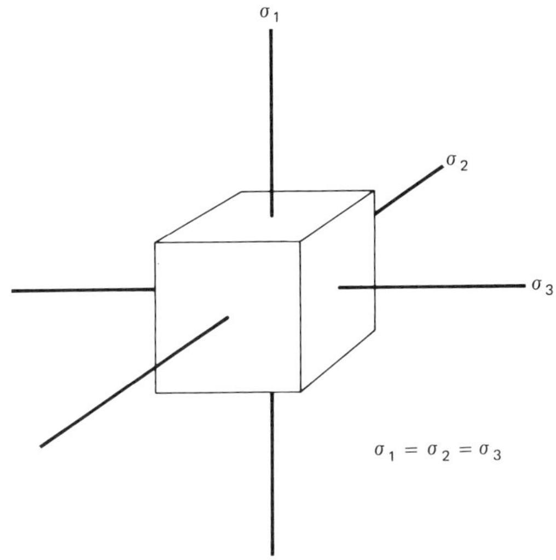

FIGURE C–2
A cube showing three reference axes corresponding to the principal stress directions.

STRESS APPLICATION AND RESULTING STRAIN

When stress is applied to a rock mass by internal processes in the earth, both movement and deformation may result. For example, convection currents in the mantle are thought to cause crustal plates to move and movement is obviously involved in the uplift of mountains composed of marine sedimentary rocks. When we deal with effects of stress applied at some time in the past, it is often impossible to obtain an accurate impression of the amount of movement involved. However, we can often describe in detail the deformational effects shown by strains, that is, the changes in shape of rock masses. The types of strain that result are directly related to the manner of stress application (Fig. C–3); *compression, dilation, tension,* and *simple shear* are examples of types of stress application.

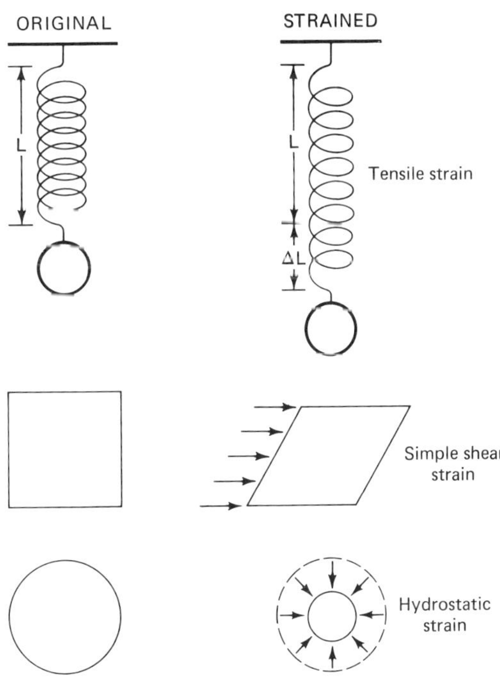

FIGURE C–3
Stress applications by tension (extension), top; simple shear (distortion), middle; and all-sided compression (dilation), bottom; and the resulting strain.

1. Compression occurs when inwardly directed stresses of unequal magnitude act on a material from different directions. A vise is designed to apply compression.
2. Dilation is a deformation involving change of volume but not shape. Dilation may occur when the three principal stresses are equal.
3. Tension is produced when a material is deformed in such a way that it is extended. A spring is extended by tensional stress application.
4. Simple shear occurs when a rotational couple (opposing forces that tend to rotate the material to which they are applied) acts so that the specimen is sheared (formerly perpendicular lines are rotated relative to one another).

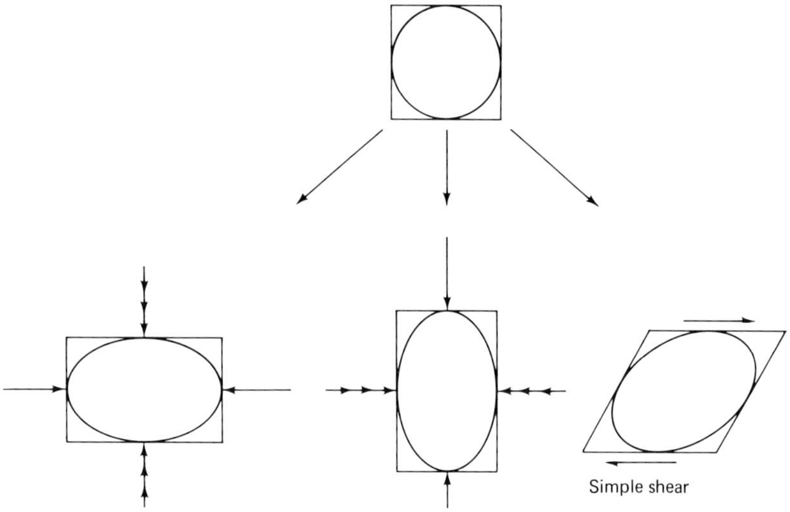

FIGURE C–4
Deformation of a circle into an ellipsoid by compression, extension, and simple shear. This illustrates that simpler strains can be produced by various stress systems.

Compression, extension, and simple shear can be easily demonstrated in terms of the effects each has in straining a reference cube and sphere (square and circle in two dimensions), as shown in Fig. C–4. The maximum and least principal stresses are illustrated in these cases with the intermediate stress direction coming out of the page. Note that the principal stresses are thought of as compressions; thus when a body is being extended, the direction of extension is the least principal stress. The same effect would be produced if the body were compressed at right angles to the extension.

A definite relationship exists in theory between the principal stresses and strains and this is confirmed by experimental evidence. Thus it is sometimes possible to postulate that certain types of stresses were acting when particular structural features were produced. Unique solutions to the type of stresses acting are not always possible because the same strains may be produced by more than one manner of stress application (Fig. C–4).

ELASTIC PROPERTIES OF ROCKS

The elastic properties of rock or other materials are readily determined by simple experimental tests in which a specimen of the material is subjected to a stress of known value and method of application, and the amount of strain is measured. When these tests are run within the elastic limit of the material, the ratio of the stress to the strain is constant. These ratios have constant values, called *elastic moduli,* which determine the velocities of the seismic waves in the earth.

Young's modulus, E, for simple tension =

$$\frac{\text{tensile stress}}{\text{change in length per unit of length}}$$

Shear modulus (rigidity), μ, for simple shear =

$$\frac{\text{shearing stress}}{\text{shearing strain}}$$

Bulk modulus, B, for hydrostatic pressure =

$$-\frac{\text{increase in stress}}{\text{change in volume per unit volume}}$$

When a body is extended in one direction, there is a concurrent shortening of its dimensions in a direction transverse to that longitudinal strain. The ratio of the transverse strain to the longitudinal strain is called *Poisson's ratio* (σ).

$$\sigma = -\frac{\text{change of width/original width}}{\text{change of length/original length}}$$

Poisson's ratio is related to the other moduli as follows:

$$\mu = \frac{E}{2(1 + \sigma)}$$

$$B = \frac{E}{3(1 - 2\,\sigma)}$$

The reciprocal of B is called the compressibility, k. The elastic moduli provide a convenient way to describe the elastic properties of a material.

D

SOURCES OF GEOLOGICAL INFORMATION

It is important for every student of geology to learn how to find desired geological information. The primary concern of the student may be finding source material for a term paper, but knowledge of how to use a geology library to find information not available in textbooks will prove a valuable asset to many businessmen, contractors, engineers, teachers, and lawyers as well; it is, of course, essential for all geologists. The primary sources of information are the following:

1. Books.
2. Publications of the federal government, for example, United States Geological Survey, Coast and Geodetic Survey, Bureau of Mines.
3. State Geological Survey publications.
4. Publications of the American Geological Institute.
5. Publications of geological societies and organizations.
6. The Library of Congress Card Catalog.

The main body of geological literature consists of short papers published in periodicals or as separates in series by private organizations and by state and federal geological surveys. Geological and topographic maps constitute a very important part of this literature for those skilled in map interpretation. Much of this literature is descriptive and regional in nature, and it often has a long, useful life expectancy.

LIST OF SELECTED REFERENCE SOURCES

Directory of Geological Material in North America (1957). This volume, published by the American Geological Institute, lists by state the names and addresses of state geological surveys and sources within each state of maps, photographs, well data, libraries, museums, and so on. The directory does not list titles or authors' papers.

United States Geological Survey (U.S.G.S.). The United States Geological Survey periodically publishes a list of its publications entitled "Publications of the Geological Survey," which is available on request from the United States Geological Survey, Washington, D.C., 20242. This booklet is cross-indexed according to subject, state, and author. Among the series of Survey publications are the following:

Professional Papers
Water Supply Papers
Geological Quadrangle Maps
Hydrological Investigation Maps
Mineral and Oil Investigation Maps
Bulletins
Coal Investigation Maps
Geophysical Investigation Maps
Topographic Maps

U.S. Bureau of Mines. The Bureau of Mines provides the "List of Publications Issued by the Bureau of Mines," with a subject and author index. It is sold through the U.S. Government Printing Office, Washington, D.C. Among the publications of the Bureau of Mines are the following:

Bulletins
Technical Papers
Information Circulars
Economic Papers
Reports of Investigation
Minerals Yearbook (a particularly valuable reference of statistics of United States mineral production)

State geological surveys. Most states have a geological survey or division of mineral resources, and all of these publish one or more series of bulletins, information circulars, and so on. Usually a publications list is published periodically and is available on request. The surveys are located in the state capital or at the state university in most states. *An Index of State Geological Survey Publications Issued in Series* (1965), compiled by J. B. Corbin, is available through the American Geological Institute. A list of survey addresses is included.

Bibliographies and abstracts. Several major sources of abstracts and cross-indexed (usually by subject, author, and region) bibliographies are available. Significant changes have recently taken place in several of these.

1. *Bibliography of North American Geology* is published as a Bulletin of the United States Geological Survey. It is generally compiled on a one-year-per-volume basis.
2. *Abstracts of North American Geology* has been published monthly by the United States Geological Survey since January 1966.
3. *Bibliography and Index of Geology Exclusive of North America* was published by the Geological Society of America annually through 1968. Since 1968 the bibliography and abstracts have been published monthly and cross-indexed in a final number each year.
4. *Bibliography and Index of Geology* is now published by the Geological Society of America in place of the former *Bibliography and Index of Geology Exclusive of North America.* The new serial covers the world and includes a cumulative index for each year as well as monthly issues.
5. *Bulletin Signaletique—Sciences de la Terre I, II* is a French abstract and cross-indexed bibliography covering the world. It is available through the ministry of National Education and has been available since 1939.
6. *Geoscience Abstracts* was published monthly by the American Geological Institute from 1958–66. It covered many North American journals.
7. *Index of Publications of the Geological Survey of Canada* is published by the Department of Mines and Technical Surveys, Ottawa. The years 1845–1958 are covered in a single volume with annual supplements.

Journals—serial publications. The number of professional journals devoted to the earth sciences has expanded greatly in recent years. One of the most complete lists of serial publications is printed in the first part of the "Cumulative Bibliography" of the *Bibliography and Index of Geology,* published by the Geological Society of America.

A partial list of some important journals published in English follows:

American Journal of Science
American Mineralogist
Bulletin of the American Association of Petroleum Geologists
Bulletin of the Geological Society of America (also publishes *Memoirs* and *Special Papers*)
Bulletin of the Seismological Society of America
Canadian Journal of Earth Sciences
Earth-Science Reviews
Economic Geology
Geological Magazine
Geophysics
Journal of Geological Education
Journal of Geology
Journal of Geophysical Research
Journal of Sedimentary Petrology
Marine Geology
Proceedings of the Geologists' Association
Science
Sedimentology
Tectonophysics

Most journals publish a cross index in the December number, and some journals compile indexes for long periods of time.

Special publications. A number of special books have been prepared to assist in information search. Among these are the following:

1. *A Basic Geographical Library* (1966), Association of American Geographers, Washington, D.C.
2. *Earth for the Layman* (1957), American Geological Institute.
3. *Directory of Geoscience Departments,* American Geological Institute.
4. *Sources of Earth Science Information* (1964), Prentice-Hall.

HOW TO COMPILE A BIBLIOGRAPHY

The main problem in compiling a bibliography is that most indexes are annual. Only in the last few years have relatively complete and up-to-date listings been available. *The Bibliography of North American Geology* and *The Bibliography and Index of Geology Exclusive of North America* provide good starting places for a year-by-year search for material pertinent to any region or topic.

It is very helpful if an existing bibliography can be found and supplemented. Many advanced textbooks provide excellent bibliographies; thus if the student can find a new book or revision that covers the topic in which he is interested, its bibliography (either at the end of chapters or in the back of the book) should be useful. It should be remembered that the most recent references listed in a book usually will be at least a year older than the date of publication of the book. Review articles found in some journals are good sources for ready-made bibliographies. If the topic is one to which a journal is entirely devoted—for example, micropaleontology—articles in recent issues of the journal often will have bibliographies that provide valuable starting points; most journals, however, are general in their coverage.

E

OIL AND GAS

The petroleum industry celebrated its first 100 years in 1959; however, oil seeps had been known for many years prior to 1859, and petroleum was used as far back as biblical times for medicine, waterproofing, and occasionally even for fuel. American Indians used it for war paint. The first real pressure for more petroleum came from its use as a light source. Colonel A. C. Ferris of New York discovered that petroleum made an excellent illuminant when run through his process for making coal oil. The first well drilled to recover oil for this purpose was put down by Colonel Edwin Drake in Pennsylvania in 1859. The new industry expanded rapidly as new uses for oil were found. One of the most important changes in the industry occurred when the automobile became widely used. Since that time more and more uses have been found, and production has expanded until now more than 12 billion barrels of oil are produced each year to meet the steadily growing demands of the world. In recent years as many as 5,000 wells have been drilled each year. About one-fifth of these are drilled in efforts to discover new accumulations. The remainder are drilled in and around areas where petroleum is already being produced.

Composition. The term *petroleum* includes natural gas, crude oil, and solid bitumins. It is a complex mixture of compounds of carbon and hydrogen called hydrocarbons. The mixture is generally made up of one or more of the following groups of hydrocarbons: paraffin, napthene, acetylene, turpene, and benzene. The solid hydrocarbons, called *bitumins,* include asphalts and waxes.

ORIGIN OF OIL

Inorganic theories of oil's origin have been discarded in favor of an organic derivation, but still no direct evidence of how decaying matter is transformed into petroleum is known. Generally accepted ideas related to the origin of oil are the following:

1. Solid organic matter found in sedimentary rocks is composed mainly of carbon and hydrogen.
2. Oil initially forms in sedimentary rocks formed in the ocean. Where oil is found in igneous or metamorphic rocks, it is believed to have migrated there from sediments.
3. Petroleums almost always differ in composition.
4. Most oil is found in younger rocks.
5. Oil has not been subjected to temperatures above 390°F. It would be highly altered at that temperature.
6. Oil forms under *anerobic* conditions, in the absence of oxygen, because carbon in it would be oxidized to form carbon dioxide in an oxidizing environment.
7. Crude oil contains porphyrin pigments, formed from the red coloring matter of blood, *hemin,* or from the green coloring matter of plants, *chlorophyll.*

The most probable source materials for petroleum are diatoms, marine fungi, bacteria, and algae, among the forms of marine plant life, and marine invertebrates. The most important marine-invertebrate sources are the protozoans, extremely abundant single-celled animals. Other potential sources among the marine invertebrates are sponges, corals, worms, bryozoans, brachiopods, crustaceans, mollusks, and echinoderms. Vertebrates may also provide lesser amounts of organic material.

The principal questions regarding the origin of petroleum deal with what happens to the organic matter to convert it into oil and gas. A number of processes may be responsible for the conversion, including distillation, cracking, hydrogenation, polymerization, bacterial action, and radioactivity. Of these, the one generally thought to be likely is bacterial action working in conjunction with one or more of the other processes. Bacteria are found universally with petroleum, and it has been definitely established that they can produce hydrocarbons from organic matter.

ENVIRONMENT OF FORMATION OF OIL

Within the bounds set by the facts stated above regarding the origin of petroleum, it is not difficult to visualize the general types of environ-

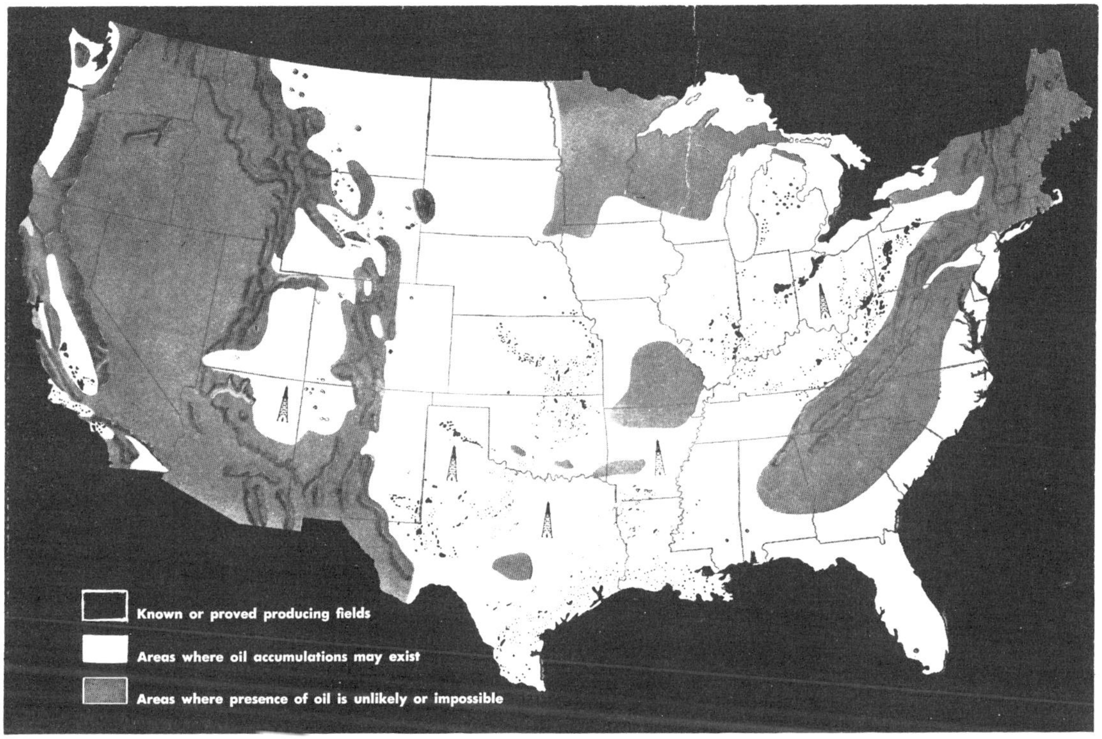

FIGURE E–1
Oil-producing areas in the continental United States (black) and potential oil-producing areas (white). Oil is less likely to occur in the highly deformed and metamorphosed rocks of the mountain belts and shield areas.
(American Petroleum Institute)

ments in which oil might begin to form. Areas similar to the shallow waters of parts of the continental shelves and the Gulf of Mexico are probable localities. Droplets of oil have been found near the top of the sediments deposited in the Gulf of Mexico. Although these droplets have had neither the time nor the necessary conditions to accumulate in quantity, their presence indicates that relatively short periods of time are necessary for oil to form. One particularly suitable area is in bodies of marine water that have shallow connections with the ocean but are otherwise cut off. Here the free-floating and swimming plants and animals are abundant at the surface, but because the body of water is enclosed, at greater depth there is poor circulation of water and thus a deficiency of oxygen. This hinders oxidation, and poisonous gases produced by the decaying organic matter form toxic conditions so that scavengers cannot roam the bottom, eating and destroying the dead animal and plant remains. Clay, mud, sand, and silt might be deposited along with the organic material. Depending on location of the area, its distance from shore, and other environmental factors, one of these sediments would probably predominate. Black shales are known to contain unusually great quantities of organic matter. Although oil is rarely found in black shale, accumulations of petroleum are found in sandstones that are overlaid or underlaid by black shales. Some think that the black shales are the principal source of oil, and almost all geologists agree that they are probably sources of at least some oil and gas.

MIGRATION

In order for oil to move through a consolidated or semiconsolidated rock, the rock must have porosity and permeability. Pore spaces may be original spaces between grains of sand or gravel, spots that have been dissolved in soluble rocks, or fractures. The first step in migration of oil is out

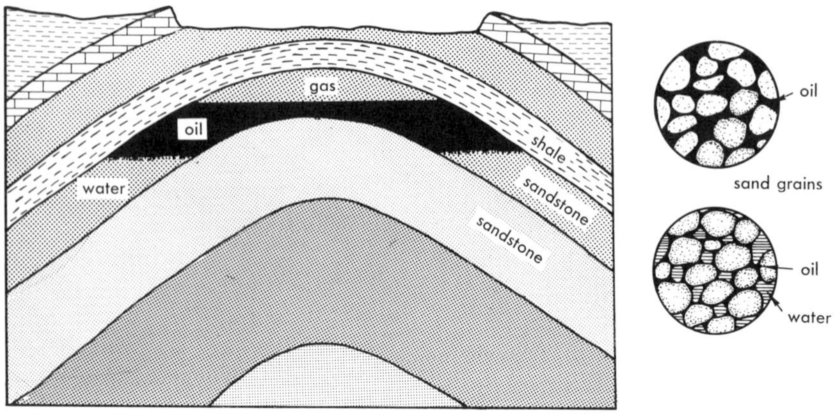

FIGURE E–2 (*above*)
A cross section showing the nature of the occurrence of oil. Oil is stored in pore spaces between sand grains as shown in the enlarged sections at right. The oil probably gets into the sandstone from black muds or shales. Because oil tends to rise through water, it moves up in sandstone units that contain water. This migration continues until the oil becomes trapped as shown in the anticlinal structure.

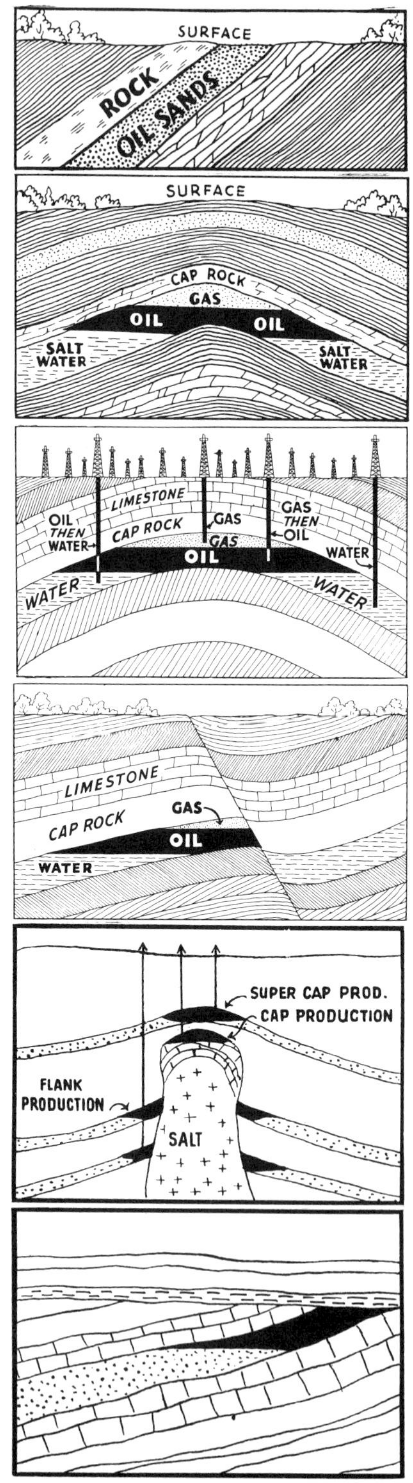

FIGURE E–3
Oil traps. These six cross sections illustrate conditions in the earth that lead to the accumulation of oil and gas. Oil from the oil sands in the top section seeps out on the ground surface. When this happens, the gases tend to escape and a tar seal forms over the top of the unit, making it a trap. The second section illustrates an anticlinal trap in a folded structure, as does section three. The last three sections show how oil accumulates along faults, over salt domes, and as a result of an angular unconformity.
(American Petroleum Institute)

of the source rock into a rock that is porous and permeable. Shales are usually impervious and not very porous. This movement occurs when accumulation of sediments on top of the oil-bearing unit becomes thick enough to begin to compress the source bed. Muds contain a great deal of water; therefore, if a mud containing oil and water were compressed, the oil and water would migrate out of the mud and into the adjacent beds of rock. If these beds are porous and permeable, the second phase of migration begins. Because oil is lighter than water, oil tends to float on top of water. Oil moves up slowly around the sand grains or through other pore spaces until it reaches the top of the unit. If it is tilted, oil continues up toward the surface until it is stopped or trapped. Where the upward movement is halted, oil accumulates. Gases that have been produced are lighter than the oil and tend to move to the top of the oil accumulation.

OIL TRAPS

Two general types of traps are recognized: (1) structural traps caused by folding or rupture and displacement of the rock units, and (2) stratigraphic traps in which some type of variation in the stratigraphy or lithology causes oil to be trapped. Combinations of these two types are known, and a few defy clear-cut classification.

Structural traps include anticlines, where the porous unit is capped by an impervious layer; fault traps, where up-dip migration of oil is halted at the fault zone; domes; synclines; and fractures.

Many of the largest oil fields are caused by stratigraphic traps. These include lenses such as buried stream-channel deposits, offshore bars and beaches, buried reef and reef-derived sediments, and porous limestones. Other traps are formed along unconformities and where sediments have been altered by solution or recrystallization to cause porosity.

GLOSSARY

GLOSSARY[1]

Ablation. The combined processes, particularly melting and sublimation, by which a glacier wastes.

Ablation till. Glacial till deposits formed from the debris that is transported within or on the ice and gradually lowered as ablation causes the glacier to shrink.

Abyssal. Used in reference to the oceans, generally in connection with depths below 1,000 fathoms (6,000 ft.).

Abyssal floors or plains. Flat areas in the deep ocean having low relief; they are irregular in shape and are found in a great range of sizes up to widths of several hundred kilometers.

Accessory minerals. Minerals commonly present in small quantities in rocks. They do not determine the classification of the rock.

Acicular crystal shape. Crystals with a needlelike shape.

Accumulator plants. Plants that accumulate and store certain elements or compounds.

Acre-foot. A measure of water volume, the amount of water contained in an area of one acre when uniformly covered to a depth of one foot; 43,560 cubic feet or 1233.5 cubic meters.

Agglomerate. Medium to coarse fragmental rocks formed by volcanic eruptions.

Aggradation. The process of building up a depositional surface.

Alkali flat. The dried up bed of a saline lake in which alkaline salts are concentrated.

Alluvial. Referring to materials (sand, gravel, etc.) transported by and deposited from streams.

Alluvial fan. A fan-shaped deposit of stream alluvium laid down where a change in stream gradient occurs.

Alluvial soil. Soil formed by material transported by streams and deposited in stream valleys.

Alluvium. Material transported by streams and deposited in stream valleys.

Alumina. Aluminum oxide.

Amorphous solid. A solid that does not possess a crystalline structure, for example, glass.

Amphibolite facies. Rocks produced by medium- to high-grade regional metamorphism.

Amphibolites. Metamorphic rocks containing mostly plagioclase and hornblende and often some biotite.

Amphidromic. The circular motion by which oscillating standing waves move around nodes in bodies of water.

Andesite line. The line around the northern and western Pacific Ocean that separates volcanoes erupting basalt from those erupting andesite or other sialic volcanic products.

Angle of repose. The steepest slope on which a uniform, uncemented, granular material can maintain stability.

Anisotropic. Having physical properties that vary in different directions.

Anomaly (gravity). A gravity measurement that is greater (positive) or less (negative) than should be observed at a locality if the locality fits an idealized model of the earth.

Anthracite coal. Dark black in color, dull to brilliant and even submetallic in luster. It is hard, burns with a short, blue flame, and emits little odor.

Anthraxylon. A vitreous component of coal. It is homogeneous and black, and may appear as a smooth black or highly lustrous material. It is derived from plant tissues.

Anticlinal valley. A valley that occupies the crest of an anticlinal structure.

Anticline. An upfold in which the limbs dip away from the fold axis. Older beds are exposed in the crest than on the limbs after erosion.

Anticlinorium. A fold system located on an uplifted area.

Aphanitic. A textural description applied to a crystalline rock in which individual crystals are too small to be seen or identified by the unaided eye.

Apophyses (*or* tongues). Dikes or veins that can be traced to larger intrusive bodies as the source of magmatic supply.

Archipelagic aprons. The broad, gently sloping surfaces that extend out from some seamounts, islands, and island groups.

Arenaceous. Sand size.

Arêtes. Sharp ridges produced by glacial erosion.

Argillaceous. Containing clay. See also *lutaceous.*

1 The *Glossary of Geology and Related Sciences*, published by the American Geological Institute, has been followed as the authority for geological definitions in the text and in this glossary. The student should refer to the A.G.I. *Glossary* for words not defined here, for various meanings used for a word, and for original sources of the terms.

Arkose. An arenaceous rock made largely of a mixture of quartz and feldspar fragments.

Artesian aquifer. A water-bearing rock unit that contains artesian water (water under pressure).

Artesian basin. A structural basin in which water is confined under artesian pressure.

Artesian water. Ground water that is under sufficient pressure to rise above the zone of saturation.

Ash. A designation in size ($\frac{1}{16}$ to 4 millimeters in diameter) given to fragmental eruptives.

Assimilation. The digestion or incorporation of materials originally in the wall rock of a magma chamber into the magma.

Asthenosphere. A zone in the earth some tens of kilometers deep in which the rock is plastic and where any stresses are removed by flowage.

Asymmetrical fold. A fold in which the limbs are not symmetrical about the axial surface.

Atolls. A ring-shaped island with coral reefs that enclose a lagoon.

Atomic number. The number of protons in the nucleus of an atom.

Attenuation. Decrease in the energy associated with a segment of a wave front as it is spread out.

Attritus. The dull grey to black material derived from plant matter that composes a large part of some coals.

Aurora. A phenomenon associated with the earth's magnetic field in which patterned and colored lights appear in the sky—usually at high latitudes.

Auroral zones. Belts located about 23 degrees from the earth's rotational axis.

Axial surface. An imaginary surface that approximately bisects a fold into symmetrical halves.

Axis. An imaginary line formed by the intersection of the axial surface of a fold with a bedding surface.

Backshore. That part of the shore that is level or sloping landward. It is covered by water only during storms.

Bajada. The gentle sloping ground surrace located between a pediment and a playa or flats in arid regions. It is often restricted to surfaces formed by deposition.

Bar. A sand or gravel deposit submerged at shallow depth.

Barchan dune. A sand dune that is curved with its convex side facing the wind and wings pointed downwind.

Barrier reef. The name given to coral reefs that run parallel to the shores of islands or continents, separated from them by lagoons.

Basalt. A dark-colored, fine-grained rock composed primarily of plagioclase and pyroxene, with or without olivine.

Base level. A level or elevation below which land cannot be eroded by streams. Sea level is referred to as the ultimate base level.

Basin (structural). A circular or elliptical downwarp or structural depression, with younger beds in center.

Batholith. A large pluton of intrusive rock, many square kilometers in areal extent.

Bathyal. Ocean depths between 200 and 2,000 meters (also sometimes applied to depths of 100–1,000 fathoms).

Bauxite. The principal ore of aluminum, composed of hydrated alumina ($Al_2O_3 \cdot 2H_2O$).

Bay-mouth beach. A beach located across the mouth of a bay.

Beach. The shore of a body of water, particularly the sea, where it has a gentle slope and is washed over by waves or tides.

Beach drifting. Movement of sediment along the shore by combined effects of incoming waves and return flow from beach.

Beheaded stream. A stream that has lost part of its drainage system as a result of stream piracy.

Bench mark. A survey marker for which location and elevation are carefully determined.

Benioff zone. See *seismic shear zone.*

Benthonic. Bottom-dwelling types of life.

Bladed crystal shape. A tabular crystal with bladelike form.

Blind valleys. Valleys that lead into a hillside or gradually lose the characteristics of a valley as water from their streams is lost to subsurface channels.

Body waves. Seismic waves that are propagated throughout a three-dimensional continuum and not related to a boundary surface.

Bog. See *swamp*.

Bog iron. Hydrated oxide of iron sometimes found in sandy layers overlain by bogs.

Bombs. Fragments (32 millimeters in diameter or larger) of material erupted from volcanoes.

Botryoidal. Minerals shaped like a bunch of grapes.

Bottom-set beds. The subhorizontal layers of sediment deposited in front of a delta.

Bouguer anomaly. Any residue when the theoretical value of gravity for a given latitude calculated by the international formula is subtracted from the observed gravity at a station corrected for free-air effect, Bouguer effect, and topography.

Bouguer effect. The attraction that the slablike mass of material between the geoid or sea level and the elevation at which a gravity reading is taken has on the value of gravity at that elevation.

Breaker. A wave breaking on the shore, formed when the velocity of the top part of the wave begins to exceed the forward velocity of the wave as a whole; the top outruns the rest of the wave and spills over in front of it.

Breccia. A fragmental rock containing angular instead of rounded pieces.

Brittle failure. Rupture or breakage characteristic of brittle substances. Failure preceded by little flow or plastic deformation.

Calcareous. Containing calcium carbonate.

Caldera. A large enclosed or partially enclosed depression caused by collapse or explosion of a volcano.

Caliche. Calcium carbonate commonly found as a layer or nodular mass in soils underlain by limestone or other rocks that contain calcium carbonate in semiarid plains or deserts.

Calving. A process of spalling off of masses of ice to form icebergs where an ice sheet extends into the seas.

Capacity (stream). A measure of the ability of a stream to transport suspended and bed load.

Capillary. Very fine; hairlike.

Capillary fringe. A thin zone immediately above the water table in which very thin tubelets of water extend up into the overlying rock or soil.

Capillary waves. Small waves, wave lengths less than 2.5 centimeters, with periods of less than 0.1 second. Their propagation is controlled primarily by surface tension.

Carbonaceous. Containing carbon.

Carbonation. Process by which carbon dioxide is added to oxides of calcium, magnesium, sodium, and potassium to form carbonates of these metals.

Carbonic acid. A weak acid formed by a combination of carbon dioxide and water.

Cataclastic deformation. Rock deformation in which brittle fracture of individual mineral grains is important.

Cave. A naturally formed underground cavity or void.

Cavitation. Corrosive effect of collapsing bubbles.

Cenozoic era. The most recent era into which geologic time is divided. It extends from the end of the Mesozoic era, about 70 million years ago, to the present.

Chalk. A fossiliferous limestone composed of the shells of protozoans and particularly fossil globigerina. It is white and soft.

Channel roughness. A measure of the resistance to flow of water in a stream due to the character of the channel.

Chatter marks. A scar caused by the chipping of a bedrock surface by rock carried at the base of a glacier.

Chemical magnetization. Magnetization of a material through chemical processes, for example, oxidation, reduction, recrystallization, precipitation, or exsolution.

Chernozem soil. A zonal group having deep, dark to nearly black surface horizons and rich in organic matter, which grades into lighter colored soil below.

Chert (flint). A dense, hard rock composed of SiO_2 for which color ranges from white through gray to black.

Chloroplasts. The part of a plant that contains chlorophyll.

Chondrite. See *stony meteorite*.

Chute (stream). A narrow passage of water between an island and the main bank of the stream. These are commonly found on the inside of curves along meandering streams.

Cinder cone. A cone-shaped hill formed from accumulation of cinders and ash around a volcanic vent.

Cirque. An amphitheater-shaped depression formed by glacial quarrying, or plucking, where the glacier starts its movement down the valley.

Clastic particles. Fragmented materials such as sand and gravel that are produced by the mechanical disintegration of all types of pre-existing rocks.

Cleavage (mineral). The property that some minerals exhibit of breaking along definite smooth planes. Mineral cleavage is a manifestation of the internal orderly arrangement of atoms in a mineral.

Cleavage (rock). The property of some rocks of breaking along definite smooth, subparallel, planes. Rock cleavage is caused by closely spaced fractures or by alignment of platy minerals.

Closed system. A system is closed if no matter is transferred either into or out of it during some process.

Coal. Vegetable matter that has been altered both physically and chemically through geological processes to a black, rocklike substance.

Coast. A broad zone directly landward from the shore.

Coccoliths. Minute calcareous plates formed on some marine flagellate organisms.

Coking coal. A variety of bituminous coal that can be used to produce coke.

Colloform. Spherical or rounded masses of various sizes and shapes resulting from the radiating growth of individual crystals.

Colloids. A particular size of matter—anything in the range of 10^{-5} to 10^{-7} centimeters in diameter.

Colluvium. A mixture of soil, stream deposits, and talus.

Cols. Passes formed where two cirques converge, cutting into the same wall and thus lowering it below the level of the remainder of the summit area.

Columnar crystal shape. A crystal that is nearly equidimensional in cross section but elongate in one direction.

Compaction. Process by which material is pressed together (e.g., as a result of the weight of sediment deposited on a given layer).

Competence (stream). The largest size particle (measured by diameter) a stream can move.

Composite cones. Volcanic cones composed of both cinders and lava flows. They are circular in plan, and the sides are concave upward in profile.

Compressional waves. See *dilational waves.*

Compressive stress. A stress applied (as in a vise) in such a way that material is forced together.

Concentric folds. Folds in which the thickness of the individual layers remains constant (measured perpendicular to the bedding) with depth.

Conchoidal fracture. A broken surface that is curved, or shell-shaped, as exhibited by glass.

Concordant injections. Intrusions emplaced along planes of stratification, layering, or schistosity so that the borders of the intrusion are parallel to pre-existing layers.

Cone (submarine). A term applied to the large accumulations of sediment in the ocean basins that have a cone shape reaching an apex on the continental shelf near the mouth of a major river, where sands, silts, and clays are carried into the ocean.

Cone sheets. A dike of arcuate outcrop and regular inclination toward a focus.

Confining pressure. Term applied to nondirected pressure due usually to depth of burial in rocks.

Conglomerate. Consolidated gravel.

Connate water. Marine or fresh water trapped in sediments when they are deposited on lake or sea bottoms.

Consequent (stream). A stream that flows in the same direction as the inclination of the slope on which it formed originally.

Conservative constituents. Certain constituents of sea water such as nitrogen, argon, helium, neon, and krypton, which have relatively constant concentrations.

Contact metamorphic aureoles. Concentric zones of alteration around plutons due to the effects of heat and chemically active fluids in the magma on the country rock.

Contact metamorphism. The alteration of country rock around an intrusion near its contact.

Continental accretion. The theory that continents have grown by incorporation of mountain belts around their margins.

Continental drift. The theory that continents have moved relative to one another.

Continental margins. The zone of transition from crustal structure of the continents to that of the ocean basins.

Continental rise. The gently sloping surface located at the base of the continental slope.

Continental shelf. A shallow, gently sloping surface of low local relief that extends from the shore line to the self break where the seaward gradient sharply increases to greater than 1:40.

Continental slope. That relatively steep portion of the sea floor that occurs at the seaward border of the continental shelf.

Convection. A slow, boilinglike motion.

Coordination number. Number of anions in contact with a cation in any given packing arrangement.

Coquina. Name given to sedimentary rocks composed of a loose aggregation—cemented together as a rock—of shell fragments.

Cordillera. A mountain chain.

Coriolis effect. The deflection of any object in motion on a rotating sphere.

Corrosion. Most often used to denote erosion accomplished by chemical processes.

Cotidal lines. Lines connecting places that experience high or low tides at the same time.

Covalent bond. A type of bonding in which elements share electrons; there is no transfer of electrons from one atom to another.

Crater. An inverted cone-shaped depression developed around a volcanic vent.

Craton. The stable portion of each continent, the shields, and the surrounding areas.

Creep (rock). Extremely slow deformation of materials subjected to stress below the elastic limit.

Creep (soil). The imperceptibly slow movement of soil downslope.

Crest. The highest part of a fold form; also the top of a wave form.

Critical temperature. The temperature above which the gaseous and liquid phases become indistinguishable.

Cross-bedding (*or* **cross laminations).** Bedding that consists of successive, systematic, internal bedding that is inclined to the principal surface of accumulation.

Cryptocrystalline. Detectably crystalline matter but so fine-grained that individual crystals are not distinguishable with a magnifying lens.

Crystal. A homogeneous body the surfaces of which have grown into smooth planes as a result of the internal orderly arrangement of atoms.

Crystal class. One of the thirty-two different combinations of the symmetry operations.

Crystal system. One of the 6 crystal systems into which all crystals are grouped depending on the length and angular relationship of the crystallographic axes.

Crystallization differentiation. Processes that bring about a separation of a magma into two phases (a solid and a liquid), resulting, after crystallization, in two or more different types of rocks.

Curie point. The temperature at which a material loses its spontaneous magnetization. It is about 570°C for ferromagnetic minerals.

Cutoff (stream). The short channel formed where a stream cuts through the neck of a meander loop.

d-spacing. The spacing between successive identical planes in a crystal structure.

Debris. A general term applied to mixtures of rock, soil, plant matter, or mud.

Declination (magnetic). The horizontal angle at a locality between the direction to magnetic north and true (geographic) north.

Décollement. "Unglueing" of a bed or zone, during which process beds above slip over those below. Usually the layer is composed of salt, gypsum, clay, or other weak rock.

Deflation. Blowing away of materials.

Degradation. The general lowering of the land surface by erosion.

Deltas. A deposit of alluvial material formed at the mouth of a stream. Some are triangular in plan.

Dendritic. A form that resembles a branchlike outline.

Denudation. Laying bare of the surface of the earth. Now applied to reduction of the level of the land by processes of erosion.

Desert pavement. A tight mosaic cover formed by lag gravels that protects underlying material from wind erosion.

Desert soil. A zonal group that have light-colored surfaces and usually are underlain by calcareous material and frequently by hard layers.

Diabase. A rock of basaltic composition, consisting mainly of labradorite and pyroxene, and with a texture characterized by euhedral or subhedral crystals of plagioclase embedded in a fine-grained matrix.

Diamagnetic. Substances that tend to move from stronger to weaker fields and, if free to rotate, will become oriented normal to the magnetic field.

Diastrophism. Deformation of the earth's crust—the processes by which the crust is deformed and the results of this deformation.

Diatom ooze. A deposit consisting of siliceous remains of diatoms.

Diatoms. Simple siliceous plants of microscopic size that have many different shapes. They may resemble a rod, sphere, or a circular disc.

Dike swarms. Term used when many dikes, often of similar trend or orientation, occur together.

Dikes. An injected body with parallel or subparallel walls that is narrow relative to its lateral extent and discordant with respect to pre-existing layers.

Dilational waves (*or* primary, P-, compressional, *or* longitudinal waves). Seismic waves propagated like sound waves with movements in the medium through which they pass parallel to the direction of propagation.

Diorite. A coarse- to medium-grained rock composed of intermediate plagioclase and biotite, hornblende, or pyroxene.

Dip. The angle between a plane and the horizon, measured in a vertical plane at right angles to the strike.

Dip needle. A magnetized needle mounted to rotate on a horizontal axis that measures inclination of the magnetic field.

Dip separation. The distance measured down the dip of a fault between two parts of a displaced layer or other marker.

Dip-slip fault. A fault in which the direction of displacement is up or down the dip of the fault.

Dipole. Object having opposite charges at two points.

Directed pressure. Application of a pressure, stress, or force in such a way that it has direction.

Discharge (stream). A measure of the quantity of water passing through a cross-sectional area per unit of time.

Disconformity. A break in a succession of sedimentary layers caused by a period of erosion or nondeposition. Layers above the break are parallel to those below.

Discontinuous reaction series. A type of reaction series (used in crystallization of magma) in which abrupt phase changes occur as reactions progress.

Discordant intrusion. Intrusion injected across planes of stratification or schistosity.

Disharmonic folds. Folds in which the geometry of the fold form varies in an irregular way with depth.

Dispersion (of seismic waves). The continual spreading out of an initial disturbance into a train of waves as a result of variation in the speed of various wave lengths.

Displacement. The straight-line distance between two points that were originally adjacent on opposite sides of the fault after the movement.

Distributaries. Branches of a stream, usually at the mouth of a river on a delta, formed by the breaking up of a single stream into smaller streams.

Disturbed days. Days when the measurements of the earth's magnetic field are erratic.

Doldrums. The equatorial belt characterized by calms and light winds and moisture-laden air bounded on either side by trade-wind belts.

Dolomite. A sugary-textured, dense rock that does not effervesce in dilute acid unless it is powdered. Dolomite is a calcium magnesium carbonate.

Dome. A circular or elliptical uplift; older beds are exposed in center after erosion.

Drainage density. Ratio of total length of all channels in a given drainage basin to the area of that basin.

Drift (glacial). Glacial deposits.

Drumlin. A streamlined hill shaped somewhat like a whaleback and generally composed of glacial till.

Dunes. Heaplike accumulations of sand that are mobile and independent of obstructions.

Dunite. A rock composed almost completely of the mineral olivine.

Duricrust. The hardened crust of soil formed by deposits in the soils of semiarid regions of salts precipitated from ground water.

Dynamic metamorphism. Metamorphism produced when deformation is significant in the transformation of rock. The term is applied when effects of directed stress are important in metamorphic processes.

Earth. A general term used to describe disintegrated rocks and loosely consolidated sediments.

Earth tides. Tidal effects (periodic movements in the earth's crust related to movement of the sun and moon) observed in the solid earth.

Earth's crust. That part of the earth above the M-discontinuity—approximately the outer 10 kilometers in ocean basins and the outer 20 to 40 kilometers under continents.

Eclogite facies. Metamorphic rocks formed at great depth in the earth's crust and mantle. Eclogite is a metamorphic rock composed primarily of a red garnet and a green pyroxene and having a bulk chemical composition that could only have formed from rocks of basaltic or gabbroic composition.

Edgewise conglomerate. A conglomerate layer in which the clasts are platelike, many oriented at a high angle to the bedding.

Effluent condition. A condition existing when water moves from beneath the water table into lakes and streams.

Elastic. Used to describe a material that returns to its original form when a deforming force is removed.

Elastic limit. The limit of stress beyond which a rock cannot fully recover its original shape after a deforming force is removed.

Elastic rebound. Recovery of elastic strain that occurs when a deforming stress is removed.

Elastic rebound theory. All theories of earthquake wave generation that rely on the concept of waves caused by sudden release of stored strain through elastic recovery of strain.

Elastico-viscous. Material behavior characterized by a combination of elastic and viscous properties.

Electron. One of the fundamental particles of which atoms are composed. The mass is 9.107×10^{-28} grams and is assigned a relative negative electrical charge of one unit, which equals 4.803×10^{-10} statcoulombs, or 1.602×10^{-19} coulombs.

Elements. Atoms and groups of atoms that are stable and cannot be broken down by ordinary chemical methods.

En echelon. An arrangement of structural features (e.g., fractures, folds, and so on) such that each feature is parallel to the others but all are offset, in effect like the edges of a shingle roof.

Entrenched meander. See *incised meander.*

Entrenched (streams). A stream that flows in a channel cut below the general level of the surrounding land and often into bedrock. Cliffs may occur along one or both banks.

Epeirogenic. A type of regional deformation characterized by broad uplifting or downwarping of a large area.

Epicenter. Point on the ground directly over the focus of an earthquake where motion is initiated.

Epicontinental marginal seas. Shallow seas on the continental block, distinguished from the continental shelves by their greater depth and possibly greater topographic irregularity.

Epithermal deposits. Ore deposits formed at shallow depths. They occur in planar veins, fractures, and usually have simple mineralogy.

Epoch. A division of geologic time. Geologic periods are often divided into epochs on the basis of unconformities.

Equant crystal shape. A shape characterized by crystals that have approximately the same diameter measured in any direction.

Equatorial tides. Tides that occur when the moon is directly over the equator. The two daily tides are about equal in range.

Era. The largest division of the geologic time scale. Time since the Precambrian is subdivided into the Paleozoic, Mesozoic, and Cenozoic eras.

Erosion. In broad context, erosion includes the effects of all those processes that loosen and remove earth materials: weathering, solution, corrosion, and transportation. Many geologists prefer to restrict its use to processes involving removal of material, such as by streams, glaciers, wind, waves, and ground water.

Erratic. A rock carried by a glacier and laid down on a different type of rock.

Eskers. Sinuous ridges composed of till and glaciofluvial deposits and thought to originate in stream channels and sometimes under glaciers.

Eugeosyncline. Many geosynclines can be subdivided into two major parts, *miogeosyncline* and *eugeosyncline*. The eugeosyncline is located generally on the oceanward side of the miogeosyncline, and is usually characterized by greater mobility and the presence of volcanic products and graywacke as well as other sediments.

Euxenic stage. The initial stage of activity in an orogenic cycle in which deeper basins having poor water circulation develop.

Evaporation. The process by which water is transferred into water vapor below the boiling point.

Exfoliation. Scaling on exposed rocks.

Extension. Stretching or pulling apart.

Fault block. A mass bounded by faults on at least two sides.

Fault plane. The zone along which movement occurs when faulting takes place.

Fault scarp. An escarpment marking the position of displacements where one side moved up in relation to the other.

Fault. A shear zone along which displacement has occurred.

Feldspars. A group of silicate minerals including potash feldspars (orthoclase and microcline) and the plagioclase feldspars that contain Ca and/or Na.

Felsite. A general term applied to igneous rocks of fine-grained texture and light color, indicating a granitic composition.

Ferromagnetic. Paramagnetic substances that reach an intensity of magnetization of the same order of magnitude as iron.

Ferruginous. Containing iron.

Festoon cross-bedding. A type of cross-bedding formed when channel filling takes place. Beds have a characteristic curved (concave upward) appearance in cross section.

Fetch. The horizontal extent of water over which the wind blows.

Fibrous crystal shape. Having individual crystals shaped like fibers.

Filter pressing. A process of magmatic differentiation in which residual liquids may be removed from a partially crystallized magma by pressure.

Fiords. Glaciated valleys, the floors of which are submerged below sea level.

Firn. Compacted, granular snow.

Fissility. A property of some materials of splitting along closely spaced parallel planes.

Flagellates. A group of microscopic-sized plants with shapes ranging from that of a balloon with a string attached to that of a pot or fancy mask; all have whiplike projections that move.

Flat-pebble conglomerate. A conglomerate or breccia layer formed where a newly deposited sediment layer is broken by wave action, producing platy pebbles.

Flexural fold. A fold formed where flow or slip is restricted by the layer boundaries. The layering exercises an active control on the deformation; the resulting folds represent a true bending of layers.

Flexure. A bend or arch that is gentle.

Flood tide. The rising, or flow, of water toward the shore as the tide approaches the coast.

Flood plains. The flat areas along a stream that are occasionally flooded by the stream.

Flow casts (*or* **load casts**). Rolls, lobate ridges, or other raised features that may be produced in a sandstone bed as a result of the flowage of sediment in an underlying bed, usually composed of some soft sediment.

Flow fold. A fold formed by rock flowage.

Flowstone. Calcium carbonate deposits formed from water in caves, often taking the shape of a cascade-like mass.

Fluidity. A measure of the ease with which a liquid flows. The inverse of viscosity.

Flysch stage. A period in an orogenic cycle. At this stage uplift in the core of the orogenic belt has started and turbidite deposits of sandstone and shale are deposited in deep water.

Focus. The point at which an earthquake originates, or where motion is initiated.

Folds. Bends or distortions of rock masses most easily recognized when bedding is deformed.

Foliation. A secondary rock fabric in which platy minerals or shear surfaces are in parallel alignment. Schistosity and rock cleavage are types of foliation.

Footwall. The block below a dipping fault.

Foraminifers. Single-celled animals of the sea that possess small shells composed of calcite.

Foreland folds. Folds formed in a belt located on the flanks of a mountain system's core.

Fore-set beds. The layers of inclined sediment on the front slope of a delta.

Foreshore. The zone between low tide and the area where the beach either becomes horizontal or slopes landward.

Fossiliferous limestone. Those limestones formed from the shells of marine animals or in which fossils make up a significant part of the rock.

Fractional crystallization. Separation of a magma into two or more phases.

Fracture. Breaks in rock along which displacements have not occurred.

Free-air gravity anomaly. A gravity anomaly that remains after the reading has been corrected for elevation above the reference surface (geoid).

Frost action. The effects caused by freezing water, particularly as an agent of mechanical weathering.

Frost heaving. The process in which the soil or rocks are lifted up as a result of the freezing of ice under them.

Fumarole. A hole or vent through which fumes or vapors issue in an area underlain by volcanic materials.

Gabbro. A rock composed of coarse- to medium-grained pyroxene, hornblende, and biotite.

Gal. A unit used to express gravitational acceleration. 1 gal = 1 cm/sec^2.

Gamma. A unit of measure of magnetic field strength. 1 gamma = 10^{-5} oersted.

Gauss (*or* oersted). A measure of magnetic field strength. 1 gauss = 1 dyne/unit pole.

Gel. The gelatinlike mass formed when colloids coagulate.

Geodesy. The field of geophysics concerned with the shape and dimensions of the earth.

Geographic poles. See *pole(s), geographic.*

Geoid. The shape of a sea-level surface on earth extended through continents.

Geophones. Instruments designed to pick up vibrations of the ground; used in seismic exploration methods.

Geostrophic currents. Currents in the oceans caused by deflective forces arising from the rotation of the earth.

Geosynclines. Long, narrow belts of long-term subsidence and sedimentation.

Geyserite. A chemical sediment composed of SiO_2 formed around the vents of geysers.

Glaciofluvial deposits. Deposits formed from water on, in, or under a glacier as well as beyond the glacier terminus by melt-water.

Glassy. A term applied to the texture of amorphous materials; such materials show no crystals and thus are glasslike.

Globigerina. Single-celled marine animals having globular-shaped shells that resemble snail shells.

Globigerina ooze. Calcareous deposits on the sea floor composed of the shells of the protozoans globigerina.

Gneiss. A medium- to coarse-grained metamorphic rock characterized by compositional layering that produces a banded appearance.

Gondwanaland. The name of the southern continental mass thought by some to be the source of Africa, India, South America, and Australia.

Gossan. Deposits of hydrated oxide of iron formed in the upper parts of mineral veins and masses containing pyrite as a result of oxidation and leeching.

Gouge. Soft powdery-textured material resulting from grinding and abrasive action along a fault.

Graben. Elongate crustal block, bounded by normal faults on either side, that has moved down in relation to these sides.

Gradation. A term applied to the combined effects of aggradation and degradation of the earth's crust. Sometimes restricted to effects of stream action.

Graded bedding. Sedimentary structure in which there is a systematic vertical change in sediment grain size.

Gradient (stream). The rate of change of elevation of the stream with distance along the stream (e.g., slope).

Granite. A coarse- to medium-grained equigranular rock composed of potassium feldspars and quartz, with some biotite mica and hornblende usually present in small amounts.

Granite pegmatite. A very coarse-grained rock with granitic composition.

Granitization. The processes by which granitic rocks are produced from sedimentary rocks by metasomatism.

Granodiorite. A medium- to coarse-grained plutonic rock that contains quartz, calcic plagioclase, and orthoclase as light-colored constituents and biotite, hornblende, or pyroxene as mafic constituents.

Granulation. The crushing of rock under such conditions that no visible openings result.

Granulite facies. Facies characterized by high-grade metamorphic rocks in which micaceous minerals are altered to garnet and kyanite or sillimanite, and in which hypersthene and diopside are produced from hornblende.

Graphic granite. Rocks of granitic composition that have a texture consisting of an intergrowth of orthoclase and quartz in such a way that cross sections resemble certain hieroglyphics.

Gravimeter. An instrument designed to measure gravity.

Gravity separation. The process by which the first crystals that form in a magma, if heavier or lighter than the melt, will sink or rise through the melt and become segregated either at the bottom or as a raft or floating minerals near the top of the magma chamber.

Graywacke. An impure sandstone consisting of quartz and feldspar fragments and small fragments of rocks.

Great circle. The circle defined by the intersection of the earth's surface and a plane passing through the center of the earth.

Greenschist facies. Metamorphic facies produced by temperatures and pressure conditions that cause the original sedimentary minerals, including clay and zeolite, to be altered. Rocks assume a greenish appearance as a result of the formation of chlorite and sericite.

Ground moraines. Glacial deposits resulting from melting so that debris on the surface, in the ice, and caught near the bottom of the ice is slowly lowered.

Ground state. The normal and lowest energy state for the electrons of an atom.

Ground-water basin. An area bounded by natural geologic conditions that cause ground water within it to be more or less separated from adjacent basins.

Group velocity. The velocity of movement of a wave train.

Guyots. Submerged flat-topped seamounts. Most are thought to be volcanoes that have been eroded by wave action.

Gyres (current). The large oval or circular current patterns found in each ocean in the surface currents.

Half life. The constant amount of time required for one-half the mass of a radioactive isotope to break down.

Hanging valley. A valley having a floor higher than the valley or shore into which it leads. Hanging valleys are usually formed by glaciation.

Hanging wall. The surface or block of rock above an inclined fault zone.

Hard pan. A hard, impervious layer of soil containing insoluble materials and clay.

Hardness. (1) In water, the amount of calcium carbonate and magnesium carbonate in solution. (2) The resistance a mineral offers to scratching.

Headlands. Those parts of the shore that protrude into the ocean.

Headward erosion. Progressive extension of the head of a stream as a result of erosion.

Homocline. General name for any block of bedded rocks that has a single, relatively uniform direction of dip (one limb of a large fold may be described as homoclinal).

Hornblendite. An igneous rock of coarse texture and nearly equigranular fabric composed mostly of hornblende.

Hornfels. Fine-grained, nonfoliated, dense, usually dark-colored rocks formed near the contacts of igneous intrusions.

Horns. Sharp peaks that project above the surrounding area and that are shaped by glacial erosion.

Horse latitude. A belt, characterized by calm or light wind, located in each hemisphere between the trade winds and the prevailing westerlies.

Horst. An elongate block, bounded by normal faults on either side, that has moved up in relation to the sides.

Hortons' Laws. A series of statements regarding relationships of streams in a drainage system.

Humic acid. The organic acid formed by partial decay of organic matter.

Hummocky. Topography characterized by small, uneven knolls.

Hydration. Absorption and combination of water with other compounds, as when anhydrite absorbs water to become gypsum.

Hydraulic factors. Characteristics of a stream that influence its behavior (e.g., velocity, width, depth, gradient, channel roughness).

Hydraulic gradient. The rate of change of pressure head per unit of distance at a given place and in a given direction.

Hydraulic head. The pressure at a point in a body of moving water expressed in terms of the height of the column of water that can be supported by the pressure.

Hydrologic cycle (*or* **water cycle).** The cycle of phenomena through which water passes.

Hydrolysis. A reaction between water and another compound.

Hydrostatic head. An expression of the hydrostatic pressure in terms of the height of a column of water that can be supported by the pressure.

Hydrostatic level. The level to which water will rise in a sealed well.

Hydrostatic pressure. The pressure exerted at a given point in a body of water at rest by the weight of the overlying body of water.

Hypothermal deposits. The deepest ore deposits associated with metamorphic rocks; they are usually lenticular in shape and include such minerals as native gold, chalcopyrite, tin, and molybdenite.

Ice sheets. Large bodies of ice not confined to valleys. Glaciers that may reach continental dimensions and bury whole mountain ranges.

Icebergs. Big blocks of ice broken off from glaciers in the sea.

Igneous. A term meaning "born from fire," applied to both volcanic rocks and rocks that crystallize slowly within the earth's crust.

Illuvial. Referring to soil in which soluble salts leached from high horizons are deposited lower down.

Impressions. Marks on sand or soft sediment on the sea floor.

Incised meander (*or* **entrenched meander).** The meandering course of a river that flows in a deep valley.

Inclination. A measure of the slope of a plane or plunge of a line.

Induced magnetism. Magnetic effects induced in a substance as a result of its presence in a magnetic field.

Inert gases. The remarkably stable unreactive gases (He, Ne, A, Kr, Xe, Rn).

Infiltration. The movement of solutions into the ground.

Influent. A condition existing when the water table is depressed below the level of stream and lake bottoms, and water moves from lake or stream toward the water table.

Influent streams. See *influent.*

Ingrown (streams). A meander that is enlarged as the stream is deepened and the channel is lowered.

Inselberg. A residual hill with steep sides that rise abruptly from adjacent plains.

Insolation. The weathering effect of the sun's heat on exposed rocks.

Internal drainage. A drainage system that does not have any streams leading out of the drainage system.

Ionic bonds. Bonds in certain compounds in which ions of the elements involved are attracted as a result of electric charges on the ions.

Ionization energy. The amount of energy needed to remove an electron from an atom.

Ions. Charged atoms.

Isograds. Lines connecting points of equal metamorphic grade. They form the boundaries between zones.

Isomorphous solid-solution series. A series of minerals in which various ions may substitute in the crystal structure without appearance of a second phase.

Isopach maps. Maps showing the thickness variation of a stratum or sequence of strata. Isopach lines connect points of equal thickness on isopach maps.

Isopiestic lines. Imaginary lines along which the static level of confined ground water is constant.

Isopor line. A line on an isoporic map that connects points of equal annual change in earth's magnetic field.

Isoporic maps. Maps that show lines of no change in the magnetic field and points, called foci, where high rates of change occur.

Isoseismal map. A map showing lines connecting points of equal intensity for an earthquake.

Isostasy. The principle that the upper parts of the earth's outer layers are in a state of flotational equilibrium on a lower level.

Isostatic equilibrium. The condition of the earth's outer layers when they are in flotational equilibrium, that is, they have no tendency to move up or down.

Isostatic gravity anomaly. A gravity anomaly that remains after the reading has been corrected for density variations thought to exist to the level of isostatic compensation.

Isotherms. Contours connecting points of equal temperature.

Isotopes. Atoms of an element having the same number of protons in the nucleus as generally found in the element, but different numbers of neutrons. Isotopes of an element have the same atomic numbers but different atomic weights.

Isotropic. Having the same properties in all directions.

Jet stream. The rapid air currents found in both hemispheres at 9,000 to 15,000 meters (30,000 to 50,000 feet). At this level air may move at several hundred kilometers per hour.

Juvenile water (*or* magmatic water). Water produced by magmatic activity.

Kame terraces. Terraces built along the sides of a glacial valley by load that is dropped by water between the glacier and the adjacent valley wall.

Kames. Glacial deposits laid down in contact with ice. Although used in various ways, the term is often applied to small conical-shaped hills formed of stratified drift.

Karst. A plateau characterized by sinks and other solution-formed features and underlain by limestone containing caves and underground streams.

Karst topography. Topography shaped in part by underground solution and diversion of surface water underground in areas of limestone and dolomite bedrock.

Karst windows. Features formed where collapse of the surface over an underground stream exposes the channel.

Kettle. An enclosed depression left in drift as the last remnants of ice melt away.

Klippe. An isolated block of material separated from underlying rocks by a fault. A feature associated with low-angle thrust faults.

Laccoliths. An igneous intrusion shaped in the form of a planoconvex or doubly convex lens, flattened in the plane of the bedding of the invaded formation.

Lag gravels. Loose gravel-size particles left in arid and semiarid regions as wind removes finer size materials.

Lagging tide. The tidal condition resulting when the arrival of high tide follows the meridian passage of the moon.

Laminae. Thin layers.

Laminar flow. Flow in which the fluid behaves as though it were made up of very thin layers that glide over one another—streamline flow.

Land form. Refers to the naturally produced shapes of the features of the earth's surface.

Landslides. A term generally applied to perceptible downslope movement of earth, rock, soil, and mixtures thereof.

Lapilli. Small fragments (4 to 32 millimeters in diameter) erupted from volcanoes.

Lateral moraine. A deposit from a valley glacier composed of debris that accumulated on the glacier surface as a result of normal downslope movement of debris from the sides of the valley.

Laterite. A special type of clay-rich soil that has a tendency to harden into a bricklike solid.

Laurasia. The name of the northern continental mass thought by some to be the source of North America, Europe, and Asia.

Lava. Magma that reaches the ground surface.

Law of faunal succession. Principle stating that the fossil fauna of rocks of different age are different and provide a key to identification of rocks of that age.

Law of superposition. Principle stating that in a normal stratified sequence each stratum lies on one of greater age.

Leveling. A technique of surveying in which the elevation of various points is determined.

Lignite. A form of coal more compact than peat. It may be woody or amorphous in texture, but twigs and branches can usually be found in it. When dried, it tends to split into slabs.

Limbs. The flanks or sides of a fold.

Liquid immiscibility. The property displayed by liquids when they will not mix with each other or will do so to only a very limited degree.

Liquidus. A curve (in a two-component system) connecting points in a temperature-composition diagram representing the saturation of the liquid phase.

Lithification. The process by which unconsolidated sediments become consolidated rocks.

Lithophysae. The collapsed remains of glass bubbles found in lava.

Lithosphere. The relatively rigid outer rind of the earth. The exact limits are not specified but it is distinguished from a plastic zone below called the asthenosphere.

Lithostatic pressure. The pressure at a point in the earth caused by the weight of the overlying body of rock.

Lithothamnium. A genus of red calcareous algae.

Littoral. The zone along the shore that lies between high and low tide.

Lodgement tills. Glacial till deposits that become pressed, plastered, or lodged in deposits under the glacier.

Loess. An unconsolidated, unstratified, and homogeneous sediment composed of small, angular particles derived from a variety of rock types.

Longitudinal dune. A sand dune that is elongate in the direction of the prevailing wind.

Longitudinal waves. See *dilational waves.*

Longshore current. A component of motion in the water directed along the shore line.

Lopoliths. A large, lenticular, centrally sunken, generally concordant intrusive mass, with its thickness approximately one-tenth to one-twentieth of its width or diameter.

Low-velocity layer (*or* Gutenberg low-velocity zone or channel). A zone in the upper mantle in which seismic waves move with reduced velocity. The depth of the zone is between 100 and 200 kilometers.

Luminescence. The emission of light from some substances when they are subjected to radiant energy.

Lunitidal interval. The interval between the tide's arrival and the meridian passage of the moon.

Luster. Appearance taken on by a mineral in reflected light.

Lutaceous (*or* argillaceous). Descriptive of fine-grained sediment in the size range of silt and clay.

M-discontinuity (*or* Moho). The major seismic discontinuity used to define the base of the earth's crust. It occurs at a depth of 20 to 40 kilometers beneath continents and 10 kilometers under oceanic crust.

Magma (*or* rock melt). Multicomponent silicate melt that usually contains some solid minerals, liquids at high temperature, and gases under pressure.

Magmatic differentiation. Processes by which a magma becomes altered and yields different rock types during crystallization.

Magmatic stoping. The removal of blocks from over the magma which then sink into the magma.

Magmatic water. See *juvenile water.*

Magnetic field. The area around a magnet within which the influence of the magnet can be detected.

Magnetic-field strength. The force (in dynes) that would be exerted on a unit pole at a point in a magnetic field.

Magnetic pole. See *pole, magnetic.*

Magnetic storms. Intervals during which the magnetic field undergoes strong fluctuations.

Magnitude (earthquake). A measure of the amount of energy released at the focus of an earthquake.

Mammillary. A crystal shape similar to mammillary glands.

Mantle. The zone in the earth's interior below the crust and above the core.

Marble. The metamorphic equivalent of calcite, limestone, or dolomite.

Marginal escarpment. An escarpment that forms the seaward slope of a marginal plateau on the sea floor.

Marginal plateau. A shelflike feature that occurs at greater depths than the continental shelf and is separated from the continental shelf by an incipient continental slope.

Marl. Mixture of shells and shell fragments with muds, clay, particles of calcite or dolomite, or sand. It is an impure limestone found in a semiconsolidated state.

Mascons. Relatively high gravity anomalies on the moon postulated as resulting from concentration of mass.

Mean velocity. Average velocity.

Meandering stream. A stream that flows in a strongly curved course consisting of loops in which the stream almost doubles back on itself.

Mechanical concentration. Concentration of minerals through mechanical processes such as gravity separation in streams, along coasts, or by wind.

Medial moraines. Ridges of till left in the middle of a valley after a glacier melts.

Mesothermal deposits. Intermediate-depth ore deposits that tend to be in continuous veins; metamorphism, if any, is low grade. Such deposits are often associated with granites and diorites.

Mesozoic era. The interval of geologic time following the Paleozoic and preceding the Cenozoic.

Metallic bonds. A type of bonding found in native metals and characterized by close packing of atoms in such a way that electrons are shared throughout the structure.

Metamorphic facies. Facies comprised of all rocks exhibiting a unique and characteristic correlation between chemical and mineralogical composition in such a way that rocks of a given chemical composition always have the same mineralogical composition. Differences in chemical composition from rock to rock are reflected in systematic differences of their mineralogical composition.

Metamorphic zone. An area that was exposed to similar metamorphic conditions. Zones are recognized by certain diagnostic minerals.

Metaquartzite. A quartzite formed as a result of metamorphism of a quartz-rich rock.

Metasomatic replacement. A process by which simultaneous solution and deposition take place as new minerals grow at the expense of existing minerals in a rock.

Metasomatism. A process of simultaneous solution and deposition that replaces old minerals.

Meteoric water. Precipitation including rain, sleet, snow, hail, as well as other forms of precipitation.

Mica. One of the most distinctive of all mineral groups, exhibiting single perfect cleavage and having a sheet structure. Included are biotite, muscovite, and other less common varieties.

Micaceous (*or* **lamellar, foliated**)**.** Descriptive of minerals that consist of thin, flat plates.

Microcrystalline. Descriptive of mineral texture when the individual grains of a rock can be distinguished only with the aid of a microscope.

Microseisms. Small-amplitude seismic disturbances caused by wind, waves, and other earth vibrations and recorded on seismograms.

Migmatite. Rock of mixed nature containing metamorphic- and igneous-appearing rock.

Mineral. A naturally occurring element or compound formed by inorganic natural processes and having generally a definite chemical composition and a characteristic atomic structure which is expressed in an external crystalline form and in other physical properties.

Mineral groups. Classification of minerals on the basis of their chemical composition.

Miogeosyncline. A long, relatively narrow belt in which subsidence and accumulation of sediment (notably shallow-water sand, mud, and limestone) occur. See also *eugeosyncline.*

Miscible. Capable of being mixed.

Mixing of magmas. The intrusion of one magma by another magma of different composition.

Mobile belts. Long, narrow belts of the crust that are affected by orogenic movements.

Moho. See *M-discontinuity.*

Mollase stage. The final stage in the sedimentary orogenic cycle. The mountains are very high and coarse sediment is shed into the sedimentary basins.

Monocline. A flexure across which the beds maintain a constant direction of dip.

Monsoon season. The time of year (May–September) when the winds blow from the southwest and the currents move in a clockwise direction in the northern Indian Ocean.

Morass. See *swamp*.

Mountain soil. Any type of soil found in mountains, but usually containing fragments of broken rock in slow transport down the mountainsides.

Mud. A mixture of water and the finer particles of earth and soil.

Mud cracks. Cracks that commonly form in mixtures of clay, sand, and silt when the sediment is dried.

Mudflows. A flow of a mixture of mud and water.

Mylonite. The name given to breccia when it is crushed so completely that it is reduced to a lithified powdery-textured material.

Nappe. A large mass of folded rock that has been moved a long distance from its point of origin.

Natural bridges. Rock bridges formed by the diversion of surface streams underground, remnants of caves, or natural tunnels.

Natural levee. A low ridge formed on either side of a stream channel from deposits laid there during flood.

Neap tides. Low tidal ranges occurring about the time of quadrature of the moon.

Neritic. The depth interval and environments in the sea between low tide and 100 fathoms (183 meters).

Net slip. The straight-line distance, measured after fault movement, between two points that were originally adjacent on opposite sides of the fault.

Neutron. One of the fundamental particles of atoms. It has no net resultant electrical charge and a mass about equal to that of a proton.

New global tectonics (*or* **plate tectonics).** A unified theory of tectonics combining the concepts of convection in the mantle, continental drift, and sea-floor spreading. Also see *tectonics*.

Nip. A curved notch formed by wave action, usually at high- or low-tide level.

Nodes. The static points about which a wave moves.

Nodules. Rounded bodies that are generally harder than the enclosing rock. They may or may not have internal structure.

Nonconservative constituents. Oxygen, carbon dioxide, and nutrients such as nitrates, phosphates, and silicates that enter into biological processes in the ocean.

Normal faults. Steeply inclined faults along which the hanging wall has moved downward in relation to the footwall.

Normal pressure surface. The upper surface of the zone of saturation in the ground under unconfined conditions.

Nunataks. Peaks or isolated hills that protrude through snow fields or glaciers.

Oblate spheroid. A solid figure resembling a sphere but flattened in one direction.

Oblique-slip fault. A fault on which the displacement is oblique across the fault plane.

Obsequent stream. A stream that flows in the direction opposite that of the dip of the underlying rocks.

Obsidian. A glassy rock compositionally equivalent to granite or felsite; a solid, natural glass containing no crystals.

Oceanic rises. Large areas in the ocean higher than the surrounding sea floor. They are neither isolated seamounts nor parts of midocean ridges.

Oersted. A magnetic force designated in terms of dynes per unit pole strength. Also see *gauss*.

Offshore. From the position of low tide seaward.

Offshore barrier beach. A beach formed by an emergent bar and separated from the mainland by a body of water or a marsh.

Olivine. A mineral series varying in composition from forsterite (Mg_2SiO_4) to fayalite (Fe_2SiO_4).

Oölites. Round, egglike bodies usually having concentric internal layering.

Oozes. Marine sediment consisting largely of shell fragments of microscopic marine animals.

Open system. A chemical system in which constituents can move in or out of the system.

Orogenic. A type of deformation of the crust; usually descriptive of mountain-building processes.

Orthogneiss. A gneiss formed as a result of metamorphism of igneous rock.

Orthoquartzite. A clastic sedimentary rock composed of sandstone cemented by silica.

Outwash plain. An area of low relief located beyond the edge of a glacier. It is the site of deposition of glaciofluvial deposits.

Overturned fold. A fold in which the axial surface is inclined so one limb is over part of the other limb.

Oxbow lake. A lake formed as a result of the cutoff of a meander.

Oxidation. The process by which oxygen combines with elements or compounds.

P-waves. See *dilational waves.*

Pack ice. Floating ice consisting of pieces of ice driven together.

Paired terraces. Terraces on either side of a stream that coincide in elevation.

Paleo-pole location. The position of the magnetic pole at some time in the past.

Paleozoic era. A division of geologic time covering that interval after the Precambrian and before the Mesozoic.

Pangaea. Name of the single protocontinent thought by some to be the original source of all continents.

Parabolic dune. A curved sand dune with its concave side facing the wind. The convex side is the slip face.

Paragneiss. A gneiss formed by metamorphism of sedimentary rock.

Paramagnetic. Weakly magnetized in the direction of the magnetic field. Paramagnetic substances are not as strongly magnetized as ferromagnetic materials.

Passive folds. Folds in which flow or slip crosses the layer boundaries, the layers exercising little or no control on the deformation. Layer boundaries serve merely as markers, parts of which are displaced relative to other parts to produce an apparent bending.

Peat. An accumulation of vegetal matter that is in the initial stages of disintegration and decomposition.

Pediment. A gently sloping surface formed by erosion of cliffs or steeper slopes. Pediments are often rock surfaces with a thin veneer of alluvium covering them.

Pedology. Study of the soil, its origin, character, and uses.

Pegmatitic. Rock composed of very coarse crystals, usually in the form of a dike or lense, formed in the late stage of crystallization of a magma.

Peneplain (almost plain). A large area reduced by erosion to a nearly featureless plain at a low level.

Perched water table. A condition found where a zone of saturation is located above an unsaturated zone and above the regional water table.

Peridotite. A dark, coarse-grained, equigranular rock containing a large quantity of olivine and/or pyroxene or hornblende, but no quartz or feldspar.

Period (wave). The time between arrival of adjacent peaks in a wave train.

Permafrost. Permanently frozen ground.

Permeability. (*Rocks*) A measure of the ability of a stratum or rock unit to transmit fluids. (*Magnetism*) The ratio of magnetic induction to magnetic intensity. It is used to characterize the magnetic properties of materials.

pH. Symbol used to indicate the acidity of a solution (the concentration of hydrogen ions). Values range from 0 to 14, with 7 neutral, less than 7 acid, and greater than 7 alkaline.

Phacoliths. Intrusions of lensoid shape in cross sections located at the hinge of folds.

Phaneritic. Crystals large enough to be seen with the unaided eye. Phaneritic textures may be subdivided into coarse-, medium-, and fine-grained subdivisions.

Phase. A part of a system that is physically different from other parts; a part that can be mechanically separated.

Phase rule. Rule stating that the number of degrees of freedom in a chemical system in equilibrium is equal to the difference between the number of components and the number of phases plus 2.

Phase velocity. The individual wave velocity.

Phenocrysts. Large crystals in a glassy or finer grained igneous rock.

Photic zone. The upper layers of the ocean penetrated by sunlight.

Phyllites. Fine-grained schistose rocks identified by the lustrous, silky sheen that characterizes light reflected from the chlorite and muscovite micas of which they are composed.

Phytoplankton. Name applied to all floating plants (e.g., diatoms, sargassum weed).

Piedmont glaciers. Glaciers that coalesce along a mountain front.

Piestic interval. The vertical distance between two isopiestic lines.

Piezometric level. The static level to which water in an artesian aquifer will rise if it is penetrated by a well.

Piezometric surface. An imaginary surface that everywhere coincides with the static level of the water in the aquifer.

Pillow lava. Lava characterized by ball- or pillow-shaped masses, formed by eruption of lava underwater.

Pingos. Small, dome-shaped hills 10 to 100 meters in diameter and sometimes over 30 meters high, formed by the growth and intrusion of ice mounds and found in regions of permafrost.

Pipe. The hole or fissure in a volcano leading to the vent from which magma escapes.

Pirate stream. A stream that diverts the drainage of another stream.

Pisolitic structure. A rock in which small (2 mm+) spherical bodies are present.

Piston corer. A coring device consisting of a long section of heavy steel pipe with a sharp rim at the bottom and a large weight attached to the top.

Pit and mound features. Features that develop from the escape of gas from sediment.

Pitchstone. A variety of obsidian containing water and having a luster like that of resin instead of glass.

Placer deposits. Accumulations of minerals that are heavy, resistant to chemical weathering, hard, malleable, and durable enough to survive the mechanical concentration processes active in streams or in waves.

Plankton. Drifting plants and animals, most of which are no more than 0.05 centimeter in diameter.

Plasma clouds. Invisible clouds of particles containing about equal quantities of positive ions and electrons.

Plastic flow. Deformation characterized by continuous nonreversible strain.

Plate tectonics. See *new global tectonics.*

Playa lake (*or* salina). A shallow lake located in a basin, usually having internal drainage. Such lakes commonly contain water only intermittently and are generally high in salt content as a result of evaporation of water from them.

Plucking (*or* quarrying). Process by which pieces of bedrock are lifted out of place by glaciers.

Plug. Crystallized lava in the volcano's pipe.

Plunge. The angle (measured in a vertical plane) between a line and the horizon.

Plunging axis. The angle of inclination measured from the horizon when the axis of a fold is not horizontal.

Pluvial. Pertaining to conditions caused by rain or to ages when patterns of atmospheric circulation were different and precipitation much increased.

Podzol (*or* podsol). A zonal soil having a thin organic layer above a leached layer that rests upon a dark brown layer.

Poise. A unit of measure of viscosity. 1 poise = 1 g/cm sec.

Polar covalent compounds. Compounds made of molecules that behave somewhat like small magnets.

Poles, geographic. The ends of the earth's axis of rotation. These rotational points are the *north* and *south* geographic poles.

Poles, magnetic. Any of several points on the surface of the earth where the lines of magnetic force are vertical. Magnetic poles do not coincide with geographic poles, though the earth has *north* and *south magnetic poles.*

Poles (of a magnet). The ends of a bar magnet where the magnetic forces appear to be concentrated. Bar magnets have *north* and *south poles.*

Porosity. The volume of pore space in a rock.

Porphyroblastic. A metamorphic texture that resembles the porphyritic texture of igneous rocks.

Porphyry. An igneous rock composed of large crystals embedded in a fine-grained groundmass.

Potential evapotranspiration. The amount of water that would be lost through evaporation and transpiration if the soil were supplied with all the moisture it could hold.

Potholes. Oval or round holes formed in the channel of a stream by concentration of abrasion, impact, and eddy current action.

ppm (parts per million). A measure of concentration. One ppm is one part by weight of an ion or other substance to a million parts by weight of water.

Precession. The generally slow movement of the axis of rotation of a spinning body. The axis defines a cone as it moves.

Precipitation. Rain, snow, sleet, and all other forms of water that fall from the atmosphere onto the land or water surface.

Preorogenic stage. The period before orogeny in the theory of sedimentary orogenic cycles when sediments deposited in the geosyncline are clean sandstones and limestone.

Pressure shadows. Shadowlike zones formed behind the edges of a vent or rotated crystal. Other mineral matter fills in the shadow.

Primary rocks. First-formed rocks.

Primary waves. See *dilational waves.*

Priming tide. The resultant tide when the sun and moon approach alignment with the earth and the tidal bulge arrives a little before the passage of the moon overhead.

Progressive wave. Wave in which the form moves laterally.

Proton. One of the fundamental atomic particles. The proton has a single positive charge, but it is much more massive than the electron, having a mass of 1.672×10^{-24} gram.

Pseudomorph. A mineral having the crystal shape of some other mineral that has been replaced.

Pseudo-plastic. Material behavior that is a combination of elastic and plastic characteristics.

Pseudoviscous flow. Type of solid flow whereby rock subjected to deformation changes shape and can undergo large-scale deformation without the development of fractures and loss of cohesion within the rock mass.

Pumice. A light-colored, porous, glassy, highly vesicular rock.

Pyrolite. A name given to mixtures of olivine, pyroxene, plagioclase, and garnet.

Pyrometamorphism. A type of alteration in which country rock may be baked without melting.

Pyroxene. A group of minerals that are single-chain silicates.

Pyroxenite. A dark, coarsely crystalline rock composed mostly of pyroxenes.

Quadrature. Resultant condition when lines from the sun to the earth and from the moon to the earth are at right angles.

Quarrying. See *plucking.*

Quartzite. The metamorphic equivalent of quartz sandstones.

Quasi-flexural (folds). Folds in which individual layers within a sequence are flexed in response to passive behavior in the associated rocks.

Radioactive isotopes. Isotopes that are unstable and break down spontaneously by the emission of radiant energy.

Radiolarian ooze. Sediment composed largely of the remains of radiolarians.

Radiolarians. Single-celled marine animals having an internal skeleton composed of radiating projections composed of silica; they are important constituents of sediment in some areas of the ocean.

Rain drop and hail imprints. Imprints of rain drops and hail on sediment, characterized by pressed center and raised edges.

Range (tide). Variation in level of water surface resulting from tidal influences.

Recessional moraines. A ridgelike mass of drift accumulation from debris dropped by the glacier when the margin of the ice remains in one position for any period of time during retreat of a valley glacier.

Recrystallization. Reorganization of the elements of the original minerals in a rock that may take place when the rock is subjected to high temperatures and pressure, particularly when water is present.

Recumbent fold. A fold for which the axial surface is nearly horizontal.

Red clay. The second most extensive deep-sea deposit, brown to reddish in color, and containing films of manganese and manganese nodules.

Reference spheroid. A mathematical model that represents an idealized geoid.

Reflection seismology. That branch of seismic exploration that deals with reflected waves.

Refraction seismology. That branch of seismic exploration that deals with refracted waves.

Regional metamorphism. The name applied to those metamorphic alterations that affect rocks over large areas and are indicative of widespread environmental changes rather than localized deformation or magmatism.

Rejuvenation. Condition whereby the behavior of a stream and its effects on the topography appear to revert to those of a more youthful stage in the erosion cycle.

Relics. A mineral that was formed earlier and survived a later alteration of the rock.

Relict structure. Preservation of an original structure or texture after replacement of the original mineral.

Remanent magnetism. Magnetic effects unrelated to the existing external field but due to permanent magnetism of the substance.

Reniform. Crystals that are kidney-shaped.

Residual concentration. A process by which some valuable mineral constituents accumulate while undesirable constituents are removed as a result of weathering and erosion.

Residual magnetic anomaly. Value left after readings for magnetic-field intensity are corrected for diurnal variations, regional effects, and temperature.

Reverse faults. Steeply inclined faults along which the hanging wall has moved up in relation to the footwall.

Rifting. Pulling apart.

Rills. Very small streams.

Ring dike. A dike of arcuate to circular outcrop.

Rip current (*or* **rip tide** *or* **"sea puss").** A strong current formed where water piled up along shore returns to greater depth.

Ripple marks. An undulating surface marking on sediment formed as a result of movement of a granular sediment in the medium of deposition.

Roches moutonnée. Rounded hills that have been streamlined by glacial abrasion and scouring.

Rock. A consolidated aggregate of minerals. Sometimes large masses of a single mineral.

Rock cleavage. The tendency of rock to break along closely spaced subparallel surfaces.

Rock flour. A powdery substance composed of fine (silt-size) particles of rock produced by the grinding and abrasive action of glacier ice moving over bedrock.

Rock flowage. Flow in which material remains solid, yet undergoes continuous deformation. Flowage of this type is accomplished by movements along planes of slip within the component crystals of the rock, by intergranular rotation, and usually it is accompanied by recrystallization of minerals.

Rock glaciers. Masses of rock fragments that assume a form similar to a valley glacier as a result of slow downslope movement.

Rock pressure. See *lithostatic pressure.*

Rotational fault. A scissorslike motion involving rotation of the fault blocks relative to one another.

Rudaceous. Sediment sizes larger than sand size.

Runoff. That portion of precipitation that flows off the ground surface in streams.

Runup. The highest elevation reached by water along the shore during a tsunami.

Safe yield. The amount of water that can be withdrawn from a ground-water basin without producing undesired results.

Salina. (See *playa lake.*)

Salinity. The total salt content per unit volume of sea water.

Salt sequence. The salts precipitated sequentially when salt water is evaporated.

Saltation. Process of transportation through a series of bounces.

Sand drifts. Sand deposits formed where a gap between obstructions acts as a funnel.

Sand shadows. Sand accumulations located downwind from and sheltered by an obstruction, such as a boulder, grove of bushes, or cliff, that interrupts the streamlined flow of wind and reduces its velocity.

Sandstone. Any rock composed of fragments of the size range $\frac{1}{16}$ to 2 millimeters, regardless of composition.

Schistosity. Ease of parting in metamorphic rocks such as schist and gneiss resulting from the parallel or subparallel alignment of platy minerals.

Schists. Strongly foliated metamorphic rocks of medium to coarse crystalline texture.

Schlieren. Streaks of minerals that become aligned in the flow pattern in intrusions.

Scoria. A vesicular volcanic rock formed by the expansion of gases in lava as it is ejected. Scoria is compositionally equivalent to basalt.

Scour and fill. A term applied to small-scale bottom scour or channels that are subsequently filled.

Scree. See *talus.*

Sea-floor spreading. The theory that the sea floor is growing by means of intrusion and formation of new crust along midocean ridges.

Sea level. The mean level of the sea after tidal and wind effects are removed.

Seamounts. Isolated mountain-shaped rises that usually protrude more than 500 fathoms (915 meters) above the ocean floor.

Secondary structural features. Features that result from deformation of rocks after deposition and often after consolidation.

Secular variations. Long-term variations that are observed as progressive changes in declination, intensity, and inclination of the magnetic field. They show rates of change that vary with time at any given point.

Sediment. Materials that settle out of suspension in water; also particles that settle out of water, wind, or ice as well as materials precipitated from solution and deposits of organic origin.

Sedimentary hematites. Hematitic deposits of sedimentary origin.

Sedimentary rocks. Hard, solid rock formed when sediments become consolidated.

Seifs. Longitudinal dunes characterized by a long, sharp ridge crest, one side of which is rounded and the other of which falls abruptly as a slip face.

Seismic sea wave. See *tsunami.*

Seismic shear zone (*or* Benioff zone). The zone that dips from deep-sea trenches under island arcs or adjacent continental margins. The zone is defined by the concentration of earthquake foci within it.

Seismograph. An instrument designed to detect and record seismic waves.

Semianthracite coal. Coal (86–92 percent fixed carbon) intermediate in properties between bituminous and anthracite, defined on the basis of volatile matter and formed by alteration of lower ranks of coal.

Semibituminous coal. Coal averaging 15–20 percent volatile matter and intermediate in properties between bituminous and anthracite coal.

Separation. The distance measured in some specified direction between the displaced parts of a marker horizon (e.g., bed).

Sessile. Organisms attached to the sea bottom.

Shadow zone (seismic). The zone between 103 and 143 degrees distance from an earthquake epicenter in which direct waves do not arrive because of refraction by the earth's core.

Shale. Consolidated mud and clay, often containing sand or lime in large amounts and having an earthy odor.

Sharp-stone conglomerates. Conglomerates or breccia layers formed where a newly deposited sediment layer is broken by wave action and moved a short distance from its point of origin.

Shear fold (*or* slip fold). A fold formed as a result of the minute displacement of beds along closely spaced fractures or cleavage planes.

Shear fractures. Breaks or fractures in rock caused by shear failure.

Shear waves (*or* secondary, S-, *or* transverse). Seismic waves that involve a shearing of the material; the waves vibrate perpendicular to the direction in which the wave front is moving.

Shearing stress. A stress applied in such a way that two adjacent parts of the material slip past one another.

Sheeting. A set of joints or fractures that may develop nearly parallel to the surface of the ground as a response to the release of confining pressure when the weight of the overlying rock is removed.

Sheets. A name given to wind-deposited masses of dust or sand that cover large areas and are relatively thin compared to areal extent.

Shield. Those areas of Precambrian rock exposed at the surface or shallowly buried that have not been folded or complexly deformed since the end of the Precambrian.

Shield volcano. The largest type of volcanic cone, which resembles a shield or low, sloping dome in profile.

Shore. The zone between mean low tide and the landward edge of wave-transported sand.

Shore line. The line along which the sea and the land meet.

Siliceous. Containing silica.

Siliceous sinter. Deposits composed of SiO_2 formed around hot springs and geysers.

Silicon-oxygen tetrahedron. The chief component of the structure of the unit cells among the silicates.

Sills. Concordant, sheetlike intrusive bodies with large lateral extent relative to their thickness.

Siltstone. A consolidated rock composed of clastic particles in the size range of $\frac{1}{16}$ to $\frac{1}{256}$ millimeter in diameter.

Sima. Refers to magnesium-rich rocks (e.g., basalt, gabbro, peridotite) of the earth's crust.

Similar folds. Folds in which the shape remains the same from one layer to another in the fold, but the thickness varies from place to place in each layer.

Simple eutectic system. A type of behavior of some two-component melts when the components do not form a solid-solution series. These systems are plotted on temperature-composition plots.

Sink holes. Enclosed depressions formed by solution, subsidence, or collapse of the surface into a cave.

Sinking creeks. Streams that disappear underground through swallow holes in the channel.

Slates. The most perfectly foliated metamorphic rocks, thought to result from dynamic metamorphism of argillaceous material.

Slaty cleavage (*or* flow cleavage). Rock cleavage like that of slate in which the micaceous minerals show a strong alignment in the plane of the cleavage.

Slickensides. Striations and grooves that appear where rocks on either side of a fault have moved across one another.

Slip face. The downwind face of a sand dune formed as a result of slip and mass movement of sand from crest of dune downslope.

Slip fold. See *shear fold.*

Slump. Mass movement characterized by downslope movement of material on a curved surface of failure.

Snow field. The area above the snow line where snow remains from year to year.

Snow line. The lower limit of the area of perennial snow.

Soil. The product of rock disintegration and decomposition by weathering modified by biological agents; it is capable of supporting plant life.

Solid-solution series. Minerals that can undergo solid-melt reactions within certain limits in such a way that the composition and physical properties vary continuously with varying amounts of the components, as exhibited by plagioclase feldspar.

Solidus. The curve in a temperature–composition diagram (for two-component systems) connecting points above which the solid and liquid are in equilibrium and below which only a solid phase exists.

Solifluction. Soil flowage, particularly flowage induced by saturation of the soil as a result of partial thawing of the ground.

Solubility. The maximum amount of an ionic compound dissolvable in a given amount of solvent under a given set of conditions of temperature and pressure.

Sorting. Separation of particles according to size, shape, or weight.

Space lattice. A three-dimensional arrangement of points representing the unit building blocks of minerals. Crystallographic axes form reference lines for each group.

Spatter cones. Mounds or towerlike structures built up by successive addition of layers of lava ejected from small openings.

Specific gravity. The ratio of a mineral's weight to that of an equal volume of water.

Spherulitic. A texture characterized by small, spherical bodies with radial structure originating at the center and extending out to the margins.

Spit. A bar and beach built out from the tip of the headland beach forming a projection out from the land mass.

Spring tides. Unusually high and low tides that result when the sun and moon lie in line with the earth and the attraction of the two bodies are cumulative.

Stacks. Columnar masses of rock standing isolated as islands just offshore.

Stacks (*or* bosses). Plutons similar to batholiths but smaller. Bosses are stacks of nearly circular ground plan.

Stalactites. Deposits of calcite that hang down from ceilings of caves.

Stalagmite. A stump-shaped cave deposit, usually of $CaCO_3$, formed on the floor of a cave.

Standing waves. Wave motion in which the form does not move laterally but the crest and troughs reverse.

Stick-slip movement. A type of fault movement characterized by sudden and intermittent movements interspersed by periods when the moving surfaces are stuck together.

Stony meteorite (*or* **chondrite**). Meteorites composed of nonmetallic materials, especially olivine.

Stoss. The side of a *roches moutonnée* that faces the advancing ice.

Strain. Change in shape and/or volume.

Strata or beds. Layers of sediment or rock of varying color, texture, or composition.

Stratification. Rock layering.

Streak. The color of the powder of a mineral.

Stream frequency. Ratio of the number of stream segments of all orders to the area of drainage.

Stream piracy. The process by which one stream extends in a headward direction, cutting into the valley of a second stream and eventually diverting the flow of the second stream.

Streaming flow. Flow of water moving on medium and low slopes. The water is smooth in some places, and exhibits mild, slow eddying in others.

Stress. Directed pressure, force per unit area.

Strike. The compass direction (bearing) of any horizontal line on a plane.

Strike separation. Separation of a marker bed measured straight along the strike of the fault.

Strike-slip fault. A fault on which the direction of movement is horizontal along the strike of the fault.

Structural terrace. A flexure that results in a flat, nearly horizontal surface.

Structure. Applies to form and orientation of parts of rock masses and their relation to the whole. Structural geology is concerned with the origin as well as the geometry of rock masses.

Structure contour map. A map showing lines connecting points of equal elevation on a marker (e.g., bed, fault, etc.).

Subbituminous coal. Coal that is usually black with a pitchy luster and parts along a surface nearly parallel to the bedding, forming slabs; ranges from a glossy black coal to lignite.

Sublimation. The processes by which a material passes from a solid to gaseous state without becoming a liquid.

Subjacent masses. Intrusive bodies that are not visibly floored. In many cases the mode of emplacement is uncertain.

Subsequent stream. A stream that flows in a channel that is adjusted to underlying rock structure (generally follows weaker rock outcrop belts, faults, etc.).

Subsoil. The B horizon. It normally contains fewer organisms than do the overlying layers, but more than the C horizon. Weathering has reduced all rock to fine material.

Sunspot activity. Phenomena and processes related to the development and decay of sunspots, particularly the effects of the magnetic storms associated with these spots.

Supercool. A condition in which a liquid is cooled below its normal freezing point.

Supergene enrichment. Processes by which near-surface ore bodies may be formed as a result of oxidation and leaching near the surface and concentration at and below the water table by means of chemical reactions.

Surface waves. Those clastic waves that are propagated near the surface of a body.

Surges (glacier). Rapid (i.e., meters/day) movement of valley glaciers.

Susceptibility. A constant (K) that depends on the nature of a material and is a measure of the degree to which the substance is attracted to a magnet.

Swallow holes. Openings in a stream channel leading into caverns and solution cavities.

Swamp. An area in which the ground is wet, usually having low relief and containing lakes or remnants of lakes.

Swash. A wave of translation occurring after a wave breaks and moving almost straight up the beach.

Swash marks. Those patterns left in sand on a beach as a result of the forward rush of water after an incoming wave breaks.

Swells. Wind-generated waves that have advanced into regions of weaker winds or calm.

Syenite. A coarse- to medium-grained equigranular igneous rock composed primarily of orthoclase with hornblende or biotite.

Symmetrical fold. A fold in which the two limbs are nearly mirror images of one another.

Symmetry. In crystallography, determined by the number of two-, three-, four-, and sixfold axes of symmetry, and the presence or absence of a center of symmetry.

Synclinal mountain. A mountain formed of rock that possesses a synclinal structure. Some are canoe-shaped mountains.

Syncline. A downfold in which the limbs dip toward the axis.

Tabular crystal shape. A crystal that has two dimensions appreciably longer than the third.

Taconite. Iron-rich chert rock.

Talus (*or* **scree).** Rock fragments that accumulate as a heap or sheet at the base of a steep rock surface.

Talus cone. A conical-shaped accumulation of rock fragments.

Tarn. A small lake in the mountains, often restricted to lakes in depressions gouged out by glaciers.

Tectogene. A downbuckle, caused by convection currents, of lower density crustal rocks into the mantle.

Tectonic. Pertaining to large-scale structural features of the earth's crust.

Tectonics. The study of the formation and deformation of the earth's crust that result in large-scale structural features.

Temporary base level. A level to which the reduction of topography of an area is restricted for a limited time.

Tensile strain. Change of shape of a deformed body as a result of application of tension.

Tensile stress. Stress applied in such a way that it tends to pull a material apart.

Terminal moraines. Ridges of drift (largely till) formed along the edge of a glacier at the position of the glacier's furthest advance by debris continually melting out of the glacier while the balance of ice supply and wastage causes the edge of the ice to remain in one location.

Ternary eutectic. The eutectic point in a three-component system.

Terra rosa. The red and clayey soil found in areas of karst topography.

Thrust fault. A fault inclined at a low angle along which reverse-type movement has occurred.

Tidal bore. A turbulent wave that moves with high velocity up streams as a result of high tides.

Tidal waves (*or* **tsunami).** The popular name applied to any gravity-wave system formed in the sea as a result of a large-scale, short-duration disturbance of the surface.

Tide. The name applied to the periodic rise and fall of sea level.

Till. Unsorted, usually unstratified or poorly stratified, glacial drift.

Tombolos. Sand bars or beaches that connect offshore islands with the mainland or another island.

Top-set bed. Beds composed of flat-lying sediments deposited on top of a delta by streams flowing across the delta, marshes, or low areas on either side of the streams.

Tracks, trails, and burrows. Impressions left in the sediment by animals.

Transform fault. A fault that terminates sharply at a place where the movement is transformed into a structure of another type.

Translational wave. A water wave in which the water moves in approximately straight lines. Waves often become translational after they break.

Transpiration. The process by which water vapor is released into the air by plants.

Transverse dune. A sand dune oriented with its long axis perpendicular to the prevailing wind.

Travel-time curve. The plot of travel time versus distance for seismic waves.

Travertine. Compact calcareous cave deposits.

Triangulation. A method of surveying in which points are located by constructing lines of intersection from points whose position is known to unknown points.

Tropical soil (latosolic). Red tropical residual soil from which silica has been leached. It contains concentrations of iron and aluminum hydroxides.

Tropical tides. Tides arising when the moon is far out of the plane of the equator and there are great differences in the range of the two daily tides.

Trough. The lowest part of a fold form; also the depression between two crests in a wave train.

Truncated spurs. Cut-back ridges between tributaries entering a major glaciated valley. These ridges have been cut back as a result of glacial erosion in the main valley.

Tsunami. A Japanese term applied to any gravity-wave system formed in the sea as a result of a large-scale, short-duration disturbance of the surface.

Tufa. Porous or cellular spring deposits; applied to spongy, fragile deposits with an earthy texture.

Tuff. Ash or dust compacted to form a rock.

Tundra soil. Black muck soil formed where there is cold weather most of the year and permafrost.

Turbidity current (*or* **density current**). A mass of water highly charged with material in solution and suspension that flows with turbulent motion down slopes through normal waters.

Turbulent flow. Fluid flow in which the paths of particle motion are irregular with eddies and swirls.

Type section. The locality where a sequence of layered rocks is first described and usually well exposed.

Ultimate base level. The elevation toward which degradational processes tend to reduce the earth's surface; the elevation the sea would have if all land areas were removed and the material deposited below water level.

Ultimate strength. The greatest stress a material can sustain without failure. The highest stress on a stress–strain curve in short-term experiments.

Underfit stream. A stream that appears too small to have formed the valley in which it flows.

Undertow. The return movement of water below the surface toward the sea after waves break.

Uniformitarianism. The theory that the operation of physical and chemical principles and processes has been unchanged throughout geologic time.

Unit magnetic pole. One pole exerting a force of 1 dyne on another like pole when the two are 1 centimeter apart in a vacuum.

Univariant equilibrium. A system in which the variance or degree of freedom is one.

Unpaired terraces. Terraces in a stream valley that do not occur at the same level across the stream.

Valley glaciers. Glaciers that are confined to valleys, generally in mountains.

Valley trains. Sand and gravel deposits laid down-valley from the end of a glacier by streams.

Van der Waals bonds. The weak attraction between molecules due to electrostatic attraction of the nuclei of one molecule for the electrons of another.

Varves. Rhythmically banded sequences of sediment in which a particular sequence of beds is deposited repeatedly. The beds are caused by seasonal variations and thus each sequence represents one year's deposit.

Vent. A hole where the pipe or feeder systems for a volcano break through to the ground surface.

Ventifacts. Smooth, polished stones resulting from repeated abrasion of sand and silt blowing over rocks for a long time.

Vertical separation. The distance measured in a vertical line between the two parts of a marker horizon (bed) offset by a fault.

Volcanic blocks. The largest masses of rock blown out of volcanoes.

Volcanic breccia. A rock composed of large, angular particles contained in fragmental eruptives.

Volcanic dust. Small fragments (less than $\frac{1}{16}$ millimeter) erupted from volcanoes.

Volcanic neck. Solid lava occupying a volcanic vent.

Volcanism. Activity, products, and features associated with volcanoes.

Washes (*or* **wadies** *or* **arroyos**). The dry beds of intermittent streams. The term is used particularly in arid regions.

Water budget. An expression of the total water resources of an area—that introduced into the area and that used or lost by any means.

Water cycle. See *hydrologic cycle.*

Water gap. A pass through a mountain ridge through which a stream flows.

Water table. The surface that separates a zone where the pore spaces in the rocks or ground are not completely filled with water (i.e., they are unsaturated) from a zone below in which they are saturated.

Wave frequency. The reciprocal of wave period.

Wave height. The vertical distance between the top of a wave crest and the bottom of an adjacent trough.

Wave length. The distance between two crests or two troughs.

Waves. The up and down movements on the surface of a body of water.

Weathering. Changes that occur where rocks and minerals come in contact with the atmosphere, surficial waters, and organic life under conditions normally found at the surface of the earth.

Welded tuffs (*or* **welded pumice** *or* **ignimbrite).** Volcanic ash deposits in which the ash and glassy particles are welded together.

Wentworth classification. A logarithmic scale that defines the limits of each type of clastic particle for the purpose of size classification of sediment.

West-wind drift. The ocean-surface currents that flow from west to east in the belts of the prevailing westerly winds.

Wetted perimeter. The distance measured along the water-channel contact in a cross section of a stream.

Wind waves. Waves generated by the wind. They are short crested; their crests are more peaked than the ideal sine waves; and they are skewed in the direction of propagation of the waves.

Xenoliths. Literally, "foreign" rock; applied to rock fragments found in an igneous rock (e.g., inclusions of wall rock).

Yield point. The stress level beyond which a material strains continuously with no additional stress.

Zeolite facies. Products of regional metamorphism at low temperatures.

Zone of aeration. That part of the soil and rock in which both air and percolating water occur.

Zone of subduction. A place in the earth's crust where crustal plates move down into the mantle.

REFERENCES CITED

REFERENCES CITED

Ahlmann, H. W., 1948, Glaciological research on the North Atlantic coasts: Royal Geogr. Soc., Research Ser. 1, 83 p.

Airy, G. B., 1855, On the computations of the effect of the attraction of the mountain masses as disturbing the apparent astronomical latitude of stations on geodetic surveys: London, Royal Soc. Trans., ser. B, v. 145.

Anderson, Don L., 1967, Latest information from seismic observations, *in* Gaskell, T. F., ed., The Earth's Mantle: London and New York, Academic Press, p. 355–420.

Anderson, E. M., 1951, The Dynamics of Faulting and Dyke Formation with Applications to Britain: Edinburgh and London, Oliver and Boyd, 206 p.

Arbenz, P., 1912, Der Gebirgsbau der Zentralschweiz: Schweizer. naturf. Gesell. Verh., 95 Jahres-Vers., Altdorf, pt. 2.

Avery, O. E., Burton, G. D., and **Heirtzler, J. R., 1968,** An aeromagnetic survey of the Norwegian Sea: Jour. Geophys. Research, v. 73, no. 14, p. 4583 ff.

Bagnold, R. A., 1941, The Physics of Blown Sand and Desert Dunes: London, Methuen and Co., 265 p.

Bailey, T. L., and **Jahns, R. H., 1954,** Geology of the Transverse Range province, southern California: California Div. Mines Bull. 170, p. 83–106.

Barazangi, M., and **Dorman, J., 1969,** World seismicity maps compiled from ESSA, Coast and Geodetic Survey, Epicenter data, 1961–1967: Seismol. Soc. America Bull., v. 59, no. 1, p. 369–380.

Barnes, V. E., and **Romberg, F., 1949,** Gravity and magnetic observations on Iron Mountain magnetite deposit, Llano County, Texas: Geophys. Case Histories, v. 1; Soc. Exploration Geophysicists, p. 400–414.

Bascom, Willard, 1964, Waves and Beaches: Garden City, N.Y., Doubleday (Anchor), 267 p.

Bateman, A. M., 1950, Economic Mineral Deposits: New York, John Wiley & Sons, 916 p.

Benioff, Hugo, 1949, Seismic evidence for fault origin of oceanic deeps: Geol. Soc. America Bull., v. 60, p. 1837–1856.

———, **1954,** Orogenesis and deep crustal structure—additional evidence from seismology: Geol. Soc. America Bull., v. 65, no. 4.

Bentz, Alfred, 1961, Lehrbuch d. angewandten Geologie, v. 1: Stuttgart, Germany, Ferdinand Enke Verlag.

Bermes, B. J., Love, G. W., and **Tarver, G. R., 1963,** Geology and ground-water resources of Flagler, Putnam, and St. Johns counties, Florida: Florida Geol. Survey Rept. Inv. 32, 97 p.

Berry, L. G., and **Mason, Brian, 1959,** Mineralogy: Concepts, Descriptions, Determinations: San Francisco and London, W. H. Freeman and Co., 630 p.

Berry, William B. N., 1968, Growth of a Prehistoric Time Scale Based on Organic Evolution: San Francisco, W. H. Freeman and Co., 158 p.

Billings, M. P., 1960, Diastrophism and mountain building: Geol. Soc. America Bull., v. 71, p. 363–397.

Bird, E. C. F., 1969, Coasts, v. 4 *of* An Introduction to Systematic Geomorphology: Cambridge, Mass., M.I.T. Press, 246 p.

Bonini, William E., 1967, The gravitational field and the shape of intrusive bodies: Geol. Eng. Rept. 67-3, Princeton, N.J., Princeton University, Dept. Geol. Eng.

———, and **Hickok, Eugene A., 1958,** Seismic-refraction method in ground-water exploration [New Jersey]: Mining Eng., v. 10, no. 4, p. 485–488.

Bott, M. H. P., 1956, A geophysical study of the granite problem: Geol. Soc. London Quart. Jour., v. 112, p. 45–68.

Bowen, N. L., 1956, The Evolution of the Igneous Rocks: New York, Dover, 332 p.

———, and **Andersen, Olaf, 1914,** The binary system $MgO-SiO_2$: Am. Jour. Sci., v. 37, p. 487–500.

Broecker, Wallace S., 1965, Isotope geochemistry and the Pleistocene climatic record, *in* The Quaternary of the United States: Princeton, N.J., Princeton University Press, p. 737–753.

———, **1966,** Absolute dating and the astronomical theory of glaciation: Sci., v. 151, no. 3708, p. 299–304.

———, and **Farrand, W. R., 1963,** Radiocarbon age of the Two Creeks forest bed, Wisconsin: Geol. Soc. America Bull., v. 74, no. 6, p. 795–802.

———, and **Kaufman, A., 1965,** The ionium method of age determination, *in* Internat. Assoc. Quaternary Research, 6th Cong., Warsaw, 1961, Rept., v. 1: Lódź, Poland, Państwowe Wydawnictwo Nauk. Oddzial w lodzi, p. 633–641.

——— and others, **1968,** Milankovitch hypothesis supported by precise dating of coral reefs and deep-sea sediments: Sci., v. 159, no. 3812, p. 297–300.

———, and **Van Donk, Jan, 1970,** Insolation changes, ice volumes and the O^{18} record in deep-sea cores:

Geophysics and Space Physics Rev., v. 8, no. 1, p. 169–198.

Bucher, Walter H., 1933, The Deformation of the Earth's Crust: An Inductive Approach to the Problems of Diastrophism: Princeton, N.J., Princeton University Press, 518 p.

Bull, William B., 1964, Geomorphology of segmented alluvial fans in western Fresno County, California: U.S. Geol. Survey Prof. Paper 352-E, p. 89–128.

Bullard, Fred M., 1962, Volcanoes, in History, in Theory, in Eruption: Austin, Tex., University of Texas Press, 441 p.

Bullen, K. E., 1961, S and the structure of the upper mantle: Royal Astron. Soc. Geophys. Jour., 4, p. 124–138.

———, 1963, An introduction to the Theory of Seismology, 3d ed.: Cambridge, England, Cambridge University Press, 381 p.

Byrne, John Vincent, and others, **1959,** The chenier plain and its stratigraphy, southwestern Louisiana: Gulf Coast Assoc. Geol. Soc. Trans., v. 9, p. 237–259.

Carey, S. W., 1958, The tectonic approach to continental drift, *in* Continental drift—a symposium: Univ. Tasmania Geology Dept., p. 177–355.

Carozzi, Albert V., 1970, New historical data on the origin of the theory of continental drift: Geol. Soc. America Bull., v. 81, no. 1, p. 283–286.

Carson, R. L., 1951, The Sea around Us: New York, Oxford University Press, 230 p.

Chapman, C. A., 1966, Paucity of mafic ring-dykes—evidence for floored polymagmatic chambers: Am. Jour. Sci., v. 264, p. 66–78.

Clark, David L., 1968, Fossils, Paleontology and Evolution: Dubuque, Iowa, Wm. C. Brown Co., 130 p.

Clark, S. P., Jr., ed., **1966,** Handbook of Physical Constants: New York, Geol. Soc. America, Inc., 587 p.

———, and **Ringwood, A. E., 1964,** Some properties of the Earth and other terrestrial planets [abs.]: Am. Geophys. Union Trans., v. 45, no. 1, p. 105.

Clark, W. E., Musgrove, R. H., Menke, C. G., and **Cagle, J. W., Jr., 1964,** Water Resources of Alachua, Bradford, Clay and Union Counties, Florida: Florida Geol. Survey Rept. Inv. 35, 170 p.

Coats, R. R., 1962, Magma type and crustal structure in the Aleutian Arc: Am. Geophys. Union Geophys. Mon. 6, p. 92–109.

Colby, B. R., 1964, Discharge of sands and mean-velocity relationships in sand-bed streams: U.S. Geol. Survey Prof. Paper 462-A. (See fig. 3—from Colby, 1961—schematic diagram showing the general reasons for variability of effect of depth on the relationship of bed-material discharge to velocity.)

———, 1964, Scour and fill in sand-bed stream: U.S. Geol. Survey Prof. Paper 462-D, 32 p.

Cole, J. W., 1965, Tarawera volcanic complex, *in* Thompson, B. N., and others, New Zealand volcanology: central volcanic region: New Zealand Dept. Sci. Industr. Research Inf. Ser. 50, p. 111–120.

Compton, Robert R., 1955, Trondhjemite Batholith near Bidwell Bar, California: Geol. Soc. America Bull., v. 66, p. 9–44.

Cook, A. H., 1963, The contribution of observation of satellites to the determination of the earth's gravitational potential: Space Sci. Rev., v. 2, pp. 355–437.

Cooper, H. H., Jr., Kenner, W. E., and **Brown, E., 1953,** Ground water in central and northern Florida: Florida Geol. Survey Rept. Inv. 10, 37 p.

Cox, Allan, and **Dalrymple, G. Brent, 1967,** Statistical analysis of geomagnetic reversal data and the precision of potassium-argon dating: Jour. Geophys. Research, v. 72, no. 10, p. 2603–2614.

———, ———, and **Doell, R. R., 1967,** Reversals of the earth's magnetic field: Sci. Am., v. 216, no. 2, p. 44–54.

———, and **Doell, R. R., 1960,** Review of palaeomagnetism: Geol. Soc. America Bull., v. 71, p. 645–768.

Crittenden, M. D., Jr., 1963, New data on the isostatic deformation of Lake Bonneville [Utah]: U.S. Geol. Survey Prof. Paper 454-E, 31 p.

Daly, R. A., 1933, Igneous rocks and the depths of the earth: New York, Hafner Publishing Co., 598 p.

Dana, J. D., and **Hurlbut, C. S., Jr., 1952,** Dana's Manual of Mineralogy, 16th ed.: New York, John Wiley & Sons.

Darcy, Henri, 1856, Les fontaines publiques de la ville de Dijon, Paris.

Davis, J. H., Jr., 1943, The natural features of southern Florida, especially the vegetation, and the Everglades: Florida Geol. Survey, Bull. 25, 311 p.

Davis, William M., 1909, Geographical Essays: New York, Dover, 777 p.

———, 1930, Origin of limestone caverns: Geol. Soc. America Bull., v. 41, p. 475–628.

Defant, Albert, 1961, Physical Oceanography, v. 1: New York, Pergamon Press, 729 p.

Denny, Charles S., 1965, Alluvial fans in the Death Valley region, California and Nevada: Geol. Survey Prof. Paper 466, 62 p.

Dickson, G. O., Pitman, W. C., III, and **Heirtzler, J. R., 1968,** Magnetic anomalies in the south Atlantic and ocean floor spreading: Jour. Geophys. Research, v. 73, no. 6, p. 2087–2100.

Dietz, R. S., 1961, Continental and ocean basin evolution by spreading: Nature, v. 189.

———**, 1963,** Wave-base, marine profile of equilibrium, and wave-built terraces: A critical appraisal: Geol. Soc. America Bull., v. 74, p. 971–990.

———, and **Menard, H. W., 1953,** Hawaiian swell, deep, and arch, and subsidence of the Hawaiian Islands: Jour. Geology, v. 61, no. 2, p. 99–113.

Dobrin, Milton B., 1960, Introduction to Geophysical Prospecting, 2d ed.: New York, McGraw-Hill Book Co., 446 p.

Donn, William L., and **Ewing, Maurice, 1968,** The theory of an ice-free Arctic Ocean: Meteorolog. Mon., v. 8, no. 30, p. 100–105.

Drake, C. L., Ewing, J. I., and **Stockard, H., 1968,** The continental margin of the eastern United States: Canadian Jour. Earth Sci., v. 5, p. 993–1010.

———**, Ewing, Maurice,** and **Sutton, G. H., 1959,** Continental margins and geosynclines: The east coast of North America north of Cape Hatteras, *in* Physics and Chemistry of the Earth, v. 3: New York, Pergamon Press, p. 110–198.

———**, Heirtzler, J.,** and **Hirshman, J., 1963,** Magnetic anomalies off eastern North America: Jour. Geophys. Research, v. 68, no. 18, p. 5259–5275.

———, and **Kosminskaya, I. P., 1969,** The transition from continental to oceanic crust: v. 7, no. 5–6, p. 363–384.

Dury, G. H., 1965, Theoretical implications of underfit streams: U.S. Geol. Survey Prof. Paper 452-C, 43 p.

Dutton, C. E., 1889, On some of the greater problems of physical geology: Washington Philos. Soc. Bull., ser. B, v. 11, p. 51–64.

Eardley, A. J., 1962, Structural Geology of North America: New York, Harper & Row, 743 p.

Eaton, A., 1818, *in* Merrill, G. P., Contributions to the History of American Geology: Dept. of U.S. Natl. Mus., 1904, p. 238.

Eaton, J. P., 1962, Crustal structure and volcanism in Hawaii, *in* The crust of the Pacific Basin: Am. Geophys. Union Geophys. Mon. 6 (Natl. Acad. Sci., Natl. Research Council Pub. 1035), p. 13–39.

———**, 1963,** Earthquakes of the 1959–1960 eruption of Kilauea [abs.]: Geol. Soc. America Spec. Paper 73, p. 37–38.

———, and **Murata, K. J., 1960,** How volcanoes grow: Sci., v. 132, no. 3432, p. 925–938.

Eckel, E. B., ed., **1958,** Landslides and engineering practice: Highway Research Board Spec. Rept. 29, NAS–NRC Pub. 544, Washington, D.C.

Egyed, L., 1956, Determination of changes in the dimensions of the earth from paleogeographic data: Nature, v. 178, 534 p.

———**, 1957,** A new dynamic conception of the internal constitution of the earth: Geol. Rundschau, v. 46, no. 1, p. 101–121.

Elasser, W. M., 1939, On the origin of the Earth's magnetic field: Phys. Rev., v. 55, p. 489–498.

Emery, K. O., 1966, Atlantic continental shelf and slope of the U.S.: U.S. Geol. Survey Prof. Paper 529-A, 23 p.

———**, Tracey, J. I., Jr.,** and **Ladd, H. S., 1954,** Geology of Bikini and nearby atolls [Marshall Islands], pt. 1, Geology: U.S. Geol. Survey Prof. Paper 260-A, 264 p.

———, and **Zarudzki, E. F. K., 1967,** Seismic reflection profiles along the drill holes on the continental margin off Florida: U.S. Geol. Survey Prof. Paper 581-A, 8 p.

Emiliani, C., 1955, Jour. Geology, v. 63, p. 538.

———**, 1964,** Paleotemperature analysis of the Caribbean cores A254–BR–C and CP–28: Geol. Soc. America Bull., v. 75, p. 129–144.

Engel, A. E. J., 1969, Time and the earth: Am. Scientist, v. 57, no. 4, p. 458–483.

———, and **Engel, C. G., 1970,** Lunar rock compositions and some interpretations: Sci., v. 167, no. 3918, p. 527–528.

Ericson, D. B., and **Wollin, G., 1964,** The Deep and the Past: New York, Alfred A. Knopf, 292 p.

Ernst, W. G., 1969, Earth Materials: New York, Prentice-Hall, 149 p.

Eskola, P., 1949, The problem of mantled gneiss domes: Geol. Soc. London Quart. Jour., v. 104, p. 461–476.

Ewing, J., and **Ewing, M., 1967,** Sediment distribution on the mid-ocean ridges with respect to spreading of the sea floor: Sci., v. 156, p. 1590–1592.

Ewing, Maurice, and **Donn, W. L., 1956,** A theory of ice ages: Sci., v. 123, no. 3207, p. 1061–1066.

———, and **Press, Frank, 1956,** The long-period nature of S waves [abs.]: Am. Geophys. Union Trans., v. 37, no. 3, p. 343.

Ewing, W. M., Ericson, D. B., and **Heezen, B. C., 1958,** Sediments and topography of the Gulf of Mexico, *in* Weeks, L. G., ed., Habitat of oil—a symposium, p. 995–1053.

———, and **Heezen, B. C., 1955,** Puerto Rico Trench topographic and geophysical data, *in* Poldervaart, A., ed., Crust of the earth—a symposium: Geol. Soc. America Spec. Paper 62, p. 255–267.

———, and ———, **1956,** Mid-Atlantic Ridge seismic belt [abs.]: Am. Geophys. Union Trans., v. 37, no. 3, p. 343.

Fahnestock, Robert K., 1963, Morphology and hydrology of a glacial stream—White River, Mount Rainier, Washington: U.S. Geol. Survey Prof. Paper 422-A, 70 p.

Fairbridge, R. W., ed., **1966,** The Encyclopedia of Oceanography, v. 1: New York, Reinhold, 1021 p.

Farrand, W. R., and **Gajda, R. T., 1962,** Isobases on the Wisconsin marine limit in Canada: Canada Dept. Mines Tech. Survey, Geogr. Branch, Geogr. Bull. 17, p. 5–22.

Faul, Henry, 1966, Ages of Rocks, Planets, and Stars: New York, McGraw-Hill Book Co., 109 p.

Ferguson, G. E., Lingham, C. W., Love, S. K., and **Vernon, R. O., 1947,** Springs of Florida: Florida Geol. Survey Bull. 31.

Ferrians, Oscar J., Jr., Kachadoorian, Reuben, and **Greene, Gordon W., 1949,** Permafrost and related engineering problems in Alaska: U.S. Geol. Survey Prof. Paper 678, 37 p.

Finn, R. S., and **Heap, W. O., 1962,** How vibratory seismic systems are performing: World Oil, v. 154, no. 5, p. 108–111.

Fisher, R. L., and **Hess, H., 1963,** Deep-sea trenches, *in* The Sea: New York, John Wiley & Sons.

Flint, N. K., and **Rubin, Meyer, 1955,** Radio-carbon dates of pre-Mankato events in eastern and central North America: Sci., v. 121, no. 3149, p. 649–658.

Ford, Arthur B., 1964, Review of Antarctic geology (repr.): Am. Geophys. Union Trans., v. 45, no. 2, 20 p.

Fraser, G. D., and others, **1965,** The Hebgen Lake, Montana, earthquake of August 17, 1959: U.S. Geol. Survey Prof. Paper 435, 242 p.

Frazier, David E., 1967, Recent deltaic deposits of the Mississippi River, their development and chronology: Gulf Coast Assoc. Geol. Soc. Trans. 17th Ann. Meeting, San Antonio, Texas, p. 287–315.

Fuglister, F. C., 1955, Alternative analyses of current surveys: Deep-Sea Research, v. 2, p. 213–229.

Fushsel, G. C., 1762, Historia terrae et maris ex historia Thuringiae permontium descriptionem erecta: Erfurt.

Fyfe, W. S., Turner, F. J., and **Verhoogen, J., 1959,** Metamorphic reactions and metamorphic facies: Geol. Soc. America Mem. 73, v. 12, 259 p.

Gaibar-Puertas, C., 1953, Varacion secular del campo geomagnetico: Observ. del Ebro, Memo. 11.

Garland, G. D., 1965, The Earth's Shape and Gravity: New York, Pergamon Press, 183 p.

Gibbs, J. W., 1928, Collected Works: New York, Longmans, Green and Co. (From Trans. Connecticut Acad., 1874–1878.)

Gordon, A. L., 1967, Circulation of the Caribbean Sea: Jour. Geophys. Research, v. 72, no. 24, p. 6207–6223.

Griggs, Robert F., 1922, The Valley of Ten Thousand Smokes: Natl. Geogr. Soc.

Grindley, G. W., 1960, Geological map of New Zealand [Taupo]: New Zealand Geol. Survey, Dept. Sci. and Industr. Research.

Gross, M. G., 1971, Oceanography: Columbus, Ohio, Charles E. Merrill, 150 p.

Gunn, R., 1947, Quantitative aspects of juxtaposed ocean deeps, mountain chains, and volcanic ranges: Geophysics, v. 12, p. 238.

Gutenberg, Beno, 1941, Changes in sea level, postglacial uplift, and mobility of the Earth's interior: Geol. Soc. America Bull, v. 50, p. 721–772.

———, **1951,** Internal constitution of the earth: Princeton, N.J., Princeton University Press.

———, **1954,** Low-velocity layers in the earth's mantle: Geol. Soc. America Bull., v. 65.

———, and **Richter, C. F., 1939,** Paths followed by waves through the earth's interior (ray diagram), *in* Richter, Charles F., Elementary Seismology: San Francisco: W. H. Freeman and Co., 1958.

———, and ———, **1941,** Seismicity of the earth: Geol. Soc. America Spec. Paper 34, 131 p.

Hack, J. T., 1960, Interpretation of erosional topography in humid temperate regions: Am. Jour. Sci., v. 258-A, p. 80–97.

Hadley, J. B., 1964, Landslides and related phenomena accompanying the Hebgen Lake earthquake of August 17, 1959, *in* The Hebgen Lake, Montana, earthquake of August 17, 1959: U.S. Geol. Survey Prof. Paper 435, p. 107–138.

Halm, J. K. E., 1935, An astronomical aspect of the evolution of the earth: South Africa Astron. Soc., v. 4, p. 1–28.

Hamilton, E. L., 1957, Marine geology of the southern Hawaiian Ridge: Geol. Soc. America Bull., v. 68, no. 8, p. 1011–1026.

Hamilton, W., and **Myers, W. B., 1967,** The Nature of Batholiths: U.S. Geol. Survey Prof. Paper 554-C, 30 p.

Hamilton, W. B., 1960, Late Cenozoic tectonics and volcanism of the Yellowstone region Wyoming, Montana, and Idaho, *in* West Yellowstone—Earthquake area: Billings Geol. Soc., 11th Ann. Field Conf. [Guidebook], p. 92–105.

Hansen, W. R., 1965, Effects of the earthquake of March 27, 1964, at Anchorage, Alaska: U.S. Geol. Survey Prof. Paper 490, 196 p.

———, 1966, Effects of the earthquake of March 27, 1964, at Anchorage, Alaska: U.S. Geol. Survey Prof. Paper 542-A, 68 p.

———, Eckel, E. B., Schaem, W. E., Lyle, R. E., George, Warren, and **Chance, Genie, 1966,** The Alaska Earthquake, March 27, 1964—field investigations and reconstruction effort: U.S. Geol. Survey Prof. Paper 541, 111 p.

Harbaugh, John W., 1968, Stratigraphy and Geologic Time: Dubuque, Iowa, Wm. C. Brown Co., 113 p.

Harr, M. E., 1962, Groundwater and Seepage: New York, McGraw-Hill Book Co., 315 p.

Harrington, H. J., 1965, Geology and morphology of Antarctica, *in* Biogeography and Ecology in Antarctica (Prepr. from Monographiae Biologicae, v. 15): The Hague, Dr. W. Junk, Publishers, 71 p.

Hayes, D. E., 1966, A geophysical investigation of the Peru-Chile trench: Marine Geology, v. 4, p. 309–351.

Heard, Hugh, 1960, Transition from brittle fracture to ductile flow in Solenhofen limestones as a function of temperature, confining pressure, and interstitial fluid pressure, *in* Griggs, D. T., and Handin, John, eds., Rock deformation—a symposium: Geol. Soc. America Mem. 79.

Heath, R. C., Foxworthy, B. L., and **Cohen, Philip, 1966,** The changing pattern of ground-water development on Long Island, New York: Geol. Survey Circ. 524, 10 p.

Heezen, B. C., 1968, The Atlantic continental margin: Univ. Missouri at Rolla Jour., no. 1, p. 5–25.

———, and **Drake, C. L., 1963,** Gravity tectonics, turbidity currents and geosynclinal accumulations in the continental margin of eastern North America: Lamont Geol. Observatory Contr. 673, 10 p.

———, and **Drake, C. L., 1964,** Grand Banks slump: Am. Assoc. Petroleum Geologists Bull., v. 48, no. 2, p. 221–225.

———, Gerard, R. D., and **Tharp, Marie, 1964,** The Vema fracture zone in the equatorial Atlantic: Jour. Geophys. Research, v. 69, no. 4, p. 733–739.

———, Hollister, C. D., an **Ruddiman, W. F., 1966,** Shaping of the continental rise by deep geostrophic contour currents: Sci., v. 152, no. 3721, p. 502–508.

———, Tharp, Marie, and **Ewing, W. M., 1959,** The North Atlantic—text to accompany the physiographic diagram of the North Atlantic [pt.] 1 *of* The floors of the oceans: Geol. Soc. America Spec. Paper 65, 122 p.

Heirtzler, J. R., Dickson, G. O., Herron, E. M., Pitman, W. C. III, and **Le Pichon, X., 1968,** Marine magnetic anomalies, geomagnetic field reversals, and motions of the ocean floor and continents: Jour. Geophys. Research, v. 73, no. 6, p. 2119–2136.

Heiskanen, W. A., 1936, Das Problem der Isostasie, chap. 45 *in* Gutenberg, B., ed., Handbuch der Geophysik, v. 1, chap. 12, p. 4: Berlin, Borntraeger.

Hely, Allen G., 1969, Lower Colorado River water supply—its magnitude and distribution: U.S. Geol. Survey Prof. Paper 486-D, 54 p.

———, Hughes, G. H., and **Irelan, Burdge, 1966,** Hydrologic regimen of Salton Sea, California: U.S. Geol. Survey Prof. Paper 486-C, 32 p.

Herbert, Don, and **Bardossi, Fulvio, 1968,** Kilauea: Case History of a Volcano: New York and Evanston, Ill., Harper & Row, 191 p.

Hess, H. H., 1938, Geological interpretation of data collected on cruise of U.S.S. Barracuda in the West Indies: Am. Geophys. Union Trans., pt. 1, pp. 69–77.

———, 1962, History of ocean basins, *in* Petrologic studies: A volume in honor of A. F. Buddington: Geol. Soc. America, p. 599–620.

———, and others, **1946,** Report of special committee on geophysical and geological study of ocean basins: Am. Geophys. Union Trans., v. 27, no. 4, p. 594–595.

Hicks, Steacy D., and **Shofnos, William, 1965,** The determination of land emergence from sea level observations in Southeast Asia: Jour. Geophys. Research, v. 70, no. 14, p. 3315–3320.

Hill, M. N., 1963, The Sea, v. 3: The Earth beneath the Sea—History: New York, Interscience (John Wiley & Sons), 962 p.

Hill, Mary R., 1964, Scenic resources of California: California Div. Mines and Geology Mineral Inf. Service, v. 17, no. 3, p. 43–51.

———, 1965, Earth hazards—an editorial: California Div. Mines and Geology Mineral Inf. Service, v. 18, no. 4, p. 57–59.

Hjulstrom, F., 1935, Studies of the morphological activity of rivers as illustrated by the River Fjris: Univ. Upsala Geol. Inst. Bull. 25, p. 221–527.

Hodgson, John H., 1964, Earthquakes and Earth Structure: Englewood Cliffs, N.J., Prentice-Hall, 166 p.

Holmes, Arthur, 1965, Principles of Physical Geology, New York, The Ronald Press, 1288 p.

Holt, C. L. R., Jr., 1965, Geology and water resources of Portage Co., Wisconsin: U.S. Geol. Survey Water-Supply Paper 1796 (text & pls.), 77 p.

Horton, R. E., 1945, Erosional development of streams and their drainage basins, hydrophysical approach to quantitative morphology: Geol. Soc. America Bull., v. 56, no. 3, p. 275–370.

Howell, Benjamin F., Jr., 1959, Introduction to Geophysics: New York, McGraw-Hill Book Co., 399 p.

Hoyt, J. H., 1967, Barrier island formation: Geol. Soc. America Bull., v. 78, p. 1125–1136.

———, and **Henry, V. J., Jr., 1967,** Influence of island migration on barrier-island sedimentation: Geol. Soc. America Bull., v. 78, no. 1, p. 77–86.

Hudson, D. E., 1963, The measurement of ground motion of destructive earthquakes: Seismol. Soc. America Bull., v. 53, no. 2, p. 419–437.

Hughes, Dudley J., 1960, Faulting associated with deep-seated salt domes in the northeast portion of the Mississippi salt basin: Gulf Coast Assoc. Geol. Soc. Trans., v. 10, p. 155–173.

Hunt, C. B., 1953, Geology and geography of the Henry Mountains region, Utah: U.S. Geol. Survey Prof. Paper 228, 234 p.

Hurley, P. M., deAlmeida, F. F. M., Melcher, G. C., Cordani, U. G., Rand, J. R., Kawashita, K., Vandoros, P., Pinson, W. H., Jr., and **Fairbairn, H. W., 1967,** Test of continental drift by comparison of radiometric ages: Sci., v. 157, no. 3788, p. 495–500.

———, and **Rand, J. R., 1969,** Pre-drift continental nuclei: Sci., v. 164, no. 3885, p. 1229–1242.

Hutton, James, 1959, Theory of the Earth: Weinheim, Engelmann Ltd., 620 p.

Idyll, C. P., 1964, Abyss, the Deep Sea and the Creatures That Live in It: New York, Thomas Y. Crowell Co., 396 p.

Inoue, Eiji, 1960, Land deformation in Japan: Geogr. Survey Inst. Bull., v. 6, pts. 2–3, 134 p.

Isacks, B., and **Oliver, Jack, 1968,** Seismology and the new global tectonics: Jour. Geophys. Research, v. 73, no. 18, p. 5855–5899.

Jackson, W. W., and **West, J., 1930,** Zeitschr. Krist, v. 83, p. 75.

Jacobs, J. A., 1963, The Earth's Core and Geomagnetism: New York, The Macmillan Co. (Pergamon Press), 137 p.

Jeffreys, Harold, 1936, The structure of the earth down to the 20° discontinuity: Royal Astron. Soc. Monthly Notices Geophys. Supp., v. 3.

———, 1937, The structure of the earth down to the 20° discontinuity: Royal Astron. Soc. Monthly Notices Geophys. Supp., v. 4.

———, 1952, The Earth, Its Origin, History and Physical Constitution: Cambridge, England, Cambridge University Press, 392 p.

Johnson, Ross B., 1961, Patterns and origin of radial dike swarms associated with W. Spanish Peak, and Dike Mountain, south-central Colorado: Geol. Soc. America Bull., v. 72, p. 579–590.

Judson, S., and **Ritter, D., 1964,** Rates of regional denudation in the United States: Jour. Geophys. Research, v. 69, p. 3395–3401.

Kachadoorian, Reuben, 1968, Effects of the earthquake of March 27, 1964 on the Alaska highway system: U.S. Geol. Survey Prof. Paper 545-C, p. C1–C66.

Kamb, Barclay, 1964, Glacier geophysics: Sci., v. 146, no. 3642, p. 353–365.

Kay, Marshall, 1951, North American geosynclines: Geol. Soc. America Mem. 48, 143 p.

———, 1967, Stratigraphy and structure of northeastern Newfoundland bearing on drift in North Atlantic: Am. Assoc. Petroleum Geologists Bull., v. 51, no. 4, p. 579–600.

Kaye, C. A., 1959, Shoreline features and Quaternary shoreline changes, Puerto Rico: U.S. Geol. Survey Prof. Paper 317-B, p. v, 49–140.

Keen, M. J., 1968, An introduction to Marine Geology: Oxford, London, New York, Pergamon Press, 218 p.

Kiersch, George A., 1965, The Vaiont reservoir disaster: California Div. Mines and Geology Mineral Inf. Service, v. 18, no. 7, p. 129–138.

King, E. R., Zietz, I., and **Alldredge, L. R., 1966,** Magnetic data on the structure of the Central Arctic Region: Geol. Soc. America Bull., v. 77, p. 619–646.

King, Lester C., 1962, The Morphology of the Earth; A Study and Synthesis of World Scenery: New York, Hafner Publishing Co., 699 p.

King, P. B., 1937, Geology of the Marathon region, Texas, U.S. Geol. Survey Prof. Paper 187, pp. 1–48.

———, 1950, Tectonic framework of the southeastern states: Am. Assoc. Petroleum Geologists Bull., v. 34.

———, 1951, The Tectonics of Middle North America: Princeton, N.J., Princeton University Press, 203 p.

———, 1964, Geology of the central Great Smoky Mountains, Tennessee: U.S. Geol. Survey Prof. Paper 349-C, p. C1–C148.

———, 1965, Tectonics of Quaternary time in middle North America, *in* Wright, H. E., Jr., and Frey, D. G., eds., The Quaternary of the United States: Princeton, N.J., Princeton University Press.

———, Neuman, Robert B., and **Hadley, J. B., 1968,** Geology of the Great Smoky Mountains National Park, Tennessee and North Carolina: Geol. Survey Prof. Paper 587, 23 p.

Kohout, F. A., 1967, Ground-water flow and the geothermal regime of the Floridian plateau: Gulf Coast Assoc. Geol. Soc. Trans. 17th Ann. Meeting, San Antonio, Texas, p. 339–354.

Krauskopf, Konrad B., 1967, Introduction to Geochemistry: New York, McGraw-Hill Book Co., 721 p.

Kuenen, Ph. H., 1936, Experiment in the formation of a tectogene, *in* Vening-Meinesz, Geol. Soc. America Spec. Paper 62, 1955.

———, 1950, Marine Geology: New York, John Wiley & Sons, 568 p.

Lachenbruch, A. H., 1968, Permafrost, *in* Fairbridge, R. W., ed., The Encyclopedia of Geomorphology: New York, Reinhold, p. 833–839.

Larmor, J., 1919, How could a rotating body such as the sun become a magnet?: British Assoc. Rept., p. 159–160.

Lehmann, Johann, 1756, Versuch einer Geschichte des Flotzgebirge: Berlin.

Lemke, R. W., 1967, Effects of the Earthquake of March 27, 1964, at Seward, Alaska: U.S. Geol. Survey Prof. Paper 542-E, 43 p.

Leopold, Luna B., and **Maddock, Thomas, Jr., 1953,** The hydraulic geometry of stream channels and some physiographic implications: U.S. Geol. Survey Prof. Paper 252, 57 p.

———, and **Wolman, M. G., 1960,** River meanders: Geol. Soc. America Bull., v. 71, no. 6, p. 769–793.

———, ———, and **Miller, J. P., 1964,** Fluvial Processes in Geomorphology: San Francisco, W. H. Freeman and Co., 522 p.

Le Pichon, X., 1968, Sea-floor spreading and continental Drift: Jour. Geophys. Research, v. 73, no. 12, p. 3661–3697.

———, and **Heirtzler, James R., 1968,** Magnetic anomalies in the Indian Ocean and sea-floor spreading: Jour. Geophys. Research, v. 73, no. 6, p. 2101–2117.

Lindsley, D. H., Andreasen, G. E., and **Balsley, J. R., 1966,** Magnetic properties of rocks and minerals: Geol. Soc. America Mem. 97 (Handbook of Physical Constants), p. 543–552.

Livesay, A., 1953, Geology of the Mammoth Cave National Park area: Kentucky Geol. Survey Spec. Pub. 7.

Lofgren, B. E., and **Klausing, R. L., 1969,** Land subsidence due to ground-water withdrawal Tulare-Wasco area California: Geol. Survey Prof. Paper 437-B, 101 p.

Lobeck, K., 1929, The geology and physiography of the Mammoth Cave National Park: Kentucky Geol. Survey.

Longwell, C. R., 1933, Eastern New York and Western New England, Internat. Geol. Congr. Guidebook 1.

Lovering, T. S., 1958, Accumulator plants and rock weathering: Sci., v. 128, no. 3321, p. 416–417.

Ludwig, W. J., Ewing, J. I., Ewing, M., Murauchi, S., Den, N., Asano, S., Hotta, H., Hayakawa, M., Asanuma, T., Ichikawa, K., and **Noguchi, I., 1966,** Sediments and structure of the Japan Trench: Jour. Geophys. Research, v. 71, no. 8, p. 2121–2137.

Lyell, Sir Charles, 1887, Principles of Geology or the Modern Changes of the Earth and Its Inhabitants, v. 1: New York, D. Appleton and Co., 671 p.

McManus, Dean A., 1967, Physiography of Cobb and Gorda Rises, Northeast Pacific Ocean: Geol. Soc. America Bull., v. 78, p. 527–546.

in batholithic rocks of Southern California: Pt. 1, structural geology of Rattlesnake Mountain pluton: Geol. Soc. America Bull., v. 75, p. 805–822.

MacDonald, Gordon A., and **Hubbard, Douglass H., 1961,** Volcanoes of the National Parks in Hawaii: Honolulu, Mid-Pacific Press, 52 p.

———, and **Kuno, Hisashi,** eds., **1962,** The crust of the Pacific Basin, Geophys. Mon. 6: Washington, D.C., Am. Geophys. Union, 195 p.

Macelwane, James B., 1947, The interior of the earth: Am. Scientist, v. 34, no. 2, p. 177–197. (Repr. *in* Science in Progress, 5th Ser., p. 24–54.)

———, **1947,** When the Earth Quakes: Milwaukee, Wisc., Bruce, 288 p.

Machado, F., Parsons, W. H., Richards, A. F., and **Mulford, J. W., 1962,** Capelinhos Eruption of Fayal Volcano, Azores, 1957–1958: Jour. Geophys. Research, v. 67, no. 9, p. 3519–3528.

MacKichan, K. A., and **Kammerer, J. C., 1961,** Estimated use of water in the U.S., 1960: U.S. Geol. Survey Circ. 456, 26 p.

Malmberg, G. T., 1965, Available water supply of the Las Vegas ground water basin, Nevada: U.S. Geol. Survey Water-Supply Paper 1780 (text), 116 p.

Mason, Brian, 1952, Principles of Geochemistry, 3d ed. 1966: New York, London, Sydney, John Wiley & Sons, 329 p.

———, and **Berry, L. G., 1959,** Elements of Mineralogy: San Francisco, W.H. Freeman and Co., 550 p.

Matthes, Francois E., 1930, Geologic history of the Yosemite Valley: U.S. Geol. Survey Prof. Paper 160, p. 94–98.

Maury, M. F., 1855, Physical Geography of the Sea and Its Meteorology: London, Sampson Low, Son and Co.

Meier, M. F., 1960, Mode of flow of Saskatchewan Glacier, Alberta, Canada: U.S. Geol. Survey Prof. Paper 351, 70 p.

Meinzer, O. E., 1923, Outline of ground-water hydrology: U.S. Geol. Survey Water-Supply Paper 494, 71 p.

Menard, H. W., Jr., 1956, Archipelagic aprons [Pacific Ocean]: Am. Assoc. Petroleum Geologists Bull., v. 40, no. 9, p. 2195–2210.

———, **1960,** The East Pacific Rise: Sci., v. 132, no. 3441, p. 1737–1746.

———, **1964,** Marine Geology of the Pacific: New York, McGraw-Hill Book Co., 271 p.

Mercalli, G., 1902, Boll. Soc. Sismologica Italiana, v. 8,

Mero, John L., 1965, The Mineral Resources of the Sea (Elsevier Oceanography Series): New York, Elsevier, 312 p.

Meyer, J. P., Steinhart, J. S., and **Bonini, W. E., 1961,** Montana 1959, *in* Steinhart, John S., and Meyer, Robert P., Explosion Studies of Continental Structure: Carnegie Inst. Washington Pub. 622, p. 305–343.

Milankovitch, M., 1938, Handbuch der Geophysik, v. 9, Koppen and Geiger, eds., p. 593–698.

Millar, C. E., Turk, L. M., and **Foth, H. D., 1966,** Fundamentals of Soil Science: New York, John Wiley & Sons.

Miller, E. W., and **Langdon, George, 1964,** Exploring Earth Environments: New York, Thomas Y. Crowell Co., 507 p.

Milliman, J. D., and **Emery, K. O., 1968,** Sea Levels during the past 35,000 years: Sci., v. 162, p. 1121–1123.

Moody, C. L., 1967, Gulf of Mexico distributive province: Am. Assoc. Petroleum Geologists Bull., v. 51, no. 2, p. 179–199.

Moore, David G., 1963, Geological observations from the bathyscaph *Trieste* near the edge of the continental shelf off San Diego, California: Geol. Soc. America Bull., v. 74, no. 8, p. 1057–1062.

Morisawa, Marie, 1968, Streams, Their Dynamics and Morphology: New York, McGraw-Hill Book Co., 175 p.

Morrison, R. B., 1965, Lake Bonneville: Quaternary stratigraphy of eastern Jordan Valley, south of Salt Lake City, Utah: U.S. Geol. Survey Prof. Paper 477.

———, **1965,** Quaternary geology of the Great Basin, *in* Wright, H. E., Jr., and Frey, D. G., eds., The Quaternary of the United States: Princeton, N.J., Princeton University Press, p. 265–285.

Murata, K. J., and **Richter, D. H., 1966,** Chemistry of the lavas of the 1959–60 eruption of Kilauea Volcano, Hawaii: U.S. Geol. Survey Prof. Paper 537-A, 26 p.

Nadia, A., 1950, Theory of Flow and Fracture of Solids: New York, McGraw-Hill Book Co., 572 p.

Nagata, T., Notes on detrital remanent magnetization of sediments: Jour. Geomagn. Geoelec., v. 14, p. 99–106.

———, **1961,** Rock Magnetism, 2d ed.: Tokyo, Maruzen Press, Ltd.

———, and **Sawada, M., 1963,** Annual mean values of geomagnetic elements since 1955: Jour. Geomagn. Geoelec. Supp. 15, v. 2, p. 1–42.

Nettleton, L. L., 1940, Geophysical Prospecting for Oil:

Neumann, Frank, 1966, Principles underlying the interpretation of seismograms: ESSA Spec. Pub. 254 (rev. ed.), 50 p.

Nordin, Carl F., Jr., and **Dempster, George R., Jr., 1963,** Vertical distribution of velocity and suspended sediment, Middle Rio Grande, New Mexico: U.S. Geol. Survey Prof. Paper 462-B, 20 p.

Oakeshott, Gordon B., 1964, The Alaskan earthquake: California Div. Mines and Geology Mineral Inf. Service, v. 17, no. 7, p. 119–125.

Oliver, J., and **Isacks, B., 1967,** Deep earthquake zones, anomalous structures in the upper mantle, and the lithosphere: Jour. Geophys. Research, v. 72, p. 4259–5275.

Parker, Garald G., and others, **1964,** Water resources of the Delaware River Basin: U.S. Geol. Survey Prof. Paper 381, 200 p.

Parker, R. H., 1959, Macro-invertebrate assemblages of central Texas coastal bays and Laguna Madre: Am. Assoc. Petroleum Geologists Bull., v. 43, no. 9, p. 2100–2166.

Pekeris, C. L., 1935, Thermal convection in the interior of the earth: Royal Astron. Soc. Monthly Notices Geophys. Supp., v. 3, p. 343–367.

Peters, Jack W., and **Dugan, Albert F., 1949,** Gravity and magnetic investigations at the Grand Saline Salt Dome, Van Zandt Co., Texas: Geophys. Case Histories, v. 1; Soc. Exploration Geophysicists, p. 105–120.

Pettijohn, F. J., 1957, Sedimentary Rocks: New York, Harper & Brothers, 718 p.

Phillips, O. M., 1968, The Heart of the Earth: San Francisco, Freeman, Cooper and Co., 236 p.

Pickard, G. L., 1968, Descriptive Physical Oceanography: Oxford, London, New York, Pergamon Press, 200 p.

Pittman, W. C., III, and **Heirtzler, J. R., 1966,** Magnetic anomalies over the Pacific-Antarctic Ridge: Sci., v. 154, no. 3753, p. 1164–1171.

———, Herron, E. M., and **Heirtzler, J. R., 1968,** Magnetic anomalies in the Pacific and sea-floor spreading: Jour. Geophys. Research, v. 73, no. 6, p. 2069–2085.

Plafker, George, 1965, Tectonic deformation associated with the 1964 Alaska earthquake: Sci., v. 148, no. 3678, p. 1675–1687.

———, and **Kachadoorian, Reuben, 1966,** Geologic effects of the March 1964 earthquake and associated seismic sea waves on Kodiak and nearby islands, Alaska: U.S. Geol. Survey Prof. Paper 543-D, 46 p.

———, and **Savage, J. C., 1970,** Mechanism of the Chilean Earthquakes of May 21 and 22, 1960: Geol. Soc. America Bull., v. 81, p. 1001–1030.

Poland, J. F. (section by Sinnott, Allen, with Poland), **1959,** Hydrology of the Long Beach-Santa Ana Area, California: U.S. Geol. Survey Water-Supply Paper 1471 (text and pls.).

Powers, M. C., 1953, Roundness of sedimentary particles: Jour. Sed. Petrology, v. 23, p. 117–119. (See fig. —comparison chart for visual estimation of roundness.)

Pratt, J. H., 1855, On the attraction of the Himalaya Mountains and of the elevated regions beyond upon the plumb-line in India: London, Royal Soc. Trans., ser. B, v. 145.

Pratt, R. M., 1968, Atlantic continental shelf and slope of the U.S.: U.S. Geol. Survey Prof. Paper 529–B, 44 p.

Press, Frank, 1956, Southern California, pt. 1 *of* Determination of crustal structure from phase velocity of Rayleigh waves: Geol. Soc. America Bull., v. 67, no. 12, p. 1647–1658.

———, 1961, The earth's crust and upper mantle: Sci., v. 133, no. 3463, p. 1455–1463.

Raff, A. D., and **Mason, R. G., 1961,** Magnetic survey off the west coast of North America 40°N latitude to 50°N latitude: Geol. Soc. America Bull., v. 72.

Rainwater, E. H., 1967, Resume of Jurassic to recent sedimentation history of the Gulf of Mexico basin: Gulf Coast Assoc. Geol. Soc. Trans. 17th Ann. Meeting, San Antonio, Texas, p. 179–210.

Ramberg, Hans, 1952, The Origin of Metamorphic and Metasomatic Rocks, a Treatise on Recrystallization and Replacement in the Earth's Crust: Chicago, The University of Chicago Press, 317 p.

Ramsey, W. H., 1948, On the constitution of the terrestrial planets: Royal Astron. Soc. Monthly Notices, v. 108, p. 406–413.

Reid, H. F., 1910, The mechanics of the earthquake, the California earthquake of April 18, 1906, Rept. State Investigation Commission: Carnegie Inst., Washington, v. 2.

Revelle, Roger, and **Shepard, F. P., 1939,** Sediments off the California coast, *in* Trask, P. D., ed., Recent marine sediments—a symposium: Am. Assoc. Petroleum Geologists, p. 245–282.

Richey, J. E., 1939, The dykes of Scotland: Geol. Soc. Edinburgh Trans., v. 13, no. 4, p. 393.

Richter, Charles F., 1958, Elementary Seismology: San Francisco, W. H. Freeman and Co.

Richter, D. H., and **Moore, J. G., 1966,** Petrology of the Kilauea Iki Lava Lake, Hawaii: U.S. Geol. Survey Prof. Paper 537-B, 26 p.

Ritter, Dale F., 1967, Rates of denudation: Jour. Geol. Education, v. 15, no. 4, p. 154–159.

Rittmann, A., 1962, Volcanoes and Their Activity: New York, John Wiley & Sons, 305 p.

Robertson, E. C., 1966, The Interior of the Earth: U.S. Geol. Survey Circ., 532, 10 p.

Rubey, W. W., 1933, Settling velocities of gravel, sand, and silt particles: Am. Jour. Sci., v. 25, p. 325–338.

Runcorn, S. K., 1959, Rock Magnetism: Sci., v. 129, p. 1002–1012.

Russell, Richard J., 1967, River Plains and Sea Coasts: Berkeley and Los Angeles, University of California Press, 173 p.

Ryba, Earle, 1968, The history of crystallography, *in* Earth and Mineral Sci., v. 37, no. 5, p. 49–54.

Scholl, D. W., and **Stuvier, M., 1967,** Recent submergence of southern Florida: a comparison with adjacent coasts and other eustatic data: Geol. Soc. America Bull., v. 78, p. 437–454.

Schott, W., 1939, Deep-sea sediments of the Indian Ocean, *in* Trask, P. D., ed., Recent marine sediments —a symposium: Am. Assoc. Petroleum Geologists, p. 396–415.

Schumm, S. A., 1966, The development and evolution of hillslopes: U.S. Geol. Survey, Short Rev.—2, p. 98–104. (Repr. Jour. Geol. Education, v. 14, no. 3.)

———, and **R. W. Lichty, 1965,** Time, space, and causality in geomorphology: Am. Jour. Sci., v. 263, p. 110–119.

Sharp, Robert P., 1960, Glaciers: Condon Lectures: Eugene, Ore., University of Oregon Books, 78 p.

Sharpe, C. F. S., 1938, Landslides and Related Phenomena: New York, Columbia University Press.

Shelton, John S., 1966, Geology Illustrated: San Francisco, W. H. Freeman and Co., 434 p.

Shepard, F. P., 1963, Submarine Geology, 2d ed.: New York, Harper & Row, 557 p.

———, 1967, Delta-front diapirs off Magdalena River, Colombia, compared with hills off other large deltas: Gulf Coast Assoc. Geol. Soc. Trans. 17th Ann. Meeting, San Antonio, Texas, p. 316–325.

———, 1967, The Earth beneath the Sea: Baltimore, Md., Johns Hopkins Press, 242 p.

Sheridan, R. E., Houtz, R. E., Drake, C. L., and **Ewing, M., 1969,** Structure of continental margin off Sierra Leone, West Africa: Jour. Geophys. Research, v. 74, no. 10, p. 2512–2530.

Shreve, Ronald L., 1966, Sherman landslide, Alaska: Sci., v. 154, no. 12, p. 1639–1643.

———, 1966, Statistical Law of Stream Numbers: Jour. Geology, v. 74, no. 1, p. 17–37.

Smith, J. G., 1965, Orogenesis in Western Papua and New Guinea: Tectonophysics, v. 2, p. 1–27.

Spencer, Edgar W., 1965, Geology: A Survey of Earth Science: New York, Thomas Y. Crowell Co., 653 p.

Stacey, Frank D., 1969, Physics of the Earth: New York, John Wiley & Sons, 324 p.

Stauder, William, 1968, Tensional character of earthquake foci beneath the Aleutian Trench with relation to sea-floor spreading: Jour. Geophys. Research, v. 73, no. 24, p. 7693–7701.

Stearns, Harold T., 1966, Geology of the State of Hawaii: Palo Alto, Calif., Pacific Books, 266 p.

———, and **Macdonald, G. A., 1946,** Geology and ground-water resources of the island of Maui, Hawaii: Hawaii Div. Hydrography Bull. 7, 344 p.

Steinbrugge, K. V., and **Cloud, W. K., 1962,** Epicentral intensities and damage in the Hebgen Lake, Montana, earthquake of August 17, 1959: Seismol. Soc. America Bull., v. 52, no. 2, p. 181–234.

Stewart, John H., and **La Marche, V. C., Jr., 1967,** Erosion and deposition produced by the flood of December 1964 on Coffee Creek, Trinity County, California: U.S. Geol. Survey Prof. Paper 422-K, 22 p.

Stommel, Henry, 1966, The Gulf Stream: Berkeley and Los Angeles, University of California Press, 248 p.

Strahler, A. N., 1950, Equilibrium theory of erosional slopes approached by frequency distribution analysis: Am. Jour. Sci., v. 248, p. 673–696; p. 800–914.

———, 1956, Quantitative slope analysis: Geol. Soc. America Bull., v. 67, p. 571–596.

———, **1963,** The Earth Sciences: New York, Harper & Row, 681 p.

Strangway, David W., 1970, History of the Earth's Magnetic Field: New York, McGraw-Hill Book Co., 168 p.

Stringfield, V. T., 1964, Relation of surface-water hydrology to the principal artesian aquifer in Florida and southeastern Georgia: U.S. Geol. Survey Prof. Paper 501-C, p. C164–C169.

Suess, Eduard, 1904, The Face of the Earth, v. 1: Oxford, England, Charendon Press, 604 p.

Sumner, John S., 1969, Geophysics, Geologic Structures and Tectonics: Dubuque, Iowa, Wm. C. Brown Co., 115 p.

Sverdrup, H. U., Johnson, M. W., and **Fleming, R. H., 1942,** The Oceans; Their Physics, Chemistry, and General Biology: Englewood Cliffs, N. J., Prentice-Hall, 1060 p.

Sykes, L. R., 1970, Earthquake swarms and sea-floor spreading: Jour. Geophys. Research, v. 75, no. 32, p. 6598–6611.

Takeuchi, H., Uyeda, S., and **Kanamori, H., 1967,** Debate about the Earth: San Francisco, Freeman, Cooper and Co., 253 p.

Talley, H. C., Jr., and **Cloud, W. K., 1962,** United States earthquakes 1960: U.S. Coast and Geod. Survey, 90 p.

Talwani, Manik, 1964, A review of marine geophysics: Marine Geol., v. 2, p. 29–80.

———, and others, **1965,** Crustal structure of the mid-ocean ridges, 2, computed model from gravity and seismic refraction data: Jour. Geophys. Research, v. 70, p. 341–353.

Taylor, F. B., 1910, Bearing of the Tertiary Mountain Belt on the Origin of the Earth's Plan: Geol. Soc. America Bull. 21, p. 179–226.

———, **1927,** The present and recent rate of land tilting in the region of the Great Lakes: Michigan Acad. Sci. Paper, v. 7, p. 145–157.

Taylor, W. H., 1933, The structure of sanidine and other feldspars: Zeitschr. Krist., v. 85, p. 425.

Terry, R. D., and **Chilingar, G. V., 1955,** Summary of "Concerning some additional aids in studying sedimentary formations," by M. S. Shevetsov: Jour. Sed. Petrology, v. 25, no. 3, p. 229–234.

Thiel, Edward C., 1962, The amount of ice on planet Earth, *in* Antarctic Research, Maury Memorial Symposium: Geophys. Mon. 7, Am. Geophys. Union, p. 172–175.

Thom, E. M., 1944, Bibliography of North American geology 1929–1939, pt. I: U.S. Geol. Survey Bull. 937, 1546 p.

Thomas, H. E., 1963, American ground water provinces (map), *in* The role of ground water in the national situation: U.S. Geol. Survey Water-Supply Paper 1800.

Thompson, B. N., and others, **1965,** New Zealand volcanology; central volcanic region: New Zealand Dept. Sci. Industr. Research Inf. Ser. 50, 211 p.

Thwaites, F. T., and **Bertrand, Kenneth, 1957,** Pleistocene geology of the Door Peninsula, Wisconsin: Geol. Soc. America Bull., v. 68, p. 831–880.

Toksöz, M. Nafi, and **Anderson, Don L., 1966,** Phase velocities of long-period surface waves and structure of the upper mantle, Jour. Geophys. Research, v. 71, no. 6, p. 1649–1658.

Tolman, C. F., 1937, Ground Water: New York, McGraw-Hill Book Co., 593 p.

Tolstoy, Ivan, and **Ewing, W. M., 1949,** North Atlantic hydrography and the Mid-Atlantic Ridge: Geol. Soc. America Bull., v. 60, no. 10, p. 1527–1540.

Tracey, J. I., Jr., Ladd, H. S., and **Hoffmeister, J. E., 1948,** Reefs of Bikini, Marshall Islands: Geol. Soc. America Bull., v. 59, p. 861–878.

Turekian, Karl K., 1968, Oceans: Englewood Cliffs, N.J., Prentice-Hall, 120 p.

Turner, Francis J., 1968, Metamorphic Petrology, Mineralogical and Field Aspects: New York, McGraw-Hill Book Co., 403 p.

———, and **Verhoogen, John, 1960,** Igneous and Metamorphic Petrology: New York, McGraw-Hill Book Co., 694 p.

Tyrrell, G. W., 1926, The Principles of Petrology; An Introduction to the Science of Rocks: New York, E. P. Dutton, 349 p.

Uchupi, Elazar, 1967, Bathymetry of the Gulf of Mexico: Gulf Coast Assoc. Geol. Soc. Trans. 17th Ann. Meeting, San Antonio, Texas, p. 161–172.

———, and **Emery, K. O., 1968,** Structure of continental margin off Gulf Coast of U.S.: Am. Assoc. Petroleum Geologists Bull., v. 52, no. 7, p. 1162–1193.

U.S. Dept. of Agriculture, 1957, Soil, the Yearbook of Agriculture: Washington, U.S. Govt. Printing Office, 784 p.

U.S. Geological Survey and Cooperating Agencies, 1964, The Hebgen Lake, Montana, earthquake of August 17, 1959: U.S. Geol. Survey Prof. Paper 435, 242 p.

Van Allen, J. A., 1957, Direct detection of auroral radiation with rocket equipment: Natl. Acad. Sci. Proc., v. 43, p. 57–62.

Van Andel, Tjeerd H., and **Bowin, Carol O., 1968,** Mid-Atlantic ridge between 22° north latitude and the tectonics of mid-ocean rises: Jour. Geophys. Research, v. 73, no. 4, p. 1279–1298.

Van Woerkom, A., 1953, *in* Shapley, H., ed., Climate Change, Cambridge, Mass., Harvard University Press, p. 147.

Varnes, D. J., 1958, Landslide Types and Processes, chap. 3, *in* Eckel, E. B., ed., Landslides and Engineering Practice, p. 20–47.

Vening-Meinesz, F. A., 1948, Gravity expeditions at sea: Netherlands Geodet. Comm. Pub., v. 4, 233 p.

Vestine, E. H., 1960, Polar auroral, geomagnetic, and ionospheric disturbances: Jour. Geophys. Research, v. 65, p. 360–362.

———, Laporte, L., Cooper, C., Lange, I., and **Hendrix, W. C., 1947,** Description of the earth's main magnetic field and its secular change, 1905–1945: Carnegie Inst. Washington Pub. 578.

Vine, F. J., 1966, Spreading of the ocean floor: New evidence: Sci., v. 154, no. 3755, p. 1404–1415.

———, 1969, Sea-floor spreading—new evidence: Jour. Geol. Education, v. 17, no. 1, p. 6–16.

———, and **Matthews, D. H., 1963,** Magnetic anomalies over oceanic ridges: Nature, v. 199, p. 947–949.

Wager, L. R., and **Deer, W. A., 1938,** A dyke swarm and crustal flexure in East Greenland: Geol. Mag. [Great Britain], v. 75, p. 39–46.

Walker, Frederick, 1940, Differentiation of the Palisade diabase, New Jersey: Geol. Soc. America Bull., v. 51, p. 1059–1108.

Weaver, K. N., 1964, The geology of Howard and Montgomery counties: Maryland Geol. Survey, 373 p.

Weeks, L. G., 1952, Factors of sedimentary basin development that control oil occurrence: Am. Assoc. Petroleum Geologists Bull., v. 36, no. 11, p. 2071–2124.

Wegener, A., 1915, Die Entstehung der Kontinente und Ozeane: Braunschweig, Sammlung Vieweg, no. 23, 94 p.

Wexler, H., Rubin, M. J., and **Caskey, J. E., Jr.,** eds., **1962,** Antarctic research; the Matthew Fontaine Maury Memorial Symposium: Am. Geophys. Union Natl. Acad. Sci., Geophys. Mon. 7, 288 p.

Williams, Howell, 1962, The Ancient Volcanoes of Oregon: Condon Lectures, 3d ed.: Eugene, Ore., University of Oregon Press.

Wilson, A. T., 1964, Origin of ice ages: An ice shelf theory for Pleistocene glaciation: Nature, v. 201, p. 147–149.

Wilson, J. T., 1965, A new class of faults and their bearing upon continental drift: Nature, v. 207, p. 343–347.

Winkler, H. G. F., 1965, Petrogenesis of Metamorphic Rocks: New York, Springer-Verlag, 220 p.

Withington, C. F., 1967, Geology—Its role in the development and planning of Metropolitan Washington: Jour. of the Wash. Acad. of Sci. 57, p. 189–199.

Witkind, I. J., 1964, The Hebgen Lake, Montana, earthquake of August 17, 1959: U.S. Geol. Survey Prof. Paper 435

———, and others, **1962,** Geologic features of the earthquake at Hebgen Lake, Montana, August 17, 1959: Seismol. Soc. America Bull., v. 52, no. 2, p. 163–180.

Woodward, John, 1695, Natural History of the Earth and Terrestrial Bodies: London.

Woollard, G. P., 1959, Crustal structure from gravity and seismic measurements, Jour. Geophys. Research, v. 64, no. 10, p. 1521–1543.

———, 1966, Regional isostatic relations in the United States: Am. Geophys. Union Geophys. Mon. 10, p. 557–594.

Wyllie, Peter J., 1966, Experimental petrology: An indoor approach to an outdoor subject: Jour. Geol. Education, v. 14, no. 3, p. 93–97.

Yoder, H. S., Jr., 1955, Role of water in metamorphism, *in* Poldervaart, A., ed., Crust of the earth—a symposium: Geol. Soc. America Spec. Paper 62, p. 505–523.

INDEX

INDEX

SIMPLIFIED PERIODIC CHART

Atomic weights are given to the nearest tenth of a unit.
(from Green, *Bulletin of the Geological Society*, Vol. 70, 1959)

Series	Group 1	Group 2	Group 3	Group 4	Group 5
1	**Hydrogen** **H** H^- (1) 1.0				
2	**Lithium** **Li** Li^+ (3) 6.9			**Carbon** **C** (6) 12.0	**Nitrogen** **N** (7) 14.0
3	**Sodium** **Na** Na^+ (11) 22.9	**Magnesium** **Mg** Mg^{+2} (12) 24.3	**Aluminum** **Al** Al^{+3} (13) 26.9	**Silicon** **Si** Si^{+4} Si^{-4} (14) 28.0	**Phosphorus** **P** P^{+3} P^{+5} (15) 30.9
4	**Potassium** **K** K^+ (19) 39.1	**Calcium** **Ca** Ca^{+2} (20) 40.0		**Titanium** **Ti** Ti^{+3} Ti^{+4} (22) 47.9	
5	**Copper** **Cu** Cu^+ Cu^{+2} (29) 63.5	**Zinc** **Zn** Zn^{+2} (30) 65.4			
6	**Rubidium** **Rb** Rb^+ (37) 85.5				
7	**Silver** **Ag** Ag^+ Ag^{+2} (47) 107.8			**Tin** **Sn** Sn^{-4} Sn^{+2} (50) 118.7	
8					
9	**Gold** **Au** Au^+ Au^{+3} (79) 197.0			**Lead** **Pb** Pb^{-4} Pb^{+2} Pb^{+4} (82) 207.2	
10					

Group

6	7		8		0
					Helium He (2) 4.0
Oxygen O (8) 16.0	Fluorine F (9) 19.0				Neon Ne (10) 20.2
Sulfur S S^{-2} S^{+4} S^{+6} (16) 32.0	Chlorine Cl Cl^{-1} Cl^{+5} Cl^{+7} (17) 35.5				Argon A (18) 39.9
	Manganese Mn Mn^{+2} Mn^{+3} Mn^{+4} (25) 54.9	Iron Fe Fe^{+2} Fe^{+3} (26) 55.8	Cobalt Co Co^{+2} Co^{+3} (27) 58.9	Nickel Ni Ni^{+2} (28) 58.7	
					Krypton Kr (36) 83.8
					[illegible] (54) 131.3
Uranium U U^{+3} U^{+4} U^{+6} (92) 238.0					

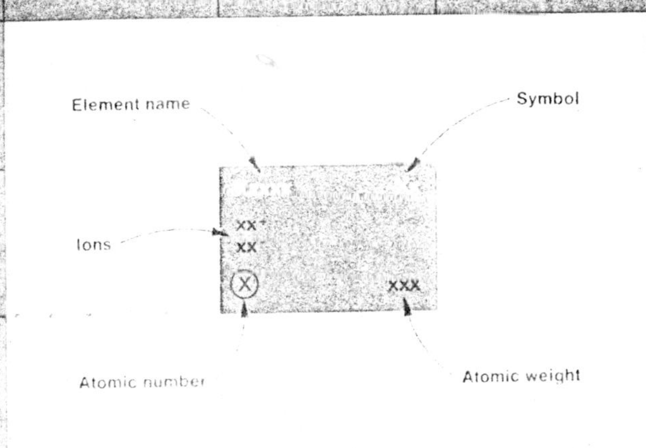

CONVERSION FACTORS METRIC AND ENGLISH MEASURE[1]

DISTANCE

1 kilometer	= 0.621 statute mile
1 meter	= 39.4 inches = 3.28 feet
1 centimeter	= 0.394 inch
1 mile (nautical)	= 1.85 kilometers
1 mile (statute)	= 1.61 kilometers
1 yard	= 0.914 meter
1 foot	= 0.305 meter
1 inch	= 2.54 centimeters
1 fathom	= 6 feet = 1.83 meters

AREA

1 square kilometer	= 0.386 square mile (statute)
1 square meter	= 10.8 square feet
1 square centimeter	= 0.155 square inch
1 square mile (statute)	= 2.59 square kilometers = 640 acres
1 acre	= 43,560 square feet
1 square yard	= 0.840 square meter
1 square foot	= 0.093 square meter
1 square inch	= 6.45 square centimeters

VOLUME

1 cubic kilometer	= 0.239 cubic mile (statute)
1 cubic meter	= 35.3 cubic feet
1 liter	= 1.06 quarts
1 cubic centimeter	= 0.061 cubic inch
1 gallon	= 0.134 cubic foot = 3.78 liters
1 quart	= .946 liters
1 cubic foot	= 7.48 gallons = 0.028 cubic meter = 28.3 liters
1 acre foot	= 43,560 cubic feet = 1,234 cubic meters
1 cubic inch	= 16.39 cubic centimeters

MASS

1 metric ton	= 2,205 pounds
1 kilogram	= 2.205 pounds
1 gram	= 0.035 ounce (avoirdupois)
1 pound (avoirdupois)	= 0.454 kilogram
1 ounce (avoirdupois)	= 28.3 grams
1 ounce (troy)	= 31.1 grams

TEMPERATURE

Celsius (°C), Kelvin (°K) and Fahrenheit (°F) conversions

$$^\circ C = \frac{(^\circ F - 32)}{1.8} \qquad ^\circ F = (1.8 \times {}^\circ C) + 32 \qquad ^\circ K = {}^\circ C + 273$$

[1]Because this table is designed for quick reference and not for extended calculations, most conversions are carried to only 3 significant figures.

MODERN GEOLOGIC TIME SCALE

Absolute time before present (in millions of years)	Eras	Periods	Epochs
	CENOZOIC	**Quaternary** (last 2–3 million years)	**Recent** (last 11,000 years)
			Pleistocene (2–3 million years)
3		**Tertiary** (67 million years)	**Pliocene** (4 million years)
			Miocene (18 million years)
			Oligocene (15 million years)
			Eocene (20 million years)
			Paleocene (10 million years)
70	MESOZOIC	**Cretaceous** (65 millions years)	
135		**Jurassic** (45 million years)	
180		**Triassic** (45 million years)	
225	PALEOZOIC	**Permian** (45 million years)	
270		**Pennsylvanian** (55 million years)	
325		**Mississippian** (25 million years)	
350		**Devonian** (50 million years)	
400		**Silurian** (40 million years)	
440		**Ordovician** (60 million years)	
500		**Cambrian** (50–100 million years)	
600	PRECAMBRIAN (to origin of earth 4.5–5 billion years ago)		
5000			

Relative time scale